# DIGITAL CONSUMER
# ELECTRONICS
# HANDBOOK

## Related McGraw-Hill Books of Interest

### Handbooks

*Avallone and Baumeister* • MARK'S STANDARD HANDBOOK FOR MECHANICAL ENGINEERS

*Benson* • AUDIO ENGINEERING HANDBOOK

*Brady* • MATERIALS HANDBOOK

*Chen* • FUZZY LOGIC AND NEURAL NETWORK HANDBOOK

*Christiansen* • ELECTRONICS ENGINEERS' HANDBOOK

*Coombs* • PRINTED CIRCUITS HANDBOOK

*Coombs* • ELECTRONIC INSTRUMENT HANDBOOK

*Fink and Beaty* • STANDARD HANDBOOK FOR ELECTRICAL ENGINEERS

*Harper* • ELECTRONIC PACKAGING AND INTERCONNECTION HANDBOOK

*Harper and Sampson* • ELECTRONIC MATERIALS AND PROCESSES HANDBOOK

*Johnson* • ANTENNA ENGINEERING HANDBOOK

*Juran and Gryna* • JURAN'S QUALITY CONTROL HANDBOOK

*Jurgen* • AUTOMOTIVE ELECTRONICS HANDBOOK

*Lenk* • MCGRAW-HILL ELECTRONIC TESTING HANDBOOK

*Lenk* • LENK'S DIGITAL HANDBOOK

*Mason* • SWITCH ENGINEERING HANDBOOK

*Schwartz* • COMPOSITE MATERIALS HANDBOOK

*Waynant* • ELECTRO-OPTICS HANDBOOK

### Other

*Inglis and Luther* • VIDEO ENGINEERING, 2/e

*Johnson* • ISO 9000

*Lenk* • MCGRAW-HILL CIRCUIT ENCYCLOPEDIA AND TROUBLESHOOTING GUIDE, VOLS. 1 AND 2

*Rohde, Whitaker, and Bucher* • COMMUNICATIONS RECEIVERS, 2/e

*Saylor* • TQM FIELD MANUAL

*Sclater* • MCGRAW-HILL ELECTRONICS DICTIONARY

*Taylor* • DVD DEMYSTIFIED

*Whitaker* • ELECTRONIC DISPLAYS

*Young* • ROARK'S FORMULAS FOR STRESS AND STRAIN

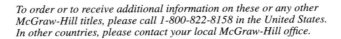

*To order or to receive additional information on these or any other McGraw-Hill titles, please call 1-800-822-8158 in the United States. In other countries, please contact your local McGraw-Hill office.*

# DIGITAL CONSUMER ELECTRONICS HANDBOOK

**Ronald K. Jurgen**   Editor in Chief

**McGRAW-HILL**

New York   San Francisco   Washington, D.C   Auckland   Bogotá
Caracas   Lisbon   London   Madrid   Mexico City   Milan
Montreal   New Delhi   San Juan   Singapore
Sydney   Tokyo   Toronto

**Library of Congress Cataloging-in-Publication Data**

Jurgen, Ronald K.
    Digital consumer electronics handbook / Ronald Jurgen.
       p.    cm.
    Includes index.
    ISBN 0-07-034143-5 (hc : acid-free paper)
    1. Household electronics.   2. Digital electronics.  I. Title.
TK7870.J87     1997
621.382—dc21
                                  97-1546
                                    CIP

# McGraw-Hill

*A Division of The **McGraw·Hill** Companies*

ISBN 0-07-034143-5

*The sponsoring editor for this book was Stephen S. Chapman, and the production supervisor was Claire Stanley. It was set in Times Roman by Publication Services.*

*Printed and bound by R. R. Donnelley & Sons Company.*

*This book is printed on recycled, acid-free paper containing 10% postconsumer waste.*

*McGraw-Hill books are available at special quantity discounts to use as premiums and sales promotions, or for use in corporate training programs. For more information, please write to the Director of Special Sales, McGraw-Hill, 11 West 19th Street, New York, NY 10011. Or contact your local bookstore.*

*This book is dedicated to my brother, Robert J. Jurgen, and to the memories of my parents, Florence T. and John H. Jurgen.*

# CONTENTS

## Chapter 4. Error Correction   *Stan Baggen*    **4.1**

## Chapter 5. Digital Modulation Techniques   *Qun Shi*    **5.1**

## Chapter 6. Fuzzy Logic   *Timothy J. Ross*    **6.1**

## Chapter 7. Channel Codes for Digital Magnetic and Optical Records   *Kees A. Schouhamer Immink*    **7.1**

# Part 3  Key Enabling Standards

## Chapter 8.  MPEG Standards

### Section 1.  Digital Video Coding Standards  *Thomas Sikora*

### Section 2.  Digital Audio Coding Standards  *Peter Noll*

### Section 3.  The MPEG Systems Layer Standard  *Qun Shi*

## Chapter 9.  The ISO JPEG and JBIG Still Image–Coding Standards

### Section 1.  The JPEG Standard  *Jechang Jeong*

# Part 4  Key Digital Delivery/Reception Systems

## Section 2.  Television Data Broadcasting
**Lynn D. Claudy**

## Chapter 16.  Interactive Television Broadcasting and Reception
**Ajith N. Nair**

## Chapter 17.  Digital Cable Systems    *Paul Moroney, Joe Waltrich,*
**Eric Sprunk, Mark Kolber**

## Part 6    Digital Audio Products

# Part 7   Digital Information/Communication Products

# CONTRIBUTORS

**Koki Aizawa**   *Pioneer Electronic Corporation* (CHAP. 22)

**Stan Baggen**   *Philips Research* (CHAP. 4)

**Bruce H. Benjamin**   *Motorola Semiconductor Products* (CHAP. 27)

**Eric Boll**   *Motorola Semiconductor Products* (CHAP. 27)

**Gordy Carlson**   *Motorola Semiconductor Products, Inc.* (CHAP. 24)

**Juin-Hwey Chen**   *Voxware, Inc.* (CHAP. 3)

**Lynn D. Claudy**   *National Association of Broadcasters* (CHAP. 15)

**Almon H. Clegg**   *CCi* (CHAP. 15)

**Chris Day**   *AuraVision Corporation* (CHAP. 25)

**Boudewijn P. Th. M. van Dijk**   *Philips Consumer Electronics* (CHAPS. 11, 12)

**E. C. Dijkmans**   *Philips Research Laboratories* (CHAP. 2)

**Rhonda Dirvin**   *Motorola, Inc.* (CHAP. 26)

**Grayson Evans**   *The Training Department* (CHAP. 29)

**Jack Fuhrer**   *Hitachi America, Ltd.* (CHAP. 20)

**Henk Hoeve**   *Philips Key Modules* (CHAP. 23)

**Robert Hopkins**   *Advanced Television Systems Committee* (CHAP. 13)

**Jae Jeong Hwang**   *Kunsan National University* (CHAP. 10)

**Tsutomu Imai**   *Sony Corporation* (CHAP. 23)

**Kees A. Schouhamer Immink**   *Philips Research Laboratories* (CHAP. 7)

**David Javelosa**   *Composer/Producer* (CHAP. 19)

**Jechang Jeong**   *Hanyang University* (CHAP. 9)

**James D. Johnston**   *AT&T Laboratories* (CHAP. 3)

**Ronald K. Jurgen**   *Editor* (CHAP. 1)

**Johannes G. F. Kablau**   *Philips Key Modules* (CHAP. 12)

**Whoi-Yul Kim**   *Hanyang University* (CHAP. 9)

**Mark Kolber**   *General Instrument Corporation* (CHAP. 17)

**David Kolkman**   *Motorola Semiconductor Products* (CHAP. 27)

**Franc Kozamernik**   *European Broadcasting Union* (CHAP. 14)

**Roger Kozlowski**   *Motorola Semiconductor Products, Inc.* (CHAP. 30)

**Beom Ryeol Lee**   *Electronics and Telecommunication Research Institute* (CHAP. 10)

**Art Miller**   *Motorola, Inc.* (CHAP. 26)

**Paul Moroney**   *General Instrument Corporation* (CHAP. 17)

**Ajith N. Nair**    *Scientific-Atlanta, Inc.* (CHAPS. 16, 18)

**Jaap G. Nijboer**    *Philips Key Modules* (CHAPS. 11, 12)

**Peter Noll**    *Technische Universität Berlin, Germany* (CHAP. 8)

**Ken Parulski**    *Eastman Kodak Company* (CHAP. 21)

**Skip Pizzi**    *BE Radio* (CHAP. 14)

**Christine Podilchuk**    *Lucent Bell Laboratories* (CHAP. 3)

**Mark Reinhard**    *Motorola Semiconductor Products* (CHAP. 27)

**Timothy J. Ross**    *University of New Mexico* (CHAP. 6)

**Jim Seymour**    *AuraVision Corporation* (CHAP. 25)

**Adnan Shaout**    *University of Michigan-Dearborn* (CHAP. 28)

**Qun Shi**    *Panasonic Technologies Inc.* (CHAPS. 5, 8)

**Kenji Shimoda**    *Toshiba Corporation* (CHAP. 20)

**Thomas Sikora**    *Heinrich-Hertz Institute* (CHAP. 8)

**Steve Sperle**    *Motorola, Inc.* (CHAP. 27)

**Eric Sprunk**    *General Instrument Corporation* (CHAP. 17)

**Mikhail Tsinberg**    *Toshiba America Consumer Products* (CHAP. 20)

**Joe Waltrich**    *General Instrument Corporation* (CHAP. 17)

# ABOUT THE EDITOR IN CHIEF

**Ronald K. Jurgen** recently retired as Senior Editor of *IEEE Spectrum.* He is the editor of the *Automotive Electronics Handbook,* published by McGraw-Hill, and an assistant editor of the *Electronics Engineers' Handbook,* also published by McGraw-Hill. He is a Life Senior member of the IEEE and a member of the IEEE Consumer Electronics Society.

# PREFACE

Consumer electronics in its early years encompassed mainly technologies and products for entertainment and communications. With the advent of digital capability, the scope of consumer electronics technologies and products has broadened considerably and rapidly. Digital consumer electronics, as we think of it today, covers not only entertainment and communications but also computers and computer-like devices, digitally controlled appliances, digital cameras, digital musical instruments, and much more.

Although the detailed engineering information content here is restricted to digital consumer electronics, with analog information covered only peripherally, it was a challenge to decide what should be included and what could be omitted without harm. I hope the decisions were wise ones. The rationale taken was to include only *key enabling* technologies, *key enabling* standards, and *key* digital delivery/reception systems that are applied broadly and then to show how those technologies, standards, and delivery/reception systems are used in *representative* digital imaging, audio, and information/communication products, as well as in digital appliances and residential automation.

The key enabling technologies described are data conversion, data reduction, error correction, digital modulation, channel coding for magnetic and optical recording, and fuzzy logic. Without those core technologies, few of the digital products and systems described later in the handbook would be possible. Similarly, the key enabling standards—the Moving Picture Experts Group (MPEG) standards, the ISO JPEG and JBIG still-image coding standards, the H.261 standard, CD-ROM standards, and Digital Versatile Disc (DVD) standards—make application of the technologies practical and the products and systems using them compatible.

The key delivery/reception systems—high-definition television (HDTV), digital audio broadcasting, the Radio Broadcast Data System, National Data Broadcasting (television), interactive television broadcasting and reception, digital cable systems, and the video dial-tone network—make it possible for service providers to send digital signals to homes either over the air or through hardwired networks.

The products and systems selected—interactive multimedia products, digital VCRs and camcorders, digital cameras, CD players, MiniDiscs, Digital Compact Cassettes, digital musical instruments, personal computers with digital video and audio capabilities, personal digital assistants, digital/video telephones, digital appliances, and the Consumer Electronics Bus—offer a thorough rundown on how digital technology is changing the look and performance of those products and systems.

Nearly every chapter has its own glossary of terms. This approach, rather than that of providing one overall unified glossary, has the advantage of allowing terms to be defined in a more application-specific manner in the context of the subject of each chapter.

I would like to thank all the contributors to the handbook, who gave generously of their time and efforts to share their expertise with users of this handbook. Two individuals deserve special mention: Gerard J. J. Vos, technology planning officer for Philips Consumer Electronics in Eindhoven, the Netherlands; and Roger Kozlowski, vice president and technical director of multimedia systems for Motorola Semiconductor Products, Inc., Maitland, Florida. Vos organized the many contributions from Philips in this handbook. Kozlowski, in addition to contributing the final chapter on the digital future, organized the other contributions from Motorola. Without their help, this handbook would not have been possible.

I also wish to thank Steve Chapman, senior editor, electronics and optical engineering, at McGraw-Hill; Donna Namorato, his assistant; Peggy Lamb and Claire Stanley, McGraw-Hill production; and Karla Scheidel and Eric Bramer at Publication Services, Inc., for their guidance and persistence in trying to keep us all on schedule.

*Ronald K. Jurgen*

# DIGITAL CONSUMER ELECTRONICS HANDBOOK

# P · A · R · T · 1

# INTRODUCTION

# CHAPTER 1
# INTRODUCTION

**Ronald K. Jurgen**
*Editor*

## 1.1  A MOMENT IN HISTORY

In June 1990 General Instrument Corp., San Diego, California, announced its all-digital Digi-Cipher system for transmitting a full high-definition television (HDTV) signal in digital format over a standard 6-MHz TV channel. That announcement stunned not only others who were developing analog HDTV systems at the time but the entire electronics community. Most researchers had felt that digital compression had not yet been developed sufficiently for HDTV application. The key to compressing the HDTV signal into a single 6-MHz channel was a proprietary algorithm developed by General Instrument. For error-free transmission of the digital data, powerful error-correction coding combined with adaptive equalization was used. With a carrier-to-noise ratio of more than 19 dB, essentially error-free reception could be achieved.

General Instrument's dramatic achievement was a major turning point in the evolution of the digital video electronics era. Arguably, it could even be called the true beginning of the overall digital electronics era. Earlier developments, such as the Compact Disc, also made use of compression and error-correction coding, but General Instrument used more advanced techniques to be able to transmit a digital signal through the air in a limited bandwidth. That had not been done previously. Within a few weeks, all of the other U.S. HDTV proponents switched over to digital systems.

## 1.2  FUTURE POSSIBILITIES

Since 1990, digital consumer electronics technology has evolved at a rapid pace, so much so, in fact, that industry seers find it difficult to restrain their enthusiasm about possible future applications. For example, Nicholas Negroponte, professor of media technology at the Massachusetts Institute of Technology and founding director of its Media Laboratory, writes in his popular book *being digital*[1] about wearable media. He states, "Computing corduroy, memory muslin, and solar silk might be the literal fabric of tomorrow's digital dress. Instead of carrying your laptop, wear it. While this may sound outrageous, we are already starting to carry more and more computing and communications equipment on our body."

Bill Gates, chairman and chief executive officer of Microsoft Corp., states in his book *The Road Ahead*[2] that

the information highway will enable innovations in the way that intellectual property, such as music and software, is licensed. Record companies, or even individual recording artists, might choose to sell music a new way. You, the consumer, won't need Compact Discs, tapes, or any other kinds of physical apparatus. The music will be stored as bits of information on a server on the highway. "Buying" a song or album will really mean buying the right to access the appropriate bits. You will be able to listen at home, at work, or on vacation, without carrying around a collection of titles. Anyplace you go where there are audio speakers connected to the highway, you'll be able to identify yourself and take advantage of your rights.

Visions of the digital future such as those of Negroponte and Gates are certainly intriguing and might lead one to believe that entirely new technologies will need to be developed in order to make those visions real. In truth, however, future digital consumer electronics products will rely, in the main, on application of the digital technologies and standards available right now and described in this handbook.

## 1.3   IS A COMPUTER A TV OR IS A TV A COMPUTER?

An important trend in the ongoing digital evolution is a blurring of the distinctions between traditional consumer electronics products, such as TV sets, and computers or computer-based products. Examples of this trend include these:

- The IEEE 1394 interface standard makes possible the availability of a video-capable, high-speed digital highway for the convergence of consumer, computer, and communication devices. It defines a method for connecting a variety of such devices, ranging from digital video cameras to printers to disk drives, over a single bus.[3]

- Web TV Networks, of Palo Alto, California, offers a set-top box or terminal for about $330 that enables consumers to access the Internet through their TV receivers and a phone line. Sony Electronics and Philips Consumer Electronics have licensed the terminals. Up to five electronic mail accounts per household are permitted. Thomson Consumer Electronics and Compaq Computer Corp. are collaborating on a proposed series of similar products.

- The Consumer Electronics Manufacturers Association, a sector of the Electronic Industries Association, and Softbank Comdex Inc. announced on March 14, 1996, the formation of a strategic partnership in response to the accelerating convergence of technologies. The two organizations also announced the creation of a new trade show and conference specifically developed to address the convergent marketplace. The first such show and conference will be held in the spring of 1997 in Atlanta, Georgia.

- Sony Corporation introduced two personal computers designed to be strong audio-video multimedia machines. They are available with either 166- or 100-MHz Intel processors, 16- or 32-Mbyte RAMs, up to 2.5 Gbyte of hard drive space, 8x CD-ROMs, and an integrated subwoofer-enhanced sound system.

- Gateway 2000 is marketing a $3800 PC and home theater system called Destination. It is a Pentium chip–based PC, with special accessory cards for high-quality sound and video, attached to a 31-in. computer monitor. Included with the system are a wireless keyboard and a remote control.

- Zenith Electronics Corp. in partnership with Diba Inc. announced plans to market a TV set equipped with a high-speed modem and an Ethernet connection so that the user can connect to the Internet without a PC.

As intriguing as "digital convergence products" may sound, they are not without their problems. Reviewing Gateway 2000's Destination in *The New York Times* of May 14, 1996, Stephen Manes said,

As a TV, the machine is not as good as a traditional model, and it is also compromised as a computer. And in an era when users can barely program their VCRs, this unit is unfortunately not a simple consumer device, but a complicated Windows machine with a confusing television built in.... The 31-inch monitor has no TV circuitry, so incoming signals must be processed within the computer. The resulting TV pictures are dimmer, fuzzier, and subtly blockier than those on a standard TV and sometimes include extraneous material on top. Since the computer has to be on for the TV to work, letting the babysitter watch it may allow her to rifle through your E-mail.

Bill Gates, in *The Road Ahead,* states,

However much like a PC the set-top box becomes, there will continue to be a critical difference between the way a PC is used and a television is used: viewing distance. Today, more than a third of U.S. households have personal computers (not counting game machines). Eventually, almost every home will have at least one, connected directly to the information highway. This is the appliance you'll use when details count or when you want to type. It places a high-quality monitor a foot or two from your face, so your eyes focus easily on text and other small images. A big-screen TV across the room doesn't lend itself to the use of a keyboard, nor does it afford privacy, although it is ideal for applications that multiple people watch at the same time.

There are others, however, who feel just as strongly that TV sets and PCs will become more and more alike and that viewing distance is not a major barrier to that result.

## 1.4  OTHER CONVERGENCES

Computers and PCs are not the only products that have convergence fever. Telephone and cable TV are also likely candidates. The new Telecommunications Act allows telephone and cable TV companies access to each other's markets so that each can offer telephone, video, and high-speed data communications. And electric utilities can seek ways to use their networks to carry communications traffic.

Telephone companies are already offering the Integrated Services Digital Network (ISDN), which combines hardware and software to allow a PC to communicate with an Internet provider at 128 kb/s instead of a maximum of 28.8 kb/s with an ordinary modem. Cable TV companies, on the other hand, are rapidly developing 10- to 32-Mb/s cable modems for downloading data into a PC, although output would still be at 28.8 kb/s. But Internet providers do not yet have server computers that operate at such high speeds.

In yet another twist on convergence, Intel Corp. has announced plans to make most new home PCs capable of making and receiving videophone calls over standard phone lines. Intel would use a 133-MHz Pentium processor and a 28.8-kb/s modem in the PC. Consumers would have to purchase a digital camera. Video would be at a slow rate of 4 to 12 frames per second. And Intel and Microsoft Corp. have worked closely together for years in creating ever more powerful chips and software and, more recently, "plotting to harness networked computers to transform everything from telephony to television."[4]

As these examples show, three industries are major candidates for convergence: consumer electronics, computers, and telecommunications. Watching the turf battles that have already begun and most assuredly will increase in the years ahead should be fascinating.

## 1.5 THE GOVERNMENT—AND THE CONSUMER— HAVE THE LAST WORD

Advances in digital electronics technology have already made possible many promising products and systems: HDTV broadcasting, digital audio broadcasting, Digital Versatile Discs, to name a few. But as history has shown, excellence in technology doesn't guarantee success in the marketplace. A classic example from the past is Sony's Betamax system for videocassette recorders. While generally acknowledged to be superior technologically to the VHS system, Betamax lost out in the marketplace. The main reason was that Sony would not license its Betamax technology to others.

While the Betamax is perhaps a classic example of misjudgment on the part of management, more often a slow-selling product is the result of a consumer perception that that product is unneeded, too expensive, too complicated, etc. For example, Laserdiscs, after a slow start, are beginning to do better but are no blockbusters in sales. More recently, Philips's Digital Compact Cassette and Sony's MiniDisc recorder–playback equipment have not caught on with the public in the United States. In March 1996 Sony announced considerably lower prices on the recorder–playback equipment and on the discs as a way to revive the technology.

HDTV in the United States could turn out to be a classic example of how state-of-the-art technology might be foiled by government and/or special interest groups. (For a fascinating behind-the-scenes account of the evolution of HDTV in the United States, read *Defining Vision* by Joel Brinkley.[5]) After the stunning technological achievement of digital HDTV, proven in field and laboratory tests, the Grand Alliance HDTV system (see Chapter 13) was endorsed by the Federal Communications Commission Advisory Committee on Advanced Television Services in late 1995. Adoption of the system would entail giving broadcasters temporary use of additional spectrum space so that they could broadcast simultaneously both analog and digital signals for a limited period of time. But subsequently the U.S. Senate Budget Committee raised the notion that allowing broadcasters temporary use of the spectrum represented a "giveaway" of the public airwaves and that perhaps the additional spectrum space should be auctioned. Adding fuel to the fire was the fear that broadcasters might not use the additional spectrum space for HDTV but instead would use it for whatever data services and programming they wanted. Subsequently there was a series of rapid-fire HDTV developments:

*March 5, 1996.* Rallying to the HDTV cause, a coalition called Citizens for HDTV was formed to fight a digital spectrum auction.

*April 3, 1996.* It was announced that TV broadcasters and equipment manufacturers had agreed to establish the first fully operational HDTV station in the United States. The model station would provide local TV stations the information and hands-on experience to move rapidly into commercial HDTV implementation. WRC-TV, owned and operated by the National Broadcasting Company, was selected as host station for the HDTV project. The David Sarnoff Research Center was contracted to implement the HDTV station.

*May 9, 1996.* Federal regulators formally proposed standards to govern the transition from analog broadcast technology to HDTV. Those standards would give broadcasters the ability to split their channel space into five or six channels of programs. They could also transmit data into homes. The standards would give broadcasters the free digital airwaves they sought with no deadline for returning analog licenses. But the chairman of the FCC reportedly had reservations about mandating those standards and was more in favor of an open, nonproprietary standard that would satisfy both broadcasters and the computer industry (interlaced scanning for broadcasters and progressive scanning for the computer industry, for example).

*June 19, 1996.* Senior leaders of the U.S. House of Representatives and U.S. Senate sent a letter to the FCC asking the agency to move forward as expeditiously as possible and, no later than April 1, 1997, to lend U.S. television stations a second channel for beginning the transition to digital broadcasting. The letter stated that Congress was no longer interested in auctioning the channels set aside for digital broadcasting. The FCC chief reportedly reacted to the letter by saying that he planned to use the time (until April 1, 1997) to lobby for having broadcasters devote at least 5 percent of the airtime on the second channel to public-interest programming. He also is reported to have said he would like the broadcast industry to give back channels 60 to 69 in the UHF band so that the FCC could auction them to the highest bidder for other purposes, such as cellular telephone service.

*July 8, 1996.* The Citizens for HDTV sent a letter to the FCC head urging him to adopt the digital television broadcasting standard and assign the broadcast channels without further delay.

*July 11, 1996.* The Electronic Industries Association and its Advanced Television Committee followed suit with a similar request.

*July 19, 1996.* The White House Office of Science and Technology Policy in a letter to the FCC head also argued for quick adoption of the standard.

*July 24, 1996.* The U.S. House of Representatives voted to reject an anti-HDTV proposal.

*July 25, 1996.* The FCC proposed to adopt a channel allocation plan for transition to the next generation of television. Stations that use channels 2 through 6 or 52 through 69 would receive a digital slot between channels 7 and 51. After the conversion from analog to digital is completed (10–20 years), those stations would return their analog channels to the FCC.

*November 16, 1996.* After months of heated discussions, representatives of the broadcasting, consumer electronics, and computer industries reached an accord on U.S. standards for advanced television systems. The main point of contention had been that the computer manufacturers, under the Grand Alliance system, would have been required to provide the capability for both progressive scanning (as used in computers) and interlaced scanning (as used in television receivers). Under the agreement, the computer industry will be able to produce computers with just progressive scanning, thereby saving hundreds of dollars per computer. The television industry, however, will be able to produce TV sets with both interlaced and progressive scanning. The agreement was subject to the approval of the Federal Communications Commission.

*December 24, 1996.* The FCC voted unanimously to approve the HDTV standard taking into account the agreement reached on November 16.

There is still no guarantee of future HDTV success. Consumers may not want HDTV sets, particularly if the prices are too high. Edmund L. Andrews, writing in *The New York Times* of November 28, 1995, summed it up this way: "The evolution of an advanced TV system illustrates both the success and the pitfalls of Government industrial policy. It suggests that changes in politics, technology, and the marketplace can confound the most concerted industrial policy, and in the end, it may be markets, not governments, that matter most."

It is also clear, whether or not specific digital consumer electronics products are successful, that there is a tremendous, and growing, core of solid digital engineering know-how upon which new products can be based. If they fail in the marketplace, it will not be for lack of engineering expertise.

## 1.6 CONCLUSION

The consumer electronics industry, from all indications, is doing well. According to the Consumer Electronics Manufacturers Association, total consumer electronic sales reached $65.7 billion in 1996. Gary Shapiro, president of the association, said at the 1997 Winter Consumer Electronics Show held in Las Vegas, Nevada, in January that factory sales of consumer electronic products will reach $69.5 billion in 1997.

Discussion of those new products and technologies, as well as more established ones, are to be found in the following pages of this handbook. Here some of the world's leading engineering experts discuss the key technologies and standards that have enabled and will continue to enable digital products to be brought to market. Major digital delivery systems; digital imaging, audio, information/communications products; digital appliances and residential automation; and the digital future round out the coverage.

## REFERENCES

1. N. Negroponte, *being digital,* Alfred A. Knopf, New York, 1995.

2. W. H. Gates III, *The Road Ahead,* Viking Penguin, New York, 1995.

3. M. Wright, "USB and IEEE 1394: Pretenders, contenders, or locks for ubiquitous desktop deployment?" *EDN,* April 25, 1996, pp. 79–91.

4. B. Schlender, "A conversation with the lords of Wintel," *Fortune,* July 8, 1996, pp. 42–44, 46, 50, 52, 56, 58.

5. J. Brinkley, *Defining Vision*, Harcourt Brace and Company, New York, 1997.

# KEY ENABLING
# TECHNOLOGIES

# CHAPTER 2
# DATA CONVERSION

**E. C. Dijkmans**
*Philips Research Laboratories, Integrated Circuit Design Section*

## 2.1 FRONTIER COUNTRY

We live in an analog world of continuous space and time, a world in which behavior is defined in terms of amplitude, frequency, and phase. The digital world, by contrast, is characterized by the processing of pure numerical values. Data conversion is the essential bridge between these worlds and must deal simultaneously with the attributes of both.

Although audio and video work in quite different frequency bands, the principles of data conversion are the same for both. In the early days, when consumer electronics data conversion was mainly audio, a large part of the necessary filtering was carried out in the analog domain, which imposed severe requirements. Moreover, digital bit values were defined by currents through switched resistor networks, calling for very high accuracy achievable only through individual component trimming. Since then, developments in converter architecture have relaxed analog filtering constraints by shifting almost all the filtering into the analog domain, while so-called bitstream technology, in combination with noise shaping, has greatly eased the demands for accuracy. All these developments are reviewed in this chapter.

## 2.2 CONVERTER BASICS

In analog-to-digital (A/D) conversion, a time- and amplitude-continuous signal is converted into a time- and amplitude-discrete signal. To transfer from time-continuous to time-discrete, the signal is sampled at regular time intervals of $1/f_s$, where $f_s$ is the sampling frequency.

### 2.2.1 Sampling

Taking samples of the input signal in the time domain is equivalent to modulation in the frequency domain of the input signal spectrum with the sampling signal spectrum. The sampled signal contains the input spectrum plus the modulation products of the input spectrum, and infinite multiples of the sampling frequency $1/T$. Various factors, both theoretical and practical, can affect the sampling rate. First among these is the Nyquist sampling theorem, which states that if an analog waveform is to be fully described by digital samples, the sampling frequency must be at least twice the highest bandwidth of the signal to be sampled. The input signal must be band-limited to half the sampling frequency; otherwise modulation products can be generated

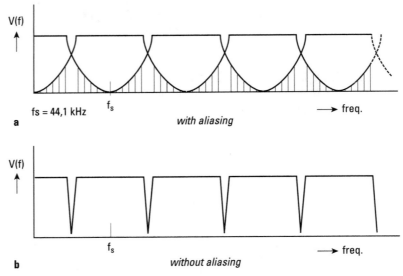

**FIGURE 2.1** Frequency spectra of a sampled analog signal: (*a*) with aliasing, (*b*) without aliasing.

in the signal passband by modulation of high-frequency, out-of-band components of the input signal with the sampling signal, called aliasing. The results of aliasing are shown in Fig. 2.1.

### 2.2.2 Choice of Sampling Frequency

For an audio signal band of 20 kHz, for example, the Nyquist theorem requires a sampling rate of at least 40 kHz. An additional factor has to be allowed for the transition band of the antialiasing filter needed in A/D conversion and of the corresponding low-pass filter needed in the reverse process.

With these constraints, the sampling rate should be as low as possible for economic reasons, particularly in consumer electronics. A practical requirement that also has to be taken into account is the high digital-signal bandwidth and the performance limitations of the equipment available to record it. In the case of the Compact Disc, the most practical solution in the early 1980s was the videorecorder. The fact that there are two video standards—525 lines at 59.94 Hz and 625 lines at 50 Hz—with no common multiple in the required sampling frequency range, was a possible source of conflict. The compromise adopted was three samples per line, 245 lines per field (half the number of active lines per frame without blanking) at 59.94 Hz, giving 44.0559 Hz, or at 60 Hz, giving the familiar CD sampling frequency of 44.1 kHz.

Two other audio sampling rates are in common use: 32 kHz and 48 kHz. The first was chosen as a satisfactory value for FM stereo broadcasting, with its bandwidth of 15 kHz; the second was originally adopted for professional audio applications and has since been applied in digital audiotape (DAT) and Digital Compact Cassette (DCC) audio recording.

### 2.2.3 Quantization

To create a digital signal, which is discrete in time as well as in amplitude, the sampled analog signal value is first stored in a hold circuit. During the hold time, the A/D converter matches the

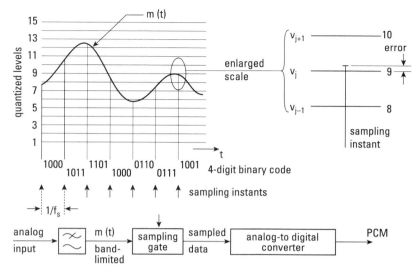

**FIGURE 2.2**    Quantization of an analog signal and the basic arrangement required.

analog signal value to the nearest integer value of a reference step $q$, which is represented by the least significant bit (LSB). The remainder is called the quantization error and will be smaller than $q/2$. The signal processing chain needed for A/D conversion is shown in Fig. 2.2. Though in theory the resolution is defined by the number of bits, several mechanisms can seriously degrade the dynamic performance to well below the theoretical value.

### 2.2.4  Specifications

The basic error in digital signal processing is the quantization error, defined as a level of uncertainty of at most $q/2$. If the input signal is sufficiently large and fast-varying and uncorrelated with the sampling process, the quantization error is random in amplitude. The error can then be treated as white noise, with a noise power of $(q \cdot 2)/12$, where $q$ is one LSB step.[1,2] It is assumed that this noise energy is spread evenly over all frequencies up to half the sampling frequency. The dynamic range is now the ratio of the power of a sine wave of maximum amplitude to the quantization noise power, and for a given number of bits

$$(S/N)_{\max} = n \cdot 6.02 + 1.76 \text{ dB} \tag{2.1}$$

The last factor depends on the signal used, in this case a sine wave, and needs correction for peak-to-peak signal-rounding effects below 7 bits.

In practice, the quantization error is not white noise—as, for instance, when a low-frequency signal is converted—but is randomized by adding a noiselike dither signal before quantizing. The minimum useful dither signal amplitude is 2 LSB,[3,4] bringing the noise power to $(q \cdot 2)/3$. This noise must be added at every point in the digital signal processing where requantization, rounding, or truncation occurs. The dynamic range of a properly dithered quantizer is now

$$S/N = (n - 1) \cdot 6.02 + 1.76 \text{ dB} \tag{2.2}$$

Sometimes the dynamic range of digital reproduction equipment is specified as the ratio between the maximum sine-wave amplitude and the noise level with no digital activity. This is done mainly to boost specification figures but is not the situation when actual signals are processed.

## 2.2.5   Linearity Errors and DC Offsets

Practical converters exhibit an additional discrimination level uncertainty that is expressed as deviation from the ideal quantizer curve. The resolution of a converter is defined by the integral linearity error, the differential linearity error (which is very important for sound quality), and the DC offset and DC gain errors (which for audio are less important). The integral linearity error is the deviation from a constant-step staircase and is usually specified to be smaller than $\frac{1}{2}$ LSB. The differential linearity error is the step size deviation; provided it is less than 1 LSB step at any point in the characteristic, monotonic behavior is guaranteed. Converters that fulfill the absolute linearity error requirement can show a maximum differential error of almost 1 LSB, as shown in Fig. 2.3.

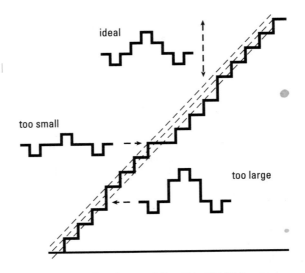

ideal

too small

too large

*conversion characteristic within 1/2 LSB IL spec.*

**FIGURE 2.3**   Differential error.

For high-level signals, the differential errors also behave as random noise, which must be added to the theoretical quantization noise. In a full-specification converter, this added noise can be equal to the theoretical noise energy. With low-level signals, however, only a few steps are used, so the error amplitude in relation to the signal amplitude is more dominant and more strongly correlated with the signal, leading to harmonic distortion. This implies, for instance, that for a 16-bit signal, the amplitude of a sine wave at $-90$ dB level (which would look like a 4-LSB-step signal when converted by a full-specification, 16-bit converter) can vary from $-85.9$ dB to 98.2 dB, depending on the DC bias level, as shown in Fig. 2.3. To reduce this phenomenon, the amplitude of the dither signal has to be increased to at least four quantization steps, which makes every step the average of a number of successive steps but also reduces the dynamic range by another 6 dB. Because a large discrepancy exists between low- and high-frequency resolution, especially in the case of high-frequency converters, the definition of effective bandwidth was introduced. This is the bandwidth at which the resolution or linearity has diminished by 3 dB ($\frac{1}{2}$ bit).

## 2.2.6  Antialiasing Filtering

For high-resolution A/D converters, the analog input filters with the required dynamic range as well as the required analog accuracy are either active filters—thus power consuming and constructed mostly of frequency-dependent, negative-resistance networks with discrete components[5]—or passive and thus bulky. The filter requirements are determined by the level of out-of-band signals that need to be suppressed to prevent aliasing. Because these levels are not known, the filter in general shows a steep roll-off and a stopband attenuation in relation to the required resolution.

For a 16-bit audio system, this means a minimum of about 100-dB attenuation. Such a specification is much easier to fulfill if the filtering can be done in the digital domain. Digital filters, however, can be implemented only up to half the sampling frequency on which the filter is operating.[5] The converter sampling frequency thus has to be increased to a multiple of the required sampling frequency. This oversampling factor allows a trade-off between analog filter, digital filter, and converter requirements. When an oversampled converter is used, a band-limited dither signal (which contains noise only in the frequency range above the signal band) can be applied to reduce the loss of dynamic range that would occur if white noise were to be applied. This technique is called out-of-band dither. It is fair to say that true high-quality audio became feasible only when oversampling converters became available.

A block diagram of an oversampling A/D conversion chain is shown in Fig. 2.4. Oversampling allows a much wider transition band of the analog filter, thereby reducing its complexity and thus the number of components contributing to the in-band noise. The amount of high-frequency noise aliased into the audio band from the input filter is also decreased. The required filtering is now done by a digital low-pass filter, after the A/D converter and prior to sample-rate reduction. One has to bear in mind, however, that though calculations can be executed with sufficient word length, the digital filter adds quantization noise due to rounding or truncation at the output where dither needs to be added before rounding.

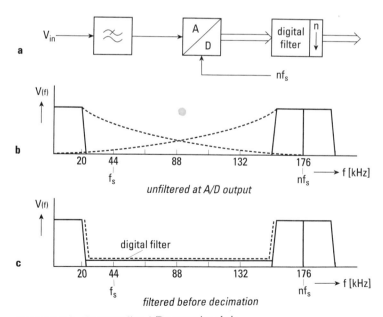

**FIGURE 2.4**  Oversampling A/D conversion chain.

## 2.2.7   Output Filtering in D/A Converter Systems

In digital-to-analog (D/A) converter systems, low-pass filtering is needed to suppress the repeated spectra around the sampling frequency and its multiples. A D/A converter system is shown in Fig. 2.5. Basically, this is the inverse function of the A/D converter of Fig. 2.4, but with different filter requirements.

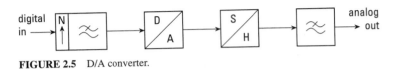

**FIGURE 2.5**   D/A converter.

At the A/D side, the filter requirements are about 100 dB of stopband attenuation. At the D/A side, however, the attenuation of the low-pass filter is determined by the ability of the associated audio equipment to handle high-frequency input signals without intermodulation distortion. A minimum attenuation value for audio applications is 50 dB. The D/A converter also performs a sample-and-hold function during the final conversion of the analog signal to an analog waveform. This zero-order hold operation is part of the filter operation and introduces extra amplitude distortion with a transfer characteristic given by

$$|H(\omega)| = \frac{\sin \omega (\tau_h/2)}{\omega (\tau_h/2)} \tag{2.3}$$

When a hold function is applied over the sampling interval at a sampling frequency of 44.1 kHz, a 3-dB loss at 20 kHz results. This loss can be reduced by shortening the hold time, which is usually done in combination with oversampling and digital filtering. The frequency spectra are shown in Fig. 2.6. At the output of the digital filter, rounding takes place to adapt to the input word length of the D/A converter. Again, quantization noise is generated, but now the noise energy is spread over the band up to half of the new sampling frequency, so that only a fraction of the noise energy is added in the audio band. Out-of-band dithering can also be applied to prevent further dynamic range reduction.

## 2.2.8   Dynamic Error Sources

***Sampling Time Uncertainty.***   Unless sampling occurs at regular time intervals, an error is introduced by time jitter that results in an amplitude error of the digital signal. The maximum allowed sampling time jitter depends on input signal bandwidth and converter resolution. For a sine-wave input signal with amplitude $A$ and frequency $f$, the maximum slope occurs at the zero crossing and is equal to $2\pi A f$. When the maximum amplitude $A$ is expressed in quantization steps, this gives:

$$\Delta t = 2^{-n}/\pi f_{\text{analog}} \tag{2.4}$$

When applied to audio signals, it follows that for a 20-kHz sine wave with 16-bit resolution, a sampling-time inaccuracy of 250 ps results in one LSB error, when sampled at the zero crossing. Because this error amplitude is signal dependent, distortion products are generated. To achieve minimum jitter, the clock signal must be obtained from a crystal-controlled oscillator (with the maximum voltage swing as allowed for aging) followed by high-speed comparators, mainly in HCMOS logic, to reduce the jitter introduced by the input noise of the comparator.[6]

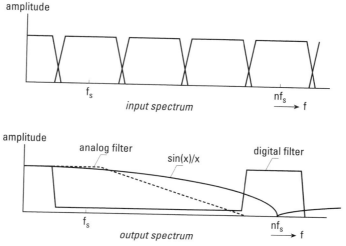

**FIGURE 2.6**   Frequency spectra.

***Switching Errors.***   Glitches are transient overshoot effects caused by time differences that occur most noticeably when different currents or voltages are switched and the delays and slopes of the "on" and "off" transitions are not equal or symmetrical. The classical example is the glitch that appears in a D/A converter using binary-weighted current sources when switching around the MSB level. The glitch energy can be reduced by an extra deglitcher, such as a sample-and-hold circuit after the D/A converter, or by segmentation of the MSB current into a number of equal currents in combination with oversampling. Low-pass filtering does not decrease the glitch energy and has no effect on audibility. The accumulated glitch energy at the output increases when dither is applied because the number of transitions increases.

## 2.2.9   Slew-Rate Distortion

A D/A converter or deglitcher sampling circuit generates current pulses with typical slopes of a few nanoseconds or less to an inverting amplifier. The current pulse amplitude varies from zero to maximum peak-to-peak level, depending on signal frequency, amplitude, and oversampling rate. As the amplifier bandwidth is always low compared with the transition speed, the error voltage at the input of the amplifier is determined by the high-frequency output impedance of the amplifier and the feedback network. If the error voltage exceeds the linear operating range of the amplifier input stage, slew-rate distortion results. By using sufficient oversampling, the difference between successive samples—and thus the input error signal—can be reduced.

## 2.3   CONVERTER ARCHITECTURES

The implementation of the D/A converter is the key factor for the performance of the complete A/D and D/A converter system. A/D conversion is always achieved by comparing the input signal with the output of a D/A converter, usually in some kind of feedback loop configuration. Different topologies have been used in the past for high-resolution, low-speed converters (audio) and for low-resolution, high-speed converters (video). Developments in silicon processes have allowed trade-offs between analog circuit complexity, required accuracy, and circuit speed.

### 2.3.1  Binary-Weighted Current Sources

The resistor-ladder converter is the best known and most widely used converter. The most common resistor network is the $R$-$2R$ ladder. In IC processing, matching limits are between 10 and 12 bits. As batch processing does not achieve highly accurate resistor ratios, some kind of trimming is used either in the factory by, for example, laser trimming or zener zapping[7] or off-line, using a small calibration D/A and a lookup table.[8] Differential nonlinearity and glitch errors are most likely to occur around the level where the output is switched from the MSB to the sum of the remaining bits. Often the MSBs are divided between a number of equal resistors and switched by a thermometer code to improve differential linearity. When MOS switches are used, the network is directly connected to the input of a summing amplifier. In a bipolar solution, the resistor network is cascaded by a row of transistors, with emitter scaling to obtain equal base-emitter voltages, as shown in Fig. 2.7.[9]

In transistor-only solutions, binary-weighted currents are obtained by grouping equal-sized transistors from a matrix. When parameter spread is random, accuracy improves with the square of the number of transistors in parallel. By combining the transistors evenly over an area, linear gradients are compensated. In this way, 10- to 12-bit resolution can be obtained. The remaining 6 to 4 bits are segmented into equal currents, again by using a thermometer code, as shown in Fig. 2.8.[10]

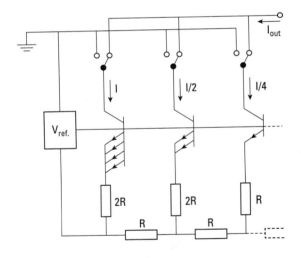

| fab. process | matching tolerance | | | |
|---|---|---|---|---|
| | σ (%) | | mean (%) | |
| | 10μ | 40μ | 10μ | 40μ |
| diffusion | 0.44 | 0.23 | −0.1 | 0.07 |
| thin film | 0.24 | 0.11 | −0.1 | −0.06 |
| ion implant | 0.34 | 0.12 | 0.05 | 0.05 |

**FIGURE 2.7**   Binary weighting: a bipolar solution.

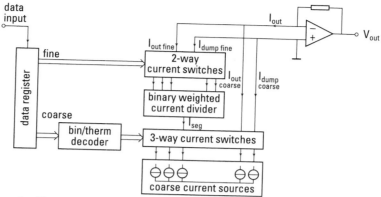

*simplified block diagram*

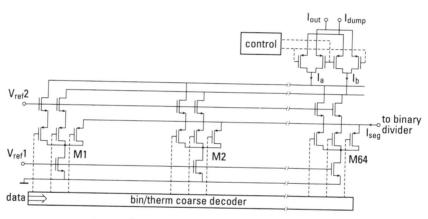

*coarse current network*

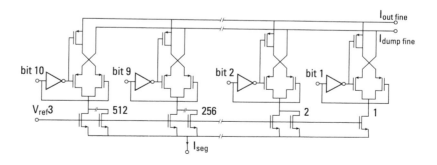

*fine binary weighted current divider*

**FIGURE 2.8**   Binary weighting: a transistor-only solution.

For small signals, sign and magnitude coding is much more attractive. With offset binary or 2's complement signals, the signal is centered around the MSB. As a result, a large current or voltage is always generated, with its relatively high thermal noise level, even when small signals are generated. The integral precision requirements of the bits used for coding low-level signals are small, and the differential linearity in this part of the characteristic is still good, deteriorating when the signal level increases. However, good integral linearity depends on the matching of positive and negative amplitudes, so the degree of distortion at full signal level is critical.

### 2.3.2   Integrating D/A Converters

An example of an integrating D/A converter is shown in Fig. 2.9. The digital input word is stored in a register and compared with the content of a counter operating from a crystal-controlled oscillator. In this part of the circuit, the input data are converted into a time signal by the digital comparator. This time signal switches a reference current into an integrating amplifier, which converts the current to a voltage across a capacitor. At the end of the integrating cycle, the output voltage is sampled in a sample-and-hold circuit. At the beginning of the following cycle, the sample-and-hold circuit is in hold mode, and the integrator is reset.

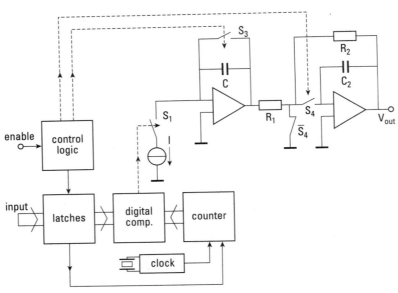

**FIGURE 2.9**   Integrating D/A converter.

This type of converter achieves high linearity but is limited in conversion speed. To increase the speed, a dual ramp converter can be used. The input data are split into coarse and fine values. At the same time, the reference current is split up into coarse and fine values. For instance, a 16-bit word is split into two 8-bit words and two currents are used with a ratio of 256:1. The conversion speed now improves 256 times, but at the cost of an increased accuracy requirement. The error between one count of the coarse current and a full 256 counts of the fine current should be smaller than $\frac{1}{2}$ LSB, giving an accuracy requirement for the current ratio of 0.2 percent. The time uncertainty with which the current can be switched can be calculated. For a sampling frequency of 44.1 kHz, 8 bits resolution in time counting, 8 bits in time accuracy, and half the sampling period devoted to generating the analog value, the time accuracy for $\frac{1}{2}$ LSB error is

$$\Delta\tau = \frac{1}{f_s \cdot 2^n \cdot 2^n \cdot 2 \cdot 2} = 86.5 \text{ ps} \tag{2.5}$$

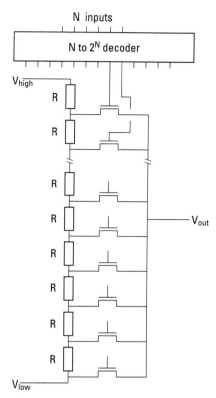

N inputs

N to $2^N$ decoder

$V_{high}$

$R$

$R$

$R$

$R$

$R$ — $V_{out}$

$R$

$R$

$R$

$R$

$V_{low}$

**FIGURE 2.10**    Voltage division converter.

### 2.3.3  Voltage-Division Converters

A tapped string of equal resistors with switches connecting the taps to a voltage follower makes an inherent monotonic D/A converter with very good differential linearity and integral linearity depending on resistor accuracy. This type of converter is mainly found in high-speed (video) applications, with moderate 10- to 12-bit integral linearity (Fig. 2.10),[11] while two-stage solutions have been used in high-resolution applications.[12]

### 2.3.4  Self-Calibrating Converters

If a D/A converter is not continuously in use, a calibration procedure can be used to eliminate inaccuracies of the elements.[8] The calibration reference must be linear in construction, like a single-ramp integrating converter. The output signal of the D/A converter is compared off-line with the reference, and a correction value is added to the output signal via a small subsidiary D/A converter. This correction value is stored for every weighting value of the main D/A converter. During calibration the converter cannot be used. To overcome this problem, an accurate converter can be constructed by calibrating the individual elements, while using spare elements to take the place of the elements under calibration.[13] In a MOS process, it is possible to use a charge storage principle in an accurate current calibration system. In Fig. 2.11 the basic calibration and the operational cycle are shown.

During calibration of a MOS current source, the MOS transistor is connected as a diode via switches $S_1$ and $S_2$ to a reference current source $I_{ref}$. When switch $S_1$ opens, the gate-source voltage of the transistor stays constant. Switch $S_2$ is then switched to the output. At that moment the drain current is equal to $I_{ref}$, and another MOS transistor can be connected to $I_{ref}$. When $N$ equal current sources are needed, $N + 1$ sources are used in the calibration sequence, so that $N$ calibrated sources are always available. A block diagram of such a continuous calibration system is shown in Fig. 2.12. The accuracy has proven to be in the 0.001 percent range. However, due to the calibration cycle, the total noise of all the sources within the calibration circuit will be subsampled, increasing the noise of a single calibrated current cell.

## 2.3.5  Error Averaging

The difference between two almost equal currents can be reduced by the chopping technique called dynamic element matching.[14] Figure 2.13*a* shows a simplified circuit diagram consisting

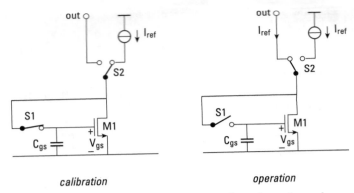

*calibration*          *operation*

**FIGURE 2.11**  Self-calibrating converter: calibration and operating cycle.

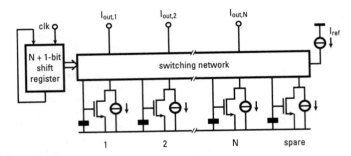

**FIGURE 2.12**  Continuous current calibration system.

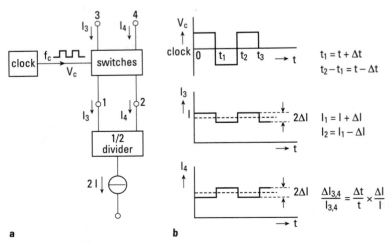

**a**                    **b**

**FIGURE 2.13**  (*a*) Dynamic element matching; (*b*) dynamic element-matching output currents.

of a passive current divider and a set of switches driven by a clock generator. The passive divider splits the total current $2I$ into two nearly equal parts, $I_1 (= I + \Delta I)$ and $I_2 (= I - \Delta I)$. $I_1$ and $I_2$ are interchanged during equal time intervals with respect to output terminals 3 and 4. The output currents at these terminals are shown in Fig. 2.13b. If there is also a slight difference in clock time periods, $t_1 (= t + \Delta t)$ and $t_2 (= t - \Delta t)$, the output current at terminal 3 over the period $2t$ will be

$$I_3 = \frac{(t + \Delta t)(I + \Delta I)}{2t} + \frac{(t - \Delta t)(I - \Delta I)}{2t} = I\left(1 + \frac{\Delta t \cdot \Delta I}{t \cdot I}\right) \tag{2.6}$$

and the output current at terminal 4 over the period $2t$ will be

$$I_4 = \frac{(t - \Delta t)(I + \Delta I)}{2t} + \frac{(t + \Delta t)(I - \Delta I)}{2t} = I\left(1 + \frac{\Delta t \cdot \Delta I}{t \cdot I}\right) \tag{2.7}$$

As can be seen from Eqs. (2.6) and (2.7), the final accuracy is determined by the product of two small errors. An overall accuracy of better than 0.001 percent can be obtained with normal parameter spreads. The ripple in the output currents is removed by (mainly external) filter capacitors.

## 2.4   APPLICATION IN A/D CONVERTERS

In the classical successive-approximation A/D converter, the output of a binary-weighted current or voltage network is compared with the input signal, starting with the MSB and adding bits until the error is smaller than 1 LSB. The block diagram of a converter and a typical decision sequence during the successive-approximation cycle are shown in Fig. 2.14.

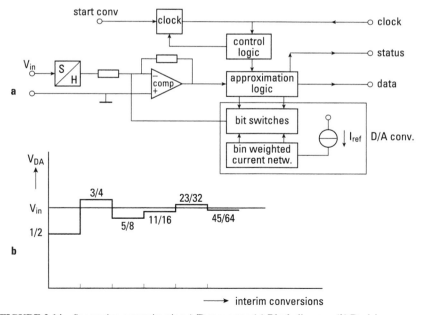

**FIGURE 2.14**   Successive-approximation A/D converter. (*a*) Block diagram. (*b*) Decision sequence.

The number of clock cycles needed for one conversion is dependent on the number of bits and the settling time of the preceding sample-and-hold circuit. For 16-bit resolution, this limits the conversion speed to about 100 kHz. The necessary bandwidth of the sample-and-hold amplifier is about 10 MHz, which means that high-frequency noise is subsampled into the baseband. The noise in the baseband now increases by a factor

$$\sqrt{N_{\text{fold}}} = \sqrt{\frac{2f_{sh}}{f_s} - 1} \qquad (2.8)$$

in which $f_{sh}$ is the bandwidth of the amplifier section that contributes to the sampled white noise, and $f_s$ is the sampling frequency. In this case, the white-noise energy of the amplifier in the baseband increases by about 23 dB.

In an integrating converter, the integrating network is also part of the input sample-and-hold circuit. Recently, however, a totally different type of converter has become increasingly popular because it combines oversampling with reduced precision requirements.

## 2.5 OVERSAMPLED NOISE-SHAPING CONVERTERS

### 2.5.1 Principles

First described in 1946,[15] and in detail in 1950 by de Jager of Philips Research Laboratories, a method called delta modulation was used in telephone transmission to extend the dynamic range of a 1-bit quantizer by applying it in a frequency-dependent feedback loop together with oversampling.[16] The frequency characteristic of the filter in the feedback loop was compensated for by placing a similar filter at the receiving side. In 1960 Inose[17] proposed sigma-delta modulation, which, by placing the filter in front of the quantizer, gave a frequency-independent transfer characteristic. The basic block diagram of sigma-delta modulation is shown in Fig. 2.15a.

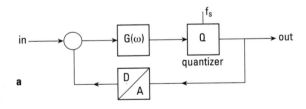

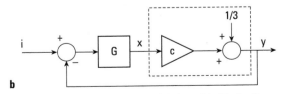

**FIGURE 2.15** Sigma-delta modulation. (a) Block diagram. (b) Linearized noise model.

The modulator consists of a feedback loop in which a quantizer reduces the word length to 1 bit and a loop filter $G$ that exhibits a low-pass characteristic. The modulator output is a stream of 1-bit words, which gave rise to the name *bitstream converter*. The feedback causes the average value of the output bitstream to follow the input signal, and a pulse-density-modulated signal is created. The integral and differential linearity of the system is independent of accurate analog components but is solved completely in the time domain.

Figure 2.15*b* shows the linearized noise model of the 1-bit quantizer. It is replaced by a gain stage with gain $c$ and a white noise source with a noise power of $\frac{1}{3}$ determined by the quantization step of 2 (the step is from $+1$ to $-1$). The average gain of the quantizer $c$ is determined by the power balance at the quantizer output. The output of the modulator is 1 and the peak level of a coded sine wave is 1, so the output noise power and the signal power are both 0.5. With smaller input signals, the noise power rises to 1 and the loop circulates almost only quantization noise. When the low-frequency gain of the loop filter-quantizer combination $cG$ is large, the transfer characteristic can be approximated by

$$Y = \frac{cG}{1 + cG} \cdot I + \frac{1}{1 + cG} \cdot N \qquad (2.9)$$

The quantization noise energy is now shaped in frequency and strongly suppressed at low frequencies by the filter. Furthermore, if the quantization noise is assumed to be white, the noise energy in the audio band is also reduced by oversampling because it is spread over a larger bandwidth. Figure 2.16*a* shows the noise density at the modulator output for loop filters from the first order to the fourth order at $128\times$ oversampling. As a result of the noise shaping, the resolution for large oversampling ratios increases by $(n + \frac{1}{2})$ bits (where $n$ is the filter order) for every factor-of-2 increase in sampling frequency. Maximum $S/N$ ratios as a function of oversampling ratio and filter order are shown in Fig. 2.16*b*.

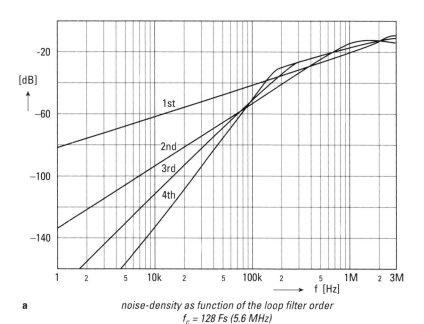

**a**          *noise-density as function of the loop filter order*
                          $f_c = 128\ Fs\ (5.6\ MHz)$

**FIGURE 2.16**   Noise shaping. (*a*) Effect of filter order on noise density.

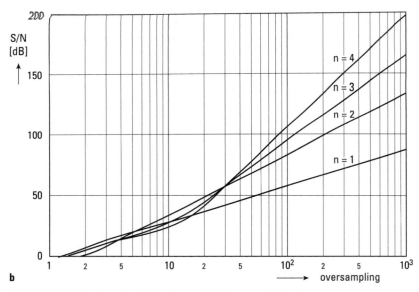

**FIGURE 2.16 (*Continued*)**   Noise shaping. (*b*) Maximum S/N ratios.

All quantization levels are obtained by averaging (low-pass filtering) over a large sequence of samples of the same step size. Also, when high oversampling ratios are used both in the A/D and D/A converter systems, most of the low-pass filtering can be done in the digital domain. But in this case too, performance in practice differs from theory.

### 2.5.2  Noise Correlation

In practice, a sigma-delta modulator cannot handle signals with an amplitude equal to the quantization step because at high signal levels the quantization error increases greatly, as shown in Fig. 2.17*a*, for a second-order modulator with a sampling frequency of 5 MHz. The quantization error rapidly becomes correlated with the signal and contains more low-frequency components. Figure 2.17*b* shows the input signal of the quantizer during one period of a 5-kHz sine wave with an input level of $-0.7$ dB. The overload area is avoided by scaling the input signal.

Correlation of the signal and the quantization error can also occur with small signals. This is easily demonstrated by analyzing a first-order loop (Fig. 2.18). When zero input signal is applied, the modulator generates a fixed pattern alternating between $+1$ and $-1$. To disturb this pattern, the input signal needs to generate at least a threshold value of 1 at the input of the quantizer. This means that for a small input signal a time $T$ elapses before the output responds. The output bitstream will contain a lot of harmonic components and discrete components with a frequency of $\frac{1}{2}T$. When the input amplitude decreases, $T$ increases and in-band harmonics and whistles are generated. By using higher-order modulators, however, the amplitudes of these harmonics and whistles are much reduced.

The application of dither improves linearity, as shown in Fig. 2.19 for a second-order modulator and a sine-wave input signal with and without dither. Provided that the dither signal is applied out of band, the dynamic range in the passband is hardly affected.

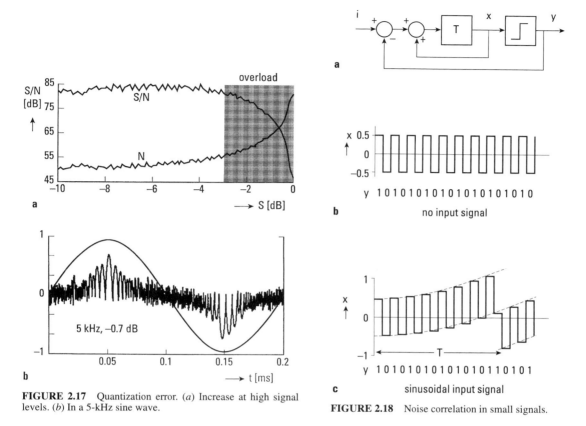

**FIGURE 2.17** Quantization error. (*a*) Increase at high signal levels. (*b*) In a 5-kHz sine wave.

**FIGURE 2.18** Noise correlation in small signals.

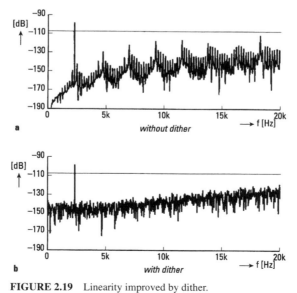

**FIGURE 2.19** Linearity improved by dither.

### 2.5.3 *Stability*

Because a sigma-delta modulator is allowed to show repetitive patterns, a different definition of loop stability is necessary. A modulator is said to be stable when it codes the input signal by means of high-frequency output patterns. If the modulator is brought into an unstable situation, it will generate longer sequences of equal sign pulses and the signal amplitude inside the loop will increase sharply until clipping occurs.

First-order loops are always stable but do not perform well. Second-order loops can be stable, depending on the filter coefficients. Higher-order loops become unstable under overload conditions and stay so after removal of the input signal. This instability can be suppressed by the application of limiters inside the loop.[18-19] Figure 2.20 shows the input and output signals of the quantizer in a third-order modulator, excited by one period of an input sine wave of too large amplitude (case A) and moderate amplitude (case B). In case A, clipping occurs at the (arbitrary) value 150 and a low-frequency pattern is generated, which stays after the input signal has returned to zero. When the clipping level is reduced to the (arbitrary) value 75 (case B), the modulator returns to its high-frequency idling patterns when the input signal returns to zero.

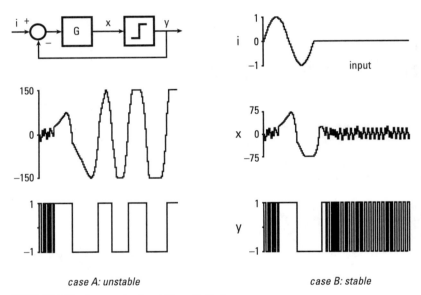

**FIGURE 2.20**   Suppressing instability with limiters.

The proper clipping levels can be calculated using the root-locus methods, well known in control theory for determining the stability of linear systems as a function of varying parameters.[18]

In a sigma-delta modulator, the varying parameter is the quantizer gain $c$. In principle, $c$ can have any positive value, depending on the value of the quantizer input signal. With high input signal levels, the quantizer input peaks and the gain $c$ drops. Because of this gain variation, higher-order modulators can become unstable due to overload or during start-up.

As an example, we may analyze a third-order modulator with loop filter transfer function $G = T/N$. The denominator of the feedback system is given by ($N = cT$), and the system is stable when the roots are within the unity circle. Figure 2.21 shows the root-loci for positive $c$ values. Note the trespassing of the unity circle for $c = 0.013$. The left root-locus through $z = -1$

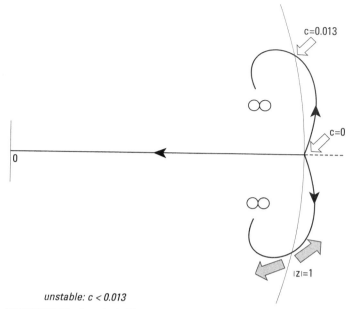

unstable: c < 0.013

**FIGURE 2.21**   Root-loci, third-order sigma-delta.

corresponds to an oscillation at half the sampling frequency, which is the oscillation required. Furthermore, when the quantizer gain falls below 0.013, a growing low-frequency oscillation is initiated. By introducing a limiter with the value $1/c$ in the last integrator, the quantizer gain is always maintained larger than the critical value, and the modulator stays stable.

Though more complex loop filters perform better with respect to quantization noise, the maximum allowable signal level also has to be reduced for stability reasons. This means that when thermal and other noise sources are taken into account, the dynamic range might well be reduced instead of increased.

### 2.5.4  Implementation

With all poles on the real axis, the noise density in the passband will rise sharply with frequency. By using complex pole pairs, the noise density in the passband can be shaped more or less flat, giving a lower integral noise level. The maximum signal level and quantizing noise level for a fifth-order coder at $64 f_s$ with different stability criteria, for all poles at the real axis and for one real and two complex pole pairs, are shown below.

| Parameter set | $S_{max}$ (dB) | Noise with real poles (dB) | Noise with complex poles (dB) |
|---|---|---|---|
| A | −3.5 | −97 | −114 |
| B | −6 | −112 | −128 |
| C | −9 | −120 | −138 |

In Fig. 2.22, the noise density spectra are shown for parameter set B.

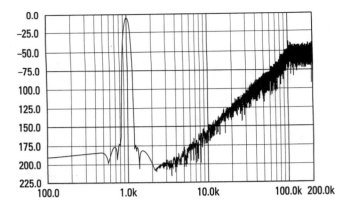

*5th order noise shaper, all poles on the real axis, 64 Fs = 2.82 MHz*

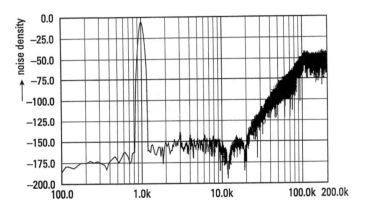

*5th order noise shaper, one real, two complex pole pairs, 64 Fs = 2.82 MHz*

**FIGURE 2.22**   Noise density spectra.

Implementation is mainly done in the switched-capacitor technique but sometimes by switched current sources. Distortion can occur from intersymbol interference if the analog values of successive samples depend in some way on the previous samples. This can occur with the switched-capacitor technique because of settling errors in the associated circuitry. When currents are switched, the difference in integral error between the rising edge and the falling edge can generate substantial distortion. If distortion occurs, this generates harmonics of the input signal and high-frequency quantization noise products, which are aliased back into the passband. This phenomenon creates extra noise as well as discrete tones. To avoid this, various types of return-to-zero pulses can be used. To reduce the high-frequency noise generated by 1-bit D/A converters at the output, some analog filtering has to be applied at the output. To reduce this to a bare minimum, the D/A conversion can be combined with digital FIR filtering.[20] One-bit data words are fed into a shift register, while at the taps the filter coefficients are implemented as an analog-weighted current source or weighted switched-capacitor array.

### 2.5.5.  Multistage (MASH) Sigma-Delta Modulators

A different method of constructing stable higher-order modulators is by using the multistage sigma-delta modulator or MASH converter.[21] Each stage of the MASH converter of Fig. 2.23 is a basically stable first- or second-order modulator. The difference between the input and output of the first quantizer is converted by the second modulator, and so on. As the output signals of the second and third modulator are taken after the integrator of the previous stage, they have to be differentiated and corrected for gain before addition. The output signal is thus a multilevel signal. The input levels of the second and third sigma-delta modulator loops have to be scaled to prevent overload. The summed output signal is an attenuated version of the input signal.

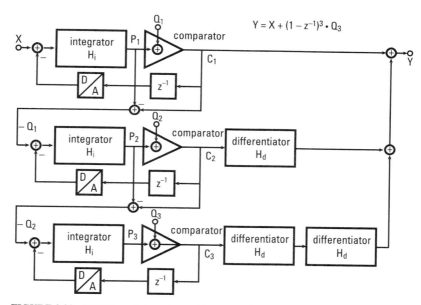

$$Y = X + (1 - z^{-1})^3 \cdot Q_3$$

**FIGURE 2.23**   MASH multistage sigma-delta conversion.

When this principle is used for D/A conversion, the coder is followed by a pulse-width modulator to obtain two-level output signals.[22] The actual D/A converter circuit is mostly implemented as a set of switches operated in a balanced configuration. By using pulse-width modulation with less than 100 percent modulation depth, a rising edge and a falling edge are always present in the output signal, thus preventing interference between successive samples.

## 2.6  HIGH-FREQUENCY CONVERTERS

The most common solution for video A/D and D/A converters is the full-flash A/D converter. This has as many input comparators or quantization levels, combined with a series resistor voltage divider, as a reference ladder or D/A converter, as shown in Fig. 2.24. The analog input signal is converted into a thermometer code by comparators and latches. A ROM structure is used to change the thermometer code into a binary-coded output word. The largest source of differential nonlinearity is now the DC offset of the comparators. Usually, an offset compensating scheme is used.[23,24]

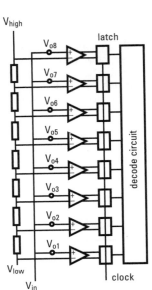

**FIGURE 2.24**    Full-flash A/D converter.

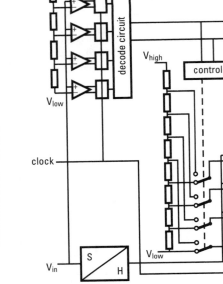

**FIGURE 2.25**    Two-stage full-flash A/D converter.

The real performance limitations are dynamic errors caused by timing errors, capacitive feedback, and delay mismatches. In coding video signals with a bandwidth of 5 MHz and 10-bit resolution, 1 LSB equals a timing error of 62 ps. When the input signal is converted directly, all delays in the comparators need to be carefully matched and optimized. This problem is commonly solved by putting a sample-and-hold operation in front of the converter, which has the additional effect of improving maximum speed. The reference signal can be disturbed by capacitive feedback of the comparator output signals, the so-called kickback noise. Because the amount of feedback depends on signal amplitude, distortion is introduced. A full-flash converter presents a considerable capacitive load at its input, thus requiring a buffer amplifier that can handle a signal-dependent load. The effects of dynamic errors such as clock skew to the comparators and comparator aperture time generate signal-dependent quantization errors.

To reduce the number of clocked comparators, which reduces such factors as the area required and the digital interference, analog encoding of the comparator input signals is used. Implementations of both Gray code[25,26] and circular code[27] have been employed. Two-stage conversion (Fig. 2.25), is also frequently applied to reduce the number of comparators. In this method, a first flash A/D converter performs the MSB conversion step and, via a D/A converter, the digital output signal is subtracted from the input signal in the analog domain. A second A/D converter then performs the LSB conversion on the remainder. The resolution of the second A/D converter is chosen to be slightly larger than necessary to cover one MSB step. The overlap in the digital output range is needed to compensate for gain mismatches. When the subtraction circuit is combined with a second sample-and-hold stage, the second conversion can be performed while the first sample-and-hold stage takes in a new sample. This pipelining compensates for the additional clock cycles required. With two-stage conversion, a reduction from 255 to 40 comparators can be obtained for an 8-bit converter.[28–30]

The most popular D/A converter configurations for video applications are the resistor string converter followed by a noninverting amplifier, and the (partially) segmented switched current sources, which are summed either into an inverting amplifier or straight into a 50- or 75-$\Omega$ termination resistor.[31,32]

## 2.7 CONCLUSION

The construction of low-frequency, high-resolution A/D converters has shifted almost without exception from mainly analog solutions toward highly oversampled noise-shaping converters with greatly reduced precision requirements and with almost all the filtering done in the digital domain. This trend has matched the developments of higher speed and increasing packing density in IC processes. On the D/A converter side, the trade-offs have been less obvious because the noise-shaping converter generates a large amount of high-frequency noise, which still requires analog low-pass filtering.

All the current D/A products developed for sound reproduction use digital filtering and up-sampling. The converters applied range from analog "20-bit resolution" to noise-shaping, single-bit converter types.

For high-speed applications, the trend of development is expected to be similar, though the accuracies required are still achievable in IC processing. Video converters have moved from bipolar to CMOS processes, although higher-bandwidth applications still rely on bipolar solutions or gallium arsenide JFET technology.

The trend in recent years has been toward integration of conversion and digital signal processing. The process developments needed for higher digital packing densities in the future, however, will make these processes less suitable for analog functions. The place of the A/D converter in consumer electronic systems is shifting to the conversion of IF and RF signals, while investigations to combine the D/A function with output power functions are reported in many publications.

## GLOSSARY

**Aliasing**   The (mostly unwanted) shifting of information from one frequency band to another one due to sampling.

**Clock skew**   Circuit- and device-dependent variable delay of a timing signal.

**Correlation**   The degree of quantitative association between two variables.

**Glitch**   Short settling error when switching from one state to another.

**Intersymbol interference**   An error in the value of a sample caused by storage effects (settling errors) of previous samples.

**Jitter**   Variation of the sampling interval due to noise or crosstalk.

**Nyquist theorem**   The frequency at which samples are taken from a signal must be at least twice the highest frequency of the signal to be sampled to allow reconstruction of the original signal from a sequence of samples.

**Oversampling**   Sampling a signal at a rate higher than the Nyquist rate.

**Quantizer**   A nonlinear system whose purpose is to transform the input sample into one of a finite set of prescribed values.

**Requantization**   Transforming a quantized value into a different (mostly smaller) finite set of prescribed values.

**Zener zapping**   A method to create a short between two conductors by applying a voltage exceeding the zener breakthrough voltage over a bipolar junction with sufficient power to melt the metal.

## REFERENCES

1. M. Schwartz, *Information Transmission, Modulation and Noise,* McGraw-Hill, New York, 1980.

2. A. B. Carlson, *Communication Systems,* McGraw-Hill, New York, 1975.

3. J. Vanderkooy and Lipshitz, S. P., "Digital dither," *Journal of the Audio Engineering Society (Abstracts),* vol. 34, Dec. 1986, p. 1030, preprint 2412.

4. J. Vanderkooy and Lipshitz, S. P., "Digital dither: Signal processing with resolution far below the least significant bit," Audio Engineering Society, 7th International Conference, "Audio in Digital Times," May 14–17, 1989.

5. M. S. Ghausi and Laker, K. R., *Modern Filter Design,* Prentice Hall, Englewood Cliffs, N.J., 1981.

6. R. J. van de Plassche, *Integrated Analog-to-Digital and Digital-to-Analog Converters,* Kluwer Academic Publishers, Norwell, Mass., 1994.

7. J. R. Naylor, "A complete high-speed voltage output 16-bit monolithic DAC," *IEEE Journal of Solid State Circuits,* vol. SC-18, no. 6, Dec. 1983, pp. 729–735.

8. K. Maio, Hotta, M., Yokozama, N., Nagata, M., Kaneko, K., and Iwasaki, T., "An untrimmed D/A converter with 14-bit resolution," *IEEE Journal of Solid State Circuits,* vol. SC-16, no. 6, Dec. 1981, pp. 616–621.

9. G. Kelson, Stelrecht, H. H., and Perloff, D. S., "A monolithic 10-bit digital-to-analog converter using ion implantation," *IEEE Journal of Solid State Circuits,* vol. SC-8, Dec. 1973, pp. 396–403.

10. H. J. Schouwenaars, Groeneveld, D. W. J., and Termeer, H. A. H., "A low-power stereo 16-bit CMOS D/A converter for digital audio," *IEEE Journal of Solid State Circuits,* vol. SC-23, no. 6, Dec. 1988, pp. 1290–1297.

11. M. J. M. Pelgrom, "A 10-bit 50 MHz CMOS D/A converter with 75 ohm buffer," *IEEE Journal of Solid State Circuits,* vol. 25, Dec. 1990, pp. 1347–1352.

12. P. Holloway, "A trimless 16-bit digital potentiometer," *ISSCC Digest of Technical Papers,* Feb. 1984, pp. 66–67.

13. D. W. J. Groeneveld, Schouwenaars, H. J., and Termeer, H., "A self calibration technique for monolithic high-resolution D/A converters," *IEEE Journal of Solid State Circuits,* vol. SC-24, Dec. 1988, pp. 1517–1522.

14. R. J. van de Plassche and Goedhart, D., "A monolithic 14-bit D/A converter," *IEEE Journal of Solid State Circuits,* vol. SC-14, June 1979, pp. 552–556.

15. E.M. Deloraine, van Miero, S., and Derjavitch, B., "Methode et systeme de transmission par impulsions," French Patent No. 932.140, 1946.

16. F. de Jager, "Delta modulation, a new method of PCM transmission using the 1 unit code," Philips Research Report, no. 7, Dec. 1952, pp. 442–446.

17. H. Inose, and Yasuda, Y., "A communication system by code modulation Delta-Sigma," *Journal of the Institute of Electrical Engineering* (Japan), vol. 44, no. 11, Nov. 1961, pp. 1775–1780.

18. S. K. Tewksbury, and Hallock, R. W., "Oversampled, linear predictive and noise-shaping coders of order $N > 1$," *IEEE Transactions on Circuits and Systems,* vol. CAS-25, no. 7, July 1978, pp. 436–447.

19. E. F. Stikvoort, "Some remarks on stability and performance of the noise shaper or sigma-delta modulator," *IEEE Transactions on Communications,* vol. 36, no. 10, Oct. 1988, pp. 1157–1162.

20. B. A. Su, and Wooley, D. K., "A CMOS oversampling D/A converter with a current-mode semidigital reconstruction filter," *IEEE Journal of Solid State Circuits,* vol. 28, Dec. 1993, pp. 1224–1233.

21. Y. Matsuya, Uchimura, K., Iwata, A., Kobayashi, T., Ishikawa, M., and Yoshitome, T., "A 16-bit oversampling A-to-D conversion technology using triple-integration noise shaping," *IEEE Journal of Solid State Circuits,* vol. 22, Dec. 1987, pp. 921–929.

22. Y. Matsuya, Uchimura, K., Iwata, A., and Kaneko, T., "A 17-bit oversampling D-to-A conversion technology using multistage noise shaping," *IEEE Journal of Solid State Circuits,* vol. 24, no. 4, Aug. 1989, pp. 969–975.

23. Y. Fujita, Masuda, E., Sakamoto, S., Sakaue, T., and Sato, Y., "A bulk CMOS 20 Ms/s 7-bit flash ADC," *ISSCC Digest of Technical Papers,* Feb. 1984, pp. 56–57.

24. M. J. M. Pelgrom, van Rens, A. C. J., Vertregt, M., and Dijkstra, M. M., "A 25 Ms/s 8-bit CMOS A/D converter for embedded application," *IEEE Journal of Solid State Circuits,* vol. 29, no. 8, Aug. 1994, pp. 879–886.

25. K. Poulton, Corcoran, J. J., and Hornak, T., "A 1 GHz, 6-bit ADC system," *IEEE Journal of Solid State Circuits,* vol. SC-22, Dec. 1987, pp. 962–970.

26. R. J. van de Plassche and van der Grift, R. E. J., "A high-speed 7-bit A/D converter," *IEEE Journal of Solid State Circuits,* vol. SC-14, Dec. 1979, pp. 938–943.

27. R. J. van de Plassche and Balthus, P., "An 8-bit 100 MHz full Nyquist analog-to-digital converter," *IEEE Journal of Solid State Circuits,* vol. SC-23, Dec. 1988, pp. 1334–1344.

28. N. Fukushima, Yamada, T., Kumazawa, N., Hasagawa, Y., and Soneda, M., "A CMOS 8-bit 40 MHz 2-step parallel A/D converter with 105 mW power consumption," *ISSCC Digest of Technical Papers,* Feb. 1989, pp. 14–15.

29. M. Ishikawa and Tsukahara, T., "An 8-bit 40 MHz CMOS subranging ADC with pipelined wideband S/H," *IEEE Journal of Solid State Circuits,* vol. SC-24, Dec. 1989, pp. 1485–1491.

30. S. H. Lewis, Fetterman, H. S., Gross Jr., G. F., Ramachandrum, R., and Viswanathan, T. R., "A 10-bit 20 Msamples/s analog-to-digital converter, *IEEE Journal of Solid State Circuits,* vol. 27, Mar. 1992, pp. 351–358.

31. T. Miki, Nakamura, Y., Nakaya, M., Asai, S., Akasaka, Y., and Horiba, Y., "An 80 MHz 8-bit CMOS D/A converter," *IEEE Journal of Solid State Circuits,* vol. SC-21, Dec. 1986, pp. 983–988.

32. P. Vorenkamp, Verdaasdonk, J., and van der Plassche, R. J., "A 1 GHz, 10-bit digital-to-analog converter," *ISSCC Digest of Technical Papers,* Feb. 1994, pp. 52–53 .

## ABOUT THE AUTHOR

Eise Carel Dijkmans joined the Philips Research Laboratories in Eindhoven, the Netherlands, in 1962 and was involved in research on telecommunications subjects such as data transmission, electronic switching networks, and companding delta-modulation converters. In 1977 he joined the Philips Research Group on Consumer Electronics, where he has been involved in research on high-resolution A/D and D/A converters. He has also worked on various audio subjects, such as bipolar integrated volume and tone controllers and integrated power amplifiers. In recent years the focus of his research has been on circuits in CMOS technology, from which emerged the bit-stream converter concepts. In 1992 he received a fellowship award from the Audio Engineering Society for his work on high-quality A/D and D/A converter systems. He holds more than 40 patents and has contributed to 15 papers on telecommunications and audio subjects.

# CHAPTER 3
# DIGITAL CODING (DATA REDUCTION) METHODS

**James D. Johnston**
*AT&T Laboratories*

**Christine Podilchuk**
*Lucent Technologies, Bell Laboratories*

**Juin-Hwey Chen**
*Voxware, Inc.*

## 3.1 WHAT IS CODING, AND WHY DO WE NEED IT?

In the field of digital signal processing, the term *coding* usually refers to the process of bit rate reduction. Coding, sometimes referred to as *compression* or *signal compression,* is necessary in applications where the storage or communications capacity available is smaller than the raw signal data rate. (*Note:* In this chapter, the word *coding* will be used in order to avoid confusion with other processing known in the audio and video arts as *compression.*)

As an example, the usual format for a digitally placed telephone call (Fig. 3.1) requires 8 bits/sample at a rate of 8000 samples/second, for a total bit rate of 64,000 bits/second (b/s). Given the continued rate of growth in telephone traffic, it is desirable to reduce the bit rate in order to allow more calls to use the same data capacity. Therefore, a compression algorithm is placed in the communications or storage path to reduce the bit rate. For instance, using various compression methods as shown in Fig. 3.2, the bit rate for the telephone call can be reduced to 32,000 b/s or 32 kb/s (G726 compression, Fig. 3.2*a*), or to 16 kb/s (G728 compression, Fig. 3.2*b*), or, alternatively, the bandwidth encoded may be increased from 4 kHz to 8 kHz (G722 compression, Fig. 3.2*c*) with a compression algorithm used to keep the total bit rate at 64 kb/s.

Table 3.1[1] lists some basic (uncoded) data rates for a variety of speech, audio, and video signals. The column labeled "Raw bit rate" shows the basic rate of the raw data signal. The column labeled "Channel capacity" shows the channel capacity of the signal as applied to that particular application. As in the case of HDTV and digital audio radio transmission, there is sometimes a large difference between the available channel or storage capacity and the actual raw data rate. For such applications, the use of compression is not merely an economic desire; it is a necessity in a practical system.

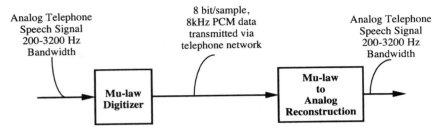

**FIGURE 3.1** Standard digital telephony without compression.

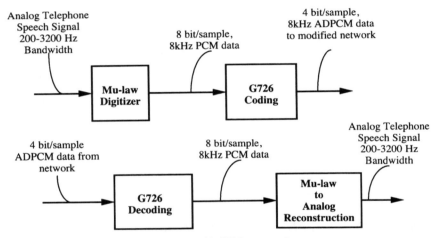

**FIGURE 3.2a** Standard digital telephony with G726 compression.

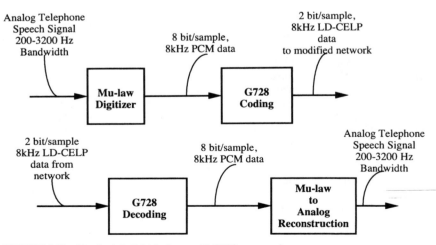

**FIGURE 3.2b** Standard digital telephony with G728 compression.

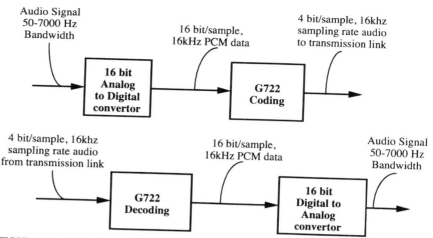

**FIGURE 3.2c**   Digital audio transmission with G722 compression.

**TABLE 3.1**   Uncoded Data Rates for a Variety of Speech, Audio, and Video Signals

| Signal application | Raw bit rate, kb/s | Channel capacity, kb/s |
|---|---|---|
| Telephone speech | 64 | 64 |
| Commentary grade speech | 192 | 64 |
| Compact Disc audio | 1411 | 1411 |
| Digital audio radio | 1411–1536 | 160–256 |
| SmartCard audio storage | 1411–1536 | 128–192 |

Some examples of established compression algorithms, showing compression ratio versus complexity, are given in Fig. 3.3a, b, and c for speech, music, and video, respectively. In Fig. 3.3, the bit rate is shown in b/s, the algorithms as data points on the graph, and the complexity in terms of operations per sample. The trade-offs between complexity, compression rate, and quality are in general quite complicated and hard to show graphically. Additionally, the kinds and amounts of coding impairments that a listener will tolerate in speech, video, and audio applications vary widely. Therefore, the wide variety of "quality" results available will not be addressed or summarized here because of the greatly different conditions, requirements, test methodologies, and subject bases involved in the various experiments. In speech, large impairments (limited bandwidth, some information loss) are very lightly penalized, while in audio a small impairment in bandwidth or nearly any audible coding artifact are grounds to rate the impairment "very annoying."

## 3.2   THE SOURCE CODER

Historically, coding methods have been based on mathematical methods using the same approach: removing redundant (or predictable) parts of the signal by creating either a mathematical or statistical model of the signal source—hence the term *source coder*. The classical forms of source-coding methods and some examples of each are listed in Table 3.2, and typical block diagrams for each of the three forms are shown in Figs. 3.4–3.6.

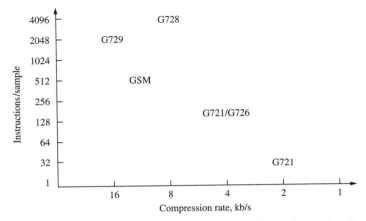

**FIGURE 3.3a**    Instructions per sample versus compression rate for speech coding.

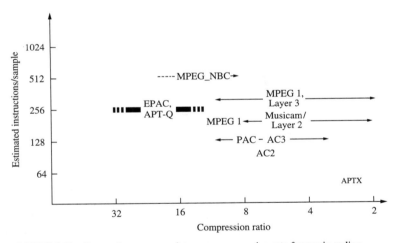

**FIGURE 3.3b**    Instructions per sample versus compression rate for music coding.

Figure 3.4 shows a generalized entropy coder. In an entropy coder, the usual method is to measure, or take from history, a statistical measure of the input signal by considering the likelihood of each individual signal value or sequences of signal values. Once this statistical measure is established or taken from history, a codeword is calculated for the input signal value, and this codeword is transmitted.

In simple entropy coders, the codebook is known and unchanging in the decoder. In more complicated entropy coders, such as Ziv-Lempel or arithmetic coders, the decoder may keep a complementary set of statistical values and continually recalculate the codewords in parallel with the encoder. Once the codeword means is established, in either case, the decoder converts the channel codeword back to the individual value (or values) of the signal.

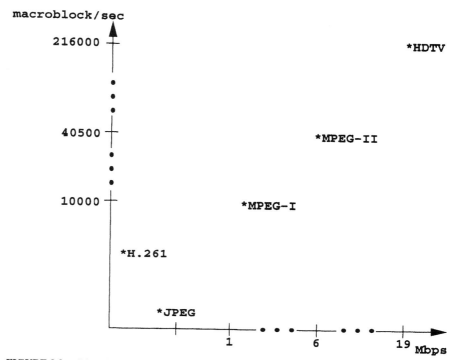

**FIGURE 3.3c**   Macroblocks per second versus compression rate for video.

**TABLE 3.2**   Classical Forms of Source Coding and Examples

| Coding method | Coding example | Reference key | Figure |
|---|---|---|---|
| Entropy | Huffman | 86 | 3.4 |
| | Ziv-Lempel | 41 | 3.4 |
| | Arithmetic | 98 | 3.4 |
| Model-based | DPCM | 3 | 3.5 |
| | ADPCM | 3 | 3.5 |
| | LPC | 3 | 3.5 |
| | APC | 5 | 3.5 |
| | CELP | 6, 9–10 | 3.5 |
| | Motion compensation | 34 | 3.5 |
| Filterbank (rate-based) | Transform | 3 | 3.6 |
| | Subband | 3 | 3.6 |
| | JPEG | 16 | 3.6 |
| | MPEG-I video | 52 | 3.6 |
| | Wavelet | 21–23 | 3.6 |

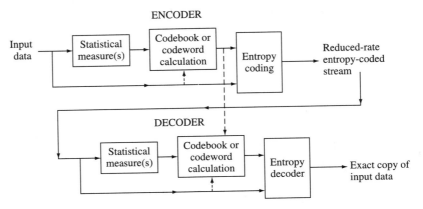

**FIGURE 3.4** A generalized entropy coder.

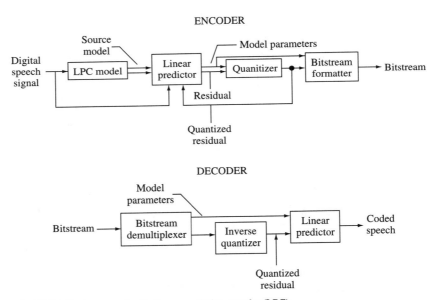

**FIGURE 3.5** An example of a linear predictive encoder (LPC).

Figure 3.5 shows an example of an LPC encoder. LPC, meaning *linear predictive coding,* is a source-modeling technique that calculates a model of the speech production mechanism and transmits the model values along with the "residual," or left-over part of the signal, or sometimes with a parametric set of values that reconstitute an approximation of the original signal.

In the code shown, the LPC model is calculated directly from the original signal and is used to set up the values in a linear predictor. This predictor calculates the residual of the coded signal versus the original signal, which is quantized (thereby adding noise) for transmission. This coder provides bit-rate reduction because the predicted signal is always smaller than the original signal and, therefore, fewer bits are required to transmit the residual.

The LPC decoder simply parses the bitstream and reverses the process, reconstituting the quantized residual and applying it to the model in order to recreate an approximation of the input signal at the encoder.

In some forms of LPC encoding, the model is calculated from the quantized residual. This is called *backward adaptation* as opposed to the method shown, called *forward adaptation.* In general, a forward-adaptive coder does not have delay in the model adjustment but requires that bits representing the model be transmitted. In contrast, the backward-adaptive coder does not have to send bits to explicitly transmit the model, but the model must always lag the signal statistics.

The filterbank coder shown in Fig. 3.6 is an example of a rate-distortion coder. Instead of using a special source model, it uses a filterbank that imposes a constant set of "models" on the source. Usually those models are frequency-like, i.e., FFT-, MDCT-, or OBT-like, to allow for simple interpretation of the filtered outputs.[3]

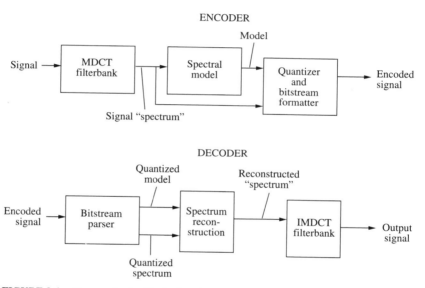

**FIGURE 3.6**    An example of a filterbank coder.

In the filterbank coder, each filterbank output is individually quantized and transmitted. In the example shown, the coder builds a spectral model and transmits that model along with the quantized filterbank outputs, allowing approximate reconstruction of the signal in the decoder.

Because most, though not all, signals have frequency shaping or a line spectrum, the filterbank coder can gain in bit rate relative to the PCM signal by isolating the energy in each frequency bin and coding it accordingly.

A source coder extracts sample-to-sample redundancy, also expressible as spectral nonflatness, to lower the energy of the signal that must be finally quantized and transmitted. In this sense, the source coder uses an analytic metric of coding quality, usually some sort of signal-to-noise ratio (SNR), and optimizes it via a mathematical process. In the forms of source coder with the most extreme compression rates, the idea of a waveform or a SNR is abandoned completely, and only the parameters of the signal model energy and pitch (in the LPC model, for instance) are transmitted. Such coders are sometimes referred to as parametric coders.

All of these coders use some kind of model or measurement of the signal source, or the statistics of the signal source, to compress the signal. It is worth noting that while entropy coding is usually "noiseless" (the signal is not modified by the coding process), the other source-coding techniques are lossy (noise is added to the signal by the coding process, in a fashion that is reduced as much as possible by the source model used in that coder). The goal of the source coder is to reduce the amount of noise by as much as possible, given the available signal redundancy and mathematical model. A good overview of source coding can be found in Jayant and Noll.[3]

## 3.3  THE PERCEPTUAL CODER

The perceptual coder is, effectively, a "destination coder," one that uses a model of what the end user (listener or watcher) can perceive, and a method of coding that removes parts of the signal that the end user cannot detect. This kind of coder is said to remove "irrelevancy" rather than redundancy and is always a lossy coder, where noise is added to the original signal in a fashion that is not noticeable by or objectionable to the user. As a matter of practice, nearly all perceptual coders are both source and destination coders, partially because the filterbanks necessary to relate the perceptual process to the signal provide some rate gain (i.e., redundancy extraction), and partially because the removal of any remaining redundancy provides an extra measure of bit-rate reduction beyond the removal of the irrelevant part of the signal. The metric used to control or evaluate coding noise in the perceptual coder is an estimate of the perceptual threshold or just noticeable difference (JND). In the literature, the JND is the lowest level of noise or distortion that can be added to a signal to provide the listener with the ability to distinguish the noisy/distorted signal from the original. In the perceptual coder, the goal is often to keep the noise and distortion below that figure.

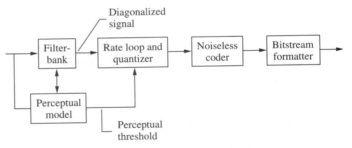

**FIGURE 3.7**    Block diagram of a typical perceptual coder.

The block diagram of a perceptual coder is shown in Fig. 3.7. The basic blocks of a perceptual coder are

- *Filterbank.* The filterbank serves to convert the input signal to a form that can be related to the perceptual model. The filterbank often provides a measure of signal diagonalization, or rate gain, as well.[2,4]
- *Perceptual model.* The perceptual model examines the signal, the filter signal, or both, to determine which parts of the signal are below the JND and can be removed without affecting the perceived quality of the signal.
- *Rate loop and quantizer.* The rate loop and quantizer is the part of the coder that actually removes the information from (adds noise to) the signal. In many coders the output rate is

required to be constant. In such cases the quantizers are adjusted by a "rate loop" that changes the quantization values, either to below threshold (increased rate) or above threshold (reduced rate, reduced quality), in order to maintain the required bit rate. In other coders, particularly those that provide data for storage media or for a packet network, it is advantageous to retain the variable-rate nature of the perceptual coder. In such coders the rate loop may be omitted or may be used only to provide control over the maximum data rate or long-term average rate. In general, a variable-rate coder with a given average bit rate can always provide better performance than a fixed-rate coder at the same rate because the bits in the variable-rate coder can be more effectively allocated over time.

- *Noiseless coder.* The noiseless coder is used to additionally compress and pack the values from the rate loop in the perceptual coder. The implementation of a quantizer in most perceptual coders is strongly controlled by the implementation of the perceptual threshold, and very often such quantizers are not efficient in terms of bit rate or redundancy removal. In such cases, it is highly advantageous to add an entropy-coding block to the coder in order to extract this additional redundancy.

- *Bitstream formatter.* The bitstream formatter packs the quantized signal data, step sizes, and other relevant signal and side information into a form suitable for transmission or storage. In most modern perceptual coders, there are two parts to the bitstream generator: a data layer formatter that packs the actual signal data into a form that can be decoded when no channel or channel errors are present, and, for transmission applications, a transmission formatter that adds redundant information allowing for detection, correction, or mitigation of difficulties in the transmission medium. This layer may also add information that makes synchronization, buffer control, or other signal recovery functions possible, faster, or easier.

## 3.4   DIFFERENCES BETWEEN SOURCE CODERS AND PERCEPTUAL CODERS

Source coders and perceptual coders address different aspects of the signal and have different characteristics. Following are some characteristics worth noting.

- A source-only coder has a higher SNR, in general, than a perceptual coder of the same complexity operating at the same bit rate. This is the case simply because the source-only coder deliberately maximizes some measure of SNR, and the perceptual coder maximizes a measure of perceived quality rather than a measure of SNR.

- Despite the lower SNR at a given bit rate, the perceptual coder usually has substantially better quality when the noise in either type of coder is perceptible.

- Most source coders implicitly incorporate some element of perceptual coding. In some cases, methods such as noise shaping may approach a simple perceptual noise allocation; in simpler cases, the coder arranges to have a noise spectrum tailored to provide good quality for the typical signal.

- Most perceptual coders, driven by the necessity to lower the noise level as much as possible, also use some aspects of source coding. The more complex perceptual coders use substantial source coding, implemented within the perceptual constraints, to provide the lowest possible bit rate.

- Source coders are often simpler and more suitable for signals that have well-understood production methods, such as the human voice.

- Perceptual coders are often more efficient and more suitable for signals that do not have well-understood, or consistent source models. Some such signals are

  *Music.* Most music signals have sudden transitions and a substantial variety of source models ranging from high-$Q$ LPC models to gated white noise sources.

*Still-frame images.* Images, unlike speech or music, can have signal content that is effectively arbitrary from the mathematical or modeling point of view.

*Motion video.* Motion video adds the complications of arbitrary rotation, translation, and scaling, as well as scene changing and other artificial constructs, to the already arbitrary image content.

## 3.5   *STANDARDIZED TELEPHONE BANDWIDTH AND WIDEBAND SPEECH-CODING ALGORITHMS*

Although speech can be digitized at many different sampling rates, in practice only two sampling rates are popular for speech-coding applications. For many decades, 8 kHz has been the dominant (and almost the only) sampling rate for speech transmission. Essentially all wire-line phone and digital cellular phone transmissions carry speech at this sampling rate. Such telephone bandwidth speech typically has a passband of 200 to 3400 Hz. The other popular sampling rate for speech is 16 kHz. The resulting speech has a nominal passband of 50 to 7000 Hz and is generally referred to as *wideband speech* in the literature. The extended bass and treble make wideband speech sound much closer to what we normally hear in face-to-face conversation. Wideband speech coding is used in some videoconferencing applications.

The most ubiquitous speech coder is the 64-kb/s $\mu$-law and A-law pulse code modulation (PCM), which was standardized as Recommendation G.711 by the CCITT (now ITU-T) in the early 1960s.[2] It can be considered a speech coder because it compresses 12-bit or 13-bit linear (uniform) PCM output into 8 bits using nonuniformly spaced quantization levels, with large quantizer step sizes for larger signal magnitudes and small step sizes for smaller magnitudes.

The next most popular speech coder is the 32-kb/s adaptive differential PCM (ADPCM) coder, which was standardized by the CCITT as Recommendation G.721 in 1984. The CCITT revised it in 1986 and also standardized a few variations of it as Recommendations G.723, G.726, and G.727. A simplified block diagram of the G.721 ADPCM is shown in Fig. 3.8. The encoder subtracts a predicted speech signal from the input speech to generate a prediction residual signal, which is quantized by an adaptive quantizer. The predicted speech is obtained by adding the outputs of two predictors, one operating on the quantized prediction residual and the other operating on the quantized speech. The decoder duplicates the part of the encoder that generates the quantized speech from the quantized prediction residual.

It can be shown that the difference between the input speech and its quantized version is the same as the difference between the prediction residual and its quantized version. The two predictors remove some redundancy in the input speech signal. The resulting prediction residual is generally smaller in magnitude than the input speech and therefore requires fewer bits to

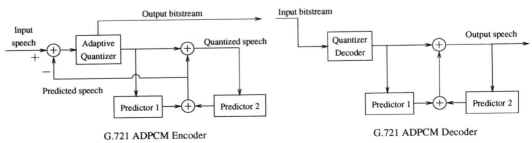

G.721 ADPCM Encoder                    G.721 ADPCM Decoder

**FIGURE 3.8**   Simplified block diagram of ITU-T 32-kb/s G.721 ADPCM coder.

quantize to the same accuracy. The 32-kb/s G.721 ADPCM is thus able to compress the bit rate of the 64-kb/s G.711 PCM by half without introducing much degradation in speech quality. For more details on ADPCM, see Jayant and Noll, Chapter 6.[2]

Multipulse linear predictive coding (MPLPC)[5] and code-excited linear prediction (CELP)[6] are two newer and fairly common speech-coding algorithms targeted at low bit rates. Their encoder and decoder structures are quite similar, as shown in Fig. 3.9. Both coders use an LPC synthesis filter, which is an all-pole filter containing an LPC predictor[7] in a feedback loop. The LPC synthesis filter has a frequency response that roughly follows the spectral envelope of the speech signal in the LPC analysis window. The LPC filter models the human vocal tract and exploits the short-term redundancy in adjacent speech samples. The pitch synthesis filter models the pitch periodicity in voiced adjacent pitch cycles. The pitch filter is sometimes implemented as an adaptive codebook[8] in the context of CELP coders. Although omitted in Fig. 3.9 for simplicity, both the LPC and pitch filter parameters are usually transmitted to the decoder periodically—about once every 20 to 30 ms for the LPC filter and once every 5 to 10 ms for the pitch filter.

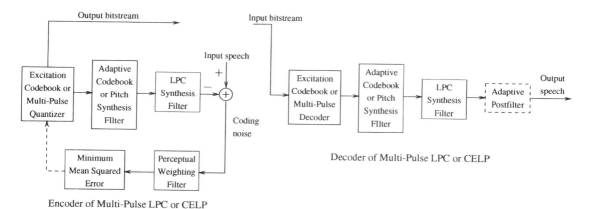

Encoder of Multi-Pulse LPC or CELP

**FIGURE 3.9**   Simplified block diagrams of multipulse LPC and CELP coders and decoders.

The MPLCP and CELP coders differ mainly in the way the excitation signal to the pitch and LPC filters is quantized. For MPLPC, the locations and amplitudes of a few pulses are encoded and transmitted, and the remaining excitation samples are set to zero. For CELP, a block of excitation samples is vector-quantized jointly using a codebook containing a collection of representative excitation patterns. As shown in Fig. 3.9, the excitation is quantized by passing each excitation codebook entry through the pitch and LPC synthesis filters, subtracting the resulting quantized speech from the input speech, passing the difference signal through a perceptual weighting filter, and then selecting the excitation codebook entry that gives the minimum mean squared error at the weighting filter output. The codebook index for this selected codebook entry is then sent to the decoder. The perceptual weighting filter shapes the coding noise to make it less audible.

The decoder in Fig. 3.9 extracts the selected excitation codebook entry for CELP, or decodes the pulse locations and magnitudes for MPLPC, and then passes the resulting excitation signal through the pitch and LPC synthesis filters. In practice, most MPLPC and CELP decoders also include an adaptive postfilter[9] to enhance the speech quality (by reducing the "perceived" coding noise).

There are several common telephone-bandwidth speech-coding standards based on CELP. The 16-kb/s low-delay CELP (LD-CELP) coder[10,11] was standardized by the ITU-T in 1992 as Recommendation G.728. Unlike traditional CELP, the G.728 LD-CELP does not use a pitch synthesis filter but uses a high-order LPC filter to exploit both the adjacent sample redundancy and the long-term (pitch cycle) redundancy. Furthermore, LD-CELP does not transmit LPC parameters but instead uses backward adaptation to reduce the buffering delay.

The 8-kb/s conjugate structure algebraic CELP (CS-ACELP) coder was standardized by the ITU-T in 1996 as Recommendation G.729. It encodes the excitation by using an algebraic code-book, where each codebook entry can have only a few pulses; all pulses have the same magnitude and a pulse can only assume its location from a regularly spaced grid.[12] In 1995 the ITU-T also standardized a 6.3-kb/s coder and a 5.3-kb/s coder as Recommendation G.723.1 for videophone applications. The 5.3-kb/s coder is a CELP coder with an algebraic codebook, and the 6.3-kb/s coder is a variation of multipulse LPC. The-13 kb/s group special mobile (GSM) coder for pan-European digital cellular telephone is another variation of multipulse LPC, where the pulses are located at fixed, regularly spaced intervals. Other CELP-based speech coding standards in-clude the 8-kb/s and 6.7-kb/s vector sum excited linear prediction (VSELP) coders for North American and Japanese digital cellular telephone, respectively; the 3.45-kb/s pitch synchronous innovation CELP (PSI-CELP) for Japanese half-rate digital cellular telephone; and the 4.8-kb/s Federal Standard 1016 CELP coder for secure voice.

At an even lower bit rate of 2.4 kb/s, a Federal Standard 1015 LPC vocoder was standard-ized in 1984 for secure voice. This coder is a parametric coder based purely on the source-filter speech production model.[7] The coder also uses an LPC synthesis filter, but the excitation signal is either a pulse train or white noise, depending on whether the current frame of speech is classi-fied as voiced or unvoiced. Unlike the other coders previously described, the LPC vocoder does not attempt to match the input speech waveform at all. Instead, it tries only to preserve some essential characteristics of speech. As a result, its output speech sounds synthetic and buzzy. A much improved new Federal Standard 2.4-kb/s coder, called mixed-excitation linear predic-tion (MELP),[13] was standardized in 1996. This coder performs as well as the 4.8-kb/s Federal Standard 1016 CELP coder.

In contrast to the numerous telephone-bandwidth speech coding standards mentioned earlier, so far there is only one international standard for 7-kHz bandwidth wideband speech coding—the ITU-T Recommendation G.722, which was standardized in 1988. G.722 is a two-band subband coder with ADPCM applied to each band. The higher subband signal is encoded at 2 bits/sample. The lower subband signal is encoded at 4, 5, or 6 bits/sample for a total bit rate of 48, 56, or 64 kb/s. The ITU-T is currently trying to standardize a new family of wideband speech coders at 16, 24, and 32 kb/s. This new standard is expected to be finalized in 1998. In addition, sev-eral of the MPEG audio algorithms have specified low-bit-rate configurations that can provide a bandwidth near 7 kHz for speech and audio signals. For more information on various speech coding standards, see Cox et al.[14]

## 3.6   SPEECH AND AUDIO PERCEPTUAL CODING CONSIDERATIONS

In speech applications, a perceptual model is rarely explicit. The handling of perception is done by a noise-shaping method,[3] and the filterbank may be replaced by an open-loop or closed-loop predictor, or an even more complicated speech-specific or parametric model. In audio applica-tions, the speech-specific model does not provide acceptable performance for many important signals because of the inapplicability of the speech generation mechanism to the general case of audio signals. Evidence of this is shown in Tables 3.3, 3.4, and 3.5, which list basic measure-ments of audio signals. In these tables, all of the data are taken on a 256-sample-shift basis, with a 512-sample overlap using a Hann window.

The first four columns, respectively, list the mean, sigma, maximum, and minimum of the LPC gain for an LPC predictor with 16 poles. This "gain" shows how much energy can be

**TABLE 3.3**  LPC Statistics for 16 Poles, 200–3200 Hz Passband

| Speaker | Mean gain, dB | Sigma, dB | Max, dB | Min, dB | Max distance, dB | Mean distance, dB | Mean SFM, dB |
|---|---|---|---|---|---|---|---|
| Male 1 | 17.7 | 5.3 | 26.4 | 2.8 | 14.2 | 2.4 | 21.8 |
| Female | 14.6 | 6.3 | 28.9 | 3.5 | 14.9 | 3.1 | 20.7 |
| Male 2 | 11.5 | 5.3 | 23.6 | 0.72 | 12.7 | 3.45 | 26.6 |

**TABLE 3.4**  LPC Statistics for 16 Poles, 50–7000 Hz Passband

| Signal | Mean gain, dB | Sigma, dB | Max, dB | Min, dB | Max distance, dB | Mean distance, dB | Mean SFM, dB |
|---|---|---|---|---|---|---|---|
| Female voice | 26.5 | 7.57 | 43.4 | 6.4 | 12.1 | 2.8 | 30.3 |
| Male voice | 26.3 | 7.3 | 41.7 | 9.9 | 16.5 | 2.4 | 30.2 |
| Female singing | 20.7 | 3.34 | 27.4 | 5.5 | 5.8 | 0.87 | 26.1 |
| Acoustic guitar | 18.5 | 2.8 | 27.0 | 9.9 | 5.9 | 0.85 | 25.8 |
| Male singing | 21.6 | 3.85 | 37.6 | 6.01 | 8.7 | 0.99 | 26.4 |
| Pipe organ | 11.8 | 2.06 | 17.4 | 7.4 | 9.5 | 1.1 | 22.2 |
| Solo piano | 23.9 | 3.77 | 34.6 | 15.5 | 8.5 | 1.4 | 30.0 |
| Solo violin | 15.4 | 3.36 | 25.0 | 7.5 | 3.2 | 0.93 | 25.1 |

**TABLE 3.5**  LPC Statistics for 16 Poles, CD Passband

| Signal | Mean gain, dB | Sigma, dB | Max, dB | Min, dB | Max distance, dB | Mean distance, dB | Mean SFM, dB |
|---|---|---|---|---|---|---|---|
| Narrowband noise | 40.5 | 5.2 | 59.2 | 25.0 | 19.1 | 1.0 | 43.5 |
| Bassoon | 40.1 | 3.8 | 23.8 | 28.8 | 10.9 | 0.68 | 45.6 |
| Viola | 30.7 | 3.7 | 43.5 | 20.9 | 11.9 | 0.92 | 34.1 |
| Castanets | 19.9 | 4.5 | 38.8 | 4.9 | 17.2 | 1.68 | 22.9 |
| Solo piano | 36.9 | 3.9 | 51.1 | 23.3 | 12.1 | 1.03 | 39.6 |
| Cabasa | 5.4 | 0.5 | 7.6 | 3.7 | 1.6 | 0.5 | 8.7 |
| Female operatic | 32.6 | 4.3 | 44.6 | 14.5 | 10.7 | 0.77 | 36.9 |
| Trumpet concerto | 36.6 | 2.8 | 47.1 | 28.2 | 6.7 | 0.64 | 39.7 |
| "Zarathustra" | 33.6 | 4.9 | 50.6 | 12.7 | 10.6 | 0.65 | 37.7 |
| Metal percussion | 23.1 | 4.2 | 35.1 | 8.2 | 26.0 | 0.84 | 29.2 |
| Suzanne Vega | 23.4 | 5.7 | 36.4 | 3.7 | 18.0 | 1.4 | 27.3 |

removed from the signal via pure mathematical redundancy extraction using a 16-pole signal model. The next two columns, "Max distance" and "Mean distance," are the LPC distance for a shift of 256 samples. These distances show the amount of loss that occurs when the predictor model lags the signal statistics by 256 samples. Finally, the column labeled "Mean SFM" shows the spectral flatness measure of the signal for a 512-point FFT with a Hann window. SFM represents the maximum prediction gain available over that data and window. Due to the differences in sampling rate, the shift of 256 samples represents a different time delay, as shown in Table 3.6.

**TABLE 3.6**   Time Strides of 3:5

| Sampling rate | Signal bandwidth, Hz | Delay, ms |
|---|---|---|
| 8,000 | 200–3,200 | 32 |
| 16,000 | 50–7,000 | 16 |
| 4,100 | 20–20,000 | 5.8 |

There are notable features in this set of tables. First, the audio signal model varies much more suddenly than the speech signal model, as shown by the LPC statistics. In the case of speech, the average LPC distance is larger, as one would expect from the much greater time delay involved in the 8-kHz sample rate case. However, the maximum LPC distances for the music signals are larger, showing that although the music has relatively long stationary parts, the model changes abruptly—for instance, at note changes or when a percussion instrument is struck. While the table does not show the sigma of the LPC distance, for music signals it is actually larger than it is for speech signals, even though the average distance is smaller. Since LPC distance must be positive and definite (the smallest possible distance is greater than zero), this suggests a pattern of very consistent periods for music signals, with sudden variations occurring at musical "events." In practice, this sudden change of the signal model creates a problem for coders using the LPC model, which is most often assumed to be slowly varying. Second, even the SFM of the audio signals, implying, effectively, a "best case" predictor, does not provide nearly enough gain to reproduce a 16-bit audio signal while providing a significant rate reduction.

While many conclusions can be drawn from these data, the most significant conclusion is that, in and of itself, redundancy removal does not provide the necessary coding gain required for audio signals. On the other hand, redundancy removal provides quite enough gain for the speech coding case and in practice is the accepted and demonstrated coding method for speech signals.

Effectively, speech coders use the speech production model in some form and manage any approach to perception via noise shaping or other implicit methods. Audio coders, on the other hand, cannot achieve anything near the required coding ratios (coding ratio is the ratio of input to output bit rate) by using source-coding methods alone, and must rely heavily on both the removal of redundancy and the removal of irrelevancy.

## 3.7   IMAGE AND VIDEO SIGNAL CONSIDERATIONS

In video, a different set of tools are used, both in the perceptual and the source coder. Typically, image coders are filterbank coders, using a discrete cosine transform (DCT), wavelet, or pyramidal filterbank in addition to either a block scalar quantizer or a vector quantizer. We begin by presenting a summary of the different image formats and target bit rates for various applications and then review the current still-image and video standards as well as some of the nonstandard technology that may be the foundation of future standards.

### 3.7.1   Image/Video Formats and Target Bit Rates

The current standard for color television in the United States and Japan is the National Television System Committee (NTSC) standard. It consists of a picture resolution of 525 lines/frame and 59.94 fields/second. The color signal is represented by three components, *YIQ,* where *Y* is the luminance signal and *I* and *Q* are the color-difference chrominance signals. Similarly, the Phase Alternation Line (PAL) system used in most of western Europe is based on 625 lines/frame and

50 fields/second. Here, the three components representing the image signal are $YUV$, where $Y$ is the luminance signal and $U$ and $V$ are the two chrominance signals. The Sequential Couleur avec Memoire (SECAM) system used in France, the Middle East, and much of Eastern Europe and Russia is very similar to PAL. The aspect ratio (visible picture width to picture height) is 4:3 for all systems. For HDTV, the number of lines per frame is 720 for progressive and 1080 for interlaced, and the aspect ratio is 5:3.

In order to provide a digital representation that is compatible with NTSC, PAL, and SECAM, the CCIR 601 international standard has been set for component video coding. The luminance and two chrominance components defined by this standard are linear combinations of gamma-corrected, normalized NTSC primaries, $\tilde{R}$, $\tilde{G}$, and $\tilde{B}$. The digital luminance signal is defined as

$$Y_D = 219Y + 16 \tag{3.1}$$

where

$$Y = 0.299\tilde{R} + 0.587\tilde{G} + 0.114\tilde{B} \tag{3.2}$$

and the digital color-difference signals are defined as

$$C_B = \frac{112(\tilde{B} - Y)}{0.886} + 128 \tag{3.3}$$

and

$$C_R = \frac{112(\tilde{R} - Y)}{0.701} + 128 \tag{3.4}$$

The range of the digital luminance signal is 16–235 while the range of the two color-difference components is 16–240. The maximum and minimum values of $C_B$ correspond to blue and yellow, respectively, while the maximum and minimum values of $C_R$ correspond to red and cyan, respectively. The CCIR 601 standard specifies a picture resolution of 720 pixels/line. The chrominance signals are subsampled by a factor of 2 in the horizontal direction.

The H.261 video-coding standard defined a *common intermediate format* (CIF) corresponding to an image size of $360 \times 288$ and a coded image size of $352 \times 288$. The value 352 was chosen to facilitate encoding blocks of size $16 \times 16$. MPEG-I extended the CIF format to accommodate the different television formats with the *standard intermediate format* (SIF), where the image size is $360 \times 240$ for NTSC and $360 \times 288$ for PAL and SECAM. MPEG also defined the coded image width as 352 to facilitate encoding blocks of size $16 \times 16$. For SIF as well as CIF, the chrominance signals are subsampled by a factor of 2 in both the horizontal and vertical directions.

For videoconferencing applications, primary rates have been defined as 1.544 Mb/s in North America and Japan and 2.048 Mb/s in Western Europe. The Integrated Services Digital Network (ISDN) allows for data transmission at multiples of 64 kb/s up to the primary rates. ISDN provides B channels with 64-kb/s rates and H0 channels with 384-kb/s rates. An interesting target rate for HDTV is 20 Mb/s, which can be transmitted on a 6-MHz channel. This would allow HDTV to be transmitted on currently vacant NTSC channels.[15] For voice-band videotelephony, target bit rates are in the range of 8 to 40 kb/s.

## 3.7.2  Standards

A brief overview will now be presented of some of the still-image and motion video standards. The compression standards described in this section have been established by the Joint Picture Experts Group (JPEG) and the Motion Picture Experts Group (MPEG); see Chapter 8. The JBIG standard is covered in Chapter 9 along with the JPEG standard.

JPEG established an international standard for color still-image coding.[16,17] The JPEG standard consists of two coding systems: a lossless codec and a family of DCT-based lossy codecs.

The JPEG lossless codec consists of a simple set of spatial predictors followed by an entropy coder. The lossless codec is based on either arithmetic or Huffman coding for the entropy coder. The lossless codec typically produces approximately 2:1 compression for color images of moderate complexity.

The lossy JPEG codecs are based on the DCT, which is the foundation of all current coding standards both for still images (JPEG) and video (H.261, MPEG-I, MPEG-II). Figure 3.10 shows a block diagram of the basic JPEG encoder. In particular, the $8 \times 8$ DCT is used as the initial step of frequency decomposition. An illustration of the basic functions for the $8 \times 8$ DCT appears in Fig. 3.11. Although the DCT provides a cost-effective and adequate solution for some bit rates, DCT-based coders are limited by the block-based nature of the transform, which does not take full advantage of the signal characteristics or provide a framework where perceptual properties can be optimally exploited. In particular, by processing each $8 \times 8$ block separately, we do not take advantage of the strong interblock correlations. This results in discontinuities along the block boundaries, which are especially noticeable at low bit rates. The end result is a strongly structured type of degradation that is very visible and perceptually objectionable. A way to avoid this type of degradation is to use a wavelet or subband framework for the frequency decomposition. This framework will be described in the next section.

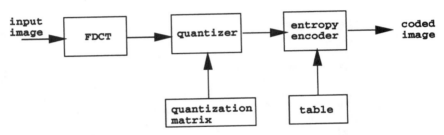

**FIGURE 3.10**    JPEG encoder.

The simplest of the JPEG codecs, the baseline sequential codec, consists of three parts: a forward transform, a quantizer, and an entropy coder. The DCT performs energy compaction on the signal and reduces the number of samples that need to be transmitted. Since most images have a low-pass spectrum, transforming the spatial-domain data into the frequency domain results in fewer significant samples, and these samples cluster at the low frequencies.

The quantizer specifications control the bit rate and output image quality. The 64 DCT coefficients are uniformly quantized, and the quantizer step for each coefficient is specified by a quantization table. It is transmitted to the decoder and may be altered based on the image or application. Later we describe perceptually based image coders that are compliant with the standards. We are able to take advantage of properties of the human visual system (HVS) within the standards framework in two ways: design of the quantization matrix and a perceptually based prequantization step.

Since image data tend to have a low-pass spectrum, the nonzero quantized DCT coefficients will cluster at low frequencies followed by zero-valued high-frequency coefficients. The last stage of the JPEG encoder consists of entropy coding using either an arithmetic or a Huffman scheme. The entropy coder provides lossless compression by exploiting the low-pass characteristic images and resulting DCT structure.

In addition to the baseline sequential codec, JPEG supports several other modes of operation, including the *progressive* and *hierarchical modes*. The key difference for the progressive mode

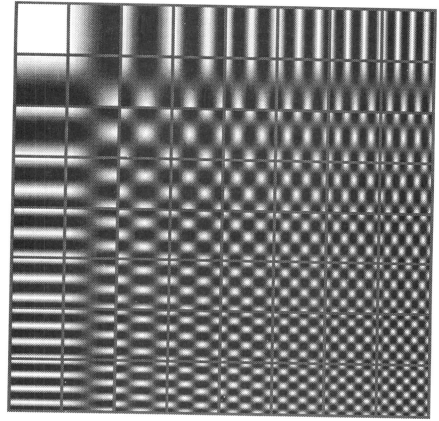

**FIGURE 3.11**  Basic functions for an 8 × 8 DCT.

is that the nonzero DCT coefficients are transmitted in several passes. The idea is to produce a recognizable, lower-quality image quickly followed by several iterations of refinement. Two techniques for JPEG progressive transmission are defined: *spectral selection* and *successive approximation*. In spectral selection bands of increasing frequency are transmitted at each iteration, and in successive approximation the most significant bits are transmitted followed by less significant bits of the nonzero coefficients at each iteration. The hierarchical mode of operation for JPEG is based on a pyramidal representation of the input image, where each layer in the pyramid is an image at half the resolution in the horizontal and vertical directions of the next layer. The image at the lowest resolution is encoded using any of the previously mentioned modes of operation. The image is decoded, up-sampled to the next resolution level, and used to predict the image at the corresponding resolution. The prediction difference is also encoded using one of the previously described modes of operation. This step is repeated at every resolution up to the original image resolution.

An interesting application of JPEG is motion JPEG, the coding of video based on intraframe JPEG coding of the frames in the sequence. Advantages of motion JPEG include low implementation complexity, predictable image quality, and flexibility in terms of image size and format. Applications include nonlinear video editing and digital video camcorders.

The current video coding standards include H.261 (see Chapter 10), MPEG-I, and MPEG-II (Chapters 8 and 9). Here we summarize some of the salient features. All of the video coding standards are based on a motion compensation (MC) DCT paradigm. Figure 3.12 shows a block diagram of such a coder. Motion compensation takes advantage of interframe correlations by providing simple yet quite effective motion estimation for interframe prediction. The current frame is predicted by displaced blocks from the previous frame. Specifically, motion vectors are calculated based on translational motion in the image plane of image blocks of a predetermined size, typically 8×8 or 16×16. The technique assumes constant lighting so that motion estimation between frames can be done by direct comparison of pixel intensities. The difference between the image block in the current frame and the estimate based on a displaced block from the previous frame is encoded using the block-based DCT.

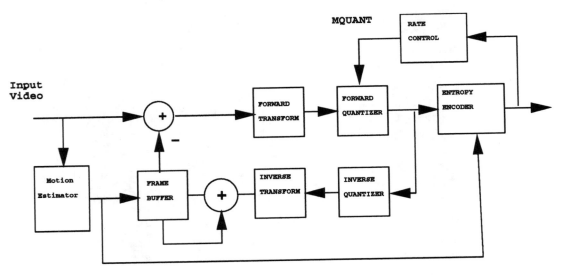

**FIGURE 3.12**   Block diagram of an MC/DCT-based encoder.

The goal of the H.261 video teleconferencing standard, also known as $p*64$, is to provide standardized video service on the ISDN. It defines a codec optimized for videophone and video teleconferencing applications that can operate at a fixed set of bit rates. H.261 operates from 64 kb/s to 1920 kb/s in multiples of the basic ISDN channel rate of 64 kb/s ($p*64$, where $p = 1, 2, \ldots, 30$).

A critical feature of H.261 is minimal delay for interactive applications. The minimal delay makes this standard suitable for videophone applications. For rates greater than 384 kb/s, this standard is suitable for videoconferencing applications. This scheme is based on the generic MC/DCT framework with the addition of a loop filter. This filter can be switched on and off and serves to attenuate high-frequency noise in the feedback loop. The codecs are defined to operate at a constant bit rate, so a buffer is provided prior to transmission to monitor the bit rate and modify the quantization step size when necessary.

The MPEG video compression standards provide a framework for coding audiovisual data for storage and transmission. Initially, the application for MPEG-I, or ISO 11172, was focused on the coding of moving pictures and associated audio for digital storage media, but the end result provides much more generic utility.

MPEG-I supports a maximum image resolution of $4096 \times 4096$ pixels and several frame rates between 24 and 60 frames/second. The most common use of MPEG-I is in the compression of SIF video for movie distribution on Compact Disc with a video coding rate of 1.1 Mb/s. The picture quality in this application is scene dependent and is in general on the level provided by consumer VHS VCRs. Many of the requirements for MPEG-I are based on having a cost-effective decoder for applications such as movies on Compact Disc.

The MPEG-I video standard provides a toolkit oriented toward the compression of progressive video for storage media. MPEG-1 provides a layered bitstream syntax that allows a great amount of flexibility in both the applications domain and encoder optimization. The top layer of the video syntax provides the ability to specify the image size, pixel aspect ratio, data frame rate, compressed bit rate, and decoder-compressed buffer size. MPEG-I contains separate quantization tables for the intraframe and interframe DCT coefficients based on uniform quantization of each coefficient. The second layer of the video syntax, the *group of pictures* (GOP) layer, provides the ability to perform "VCR functions" such as fast forward and fast reverse. The frames in a GOP can be encoded in one of three ways: intracoded (I), forward predictively coded (P), or bidirectionally predictively coded (B). The first frame in a GOP is always an I frame, and encoder complexity limits the number of consecutive B frames allowed in a GOP. In general, P frames provide better compression than I frames, and B frames provide the best compression. The next layer of the bitstream, the *picture layer,* is used to describe a single frame of video. It contains information about the picture type (I, P, or B), a temporal reference, buffer occupancy, and, in the case of P and B frames, the size of motion vectors. The next layer, the *slice layer,* contains the starting location of the slice within the frame and a value of the quantizer scale parameter that is used to adjust the output bit rate. The final layers, the *macroblock* and *block* layers, transmit the motion vectors and quantized DCT coefficients in a manner similar to H.261.

The purpose of MPEG-II was to address the limitations of MPEG-I, namely, the ability to deal with higher-bit-rate applications and interlaced source material. Originally, MPEG-II was developed for the coding of movie sequences in the range of 4 to 10 Mb/s and CCIR 601 image resolution. MPEG-II has also been proposed for the U.S. standard for HDTV with a target transmission rate of 20 Mb/s (see Chapter 13). One significant feature of MPEG-II is the ability to perform MC and DCT operations on field as well as frame data in order to be able to handle broadcast-quality interlaced video sources. MPEG-II provides hierarchical coding using either *SNR scalability* or *spatial scalability.*

MPEG-III, initiated for high-definition television applications, was dropped when it was realized that MPEG-II was able to handle the applications targeted for MPEG-III.

### 3.7.3  Wavelet/Subband Coders

Currently, wavelet- and subband-based coders have not been incorporated into any of the major standards for still-image or video coding. One exception is an FBI standard for fingerprint compression based on wavelet coding. Although much of the research in recent years shows an obvious advantage in going from a block-based transform such as the DCT to a wavelet- or subband-based scheme, the resistance in adopting a framework different from the DCT or MC/DCT model in the standards stems from both historical and practical factors. These include the unavailability of mature subband/wavelet-based schemes at the time the initial standards were set and the initial investment in optimizing DCT- and MC/DCT-based schemes. As future standards become more demanding and as new applications emerge, it may be necessary to adopt new coding frameworks.

This section covers coding schemes that fall under the common category of a multiresolution framework. This includes what is commonly referred to as subband coding in the engineering literature and wavelet-based coding in general. The idea of composing an image signal into different frequency channels prior to compression can be justified on the basis of both statistical and psychovisual studies. Multifrequency channel decompositions have found many applications in computer vision and image-processing applications and are valuable in understanding some low-level processes in the HVS. The advantage of using subband-based coding over block-based transforms is that by decomposing the signal into a multiresolution representation, we avoid a block-based coding scheme and the associated block-based artifacts. The multiresolution representation also lends itself naturally to the incorporation of perceptually based bit allocation and quantization. In a subband-based scheme, lower frequency bands are given higher priority, and at low bit rates the effect of not being able to accurately represent the high-frequency components of the image signal manifests itself in a loss of high-frequency details over the whole image. This distortion is quite different from the tiling effect of DCT-based algorithms.

The basic subband or wavelet encoder consists of three parts: frequency analysis, quantization, and entropy coding. The frequency analysis could consist of equal-band decomposition,[18] octave-band decomposition,[19] or image-dependent decompositions as offered by wavelet packets.[20] The basic idea of an octave-band representation is to provide good frequency resolution at low frequencies and good time (space) resolution at high frequencies. The flexibility of trading off time (space) and frequency resolution is desirable for analyzing nonstationary signals such as still images and video. Intuitively, this type of decomposition is most useful for signals that are composed of low-frequency components of long duration and high-frequency components of short duration—a situation that generally holds for many signals in practice.[21] There is also reason to use an octave-band decomposition to take advantage of limitations of the HVS in the coding scheme. Figure 3.13 illustrates the difference between a uniform-band and an octave-band decomposition for a one-dimensional signal in terms of time frequency resolution.

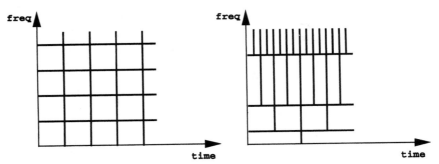

**FIGURE 3.13**   Time frequency resolution for uniform-band (*left*) versus octave-band (*right*) decompositions.

Wavelet theory provides a unifying theory for multiresolution representations of signals and for the design of the filters to be used for the decomposition.[19,22–24] Different filter design characteristics and their relevance to image compression are reviewed in ref. 25. In general, filter properties that are often considered in the design of filterbanks for signal compression include perfect reconstruction, regularity, orthogonality, frequency response, and linear phase. The filters described in ref. 26—which have almost perfect reconstruction, are approximately regular, and have linear phase—are widely cited in the subband-based image coding schemes. So far, much of the literature shows that decomposition structure and quantization strategy are much more critical in coder performance than particular filter design. Of course, this assumes that the filter decomposition is not block-based as in the DCT-based coders.

The theory of subband decomposition has been extended to multidimensional signals in ref. 27. The first subband-based coder for still-image compression was proposed in ref. 28 and was followed by many different subband-based coding strategies for still-image data. A wavelet-based coding scheme introduced in ref. 29 presents the concept of the *embedded zerotree wavelet* (EZW). This scheme takes advantage of the low-pass nature of the image spectrum and the likelihood that most of the signal energy will reside in the lower frequency bands followed by a string of low-energy, insignificant high-frequency components. In addition to providing coding results that are significantly better than JPEG, especially at low bit rates, this technique has the advantage that it does not require training or a priori knowledge about the image statistics.

For video coding, a commonly used paradigm is two-dimensional spatial subband decomposition with motion compensation.[30,31] The original work on decomposing and encoding a video sequence using a three-dimensional subband decomposition appears in ref. 32. For image and video coding, it is common to decompose the images spatially into subbands using separable one-dimensional filters in the horizontal and vertical directions. This idea can be easily extended to three dimensions—horizontal, vertical, and temporal—for the encoding of a video sequence.

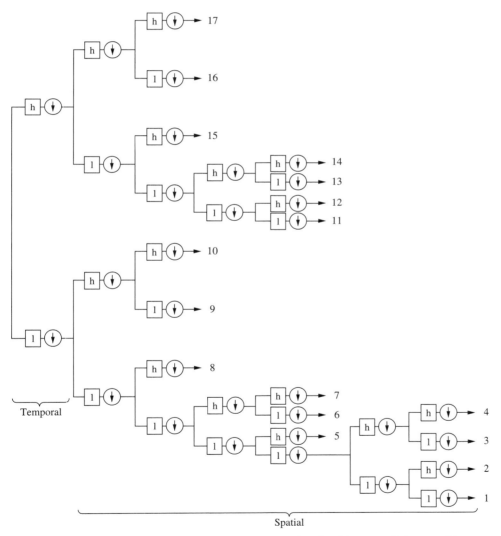

**FIGURE 3.14**  Seventeen-band spatio-temporal decomposition, where *h* and *l* represent high-pass and low-pass filtering, respectively, and ↓ represents down-sampling by a factor of 2.

As an example, the work presented in ref. 33 decomposes an original image sequence into 17 frequency bands as shown in the tree structure of Fig. 3.14. Some examples of a video sequence decomposed into the frequency bands illustrated in Fig. 3.14 are shown in Fig. 3.15, where the left-hand side represents the high temporal frequency band, the right-hand side represents the low temporal frequency band, and the lower right corner represents the lowest spatial frequency band. Other work based on a three-dimensional decomposition include refs. 34–36. In refs. 33 and 37 the video sequence is decomposed into spatiotemporal frequency bands, and the idea of geometric vector quantization (GVQ) is introduced to encode the high-frequency data. GVQ takes advantage of the sparse, structured nature of the data in the higher frequency bands. A spatiotemporal decomposition presents a framework in which bits can be adaptively allocated between spatial and temporal information.

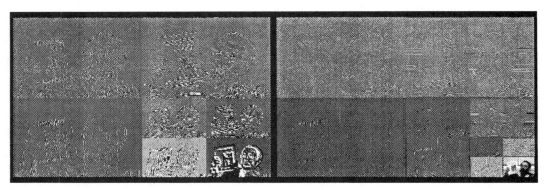

**FIGURE 3.15** Example of a spatiotemporal decomposition.

Subband coding for speech and image compression is based on critical sampling; that is, the number of samples after the analysis stage is equal to the number of original samples. The critical-sampling property of a subband system imposes strict constraints on the design of the filters. Another approach for multiresolution representation of a signal is to allow for redundancy or oversampling of the signal in the analysis stage, which reduces the requirements on the design of the analysis/synthesis filters. This is the approach used in the pyramid decomposition introduced in ref. 38. Such a framework could be used for coding, although most coding techniques are based on critically sampled frequency decompositions to avoid an initial data expansion.

Many books (e.g., refs. 39 and 40) have been written about wavelet theory and subband coding for those seeking more details.

## 3.8 THE HUMAN AUDITORY SYSTEM

In order to understand the requirements of a perceptual audio coder, it is necessary to understand some of the functions of the human auditory system (HAS). The peripheral HAS is usually separated into three parts, listed in order from the "outside" to the "inside" of the body: the head, pinna, and ear canal; the middle ear; and the inner ear. In addition, the central nervous system (CNS), specifically the brain, is heavily involved, in terms of recognizing and attaching meaning to the features extracted by the peripheral part of the HAS.

In this section only the issues involving the non-CNS part of audition will be addressed, up to and especially including the inner ear, in the light of how those issues relate to the basic detection abilities—both the detection of one signal as a function of level and frequency and the detection of one signal in the presence of others. The level of detection of one signal as a function of frequency and level is often called the "absolute threshold" question or, more completely, the "loudness contour" issue.[41] The question of the detection of a signal or signals in the presence of other signals is called the question of "auditory masking."[42-44] Both of these phenomena are used in audio coding algorithms, but in quite different ways.

As a practical matter, since most people have volume controls on their audio equipment, and the audio coding algorithm can neither know nor control the volume control position, the levels demonstrated by the absolute threshold of hearing must be handled in a very conservative fashion. The auditory masking phenomenon, on the other hand, is of great use in the coding of audio signals, because it can be shown to be a ratelike[2] function that varies with the level of the signal, rather than an absolute lower limit to the quantization levels that are perceptually tolerable. Because of this variation with signal energy, the masking phenomenon can allow much more irrelevancy removal under most circumstances.

Here the peripheral parts of the HAS are described, and then their actual masking and frequency analysis performance are discussed in detail. First the functions of the peripheral hearing apparatus, partitioned into its usual three parts as mentioned previously, will be discussed, and then the ear's frequency partitioning, masking ability, and ability to detect changes in different types of signals will be addressed.

### 3.8.1    The Head and Outer Ear

The external hearing apparatus consists of the head, the pinnae (outer ears), and the ear canals.[46] The primary effects of the external parts of the ear are to impart directionality information individually to each ear. While these effects are important to binaural hearing, they do not significantly affect the coding process because the coder designer must assume that the head is oriented in a fashion that will provide the most sensitive listening situation as far as coding noise is concerned. In practice, this is often the case. Although the head-related issues do not directly affect most coding problems, they are quite important in other settings, involving imaging, front/back discrimination, and 3D localization.

### 3.8.2    The Middle Ear

The middle ear consists of the eardrum (tympanum) and the three small bones that connect the eardrum to the cochlea or inner ear. The middle ear does have a strong effect on the audition process, but one that relates mostly to minimum audible levels and frequency shaping. A good description of the general effects of the middle-ear processes can be found in Yost.[45]

### 3.8.3    The Inner Ear

The part of the HAS of primary concern to the perceptual coder designer is the inner ear, specifically that part of the inner ear directly involved in audition: the cochlea and organ of Corti. In this part of the HAS are located the mechanical and neural components that correspond to the frequency analyzer and detectors that provide the input to the CNS. It is in the inner ear, in the cochlea, that frequency analysis in human hearing takes place, and in the organ of Corti, in the basilar membrane, where this frequency-analyzed sound is actually neurally detected. The rest of this section will deal with effects attributable to the frequency analysis and detection abilities of the inner ear—more specifically, the basilar membrane.

### 3.8.4    More Reading

Both William Yost[45] and Brian C.J. Moore[46] do a very good job of covering various parts of the hearing chain—the physiology, psychology, and more. The reader wishing to go beyond the surface treatment given here is advised to consult one or both of these books for more information. In addition, Allen's reprint of Fletcher's book[47] is very good at demonstrating the rather advanced state of psychoacoustics in the early Bell Labs era. The telling part of Fletcher's book is that the data and conclusions reached in his book are still relevant and useful today.

### 3.8.5   Critical Bands, Masking, Tonality, and Just Noticeable Difference

Many people have researched the performance of the human ear. Harvey Fletcher, in his early work, documented frequency ranges over which the signals interacted in certain fashions—in particular, adding arithmetically over some frequency range and in power beyond that range—and pointed out that this frequency scale was similar to the width of the basilar membrane, according to von Bekesy's[48] and Fletcher and Allen's[47] experiments. Scharf,[43] in his work, identified a similar set of frequency ranges based on the tone-masking-noise result with varying bandwidth and power of noise probe versus the tone masker. The term *critical band* was coined to represent the frequency ranges over which loudness[42] or the masking phenomenon[43] showed a more or less fixed behavior. Greenwood[49] has related these critical bands to distance along the basilar membrane. All of these measurements, within the limits and roll-off constraints used in the particular calculations, show the same results. The "critical band" or "Bark" scale (after Barkhausen) has been established to standardize this scale of auditory phenomena. A tabulation of this scale, according to Scharf,[43] is shown in Table 3.7, along with a good approximation of Bark value as a function of frequency.

**TABLE 3.7**   Critical Band Centers and Edge Frequencies[44]

| Band number, Hz | Lower edge, Hz | Center, Hz | Upper edge, Hz |
|---|---|---|---|
| 1 | 0 | 50 | 100 |
| 2 | 100 | 150 | 200 |
| 3 | 200 | 250 | 300 |
| 4 | 300 | 350 | 400 |
| 5 | 400 | 450 | 510 |
| 6 | 510 | 570 | 630 |
| 7 | 630 | 700 | 770 |
| 8 | 770 | 840 | 920 |
| 9 | 920 | 1,000 | 1,080 |
| 10 | 1,080 | 1,170 | 1,270 |
| 11 | 1,270 | 1,370 | 1,480 |
| 12 | 1,480 | 1,600 | 1,720 |
| 13 | 1,720 | 1,850 | 2,000 |
| 14 | 2,000 | 2,150 | 2,320 |
| 15 | 2,320 | 2,500 | 2,700 |
| 16 | 2,700 | 2,900 | 3,150 |
| 17 | 3,150 | 3,400 | 3,700 |
| 18 | 3,700 | 4,000 | 4,400 |
| 19 | 4,400 | 4,800 | 5,300 |
| 20 | 5,300 | 5,800 | 6,400 |
| 21 | 6,400 | 7,000 | 7,700 |
| 22 | 7,700 | 8,500 | 9,500 |
| 23 | 9,500 | 10,500 | 12,000 |
| 24 | 12,000 | 13,500 | 15,500 |
| 25 | 15,500 | 19,500 | |

Bark frequency $= 13 \cdot \text{atan}(0.76 \cdot f/1000) + 3.5 \cdot \text{atan}[(f/7500) \cdot (f/7500)]$, where atan is the radian arctangent function and $f$ is the frequency in Hz

Allen and others have shown that at a given point along the basilar membrane, one can determine a cochlear filter impulse response (in either frequency or time) of the waveform at that point from an impulsive excitation. Example impulse responses for two points on the cochlea are shown in Fig. 3.16a (the frequency and time responses). Allen notes that the responses are nearly minimum-phase, so the change in bandwidth of the critical bands results in a similar, inverse change in the length of the appropriate time or impulse response, resulting in a difference in time/frequency resolution along the cochlea of about 40:1. The frequency resolution is best at low frequencies, where the cochlear filters have narrow bandwidths. Conversely, the time resolution is best at higher frequencies, where the cochlear filters have very wide bandwidths and therefore narrower impulse responses.

One of the effects that helped to establish the critical band scale was the hiding of one signal at a given frequency by another signal at or near that frequency; this effect is called "masking." Fletcher, Scharf, and others have reported on this effect. To test for masking, one signal, called a masker, is presented to the listener with and without another signal, called the probe. If the listener cannot hear the difference between the signal with and without the probe, then the probe signal is "masked." The level at which the probe is just audible is called the "just noticeable difference" (JND). Section 3.8.4 lists reports of a multitude of such experiments.

The two kinds of masking experiments most interesting to the coder designer are those with noise as the probe, i.e., *tone masking noise* (TMN), and *noise masking tone* (NMT), where either a tone or a critical bandwidth of noise is the masker. What is interesting about the two tests is that they show a strong difference in masking ability between tones and noise; a noise probe at $-3.5$ to $-5.5$ dB is reported as masked for a noise masker while, depending on frequency, a noise probe under a tone is masked between $-16$ and $-40$ dB. Schroeder et al.[50] report that $15.5 + i$dB is a good estimate for the masking ratio of TMN, where $i$ is the actual Bark frequency of the tone. Other sources suggest different numbers, but the difference in the tests used to establish the JND

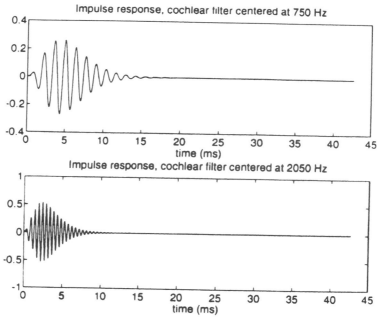

**FIGURE 3.16a**    Impulse response (amplitude versus time) for two example cochlear filters.

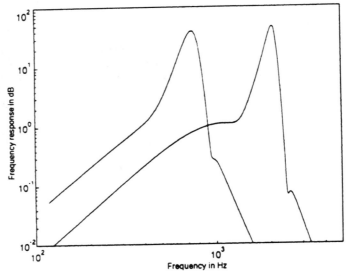

**FIGURE 3.16***b* A plot of log amplitude versus frequency for two example cochlear filters.

create some uncertainty. The asymmetry between TMN and NMT leads to the next problem in the audio perceptual coder, that of *tonality.*

A practical perceptual coder must be able to distinguish in some way a tonal input signal from a noiselike one. This distinction has been referred to in the literature as the "tonality" problem,[51] and several metrics, each with its own advantages and disadvantages, have been proposed.[51,52] A complete description of tonality metrics is best left to the literature, and to new literature as it becomes available. In any case, accounting for tonality is a necessary part of any masking model used for audio signals.

### 3.8.6 Time and Frequency Considerations in Masking

As suggested in the previous section, masking is the primary source of irrelevancy extraction in the perceptual coder. There are an enormous number of masking experiments reported in the literature, and they fall in two groups: "nature of masker/maskee" and "time duration/relation between masker and maskee." For a perceptual coder to successfully hide the added coding noise, both sets of considerations are important. As mentioned earlier, the tonality measure can be used to adjust a coder to discriminate between a tonelike and a noiselike signal and adjust the masking level accordingly. As also noted, the measurements for this sort of masking are made with the masker and the probe signal both present at the same time, called *simultaneous masking.* There is also evidence that masking occurs when the probe comes after (forward masking) or before (backward masking) the masking signal. These two situations arise as a result of both neural/CNS effects and the fact that the bandwidth of the cochlear filterbank establishes a minimum length to the time response for the cochlear filterbank.

Two components contribute to the length and shape of the forward masking phenomenon: the length of the impulse response of the relevant cochlear filter and the neural "overhang" due to higher-order processing in the CNS. Since the length of the impulse response of the cochlear filter is a strong function of the frequency involved, this effect can vary from 1 to 25 ms or so.

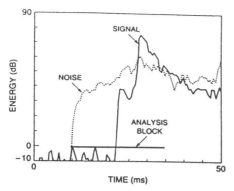

**FIGURE 3.17**    An example of pre-echo, where noise occurs before the transient signal.

Additionally, some forward masking due to neural effects has been reported for up to 50 to 200 ms following the masker. Both effects (neural and mechanical) diminish with time.

The extent of backward masking is much shorter. While some authors report some effect at up to 50 ms, there are also well-established reports of unmasking on an attack at 1 ms or less in advance of the probe. In practice, pre-echo is a real problem for audio coders that use long analysis windows, because the delay in the filterbank can allow noise to occur before, as well as after, the transient signal. This pre-echo is very perceptible under some circumstances and creates a distortion that some listeners find quite objectionable. For an example of a pre-echo situation, see Fig. 3.17.

The time duration of both forward and backward masking affects the question of audio coding substantially, as will be shown in the next section.

### 3.8.7   Audio Signal Statistics

As mentioned in the introduction, a good coder must do a good job of both satisfying the perceptual processes (irrelevancy) and the mathematical processes (redundancy). The previous section suggests some mechanisms (simultaneous masking, forward masking) in the HAS that can be exploited to substantially reduce the irrelevancy of an audio signal. In this section, the redundancy present in a wide variety of audio signals and the way in which the perceptual processes permit the removal of that redundancy are examined. The primary measure of redundancy used in this section is the SFM,[3] which measures the redundancy available to a "perfect" predictor operating over the time span of the SFM measurement.

The SFM was calculated for a large variety of music signals, from those known to be easily coded to those known to be difficult outliers (out of the main stream), for current audio coders, totaling about 2.5 hours of 16-bit stereo audio signal. These measures were calculated separately for each channel and do not reflect any additional redundancy that may be available by using interchannel processing. The number of samples was varied in powers of 2 from $2^5$ to $2^{13}$, representing a time span, at 44,100 or 48,000 samples per second, of 0.666 ms to 170.66 ms at a 48-kHz sampling rate. The results of this set of calculations are shown in Fig. 3.18,[53] where the SFM is calculated in decibels of available prediction gain. Each "+" represents the mean SFM over an entire song or music sample at a given time duration. The three lines represent the mean and $\pm 1$ sigma of the data values shown at a given time duration.

While this measure of redundancy is revealing and shows that music signals vary substantially in their redundant content, what is more revealing is Fig. 3.19,[53] which shows the SFM and the actual redundancy extraction available after considering the time constraints of the perceptual process for two signals, one with substantial time-domain stationarity and one with much less time-domain stationarity. It is clear from these figures that a great deal of redundancy extraction may be offset by perceptual requirements if the analysis window does not vary according to the perceptual constraint. In fact, modern low-rate audio coders have a "window-switching" or "block-switching" filterbank that allows the coder to use a psychoacoustically appropriate filterbank during situations (transients, some pitchlike signals) where short filterbanks are appropriate. The method and exact details of this switching vary between coders.

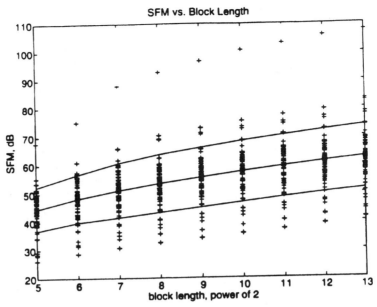

**FIGURE 3.18** Scatter plot of SFM versus block length for 44.1- and 48-kHz sampled signals. Each + represents an observation of one sample signal at one block length. The middle curve is the mean.

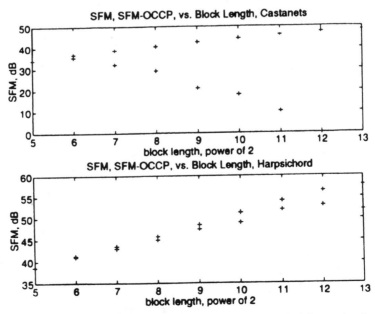

**FIGURE 3.19** Plot of SFM and SFM-OCP (over coding protection) for two signals: a harpsichord and castanets. The upper + in each plot is the SFM for the given signal and block length. The lower + is the SFM-OCP for that signal at that block length.

## 3.9  CURRENT TECHNOLOGY

A variety of audio coders are currently in use or proposed for the marketplace. In general, most of the audio coders—and all of the low-bit-rate coders—are "perceptual" coders, with high-resolution switched filterbanks, active psychoacoustic models, and noise-masking controlled noise or bit allocation across the frequency bands. In this section, short descriptions are presented of the basic audio coder algorithms that are, or are likely to soon be, available in the consumer marketplace. The algorithms are organized by complexity, and the least complex algorithms will be discussed first. Basically, the classes or algorithms are

- Simple subband coders
- Perceptually controlled short-filterbank subband coders
- Perceptually controlled switched-filterbank subband coders

  Coders without noiseless compression facilities

  Coders with noiseless compression facilities

  Coders using hybrid filterbanks

  Coders using transform-form filterbanks

- Perceptually controlled, switched-filterbank, linear predictor subband coders

### 3.9.1  Simple Subband Coders

There are two examples of a simple subband coder in current use. The least complex is the G722 7-kHz bandwidth algorithm discussed in the speech coding section. The other is the APTX coder,[55] which breaks a signal into four bands and handles each band independently. This coder is specified to provide a bit rate of 192 kb/s per channel and offers good, but not nearly transparent, coding on typical audio signals, and it has been known to cause artifacts on unusual audio signals. Despite the performance issues, this algorithm, which is simple, transmission-error tolerant, and easily implemented, is in place for many radio station studio-transmitter links and some other applications, including motion picture theater storage and reproduction.

### 3.9.2  Perceptually Controlled Short-Filterbank Coders

There is one perceptually controlled short-filterbank coder, available in several forms and marketed for a variety of purposes. This coder is variously known as Musicam, PASC, or MPEG-I Audio Layer 1/Layer 2.[52,55] This coder uses a 32-band pseudo-exact reconstruction filterbank, PCM quantizers in each of the 32 bands with the step size of the quantizer adjusted every 8 ms according to psychoacoustic constraints, and a simple bitstream generation method that uses either straight PCM codewords or a radix encoding scheme to combine several small codewords. Layers 1 and 2, sometimes referred to as PASC and Musicam, respectively, have different block structures and bit rate overheads. Layer 1 uses a simple bitstream format with an 8-ms block size, while Layer 2 uses a 24-ms block size, with three 8-ms blocks per macro block and shared information between the 8-ms frames.

PASC is used in the Digital Compact Cassette format sold by various manufacturers (see Chapter 23), and Musicam is used for the Eureka 147 digital audio broadcast standard (see Chapter 14). Although PASC and MPEG-I Layer 1 are the same in principle, they have slightly differently defined bitstreams upon application. The same is true of Musicam and MPEG-I Layer 2.

### 3.9.3 Perceptually Controlled Hybrid-Filterbank Subband Coders

The currently prominent hybrid coder is the ATRAC coder used in the Sony MiniDisc system (Chapter 23). While the exact details of the ATRAC coder are not available, it is known to be a coder that uses an octave-band split followed by an MDCT[3,57] of different lengths, resulting in differing lengths of impulse response at different frequencies. While this coder is said to be hardware efficient, its audio quality performance has been questioned.

### 3.9.4 Perceptually Controlled Switched-Filterbank Subband Coders

There are several groups of switched-filterbank subband coders. The one that is the least complex is the switched-filterbank coder without noiseless compression facilities. The one notable coder in this group is the Dolby AC2/AC3 family of audio coders, which have not used Huffman or other noiseless coding. The AC2 and AC3 audio coders have been proposed or adopted for many uses from radio station studio-transmitter links to the five-channel audio transmission for the proposed new U.S. HDTV standard.

***Coders with Noiseless Compression Filterbanks.*** There are two subclasses of this variety of coder: that with the hybrid filterbank and that with the structured filterbank. The first variety is exemplified by the MPEG-I Layer 3 audio coder. In this coder, the 32-band filterbank used in layers 1 and 2 is further separated in frequency by a factor of either 6 or 18 to provide additional frequency separation at low frequencies as well as additional coding gain at all frequencies. This coder also uses a Huffman coding back end to get the most out of the perceptually appropriate (but information-theoretically inefficient) quantizers. This coder is the first very-low-rate audio coding standard to provide good performance at bit rates of 64 kb/s or below. Its parent, ASPEC, was a transform-structure switched-filterbank coder with Huffman coding and was derived from two earlier coders called PXFM and OCF.

Finally, there are the current coders that use noiseless compression on a transform-structure filterbank. While there are several such coders, the current coder with that structure is the PAC coder,[58] which is a direct stereo or stereo pair coder that degenerates in an efficient way to monophonic encoding. Relevant features of the PAC coder beyond the use of a transform structure, the perceptual encoding, and the noiseless coding include multichannel coding and the use of wavelet filterbanks in addition to MDCT filterbanks for transient signals.

***Perceptually Controlled, Switched-Filterbank, Linear-Predictor Subband Coders.*** The MPEG-NBC coder is currently proposed as an MDCT filterbank coder with block lengths of 1024 and 128 lines for stationary and nonstationary signals, respectively; line-by-line backward adaptive linear predictors for long blocks; Huffman coding; and several multichannel techniques. The multichannel techniques proposed (but not yet accepted) for this standard include a stereo pair structure like the PAC coder, an intensity stereo encoding like MPEG-I, and a "coupling channel" that is a multichannel version of an intensity stereo coder, whereby more than two channels involving signals other than a stereo pair are coded jointly, using one waveform.

Each of the preceding audio coding algorithms has been designed with a particular target result, e.g., low complexity, high quality, low bit rate, or high bit rate. As a result, the various combinations of filterbank, perceptual control, and entropy coding have been explored fairly thoroughly and, in general, the more complex algorithms provide a better compression/quality trade-off. There are some exceptions to this rule, of course.

The MPEG-NBC coder has recently been tested (fall 1996) in the five-channel format at a bit rate of 320 kb/s and demonstrated coding results above 4 on the CCIR scale for all signals tested. This performance occurred both for the full NBC coder (called the main profile) and for a reduced-complexity version that does not use the LPC predictors (called the low complexity profile).

## 3.10   *VIDEO CODING AND THE HUMAN VISUAL SYSTEM*

This section begins with a brief overview of the human visual system (HVS) and the underlying properties that form the basis for the visual models that have been successfully incorporated into image coding systems. Excellent references are provided for those interested in more details. Several perceptually based coders are described, as are techniques that effectively incorporate perceptual models into several of the current standards for image compression.

### 3.10.1   The Human Visual System

Light from an object is focused by the cornea and lens to form an image on the retina, a thin layer of neural tissue at the back of the eyeball. The retina contains photoreceptors composed of photosensitive pigments that sample the image and encode the visual information into different intensity ranges. The human eye consists of two basic types of photoreceptors: cones and rods. The cones are responsible for spatial acuity and color vision under photopic (high) light levels, and the rods are responsible for vision under scotopic (low) light levels. The fovea is the region of highest visual acuity in the retina and consists of densely spaced cones and no rods. It has been established that there are three types of cones, each containing a different absorption spectrum. Although there are many more rods than cones, and the rods are densely spaced in the retina, many rods converge to one neuron, resulting in very poor spatial resolution but increased light sensitivity. Each cone, on the other hand, excites many neurons, resulting in high spatial resolution. The visual data from the retina is output through the retinal ganglion cells and carried through the lateral geniculate nucleus (LGN) to the visual cortex.

Most of the visual signal from the retina arrives at a single area within the occipital lobe of the cortex called the *primary visual cortex*. The data collected by vision scientists leads us to believe that visual information is split into different paths, specialized for different tasks. Within each path, visual data seem to be further split based on local orientation, spatial scale, and color. Although tremendous progress has been made in understanding the HVS over the last several decades, there is still much to be learned, especially concerning the processes occurring in the visual cortex. Excellent books detailing the progress made in vision science are available.[57,58]

### 3.10.2   Visual Masking

As mentioned earlier, a perceptual coder can be thought of as a *destination coder* rather than a *source coder.* The idea behind a destination coder is to use a model of what the end user can perceive in order to remove "irrelevancy." This is in contrast to source coding, which focuses on removing redundancy. In practice, an effective coder is both a source and a destination coder. Here we examine some schemes that try to model characteristics of the human receiver, the HVS, in the design of image and video compression systems. Many of the source coders are motivated by mean square error minimization. If the HVS utilized mean square error as a distortion measure, the task would be easy. Unfortunately, the visual system is not that simple.[59]

Over the years there has been a steady increase in the extent to which knowledge about human perception has been incorporated into image compression. An overview of the current state of the art in perceptually based compression can be found in the literature.[57]

*Frequency.*   Most of the early work on perceptual coders has used some form of *frequency sensitivity* of the HVS as described by the *modulation transfer function* (MTF).[58] This function describes the human eye's sensitivity to sinewave gratings at various frequencies. Some of the literature refers to the spatial frequency sensitivity as a *contrast sensitivity function* (CSF). A commonly used model for the MTF can be found in the literature.[61,62] To apply properties of the MTF to differential cosine transform (DCT)–based coders, Nill[63] has developed a transformation that accounts for the difference between the Fourier and DCT bases. Similar procedures could be used to transform the model to other frameworks for frequency analysis. From these frequency sensitivity–based models, given a fixed minimum viewing distance, it is possible to determine a *static JND* threshold for each frequency band. These thresholds can be used for both quantization[64] and bit allocation.[65] Many of the perceptual coding models are based on finding a JND threshold for each image location. Frequency sensitivity is a good starting point for building a JND threshold profile; however, to have a more effective coder, we must consider image-dependent, local visual masking properties.

*Lightness.*   The perceptual models can be extended to include lightness or sensitivity to different gray levels as well. Lightness sensitivity refers to the threshold of detectability of noise on a constant background. This phenomenon depends on the gray level of the background as well as the gray level of the noise. For the HVS, this is a nonlinear function. Lightness sensitivity can be included either implicitly, by developing functions of masking threshold versus local lightness, or explicitly, by using homomorphic systems or so-called perceptually uniform or equi-luminant color spaces.[66]

Although the preceding methods are a good starting point for utilizing properties of the HVS for image compression, they do not fully utilize the masking properties of the HVS. Frequency sensitivity is a global property dependent only on the image size and viewing conditions. Lightness sensitivity has been exploited mainly via pre- and postprocessing of the entire image. To fully exploit masking properties, we need a more dynamic model that allows for finer control of the quantization process.

*Contrast.*   Another useful perceptual property for coding is known in the vision literature as *contrast masking.*[67] Contrast masking refers to the ability to detect a signal component in the presence or absence of other signal components in the same locality. Contrast masking is strongest when the two signals are of the same spatial frequency and orientation, so choosing an appropriate frequency decomposition is critical in being able to effectively utilize contrast masking in coder design. We can best illustrate this effect with an example. A single pixel with a value above the lightness detection threshold will be quite visible against a flat-field background but will become imperceptible against a random-noise background with the same mean value as the flat field. In other words, the presence of the noiselike background signal *masks* the presence of the second signal. Contrast masking depends on the local scene content or texture and has the desired property of *local control.*

In summary, in order to maximize perceptual coding effectiveness, the JND profiles should be functions of the MTF of the human visual system, the local lightness sensitivity, and local contrast. The MTF provides the most conservative estimate of JND and is image independent. Lightness provides additional perceptual gain but is based on a constant gray-level background. Finally, contrast masking adds additional perceptual gain over frequency and lightness due to the ability to hide distortion locally in highly textured areas.

### 3.10.3   Perceptual Image Quality Models

Signal-to-noise ratio, or mean square error, although widely used in source-coding design, is not an accurate predictor of subjective image quality.[68–73] This becomes especially obvious at low bit rates.[74] One way of addressing this problem is to transform the errors into a "human visual space" where mean square error could be used. This is the approach used in refs. 63 and 75, where original and coded images are transformed into a human visual space by utilizing Weber's law and the MTF. The mean squared error between the transformed images provides a weighted distortion that produces better correlation to subjective evaluation than the unweighted mean squared error but is limited by the fact that the visual model is global in nature.

The visible differences predictor (VDP) has been introduced[76] to provide a tool for evaluating the perceptual image quality of monochrome still images in a more effective way than the global approach provides. A detailed visual model is applied to the original image, and local JND thresholds are computed for each pixel in the image. The visual model is based on spatial frequency sensitivity, lightness sensitivity, and contrast masking. A frequency decomposition based on a modified version of the cortex transform[77] is used for the contrast masking stage. The differences between the original and compressed images are computed and transformed into JND units and then displayed as an output image indicating the location and magnitude of the visible differences between images. The VDP yields very good results in assessing image quality and provides a valuable tool in evaluating and designing imaging systems. However, since such a technique avoids providing a global metric, it does not provide a direct way to rank suprathreshold image coding systems—an important consideration, especially for low-bit-rate applications.

### 3.10.4   Perceptual Image/Video Coders

Here we describe several perceptually based image coders that, as mentioned before, can be thought of as *destination coders*. The perceptual coding paradigm for these codes is based on using an HVS model to compute a JND profile, which in turn is used directly for coder design. The coders are separated into two categories: ideal perceptual coders and standards-based perceptual coders. The so-called ideal perceptual coders are not constrained to any particular frequency decomposition or predefined bitstream syntax, whereas the standards-based perceptual coders are constrained to meet specified standards requirements. The unconstrained ideal coders are able to provide more perceptual gain than the standards-based coders. However, since standards-based coders are prevalent in today's technology, it is important to see how visual properties can be exploited within such frameworks.

*Ideal Perceptual Coder.*     Several coders have been proposed that utilize perceptual models to achieve higher perceived quality for a given bit rate. Chen and Pratt[78] provide a fixed quantization matrix accounting for the MTF of the visual system, in conjunction with the concept of encoding only the portion of the signal that exceeds a local threshold. More recent systems[79–81] utilize frequency and lightness sensitivity and, in addition, use some form of contrast masking. In these methods, instead of computing JND thresholds based on an explicit visual model, each DCT block is classified according to its energy and quantized appropriately. The perceptual coder proposed by Xie and Stockham[82] is based on a vector quantization coding scheme with a perceptually weighted distortion metric.

Several coders will now be described that use visual models to generate JND profiles that are directly used in designing the coding system.

*The Cortex Coder.*     The choice of the analysis filterbank used in a coder can affect the performance of the coding system both from a source-coding and destination-coding point of view. Perceptually, the filterbank should be chosen so that the spatial frequency location of the quantization distortion can be controlled. Ideally, the addition of quantization distortion to one coefficient should not affect coefficients that are not adjacent to the one perturbed. The analysis filterbank should also mimic the visual system's structure. For a human, this structure is a set of filters with frequency spacing of about 1.5 octaves and an angular width of about 40 degrees.

The DCT (and other uniform filterbanks) meet the first criterion,[83] but not the second. This leads to difficulty in creating masking models since there is a mismatch between the underlying structure of the model and the structure of the transform being implemented. One means of implementing a transform that mimics the visual system is the *cortex transform.*[77] The cortex transform provides a signal decomposition ideally suited to take advantage of contrast-masking effects. This is due to the fact that masking effects are strongest when the signals are of the same spatial frequency and orientation. These ideas provide the basic structure for an image coder.[84] The transformed data are quantized using a scheme based on contrast-masking results. The main stumbling block in using this approach is that the cortex transform is not a maximally decimated filterbank. This results in an expansion in the amount of data that need to be encoded. At present, visual masking models have not been able to provide enough coding gain to overcome this disadvantage.

*The Perceptual Image Coder (PIC).*   The perceptual image coder (PIC), first presented by Safranek and Johnston,[18] is based on a uniform subband decomposition and a visual model for generating a JND profile to efficiently quantize the image data. A block diagram of the PIC coder is given in Fig. 3.20. To take advantage of visual masking as well as the characteristically low-pass nature of image spectra, it is useful to perform some sort of frequency decomposition as a first step to an effective image coding system. The PIC coder is based on decomposing the original image into 16 uniform frequency bands using separable generalized quadrature mirror filterbanks (GQMFs).[85] The 16-band decomposition was chosen to allow a reasonable trade-off between spatial and frequency localization.

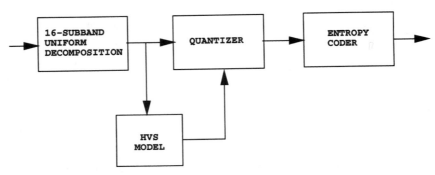

**FIGURE 3.20**   Block diagram of the PIC coder.

The visual masking model introduced by Safranek and Johnston[18] determines a distortion sensitivity profile based on frequency, lightness, and contrast masking. This profile provides a measurement of the JND and the minimal noticeable different (MND), which can be used to determine an optimal quantizer step size for either a minimum-rate perceptually transparent coder or a fixed-rate minimum-distortion coder.

The frequency sensitivity and lightness sensitivity are determined using the basic experimental paradigm in which a subject views a variable-power noise square at a specified distance and determines the noise variance for which the noise square is *just noticeable* against a specified flat-field background. The frequency sensitivity is experimentally measured by repeating the experiment for each of the 16 subbands at the most sensitive level of a mid-gray background. This is referred to as the *base sensitivity* by Safranek and Johnston[18] and is a measurement of the MTF within this particular subband framework. The experiment is repeated at various background lightness levels to determine the lightness sensitivity. The *texture sensitivity,* or contrast-masking portion of the visual model, is determined from an empirical calculation in which texture is defined as a local intrablock variance with a local dominant frequency and the MTF is used to establish a relation between visibility of distortion and texture. Note that the base sensitivity assumes a flat, mid-gray area and gives a very conservative estimate of the JND. By adding lightness, the JND is modulated according to local lightness but still assumes a flat image area. The texture model further modulates the JND by taking advantage of contrast-masking properties in local nonflat areas.

The JND estimates are used to generate spatially adaptive bit allocation in each subband. Each subband is coded using a differential pulse code modulation (DPCM) coder with a variable uniform midriser quantizer.[86] For a perceptually lossless coder, the quantizer step size is adjusted so that the perceptual criterion is met at the most sensitive area in the subband. Quantization is followed by Huffman coding using $1 \times 1$, $2 \times 2$, or $4 \times 4$ blocks.

The PIC coder was one of the early attempts to use a perceptual model in the design of a practical coding system. Although the original coder could be improved upon, it produces very good, competitive results and often outperforms more current coders that use more sophisticated frequency decompositions and entropy coding schemes. A recent study of several coders operating in the suprathreshold range[74] shows that a modified version of the PIC coder (with an $8 \times 8$ uniform subband decomposition) does very well at low bit rates, outperforming JPEG considerably and competing with a wavelet-based scheme designed to operate at lower bit rates. This study also shows that the PIC coder (based on the original $4 \times 4$ decomposition) outperforms other coders, including JPEG, at higher bit rates. The authors show that the metric provided by the visual model used in the PIC coder provides a more accurate estimate of image quality, based on subjective testing of several coders operating over a wide range of bit rates in the suprathreshold range. The authors also propose that subjective image quality depends very much on *image content*. In other words, coder performance depends on particular image characteristics. For example, a coder that yields the best results for portrait-type images at low bit rates may not perform as well for images with strong structured edges. Steps for improving the performance of PIC include a better-suited frequency decomposition, as in the coder recently proposed in Watson et al.[87] A better model for contrast masking should also yield improved results over the current PIC coder.

Other work related to incorporating a visual model includes the coding of gray-level originals for the transmission of facsimile images.[88-90] In one proposition,[88] instead of halftoning the image at the transmitter and encoding the binary information, a perceptually coded gray-level image is sent and the image is halftoned at the receiver. The proposed algorithms[88-90] result in a greatly reduced bit rate over the standard approach and provide a framework for incorporating model-based methods for efficient halftoning at the receiver. Models of the HVS as well as the printer have been examined, and the results show a dramatic improvement in image quality over the standard techniques.

***Standards-Based Perceptual Coders.***    An interesting question is if any coding gain can be realized by taking advantage of visual models within the framework of current compression standards. Although compliance with the standards limits the amount of perceptual gain that can be incorporated into the coders, a significant improvement in coder performance could be realized by using, for example, perceptually designed quantization matrices. Here we describe several algorithms developed for both still-image and video coding that take advantage of the properties of the HVS within the standards framework.

*Still-Image Coders.*    Perceptual prequantization[91] applied within the framework of the standards (JPEG and MPEG) has provided significant coding gain over the standard implementation. Perceptual coding has been applied to the baseline lossy JPEG still-image compression standard.[91,92] The challenge is to find a means of incorporating visual properties within the existing bitstream syntax.

Since the quantization table is not fixed for JPEG, visual models could be used to derive perceptually optimal tables. In fact, perceptually optimal quantization matrices based on image size, monitor white point, and viewing distance have been presented.[93] An image-independent model for predicting the detection thresholds based on the viewing conditions and global properties of the visual system (MTF) has been derived. This work provides a strong start toward a perceptually based coder but does not take advantage of local, image-dependent masking properties. This has been addressed,[94] and the approach has been extended to determine an image-dependent quantization table that not only incorporates the global conditions, but also accounts for local contrast masking. An iterative approach is proposed that determines an image-dependent quantization table providing a specified level of visual distortion. This offers significant improvement over the global approach, but the quantization table adapted for the whole image does not take advantage of local adaptation. It can, however, be applied in a noniterative manner. The local

JND value for each coefficient can be computed as a function of the global quantization table and the value of the DCT coefficients of the local block. This results in a perceptually optimal quantization table for each block. Unfortunately, JPEG allows only one quantization table for each image, and the overhead needed to send a different quantization table for each block is impractical. However, some local control through adaptive quantization can be incorporated into JPEG by zeroing out all input coefficients that fall below their corresponding perceptual quantization thresholds while maintaining perceptually transparent image quality. This produces a result that complies with the JPEG bitstream specification while reducing the bit rate required to encode the image. A block diagram of the perceptually based JPEG encoder is presented in Fig. 3.21. The perceptually based encoder results in a bit rate savings of 10–40 percent. Figures 3.22 and 3.23 illustrate the perceptual gain of perceptual JPEG (PJPEG) over JPEG for an image; Fig. 3.22 represents a quality factor of 75 for both images, and Fig. 3.23 represents a quality factor of 20 for both images.

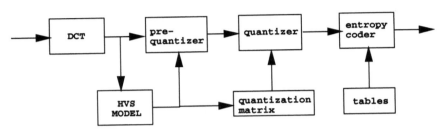

**FIGURE 3.21**   JPEG encoder with perceptually based prequantization.

*Motion Video Coders.*   In applying visual models to motion video coding, we add the dimension of temporal information. Understanding the complexities of temporal masking and how it can be used effectively for coding of video remains a challenging area of research. On a different level, even prior to quantization, we still use very informal criteria for fundamental designs, such as choice of temporal (and spatial) resolutions and the degree of chrominance subsampling.

The framework for all current video coding standards, the MC/DCT approach described in brief previously, consists of motion compensation to take advantage of interframe correlations followed by frequency-domain coding of the interframe error. One approach to incorporating perceptually based quantization into motion-compensated DCT coders is described by Puri and Aravind.[95] This approach modifies the quantizer scaling parameter based on macroblock classification of texture and global scene complexity. This differs from the strict JND/MND approach in that perceptual thresholds are not specifically determined but are implicit in the adjustment rules.

A prequantization technique as presented earlier for JPEG can also be adapted to MPEG and CCITT systems. The MND thresholds provided by the visual model could be used to zero out local differences that are smaller than their corresponding thresholds. For MPEG-type encoders running in a constant quality mode, a rate reduction of 5–35 percent is achievable with no visible loss in picture quality.

A low-bit-rate HDTV coder based on perceptual coding of the interframe error has also been described recently. The methodology used is similar to that of the PIC coder, but the extension is approximate in that the structure of the interframe error is very different from the uniform-noise model used in other JND experiments. Further, the HDTV viewing distance is smaller than the 6×-picture-height model of the original PIC experiment. These effects are offset by the fact that the JND can now be a function of the time dimension as well, and the benefits of temporal masking are now available. In the context of newly uncovered background, there is a small latency period during which the newly exposed areas do not have to be perfectly encoded.

**FIGURE 3.22**   JPEG *(top)* versus PJPEG *(bottom)*. The quality factor is 75 for both images, but the PJPEG bit rate is 10 percent less than that for JPEG.

**FIGURE 3.23** JPEG *(top)* versus PJPEG *(bottom)*. The quality factor is 20 for both images, but the PJPEG bit rate is 17 percent less than that for JPEG.

Using this procedure, HDTV images ($1280 \times 720$ pixels, 60 frames per second) are compressed to 17 Mb/s to facilitate simulcasting over 6-MHz NTSC channels.[96] The rate of 17 Mb/s is not adequate to provide perceptually transparent coding of all HDTV inputs, and the result is a constant-bit-rate system that oscillates between the JND and MND modes. More sophisticated visual models could improve performance for HDTV as well as other video coding applications.

## 3.11  SUMMARY

In this chapter we have tried to cover the basic coding models in a conceptual fashion and have provided the reader with a variety of specific and technological references for further investigation. We would like to close with a few words on the use of compression algorithms and on perceptual compression algorithms in particular.

First, most compression is lossy. This means that with multiple generations of encoding/decoding, an additional quality loss is introduced at each generation. This is particularly true of perceptual coders, because they eliminate the most information and come closest to the perceptual limits on the first encoding. Herb Squire[97] has shown some very enlightening results on the tandeming of audio coders, and they are required reading for anyone considering such uses for perceptual audio coders.

Second, perceptual coding in particular is intended for *final delivery* only. It is not, despite some claims to the contrary, a wise choice for data acquisition or perhaps even for archival storage, because once the perceptual encoding process is completed, any substantial reprocessing may create unmasked noise that will be quite objectionable.

Finally, in compression, as in most things, you get what you pay for, both in terms of compression ratio and encoder/decoder complexity. As you go to lower and lower bit rates, the quality goes down, and as you proceed to encoders and decoders with more complexity, the threshold for audible degradation as a function of compression ratio also goes down. This does not imply that with sufficient complexity one can reduce the bit rate to an arbitrary level. However, it is always possible to reduce the total rate to a level below that at which the perceptual system involved (audition, vision) can receive information from a given signal. At such an information rate, "transparent" or "unchanged" signal quality is very likely to be impossible for some signals.

Within the intended limits of coding algorithms, the use of a signal coder is well suited to situations where the raw data rate of the signal creates either economic or technological difficulties in a system. In the case of economics, the use of compression has directly affected the cost of the telephone call. As for technological issues, the use of compression algorithms makes such devices as the MiniDisc, the Digital Compact Cassette, digital audio radio, and Internet audio distribution possible. It is often the case that without enabling compression technology, a service may be not only uneconomical, but impossible.

## GLOSSARY

**Audio bandwidth**   The signal range from 20 Hz to 20 kHz, according to the standard specification, although there is some evidence of acoustic perception outside that bandwidth in exceptional individuals, particularly young children.

**Brightness**   The aspect of perception that changes as the intensity of a patch of light is varied.

**Coding**   Also called compression or bit rate reduction, the process of taking a signal and reducing the number of bits, energy, time, or bandwidth necessary for its transmission. In the context of this chapter, coding refers specifically to bit rate reduction of a digitally captured or encoded signal.

**Contrast**   A measure of the difference in intensity between two adjacent patches of light.

**Human auditory system (HAS)**   The outer, middle, and inner ear along with the parts of the central nervous system that are involved in hearing, speech understanding, spatial localization, and other auditory functions.

**Human visual system (HVS)**   The eye, optic nerve, visual cortex, and other parts of the central nervous system involved in seeing and interpreting images.

**Irrelevancy**   The part of a signal that is not perceptible to the human user under the normal circumstances of use. In other words, irrelevancy is the part of the signal that may be removed at the encoder, and removed at the decoder, with no loss of content.

**Just noticeable difference (JND)**   The level at which an observer can begin to distinguish between an original signal and a noisy/corrupted/distorted signal. The JND is sometimes referred to as a threshold or perceptible threshold.

**Linear predictive coding (LPC)**   A particular form of redundancy removal, where a "linear predictor" is created to remove the redundancy due to vocal tract (or other signal generation) effects. LPC is commonly used in a speech coder because this method allows for good tracking of the redundancy introduced by the human vocal tract.

**Modulation transfer function (MTF)**   A representation of the human visual system's response to visual stimuli at various frequencies.

**Noiseless coding**   Coding methods that do not remove any information present in the signal, i.e., that allow for perfect reconstruction of the original digital input signal. The classic set of noiseless coders are the information-theoretic coders such as Huffman coders, Ziv-Lempel coders, and arithmetic coders. However, other coders—including subband, transform, and other source coders—can be used as noiseless coders in some circumstances. A particular example is lossless JPEG.

**Perceptual coding**   A lossy form of coding where parts of the signal that are not perceptible to the end user are removed. Perceptual coding is always lossy and is in some sense the dual of source coding, i.e., "destination coding," where information important to the destination (human being) is preserved and information not important is removed.

**Quantization**   A process by which information, redundancy, and irrelevancy are removed from a signal. A quantizer is a construct that takes an amplitude-continuous signal and converts it to discrete values that can be reconstructed by a decoder.

**Redundancy**   Parts of a signal that are, because of the process(es) that generated it, predictable or in some way related to other parts of the same signal. In a strict sense, redundancy in a signal is not part of the information content of that signal, as it may be removed at the encoder, and restored at the decoder, with no loss of content.

**Signal-to-noise ratio (SNR)**   The measure of the energy in the signal, within a certain bandwidth and/or time, versus the energy of the noise that was added to the signal by some process. In this context, *noise* refers not only to added random noise but to any other difference, other than simple scaling, present in the reconstructed signal. In video, a variant of SNR called PSNR (peak signal-to-noise ratio) is often used; this corresponding to the energy of the noise versus the maximum possible signal energy.

**Simultaneous contrast**   A measure of the perceived difference between two adjacent patches of light.

**Source coding**   Techniques that use the redundancy in a signal to reduce the bit rate. Some source coders (noiseless coders) remove only redundancy and provide perfect reconstruction of the decoded signal. Others remove redundancy and some information as well and provide a noisy decoded signal, but usually with unobjectionable noise characteristics. Typical kinds of source coding are LPC, ADPCM, CELP, subband, and transform.

**Telephone bandwidth**   Usually 200 Hz to 3200 Hz, although some algorithms and transmission facilities provide signals from 50 Hz to 3600 Hz, depending on the situation and equipment involved.

**Wideband speech**   Speech having a bandwidth of 50 Hz to 7 kHz, also referred to as *commentary* grade or *remote contribution* quality.

## REFERENCES

1. J. D. Johnston and Brandenburg, K., "Wideband coding: Perceptual considerations for speech and music," in Sadaoki Furui and M. Mohan Sondhi (eds.), *Advances in Speech Signal Processing,* Dekker, New York, 1992, pp. 109–139.

2. H. S. Malvar, *Signal Processing with Lapped Transforms,* Artech House, New York, 1992.

3. N. S. Jayant and Noll, P., *Digital Coding of Waveforms,* Prentice Hall, Englewood Cliffs, N.J., 1984.

4. N. Morrison, *Introduction to Fourier Analysis,* John Wiley & Sons, New York, 1994.

5. B. S. Atal and Remde, J. R., "A new model of LPC excitation for producing natural-sounding speech at low bit rates," *Proceedings of IEEE International Conference on Acoustics, Speech, and Signal Processing,* Paris, April 1982, pp. 614–617.

6. M. R. Schroeder and Atal, B. S., "Code-excited linear prediction (CELP): High quality speech at very low bit rates," *Proceedings of IEEE International Conference on Acoustics, Speech, and Signal Processing,* March 1985, pp. 937–940.

7. L. R. Rabiner and Schafer, R. W., *Digital Processing of Speech Signals,* Prentice Hall, Englewood Cliffs, N.J., 1978.

8. W. B. Kleijn, Krasinski, D. J., and Ketchum, R. H., "Improved speech quality and efficient vector quantization in self," *Proceedings of IEEE International Conference on Acoustics, Speech, and Signal Processing,* April 1988.

9. J.-H. Chen and Gersho, A., "Adaptive postfiltering for quality enhancement of coded speech," *IEEE Transactions on Speech and Audio Processing,* Jan. 1995, pp. 59–71.

10. J.-H. Chen, Cox, R. V., Lin, Y. -C., Jayant, N. S., and Melchner, M. J., "A low-delay CELP coder for the CCITT 16 kb/s speech coding standard," *IEEE Journal on Selected Areas in Communications,* June 1992, pp. 830–849.

11. J. -H. Chen, Kleijn, W. B., and Paliwal, K. K., "Low-delay coding of speech," *Journal of Speech Coding and Synthesis,* Oct. 1995, pp. 209–256.

12. R. Salami, et al., "Description of the proposed ITU-T 8 kb/s speech coding standard," *Proceedings of the 1995 Speech Coding Workshop,* Annapolis, Md., Sept. 1995, pp. 3–4.

13. A. V. McCree, et al., "A 2.4 kbit/s MELP coder candidate for the new U.S. Federal Standard," *Proceedings of IEEE International Conference on Acoustics, Speech, and Signal Processing,* May 1996, pp. 200–203.

14. R. V. Cox, Kleijn, W. B., and Paliwal, K. K., "Speech coding standards," *Journal of Speech Coding and Synthesis,* Oct. 1995, pp. 49–78.

15. A. N. Netravali, Petajan, E., Knauer, S., Mathews, K., Safranek, R. J., and Westerink, P., "A high quality digital HDTV codec," *IEEE Transactions on Consumer Electronics,* Aug. 1991, pp. 320–330.

16. W. B. Pennebaker and Mitchell, J. L., *JPEG Still Image Data Compression Standard,* Van Nostrand Rinehold, New York, 1993.

17. G. K. Wallace, "The JPEG still picture compression standard," *Communications of the ACM,* Apr. 1991, pp. 31–43.

18. R. J. Safranek and Johnston, J. D., "A perceptually tuned sub-band image coder with image-dependent quantization and post-quantization data compression," *Proceedings ICASSP,* 1989.

19. S. Mallat, "A theory of multiresolution signal decomposition: The wavelet representation," *IEEE Transactions on Pattern Recognition and Machine Intelligence,* 1989, pp. 674–693.

20. R. Coifman and Wickenhauser, V., "Entropy-based algorithms for best basis selection," *IEEE Transactions on Information Theory,* vol. 38, 1992, pp. 713–718.

21. O. Rioul and Vetterli, M., "Wavelets and signal processing," *IEEE Signal Processing Magazine,* 1991, pp. 14–38.

22. I. Daubechies, "Orthonormal bases of compactly supported wavelets," *Communications on Pure and Applied Mathematics,* 1988, pp. 909–996.

23. I. Daubechies, "The wavelet transform, time-frequency localization and signal analysis," *IEEE Transactions on Information Theory,* 1990, pp. 961–1005.

24. S. Mallat, "Multiresolution approximations and wavelet orthonormal bases of $l^2(r)$," *Transactions of the American Mathematical Society,* 1989, pp. 69–87.

25. C. I. Podilchuk, Jayant, N. S., and Farvardin, N., "Three-dimensional subband coding of video," *IEEE Transactions on Image Processing,* vol. 4, Feb. 1995, pp. 125–139.

26. J. D. Johnston, "A filter family designed for use in quadrature mirror filter banks," *Proceedings ICASSP,* 1980.

27. M. Vetterli, "Multidimensional subband coding: Some theory and algorithms," *Signal Processing,* 1984, vol. 6, pp. 97–112.

28. J. W. Woods and O'Neil, S. D., "Sub-band coding of images," *IEEE Transactions on Audio, Speech, and Signal Processing,* 1986, pp. 1278–1288.

29. J. M. Shapiro, "Embedded image coding using zerotrees of wavelet coefficients," *IEEE Transactions on Signal Processing,* vol. 41, 1993, pp. 3445–3462.

30. P. H. Westerink, Biemond, J., and Muller, F., "Subband coding of image sequences at low bit rates," *Signal Processing; Image Communication,* 1990, pp. 441–448.

31. Y.-H. Kim and Modestino, J. W., "Adaptive entropy-coding subband coding of images," *IEEE Transactions on Image Processing,* 1992, pp. 31–48.

32. G. Karlsson and Vetterli, M., "Three-dimensional subband coding of video," *Proceedings ICASSP,* 1988.

33. C. I. Podilchuk, "Low bit rate subband video coding," *Proceedings ICIP 94,* 1994.

34. J.-R. Ohm, "Three-dimensional subband coding with motion compensation," *IEEE Transactions on Image Processing,* vol. 3, no. 5, Sept. 1994.

35. F. Bosveld, Lagendijk, R. L., and Biemond, J., "Hierarchical video coding using a spatio-temporal subband decomposition," *Proceedings ICASSP 92,* 1992.

36. A. Lewis and Knowles, G., "Video compression using 3D wavelet transforms," *Electronic Letters,* vol. 26, 1990, pp. 396–398, 1184–1185.

37. C. I. Podilchuk and Farvardin, N., "Perceptually based low bit rate video coding," *Proceedings ICASSP,* 1991.

38. P. J. Burt and Adelson, E. H., "The Laplacian pyramid as a compact image code," *IEEE Transactions on Communications,* 1983, pp. 532–540.

39. M. Vetterli and Kovacevic, J., *Wavelets and Subband Coding,* Prentice Hall, Englewood Cliffs, N.J., 1995.

40. P. P. Vaidyanathan, *Multirate Systems and Filter Banks,* Prentice Hall, Englewood Cliffs, N.J., 1993.

41. J. Ziv and Lempel, A., "Compression of individual sequences via variable rate coding," *IEEE Transaction on Information Theory,* Sept. 1978, IT-24, no. 5, pp. 530–536.

42. H. Fletcher, "Auditory patterns," *Review of Modern Physics,* 1940, pp. 47–65.

43. B. Scharf, "Critical bands," in J. Tobias, ed., *Foundations of Modern Auditory Theory,* Academic Press, New York, 1970, pp. 159–202.

44. R. P. Hellman, "Asymmetry of masking between noise and tone," *Perception and Psychophysics,* vol. 11, 1972, pp. 241–246.

45. W. Yost, *Fundamentals of Hearing: An Introduction,* Academic Press, New York, 1994.

46. B. C. J. Moore, *An Introduction to the Psychology of Hearing,* 3rd ed., Academic Press, New York, 1989.

47. H. Fletcher, *The ASA Edition of Speech and Hearing in Communication,* edited by J. B. Allen, Acoustical Society of America, Woodbury, N.Y., 1995.

48. G. von Bekesy, *Experiments in Hearing,* translated by E. G. Weaver, McGraw-Hill, New York, 1960.

49. D. D. Greenwood, "Critical bandwidth and frequency coordinates of the basilar membrane," *JASA,* vol. 33, 1961, pp. 1344–1356.

50. M. R. Schroeder, Atal, B. S., and Hall, J. J., *JASA,* no. 6, Dec. 1979, pp. 1647–1651.

51. J. D. Johnston, "Transform coding of audio signals using perceptual noise criteria," *IEEE Journal of Selected Areas in Communications,* Feb. 1988, pp. 314–323.

52. MPEG-I, *Information Technology: Coding of Moving Pictures and Associated Audio for Digital Storage Media at up to about 1.5 Mbit/s,* CD 11172, Part 3.

53. A. N. Akansu and Smith, M. J. T., "Subband and wavelet transforms, design and applications," in *Filter Banks in Audio Coding,* Kluwer, Boston, 1996.

54. M. Smyth and Smuth, S., "apt-X100: A low-delay, low bit-rate, sub-band ADPCM coder for broadcasting," *Proceedings of the 10th International AES Conference on Images of Audio,* Audio Engineering Society, London, 1991, pp. 41–56.

55. G. Stoll and Dehery, F. "High quality audio bit-rate reduction family for different applications," *Proceedings of IEEE International Conference on Communications,* 1990, pp. 937, 941.

56. J. Princen, Johnson, A., and Bradley, A., *Proceedings of IEEE Conference on Acoustics, Speech, and Signal Processing,* 1987, pp. 2161–2164.

57. T. N. Cornsweet, *Visual Perception,* Academic Press, New York, 1970.

58. B. A. Wandell, *Foundations of Vision,* Sinauer Associates, Inc., 1995.

59. B. Girod, "Psychovisual aspects of image communication," *Signal Processing,* vol. 28, 1992, pp. 239–251.

60. N. S. Jayant, Johnston, J. D., and Safranek, R. J., "Signal compression based on models of human perception," *Proceedings of the IEEE,* Oct. 1993.

61. F. W. Campbell and Robson, J. G., "Application of Fourier analysis to the visibility of gratings," *Journal of Physiology,* 1968, pp. 551–566.

62. J. L. Mannos and Sakrison, D. J., "The effects of a visual fidelity criterion on the encoding of images," *IEEE Transactions on Information Theory,* vol. IT-20, no. 4, July 1974.

63. N. B. Nill, "A visual model weighted cosine transform for image compression and quality assessment," *IEEE Transactions on Communications,* June 1985, p. 551.

64. G. K. Wallace, "The JPEG still picture compression standard," *Communications of the ACM,* Apr. 1991, pp. 31–43.

65. M. G. Perkins and Lookabaugh, T., "A psychophysically justified bit allocation algorithm for subband image coding systems," *Proceedings ICASSP,* 1989.

66. R. E. Van Dyck and Rajala, S. A., "Subband/VQ coding in perceptually uniform color spaces," *Proceedings ICASSP,* 1992, pp. III-237–III-240.

67. C. F. Stromeyer and Julesz, B., "Spatial frequency masking in vision: critical bands and spread of masking," *Journal of the Optical Society of America,* Oct. 1972, pp. 1221–1232.

68. A. J. Ahumada, Jr., "Putting the visual system noise back in the picture," *Journal of the Optical Society of America A,* Dec. 1987, pp. 2372–2378.

69. Z. L. Budrikis, "Visual fidelity criterion and modeling," *Proceedings of the IEEE,* vol. 60, no. 7, July 1972, pp. 771–779.

70. F. J. Kolb, Jr., "Bibliography: Psychophysics of image evaluation," *SMPTE Journal,* Aug. 1989, pp. 594–599.

71. J. O. Limb, "Distortion criteria of the human viewer," *IEEE Transactions on Systems, Man, and Cybernetics,* Dec. 1979.

72. A. Netravali and Haskell, B., *Digital Pictures, Representation and Compression,* Plenum, New York, 1988.

73. L. G. Roberts, "Picture coding using pseudo-random noise," *IRE Transactions on Information Theory,* Feb. 1962, pp. 145–154.

74. T. N. Pappas, Michel, T. A., and Hinds, R. O., "Supra-threshold perceptual image coding," *Proceedings of ICIP 96,* 1996.

75. J. A. Saghri, Cheatham, P. S., and Habibi, A., "Image quality measure based on a human visual system," July 1989, pp. 813–818.

76. S. Daly, "The visual difference predictor: An algorithm for the assessment of image fidelity," *Proceedings of SPIE Conference on Human Vision, Visual Processing, and Digital Display III*, 1992, pp. 2–15.

77. A. B. Watson, "The cortex transform: Rapid computation of simulated neural images," *Computer Vision, Graphics, and Image Processing*, 1987, pp. 311–327.

78. W. H. Chen and Pratt, W. K., "Scene adaptive coder," *IEEE Transactions on Communication*, Mar. 1984, pp. 225–232.

79. B. Chitprasert and Rao, K. R., "Human visual weighted progressive image transmission," *IEEE Transactions on Communication*, July 1990, pp. 1040–1044.

80. D. L. McLaren and Nguyen, D. T., "Removal of subjective redundancy from DCT-coded images," *Proceedings IEE: Part I*, Oct. 1991, pp. 345–350.

81. K. N. Ngan, Leong, K. S., and Singh, H., "Adaptive cosine transform coding of images in perceptual domain," *IEEE Transactions on ASSP*, Nov. 1989, pp. 1743–1750.

82. Z. Xie and Stockham, T. G., Jr., "Previsualized image vector quantization with optimized pre- and postprocessors," *IEEE Transactions on Communication*, vol. 39, 1991, pp. 1662–1671.

83. S. A. Klein, Silverstein, A. D., and Carney, T., "Relevance of human vision to JPEG-DCT compression," *Proceedings of SPIE Conference on Human Vision, Visual Processing, and Digital Display III*, 1992, pp. 200–215.

84. A. B. Watson, "Efficiency of an image code based on human vision," *Journal of the Optical Society of America A*, 1987, pp. 2401–2417.

85. R. V. Cox, "The design of uniformly and nonuniformly spaced pseudoquadrature mirror filters," *IEEE Transactions on ASSP*, 1986, pp. 1090–1096.

86. D. Huffman, "A method for the construction of minimum redundancy codes," *Proc. IRE Transactions on Communication Systems*, vol. CS-11, 1952, pp. 289–296.

87. A. B. Watson, Yang, G. Y., Solomon, J. A., and Villsenor, J., "Visual thresholds for wavelet quantization error," *Proceedings of SPIE Conference on Human Vision and Electronic Imaging*, 1996, pp. 382–392.

88. T. N. Pappas, "Perceptual coding and printing of gray-scale and color images," *SID-92 Digest of Technical Papers*, 1992.

89. D. L. Neuhoff and Pappas, T. N., "Perceptual coding of images for halftone display," *IEEE Transactions on Image Processing*, vol. 3, July 1994.

90. T. N. Pappas and Neuhoff, D. L., "Printer models and error diffusion," *IEEE Transactions on Image Processing*, vol. 4, Jan. 1995.

91. R. F. Safranak, "Perceptually based prequantization for image compression," *Proceedings of SPIE Conference on Huam Vision, Visual Processing, and Digital Display V*, 1994.

92. R. J. Safranak, "A comparison of the coding efficiency of perceptual models," *Proceedings of SPIE Conference on Human Vision, Visual Processing, and Digital Display VI*, 1995.

93. H. A. Peterson, Ahumada, A. J., Jr., and Watson, A. B., "Improved detection model for DCT coefficient quantization," *SPIE Conference on Human Vision, Visual Processing, and Digital Display IV*, Feb. 1993, pp. 191–201.

94. A. B. Watson, "DCT quantization matrices visually optimized for individual images," *Proceedings of SPIE Conference on Human Vision, Visual Processing, and Digital Display IV*, 1992, pp. 202–216.

95. A. Puri and Aravind, R., "Motion-compensated video coding with adaptive perceptual quantization," *IEEE Transactions on Circuits and Systems for Video Technology*, vol. 1, no. 4, Dec. 1991.

96. A. N. Netravali, Petajan, E., Knauer, S., Mathews, K., Safranek, R. J., and Westerink, P., "A high quality digital HDTV codec," *IEEE Transactions on Consumer Electronics*, Aug. 1991, pp. 320–330.

97. Herb Squire, private communication.

98. G. G. Langdon, "An introduction to arithmetic coding," *IBM Journal of Research and Development*, 28(2), March 1984, pp. 135–149.

99. J. D. Johnston, Sinha, D., Dorwward, S., and Quackenbush, S., "AT&T perceptual audio coding (PAC)," in N. Gilchrist and C. Grewin (eds.)," *Collected Papers on Digital Audio Bit-Rate Reduction,* Audio Engineering Society, New York, 1996.

## ABOUT THE AUTHORS

James D. Johnston is employed by the newly formed AT&T Laboratories–Research, Murray Hill, N.J., where he is working on perceptual coding of audio. He is a co-inventor and standards proponent of the ASPEC algorithm for MPEG-I audio. He is the primary researcher and inventor of AT&T's PAC audio coding algorithm. Previously he was employed by AT&T Bell Laboratories, first in the Acoustics Research Department and then in the Signal Research Department. He also represents AT&T in the ANSI-accredited group X3L3.1 and represents that group in the ISO-MPEG-AUDIO (NBC) arena in support of the PAC algorithm. Johnston received the B.S.E.E. and M.S.E.E. degrees from Carnegie-Mellon University, Pittsburgh, in 1975 and 1976, respectively.

Juin-Hwey Chen has been, since October 1996, director of research at Voxware Inc., Princeton, N.J., where he is responsible for speech and audio processing research. He joined AT&T Bell Laboratories, Murray Hill, N.J., in 1988 and first worked in the Signal Processing Research Laboratory and then in the Speech Coding Research Department. He became a Distinguished Member of the Technical Staff in 1994. Due to the AT&T split, he became part of AT&T Research in 1996. While working at AT&T Bell Laboratories, he created several speech coding algorithms for use in various AT&T products and services. His research there also led to the ITU-T (formerly CCITT) G.728 speech coding standard (16-kb/s low-delay CELP). Besides speech processing, he is also interested in image processing and digital communications. He is a Fellow of IEEE and serves on IEEE's Speech Processing Technical Committee. Chen received the B.S.E.E. degree from National Taiwan University in Taipei in 1980 and the M.S. and Ph.D. degrees in electrical engineering from the University of California, Santa Barbara, in 1983 and 1987, respectively. At that university, he was a teaching assistant from 1983 to 1984 and a research assistant from 1984 to 1987. He was a senior engineer at Codex Corp., Mansfield, Mass., from 1987 to 1988.

Christine Podilchuk is with the Image Synthesis and Recognition Research Department at Lucent Technologies, Bell Laboratories, Murray Hill, N.J. Her research interests are in signal processing, particularly for image and video applications. Recently she has been working on video compression, video over wireless, image recognition, and digital watermarking of images. She received the B.S., M.S., and Ph.D. degrees in electrical engineering from Rutgers University, New Brunswick, N.J., in 1984, 1986, and 1988, respectively.

# CHAPTER 4
# ERROR CORRECTION

**Stan Baggen**
*Philips Research Laboratories*

## 4.1  INTRODUCTION

Error correction is a technique for enhancing the reliability of digital information to be stored or transmitted.

With the advance of storage technologies, bits are becoming ever smaller on the media. Apart from increased storage densities, this generally leads to an increased sensitivity to physical disturbances, resulting in more errors. Likewise, the minimization of energy and bandwidth for transmitting digital information results in a large sensitivity to channel disturbances and noise, again leading to errors. Error correction can be used to achieve robust communication, in terms of both time (storage) and space, in spite of unreliable individual bits. Advances in IC technology and the resulting fall of implementation costs have opened up the possibility of using advanced digital signal-processing techniques such as error correction even in consumer products.

Figure 4.1 is a block diagram of a general communication system, from source to user, for clarifying the functional position of error correction. The functional blocks to the right of the dotted line in Fig. 4.1 serve to create a reliable digital communication channel from point A to B. Current source-coding techniques and data applications often require an error rate of no more than one error event in $10^{-12}$ at these points. The error correction encoder and decoder together form the outer shell of such a digital communication channel. As will be shown later, an error correction system usually works with fairly large groups (hundreds or thousands) of bits called *codewords*. The interfaces at points A and B in Fig. 4.1 are then usually defined in terms of these codewords. The bits at these points are called *user bits*.

The interfaces at points C and D must be matched to the modulation process. Also, these functional blocks usually use groups of bits (called *channel bits* at these points) for a mapping of digital information onto waveforms that can be efficiently transmitted through the physical channel. These groups typically contain fewer than 10 bits (for instance, 2 bits in the case of 4-PSK for digital satellite communication, 6 bits in the case of 64-QAM for cable transmission, and 8 bits in the case of EFM, the modulation code used in the Compact Disc). Often the error correction code operates on symbols that correspond to the small groups of channel bits used by the modulator. By the modulation/demodulation processes, the physical channel is transformed into a channel having discrete inputs and outputs at points C and D, respectively. Some discrete channels also give reliability information at point D. Depending on the underlying physics, the error behavior of the discrete channel at point D can be random (e.g., the satellite channel), bursty, or a combination of both (e.g., optical recording).

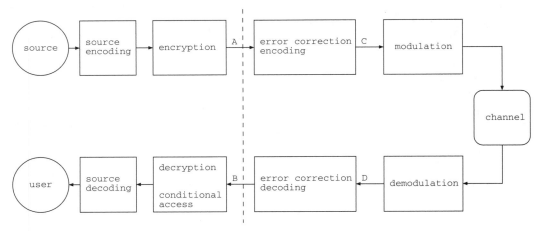

**FIGURE 4.1**   Block diagram of a general communication system.

Note that in most textbooks on error correction, communication theory, and information theory, error correction coding is called *channel coding*. In the field of magnetic and optical storage, however, the term *channel coding* is often used for denoting the modulation process, such as EFM in the case of Compact Discs.

## 4.2   *GENERAL CHARACTERISTICS OF ERROR CORRECTION CODES*

The basic idea behind error correction is that, although individual bits might be unreliable, information conveyed by using groups of bits can be more reliable, provided that information is transmitted in a redundant manner. For example, consider the following code, called a repetition code, where each "1" user bit is replaced by a codeword of three consecutive 1s and, likewise, each 0 is replaced by three consecutive 0s in the encoder:

$$1 \rightarrow 111$$
$$0 \rightarrow 000$$

(4.1)

If the channel now creates an error in at most one of the three transmitted channel bits of a codeword, a decoder can find the original transmitted codeword by looking for the closest (most similar) candidate; i.e., a decoder can correct at most one channel bit error and find the correct original user bit.

### 4.2.1   Hamming Distance and Error Correction

The error correction capability of code (4.1) results from the two codewords differing in three places. The number of places in which two strings differ is called their *Hamming distance d*. If another repetition code were used by replacing each user bit by five copies, a Hamming distance $d = 5$ would be obtained. Such a code is able to correct at most two errors by looking for the closest candidate. In general, if a decoder has to choose between two codewords that differ in $d$ symbol positions, it can correct any $t$ symbol errors by looking for the closest candidate (in the Hamming distance sense) if $2t < d$.

If a code $C$ consists of more than two codewords and each pair of codewords still differ in at least $d$ positions, the code $C$ is said to have minimum Hamming distance $d$. By comparing a mutilated word with all possible alternatives of $C$ and choosing the closest, any $t$ errors can be corrected if $2t < d$. The proof goes as follows. Let $t^*$ be the largest $t$ that satisfies $2t < d$. Suppose the codeword $c_1 \in C$ is transmitted, which the channel modifies to $r$ by creating $t^*$ or fewer errors. Suppose that a decoder, looking for the codeword closest to $r$, finds $c_2 \neq c_1$ by changing at most $t^*$ symbols in $r$. In that case we can obviously change $c_1$ via $r$ into $c_2$ by modifying $2t^*$ or fewer symbols, which contradicts the assumption that any two codewords have a Hamming distance larger than $2t^*$.

### 4.2.2    Detection and Erasure Correction

If a code is used for error detection, only valid codewords are accepted at the receiver side; i.e., no attempt at correction is made. Using the simple code (4.1) for error detection, it is immediately obvious that any two errors will be detected because we need at least three errors to change 000 to 111, or vice versa, thus creating an undetectable error. In general, a code $C$ can certainly detect $d - 1$ or fewer errors.

An erasure is a symbol at point D in Fig. 4.1 that is known to be unreliable and that the decoder has to guess using the other symbols of the code. Such erasure information may be provided by the channel decoder. Suppose that erased symbols are indicated by an "x" and that there are no errors outside the erasures. If we now use code (4.1) for such an erasure correction, we can see that the word x0x must have originated from 000 since otherwise we would have seen a 1 in the middle of the received word. In general, a code $C$ can correct $d - 1$ or fewer erasures. Note that a given code (with a fixed $d$) can correct about twice as many erasures as the number of errors. The performance of a code can therefore be enhanced if reliability information is available and used by the decoder.

It also turns out that a decoder for $C$ can perform a combination of error and erasure correction. Any error pattern simultaneously having $t$ errors and $e$ erasures is jointly correctable if $2t + e < d$. Similar expressions exist for joint detection and correction capabilities.[1,2]

### 4.2.3    Code Rate

Hamming distance is created by transmitting or storing extra redundant symbols. For the simple code of (4.1), three transmitted channel bits contain only 1 bit of useful user information. It turns out that one can reduce the relative amount of redundant symbols, while keeping $d$ fixed, by using larger codes. Consider the code that replaces user bits by channel bits as follows:

$$
\begin{aligned}
000 &\rightarrow 000000 & 110 &\rightarrow 101110 \\
100 &\rightarrow 110100 & 101 &\rightarrow 001101 \\
010 &\rightarrow 011010 & 011 &\rightarrow 100011 \\
001 &\rightarrow 111001 & 111 &\rightarrow 010111
\end{aligned}
\tag{4.2}
$$

By inspection, one can see that each pair of words from code (4.2) has a Hamming distance of at least 3; i.e., this code can correct one error just like code (4.1). Code (4.2) has eight different codewords, so each choice corresponds to 3 bits of information. Actually, the rightmost three channel bits of each codeword can be taken as the user information bits. The leftmost three channel bits of each codeword then serve as redundancy (generally called *parity bits*) for creating additional Hamming distance between codewords. Note that the number of channel bits of code (4.2) is only twice the amount of user bits, which is less than for code (4.1).

In general, we find that words of systematic block codes can be represented as strings of $n$ channel symbols, where the rightmost $k$ symbols contain the user information and the leftmost $n - k$ symbols form the parity (Fig. 4.2). The cost of redundancy is expressed by the rate $R$, where $R = k/n$, i.e., the fraction of transmitted channel symbols containing user information. For code (4.1) $R = \frac{1}{3}$, while for code (4.2) $R = \frac{1}{2}$.

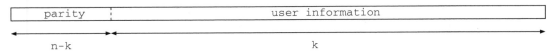

**FIGURE 4.2**   Codeword of a block code.

## 4.3   BLOCK CODES

An $[n, k, d]$ block code $C$ replaces fixed-sized blocks of $k$ user symbols by $n$ channel symbols, like the examples in the previous section, thereby creating a Hamming distance of $d$.[1-4] We need methods of designing codes that offer a large Hamming distance for a given rate $R$ and codeword length $n$. Furthermore, because $C$ usually consists of a very large set of codewords, we need an efficient algorithm for mapping the user information onto codewords at the transmitter side. Finally, we need efficient procedures for finding the closest codeword at the receiver and recovering the user information from the received, possibly mutilated transmitted word. Algebraic structures greatly simplify these tasks.

### 4.3.1   Algebraic Structures

***Fields.***   A code consists of strings of symbols. For a binary code these symbols are bits, and for other codes these symbols may consist of groups of bits (e.g., bytes in the case of CD). To be able to compute with such symbols, we impose on them the structure of a finite field (Galois field).[5] A finite field is a finite set of elements on which the operations of addition, subtraction, multiplication, and division can be performed in a mathematically well-defined manner. If a code uses bytes as symbols, the Galois field $\mathbb{F}_{2^8}$ can be used as its underlying field. It has 256 elements that can be brought into one-to-one correspondence with all possible 8-bit patterns. For a binary code, $\mathbb{F}_2$ is used. $\mathbb{F}_2$ can be represented by the set $\{0, 1\}$ using arithmetic modulo 2; i.e., addition of any two elements in $\mathbb{F}_2$ corresponds to the exclusive OR operation ($1 + 1 = 0$), and multiplication corresponds to the logical AND operation. Also note that subtraction and addition of any two elements lead to the same result in this field.

***Vectors over Fields.***   Mathematically, a string of $n$ symbols can be represented by a vector over the corresponding field:

$$\bar{c} = (c_0, c_1, c_2, \ldots, c_{n-1})$$

Two vectors $\bar{c}$ and $\bar{c}'$ can be added by componentwise addition in the field:

$$\bar{c} + \bar{c}' = (c_0 + c_0', c_1 + c_1', c_2 + c_2', \ldots, c_{n-1} + c_{n-1}')$$

So for binary vectors we obtain $(111001) + (011101) = (100100)$. Note that the sum (or difference) of two binary vectors has 1s at exactly those places where the two vectors differ because the additions of the components are in $\mathbb{F}_2$.

In the language of vectors, a code $C$ consists of a set of $n$-vectors, each corresponding to a codeword. If a code can represent $k$ user bits, there must be at least $2^k$ different code vectors.

It turns out that the codes that are mostly used are so-called linear codes; i.e., the set of vectors that constitute the code form a $k$-dimensional linear subspace of the $n$-space. For such a code any linear combination of codewords is also a codeword. Such a structure enormously facilitates representation of the code, its encoding, and finding its minimum distance. Instead of keeping all $2^k$ different code vectors, we only need to know $k$ code basis vectors, from which the rest can be computed. For instance, in code (4.2), we need to know only the matrix $G$ consisting of the three bottom codewords of the left column, which form a basis of this code:

$$G = \begin{pmatrix} 110100 \\ 011010 \\ 111001 \end{pmatrix} \tag{4.3}$$

All code vectors from (4.2) can be generated by taking the eight possible linear combinations over $\mathbb{F}_2$ of the rows of $G$.

If a code $C$ forms a $k$-dimensional subspace, there is also an $n - k$-dimensional subspace (generated by a parity check matrix $H$) that is orthogonal on $C$. For instance, for code (4.2), a parity check matrix is given by

$$H = \begin{pmatrix} 100101 \\ 010111 \\ 001011 \end{pmatrix} \tag{4.4}$$

The inner product between each row of $H$ and any code vector is zero. This property is useful for decoding, as we shall see in Section 4.3.3. In specifications for a standard (e.g., the CD standard), codes are often defined by their parity check matrix $H$. A parity check matrix allows us to build a kind of filter at the receiving end, the output of which depends only on the errors that have occurred, and not on the particular codeword that has been sent.

*Polynomials over Fields.*    Mathematically, a string of symbols also can be represented as a polynomial over the corresponding field $\mathbb{F}$:

$$(c_0, c_1, c_2, \ldots, c_{n-1}) \leftrightarrows c(x) = \sum_{i=0}^{n-1} c_i x^i = c_0 + c_1 x + c_2 x^2 + \cdots + c_{n-1} x^{n-1}$$

Terms with coefficient zero are usually not shown. Two polynomials can be added by adding (over $\mathbb{F}$) the coefficients of the corresponding powers of $x$:

$$c(x) + c'(x) = \sum_{i=0}^{n-1} (c_i + c_i') x^i$$

Polynomials also can be multiplied. Let $g(x) = 1 + x + x^3$ and $f(x) = 1 + x$; then multiplication over $\mathbb{F}_2$ results in

$$f(x)g(x) = (1 + x)(1 + x + x^3) = 1 + x^2 + x^3 + x^4 \tag{4.5}$$

Note that the coefficient of $x^1$ equals $1 + 1$, which is zero in $\mathbb{F}_2$. Likewise, polynomials can be divided,

$$\frac{1 + x^2 + x^3 + x^4}{g(x)} = 1 + x$$

by using long division. Polynomial division is easily implemented in hardware by using recursive filter structures, as shown in the next section.

### 4.3.2   Code Constructions

***Generator Polynomials.***    Most practical block codes are characterized by their generator polynomial $g(x)$; such codes are called *cyclic codes*. All codewords of such a code (in polynomial notation) are multiples of $g(x)$. For instance, code (4.2) has a generator polynomial $g(x) = 1 + x + x^3$. Each of its codewords corresponds to some $f(x)g(x)$, with an $f$ of degree at most 2 (there are eight different polynomials $f$ of degree at most 2 over $\mathbb{F}_2$). For instance Eq. (4.5) corresponds to the top word in the right column of (4.2). Since each codeword is divisible by $g(x)$, we can check whether a received binary polynomial $r(x) = r_0 + r_1 x + r_2 x^2 + \cdots + r_{n-1} x^{n-1}$ is a codeword by feeding it in a division circuit. Figure 4.3 shows a division circuit corresponding to $g(x) = 1 + x + x^3$. It is a synchronous circuit. The square boxes correspond to 1-bit delay elements, and the additions are mod 2 (XOR). The feedback taps correspond to $g(x)$. A polynomial $r(x)$ to be divided by $g(x)$ is shifted in; the contents of the memory cells correspond to the remainder. This procedure is called *parity checking*. For a codeword, the remainder is zero.

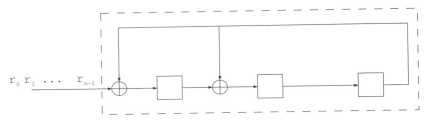

**FIGURE 4.3**    Circuit for dividing by $g(x) = 1 + x + x^3$.

For encoding, the same circuit can be used. As we saw in Section 4.2.3, code (4.2) is systematic; i.e., the three rightmost channel bits of each codeword are identical to the user bits. The leftmost bits forming the parity can be found by feeding the polynomial corresponding to the user bits followed by three trailing zeros into the circuit of Fig. 4.3. The parity bits then correspond to the contents of the memory cells.

For a general cyclic $[n, k, d]$ code, the degree of the $g(x)$ equals $n - k$, i.e., the number of parity symbols of the generated codewords. Also, the number of delays for the corresponding division circuit with feedback polynomial $g(x)$ equals $n - k$. The parities for systematic encoding can be found by feeding the polynomial that contains the user symbols, followed by $n - k$ trailing zeros, into the division circuit.

A good code with a large Hamming distance is obtained by properly choosing the factors of $g(x)$. This is the art of code design, a branch of discrete mathematics that falls beyond the scope of this book. For each fixed $g(x)$, there is a maximum codeword length $n_{max}$, called the primitive block length, for which the distance properties hold. If $n < n_{max}$, we say that the code is shortened. Shortening (reducing the number of user bits per codeword) lowers the rate of the code, but it does not compromise its error-correcting capability. In practice, a code is almost always shortened to tailor its length to an application. However, if $n$ is made larger than $n_{max}$, the Hamming distance of the resulting code suddenly reduces to 2.

***CRC Codes.***    Cyclic redundancy check (CRC) codes are used for error detection. A CRC code is a binary code characterized by its generator polynomial $g(x)$. Some well-known CCITT-recommended CRC generator polynomials are shown in Table 4.1. All of these CRC codes have $d = 4$. Checking a CRC code at the receiver corresponds to checking whether the received polynomial is divisible by $g(x)$. This can be done efficiently by circuits similar to that of Fig. 4.3.

**TABLE 4.1**   CCITT CRC Generator Polynomials

| | |
|---|---|
| $x^8 + x^2 + x + 1$ | $n_{max} = 127$ |
| $x^{12} + x^{11} + x^3 + x^2 + x + 1$ | $n_{max} = 2047$ |
| $x^{16} + x^{12} + x^5 + 1$ | $n_{max} = 32,767$ |

Also, the encoding, i.e., creating the parities at the sender side, can be done efficiently by using such circuits.

Because the Hamming distance is 4, these CRC codes detect all single, double, and triple bit errors. CRC codes also efficiently detect error patterns of greater weight, for instance, burst errors. For error patterns of large weight, the remainders after division of such mutilated codewords by $g(x)$ behave like randomly drawn patterns of $n - k$ bits. The probability that such a pattern consists of $n - k$ zeros equals $2^{-(n-k)}$, which is the probability that the decoder does not detect the presence of such large errors.

The use of error-detecting codes such as CRC codes assumes the possibility of alternative action if an error is detected. In automatic repeat request (ARR) schemes, a feedback channel exists so that a retransmission or reread of the faulty data can be requested. For audio and video applications, concealment techniques can be used to mask the presence of unreliable user bits.

***Hamming Codes.***   For each positive integer $r$, $r > 1$, there exist generator polynomials $g(x)$ of degree $r$ that generate a Hamming code such that $n_{max} = 2^r - 1$ and $d = 3$. Hamming codes form a class of $[2^r - 1, 2^r - r - 1, 3]$ binary single-error-correcting codes. The amount of redundancy is $\log(n_{max} + 1)$, which is the minimum required redundancy for recognizing and pointing out one erroneous position among $n_{max}$.

The simplest Hamming code (for $r = 2$) has $g(x) = 1 + x + x^2$ and $n_{max} = 3$. It corresponds to the simple repetition code (4.1). Code (4.2), generated by $g(x) = 1 + x + x^3$, actually is a shortened $[7, 4, 3]$ Hamming code with $r = 3$.

***Binary BCH Codes.***   If single error correction is not sufficient, one can use multiple-error-correcting BCH codes as in digital satellite radio (DSR), where a binary $[63, 44, 8]$ BCH code is used for jointly correcting two errors and detecting five errors. Also, the triple-error-correction $[23, 12, 7]$ Golay code is a special member of this class.

Like Hamming codes, these codes are characterized by generator polynomials that can be used in shift register implementations for encoding and parity checking as described before. It turns out that more polynomial factors must go into the construction of $g(x)$ if the generated code must have a larger distance.[2] Therefore, the degree of $g(x)$ for a fixed $n_{max}$ generally increases if more errors are to be corrected, i.e., more redundancy is required, and the rate becomes lower for a stronger code. Of course, these codes are also more difficult to decode.

***Reed-Solomon Codes.***   Reed-Solomon (RS) codes are a class of very efficient symbol error-correcting codes.[6] The code symbols of an RS code do not consist of bits, as in all previous examples, but usually groups of bits—for instance, bytes. For an $[n, k, d]$ RS code using bytes, the parameters length $n$, dimension $k$, and distance $d$ are counted in bytes (Fig. 4.4). Such an RS code is therefore a $t$-byte error-correcting code if $2t < d$. The exact number of bit errors in a faulty byte is not important. On the level of bits, a multiple-byte error-correcting RS code can be viewed as a multiple-burst error-correcting code. RS codes offer the maximum theoretically obtainable Hamming distance, given the number of redundant symbols:

$$d = n - k + 1 \tag{4.6}$$

If the demodulation process (because of its mapping using groups of bits) leads to symbol errors, or maybe even large bursts, it is efficient to correct these errors symbolwise. Typical applications are in the area of digital storage, where the media errors and the (de)modulation

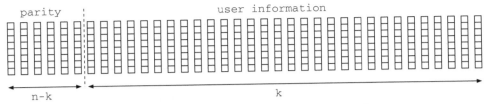

**FIGURE 4.4**   $[36, 30, 7]$ 3-byte error-correcting Reed-Solomon codeword over $\mathbb{F}_2.^8$

processes give rise to such symbol errors. Another application is in satellite communication for correcting the residual burst errors created by another code (Section 4.5.4).

### 4.3.3   Decoding Algorithms

We explain the general concept of decoding block codes by considering the binary code (4.2). If an error occurs in the $i$th position of a binary codeword $c(x)$, the coefficient $c_i$ of $x^i$ is flipped. Mathematically, we can represent the creation of errors by a channel, by adding (over $\mathbb{F}_2$, i.e., mod 2) a polynomial $e(x)$ to the transmitted codeword $c(x)$, where $e(x)$ has 1s in the error positions and 0s elsewhere. So the received mutilated codeword $r(x)$ can be written as

$$r(x) = c(x) + e(x) \tag{4.7}$$

where $e(x)$ is called the *error polynomial.*

Algebraic decoding algorithms generally first try to recover $e(x)$ from the received word $r(x)$. Next they recover the transmitted codeword by subtracting the recovered error from the received word, thus undoing the effect of channel errors as given by Eq. (4.7). Finally, the user information is extracted from the corrected $c(x)$.

***Syndromes.***   In decoding linear block codes, the receiver first divides the received polynomial $r(x)$ by the generator polynomial $g(x)$ (Section 4.3.2). Since the division circuit is a linear (over $\mathbb{F}_2$) feedback shift register, the remainder of a sum of polynomials equals the sum of the remainders. Let $r(x)\,|_{g(x)}$ denote the remainder after dividing $r(x)$ by $g(x)$; then

$$r(x)\,|_{g(x)} = \{c(x) + e(x)\}\,|_{g(x)} = c(x)\,|_{g(x)} + e(x)\,|_{g(x)} = e(x)\,|_{g(x)} \tag{4.8}$$

The last equality results from $c(x)\,|_{g(x)}$ being zero, since $c(x)$ is divisible by $g(x)$. We thus find that the remainder is a function only of the errors and not of the particular codeword that has been sent. This remainder is called a *syndrome,* because it is a symptom of what has gone wrong during transmission. The syndrome is zero if there were no errors (cf. Section 4.3.2). Note that a zero syndrome does not imply that there were no errors, as the error pattern $e(x)$ might be a codeword. As an example, if we change the second bit of each codeword of list (4.2), it turns out that the division of all these mutilated codewords by the circuit of Fig. 4.3 always leads to the same result, 010, which is called the syndrome of this error. So a receiver for code (4.2), finding the remainder 010 after dividing a received word by $g(x)$, should simply revert the second bit to correct the error (independent of the particular transmitted codeword).

***Finding Error Patterns from Syndromes.***   The second step in decoding a block code consists of finding the error pattern from the syndrome. Each of the correctable error patterns has its unique syndrome. For instance, for code (4.2), each error position leads to a syndrome that equals the corresponding column of the parity check matrix $H$; i.e., if the syndrome is equal to the leftmost column of $H$, change the leftmost bit of the received word. For a Hamming code it is thus simple to recover the error pattern from the syndrome.

In the case of multiple-error correction, this is more difficult because the syndrome is a linear combination of the syndromes of each individual error. Decomposing a multi-error syndrome into individual errors turns out to be a nonlinear problem that falls beyond the scope of this chapter.[7,8] For the class of BCH codes and RS codes, relatively efficient algorithms exist; their complexity depends only quadratically on the number of errors to be corrected. For codes such as RS codes, which have symbols consisting of groups of bits, the decoder first finds the erroneous symbol positions. Next the decoder finds the "error values," i.e., the exact bit patterns for correcting the faulty symbols.

For codes with little redundancy, one can opt for the use of a table that links each syndrome to its corresponding error pattern.

### 4.3.4 Performances

To express the improvement in reliability resulting from the use of a code, one calculates its performance. The performance $P_e$ of a code gives the residual error event rate at the output of the decoder. We define the residual error event rate as the probability (per user symbol) that one enters a group of unreliable symbols. Note that, because of the use of coding, unreliable symbols are no longer independent. They occur in groups that are related to uncorrectable codewords, each uncorrectable codeword having at least $t + 1$ errors.

If one uses an $[n, k, d]$ code that corrects $t$ symbol errors, the residual error rate is related to the probability that there are more than $t$ errors in a codeword of length $n$. For instance, if at point D in Fig. 4.1 each of the symbols independently has a probability $p$ of being erroneous, we define $P_e$ as

$$P_e = \frac{1}{k} \sum_{i=t+1}^{n} \binom{n}{i} p^i (1 - p)^{n-i} \tag{4.9}$$

In Fig. 4.5 we have plotted the performances of code (4.2) and an eight-error-correction [204, 188, 17] RS code over 8 bit symbols, as a function of the bit error rate $p$ on the binary memoryless channel. For comparison, we also plotted $P_e$ if no code is used (curve labeled "uncoded"). Note that the higher $t$ is, the steeper the slope of the curve becomes for low $p$. Also note that codes tend to become less effective at high bit error probabilities.

In a systems application, one is often interested in the mean time between failure (MTBF). The inverse error event rate, $1/P_e$, multiplied by the number of user symbols per second is the MTBF for the error-correcting decoder due to channel errors.

## 4.4 CONVOLUTIONAL CODES

Convolutional codes generate their redundancy in a completely different manner.[1,2,9,10] As shown in the example of Fig. 4.6, the information bits $i_0, i_1, \ldots$ are shifted into a shift register. At the output there are two streams of channel bits, $c_0^1, c_1^1, c_2^1, \ldots$ and $c_0^2, c_1^2, c_2^2, \ldots$. Each of the channel bitstreams is a linear combination of the input bits as determined by the taps of the shift register (calculations modulo 2). The encoder can be seen as two different FIR filters (over $\mathbb{F}_2$) acting on the input bitstream. By using a multiplexer, one can serialize the two output bitstreams into one bitstream.

Mathematically, we can describe the creation of the output as a convolution of the input power series by the connection polynomials (also called *generator polynomials*) of the FIR filter—hence the name *convolutional code*. The operation of such a convolutional encoder is often described by a state transition diagram (Fig. 4.7). Each state corresponds with the particular contents of the shift register. For each state, a new input determines the next state and both output

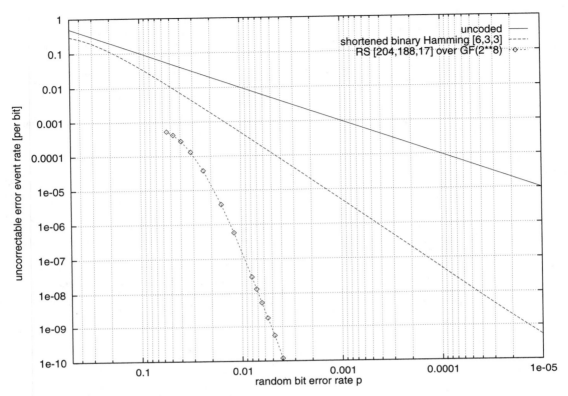

**FIGURE 4.5**    Performance of error-correcting block codes.

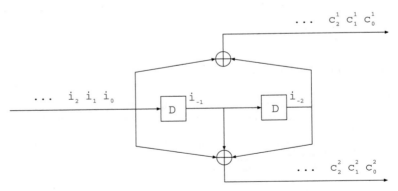

**FIGURE 4.6**    Convolutional encoder.

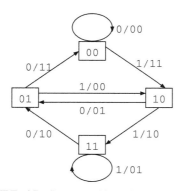

**FIGURE 4.7** State transition diagram of $\nu = 2$ encoder.

channel bits. Alongside each branch, the input causing that transition (before the slash) and the resulting output (after the slash) is given. The encoding of an information sequence corresponds to a particular path in the state transition diagram.

Unfortunately, a different notation is used with convolutional codes than with block codes, as can be seen from the ordering of the indices. With convolutional codes, the delay operator $D$ is used; i.e., binary sequences are represented as power series in $D$. However, as a function of time, the $i$th coefficient appears at discrete time $i$, contrary to block codes, where the high-order coefficients come first in time.

With block codes, each pair of finite-length codewords has a minimum Hamming distance $d$. With convolutional codes, we have to compare sequences that in principle can become infinitely long. It turns out that any two sequences of a good convolutional code also have a large Hamming distance $d_{\text{free}}$, which allows a decoder to correct errors.

### 4.4.1  Code Construction

Most convolutional codes used in practice can be represented by an encoder structure as shown in Fig. 4.6. Since two channel bits are transmitted for each incoming information bit, the rate of such a code is $R = \frac{1}{2}$. Other rates exist, as we shall see later.

The number $\nu$ of the delay elements of an encoder is an important design parameter. It is called the *constraint length* of the encoder. The higher $\nu$, the larger $d_{\text{free}}$ can be for a given rate. In some literature, the span $K$ of the generator polynomials is taken as the constraint length: $K = \nu + 1$.

***Generator Polynomials.***    By choosing proper generator polynomials $g_1(D)$ and $g_2(D)$, we can guarantee a minimum distance between its output sequences, just as with block codes. For example, the $\nu = 2$ code in Fig. 4.6, with generator polynomials $g_1(D) = 1 + D^2$ and $g_2(D) = 1 + D + D^2$, has $d_{\text{free}} = 5$, so it can correct at least two errors. Often, generator polynomials are given in binary or octal form, e.g., the $\nu = 2$ [101,111] or [5,7] code, respectively, for the code above.

It turns out that for maximizing $d_{\text{free}}$ for a given $\nu > 1$, both generator polynomials should be nontrivial. Therefore, these codes are usually not systematic; i.e., the user bits are not directly observable in the channel bit stream. A code that is often used in satellite communication is the $\nu = 6$ ($K = 7$)[171,133] code. It has generator polynomials $g_1(D) = 1 + D + D^2 + D^3 + D^6$ and $g_2(D) = 1 + D^2 + D^3 + D^5 + D^6$ and a $d_{\text{free}} = 10$. Its state transition diagram contains $2^6 = 64$ states.

***Puncturing.***    If the rate of a code must be enhanced, e.g., because the channel is relatively good and one wants to send more user data, a so-called punctured code can be obtained from a parent code. In puncturing a code, some of the channel symbols (in a regular pattern also known to the receiver) are not transmitted. In Fig. 4.8, we have shown the puncturing of the $\nu = 2[5, 7]$ parent code. The output channel bits are created pairwise on both outputs, as in Fig. 4.6. Of the lower branch, all bits are transmitted. Of the upper branch, only the even-numbered bits are transmitted while the odd-numbered bits are deleted, as indicated by the cross in the delete map in Fig. 4.8. Since we now have on average three transmitted channel bits for every two user bits, the rate

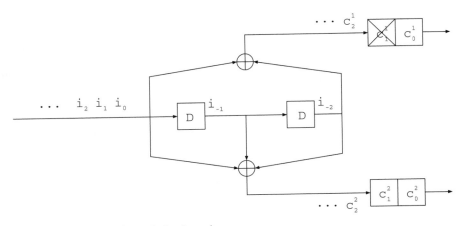

**FIGURE 4.8** Punctured convolutional encoder.

of the punctured code equals $\frac{2}{3}$. Since we have deleted channel symbols that are important in creating distance between certain channel sequences, the free distance of the code in this example has been reduced to 3. So rate can be exchanged for error-correcting power by a puncture process. In Table 4.2, we show the effect on $d_{\text{free}}$ of puncturing the $\nu = 6[171, 133]$ code.[11]

**TABLE 4.2**  Puncturing the $\nu = 6$ [171, 133] Code

| $R$ | 1/2 | 3/4 | 4/5 | 5/6 | 6/7 | 7/8 |
|---|---|---|---|---|---|---|
| $d_{\text{free}}$ | 10 | 6 | 5 | 4 | 4 | 3 |

A major advantage of enhancing the rate by puncturing is that the decoder of a punctured convolutional code is almost equal to that of the parent code. The decoder has to find out the locations of the symbols that have been deleted during the puncturing process. By inserting erasures at the locations of the nontransmitted bits at the receiver side, a virtual nonpunctured symbol stream is created that can be handled by the decoder of the parent code.

### 4.4.2 Viterbi Decoding Algorithm

The decoding of a convolutional code usually is completely different from the decoding of an algebraic block code. Mostly, one uses a maximum likelihood (ML) decoder. An ML decoder essentially compares the received waveform with all possible transmitted waveforms and chooses the one that is most likely given the noise statistics on the channel. Such an ML decoder gives the lowest error event probability if all transmitted codewords are equally likely (which is almost true in practice). Such a comparison is exponentially complex in the size of the code. The Viterbi algorithm (essentially dynamic programming) is a particular implementation of an ML decoder that keeps the complexity only exponential in $\nu$, the constraint length of the encoder, by disregarding at the earliest possible stage those codewords that can never be the most likely ones.

***Trellis.*** A Viterbi decoder cleverly generates all possible code sequences that are to be compared with the received sequence. By weeding out the unlikely sequences, it tries to keep track of the encoder state as a function of discrete time. A trellis is a way of visualizing a path through

STATE

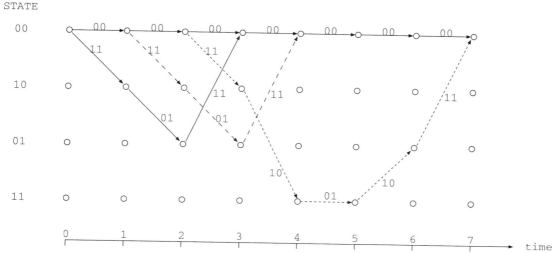

**FIGURE 4.9**    Trellis of $\nu = 2[3, 5]$ code.

the encoder state transition diagram as a function of time. In Fig. 4.9, we have shown the trellis of the $\nu = 2[3, 5]$ code. At each time $t = 0, 1, \ldots$, there are four states corresponding to the four possible contents of the $\nu = 2$ encoder. In the trellis we have shown four different paths, each of which corresponds to a different input sequence. Alongside each branch, we depict the pair of output channel bits that are generated at those time instances. The first path, the upper solid line, corresponds to the all-0 input. The state remains 0, and all generated outputs are 0. The second path (also solid) corresponds to the input $i(D) = 1$, which leads to the state sequence $00, 10, 01, 00, \ldots$, as the 1 ripples through the encoder. The third path (long dash) corresponds to $i(D) = D$, and the fourth path (short dash) corresponds to $I(D) = D^2 + D^3 + D^4$. As shown in the trellis, the three nonzero paths merge with the all-0 path at times $t = 3, 4$, and 7. At the time of merging, the decoder can decide which of the two merging codewords is more likely and throw the other one away. In general, the decoder has to choose between two possible merging paths for all states at each time. Therefore the Viterbi decoder keeps track of $2^\nu$ codewords at each time.

In the comparison between the received waveform and possible transmitted ones, one typically considers detailed amplitude information of the received waveform. Such detailed amplitude information is called *soft decision information*.

### 4.4.3  Performance

The strength of a convolutional code lies in the small complexity of its corresponding maximum likelihood decoder. It turns out that maximum likelihood decoding is of paramount importance in obtaining a significant reliability improvement on very noisy channels. A maximum likelihood decoder such as a Viterbi decoder finds the nearest codeword for any received channel output, while making use of the soft decision information of the channel; i.e., it squeezes the last juice out of the code. In this way, very noisy channels can be transformed into channels having a reasonable residual error rate. In Fig. 4.10, the curve labeled "uncoded" indicates the bit error rate (BER) for an uncoded system, using 4-PSK modulation on a Gaussian channel, as a function of the signal-to-noise ratio (SNR). By applying a convolutional code, the BER for each fixed SNR improves, as shown by the curve labeled "coded." One can also fix the BER and look for how

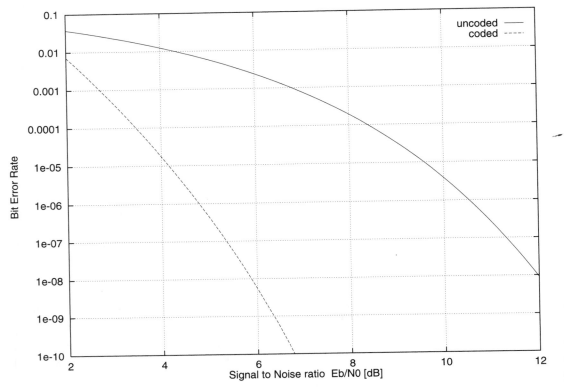

**FIGURE 4.10**  Performance of $\nu = 6[171, 133]$ convolutional code.

much less SNR is required by applying coding. This is called the *coding gain* of a code (e.g., the code in Fig. 4.10 has about 5 dB coding gain at a BER of $10^{-4}$).

Because of the exponential rise of complexity of a Viterbi decoder in terms of code parameters, we cannot apply codes as powerful (in terms of rate and distance) as the algebraic decodable block codes.

## 4.5  COOPERATING CODES

For channels that are very bad or create long bursts of errors, it can be beneficial to enhance the reliability by not using a single code but by using a set of cooperating codes, a kind of super code. In the sequel, we introduce the most common ones.

### 4.5.1  Interleaving and Burst Error Correction

A channel is called *bursty* if errors tend to occur in groups, i.e., if errors are not independent. The simplest solution for correcting those errors is to use some form of interleaving. An interleaved code consists of several codewords of a code $C$, whose symbols are interwoven on the bursty channel. In Fig. 4.11, we have shown an example of a depth $I = 4$ interleaved code. On the receiver side, the stream of demodulated symbols is deinterleaved into its constituent codewords $C_1, \ldots, C_4$. In this way a burst error is divided among the interleaved codewords in such a way that each codeword only needs to correct a fraction $1/I$ of the total burst.

Interleaving is used in all the following cooperating coding schemes.

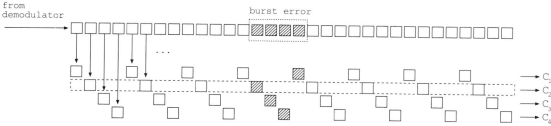

**FIGURE 4.11**  Interleaving of depth $I = 4$.

## 4.5.2  Product Codes

A simple product code can be envisioned as a two-dimensional array in which each column belongs to a column code $C_1$ and each row to a row code $C_2$. In the case of systematic codes and $C_2$, the user information can be in the upper left corner, while the right and lower edges of the array contain the parity symbols as shown on the left-hand side of Fig. 4.12. On the right-hand side of Fig. 4.12, we show a decoder for the product code. First all columns of the array are decoded using $C_1$. Next all rows are decoded using $C_2$. Assuming that the information on the channel is transmitted columnwise, the first decoder (for $C_1$) corrects most of the random errors and indicates the large bursts and concentrations of uncorrectable random errors. Next the $C_2$ decoder can apply error and erasure decoding on the deinterleaved indicated burst errors that remain after $C_1$ decoding.

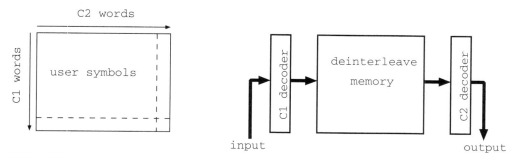

**FIGURE 4.12**  A product code and its decoder.

Such a cooperating code with its corresponding decoder has a surprisingly good performance, especially for bursty channels. The codes used in the Digital Compact Cassette Recorder and the Digital Versatile Disc player are based on this concept, with Reed-Solomon codes for $C_1$ and $C_2$ (cf. Section 4.3.2). The concept of a product code can be extended to more than two dimensions.

### 4.5.3    Cross-Interleaved Schemes

A cross-interleaved code can be envisioned as a bi-infinite strip, where each column of height $n$ belongs to a code $C_1$ having $p$ parity symbols in each codeword. The upper $n - p$ symbols of each diagonal belong to a code $C_2$, as shown on the left-hand side of Fig. 4.13. A decoder for this format is shown on the right-hand side of Fig. 4.13. First all columns of the strip are decoded using $C_1$. The upper $n - p$ symbols of each column are "tilted" by the deinterleave memory so that they can be corrected by $C_2$. If the information on the channel is transmitted columnwise, the performance of a cross-interleaved scheme is very similar to the performance of a decoder for a product code using the same component codes. The advantage lies in the reduced memory requirements for the deinterleaver. A disadvantage is the difficulty in making independent blocks for recording applications.

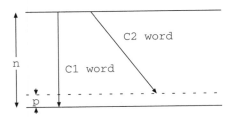 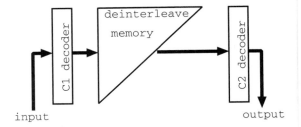

**FIGURE 4.13**    A cross-interleaved code and its decoder.

The codes that are used in the Compact Disc (and all its derivatives, such as the CD-ROM) and the MiniDisc utilize this concept. Again, the component codes $C_1$ and $C_2$ are both Reed-Solomon codes in those cases.

### 4.5.4    Concatenated Codes

For satellite communication and other rather noisy channels, one often uses a concatenated coding system, of which the decoder is shown in Fig. 4.14. The idea is that the very noisy bits at the input are first decoded by applying maximum likelihood decoding of a convolutional code, making use of the soft decision information that is available at the input. In this way, a bad channel at the output of the demodulator, as given by the $I$ and $Q$ symbols, is transformed into a reasonable bursty channel at the output of the Viterbi decoder. Next, taking bits together in bytes, using a deinterleaver (for spreading the bursts on byte level) and a Reed-Solomon code, the reasonable bursty channel is transformed into a very reliable channel at the output of the RS decoder. Since each of the codes can operate in a region where it performs best, the combined approach leads to

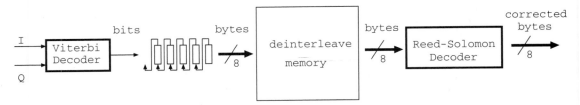

**FIGURE 4.14**    A decoder for a concatenated code.

very good overall performance in terms of low error rate for a weak signal-to-noise ratio on the channel.

The Digital Video Broadcast standard uses a concatenated coding scheme for satellite transmission.

## REFERENCES

1. P. Sweeney, *Error Control Coding: An Introduction,* Prentice Hall, Englewood Cliffs, N.J., 1991.
2. G. C. Clark Jr. and Cain, J. B., *Error-Correction Coding for Digital Communications,* Plenum Press, 1986.
3. W. Wesley Peterson and Weldon, E. J. Jr., *Error-Correcting Codes,* MIT Press, Cambridge, Mass., 1972.
4. F. J. MacWilliams and Sloane, N. J. A., *The Theory of Error-Correcting Codes,* North-Holland, 1977.
5. R. J. McEliece, *Finite Fields for Computer Scientists and Engineers,* Kluwer Academic Publishers, Norwell, Mass., 1987.
6. S. B. Wicker and Bhargava, V. K., eds., *Reed-Solomon Codes and Their Applications,* IEEE Press, 1994.
7. R. E. Blahut, *Theory and Practice of Error Control Codes,* Addison-Wesley, Reading, Mass., 1983.
8. E. R. Berlekamp, *Algebraic Coding Theory,* McGraw-Hill, New York, 1968.
9. S. Lin and Costello, D. J. Jr., *Error Control Coding: Fundamentals and Applications,* Prentice Hall, Englewood Cliffs, N.J., 1983.
10. A. J. Viterbi and Omura, J. K., *Principles of Digital Communication and Coding,* McGraw-Hill, New York, 1979.
11. Y. Yasuda, Kashiki, K., and Hirata, Y., "High-rate punctured convolutional codes for soft decision Viterbi decoding," *IEEE Transactions on Communications,* COM-23(3), March 1984, pp. 315–319.

## ABOUT THE AUTHOR

C. P. M. J. (Stan) Baggen is a senior scientist at the Philips Research Laboratories, Eindhoven, The Netherlands. His main research interests are in the areas of algebraic coding, communication, and information theory, and applications thereof to storage and transmission of digital information. After joining Philips in 1979, he spent seven years in the optical recording group, where he designed the error correction system of the first generation of 12-inch optical recorders. He was involved in the design of error correction decoders for CD players and in standardization discussions of CD-ROM third-layer error correction. In 1986 he moved to the digital signal processing group, where he became responsible for research on channel coding. Recently he has been involved in the design of channel coding systems for Digital Video Broadcast (DVB) and for the Digital Versatile Disc (DVD) systems. He received the Ingenieurs (ir.) degree from Eindhoven University of Technology in 1979. He spent the academic year 1989–90 at the Center for Magnetic Research at the University of California, San Diego, where he started working toward a Ph.D. He received the Ph.D. in Electrical Engineering (Communication Theory and Systems) from the University of California in 1993.

# CHAPTER 5
# DIGITAL MODULATION TECHNIQUES

**Qun Shi**
*Communication Systems Technology Laboratory*
*Panasonic Technologies Inc.*

## 5.1 FUNDAMENTALS OF DIGITAL MODULATION[1-6]

Compared to traditional analog modulation, digital modulation offers many advantages in today's consumer communications systems and applications. Some advantages include greater noise immunity and robustness to channel impairments, easier multiplexing of various forms of information (e.g., data, voice, and video), higher system capacity, greater error protection, and greater security. Further, advances in silicon and digital signal processor (DSP) technologies have made digital modulation more cost-effective than its analog counterpart.

### 5.1.1 Basic Signal-Processing Elements in a Digital Communication System

A digital communication system, shown in Fig. 5.1, typically encompasses three basic signal processing elements: source coding, channel coding, and digital modulation. The coding operation involves providing efficient signal representation and bandwidth (source coding) and reliable communication over a noisy channel (channel coding). The digital modulation operation involves the shaping and mapping of the digital data symbols into waveforms that are suitable for transmission over the channel. At the receiving site the digital demodulation performs the inverse process. The receiver also contains a timing and synchronization element, which provides receiver clock and symbol timing.

### 5.1.2 Digital Modulation Formats

Modulation may be viewed as changing certain characteristics of a carrier in accordance with a modulating wave. In digital modulation the modulating wave is either binary or *M*-ary coded data. The carrier signal is usually a sinusoidal signal. The digital modulation process then involves mapping a sequence of binary digits into a corresponding set of discrete amplitudes, phases, or frequencies of the carrier to distinguish one signal from another. This corresponds to the amplitude, phase, and frequency modulation in analog communications. For the binary case

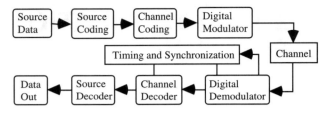

**FIGURE 5.1** Basic signal-processing elements of a digital communication system.

the modulated signal takes one of the three common formats as amplitude-shift keying (ASK), frequency-shift keying (FSK), or phase-shift keying (PSK). For the $M$-ary case, the modulation produces one of an available set of $M = 2^m$ distinct signals in response to $m$ bits of source data at a time; each modulated signal takes one of the above three formats. In addition, hybrid amplitude and phase or frequency modulation formats may be used in communication systems.

### 5.1.3 Coherent and Noncoherent Modulation and Demodulation

Coherent modulation and demodulation refer to the case where the demodulator in a digital communication system is time-synchronized with the modulator; i.e., the demodulator maintains the exact timing or phase of the modulated carrier. In many cases the coherent demodulator, as shown in Fig. 5.2, uses a locally generated carrier signal that is phase-locked to the transmitted carrier.

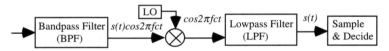

**FIGURE 5.2** Generic coherent demodulator.

Noncoherent modulation and demodulation refer to the case where no phase synchronization between the modulator and demodulator is necessary. The demodulator in this case performs nonlinear operations on the modulating signal to retrieve the desired information. The most common nonlinear operation is the envelope or square-law detector, as shown in Fig. 5.3, taking either the absolute value or the amplitude square of the signal.

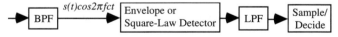

**FIGURE 5.3** Generic noncoherent demodulator.

### 5.1.4 Quadrature Modulation

Quadrature modulation involves multiplexing two data-bearing signals on the carrier frequency, $f_c$, in phase quadrature; i.e., the modulated signal is given by

$$s(t) = s_I(t) \cos 2\pi f_c t - s_Q(t) \sin 2\pi f_c t \tag{5.1}$$

where $s_I(t)$ and $s_Q(t)$ are the orthogonal in-phase (I) and quadrature (Q) components of the modulated signal and are low-pass processes with bandwidth $B \ll f_c$. Many digital modulation schemes can be represented by Eq. (5.1), with I and Q components taking the waveforms of the form $s_I(t) = \sum a_k p_I(t - kT_s)$ and $s_Q(t) = \sum b_k p_Q(t - kT_s)$, where $p_I(T)$ and $p_Q(t)$ are finite energy pulses, and $a_k$ and $b_k$ are sequences of discrete random variables with symbol rate $1/T_S$. For example, in quadrature phase-shift keying (QPSK) modulation, $(a_k, b_k) = (\pm 1, \pm 1)$ (i.e., the carrier phase takes one of four possible values: $\pi/4, 3\pi/4, 5\pi/4,$ and $7\pi/4$), and $p_I(t) = p_Q(t) = 1$, for $1 \leq t \leq T_s$.

### 5.1.5  Shannon's Limit, Channel Capacity, and Bandwidth Efficiency

Shannon has proved that for a communication channel, embedded in additive white Gaussian noise (AWGN) and limited in bandwidth to $B$ Hz, there exists a maximum rate at which information can be transmitted error-free with the employment of intelligent signal coding schemes.[1] The maximum rate, denoted $C$, is known as the *channel capacity,* in b/s, and is given by the following:

$$C = B \cdot \log_2(1 + P_s/N_0 B) \tag{5.2}$$

where $P_s$ is the received signal power and $N_0$ is the power spectral density of the AWGN in watts per Hz. Equation (5.2) provides an important relation between the system parameters channel bandwidth and signal power. In particular, as the bandwidth $B$ approaches infinity, the channel capacity approaches its limit (Shannon limit): $C = \log_2 e \cdot P_s/N_0 = 1.44 P_s/N_0$.

Alternatively, by expressing $P_s = E_b C$, where $E_b$ denotes the received signal energy per bit of data, we can rewrite Eq. (5.2) for $E_b/N_0$ in terms of the ratio $C/B$ as follows:[2]

$$E_b/N_0 = (2^{C/B} - 1)(C/B) \tag{5.3}$$

The ratio $C/B$ is defined as the bandwidth efficiency in b/s/Hz. As $B \to \infty$, $E_b/N_0 \to \ln 2 = (-1.6$ dB$)$, implying that error-free transmission is possible if $E_b/N_0 \geq \ln 2$. Equation (5.3) further implies that for a practical communication system with information bit rate $R_b$, error-free transmission is possible if $R_b < C$, and not possible if $R_b > C$. Note that the region $R_b/B < 1$ is called the *power-limited region,* whereas $R_b/B > 1$ is defined as the *bandwidth-limited region.*[1–2]

### 5.1.6  Signal Space Representation and Model in Noise

Vector space representation of signals and noise provides a convenient way for characterization and design of an optimum receiver for *M*-ary modulation. The basic model, illustrated in Fig. 5.4, is as follows. The signal source at the transmitter emits a sequence of messages with one symbol per $T_s$ seconds, each symbol being taken from a set of $q$ symbol alphabet $(m_1, m_2, \ldots, m_q)$. For each symbol $m_i$ the modulator generates and transmits a corresponding waveform signal $s_i(t)$ every $T_s$ seconds. The receiver receives the signal $s_i(t)$ corrupted by additive noise $n(t)$. To retrieve as to which of the $q$ message symbols is being sent, the receiver makes a best estimate of the signal $s_i(t)$ using the observation of the received signal $y(t) = s_i(t) + n(t)$ for a duration of $T_s$ seconds.

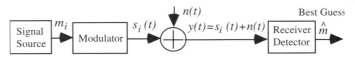

**FIGURE 5.4**  Basic digital communication model.

The modulated signals, $s_i(t)$, $i = 1, \ldots, q$, can be represented by a set of $p$ orthonormal basis functions according to the Gram-Schmidt orthogonalization principle:[1]

$$s_i(t) = \sum_{j=1}^{p} s_{ij}\phi_j(t), \quad i = 1, 2, \ldots, q; \quad 0 < t < T_s \tag{5.4}$$

where the coefficients

$$s_{ij} = \int_0^{T_s} s_i(t)\phi_j(t) \, dt, \quad i = 1, \ldots, q; \quad j = 1, \ldots, p \quad (p \le q) \tag{5.5}$$

and $\int_0^{T_s} \phi_i(t)\phi_j(t) \, dt = 1$ if $i = j$, and $0$ if $i \neq j$. Thus, each signal in the signal set $\{s_i(t)\}$ is completely specified by the $p$-dimensional vector $\underline{s}_i = (s_{i1}, s_{i2}, \ldots, s_{ip})^*$, $i = 1, \ldots, q$, where the asterisk denotes the vector transpose. The space spanned by the $p$-dimensional vector $\phi = (\phi_i, \ldots, \phi_p)$ is called the *signal space*, the vector $\underline{s}_i$ is called the signal vector, and the signal $s_i(t)$ denotes a point in the signal space. Hence, the $q$ message symbols or the signal set $\{s_i(t)\}$ denotes a point in the signal space. Hence, the $q$ message symbols or the signal set $\{s_i(t)\}$ represents a collection of $q$ points in the signal space referred to as the *signal constellation*.

Using the signal space representation, the basic signal plus noise model as shown in Fig. 5.4 can be expressed by the equivalent vector form:

$$\underline{y} = \underline{s}_i + \underline{n} \tag{5.6}$$

where the components of the vectors $\underline{y}$ and $\underline{n}$ are similarly expressed by Eq. (5.5). Because of the presence of the noise, the received signal point in the signal constellation deviates from the transmitted signal point; the deviation equals the length of the noise vector.

### 5.1.7   Performance Measure of Digital Communication Systems: Symbol Error Rate (SER), Bit Error Rate (BER), Eye Diagrams, and Signal Constellations

In digital communications an important figure of merit for performance assessment is a measure of the system's error-producing behavior. In general, the performance measure can be divided into two broad categories: statistical measure and visual measure. The statistical measure deals with the error-producing behavior of a digital transmission in an average sense; it includes symbol error probability, bit error probability, error-free intervals, etc. The visual measure, on the other hand, involves using instruments or simulation tools to give a qualitative "visual" indication of the performance of a digital system. The measure can be an eye diagram and/or a signal constellation.

***Symbol Error Rate (SER).***   The symbol error rate, or the average symbol error probability, is defined as the probability that the observed symbol is not equal to the transmitted symbol. From a signal space viewpoint the SER may be written as $P_{es} = \sum P_i P(\hat{m} \neq m_i)$, where $P_i$ is the *a priori* probability that the symbol $m_i$ is sent. If one assumes that all $q$ symbols of the alphabet are equally likely and lets $R_i$ be the *decision region* in which the receiver estimates $m_i$ if the observation vector point falls inside it, then the SER becomes the following:

$$P_{es} = \frac{1}{q}\sum_{i=1}^{q} P(\underline{y} \text{ not in } R_i | m_i \text{ sent}) = 1 - \frac{1}{q}\sum_{i=1}^{q} P(\underline{y} \text{ in } R_i | m_i \text{ sent}) = 1 - \frac{1}{q}\int_{R_i} f(\underline{y}|m_i)d\underline{y} \tag{5.7}$$

where $f(\underline{y}|m_i)$ denotes the probability density function of the observation vector $\underline{y}$ given that the message $m_i$ was sent. For example, consider binary transmission with $s_1$ and $s_2$

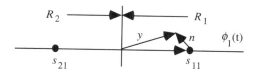

**FIGURE 5.5**   Signal-space diagram of a binary modulation system.

representing symbols 1 and 0, respectively. In this case the signal space is one-dimensional spanned by the basis function $\phi_1$. The signal space diagram is shown in Fig. 5.5. Assuming an AWGN with zero mean and variance $\sigma_n^2$ being present, the observed signal $y = s_i + n$ is also Gaussian with mean $s_i$ and the same variance. From Fig. 5.5 and Eq. (5.7), one can obtain the following:

$$P_{es} = 1 - \frac{1}{2} \int_{R_1} \frac{1}{\sqrt{2\pi}\sigma_n} \exp[-(y - s_{11})^2/2\sigma_n^2]dy = \frac{1}{2}\text{erfc}\left(\frac{s_{21} - s_{11}}{2\sqrt{2}\sigma_n}\right) \tag{5.8}$$

where $s_{11}$ and $s_{21}$ are the components of $s_1(t)$ and $s_2(t)$, respectively, and erfc($\cdot$) is the complementary error function.

***Bit Error Rate (BER).***   The bit error rate, or the average probability of bit error, $P_{eb}$, may be defined as the ratio of the expected number of bit errors per symbol block to the total number of the information bits transmitted per symbol block. In $M$-ary modulation the BER and the SER are related in one of the following two ways:[2]

1. For $M$-QAM using the *Gray code* mapping of binary bits to $M$-ary symbols, then

$$P_{eb} = \frac{P_{es}}{\log_2 M}, \quad M \geq 2 \tag{5.9}$$

2. For $M$-ary orthogonal modulation (e.g., $M$-FSK) with $M = 2^K$ equiprobable symbols, then

$$P_{eb} = \frac{2^{K-1}}{2^K - 1}P_{es} = \frac{M/2}{M - 1}P_{es} \tag{5.10}$$

***Eye Diagram.***   The eye diagram is commonly used to provide a qualitative performance indication of a digital communication system. An eye diagram is formed by simultaneously displaying the received signal waveform and its delayed versions (the delays being in multiple of symbol period) over symbol intervals. Experimentally, this is done by connecting the received signal to the vertical input of an oscilloscope and a clock signal to the horizontal input with clock rate equal to the symbol rate of the received signal. The pattern seen on the oscilloscope is then the eye diagram mimicking a person's eye. Figure 5.6 shows a typical eye diagram, from which the effects of noise and distortion (and in particular the intersymbol interference [ISI] on a digital modulation system can easily be observed as follows:[2]

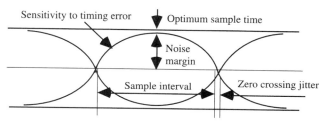

**FIGURE 5.6**   A typical eye diagram.

1. The width of the eye opening shows the ISI-free sampling time interval. Hence, the optimum sampling time is the instant at which the eye is widest open.

2. The width of zero crossings shows distortions incurred as a result of sampling clock jitter. The rate of the eye closure (slope) indicates the sensitivity to timing error.

3. The nonsymmetrical shape of the eye pattern may indicate the presence of nonlinearities.

4. The height of the eye opening indicates the noise margin, useful for estimating the SER.

***Signal Constellation.*** Signal constellation represents sampled signal values in signal space. For quadrature modulations the signal constellation is a two-dimensional vector space where each of the signal points (symbols) represent the sampled in-phase and quadrature signal values at the output of the demodulator. For an ideal communication system with no noise and distortion, the constellation would show distinct points, with each point corresponding to a certain amplitude and phase of the carrier. Practically, due to the presence of the noise and distortion, each sampled value of the received signal deviates from that of the transmitted signal. As a result, in the signal constellation, scattered points corresponding to all sampled values appear around the area. Thus, the signal constellation also provides a qualitative indication of the behavior of a digital modulation. As an example, Fig. 5.7 shows the signal constellation of a QPSK system with (*a*) the ideal constellation and (*b*) the nonideal constellation. The scatter around the ideal points indicates the effects of noise and ISI. The rotation and the compression of the constellation shown in Fig. 5.7*b* indicate the effect of phase and amplitude distortions.

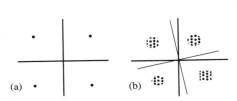

**FIGURE 5.7** Signal constellation of a 4-QAM system.

### 5.1.8 Optimum Demodulation and Matched Filter Receiver

For *M*-ary transmission in additive noise, the demodulator upon receiving the signal must decide which one of the *M* possible signal waveforms was sent. It is desirable that the decision be made to minimize the average SER. The correspondent demodulator is then said to be optimal in the minimum SER sense.

One decision criterion is the maximum *a posteriori* probability (MAP) criterion.[1-2] That is, given the received signal $y(t)$, the demodulator minimizes the SER by selecting the message signal $\hat{m}_i$ that has the largest *a posteriori* probability $P(m_i$ transmitted $\mid y(t), 0 < t < T_s), i = 1, \ldots, M$. When all the symbols are sent with equal *a priori* probability, the MAP criterion can be translated into the maximum-likelihood (ML) criterion. That is, the demodulator chooses the signal that has the maximum-likelihood function $f(y \mid m_i)$, which is the probability density function of $y$ conditional on $m_i$. Further, in the presence of a zero mean AWGN with variance $\sigma_n^2$, the received signal $y$ is also Gaussian, with the following likelihood function:[1-2]

$$f(\underline{y} \mid m_i) = (2\pi\sigma_n^2)^{-p/2} \exp\left[-\frac{1}{2\sigma_n^2}\sum_{j=1}^{p}(y_j - s_{ij})^2\right] = \frac{\exp(-\|\underline{y} - \underline{s}_i\|^2/2\sigma_n^2)}{(2\pi\sigma_n^2)^{p/2}}, \quad i = 1, \ldots, q$$

(5.11)

where $\|y - \underline{s}_i\|$ is the distance between the received signal vector and the *i*th transmitted signal vector. Hence, the optimum demodulator is the one that selects the signal $\hat{s}_i$ for which the quantity $\|y - \underline{s}_i\|^2$ is a minimum. The optimum demodulator may be implemented by either a correlation receiver or a *matched filter receiver* structure.

***Correlation Receiver.***   The structure of the correlation receiver can be obtained through the signal space representation of the received signal $y$ and the expansion of the quantity $\|y - s_i\|^2$. The resulting structure is shown in Fig. 5.8.

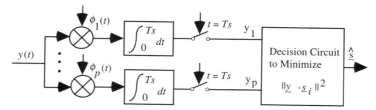

**FIGURE 5.8**   Correlator receiver structure.

***Matched Filter Receiver.***   Because the orthonormal basis functions $\phi_1, \ldots, \phi_p$ are zero outside the symbol interval $0 \le t \le T_s$, the integrators in the correlation receiver can be replaced by a bank of filters with impulse responses $h_i(t) = \phi_i(T_s - t)$, $i = 1, \ldots, p$. The filters are then said to be matched to the basis functions representing the signal waveforms, and the resulting demodulator structure is called a *matched filter receiver*, which is shown in Fig. 5.9. Note that the matched filter receiver may also be obtained by maximizing the filter output signal to noise ratio (SNR).[1]

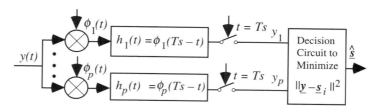

**FIGURE 5.9**   Matched-filter receiver structure.

### 5.1.9   Band-Limited Transmission and Pulse Shaping

The optimum demodulator in AWGN reflects the ideal situation, i.e., no bandwidth constraint on the system is imposed. In practice, filters with finite bandwidth are used to select desired signals, reduce noise and interference, and shape the signal waveforms to meet the system requirements. In addition, the transmission media have bandwidth limitations as well. Figure 5.10 shows a block diagram modeling the band-limited digital transmission system.

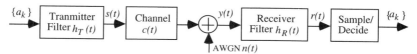

**FIGURE 5.10**   Band-limited digital communication system model.

Consider that the input to the transmitter filter is a message sequence $d(t) = \sum a_k \delta(t - kT_s)$, where $\delta(t)$ is the unit impulse function and $a_k$ are message symbols (for binary, $a_k = 0$,

1 or $\pm 1$). Then the output of the receiver filter, $r(t)$, at the sampling times $mT_s + t_d$ is given by the following:

$$r(mT_s + t_d) = a_m g_R(t_d) + \sum_{k \neq m} a_k g_R[(m - k)T_s + t_d] + n_R(mT_s + t_d) \qquad (5.12)$$

where $g_R(t) = h_T(t) \otimes c(t) \otimes h_R(t)$ ($\otimes$ being the convolution symbol), $n_R(t) = n(t) \otimes h_R(t)$, and $t_d$ denotes the channel delay time. The first term in Eq. (5.12) is the desired signal, whereas the second and third terms denote ISI and noise, respectively. Thus, it is required that the transmitter and receiver filters be designed to minimize the ISI and noise. For zero ISI in the absence of noise, we have the following criterion.

***Nyquist Zero ISI Criterion.*** If $G_R(f)$—the Fourier transform of $g_R(t)$—satisfies the following condition,

$$\sum_{k=-\infty}^{\infty} G_R(f + k/T_s) = T_s \qquad |f| \leq 1/2T_s \qquad (5.13)$$

and zero otherwise, then $g_R(mT_s - kT_s) = 1$ for $m = k$, and zero for $m \neq k$.

***Raised Cosine Pulse Shaping.*** A family of the transfer function $G_R(f)$ that satisfies Eq. (5.13) is the so-called *raised cosine* family, with the following,[2]

$$G_R(f) = \begin{cases} T_s, & |f| < (1 - \alpha)/2T_s \\ T_s \cos^2 \frac{1}{4\alpha}[2T_s|f| - (1 - \alpha)], & (1 - \alpha)/2T_s \leq |f| < (1 + \alpha)/2T_s \\ 0, & |f| \geq (1 + \alpha)/2T_s \end{cases} \qquad (5.14)$$

and its time domain waveform is given by

$$g_R(t) = \frac{\sin(\pi t/T_s)}{\pi t/T_s} \frac{\cos(\pi \alpha t/T_s)}{1 - 4\alpha^2 t^2/T_s^2} \qquad (5.15)$$

where $\alpha(0 \leq \alpha \leq 1)$ is the roll-off factor, playing a critical role in the filter design. From Eqs. (5.14) and (5.15), it can be seen that low $\alpha$ values provide bandwidth efficiency that closes to the ideal low-pass filter ($\alpha = 0$), but have slow decaying impulse responses that are more sensitive to sampling errors and ISIs. On the other hand, large $\alpha$ (e.g., $\alpha = 1$) yields fast decay at the price of increasing the bandwidth. The bandwidth of the raised cosine filter is $B = B_o(1 + \alpha)$, where $B_o = 1/2T_s$ is the minimum bandwidth corresponding to the ideal low-pass filter.

It is noted that for ideal channels with constant impulse response [$c(t) =$ a constant] and the AWGN, the optimum transmitter and receiver filters (in the sense of minimum BER or maximum SNR) are such that they are matched and identical,[1] i.e., $H_T(f) = H_R(f) = G_R(f)^{1/2}$. This is the so-called square-root raised cosine filters.

## 5.1.10 Channel Characteristics and Impairment Models

In digital communications, channel characterization can be broadly classified as the discrete channel and the *waveform channel*.[1]

***Discrete Channel.*** The discrete channel combines the signal-processing blocks, such as modulator and demodulator, and the physical channel. The channel provides a probabilistic model for describing the statistical behavior of a digital communication system.

A discrete channel is memoryless if the value of the output symbol depends only on the current value of the input symbol and not on any of previous ones. Denote the input symbols as

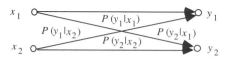

**FIGURE 5.11**   Transition diagram of a binary DMC.

$x_1, x_2, \ldots, x_q$ and the output symbols as $y_1, y_2, \ldots, y_p$. Then the discrete memoryless channel (DMC) is solely represented by the transition (conditional) probabilities $P(y_i|x_j)$ for all $j$ and $i$. Conveniently, this can be written as a probability transition matrix $\boldsymbol{P} = [P(y_i|x_j)]$ or arranged in a probability transition diagram, as shown in Fig. 5.11 for a binary DMC. Given $P(y_i|x_j)$, the probability of an output $y_k$ can be expressed as follows:[1]

$$P(y_k) = \sum_{j=1}^{q} P(y_k|x_j)P(x_j) \tag{5.16}$$

where $P(x_j)$ is the probability of input $x_j$.

***Waveform Channel.***   The waveform channel is the physical medium between transmitter and receiver, such as telephone lines, coaxial cables, optical fibers, and microwave and satellite links.

The waveform channels often experience some or all of the impairments such as amplitude and phase distortion, noise and interference, and multipath fading. Figure 5.12 shows a waveform channel model including some of these impairments.

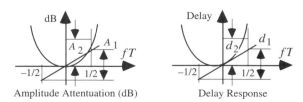

**FIGURE 5.12**   Generic waveform channel mode.

In many situations, signals are sent through conducting or guided media such as telephone lines, coaxial cables, or optical fibers. In these cases the waveform channel can be modeled[2] as a "band-limited linear filter" with a frequency response $C(f) = |C(f)| \exp[j\theta(f)]$ ($|f| \leq B$, bandwidth of the channel). The channel is characterized by the amplitude response $|C(f)|$ and phase delay response $t_p = -\theta(f)/2\pi f$ or group delay $t_g = -d\theta(f)/2\pi df$. Nonconstant amplitude response and group delay introduce linear distortions, which in turn result in ISI to the system. To a first order, the amplitude and delay responses can be modeled in terms of linear and quadratic variations in frequency $f$. Figure 5.13 displays such a model with the linear and quadratic parameters ($A_1, A_2, d_1, d_2$) representing the two responses.[3] Note that each parameter is defined to be the maximum change in the response (amplitude attenuation or group delay) over all frequencies within $\pm 1/2T_s$ of the channel center.[3]

**FIGURE 5.13**   Channel response model of linear distortions. (From Noguchi et al.,[3] ©1986 IEEE, with permission.)

In other cases, signals travel through the air or atmosphere from the sender to the destination. Typical examples are terrestrial satellite and microwave links. Here, the transmitted signals are exposed to atmospheric effects and multipath fading perturbations.[2] The waveform channel in this case has randomly time-variant impulse responses due to time-varying media, and may be best characterized by statistical models. A multipath channel is one in which the receiver sees the transmitted signal (direct path) plus its delayed versions (indirect path) due to reflections. Figure 5.14 shows a two-path phenomenon. The received signal in a multipath channel can be written as follows:[2]

$$y(t) = \sum_n a_n(t)s[t - \tau_n(t)] \tag{5.17}$$

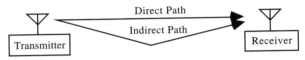

**FIGURE 5.14**   A two-path multipath scenario.

where $a_n(t)$ and $\tau_n(t)$ denote the attenuation and propagation delay along the $n$th path. Let the input signal $s(t) = \text{Re}\{\tilde{s}(t)\exp(j2\pi f_c t)\}$. Then, the equivalent time-varying impulse response of the channel is given by the following:[2]

$$\tilde{h}(\tau, t) = \sum_n \alpha_n(t)\exp(-j2\pi f_c t)\delta[t - \tau_n(t)] \tag{5.18}$$

To give an example, a group of multipath models incorporating typical delays and attenuations for off-the-air broadcast channels is given in Table 5.1.[4]

**TABLE 5.1**   Multiecho Multipath Model for Terrestial Broadcast Channels

| | Model | | | |
|---|---|---|---|---|
| | A | B | C | D |
| Path | Delay/Att. | Delay/Att. | Delay/Att. | Delay/Att. |
| Path 1 | 0.45 μs/−19 dB | 0.2 μs/−14 dB | −0.7μs/−26 dB | 0.1 μs/−9 dB |
| Path 2 | 2.3 μs/−24 dB | 1.9 μs/−18 dB | 0.1 μs/−26 dB | 0.25 μs/−17 dB |
| Path 3 | | 3.9 μs/−24 dB | 0.15 μs/−31 dB | 0.6 μs/−14 dB |
| Path 4 | | 8.2 μs/−12 dB | 0.4 μs/−28 dB | 1.1 μs/−11 dB |

*Source*: From Wu et al.[4] ©1994 IEEE, with permission.

*Fading* refers to the amplitude variations in the received signal as a result of time-variant attenuation and delay. If the envelope $|\tilde{h}(\tau; t)|$ follows a Rayleigh distribution at any instant $t$, we have a Rayleigh fading channel. Statistically, the multipath fading can be characterized by the scattering function,[2] $S(\tau, f_d)$, which is the power spectrum of the channel as a function of the time delay $\tau$ and the Doppler frequency $f_d$. Mathematically, the scattering function is the Fourier transform of the multipath intensity profile, $\phi_h(\tau, \Delta t)$, defined by the following:[2]

$$\phi_h(\tau, \Delta t) = E\{\tilde{h}^\dagger(\tau, t)\tilde{h}(\tau, t + \Delta t)\}/2 \tag{5.19}$$

where $E\{\cdot\}$ is the ensemble average. Based on the scattering function, two system parameters are often used to measure the time and frequency dependencies of a channel:[2]

1. The multipath or delay spread $T_m$, the nonzero range of $\phi_h(\tau, 0)$ over $\tau$, determines the frequency selectivity of a channel. A channel is frequency-selective if $1/T_m \ll B$ (signal bandwidth). Note that $1/T_m \approx B_c$ is often called the *coherence bandwidth* of the channel.

2. The Doppler spread $B_d$, the nonzero range of the Doppler power spectrum $S_D(f_d)$ over $f_d$, determines the time variation of a channel, where $S_D(f_d)$ is obtained by the Fourier transform of $S(\tau, f_d)$ over the delay $\tau$. The reciprocal of $B_d$ defines the coherence time of the channel; i.e., $T_c \approx 1/B_d$. Thus, if $(\Delta t)_c \gg T_s$ (signal symbol period), a multipath channel is said to be slowly fading; i.e., the channel response is essentially flat over the signal symbol duration.

In effect, the product $T_m B_d$ gives the channel spread condition in frequency and time. For channels $T_m B_d < 1$, we say that the channels are essentially frequency-nonselective and slowly fading. In terrestrial broadcast (VHF/UHF) and mobile radio,[2] for example, $T_m B_d$ is on the order of $10^{-3}$ to $10^{-5}$.

*Noise Sources and Models.*    Ideal transmission of digital data is impeded, in addition to channel distortions, by noise sources and interferences that exist in the communication link.

Thermal noise is the most common noise source in a communication channel, and its characteristics are well known. In practice, thermal noise is modeled as AWGN with power spectral density $N_o$. In band-limited channels, the AWGN is also band-limited and can be represented in quadrature form as $n(t) = n_1(t) \cos 2\pi f_c t - n_Q(t) \sin 2\pi f_c t$, where $n_1(t)$ and $n_Q(t)$ are in-phase and quadrature components; each is a zero-mean AWGN with variance $N_o B$.

Impulse noise (for which shot noise is a special case) is another common noise source in a communication channel, which appears like spike pulses occurring at random times. Impulse noise may arise from a variety of natural and human-made noise sources (e.g., atmospheric noise, ignition noise, switching, and signal clipping) and can cause severe damage to digital modulation systems if not accounted for. The model of the impulse noise may be derived based on a generalized shot noise model:

$$n_i(t) = \sum_k I_k \delta(t - t_k) \tag{5.20}$$

where $t_k$ are the random arrival times of the impulses modeled as Poisson points and $I_k$ are the random "tail" areas of the impulses with independent and identical distributions. The distribution of $I_k$ may be modeled as power Rayleigh, exponential, etc.[5] The mean and variance of the impulse noise can be obtained from Eq. (5.20) as $\eta_i = \lambda$, and $\sigma_i^2 = \lambda \langle I_k^2 \rangle$, $\lambda$ being the impulse-occurring rate. One important parameter of the impulse noise is the impulsive index, $\gamma$, defined as the product $\gamma = \lambda \bar{\tau}$,[5] where $\bar{\tau}$ is mean impulse duration. The smaller the $\gamma$ is, the smaller the impulse rate $\lambda$ and/or the mean impulse duration $\bar{\tau}$ and, therefore, the more impulsiveness of the noise. Conversely, as $\gamma \rightarrow \infty$ the impulse noise approaches Gaussian noise.

*Interferences and Nonlinear Impairments.*    *Interference* may refer to any unwanted signal(s) residing in the desired signal band. Intentional interferences such as jammers tend to mask the desired signal by having the same or higher power. Unintentional interferences, relatively low-level in power, are present as a result of the nonlinear devices in the system and/or sharing a communication resource, where the signals are not perfectly isolated from one another. In mobile radio and terrestrial broadcast systems, for example, cochannel and adjacent-channel interferences are commonplace.

*Nonlinear channels* are nonlinear devices in a communication system (e.g., amplifiers, limiters, diodes, etc.). In general, a nonlinear channel with memory can be modeled by the Volterra series,[1] which relates the input and output via higher-order impulse responses—the Volterra kernels. For a memoryless nonlinear system with an input-output relation $y(t) = g[x(t)]$, the Volterra model can be simplified to a power series,[1] $y(t) = g[x(t)] = \sum a_n x^n(t)$, where the $a_n$ are the coefficients. If the input $x(t) = \sum a_i \cos(2\pi f_i t + \theta_i)$ is a sum of multiple carrier signals (e.g., in a multichannel satellite or cable environment[1,6]) the output $y(t)$ will consist of the desired signal and a plural number of "beat" interferences, known as the *intermodulation* (IM) *products*, in the signal band with beat frequencies $w_i \pm w_j$, $w_i \pm w_j \pm w_k$, etc.

Bandpass, memoryless, nonlinear devices may also be characterized by their amplitude-gain (AM/AM) and amplitude-phase (AM/PM) transfer functions,[6] $A(r)$ and $\phi(r)$, where $r(t)$ is the input envelope. The functions can be readily obtained from single-tone measurements—that is, letting the input $x(t)$ equal $r\cos(2\pi f_o t + \theta)$, the output $y(t)$ is equal to the following:

$$A(r)\cos[2\pi f_o t + \theta + \phi(r)] = y_I(t)\cos(2\pi f_o t + \theta) - y_Q(t)\sin(2\pi f_o t + \theta) \quad (5.21)$$

where $y_I(t) = A(r)\cos\phi(r)$ and $y_Q(t) = A(r)\sin\phi(r)$ are the quadratic nonlinearity functions.

## 5.2 M-ARY QUADRATURE AMPLITUDE MODULATION (M-QAM)[1-27]

Since its first publication, the *M*-ary quadrature amplitude modulation (*M*-QAM) has become a digital modulation technique and a de facto standard that has a broad spectrum of applications because of its highly spectral efficiency. To date, *M*-QAM has been successfully employed in telephone modems,[7] satellite links,[8] microwave and mobile radios,[4,7] as well as cable and terrestrial advanced digital television systems.[9-11]

### 5.2.1 Modulation and Demodulation Processes

QAM is effectively a double-sideband suppressed carrier amplitude modulation scheme. In general, there are several ways to generate QAM signals. These methods are described in the following sections.

***Standard QAM.*** This method is by far the most common way of generating QAM signals. As Fig. 5.15 shows, the method involves (*a*) dividing the input binary stream into two independent data streams, one in-phase to the carrier (the *I*-data) and one quadrature to the carrier (the *Q*-data); (*b*) converting or mapping the I/Q data from binary level to *L* level ($L = \sqrt{M}$); (*c*) pulse shaping, filtering, and converting the data to analog waveform; and (*d*) modulating the signal onto an intermediate frequency (IF) and combining the I/Q components to form a QAM signal.[7]

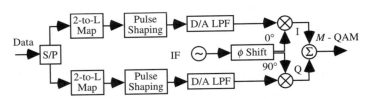

**FIGURE 5.15** Standard *M*-QAM modulator block diagram.

***Offset QAM.*** This method is developed to reduce the envelope variations of the transmitted signal. The structure of offset-QAM is similar to that of the standard QAM, except that one quadrature arm is delayed by half a symbol period relative to the other arm.[7] Thus, when the maximum rate of level change occurs in one arm (in the transition center between levels), the other arm is near stationary. It is reported that the offset-QAM creates less out-of-band power spillage and higher spectral efficiency than the standard QAM.[7]

***Superposed QAM.*** This method is basically constructed by using a number of QPSK modulators in tandem.[7] Figure 5.16 shows the structure of a superposed 16-QAM. This method has

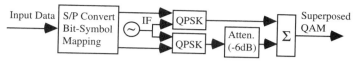

**FIGURE 5.16** Structure of superposed 16-QAM.

inherent advantages for power efficiency and simple modular implementation because it makes use of available QPSK circuitry.

***Nonlinear Amplified (NLA) QAM.*** NLA-QAM is designed for the use in power-dominated channels such as satellite links.[7] The structure of NLA-QAM is similar to that of the superposed QAM, except that nonlinear amplifiers are here used to boost the QPSK signals before I/Q combination. It has been claimed that the NLA-QAM can offer a 5-dB gain compared to the standard QAM.[7] However, the hardware complexity, especially for higher-order QAM, has hindered the practical application of the technique.

***Demodulation Process.*** The demodulation of QAM performs the inverse functions of the modulation, coherent carrier/timing recovery, and equalization if required. Figure 5.17 shows the generic QAM demodulation structure. The matched filters are practically roll-off filters. In addition, the timing recovery circuit, not shown, supplies the symbol clock for the demodulator (i.e., analog-to-digital converters [A/D], samplers, decision circuits, etc.).

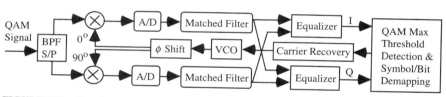

**FIGURE 5.17** Generic structure of *M*-QAM demodulator.

## 5.2.2 QAM Signal Representation and Characteristics

The signal waveform of *M*-QAM in one symbol interval can be expressed in quadrature form as follows:

$$s_i(t) = a_i \sqrt{2E_o/T_s} \cos 2\pi f_c t - b_i \sqrt{2E_o/T_s} \sin 2\pi f_c t, \quad i = 1, 2, \ldots, M; \quad 0 \le t \le T_s$$

$$(5.22)$$

where $E_o$ is the energy of the signal with the lowest amplitude, and $a_i$ and $b_i$ are independent information-bearing signal amplitudes of the in-phase and quadrature carriers. The *M*-QAM signal $s_i(t)$ can also be expanded in terms of the two basis functions $\phi 1(t) = \sqrt{2/T_s} \cos 2\pi f_c t$ and $\phi 2(t) = \sqrt{2/T_s} \sin 2\pi f_c t$ for $0 \le t \le T_s$, and represented by the signal vector $s_i = (a_i \sqrt{E_o}, b_i \sqrt{E_o})$. The distance between two signal vectors $d_{ij} = \|s_i - s_j\| = \sqrt{E_s[(a_i - a_j)^2 + (b_i - b_j)^2]}$.

***Bandwidth Efficiency of M-QAM.*** Compared to binary modulation, *M*-QAM is bandwidth-efficient. According to Section 5.1.5, the bandwidth efficiency of *M*-QAM with a roll-off factor $\alpha$ can be expressed by the following:

$$R_b/B = \log_2 M/(1 + \alpha) \tag{5.23}$$

The ideal bandwidth efficiency of $M$-QAM is the case of $\alpha = 0$. The practical roll-off values for $M$-QAM are $\alpha = 0.2$–$0.3$, which is about 77 percent to 83 percent of the ideal case.

***QAM Constellation.*** The constellation is a two-dimensional representation of the QAM signal. An optimum constellation provides a good trade-off between noise immunity and required signal power. In general, an optimum constellation maintains minimum distance and phase rotation between points (for noise and phase jitter immunities), and lowest peak-to-average power ratio (for nonlinear distortion immunity).[7] Various optimum constellations have been proposed for QAM. The preferred ones are square and star or ring constellations due to the ease of implementation and their robustness to channel impairments. These constellations, however, are optimal only for point-to-point communications. For broadcast channels (i.e., a point-to-multipoint communication scenario) a multiresolution (MR) constellation recently proposed has been shown to be optimal.[12] Figure 5.18 shows the above three constellation types for 16-QAM. Note that for square QAM the constellation points are specified by only one parameter—the distance between adjacent points (i.e., the minimum distance $d$). For star and MR QAM, on the other hand, two parameters, $d_1$ and $d_2$, specify the constellation points.

Square 16-QAM
d
Star 16-QAM
d1
d2
MR 16-QAM
d1
d2

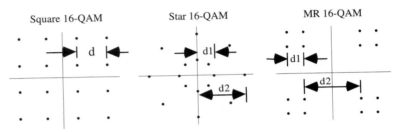

**FIGURE 5.18** Constellation shapes of 16-QAM.

***Bit-to-Symbol Mapping.*** The mapping of data bits (say, $k$ bits) to the $M$ ($M = 2^k$) possible symbols (or constellation points) can be done in a number of ways. The preferred assignment is the one in which adjacent points differ by one binary digit, as illustrated in Fig. 5.19*a*. This is called *Gray coding*. For demodulation, because the most likely errors caused by noise involve the erroneous selection of an adjacent point to the transmitted point, only a single bit error occurs in the $k$-bit sequence. Thus, the Gray coding minimizes the error probability.

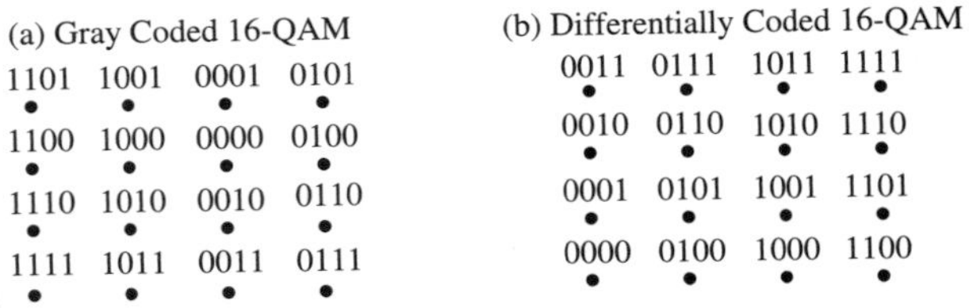

(a) Gray Coded 16-QAM
1101 1001 0001 0101
1100 1000 0000 0100
1110 1010 0010 0110
1111 1011 0011 0111
(b) Differentially Coded 16-QAM
0011 0111 1011 1111
0010 0110 1010 1110
0001 0101 1001 1101
0000 0100 1000 1100

**FIGURE 5.19** Bit-to-symbol mapping examples of 16-QAM.

For coherent demodulation an inherent problem is the phase ambiguity; the carrier recovery loop cannot determine an absolute value of the carrier phase due to the nonlinear process involved. In general, $M$-QAM has an $L$-fold phase ambiguity ($L = \sqrt{M}$).[7] Hence, in practice, differential coding is often incorporated into the bit-to-symbol mapping to resolve the phase ambiguity.[2,7] The differential encoding operation may be generally described by the following logic expression,[2]

$$b_k = a_k \oplus b_{k-1} \tag{5.24}$$

where $\{a_k\}$ and $\{b_k\}$ are the input and output bit sequences, and $\oplus$ denotes the modulo-2 addition. Figure 5.19b illustrates a constellation mapping based on differentially encoded 16-QAM.

***QAM Power Spectrum and Pulse Shaping.*** In the ideal case each $M$-QAM symbol is a rectangular pulse. Thus, the power spectrum of $M$-QAM is of $\sin^2 x/x^2$ form. Denoting $T_s$ as symbol period and $E_s$ as symbol energy, the baseband power spectrum of QAM is given by the following:

$$S_B(f) = 2E_s \sin^2 \pi T_s f/(\pi T_s f)^2 \tag{5.25}$$

In practice, when pulse shaping is involved (e.g., with raised cosine pulse), the power spectrum of QAM will be the squared raised cosine waveform given by Eq. (5.14) of Section 5.1.

***Peak-to-Average Power Ratio (PAR).*** An important design factor for QAM is the PAR, which determines the robustness of the QAM signal against nonlinear distortions. Because the digital signals are random in nature, the PAR is often expressed in terms of the percentage of time in which a transient peak exceeds the average power by a stated value.[13] For unfiltered QAM the signal envelope would not have transient peaks. In this case the PAR is the ratio of maximal symbol power to the average power, dependent on the constellation only. For square constellation, for instance, the PAR for even $k$ bit mapping ($M = 2^k$) can be obtained by the following:

$$PAR^{(M)} = 3(\sqrt{M} - 1)^2/(M - 1) \tag{5.26}$$

In filtered QAM, a transient response occurs whenever there is a phase change during symbol transition, due to the "Gibbs" phenomenon. The characteristics of the band-limiting filter, such as the roll-off factor, determine the transient response of the system. As a result, the actual PAR of QAM is higher than the ideal case. Table 5.2 illustrates some PARs of QAM with various levels, with and without filtering.

**TABLE 5.2** The PARs of QAM at Different Levels with and without Filtering

|  | QAM Level | | |
| --- | --- | --- | --- |
|  | 16-QAM | 64-QAM | 256-QAM |
| PAR@ $\alpha = 0$ (no filtering) | 2.55 dB | 3.67 dB | 4.23 dB |
| PAR@ 99.9%, $\alpha = 0.1$–0.2 | 5.50 dB | 6.00 dB | 6.40 dB |

### 5.2.3 Synchronization

The coherent reception of an $M$-QAM signal requires that the receiver be synchronized to the transmitter. First, accurate carrier frequency and phase must be known—the carrier recovery. Secondly, accurate symbol timing must be known to align transmitter and receiver clocks and give correct sample instants—the symbol timing recovery. These two modes of synchronization can be coincident with each other or they can occur sequentially. Note that for noncoherent systems, carrier recovery is of no concern.

***Carrier Synchronization.*** There are two basic approaches for carrier recovery. One approach is to send a pilot carrier along with the data-bearing signal, allowing the receiver to extract and synchronize its local oscillator to the carrier frequency and phase of the received signal.

The second method is to derive the carrier information directly from the modulated signal. This approach has the advantage that the total transmission power can be allocated to the data-bearing signal. For this case there are essentially three ways for deriving the carrier:[7,14] (1) times-$n$, or $n$th power; (2) Costas loop; and (3) decision-directed (DD).

The times-$n$ method employs an $n$th power-law device to raise the received signal to its $n$th power and extract the spectral line component at $n$ times the carrier frequency using a phased-lock loop (PLL). Figure 5.20 shows the block diagram of this method. For QAM, however, the times-two (i.e., the squaring approach) is not suitable. Thus, in practice, a times-four method has to be used.[7] Note that the times-$n$ method enhances the noise due to using the $n$th power device and has phase ambiguity, which requires differential coding.[2] However, the method may be suited to certain situations without the need of decision direction.

**FIGURE 5.20**    Times-$n$ carrier recovery.

A Costas loop is another tool for generating a properly phased carrier. The method is similar in concept to the squaring approach, but instead makes use of both the I and Q arms with a common PLL feedback.[2] The method also has the phase ambiguity problem (ambiguity of 180°), necessitating differential encoding/decoding.

In contrast to times-$n$ or Costas loop, the decision-directed carrier recovery (DDCR) estimates the carrier frequency and phase, based on the decision that chooses the constellation point closest to the received symbol as the transmitted symbol, and assuming that any phase difference between the received symbol and the chosen constellation point is due to carrier drift. Figure 5.21 shows the generic structure of the DDCR and, by example, a 16-QAM constellation (first quadrant only) template used for carrier phase correction.[3] When a constellation point moves from its ideal position to the shaded square area due to noise and/or impairments, the point needs to be rotated clockwise to correct its position. Similarly, counterclockwise rotation would be needed if points fall into the slashed areas. Figure 5.22 shows a practical implementation of the DDCR using only slicers and logic gates.[14] The error signal before the loop filter (also called *loop S-curve or phase detector characteristics*) can be expressed as follows:[14]

$$e_t = \text{sgn}\{Q - \hat{Q}\} \cdot \text{sgn}\{I\} - \text{sgn}\{I - \hat{I}\} \cdot \text{sgn}\{Q\} \tag{5.27}$$

where sgn$\{\cdot\}$ is the sign function. It is shown that this circuit leads to good loop performance and does not exhibit any false lock points compared to earlier carrier recovery methods.[14]

The DDCR method has drawbacks, however. First, it may fail if the BER is higher than a certain threshold, and, secondly, it has low acquisition speed due to the use of narrowband PLL for phase jitter reduction.[7] Possible solutions include (1) using two loop filters—one wide-band for initial acquisition and one narrow-band for tracking; (2) using nonlinear elements in the loop filter; (3) using frequency sweeping; and (4) using frequency detectors. In particular, a mixed-phase and frequency detector (PFD) circuit was proposed for use in DDCR.[7] The circuit uses

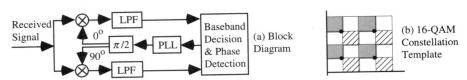

**FIGURE 5.21**    Decision-directed carrier recovery.

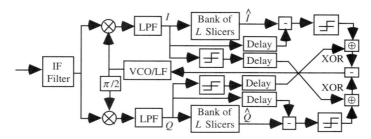

**FIGURE 5.22**  Practical implementation of *M*-QAM DDCR. (From Leclert,[14] ©1983 IEEE, with permission.)

the diagonal points (QPSK states) of the constellation with a preset window to determine the carrier phase rotation. If a signal point is inside the window, the PFD output is used for feedback; otherwise, the previous output is maintained. It is reported that the PFD approach significantly improves the acquisition range and speed as compared to the phase detector–only approach, albeit some penalty on increased phase jitter.[7]

***Symbol Synchronization.***    Timing or clock recovery for symbol synchronization may also be achieved by using pilot symbols or deriving the clock signal from the data-bearing signal. For *M*-QAM the preferred method is still the latter approach, which is described in the following paragraphs.

A simple and effective method is the squaring or times-two clock recovery,[2,7] as shown in Fig. 5.23. The I and Q signals are first squared, combined, and then filtered (tuned to the symbol rate), producing an appropriate clock signal. A PLL circuit may be used in practice to improve the clock recovery accuracy.

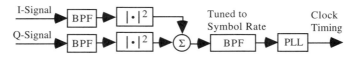

**FIGURE 5.23**  Square-law clock recovery.

Another widely used method is the early-late clock recovery,[2,7] which extracts the clock signal by processing demodulated (not necessarily detected) baseband waveforms. The method works on the peaks of the received waveform (i.e., the matched filter output) and assumes that the peaks are at the correct sampling times. Figure 5.24 shows one form of the early-late scheme.[2] The demodulated baseband signal passes through two correlators with the reference symbol and are sampled at two instants—one early and one late—equi-spaced around the predicted sampling

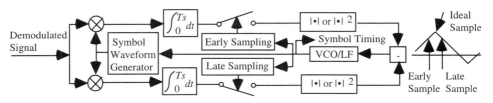

**FIGURE 5.24**  Early-late clock recovery.

instant (i.e., the peak of the correlator output). Ideally, because the autocorrelation function of any signal pulse is even, the difference (error signal) between the two correlators should be zero. In practice, when there is jitter or sampling offset, the error signal will be nonzero and drive the VCO loop to delay or advance the clock accordingly.

***Pilot-Aided Synchronization.***   Although pilot-aided carrier or symbol timing recovery (TR) techniques require extra bandwidth and/or transmitted power, they are particularly attractive to ensure coherent QAM operation in time-dispersive or fading environments.[7] This is because the pilot-aided techniques can also deliver channel conditions in terms of amplitude and phase variations, in addition to the carrier or symbol information.

One pilot-aided carrier recovery is known as the *transparent tone in band* (TTIB).[7] Its basic idea (Fig. 5.25) is to split the baseband spectrum into a lower subband (L) and an upper subband (U), creating a spectral gap in the center of the spectrum to allow the pilot insertion. The L- and U-band filters can be practically implemented using a subband splitting technique known as quadrature mirror filter (QMF).[7] For flat fading, or the total bandwidth (signal plus spectral gap) less than the channel coherent bandwidth, the data signal and the pilot suffer the same attenuation and phase shift, so the channel information derived from the pilot can be used for gain/phase correction. Note that the signal spectrum may also be split into multiple subbands incorporating multiple pilots, such that the channel is effectively sounded across the signal spectrum, making it possible to dispense with a channel equalizer.[7] However, the drawback is that the system is more complex and the spectral efficiency is reduced.

For pilot symbol–aided TR, the transmitter periodically inserts a known sequence or symbol into the data stream (Fig. 5.26). The receiver extracts the pilot symbol, usually by means of autocorrelation.[7] Again, the channel information carried by the pilot symbols can be used for compensation. Compared to TTIB, the pilot symbol–aided scheme has no change in signal peak power but still requires excess bandwidth.[7]

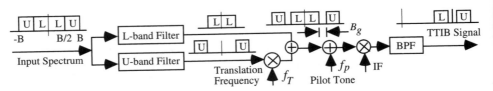

**FIGURE 5.25**   Block diagram of TTIB generation.

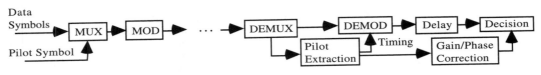

**FIGURE 5.26**   Pilot symbol–aided TR scheme.

***Maximum Likelihood Estimation and Joint Carrier/Timing Recovery.***   The carrier and symbol-timing recovery problems can be treated as a joint parameter estimation problem in which the carrier phase and time delay, denoted $f$ and $\tau$, are two parameters to be estimated.[2] The problem can be formulated by the following steps. Denote the received signal as follows:

$$y(t) = s(t; \phi, \tau) + n(t) \tag{5.28}$$

The estimation of the parameter set $\{\phi, \tau\}$ utilizes the popular maximum-likelihood (ML) criterion, which maximizes the conditional probability density $f(y)\{\phi, \tau\}$. It can be shown that the ML estimates $\{\phi_{\mathrm{ML}}, \hat{\tau}_{\mathrm{ML}}\}$ are the values of $\{\phi, \tau\}$ that maximize the following,[2]

$$\Lambda_{\mathrm{L}}(\phi, \tau) = \frac{2}{N_o} \int_{T_s} y(t)s(t; \phi, \tau)\, dt = \mathrm{Re}\{\exp(j\phi) \sum_n I_n^* y_n(\tau)/N_o\} \tag{5.29}$$

where $s(t; \phi, \tau) = \exp(-j\phi) \sum I_n u(t - nT_s - \tau)$—$I_n$ being the data sequence, $u(t)$ the data waveform, and $y_n$ the output of the matched filter. The ML estimates are then obtained by setting the derivatives of $\Lambda_{\mathrm{L}}(\phi, \tau)$ with respect to $\phi$ and $\tau$ to zero, respectively:

$$\hat{\phi}_{\mathrm{ML}} = -\tan^{-1}\left[\mathrm{Im}\{A(\hat{\tau}_{\mathrm{ML}})\}/\mathrm{Re}\{A(\hat{\tau}_{ML})\}\right] \tag{5.30a}$$

$$\mathrm{Re}\{A(\tau)\}\mathrm{Re}\left\{dA(\tau)/d\tau\right\} + \mathrm{Im}\{A(\tau)\}\mathrm{Im}\left\{dA(\tau)/d\tau\right\}\big|_{\tau = \hat{\tau}_{\mathrm{ML}}} = 0 \tag{5.30b}$$

where $A(\tau) = \sum_n I_n^* y_n(\tau)/N_o$. Note that these estimates are also decision-directed estimates. In effect, the PLL-based implementation can be shown as approximations to the ML estimates.[1]

### 5.2.4  Equalization

The aim of channel equalization is to compensate for the ISI distortions introduced in the channel. The equalization can be carried out either in the frequency domain or in the time domain. In practice, equalizers must be made to automatically adapt their parameters to reflect the varying channel characteristics, i.e., the adaptive equalization.[2] Figure 5.27 illustrates the basic structure of an adaptive equalizer. Usually, a training signal is applied to initially adjust the equalizer coefficients and stabilize the error signal, after which the equalizer's adaptation becomes decision-directed.

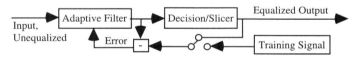

**FIGURE 5.27**    Basic structure of adaptive equalizer.

In practice, the QAM signal can be equalized either at baseband after demodulation or at passband before demodulation (see Fig. 5.28).[2] Note that for QAM the complex equalizer effectively consists of four real-valued equalizers—one for the I-signal, one for the Q-signal, and two for the cross-talks from I into Q and vice versa.

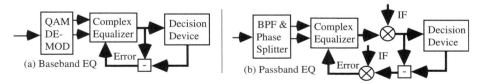

**FIGURE 5.28**    QAM equalization at passband and baseband. (From Proakis,[2] ©1989 McGraw-Hill, with permission.)

***Linear Equalizer (LE).***   The simplest LE is a linear tapped-delay line filter, as shown in Fig. 5.29, in which the tap spacing is equal to symbol period ($T$-spaced). In the figure, $\{h_j\}$ are the tap coefficients of the filter. The equalized output $\{\hat{I}_k\}$, the estimate of the data symbols, can be obtained by convolving the received signal $y_k$ with the equalizer's impulse response:

$$\hat{I}_k = \sum_{j=-N}^{N} c_j y_{k-j} \tag{5.31}$$

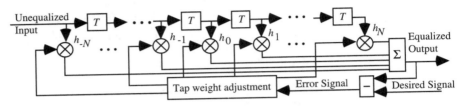

**FIGURE 5.29**   Linear transversal equalizer.

The filter taps may also be spaced in a fraction of the symbol interval, which is often called the *fractionally spaced equalizer* (FSE). In practice, $T/2$-spaced equalizers are often used.[2] In effect, the FSE can alleviate the aliasing problem, which is often associated with a $T$-spaced LE, which will likely produce sampling errors in the unknown channel conditions.[2] This is because the FSE can choose a sampling rate beyond the Nyquist frequency so that there is no aliasing before filtering. It can be shown that an optimal FSE is equivalent to an optimal LE consisting of a matched filter followed by a symbol rate equalizer, and yet performs better than the $T$-spaced LE in amplitude- and phase-distorted channels.[2]

***Decision Feedback Equalizer (DFE).***   The LE is not optimal in ISI, and its performance can be improved by introducing nonlinearity. The DFE is just such an equalizer.[2] The DFE, depicted in Fig. 5.30, consists of two filters, a feed-forward (FF) filter and a feedback (FB) filter. The FF filter takes as input the received sequence and is identical to an LE or FSE with a finite-impulse-response (FIR) structure. The FF filter removes the precursor ISI caused by current symbols not yet reaching the decision circuit. The FB filter has its input from the decision detector and is similar to an infinite-impulse-response (IIR) structure. The FB filter removes the postcursor ISI caused by previously detected symbols. From the DFE structure the DFE output can be expressed as follows:[2]

$$\hat{I}_k = \sum_{j=-K_1}^{0} c_j y_{k-j} + \sum_{j=1}^{K_2} c_j \tilde{I}_{k-j} \tag{5.32}$$

where $\{\tilde{I}_{k-1}, \ldots, \tilde{I}_{k-K_2}\}$ are previously detected symbols. The equalizer is assumed to have $K_1 + 1$ taps in its FF section and $K_2$ taps in its FB section.

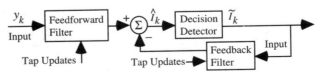

**FIGURE 5.30**   Block diagram of a decision feedback equalizer.

Because the FB filter takes the input from the decision circuit and thus contains no noise, the DFE can increase the accuracy of the interference cancellation and yield lower noise enhancement and mean-square errors than an LE.[7] However, a DFE may suffer from the impairment caused by any decision errors that will propagate through the FB section.

*Equalization Criteria.*     Two criteria have found widespread use in optimizing the equalizer coefficients: (1) peak distortion, and (2) minimum mean-square error (MMSE).

The peak distortion criterion minimizes the peak value of the ISI, assuming infinite number of taps in an LE.[2] Denote the channel transfer function as $C(z)$. The criterion leads to the inverse filter or zero-forcing (ZF) solution:[2,7]

$$H(z) = C(z)^{-1} \tag{5.33}$$

where $H(z)$ is the z-transform of the coefficients $\{h_j\}$. In practice, the channel response has finite length, i.e., $C(z) = c_0 + c_1 z^{-1} + \cdots + c_N z^{-N}$. In this case the equalizer coefficients are given by the following:[7]

$$h_j = -c_0^{-1} \sum_{i=0}^{n-1} h_i c_{j-i}, \quad h_0 = c_0^{-1}; \quad j = 1, \dots, N. \tag{5.34}$$

Note that the ZF solution works only for mild ISI conditions (i.e., the eye pattern still opens), and it neglects the effects of the additive noise.[2]

In the MMSE criterion the equalizer coefficients are adjusted by minimizing the mean-square value of the error $e_k = I_k - \hat{I}_k$, where $I_k$ is the transmitted information symbol. Again, for an LE with infinite taps, by means of the orthogonality principle, the MMSE solution gives the following:[2]

$$H(z) = H^*(z^{-1})[H(z)H^*(z^{-1}) + N_0]^{-1}. \tag{5.35}$$

Note that the noise $N_o$ appears in Eq. (5.35). This means that the MMSE criterion minimizes both the ISI and noise. For finite taps the MMSE solution may be obtained by solving the following,[2,7]

$$\sum_{i=-N}^{N} h_i \Gamma_{ij} = \sum_{i=-N}^{N} I_n y_{n-j}, \quad j = -N, \dots, N \tag{5.36}$$

where $\Gamma_{ij} = \sum y_{n-i} y_{n-j} + N_o \delta_{ij}$. For DFE the optimum coefficients are the solution of the following:[2]

$$\sum_{j=-K_1}^{0} \sum_{m=0}^{-i} (c_m^* c_{m+i-j}^* + N_0 \delta_{ij}) h_j = c_{-i}, \quad i = -K_1, \dots, -1, 0; \tag{5.37a}$$

$$h_k = \sum_{j=-K_1}^{0} h_j c_{k-j}, \quad k = 1, 2, \dots, K_2; \tag{5.37b}$$

where $c_m$ is the channel response and is given by $y_k = \sum_{m=0}^{L} c_m I_{k-m} + n_k$, where $n_k$ is the noise.

*Adaptive Equalizer—The LMS Algorithm.*     In general, the solution of Eq. (5.36) or (5.37) for the coefficients $\{h_j\}$ requires solving $(2N + 1)$ or $(K_1 + K_2 + 1)$ simultaneous, linear equations, and it can be quite cumbersome. In practice, computationally efficient iterative solutions are usually preferred. Among various iterative algorithms, a frequently used algorithm is the LMS (least-mean-square) algorithm or its signed version, which adapts the tap coefficients by the following:[2,9]

$$\underline{h}_{k+1} = \underline{h}_k + \Delta\varepsilon_k\underline{y}_k^* \tag{5.38}$$

$$\underline{h}_{k+1} = \underline{h}_k + \Delta\text{sgn}\{\varepsilon_k\}\text{sgn}\{\underline{y}_k^*\} \tag{5.39}$$

where $\underline{y}_k$ is the vector of the received signal sequence at the $k$th iteration and $\Delta$ is the LMS step size that regulates the speed and stability of the algorithm. These equations apply to the LE or the FF part of the DFE. For the DFE's FB part the equations also apply if $\underline{y}_k$ is replaced by the vector $\tilde{\boldsymbol{I}}_k$, which has components $(\tilde{I}_{k-1}, \ldots, \tilde{I}_{k-K_2})$.

In practice, the LMS algorithm often exhibits slow convergence. Hence, faster convergence algorithms have been sought. One popular algorithm, for example, is the recursive least-squares (RLS), or Kalman, algorithm which minimizes the error signal in terms of its time average rather than the expected value as in LMS.[2] Compared to LMS, the fast RLS can achieve an order of magnitude increase in converging speed, with slightly more computations.[2]

***Blind Equalizer.*** In some applications, such as multipoint communication networks,[2] it is desirable for the receiver to adjust the equalizer without benefit of a training sequence. Such an equalization is called the *blind equalization*. In effect, a blind equalizer may be used as a bootstrap to bring the receiver to the point (e.g., eye reopens) where it can reliably switch to a steady-state decision-directed mode. For higher-level QAM, a blind equalizer based on a constant-modulus algorithm (CMA) has been most suitable.[2,9] The CMA equalizes the QAM signal by minimizing the MSE, $\text{E}\{(|\hat{I}_k|^2 - R)^2\}$, where $R$ is a positive real constant. The minimization leads to the tap update recursion, with $R = \text{E}\{|I_k|^4\}/\text{E}\{|I_k|^2\}$, as follows:

$$\underline{h}_{k+1} = \underline{h}_k - \mu\hat{I}_k(|\hat{I}_k|^2 - R)^2\text{Re}\{\underline{y}_k\}. \tag{5.40}$$

***Frequency Domain Equalizer.*** Alternative to the time-domain implementation, equalization can be also realized in the frequency domain using the discrete Fourier transform (DFT).[15] The basic structure of the frequency domain equalizer is shown in Fig. 5.31, where the time-domain convolution of the equalizer coefficients with the received signal is translated to frequency multiplication. Note that the tap weight updates can be done either in the time or frequency domains. Although the frequency domain equalizer has not been popular for $M$-QAM, it shows superior performance over the time-domain equalizers for amplitude distortion channels.[15]

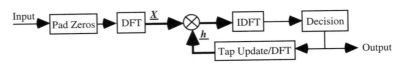

**FIGURE 5.31** Basic structure of a frequency-domain equalizer.

***Joint Equalization and Carrier/Timing Recovery.*** In practice, equalization can also be used jointly with carrier/timing recovery, as joint parameter estimation, to reduce "tap rotation" and phase jitter. The joint estimation has been proposed using either using a baseband equalizer[2] or a passband equalizer (Fig. 5.32).[16] For example, in the passband case, a joint DD-based equalizer/phase-tracking algorithm suitable for $M$-QAM can be expressed as follows:[16]

$$\hat{\underline{h}}_{k+1} = \hat{\underline{h}}_k - \beta\underline{y}_k[I_k^*\exp(-j\phi_k) - \tilde{\boldsymbol{I}}_k^*\exp(-j\hat{\phi}_k)]\exp(-j2\pi f_c kT_s)/ < |I|^2 > \tag{5.41a}$$

$$\hat{\phi}_{k+1} = \hat{\phi}_k + \alpha\text{Im}\{\hat{I}_k\tilde{I}_k^*\exp[j(\phi_k - \hat{\phi}_k)]\}/|\hat{I}_k|^2 \tag{5.41b}$$

where $\beta$ and $\alpha$ are positive constants, $\phi_k$ is the channel phase shift, and $f_c$ is the carrier frequency.

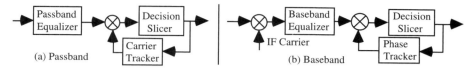

**FIGURE 5.32** Joint equalization/carrier phase tracking.

For joint baseband (blind) equalizer/carrier phase tracking, the Godard algorithm provides the tap coefficient update by (5.40) and phase estimate update by the following:[2]

$$\hat{\phi}_{k+1} = \hat{\phi}_k + \mu_\phi \text{Im}\{\tilde{I}_k \hat{I}_k^* \exp(j\hat{\phi}_k)\} \tag{5.42}$$

### 5.2.5 Coded QAM

In almost all practical QAM transmissions, channel coding is used with QAM modulation for error detection and correction. Particularly since the advent of Trellis coding,[2,7] coded modulation has become a widespread usage in many applications.

*Trellis-Coded Modulation (TCM).* The TCM was developed to improve the performance without bandwidth expansion by integrating the coding and modulation into a single entity. The expense is, however, the expansion of the constellation size. For example, a 32-TCM/QAM would actually convey an uncoded 16-QAM information; the extra data bits are used for error correction. The idea of the TCM is based on a "set partition" principle,[2,7] which involves partitioning the QAM constellation successively into 2, 4, 8, etc. subsets, and maximizing the minimum Euclidean distance between their respective signal points. As Fig. 5.33a shows, $K$ out of $N + K$ input data bits are (convolutionally) encoded into $K + 1$ bits, which are then used to partition the subsets, whereas $N$ bits are uncoded and used to select the signal points in each subset. Shown in Fig. 5.33b is an example of 16-TCM/QAM with rate 3/4 ($N = 2$, $K = 1$).

The decoding of TCM is typically performed via the Viterbi algorithm, which involves (1) determining the best signal point within each subset, and (2) selecting the trellis path that has the minimum sum of square distances from the chosen signal points. The TCM/QAM can improve the performance with an SNR (coding) gain of 3 to 6 dB, compared with the uncoded QAM.[2]

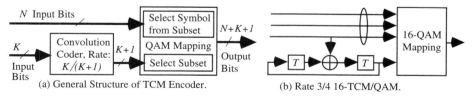

(a) General Structure of TCM Encoder.          (b) Rate 3/4 16-TCM/QAM.

**FIGURE 5.33** TCM/QAM encoder structure.

*Forward Error Correction (FEC).* In conventional digital communications the FEC is done separately with the modulation, as shown in Fig. 5.34. Basically, the FEC involves the systematic addition of redundancy to the transmitted symbols to facilitate error detection and correction at the receiver. The FEC can generally be classified into block or convolutional codes.

**FIGURE 5.34**   FEC and modulation.

In block code, blocks of $k$ information bits are encoded into blocks of $n$ bits $(n > k)$; the $n - k$ bits are redundant bits. The code rate is $R_c = k/n$. For $M$-QAM a widely used block code is the Reed-Solomon (RS) code, which is nonbinary and works on symbols instead of bits.[2] Specifically, an $(N, K)$ RS code has a total of $N (= 2^m - 1)$ $m$-bit symbols, in which $K$ symbols are information symbols, and can correct up to $t = [(N - K)/2]$ symbol errors.

In contrast, for an $(n, k)$ convolutional code, coding may be viewed as passing the $k$-bit information sequence serially through a linear finite-state shift register with connections to $n$ modulo-adders and a multiplexer that serializes the outputs of the adders.[2] The code rate is also $R_c = k/n$. The error-correcting capability of a convolutional code depends on the decoding algorithms used and the distance properties of the code.

Both the block- and convolutional-code FEC may, however, introduce bandwidth expansion if the information rate is kept the same after the coding. The expansion factor is $1/R_c$.

***Concatenated Codes.***   The error-correcting capability of FEC can be enhanced by cascading two FEC codes, interconnected by interleaving. Although the inner/outer codes can be a combination of any coding schemes, in typical QAM systems the inner code is usually a convolutional code or bandwidth-efficient trellis code, whereas the outer code is a powerful RS block code.[10]

In concatenated as well as nonconcatenated coding, interleaving is often used to randomize any burst errors and perform block/sequential conversions. This is done by rearranging or permuting the order of a sequence of symbols in a deterministic manner; the deinterleaving does the inverse functions. Interleaving can also be block- or convolution-based.[10] For block interleaving, for example, $k$-bit encoded symbols are written by column into a memory matrix of $I$ rows and $N$ columns, where $N$ is the code-word length and $I$ is the interleaving depth. When the memory is full, the symbols are then read out by rows. An $I$-depth block interleaver can handle bursts of length $I[(d_m - 1)/2]$, where $d_m$ is the minimum distance of the $(N, K)$ block code.

### 5.2.6   Performances of *M*-QAM in Noise and Channel Impairments

The performances and trade-offs of $M$-QAM in various noise and channel impairments are described in the following sections.

***AWGN.***   Using matched filter reception, the $M$-QAM is optimal for AWGN in terms of the maximum SNR or minimum BER. For square constellation and equal probable signal points, the BER of $M$-QAM can be readily derived, with $M = 2^k$, as follows,[2]

$$P_{eb} = \frac{2}{\log_2 M}(1 - 1/\sqrt{M}) \, \text{erfc}\left\{ \sqrt{1.5\gamma_{av}/(M-1)} \right\} \tag{5.43}$$

where $\gamma_{av}$ is the average SNR per $k$-bit symbol. Figure 5.35 shows the BERs for QAM levels $M = 4$ through 256. Note that doubling $M$ requires a 3-dB increase in SNR to give the same BER. In some cases it is desirable to express the BER as a function of the ratio $E_b/N_o$ ($E_b$ is energy per bit), which is given by $E_b/N_o = \gamma_{av} B/R_b$, where $B$ is the bandwidth and $R_b$ is the bit rate.

***Impulse Noise.***   The matched filter–based QAM can be vulnerable to non-Gaussian impulse noise, which may occur infrequently in time with large amplitudes.[5] The model for assessing the QAM performance for a mixture of AWGN and impulse noise is shown in Fig. 5.36$a$. As

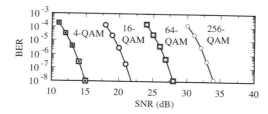

**FIGURE 5.35**    BER vs. various QAM levels under AWGN.

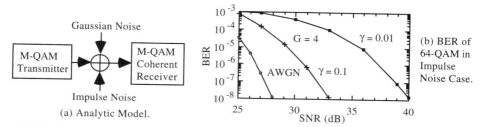

**FIGURE 5.36**    Performance of *M*-QAM in impulse noise environment.

described in Section 5.1, various impulse noise models have been developed; among them, the Middleton's impulse noise model is most attractive because it fits to a variety of non-Gaussian noises.[17] Following Figure 5.36a, the BER of *M*-QAM can be derived as follows:[17]

$$P_{eb} = \frac{2}{\log_2 M}(1 - 1/\sqrt{M})e^{-\gamma}\sum_{j=0}^{\infty}\frac{\gamma^j}{j!}\text{erfc}\left\{\sqrt{1.5\gamma_{av}(1 + G)[(M - 1)(j\gamma^{-1} + G)]^{-1}}\right\} \quad (5.44)$$

which is a function of the impulsive index *g* and the variance ratio, *G*, of the Gaussian noise to the impulse noise. Figure 5.36b shows the case of 64-QAM for $G = 4$, and $\gamma = 0.1$ and 0.01. As shown, the BER of 64-QAM degrades significantly as the noise becomes more impulsive.

*Cochannel and Adjacent Channel Interferences.*    Cochannel and adjacent channel interferences are typical in microwave, mobile, and terrestrial communication links. When the overall interference has Gaussian statistics, the performance of *M*-QAM can be estimated using Eq. (5.43) with SNR replaced by SIR (signal-to-interference ratio). However, the more severe case may be the presence of a single interferer. For this case as high as 30 dB SIR would be needed to give a negligible BER degradation.[7,17] Cochannel interference can be reduced by equalization, notch filtering, or spectrally shaped modulation, by creating a notch or gap at the interferer location.[4,10] For example, Table 5.3 shows the SIRs needed for a 32-TCM/QAM to achieve a BER of $10^{-4}$ in the presence of an NTSC cochannel interference and AWGN, with and without equalization.[4] The equalizer is a *T*/2-spaced FSE with 256 taps.

**TABLE 5.3**    SIRs of 32-TCM/QAM with NTSC Cochannel
Interference with and without Equalization

| 32-TCM/QAM@BER= $10^{-4}$ | SIR w/EQ | SIR w/o EQ |
|---|---|---|
| SNR = 40 dB | 7.8 dB | 1 dB |
| SNR = 20 dB | 12 dB | 6 dB |

***Carrier Offset and Phase Jitter.***   Let $\phi$ be the carrier phase shift at the receiver input and $\hat{\phi}$ be the receiver's phase estimate. Then, the demodulated I and Q signals can be written as follows:[2]

$$\hat{s}_{I,Q}(t) = s_{I,Q}(t)\cos(\phi - \hat{\phi}) - s_{Q,I}(t)\sin(\phi - \hat{\phi}) \qquad (5.45)$$

It is clear that the phase error $\phi \neq \hat{\phi}$ for QAM not only introduces a power penalty of the desired signal by the factor $\cos^2(\phi - \hat{\phi})$, but also a cross-talk interference into the I and Q components. In terms of constellation effect, the following observations may be made:[9] (1) a constant phase error causes a tilt or rotated constellation, (2) a carrier offset causes a constellation spin, and (3) a random phase jitter causes a circular arc displacement of the constellation points centered at the nominal position. If the phase error $\theta = \phi - \hat{\phi}$ can be modeled as a Gaussian random number with zero mean and variance $\sigma_\theta^2$, the degraded BER can be approximated by the following:

$$P_{eb} = \frac{2}{\log_2 M}(1 - 1/\sqrt{M})\int_{-\infty}^{\infty} \frac{1}{\sqrt{2\pi}\sigma_\theta}e^{-\theta^2/2\sigma_\theta^2}\mathrm{erfc}\left\{\sqrt{1.5\gamma_{av}\cos^2\theta/(M-1)}\right\} \qquad (5.46)$$

which can be evaluated numerically. Additionally, Table 5.4 lists the power penalties (i.e., the increase in SNR) required to achieve BER $= 10^{-6}$ with respect to RMS phase jitter for various QAM levels.[3] It is seen that as the phase jitter increases the power penalty increases significantly. Moreover, the higher the modulation level is, the smaller the jitter tolerance is.

The carrier phase errors can be reduced by employing effective carrier recovery and phase-tracking techniques and/or jointly with equalization.[14,16]

**TABLE 5.4**   Power Penalties of QAM in the Case of Carrier Phase Jitter (BER $= 10^{-6}$)

| Carrier phase jitter (RMS degree) | Power penalty (dB) | | | |
|---|---|---|---|---|
| | 4-QAM | 16-QAM | 64-QAM | 256-QAM |
| 0.5 | 0.0 | < 0.1 | 0.2 | 1.0 |
| 0.8 | 0.0 | < 0.2 | 0.4 | > 3.0 |
| 1.0 | 0.0 | 0.25 | 1.0 | > 6.0 |
| 1.5 | < 0.1 | 0.4 | 3.0 | ∞ |
| 2.0 | < 0.2 | 1.0 | > 6.0 | ∞ |
| 3.0 | < 0.4 | > 3.0 | ∞ | ∞ |

***Linear Distortion.***   Typically occurring in band-limited channels, linear distortion is a major source for ISI-causing SNR degradations. Figure 5.37 shows the SNR degradations for various QAM signals due to amplitude and delay distortions, respectively, based on the model given

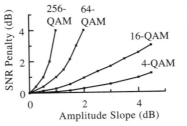
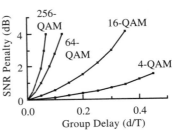

**FIGURE 5.37**   SNR penalties of QAM due to linear distortions (BER $= 10^{-6}$). (From Noguchi, et al.,[3] ©1986 IEEE, with permission.)

in Section 5.1.[3] The sensitivity to the level of modulation is evident. Linear distortions can be effectively reduced by adaptive equalization.

***Multipath and Fading.***    Multipath induces time-dispersive impairments for QAM, which are typical in terrestrial and mobile environments. In cable, multipath is a result of microreflections due to impedance mismatches. For QAM, multipath distortion induces ISI and causes eye closure or constellation smearing. For further understanding, consider a single echo with attenuation $\alpha$ and delay $\tau$. Then the demodulated I and Q signals can be written as follows:[18]

$$\hat{s}_{I,Q}(t) = s_{I,Q}(t) + \alpha[s_{I,Q}(t - \tau)\cos(2\pi f_c\tau) - s_{Q,I}(t - \tau)\sin(2\pi f_c\tau)] \tag{5.47}$$

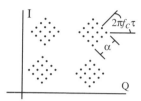

**FIGURE 5.38**   16-QAM constellation with single echo.

If the equations are plotted as depicted in Fig. 5.38 for a 16-QAM (first quadrant), the result is a rotated miniature replica of the original constellation,[18] which is $\alpha$ times smaller in size and has a rotating angle of $2\pi f_c\tau$. For multiple echoes the effect will be a super-position of all the miniature replicas of the original constellation, producing a smearing of the constellation points.

For nonfrequency selective fading, in which all frequency components of the signal undergo the same attenuation and phase shift, the Nakagami's fading model for a family of distribution on signal envelope, denoted $S$, can be used to assess the QAM BER:[17]

$$P_{eb} = \int_0^\infty P(S)P_{eb}^{(G)} \, dS = \frac{2(1 - 1/\sqrt{M})}{\log_2 M} \int_0^\infty \frac{m^m S^{2m-1}}{\Gamma(m)\Omega^m} e^{-mS^2/\Omega} \mathrm{erfc}\left\{ \frac{S}{2(\sqrt{M} - 1)\sqrt{N_o}} \right\} dS \tag{5.48}$$

where $\Gamma(\cdot)$ is the gamma function, $\Omega/2 = P_{av}$ (average signal power), and $m = \Omega^2/(S^2 - \Omega)$ is the fading figure, for which $m = 1$ or $m = \infty$ represents Rayleigh fading or no fading, respectively. For $m = 1$, the Rayleigh fading case, the integral of (5.48) has a close expression:

$$P_{eb} = \frac{2}{\log_2 M}\left[1 - 1/\sqrt{M}\right]\left[1 - \sqrt{0.5\gamma_{av}}\left(\sqrt{M} - 1\right)^{-1}\left(1 + 0.5\gamma_{av}(\sqrt{M} - 1^{-2}\right)^{-1/2}\right] \tag{5.49}$$

where $\gamma_{av} = \Omega/2N_o$. For 16-QAM, for example, calculation of Eq. (5.49) would show that a BER of $10^{-6}$ would need a 60-dB SNR, an almost 40 dB increase in power compared to no fading.

The multipath/fading distortions can be compensated for via diversity, adaptive equalization, pilot transmission,[38] or differentially coded star-QAM.[2,7]

***Nonlinear Distortions.***    $M$-QAM is also susceptible to nonlinear distortions—the cause of the nonlinear devices existing in a communication link. For AM/AM and AM/PM nonlinear distortions, the effect is the spectral spreading or sidelobe regrowth, constellation cluster, and/or warping, which in turn causes severe BER degradations.[3,6–7] In multichannel transmission, intermodulation (IM) products are generated by nonlinear distortion, further impairing the QAM quality. Also in the multichannel case, due to the sum of a large number of random signals, the composite signal peaks may sometimes drive the nonlinear device(s) over its operating limit, and thereby *clipping* results. The clipping distortion introduces impulse noiselike effects, which can result in a BER floor for $M$-QAM.[19]

To avoid the nonlinear distortion, the simplest way is to back off the power or use a linear class-A amplifier.[7] The trade-off is, however, the power efficiency. Alternatively, linearization

**TABLE 5.5** Lab/Field Test Results for 64- and 245-QAM in Cable Impairments

| Cable impairments | 64-QAM | 256-QAM | Cable impairments | 64-QAM | 256-QAM |
|---|---|---|---|---|---|
| SNR @ TOV | 21.2 dB | 29 dB | 2.5 $\mu$s echo @ $-20$ dB (+noise @ $-3$ dB TOV) | Acq. time = 0.28 s | Acq. time = 0.54 s |
| CSO/CTB @ TOV | $-28/-41$ dBc | $-38/-48$ dBc | 0.5/1.5 $\mu$s Delays | C/I = 3/5 dB | C/I = 4.5/11.5 dB |
| Phase noise @ 20 kHz | $-78.2$ dBc/Hz | $-85$ dBc/Hz | Fiber link (22 km) | BER = $1.85 \times 10^{-8}$ | BER = $4.7 \times 10^{-5}$ |
| Residual FM @ 120 Hz | > 99 kHz | 66 kHz | Coax link (23 amp) | BER = $1.85 \times 10^{-8}$ | BER = $4.3 \times 10^{-5}$ |
| AM hum @ 120 Hz | 13.8% mod. | 5.7% mod. | Subscriber | — | BER = $1.2 \times 10^{-5}$ |
| CW interference (S/I) | 23.65 dB | — | | | |

techniques such as predistortion or postdistortion may be employed.[7] Both the predistortion and postdistortion techniques are designed to cancel IM products and minimize the out-of-band power spillage by compensating the nonlinear transfer characteristics.

***Cable Impairments.***    Cable may be a good example of a benign channel environment that encompasses various noise sources, linear and nonlinear distortions, and multipath. The performance of QAM over cable has been tested extensively involving different modulation levels.[20–21] Table 5.5 provides some of the reported test results for 64- and 256-QAM.[20] In the tests the (coded) BER was equal to $3 \cdot 10^{-6}$, designated as the threshold of visibility (TOV).[21] The equalizer used is $T/2$-spaced with 64 taps for 64-QAM and 32 taps for 256-QAM. The field test was done at channel 41 of a New York City cable system, where the measured BERs are uncoded.

## 5.2.7  Advanced QAM Techniques

To cope with hostile environments and provide graceful degradations, researchers have proposed advanced QAM signal designs based on variable constellation and adaptive bit-to-symbol mapping, in addition to channel coding and equalization.

***Variate QAM.***    This scheme basically changes the constellation levels according to the channel condition;[7] for example, when the channel is subject to a fade the constellation size is decreased until a specified BER is met, and when there is no fading the constellation is increased. Alternatively, if a constant average bit rate is required, the QAM so designed provides an acceptable variable BER adapting to the varying channel conditions. Figure 5.39 illustrates the basic block diagram of the variate QAM. The system has been applied for star-QAM in mobile applications.[7] However, the system currently works only for duplex transmission (e.g., a time division duplex), because some means of informing the transmitter of the quality of the channel as perceived by the received signal is required.[7]

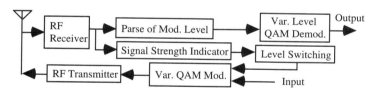

**FIGURE 5.39**    Basic block diagram of variate QAM.

***Multiresolution (MR) QAM.***    This scheme is developed to overcome the "cliff" threshold effect associated with the conventional QAM from reliable to unreliable reception, which is crucial for broadcast systems in the fringe coverage areas.[12] The idea is based on dividing the broadcast channel into fine (strong) and coarse (weak) resolution channels according to receiver distances or channel SNRs (see Fig. 5.40). The receiver closer to the emitter may access the full-quality

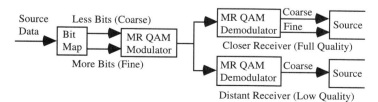

**FIGURE 5.40**    MR-based QAM system concept.

signal, whereas the distant user gets the lower-quality signal, providing a stepwise graceful degradation. The MR QAM design is then to match the different resolution levels with different bit mapping and assignment. The resultant constellation (e.g., see Fig. 5.18 for MR 16-QAM) consists of "clouds" of mini-constellations or "satellites," where the clouds and satellites carry the coarse and fine information, respectively.[12]

***Orthogonal QAM.***   Instead of dividing the source data stream into the two streams as in standard QAM, the data stream may be divided into a number of streams, each orthogonal to one another and modulated separately. The scheme is called the orthogonal multiplex QAM.[7] Because each stream can be modulated independently, the scheme has the freedom to map and assign bits to individual streams according to the channel condition. Because the scheme leads to another popular digital modulation to be described later, further detail is deferred to Section 5.4.

### 5.2.8   QAM Applications

QAM is now a mature technology that has enjoyed widespread usage in almost all consumer areas that involve the digital transmission. Described below are some of the applications.

***Telephone Modems.***   One early QAM application is for telephone modems. The CCITT V-series modem standards (V.29–V.33) adopted 16-QAM to 128-TCM/QAM.[7] Table 5.6 provides a summary of the modem features. The modems all employ scrambler, differential coding, and adaptive equalizer with initial training in order to aid synchronization and channel protection. In specific, the V.29 modem, for four-wire telephone lines, uses a 16-QAM with constellation points on I/Q coordinates and diagonal lines. The modem also has fall-back rates of 7200/4800 b/s, with reduced constellation. The V.32 modem, for two-wire duplex telephone circuits, uses either a square 16-QAM or a 32-TCM/QAM with 2/3 code rate. The modem supports the data rate up to 9600 b/s. The V.33 modem, again for four-wire lease lines, applies a 128-TCM/QAM with 2/3 code rate. The modem supports up to 14400 b/s rates and has a fall-back rate of 12000 b/s using 6 bits/symbol and 64-QAM constellation.[7]

**TABLE 5.6**   V-Series Telephone Modem Features

| Modem/date | Modulation | Baudrate | Data rate | Usage |
|---|---|---|---|---|
| V.29/1976 | 16-QAM | 2400 | 9600 b/s | 4-wire |
| V.32/1984 | 32-TCM/QAM | 2400 | 9600 b/s | 2-wire |
| V.33/1988 | 128-TCM/QAM | 2400 | 14,400 b/s | 4-wire |

***High-Speed Digital Subscriber Lines (HDSL).***   QAM has been proposed for HDSL.[22] The system intends to provide full-duplex transmission of 1.6 Mb/s for $T1$, fractional-$T1$, and primary rate integrated services digital network (ISDN). The modulation combines a precoding scheme with trellis coding and DFE to combat the near-end-talk (NEXT) impairments. Under the acronym "CAP" (carrierless AM/PM), the QAM was also used in an early asymmetric digital subscriber line (ADSL) system (i.e., ADSL-1) providing 1.5-Mb/s data rates.[22–23]

***Satellite Transmission.***   Because of severe power limitations, satellite digital transmission typically employs low-level QAM.[8,10] Recently, 16-QAM–based satellite modems were developed for field testing, providing 1.5-Mb/s (DS-1) to 54.84-Mb/s (SONET) data rates.[8] But 4-QAM (or QPSK) is mostly used in current satellite systems, such as Direct Broadcast Satellite (DBS). In a typical C-band transponder (24–36-MHz bandwidth), for example, QPSK coupled with a 3/4-rate FEC can provide a maximum transmission data rate of 40 Mb/s.[10]

***Digital Microwave Radio (DMR).*** Another widespread application for *M*-QAM is the DMR or point-to-point line-of-sight digital radio links. In DMR the deployment of higher-level QAM, such as 64/256-QAM, is commonplace, with data rates of 90–300 Mb/s.[3,24] For example, Fig. 5.41 shows a 140-Mb/s 64-QAM DMR block diagram, and Table 5.7 lists typical QAM schemes used in some North American, European and Japanese DMR systems.[24]

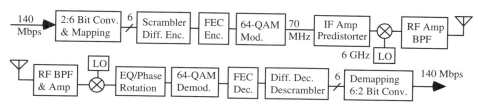

**FIGURE 5.41**   Block diagram of a 140-Mb/s 64-QAM DMR system.

**TABLE 5.7**   QAM Schemes Used in Global DMR Systems

| DMR | Freq. band/BW | QAM level/roll-off | Symbol rate/bit rate |
|---|---|---|---|
| North America | 4 GHz/20 MHz | 64-QAM/30% | 16 Mbaud/90 Mb/s |
|  | 6 GHz/30 MHz | 64-QAM/40% | 24 Mbaud/144 Mb/s |
| Europe | 4,8 GHz/32 MHz | 64-QAM/30% | 25 Mbaud/150 Mb/s |
| Japan | 4–6 GHz/20 MHz | 256-QAM/40% | 14 Mbaud/335 Mb/s |

***Mobile Cellular Radio (MCR).*** Since the early 1990s QAM has found applications in digital MCR.[7] However, due to the low-data-rate service and hostile fading conditions, the usage has been limited to mainly low-level QAM (e.g., 4/16-QAM). In addition, the star (ring-configured) QAM, in combination with differential coding, trellis coding, and block channel coding, has been found promising for MCR, particularly for the use in microcells.[7]

***Digital Terrestrial Broadcast and HDTV.*** A landmark application of the QAM technology is the development of digital high-definition television (HDTV) systems for terrestrial broadcasting.[10–11] In late 1990 General Instrument introduced the first (QAM-based) all-digital HDTV system—the Digicipher,[10] which spearheaded the race of digital HDTV for North American terrestrial broadcasting. Soon after, several QAM-based digital HDTV systems were proposed, including the ADTV[10] (advanced digital television) and CCDC (channel-compatible Digicipher),[11] which were selected for FCC testing along with Digicipher and two other non-QAM systems. The three QAM-based HDTV systems (Digicipher, ADTV, and CCDC) all employ an automatic switched 16/32-QAM mode for different data rates, concatenated FEC (RS and Trellis), adaptive equalizer, and DDCR. Table 5.8 gives a summary of modulation schemes and properties used in Digicipher and ADTV (the CCDC has essentially the same modulation

**TABLE 5.8**   Modulation Characteristics in the QAM-Based HDTV Systems

| HDTV | Modulation | FEC | Synchronization | Equalization | Data rate |
|---|---|---|---|---|---|
| Digicipher | 16-QAM | (116,106) RS, 3/4-TCM | DDCR | 256-tap LE/LMS | 4.88 baud |
|  | 32-QAM | (155,145) RS, 4/5-TCM | Transit.-det. TR | Blind CMA | 19.5/24.4 Mb/s |
| ADTV | SS-based | (147,127) RS, 0.9-TCM | DCCR | 64-Tap FSE | 4.8 baud |
|  | 16/32-QAM | Cell interleaving | Times-4 TR | Blind Update | 19.2/24 Mb/s |

structure as Digicipher but with different parameters).[10–11] Note that in ADTV, a spectrally shaped QAM (SS-QAM) is used to remove the NTSC cochannel interference; i.e., a spectral gap is created at the NTSC picture carrier location by using two separate modulators—one narrowband ($B = 1.125$ MHz) and one wide-band ($B = 4.5$ MHz).[10]

***Digital Cable Television, Set-Top Box, and Cable Modem.***   Concurrent with the terrestrial HDTV development, QAM-based digital cable television systems have also been developed to deliver high-quality video signals over a hybrid fiber/coaxial (HFC) network to subscriber homes, where the signals are received via set-top boxes.[9] Typically, the downstream transmission from cable head-end to users utilizes 64-QAM to provide high-bandwidth efficiency, whereas the upstream transmission may use QPSK for low-speed data services. As an example, Fig. 5.42 shows a cable QAM receiver architecture incorporating a DFE and DDCR.[9] The DFE initially uses a blind mode (CMA) to update the coefficients and then switches to the DD mode for normal tracking using LMS or sign-LMS. Recently, a single-QAM receiver chip has been developed for cable systems, paving the way for 500-channel television capacity and for two-way traffic on HFC networks.[25] The receiver chip can support up to 40 Mb/s. Currently, QAM-based cable modems are also being developed for two-way Internet access and services.

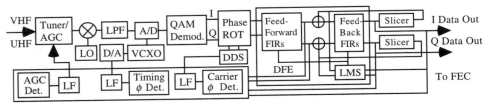

**FIGURE 5.42**   Cable QAM receiver architecture. (From Bryan,[9] ©1995 IEEE, with permission.)

In addition to hardware developments, standards have also been specified for using QAM over cable. For example, in the digital video broadcasting (DVB) over cable standard, in Europe, up to 64-QAM is specified for the modulation format.[26] In the physical layer specification of the Digital Audio-Visual Council (DAVIC), up to 256-QAM is allowed for downstream transmission (16-QAM for short distance, less than 1000 feet), whereas QPSK is used for upstream.[27]

***Multichannel Multipoint Distribution Service (MMDS).***   MMDS, also known as *wireless cable,* is a local broadcast distribution system providing multiple leased programming channels to subscribers via the microwave link at frequencies 2.5–2.7 GHz. At present, up to 33 channels may be leased using analog modulation methods. However, the proposed digital QAM-based systems for terrestrial HDTV have all claimed to be applicable to MMDS, which when deployed may provide about a fivefold increase in capacity.[10,23]

## 5.3   MULTILEVEL VESTIGIAL SIDEBAND (VSB) MODULATION[10–11,28–35]

The VSB modulation, alternative to the QAM described in Section 5.2, is another bandwidth-efficient modulation technique suitable for various applications. Recently, the 8/16 level VSB developed by Zenith/AT&T, which is an outgrowth of the 4-VSB used in the digital spectrum–compatible HDTV system (DSC-HDTV),[10–11] was adopted for the U.S. HDTV standard[28] by the Grand Alliance (GA), a consortium of companies and one university formed to develop a single, unified HDTV system for North American terrestrial and cable broadcasting (see Chapter 13).

### 5.3.1   Modulation and Demodulation Processes

The multilevel VSB is in principle a pulse amplitude modulation (PAM) with a suppressed-carrier vestigial sideband. The VSB signal can be generated by either amplitude modulation and passband filtering or baseband filtering and quadrature modulation, as shown in Fig. 5.43. The vestigial sideband is obtained in practice through root raised-cosine VSB filter.

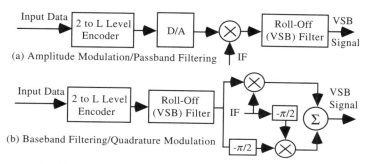

**FIGURE 5.43**   *L*-level VSB modulator.

***Bandwidth Efficiency of L-VSB.***   Because *L*-level VSB is based on PAM, it ideally has the same bandwidth efficiency as that of *L*-ary PAM;[2] i.e., $R_b/B = 2 \cdot \log_2 L$. In practice, due to the raised-cosine filtering with a roll-off factor $\alpha$, we have the following:

$$R_b/B = 2 \cdot \log_2 L/(1 + \alpha) \tag{5.50}$$

Hence, the *L*-VSB has the same bandwidth efficiency as *M*-QAM ($M = L^2$), with the same roll-off. For example, a 16-VSB with 11.5 percent roll-off has a bandwidth efficiency of 7.175 b/s/Hz.

***VSB Signal Representation.***   Similar to QAM, a VSB signal can also be represented by in-phase and quadrature components:

$$s_{\text{VSB}}(t) = s_I(t) \cos(2\pi f_c t + \theta) - s_Q(t) \sin(2\pi f_c t + \theta). \tag{5.51}$$

The quadrature signal, $s_Q(t)$, approximates the Hilbert transform of the in-phase signal, $s_I(t)$, for an upper sideband. Note that for VSB only the in-phase signal needs to be demodulated.

***L-VSB Signal Spectrum.***   The frequency spectrum of an *L*-VSB signal is shown in Fig. 5.44*a* in a 6-MHz channel. The spectrum is flat throughout most of the band, due to the noiselike attributes of the randomized data, and has two steep transition regions at each band edge due to roll-off. VSB also employs a small pilot at the suppressed carrier position near the lower band edge. In comparison to the conventional NTSC spectrum (Fig. 5.44*b*)—where most of the energy is carried by the visual, chroma, and aural carriers—the digital VSB system makes efficient use of the channel in terms of bandwidth and power.

***Demodulation Process.***   The VSB system uses a coherent demodulation process. The block diagram of a VSB demodulator is illustrated in Fig. 5.45.[29] The matched filter is a square-root raised-cosine filter matched to the transmitter filter. The synchronous detector performs the IF demodulator and the carrier recovery using the transmitted pilot tone. Additionally, a channel equalizer and a phase tracker are used to reduce linear distortions and phase noise. Finally, the slicer performs the demapping of the VSB symbols to the original data bits.

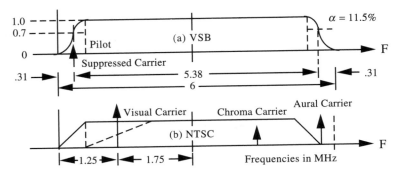

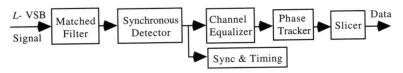

**FIGURE 5.44**   VSB and NTSC spectra. (From Sgrignoli et al.,[29] ©1995 IEEE, with permission.)

**FIGURE 5.45**   *L*-VSB demodulator block diagram.

## 5.3.2  Synchronization

The synchronization of the VSB system, i.e., carrier recovery and timing recovery, is achieved by means of pilot-aided and repetitive sync symbol insertion techniques.

*Carrier Recovery.*   At the VSB transmitter a DC bias (pilot) is inserted to the data stream and transmitted in-phase at the carrier frequency. The pilot adds the total signal power by 0.3 dB but gives a reliable carrier recovery with SNR down to 0 dB.[29] The VSB receiver coherently detects the pilot using a frequency- and phase-locked loop (FPLL), which exhibits good wide-band carrier frequency acquisition (±100 kHz) and narrow-band carrier phase tracking (±2 kHz).[28–29] Figure 5.46 depicts the block diagram of the FPLL-based carrier recovery. The FPLL utilizes the I and Q signals and an automatic frequency control (AFC) loop to act on the frequency difference between the VCO output and the incoming pilot signal and reject all other signals. Once the FPLL is phase-locked, the operation of the nominal PLL takes over, tracking the carrier phase changes. In addition, the PLL removes the effect of the NTSC cochannel picture carrier located at 900 kHz offset from the pilot and can be combined with a wide-band phase tracker to give excellent phase noise immunity.[29]

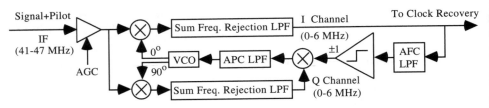

**FIGURE 5.46**   FPLL carrier recovery circuit. (From Sgrignoli et al.,[29] ©1995 IEEE, with permission.)

***Symbol Clock Recovery.***    A repetitive 4-symbol sync pattern, called *data segment sync,* is sent in binary form, along with the data symbols, to aid in clock acquisition.[28–29] Figure 5.47 shows the data segment format used for 8-VSB[28] and Fig. 5.48 shows the clock recovery block diagram.[29] The segment sync is detected in an integrate-and-dump circuit, containing a 4-symbol sync correlator and an integrator, that rejects random data input but enhances the repetitive syncs. Upon reaching a predefined confidence level (via confidence counter), which indicates that the segment sync has been found, clock phase tracking can begin. The symbol clock is then locked to the incoming data clock frequency with the proper sampling time and VCXO adjustments.

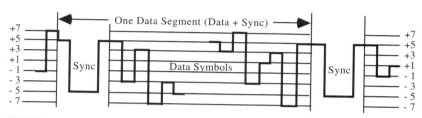

**FIGURE 5.47**    Data segment format of 8-VSB.

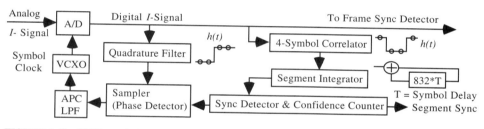

**FIGURE 5.48**    VSB's symbol clock recovery. (From Sgrignoli et al.,[29] ©1995 IEEE, with permission.)

### 5.3.3  Equalization

The channel equalization of the VSB system is based on the DFE structure.[28] The LMS algorithm is used to adjust the tap weight coefficients. Figure 5.49 shows the block diagram of the channel equalizer used in the Grand Alliance system.[28] In the adaptive mode the equalizer may operate based on three methods:[29]

**1.** Repetitive training with a known sequence regularly sent as data frame sync.

**2.** Switching to data-directed equalization once trained (eyes are open).

**3.** Using blind equalization for fast initial acquisition, adapting on data with closed eyes.

### 5.3.4  Coded *L*-VSB

The 8-level VSB designed for terrestrial broadcasting by GA employs a concatenated coding scheme, with 2/3-rate trellis coding as the inner code and 187/207-rate Reed-Solomon (RS) coding as the outer code. The RS code adds 20 parity bytes to the data and can correct up to 10 bytes of errors.

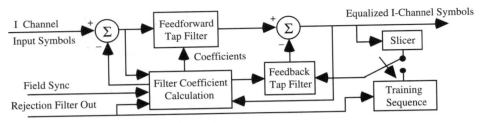

**FIGURE 5.49**    The Grand Alliance VSB equalization system.

The inner and outer codes are separated by a convolutional byte interleaver, providing inter-segment interleaving and protecting noise bursts lasting about 170 ms.[28] The interleaver matrix is a 52-data segment with a depth of 1/6 of a data field (4 ms deep).

### 5.3.5    Performances of VSB in Noise and Channel Impairments

The performances and trade-offs of multilevel VSB in various noise and channel impairments are given in the following sections.

***AWGN.***    In AGWN, VSB has essentially the same performance as PAM, with corresponding amplitude levels. Hence, the BER of $L$-level VSB is given by the following:

$$P_{eb} = \frac{1}{\log_2 L}(1 - 1/L) \operatorname{erfc}\left\{ \sqrt{1.5 \cdot \text{SNR}/(L^2 - 1)} \right\}. \tag{5.52}$$

For example, Table 5.9 lists the SNRs required to have $P_{eb} = 10^{-6}$ for various VSB levels.

**TABLE 5.9**    SNR Requirements for Multilevel VSB

| VSB level | 2-VSB | 4-VSB | 8-VSB | 16-VSB |
|---|---|---|---|---|
| SNR at BER $= 10^{-6}$ | 13.54 dB | 20.42 dB | 26.55 dB | 32.54 dB |

***Time-Domain Short Bursts and Impulse Noise.***    Multilevel VSB, like amplitude modulation techniques, is vulnerable to time-domain short bursts and impulse noise. The burst noise, if uncorrected, can cause a significant increase in BER or even a BER floor. Channel coding, inter-leaving, as well as equalization help to reduce the burst errors. Table 5.10 lists the performances of several VSB modes for a burst noise with a 10-Hz repetition rate.[21]

**TABLE 5.10**    Performance of VSB Systems under Burst Interference

| VSB modes | 4-VSB | 8-VSB | 16-VSB |
|---|---|---|---|
| Burst length @ 10 Hz, BER $= 10^{-4}$ | 21 μs | N/A | 47 μs |
| Burst length @ 10 Hz, BER $= 3 \times 10^{-6}$ | N/A | 190 μs | 150 μs |

***Frequency Domain and Cochannel Interferences.***    Simulcasting HDTV with existing NTSC services requires that the HDTV signal must work under the strong NTSC cochannel interference. In the VSB system a one-tap feed-forward rejection filter is used,[29] as shown in Fig. 5.50, to reject the three principle components of the NTSC signal; i.e., the filter's frequency response is such that the three NTSC carriers fall close to the filter nulls and are significantly reduced in strength. In addition, the VSB signal is offset in frequency by 28.615 kHz, which has insignificant consequence on the upper adjacent channel.[29]

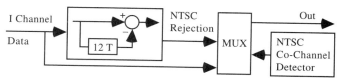

**FIGURE 5.50**   NTSC cochannel rejection block diagram. (From Sgrignoli et al.,[29] ©1995 IEEE, with permission.)

However, the filter introduces a 3-dB noise penalty; thus, it is only used when significant NTSC cochannel interference is detected.[29] The rejection filter also produces burst errors and affects the VSB constellation (e.g., eight input levels mapped to 15 output levels).[29] However, the burst errors can be mitigated by interleaving, and the constellation changes can be compensated by transmitter precoding and receiver postcoding (trellis) processing. Test results[4,30] report that the VSB signal (e.g., 4 and 8 levels) can operate with NTSC cochannel desired-to-undesired (D/U) ratios of 2 dB, for a threshold of visibility (TOV) BER of $3 \times 10^{-6}$.

***Carrier Phase Errors and Noise.***   Carrier phase noise is introduced by local oscillators in frequency conversion and receiver tuners. For VSB the phase noise results in SNR loss and constellation rotation.[29,31] Figure 5.51 shows the VSB constellation rotation due to phase error (4-VSB is illustrated). With carrier phase error $\hat{\theta}$, the demodulated VSB in-phase signal is given by the following,[31]

$$s_i(t) = s_i(t) \cos(\theta - \hat{\theta}) + s_q(t) \sin(\theta - \hat{\theta}) \qquad (5.53)$$

indicating the rotation of the I-signal and the interference of the Q-signal, resulting in an SNR loss,[31]

$$D_\varepsilon(dB) = 10 \cdot \log_{10}[1 + \sin^2(\theta - \hat{\theta}) P_{ISI}/N_o] \qquad (5.54)$$

where $P_{ISI} = (L^2 - 1) \sum [p_q(nT_s)]^2/3$, $p_q(t)$ being the pulse shape of $s_q(t)$ and $T_s$ the symbol period. For example, with 16-VSB and a carrier phase error of 0.85 degree, $D_\varepsilon = 0.6$ dB.

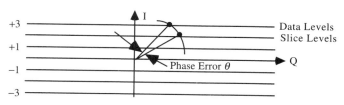

**FIGURE 5.51**   4-VSB constellation rotation due to phase error. (From Sgrignoli et al.,[29] ©1995 IEEE, with permission.)

To reduce the phase noise, the VSB system uses a decision-directed feedback loop for phase tracking.[28–29] The loop utilizes the I and Q signals (the Q-signal is derived from the I-signal via the Hilbert transform) to drive a phase de-rotator, which is also controlled by decision feedback of the data taken from the de-rotator output and estimated by the slicers.[29] It is reported that 8-VSB (trellis-coded) can correct phase errors up to $-77$ dBc/Hz, whereas 16-VSB endures the errors up to $-84$ dBc/Hz—both at a 20-kHz offset from the carrier.[29]

***Multipath and Fading.*** The multipath distortion can cause ISI and eye closure for $L$-VSB. Adaptive equalizers are effective for multipath correction. For example, 8-VSB can correct an echo range of 48 ms using an adaptive equalizer with a 511 symbol-training sequence.[29] Table 5.11 shows some of the testing results for the static and timing-varying (e.g., airplane flutter) multipath cases.[30,32]

**TABLE 5.11** Multipath Impairment into the VSB System

| 4-VSB | | | |
|---|---|---|---|
| Multipath cases | 0.32-μs echo @ 0 Hz | 2.56-μs echo @ 0 Hz | Flutter @ 2 Hz | Flutter @ 5 Hz |
| D/U @ TOV | 3.3 dB | 5.5 dB | 12.6 dB | 17 dB |

| 8-VSB | | | |
|---|---|---|---|
| Multipath cases | 1-μs echo @ 0 Hz | 15-μs echo @ 0 Hz | 1-μs echo @ 2 Hz | Flutter @ 5 Hz |
| D/U @ TOV | 1 dB | 2.9 dB | 10 dB | 11.48 dB |

***Nonlinear Distortion.*** For $L$-VSB the nonlinear distortion may arise from two sources. First, the VSB signal transient peaks may drive the transmitter to its maximum output or saturation, resulting in both the in-band and out-of-band distortion products, which may be perceived as impulsive noise. The peak-to-average power ratio (PAR) that characterizes the signal transient peaks for L-VSB depends on the VSB levels and the roll-off factor. Table 5.12 lists the PARs and also the peak symbol to average power ratio (PSAR) for several VSB levels.[30,32]

Secondly, due to the coexistence of analog and digital services, the VSB system may experience the IM products (e.g., CSO/CTB) and the clipping distortions generated by AM channels.[19,21] Table 5.13 shows a sample of threshold D/Us and SNR penalties of 16-VSB (with error correction applied) against CSO, CTB, and clipping distortion falling in-band.[21] The TOV BER is set at $3 \cdot 10^{-6}$ with SNR = 28 dB.

***Field Trials.*** Following the extensive laboratory tests, the VSB subsystem used in the GA's HDTV prototype was subjected to field testing.[21,30] The testing results for both over-the-air and

**TABLE 5.12** PARs and PSARs for VSB Modulation Levels

| VSB levels | 4-VSB | 8-VSB | 16-VSB |
|---|---|---|---|
| PAR @ 99.9% ($\alpha = 11.5\%$) | 7.6 dB | 6.2 dB | 6.8 dB |
| PSAR ($\alpha = 0$) | 2.55 dB | 3.68 dB | 4.23 dB |

**TABLE 5.13** Performance of 16-VSB in Nonlinear Impairments

| Impairments | CSO | SNR @ CSO-1 dB | CTB | SNR @ CTB-1 dB | Clipping distortion |
|---|---|---|---|---|---|
| D/U @ TOV | 33.4 dB | N/A | 44 dB | N/A | N/A |
| SNR penalty @ TOV | N/A | 6.1 dB | N/A | 0.4 dB | 2.4 dB |

cable transmissions showed that the system performed above the TOV level at the majority of test sites and worked well at the sites where NTSC signals are deemed unacceptable.[21]

The over-the-air field testing of GA's trellis-coded 8-VSB HDTV system was conducted in Charlotte, N.C., in 1994 and 1995. The tests were performed at 199 sites in 1994 and at 56 sites in 1995. The HDTV signal power is 12 dB below NTSC peak power. Table 5.14 gives a summary of the field test results.[30] In all, the test results reveal 30 to 40 percent higher satisfactory reception of the HDTV than the NTSC and has ample margin to deal with signal level variations from location to location (7–15 dB). Note that the TOV for 8-VSB is at SNR = 15 dB.

**TABLE 5.14**   Summary of Over-the-Air Field Test Results

| Satisfactory reception @ TOV | | Average SNR (dB) | | Cochannel interference D/U (dB) | |
| --- | --- | --- | --- | --- | --- |
| VHF (ch. 6) | UHF (ch. 53) | VHF (ch. 6) | UHF (ch. 53) | VHF (ch. 6) | UHF (ch. 53) |
| 82% | 92% | 22 | 30 | 0 | 0 |

The cable field tests for GA's 16-VSB system were performed in more than 40 remote sites and 8 cable systems during 1994 and 1995.[30] In 1995's field trial, for example, the tests were performed at 46 sites, over both the coax and fiber links.[30] In addition, the measurements were carried out through a sample house-in-a-box (HIAB). The 16-VSB has a data rate of 38.6 Mb/s in 6 MHz and a signal level 6 dB below the NTSC peak power. In summary, a majority of the measured SNRs were above the TOV (28-dB SNR).[30] The average SNRs are 35 dB for remote sites and 32 dB for HIAB, giving 7-dB and 4-dB headrooms to TOV, respectively.

### 5.3.6   VSB Applications

VSB modulation was first developed in the 1960s, when the 9.6-kb/s AT&T 203 telephone modem incorporated a VSB format.[33] However, it is perhaps Zenith's proposal of 2/4-VSB for DSC-HDTV in 1991[11] that has led to a sequence of VSB developments for advanced digital terrestrial and cable television systems and set-tops. Currently, VSB is also being applied to MMDS systems.[34]

***DSC-HDTV.***   The DSC-HDTV developed by Zenith/AT&T was one of the four all-digital HDTV systems proposed for digital terrestrial broadcasting in the United States.[10–11] The system employs a 2/4-VSB modulation based on service coverage and graceful picture degradation requirements. Table 5.15 shows the transmission parameters used for DSC-HDTV.[10–11]

**TABLE 5.15**   Transmission Parameters of the DSC-HDTV System

| Symbol rate: | 10.76 MByte | Synchronization: | DC pilot, FPLL, Segment sync |
| --- | --- | --- | --- |
| Data bit rate: | 21 Mb/s | DFE equalizer, LMS: | 80-tap feed-forward, 200-tap feedback |
| Bandwidth | 6 MHz | Channel coding: | RS(167,147,10), Convolutional interleaving |
| Roll-off | 11.5% | | |
| SNR@TOV: | 16 dB | Other: | Precoder/Post–comb filter |

***Grand Alliance HDTV.***   The GA HDTV system developed for the U.S. HDTV simulcast service employs a trellis-coded 8-VSB modulation. The GA system has been successfully tested in the laboratories and in the field[30] and has now been approved by the FCC as the HDTV standard for North America. Figure 5.52 shows the block diagram of the GA system, and Table 5.16 lists the system transmission parameters.[28,29]

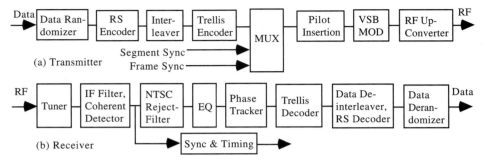

**FIGURE 5.52**   Block diagram of the GA transmission system.

**TABLE 5.16**   Transmission System Parameters of the GA HDTV System

| | | | |
|---|---|---|---|
| Symbol rate: | 10.76 MByte | Synchronization: | Pilot; FPLL; segment and frame sync |
| Total data rate: | 21.36 Mb/s | Segment, frame: | 4 sync symb./seg., 1 frame sync/313 seg. |
| BW, roll-off: | 6 MHz, 11.5% | DFE equalizer, LMS: | 64-tap feed-forward, 192-tap feedback |
| BW efficiency: | 3 bits/symbol | Channel coding: | 2/3-rate trellis code, RS(207,187,10) |
| Pilot Power: | 0.3 dB | Byte interleaving: | 52 data segments, 1/6 frame depth |
| SNR@TOV: | 15 dB | Other: | NTSC rejection filter, phase tracker |

**TABLE 5.17**   Transmission System Parameters of the Multimode USB Cable System

| | | | |
|---|---|---|---|
| Symbol rate: | 10.76 Ms/s | Synchronization: | Pilot; FPLL; segment and frame sync |
| Total data rate: | 42.7 Mb/s | Segment sync: | 836 (4 sync) symbols/segment, 77.3 $\mu$s |
| BW, roll-off: | 6 MHz, 11.5% | Frame sync: | 313 (1 sync) segments/frame, 24. 2 ms |
| BW efficiency: | 4 bits/symbol | Channel equalizer: | LMS, 63-tap feed-forward |
| VSB mode: | 2, 4, 8, 16 levels | Channel coding: | RS(207,187,10) |
| Pilot power: | 0.3 dB | Byte interleaving: | 26 segments, 1/12 frame depth (2 ms) |
| SNR@TOV: | 28 dB | Other: | Phase tracker |

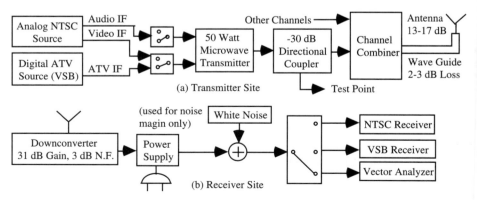

**FIGURE 5.53**   MMDS test site transmitter/receiver diagram. (Reprinted from Citta et. al.,[34] with permission.)

***Cable Television, Set-Tops, and Modems.***    The GA system has a high-speed 16-VSB cable mode providing a transmission data rate of 43 Mb/s, which can send two HDTV programs over one 6-MHz cable channel.[28] In addition, a multimode VSB cable modem, supporting 2-VSB to 16-VSB, has also been developed.[29,34] Most parts of the 16-VSB cable system are similar to the 8-VSB broadcast system, except that no trellis coding and NTSC interference rejection filtering are employed because the cochannel interference is not severe in the cable environment. Table 5.17 summarizes the system characteristics of the multimode-VSB cable modem.[28] Additionally, chip implementation for low-cost VSB demodulators (based on integrated circuits) has been reported,[35] paving the way for practical HDTV receiver and cable set-top developments.

***Multichannel Multipoint Distribution Service (MMDS).***    The field trials of using the VSB-based digital MMDS transmission for wireless cable have been underway under the auspices of the Wireless Cable Digital Alliance.[34] The trial systems for MMDS utilize 2/4/8-VSB, without trellis coding and NTSC rejection filtering. The 4-VSB and 8-VSB systems, respectively, offer date rates of 19.3 Mb/s and 29 Mb/s. Figure 5.53 shows the sample block diagrams of transmitter site [part (*a*)] and receiver site [part (*b*)] for the MMDS field test.[34] The success of the field trials will greatly increase the potential of realistic (VSB-based) digital MMDS.

## 5.4   ORTHOGONAL FREQUENCY DIVISION MODULATION (OFDM)[11,15,22–23,36–46]

After three decades of research and development, orthogonal frequency division modulation or multiplexing (OFDM) has now emerged as an alternative digital modulation method in various consumer electronics areas.[15,36–39] This can be attributed to the recent advances in VLSI and DSP technologies, so that OFDM implementation has now become realistic. Another reason for the growing popularity of OFDM is that its optimal performance has only recently been proven theoretically.[37] To date, the OFDM (particularly, the coded OFDM,[37–38]) has been widely applied to high-speed data transmission,[36,39] digital subscriber lines,[22–23] digital mobile radio,[37] and digital audio and television terrestrial broadcasting.[40–42]

### 5.4.1   Modulation and Demodulation Processes

The OFDM is a form of multicarrier modulation (MCM).[36–37] The basic principle of OFDM is described by Fig. 5.54. The input data stream is first grouped into $N$ parallel blocks or symbols, each having $m_n$ bits. Each block of data is used to independently modulate a carrier. The modulated carriers, which are equally spaced across the signal spectrum, are then summed (frequency-multiplexed) for transmission. In the receiver the carriers are first separated (demultiplexed) and then demodulated into individual but parallel data streams. The parallel data are then converted to the original serial format. Hence, the OFDM is essentially a parallel data modulation scheme, as opposed to the conventional serial data modulation scheme associated with single-carrier modulation (SCM), such as QAM. Parallel modulation can alleviate many of the problems encountered with serial systems, such as in the fading and impulse noise cases.[37]

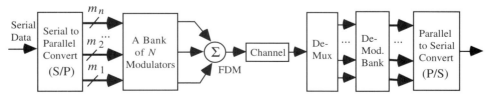

**FIGURE 5.54**    Basic principle of OFDM and MCM.

*OFDM Structure and Implementation.*   There are three MCM or OFDM structures:[36–37]

1. *Nonoverlapping frequency MCM.*   This is borrowed from the classical FDM idea, which divides the total signal spectrum into $N$ nonoverlapping frequency sub-bands and uses filters to completely separate the sub-bands. Each sub-band then employs an independent modulation and demodulation process. In practice, the limitation of filter implementation forces this classical MCM to have excessive bandwidth (with certain roll-off). Furthermore, the complexity of this system increases significantly as the number of the sub-bands increases.

2. *Staggered QAM.*   This is based on orthogonal multiplexing of the QAM signals by staggering the data (offset by half a symbol) onto alternate in-phase and quadrature sub-bands.[39] The individual spectra of the modulated carriers still use an excess roll-off bandwidth but overlap at the −3-dB frequency. The composite spectrum is flat, and the bandwidth efficiency is increased. Further, the filtering requirement is less critical than that of the classical MCM. However, the amount of filtering is still considerable when the number of carriers is large.

3. *Discrete Fourier transform (DFT)–Based OFDM.*   This is by far the most efficient OFDM scheme. The spectra of the individual sub-bands do overlap, but the individual carriers are orthogonal to one another so that they can still be separated in the receiver. The orthogonality is achieved by spacing the carriers by the reciprocal of the symbol interval.[37] Further, the FDM is obtained not by passband filtering, but by baseband processing. The composite spectrum is again flat, and, as we shall see, the band usage will be as efficient as that of the SCM. The big advantage is that both transmitter and receiver can be implemented using efficient fast Fourier transform (FFT) techniques.

Figure 5.55 shows the basic structure of the FFT-based OFDM. The modulation is effectively an inverse FFT (IFFT), and the demodulation is performed via FFT. Note that in this structure a guard interval is usually inserted after the IFFT between successive symbols, in order to ensure that the structure is free from ISI distortions.[37] Note also that the FFT-based OFDM may be further extended to an optimal structure in terms of an orthogonal transform that gives both the time and frequency localization.[38]

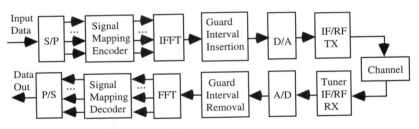

**FIGURE 5.55**   The FFT-based OFDM structure.

## 5.4.2   OFDM Signal Representation and Characteristics

In an $N$-carrier OFDM system centered at the frequency $f_c$, the transmitted OFDM signal can be expressed as follows:

$$s(t) = \mathrm{Re}\left\{ \sum_{n=-\infty}^{\infty} x_n(t)u(t - nT_s)\exp[j(2\pi f_c t + \phi)] \right\} \tag{5.55}$$

where $\phi$ is an arbitrary phase; $T_s$ is the OFDM symbol duration ($T_s = N\Delta t$, $\Delta t$ being the symbol time of the serial data); $u(t)$ is the transmitter filter impulse response and is of a rectangular shape with length $T_s$; and $x_n(t)$ is the OFDM signal, which can be written, in complex form, as follows:

$$x_n(t) = \sum_{k=0}^{N-1} X_n(k) e^{j2\pi f_k t}, \quad 0 \le t \le T_s \tag{5.56}$$

where $f_k = k/T_s$ is the $k$th carrier frequency and $X_n(k) = A_{k,n} + jB_{k,n}$ is the $n$th complex symbol for the $k$th carrier. Letting $t = m\Delta t$, the sampled sequence $x_n(m)$ of Eq. (5.56) is effectively the inverse DFT (IDFT) of the input sequence; i.e., $x_n(m) = \text{IDFT}\{X_n(k)\}$.

At the receiver the received OFDM signal is down-converted to baseband, sampled at the intervals of $\Delta t$, and passed to the DFT demodulator. If the transmission channel is distortionless and noiseless, the OFDM symbols are demodulated without error:

$$X_n(k) = \frac{1}{N} \sum_{m=0}^{N-1} x_n(m) \exp(-j2\pi km/N) = \text{DFT}\{x_n(m)\}, \quad k = 0, 1, \ldots, N - 1 \tag{5.57}$$

***OFDM Spectrum and Pulse Shaping.***   Figure 5.56 shows the power spectrum of an OFDM signal. Each subchannel has a $\sin^2 x / x^2$ spectral shape with zeros at multiples of $1/T_s$. The carriers are spaced by $1/T_s$, yielding an overall rectangular spectrum and the spectral orthogonality. Because of the IDFT, the OFDM spectrum is periodic with a period of $1/\Delta t$ Hz. Because an ideal rectangular low-pass filter is not realizable, some spectral shaping is used in practice. In this case the carriers in the roll-off region may be turned off (set to zero and referred to as "virtual carriers") to provide a flat spectrum in the non-roll-off region.[15]

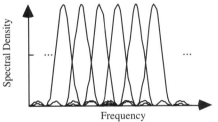
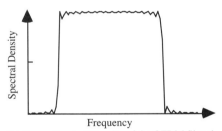

(a) Power Spectral Density of Individual Subcarriers    (b) Power Spectral Density of the OFDM Signal

**FIGURE 5.56**   Power spectrum of an OFDM signal.

***Bandwidth Efficiency of OFDM.***   For classical $M$-ary MCM it is readily shown that the bandwidth efficiency is $\log_2 M$ b/s/Hz without roll-off. For DFT-based $M$-ary OFDM systems, the bandwidth efficiency will asymptotically approach that of the $M$-ary SCM. This is seen as follows. Note that the total bandwidth of an $N$-carrier OFDM is $B = (N - 1)/T_s + 2d$, where $d = 1/T_s$ is the one-sided bandwidth of the subchannel (defined as the distance from the center of the spectrum to the first null). Assuming that each sub-band takes $M$-QAM, then the transmitted bit rate $R_b$ is equal to $N \log_2 M/T_s$. Thus, the bandwidth efficiency is given by the following:

$$R_b/B = \frac{N}{N+1} \log_2 M \quad \text{b/s/Hz} \tag{5.58}$$

For large $N$, $R_b/B \to \log_2 M$, the single-carrier efficiency. In practice, due to the guard interval insertion ($T_g$) and spectral shaping [$\delta = (1 + \alpha)/T_s$], the bandwidth efficiency reduces to the following:

$$R_b/B = [1 - (1 + \alpha)/(N + 1 + \alpha)][1 - T_g/(T_s + T_g)] \log_2 M \quad \text{b/s/Hz} \tag{5.59}$$

Thus, to obtain high-bandwidth efficiency in OFDM, $N$ must be large and $T_g$ must be small.

Note that Eqs. (5.58) and (5.59) apply to the equal sub-band bit assignment case. In general, if $\log_2 M_k$ bits are assigned to sub-band $k$, then $R_b = \sum_{k=0}^{N-1} \log_2 M_k / T_s$ should be used.

***Guard Interval.***    A guard interval $T_g$ is inserted at the IDFT output between successive symbols to eliminate the ISI caused by the multipath delays.[37,41] The duration $T_g$ is chosen long enough so that the channel transient responses induced by the multipath have settled before the signal is fed to the DFT. Figure 5.57 illustrates the concept of the guard interval, where each symbol is preceded by a periodic extension of the signal itself. The total symbol time $T$ is equal to $T_s + T_g$, where $T_s$ is now the useful symbol interval (the carriers are still spaced by $1/T_s$).

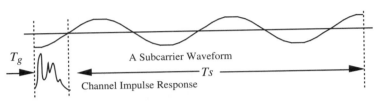

**FIGURE 5.57**    Concept of guard interval.

Because of the periodic property of the OFDM, the guard interval is a *cyclic* extension or prefix of the DFT output blocks, simply repeating $N_g$ samples of the useful symbol block (the prefix length $N_g$ corresponds to the duration $T_g$).[15]

The insertion of a guard interval, however, reduces bandwidth efficiency [see Eq. (5.59)] and power efficiency (i.e., a loss of $10 \log[(T_s + T_g)/T_s]$ dB). In practice, $T_g < T_s/4$ is often chosen to keep bandwidth expansion below 20 percent and power loss below 1 dB.[37]

***Modulation Format.***    One of the attractive features for OFDM is that different modulation types can be applied to different carriers depending on the service and channel requirements. The modulation levels of each subchannel can be specified by the complex symbols $X(k) = A_k + jB_k$ as defined in Eq. (5.56). In $M$-QAM, for example, $A_k$ and $B_k$ take on values of $\pm 1, \pm 3, \ldots, \pm(\sqrt{M} - 1)$.

***Power Division and Bit Allocation.***    Ideally, the optimum power distribution and bit allocation for OFDM can be achieved by a "water-filling" procedure in information theory.[36] In practice, in the Gaussian channel the number of bits $b_k$ assigned to the $k$th carrier can be obtained from the desired BER $P_{eb}$ and the estimated or measured $\mathrm{SNR}_k$ via the following,

$$P_{eb} = \frac{2}{b_k}\left(1 - 1/\sqrt{M_k}\right)\mathrm{erfc}\left(\sqrt{\frac{3\mathrm{SNR}_k}{2(M_k - 1)}}\right) \tag{5.60}$$

where $b_k = \log_2 M_k$. The solution of Eq. (5.60) for $M_k$ can be sought numerically.

Alternatively, iterative bit allocations can be applied. The following scheme can be viewed as a discrete approximation to the water-filling procedure.[36] The scheme maximizes the transmitted bit rate given the available total transmitted power or SNR and assuming equal BERs for the carriers. The iterative steps are summarized here:

1. Estimate the noise power $(\sigma_n^2)_k$ and channel response $H_k$ of subchannel $k$ for $k = 1, \ldots, N - 1$.

2. For all $k$, set initial values of power and bits per carrier to zero; i.e., $P_k = 0$ and $b_k = 0$.

3. Let $b_k = 2$ and calculate the power $\Delta P_k$ needed to satisfy the desired BER from Eq. (5.60), with $\mathrm{SNR}_k = (P_k + \Delta P_k)|H_k|^2/(\sigma_n^2)_k$. Then search for the smallest $\Delta P_k$.

**4.** Increment $P_k$ and $b_k$; i.e., let $P_k = P_k + \Delta P_k$ and $b_k = b_k + 2$ (or $b_k = b_k + 1$).

**5.** Repeat from step 3 with an update $\Delta P_k$ and $b_k$ until all power is allocated (i.e., $P_k \geq P_{tot}$, the available power) or all bits are allocated (i.e., $b_k \geq b_{max}$, the maximum bits per symbol).

***Peak-to-Average Power Ratio (PAR).***    Because the OFDM signal is a sum of a large number of carriers with random phase and amplitude, it may exhibit high PAR.[37,41] Because the OFDM signal tends to be Gaussian-distributed for large DFT size, the PAR may be solely determined by the probability of a transient peak exceeding the average, independent of the DFT size and subchannel constellation. A PAR of 10 dB, for example, corresponds to 0.1-percent probability that a transient peak exceeds the average, or transmitting the signal 99.9 percent of the time. As will be shown later, the high PAR requires the power amplifier to operate at high-output back-off (OBO), so as to avoid performance degradation and the adjacent channel interference.

### 5.4.3  Synchronization

The coherent demodulation of the OFDM signal (via DFT) requires the receiver to be perfectly synchronized with the transmitter. That is, an accurate carrier recovery and symbol timing must be provided in order to maintain the orthogonality of the OFDM signal and hence avoid the ISI and intercarrier interference (ICI).

***Carrier Synchronization.***    The carrier synchronization of OFDM is usually achieved by a pilot-based technique.[15,37] The pilot may be an unmodulated carrier or a known sequence. The receiver may use an AFC loop at the tuner stage, which consists of a narrow-band filter and a PLL synthesizer, to track the carrier frequency offset and phase errors obtained from the pilot signal. Figure 5.58 depicts the block diagram of the pilot-based carrier recovery technique.

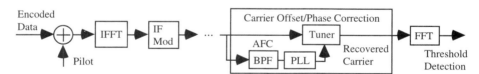

**FIGURE 5.58**    Pilot-based OFDM carrier recovery.

One drawback of the pilot-aided technique is the reduction of the OFDM's bandwidth and power efficiencies. An alternative method may use the decision-directed feedback method, similar to that of QAM as described in Section 5.2. However, the excessive delay introduced by the FFT in the loop may render the solution to be impractical.[15]

***Symbol Synchronization.***    For symbol timing recovery in OFDM, test or pilot symbols are inserted regularly in the time domain.[37,41] A test symbol may be a chirp pulse or a pseudorandom sequence. The pilot symbols may be placed at fixed positions, such as at a preamble, in an OFDM framing structure. At the receiver, the pilot symbols are captured during the frame synchronization. Figure 5.59 shows an exemplified OFDM frame structure in which the pilot symbols are preambles.

With the aid of guard interval, coarse symbol synchronization may be realized, from the test symbol, by estimating the ISI-free useful symbol boundaries. Let $s(t)$ be the transmitted OFDM signal. Because the guard interval is periodic, the difference signal $s(t + T - T_g) - s(t)$ will be zero in the guard interval and nonzero otherwise. Actually, to minimize the noise effect, an integration of the absolute value $|s(t + T - T_g) - s(t)|$ over the interval $T_s = T - T_g$ gives a coarse indication of the beginning of the symbol interval.[41] A finer synchronization can further be achieved by examining the phase rotation of the carriers based on the pilot symbols.

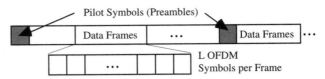

**FIGURE 5.59**    A frame structure using OFDM and pilot symbols.

### 5.4.4  Channel Equalization

In OFDM the division of the signal band into many narrow sub-bands effectively turns the channel equalization to a complex multiplier correction in frequency domain on individual subchannels.[41] This is seen as follows.

Let $C(f)$ be the channel transfer function. When the number of carriers is large, the function $C(f)$ does not change significantly over a sub-band. Hence, the received signal over a symbol duration can be approximated in complex form by the following:

$$y(t) = \sum_{k=0}^{N-1} C(k)X(k)\exp[j2\pi k(t - T_g)/T_s], \quad 0 \le t \le T_s \tag{5.61}$$

where $X(k)$ is the transmitted signal and $C(k)$ is the equivalent channel response of the $k$th subchannel. Assume that there is no ISI over $[T_g, T]$. The sampling time is at $t_n = T_g + nT_s/N$. Then the DFT performed on $y(t_n)$ yields $\hat{X}(m) = X(m)C(m)$, $m = 1, 2, \ldots, N - 1$.

Under the ZF criterion,[2] the equalization coefficient $H(m)$ is equal to $1/C(m)$; i.e., multiplying the DFT output $\hat{X}(m)$ by $H(m)$ gives the original signal $X(m)$. When additive noise is also considered, the MMSE criterion is usually used, which yields the following,[15]

$$H(m) = C(m)^* \left[ |C(m)|^2 + \sigma_n^2/\sigma_s^2 \right]^{-1} \tag{5.62}$$

where $\sigma_n^2$ is the variance of the noise and $\sigma_s^2$ is the variance of the transmitted symbols. Clearly, the MMSE solution reduces to the ZF solution if $\sigma_n^2 = 0$. MMSE is also more attractive than ZF for channels with spectral nulls in the signal band, because the latter requires infinite gain and leads to infinite noise enhancement at the correspondent subchannels.

Note that if the modulation format used is a PSK, no amplitude equalization is then needed. Further, if differential phase transmission between subchannels is used, phase equalization can also be eliminated or made differential [e.g., $Z(m) = Y(m)Y(m - 1)^*$].[15]

***Estimating Equalizer Coefficients.***    The channel response can be estimated using pilot symbols. Let $V(k)$ be the pilot symbol, known to the receiver. Then $C(k)$ is equal to $\hat{V}(k)/V(k)$, where $\hat{V}(k)$ is the received symbol. Hence, during the useful symbol interval, the estimated information symbol is $\tilde{X}(m) = \hat{X}(m)/H(m)$. In practice, to cope with a slowly time-varying channel, the equalization may need to be adaptive, with recursive estimations of the channel response and equalization coefficients. For example, weighted averages of the equalization coefficients may be used to give better estimates.

In the general time-varying case the equalization may take the convolutional form in the frequency domain. The well-known LE and DFE can be applied:[43]

$$\hat{X}_{\text{LE}}(m) = \sum_{k=-K}^{K} H_{\text{LE}}(k)Y(m - k),$$

or

$$\hat{X}_{\text{DFE}}(m) = \sum_{k=-K_1}^{0} H_{\text{FF}}(k)Y(m - k) + \sum_{k=1}^{K_2} H_{\text{FB}}(k)\tilde{X}(m - k) \tag{5.63}$$

where $H_{LE}(k)$ is the tap coefficients of the LE and $H_{FF}(k)$ and $H_{FB}(k)$ are the feed-forward and feedback tap coefficients of the DFE.

### 5.4.5  Coded OFDM (COFDM)

In practical OFDM transmission, channel coding is often used to combat channel impairments and improve performance. In certain cases (for example, in multipath fading channels with spectral nulls in the signal band destroying relevant carriers) OFDM cannot perform reliably without channel coding.

Among various codes, TCM has been a favorable choice for obvious reasons. In more severe cases, concatenate coding with TCM as inner code and RS code as outer code may further improve performance with moderate decoding complexity. Alternatively, a parallel concatenated convolutional code, the turbo code,[38] can be used in COFDM. It is claimed that the turbo-COFDM provides better performance than the TCM/OFDM in some Rayleigh fading cases.[38]

In COFDM both time- and frequency-domain interleaving may be applied. In particular, the frequency interleaving is essential to reduce the burst errors and frequency interferences over the signal spectrum and scatter the signal samples that may have fallen into deep notches. The interleaving matrix design now depends on the code length and the DFT size of the OFDM.

### 5.4.6  Performances of OFDM in Noise and Channel Impairments

The performances and trade-offs of OFDM in various noise and channel impairments are given in the following sections.

*AWGN.*    In AWGN the optimal demodulation of OFDM can be achieved, in principle, by using a parallel bank of matched filters. (Note that the FFT realization is equivalent to a bank of matched filters.[36–37]) Moreover, there is no ISI and ICI at the matched filter outputs. Assume each subchannel uses $M$-QAM. Then, the BER of each subchannel can be calculated by Eq. (5.60) with $M_k = M$. The total BER will be an average of the subchannel BERs.

*Time-Domain Short-Duration Bursts and Impulse Noise.*    Because in OFDM the symbol period is quite long (e.g., a few hundred microseconds), any time-domain short-duration bursts and impulse noise can be averaged out by the FFT process in the receiver over the entire OFDM symbol period. If the impulsive interference occurs during a small fraction of the OFDM symbol time, its impact is then negligible; the subchannels would be only slightly affected but still decodable. Thus, OFDM is inherently immune to this type of impairment.

*Frequency-Domain and Cochannel Interferences.*    Uncoded OFDM systems are susceptible to frequency-domain interferences, such as cochannel interference in terrestrial broadcast and cellular mobile systems. For simplicity consider a tone interferer, which may represent an NTSC signal component (e.g., the visual or aural carrier) co-located with the OFDM signal. Then during the useful symbol duration $T_s$, the received signal $y(t) = \text{Re}\{x(t)\} + A\cos(2\pi f_I t + \varphi)$, where $x(t)$ is given by (5.56). The demodulation (DFT) on the received signal yields the following:

$$\hat{X}(m) = X(m) + \frac{2A}{N}\sum_{n=0}^{N-1}\left\{ e^{j[\varphi+2\pi(f_I T_s - m)n/N]} + e^{-j[\varphi+2\pi(f_I T_s + m)n/N]} \right\} \qquad (5.64)$$

Two cases arise:[42] (1) $f_I$ is equal to a carrier frequency ($f_I = k_o/T_s$), and (2) $f_I$ falls between two carriers. It is easy to verify that in case 1, only that carrier is corrupted by the interferer, i.e., $\hat{X}(m) = X(m)$ for $k \neq k_o$, and $\hat{X}(m) = X(m) + A\exp(j\varphi)$ for $k = k_o$; whereas in case 2 all carriers will be corrupted, with the carriers closer to the interferer being most affected.

The cochannel interference may be mitigated by spectral shaping, i.e., offsetting the carrier locations or switching off the carriers close to the interferers.[39,41] To give an example, Figure 5.60 shows simulated BERs of a 512-carrier OFDM for case 2 and improved BERs with 10 carriers switched off.[41] Other techniques may also be used to provide better performance, such as trellis coding or the optimal power and bit allocation scheme as described earlier.

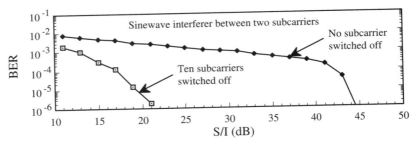

**FIGURE 5.60**    OFDM in cochannel interference. (From Tourtier et al.,[41] ©1993 Elsevier, with permission.)

***Carrier Frequency Offsets and Phase Noise.***    The long symbol duration and narrow sub-band spacing make the OFDM very sensitive to carrier frequency offset and/or phase noise. Considering the case of phase error, the DFT output for the $m$th carrier may be written as follows:[44]

$$Y(m) = X(m)I_o(\varepsilon) + \sum_{k \neq m} X(k)I_{k-m}(\varepsilon) + W(m) = \hat{X}(m) + V(m) + W(m) \quad (5.65)$$

where $\varepsilon$ is the phase error caused by the carrier offset and/or phase noise, and

$$I_n(\varepsilon) = \exp[j2\pi(N-2)(n+\varepsilon)/N] \cdot \sin[\pi(N-1)(n+\varepsilon)/N]/\sin[\pi(n+\varepsilon)/N]. \quad (5.66)$$

The first term in Eq. (5.65) is the signal term, which has an amplitude reduction and phase shift, and the second term is the ICI. Assume that $E\{X(k)X^x(l)\} = E_k\delta_{kl}$ (i.e., the signals are uncorrelated) and $E_o = E\{|I_o|^2\}$. Then, for large $N$, $V_o = E\{|V|^2\} = E_m \sum_k E\{|I_{m-k}|^2\} = E_m(1 - E_o)$. The SNR degradation for the $m$th subchannel due to the phase error can be defined by the following:[44]

$$D_\varepsilon = -10 \log[E_o/(1 + V_o/N_o)] \approx (10/\ln 10)(1 - E_o)E_m/N_o \quad \text{(dB)} \quad (5.67)$$

where the approximation is valid for small phase errors.

For carrier frequency offset, $\varepsilon = 2\pi\Delta ft + \varepsilon_o$ and $E_o = [\sin\pi\Delta fT_s/\pi\Delta fT_s]^2 \approx 1 - (\pi\Delta fT_s)^2/3$. Thus, $D_\varepsilon(dB) \approx 1.45(\pi\Delta fT_s)^2 E_m/N_o$, proportional to the square of the carrier offset. For example, with $\Delta fT_s = 0.01$ and OFDM/16-QAM at BER $= 10^{-4}(E_m/N_o = 19$ dB), $D_\varepsilon$ is equal to 0.1 dB, but with $\Delta fT_s = 0.1$ and the same $E_m/N_o$, $D_\varepsilon$ is equal to 11 dB, an intolerable degradation.

For phase noise resulting from local oscillators, if modeled as a wiener process, then $D_\varepsilon(dB) \approx 3.2\pi\beta T_s E_m/N_o$,[44] where $\beta$ is the 3-dB oscillator line width.

***Multipath and Fading.***    Following the multipath model given in Section 5.1 and the OFDM process with a proper guard interval, the DFT output at the OFDM receiver can be written as follows:

$$Y(m) = \text{DFT}\{y(n)\} = X(m) \cdot \sum_{q=0}^{Q} \alpha_q e^{-j2\pi m\tau_q/T_s} + W(m) = X(m)C(m) + W(m) \quad (5.68)$$

where $W(m)$ denotes the AWGN. Thus, the multipath distortion in an OFDM system transforms to a complex multiplicative noise in the frequency domain. To gain further insights, consider the single multipath case $(Q = 1)$. Then we have the following:

$$Y(m) = C(m)X(m) + W(m)$$
$$= [1 + \alpha \exp(-j2\pi m\Delta)]X(m) + W(m), \quad m = 1, \ldots, N - 1 \quad (5.69)$$

where $\Delta = \tau/T_s$. The transmitted symbols are recovered by using the equalizer $H(m) = C(m)^{-1}$ in the absence of noise. When noise is considered, applying equalizer $H(m)$ to Eq. (5.69) yields a new noise term, $W(m)C(m)^{-1}$, with a variance $(N_o)_m = N_0|C(m)|^{-2}$. It is easily verified that for a fixed delay $\Delta$, the noise is suppressed at some carriers $(\cos 2\pi m\Delta > 0)$ and enhanced at other carriers $(\cos 2\pi m\Delta < 0)$. If the number of carriers is large, the amounts of the noise suppression and enhancement will be roughly equal. Using Eq. (5.60), the total SER can be written as follows:

$$P_{es} = \frac{1}{N} \sum_{m=0}^{N-1} (P_{es})_m = \frac{2}{N}(1 - 1/\sqrt{M}) \sum_{m=0}^{N-1} \mathrm{erfc}\left(\sqrt{\frac{3E_s C(m)|^2}{2(M-1)N_0(1 + T_g/T_s)}}\right) \quad (5.70)$$

where the term $(1 + T_g/T_s)$ accounts for the symbol energy reduction by the guard interval. As an example, Fig. 5.61 shows the SER curves of OFDM/64-QAM for different echo amplitudes. The OFDM uses 1400 carriers over a 6-MHz channel and $T_g = 32$ $\mu$s (i.e., $T_g/T_s = 0.137$)[45]. The relative delay $\Delta$ is equal to 0.06 (in fact, the delay has negligible effect on SER as long as $\Delta < T_g/T_s$). Additionally, a SER degradation factor due to the noise enhancement may be defined as follows:[45]

$$D_e = \int_{\cos\theta<0} (1 + \alpha^2)|C(\theta)|^{-2}d\theta = \frac{4}{\pi}(1 + \alpha^2)(1 - \alpha^2)^{-1} \arctan\{(1 + \alpha)(1 - \alpha)^{-1}\} \quad (5.71)$$

For a 9-dB echo $(\alpha = 0.355)$, for instance, $D_e = 2.6$ dB, roughly equal to that given in Fig. 5.61.

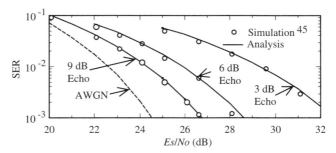

**FIGURE 5.61** SER performance of OFDM/64-QAM in single-echo multipath.

For the case of multiple echoes, a similar approach can be used; however, no closed-form solution may be obtained. Instead, if the multipath delays fall into the guard interval and the echo phases are uniformly distributed over $[0, 2\pi]$, an equivalent average echo amplitude[45] can be defined as $\alpha_{av} = (\sum a_q^2)^{1/2}$. Then, the average SER or performance degradation can be estimated using Eq. (5.70) or (5.71) with $\alpha$ replaced by $\alpha_{av}$.

The effect of a fading channel on the transmitted OFDM signal depends on the choice of the useful symbol duration $T_s$ and guard interval $T_g$. If $T_s$ is chosen so that $T_s \gg T_m$ (delay spread), the OFDM effectively turns the frequency-selective fading into a frequency-nonselective fading. Similarly, if $T (= T_s + T_g) \ll 1/B_d$ (Doppler spread), the fading is essentially a slowly fading channel over one OFDM symbol period. In practice, one may choose $T_m/T_s < 0.1$ and $TB_d < 0.02$,[40] so that $T_m B_d < 10^{-3}$; i.e., the channel is essentially nonselective and slowly fading.

For flat Rayleigh fading the transmitted signal suffers from a time-varying multiplicative distortion.[43] That is, the received signal is, in discrete form, $y(n) = x(n)z(n) + w(n)$, where $z(n)$ denotes the complex fading envelope. After the DFT the output signal is given by the following:

$$Y(m) = \sum_k X(k)Z(m - k) + W(k)$$

$$= X(m)Z(0) + \sum_{k \neq m} X(k)Z(m - k) + W(k), \quad m = 0, 1, \ldots, N - 1 \qquad (5.72)$$

The second term on the right of the second equality represents the ISI/ICI caused by the loss of orthogonality due to fading. Because the ISI/ICI is a sum from independent carriers, it can be approximated as an additive Gaussian noise. Flat fading can be effectively corrected by pilot-based method, equalization in frequency domain, and/or channel coding.

***Nonlinear Distortion.***    Because OFDM has a high PAR, it is susceptible to nonlinear or clipping distortions, since the signal peaks may occasionally thrust into the saturation region of a transmitter power amplifier. Thus, certain OBO is required to prevent BER degradation and IM products falling into adjacent channels. To illustrate, consider a soft limiter modeling the solid-state power amplifiers:

$$y = g(x) = \begin{cases} A \, \text{sgn}\{x\}, & |x| > A \\ x, & |x| \leq A \end{cases} \qquad (5.73)$$

Table 5.18 lists the in-band BER degradations and interfering powers into an adjacent channel of an OFDM/64-QAM for various OBOs. Here, the BER degradation is the difference of the signal power required to have a BER of $10^{-4}$, with and without the soft limiter. The OBO is the difference between the limiter's clipping power and the average signal power. Table 5.18 indicates that an OBO of 6.5 dB would produce a negligible BER degradation (at $10^{-4}$ BER) and lower than $-40$ dB interfering power to the adjacent channel.

**TABLE 5.18**  BER Degradation and Adjacent Channel SIR of OFDM for Various OBOs

| OBO (dB) | BER degradation (dB) | Adjacent channel SIR (dB)* |
|---|---|---|
| 3.5 | 8.18 | $\Delta + 31.2$ |
| 4.5 | 2.11 | $\Delta + 34.4$ |
| 5.5 | 0.65 | $\Delta + 38.3$ |
| 6.5 | 0.16 | $\Delta + 43.4$ |

* $\Delta$ denotes the power ratio of the adjacent channel to the in-band OFDM signal.

### 5.4.7  OFDM Applications

Owing to its attractive properties, OFDM or COFDM has been applied to a variety of digital transmission systems from data modem to digital terrestrial audio/video broadcasting. Described in the following sections are some of the applications.

***Data Modem.*** An early OFDM application was the AN/GSC-10 (KATHRYN) variable-rate data modem built for the high-frequency radio.[37] The system uses up to 34 parallel low-rate channels, each operating at 75-b/s data rate and using PSK modulation. The subchannels are spaced by 82 Hz to provide guard time between successive signaling elements.

A high-speed group-band data modem was developed by Hirosaki et al.[39] using a staggered QAM multicarrier structure. The modem can provide a maximal 256-kb/s data rate. Operating in the frequency range 63.2 kHz to 103.2 kHz, the system uses 12 staggered QAM signals, each having a 3.2-kHz baud rate and an 18-point DFT processor. In addition, a bit assignment strategy is used to optimize the system immunity against Gaussian noise.

***Digital Subscriber Lines (DSL).*** Under the name of discrete multitone (DMT), OFDM has been applied to various DSL services, such as ADSL (asymmetric DSL) and HDSL (high-speed DSL).[22–23] Recently, the DMT method was adopted by ANSI for the ADSL standard, providing up to 6.3-Mb/s data rates to the subscribers over the existing telephone lines.[23]

The basic components of a DMT ADSL transceiver is similar to those shown in Fig. 5.55, except that the system has an *L*-tap time-domain LE to reduce the channel distortion.[22] In the frequency domain a one-tap equalizer is used for each subchannel to provide a uniform symbol detection on the sub-bands. The DMT employs a 512-point FFT and a bit-allocation scheme for each of the carriers. The sub-bands with the better SNRs are assigned with more bits of data; each carrier uses QAM to encode the allocated bits. Up to 15 bits of data may be assigned among each of 512 carriers, depending on the conditions detected on each subchannel.[23] The sub-band bit allocation and equalization are all trained initially using test sequences.

***Digital Mobile Radio.*** In wireless applications OFDM/QPSK and OFDM/FM have been applied for mobile radio systems.[37] In the OFDM/QPSK system, pilot carriers are sent to provide gain/phase corrections for Rayleigh fading and cochannel interference. In the OFDM/FM system each carrier is QAM-modulated and the baseband OFDM signal is then modulated at RF by a conventional FM transmitter. The system, as claimed, may be implemented inexpensively by retrofitting existing FM radio stations.[37]

***Digital Audio Broadcast (DAB).*** A recent successful implementation of OFDM is in the DAB project developed under Eureka147[40] (see Chapter 15). In its first public demonstration the DAB system used a COFDM with system parameters as listed in Table 5.19. The COFDM encoder/decoder subsystem is shown in Fig. 5.62. The DAB trial's success has since stimulated the exploration of COFDM for digital television broadcasting.

**TABLE 5.19**   COFDM Parameters Used in the First DAB Demonstration System

| | | | |
|---|---|---|---|
| FFT size/useful subcarriers: | 512/448 | Modulation/demodulation: | DQPSK |
| Subcarrier spacing: | 15,625 Hz | Interleaving (frequency): | 448 subcarriers |
| Total symbol time/guard time: | 80 μs/16 μs | Channel coding: | R = 1/2, K = 7, Dfree = 10 |
| Total bandwidth: | 7 MHz | Channel decoding: | Viterbi decoder |
| Total useful bit rate: | 5.5 Mb/s | Synchronization: | 3 pilot symbols |

***Digital Terrestrial Television Broadcasting (dTTb).*** Based on DAB's success, OFDM/COFDM has become a favored choice in Europe for digital television and HDTV broadcasting. Various projects and field trials employing OFDM/COFDM have been engaged, including RACE dTTb, HD-DIVINE, SPECTRE, STERNE, and several others.[11,37,41–42] The goals have been not only for fixed reception but also for potential mobile and portable receivers.

Table 5.20 lists the COFDM parameters used in several field trial systems,[11,41–42] and Fig. 5.63 shows one of the experimental OFDM systems set up as given by Tourtier et al.[41] The trial results have shown the robustness of COFDM against multipath, impulse noise, cochannel interference, and other channel distortions.

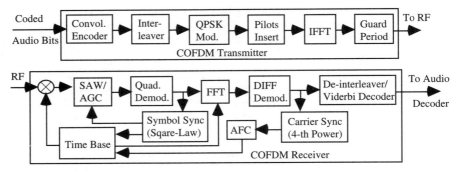

**FIGURE 5.62**   The OFDM subsystem in the DAB system.

**TABLE 5.20**   COFDM Modem Parameters Used in Various Field Trial Systems

| Field Trial System | | | |
|---|---|---|---|
| HD-DIVINE<br>Scandinavia, 1992 | SPECTRE<br>U.K., 1991 | T. De Couasnon et al.<br>Germany, 1992 | Tourtier et al.<br>France, 1990, 1992 |
| OFDM/16-QAM<br>512 FFT/448 useful<br>$T_s = 64\ \mu s$<br>$T_g = 2\ \mu s$<br>Freq. equalization<br>RS (208, 224)<br>25-Mb/s bit rate | DSP-based OFDM<br>400 subcarriers<br>QPSK, 8-PSK,<br>16-QAM, etc.<br>AFC-aided sync. | OFDM/64-QAM<br>512 FFT/7-MHz BW<br>$T_s = 70.4\ \mu s$<br>$T_g = 8.8\ \mu s$<br>RS FEC<br>34-Mb/s bit rate | OFDM/64-QAM<br>1024 FFT/982 useful<br>$T_s = 70.4\ \mu s$<br>$T_g = 8.8\ \mu s$<br>Pilot symbol<br>sync. (every 15<br>symbols) 70-Mb/s<br>bit rate |

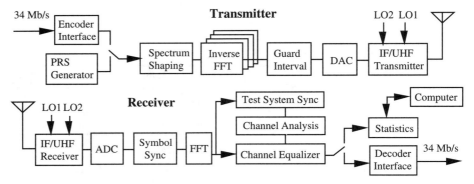

**FIGURE 5.63**   An OFDM transmitter-receiver system block diagram in field trials. (From Tourier et al.,[41] ©1993 Elsevier, with permission.)

***Single-Frequency Network (SFN).***    Owing to its multipath handling capability, COFDM has been considered an ideal solution for SFN, which intends to cover an entire service area (e.g., the entire country) by a number of terrestrial transmitters, all of them time-synchronized and broadcasting the same program, using only one frequency.[37–38] This is based on a constructive addition of the signals from all the transmitters belonging to the network, when the signal delays are within the guard interval. This, in effect, brings spatial diversity into the transmission, in addition to the time and frequency diversities provided by the COFDM, and therefore increases both the power and frequency efficiencies. COFDM is also attractive for other distributed emission networks such as dense networks in cities and on-channel relays or gap fillers.[37]

***Hybrid Fiber and Coax (HFC) Networks.***    More recently, OFDM, in terms of DMT, has been evaluated for its potential use in HFC CATV networks, especially for upstream or return-channel transmission.[46] A synchronized DMT (SDMT) was proposed for the IEEE 802.14 cable TV protocol standard.[46] The system is essentially based on the DMT ADSL structure but requires that all remote transmitters located at the subscriber premises be synchronized to the cable head-end (the receiver) for carrier synchronization. This can be achieved by sending the master clocks from the head-end downstream to all remote sites. The basic proposed SDMT system consists of a 512-carrier FFT with a guard period (cyclic prefix) of 32 samples (about 2 ms). The symbol rate is 32 kHz, and the carrier separation is 34.5 kHz.

## 5.5   COMPARISONS BETWEEN QAM, VSB, AND OFDM TECHNIQUES[15,28,30–37,47–49]

The three digital modulation techniques (QAM, VSB, and OFDM) discussed so far are all shown to be feasible for digital terrestrial and cable transmissions. The debate on choosing the best technology will be most likely based on implementation and cost issues, because the techniques theoretically provide equivalent performance. In brief, in the United States, the FCC ACATS has officially adopted VSB for the U.S. HDTV standard, which has been approved by the FCC.[30] In Europe the DVB group standardized 64-QAM for cable transmission,[26] and is in favor of COFDM for digital terrestrial transmission.[15,37–38] Globally, both the International Telecommunications Union (ITU) and DAVIC endorsed QAM for digital video transmission.[27]

### 5.5.1   QAM Versus VSB

Despite test results and implementation differences, it can be shown that the QAM and VSB techniques provide essentially the same theoretical performance and channel capacity.[35] In basic principle QAM involves a double-sideband amplitude modulation, multiplexing two independent in-phase and quadrature signals, whereas VSB involves a single-sideband vestigial amplitude modulation. Both methods give equivalent bandwidth efficiency. A summary of comparisons between QAM and VSB is provided as follows, with regard to transmission characteristics, implementation, and system performance.

***Transmission Characteristics.***    Table 5.21 compares some characteristics (e.g., data rate and bandwidth efficiency) of QAM and VSB at various encoding levels.[28,33,35,47]

**TABLE 5.21**   Transmission Characteristics of QAM and VSB

| Modulation | VSB (4, 8, 16) | QAM (16, 64, 256) |
|---|---|---|
| (Ideal) practical symbol rate | (12) 10.76 Mbaud | (6) 4.88–5 Mbaud |
| (Ideal) practical bit rate @ 6 MHz | (24) 21.5, (36) 32.3, (48) 43 Mb/s | (24) 19.5, (36) 29.3, (48) 40 Mb/s |
| (Ideal) practical BW efficiency | (4) 3.6, (6) 5.4, (8) 7.2 b/Hz/s | (4) 3.3, (6) 4.9, (8) 6.7 b/Hz/s |

***Implementation.***    Table 5.22 lists the major implementation similarities and differences between QAM and VSB systems.[31,35,47] With respect to equalization, VSB uses a real $T$-spaced equalizer running at symbol rate and adapting on a training sequence. For QAM, different equalizers may be used either at passband or baseband using LE or FSE with blind acquisition.[31] However, as described in Section 5.2, a QAM equalizer in general consists of four FIR filters, though they can be eliminated by design.

**TABLE 5.22**    Implementation Similarities and Differences Between QAM and VSB

| Modulation | VSB | QAM |
|---|---|---|
| Mod. format | Vestigial sideband, suppressed carrier | Double sideband, suppressed carrier |
| Matched filter | 11.5% roll-off | 20% roll-off |
| Carrier recovery | Pilot tone, FPLL (analog) | Data-directed loop (digital), blind |
| Timing recovery | Segment sync (sync symbol) | Data-directed loop (digital), blind |
| Acq. speed, threshold | $\sim$1 s (improving), SNR > 0 dB | 0.1–0.2 s, SNR > 18 dB |
| Equalizer | Transversal FIR, LMS | Transversal FIR, LMS |
| Adaptation | Training sequence | Blind adaptation |
| Number of taps, time span | $N$ taps, $T/2$ baud-spaced* ($NT/2$) | $N/2$ taps/carrier, $T$ baud-spaced, $NT/2$ |
| Multiplications per second | $2N/T$ | $N/T$ (w/o cross-coupled filters) |
| | | $NT/2$ (with cross-coupled filters) |
| Reciever | Analog front end demodulator | All digital |
| Interleaving and error protection | Convultional byte de-interleaving | Block byte de-interleaving |
| | Reed-Solomon and (trellis) codes | Reed-Solomon and (trellis) codes |

*$T$ = QAM baud period

***Theoretical Performance.***    Figure 5.64 compares the BER performance of QAM and VSB under AWGN.[47] Note that the steep curves in the figure are for the error correction case. The figure indicates that QAM and VSB give equivalent BER performance in the AWGN channel.

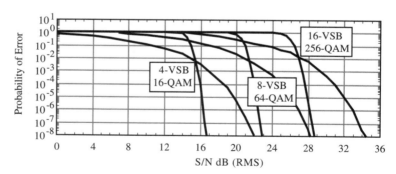

**FIGURE 5.64**    Error probabilities of VSB and QAM. (From Citta,[47] ©1993 NCTA, with permission.)

***Jitter and Carrier Phase Error Performance.***    Simulation studies on evaluating the effects of symbol timing jitter and carrier phase errors on QAM and VSB systems have been reported by Kerpez for the case of hybrid fiber/coax network.[31] The results reveal similar performances of QAM and VSB in these cases. The transmission and simulation parameters are given in Table 5.23, and the results reported based on Kerpez[31] are summarized in Table 5.24, which shows the

**TABLE 5.23**  Transmission Parameters Used in Kerpez's Simulations

| Transmit filter: | 11.5% roll-off | SNR at transmitter: | 63 dB |
|---|---|---|---|
| No. of equalizer taps: | 63 | Symbol timing jitter: | 0.01 UI RMS |
| Coaxial cable: | 2750 feet, 3/4" | | |

**TABLE 5.24**  Effects of Jitter and Phase Error on QAM and VSB

| SNR | No jitter | | 0.01 RMS jitter | | | SNR loss | |
|---|---|---|---|---|---|---|---|
| IF Freq. | 256-QAM | 16-VSB | 256-QAM | 16-VSB | Carrier phase error | 256-QAM | 16-VSB |
| 300 MHz | 36.8 dB | 36.8 dB | 33.0 dB | 32.8 dB | 0.25 deg. | 0.05 dB | 0.05 dB |
| 400 MHz | 32.8 dB | 32.8 dB | 31.0 dB | 31.0 dB | 0.45 deg. | 0.17 dB | 0.19 dB |
| 500 MHz | 29.0 dB | 29.0 dB | 27.0 dB | 27.0 dB | 0.65 deg. | 0.35 dB | 0.40 dB |
| 600 MHz | 25.2 dB | 25.2 dB | 25.0 dB | 25.5 dB | 0.85 deg. | 0.60 dB | 0.67 dB |
| 700 MHz | 22.2 dB | 22.2 dB | 22.0 dB | 22.5 dB | 0.95 deg. | 0.73 dB | 0.83 dB |

received SNRs of QAM and VSB with and without jitter influence as well as the SNR losses due to the carrier phase error. The received SNR for carrier phase simulation is 32.5 dB. The symbol timing jitter is 0.01 unit interval (UI) RMS—about 1 percent of the symbol period. Additionally, the simulations also indicate, not shown, similar adaptation or acquisition speeds of the carrier recovery loop for both QAM and VSB systems.[31]

***Laboratory and Field Testing.***    Comparative performances of VSB and QAM under various channel impairments were tested at the CableLabs and the Advanced Television Testing Center (ATTC) under the auspices of the ACATS and GA.[30,32] Table 5.25 shows the testing data of 16-VSB and 256-QAM under single impairment, and Fig. 5.65 illustrates two multiple impairment cases based on the CableLabs tests.[35] All the test data are measured at or below $3 \times 10^{-6}$ BER (TOV). In these tests VSB outperformed QAM in most impairment categories.

**TABLE 5.25**  Performances of 16-VSB and 256-QAM in Single Impairment Tests

| | 16-VSB | 256-QAM | | 16-VSB | 256-QAM |
|---|---|---|---|---|---|
| SNR: | 27.6 dB | 29.3 dB | Residual FM: | 4.7 kHz | 70 kHz |
| CTB/CSO: | 44/33 dB | 47/37 dB | LO pull-in range: | $> \pm 100$ kHz | $> \pm 100$ kHz |
| Fiber depth of mod.: | 4.7% | 3.8% | Burst error @ 10-Hz rate: | 150 μs | 27 μs |
| Fiber C/NLD: | 38.9 dB | 48.6 dB | Burst error @ 20 μs: | 2.4 kHz | 30 Hz |
| Phase noise @ 20 kHz: | −83.0 dBc | −84.2 dBc | Av. Ch. change time: | 0.54 s | 0.55 s |

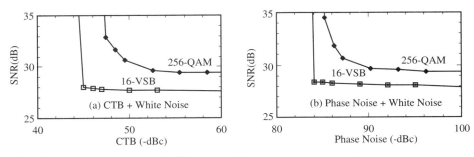

**FIGURE 5.65**  Performances of 16-VSB and 256-QAM in multiple impairments.

Over-the-air comparative testing between 8-VSB supplied by Zenith and 32-QAM supplied by General Instrument was conducted at ATTC for typical terrestrial broadcast conditions.[48] Tables 5.26 and 5.27 provide some of the test results for regular and taboo channels.[48] Also included are the first-round measurements of 2/4-VSB for DSC-HDTV and 16/32-QAM for CCDC and ADTV systems.[32,48] The results are again measured at TOV BER.

**TABLE 5.26** Over-the-Air Impairment Performances of QAM and VSB

| Modulation | 8-VSB | 32-QAM | 4-VSB | CCDC | ADTV |
|---|---|---|---|---|---|
| AWGN D/U | 14.9 dB | 15.0 dB | 16.0 dB | 15.4 dB | 18.1 dB |
| Cochannel D/U (ATV into NTSC) | 15.9 dB | 16.4 dB | 18.2 dB | 16.6 dB | 18.4 dB |
| Cochannel D/U (NTSC into ATV) | 2.07 dB | 5.87 dB | 3.47 dB | 8.05 dB | 0.50 dB |
| UAdj.-ch. D/U (ATV into NTSC) | −5.65 dB | −7.81 dB | −1.14 dB | −8.90 dB | — |
| UAdj.-ch. D/U (NTSC into ATV) | −47.1 dB | −36.3 dB | −42.1 dB | −37.2 dB | −36.0 dB |
| Echo D/U @ 0 Hz, 1-$\mu$s delay | 1.00 dB | 2.80 dB | — | — | — |
| Echo D/U @ 0 Hz, 15-$\mu$s delay | 2.90 dB | 6.90 dB | — | — | — |
| Flutter D/U @ 5 Hz, 1-$\mu$s dealy | 11.5 dB | 8.27 dB | 17.0 dB | 11.4 dB | 17.6 dB |

**TABLE 5.27** Taboo Interferences for QAM and VSB

| Modulation/taboo (ATV into NTSC) | 8-VSB | 32-QAM | 4-VSB | CCDC |
|---|---|---|---|---|
| $n − 2, n + 2$ channel D/U @ TOV | −24, −29 dB | −24, −29 dB | −24, −28 dB | −23, −30 dB |
| $n + 4$ channel D/U @ TOV | −24 dB | −27 dB | −25 dB | −27 dB |
| $n − 7, n + 7$ channel D/U @ TOV | −35, −36 dB | −36, −35 dB | −35, −34 dB | −35, −34 dB |
| $n − 8$ channel D/U @ TOV | −31, 40 dB | −31, −40 dB | −34, −36 dB | −30, −43 dB |
| $n + 14$ channel D/U @ TOV | −28 dB | −26 dB | −26 dB | −27 dB |
| $n + 15$ channel D/U @ TOV | −16 dB | −15 dB | −17 dB | −18 dB |

### 5.5.2 Single-Carrier Modulation (SCM) versus Multicarrier Modulation (MCM)

Both QAM and VSB belong to SCM, in which a single carrier is modulated onto in-phase/quadrature signals or the in-phase signal only. OFDM, on the other hand, belongs to MCM, in which a plurality of carriers are simultaneously modulated. The fundamental difference between SCM and MCM is that SCM (such as QAM) is a time-domain technique and involves serial symbol-by-symbol transmission, whereas MCM (such as OFDM) is a frequency-domain technique and involves parallel transmission of symbols. As an example, Fig. 5.66 shows the difference in influence of a fade on serial and parallel transmissions.[7] It is evident that a serial system can suffer from fading due to several symbols getting hit simultaneously, whereas in a parallel system the effect of fading is minor because the fade only affects a small portion of the symbol duration.

The following sections describe similarities and differences between QAM (or VSB) and OFDM, in terms of modulation principle, transmission characteristics, and performance.

***Time and Frequency Dualities.***  There are interesting time- and frequency-domain dualities between MCM and SCM in terms of modulation processing, equalization, synchronization, etc. Table 5.28 gives a summary of the dualities.

***Transmission Characteristics.***  Transmission properties for MCM and SCM are as follows:

1. Regarding bandwidth efficiency, OFDM and QAM (or VSB) are asymptotically equivalent with ideal Nyquist filtering. In practice, the bandwidth efficiency is reduced by guard interval

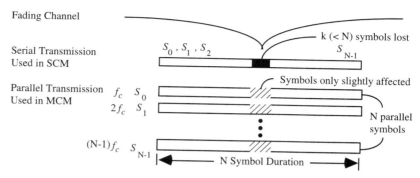

**FIGURE 5.66** Effect of a fade on SCM and MCM systems. (From Webb and Hanzo,[7] with permission of John Wiley & Sons, Ltd.)

**TABLE 5.28**  Time/Frequency-Domain Dualities Between MCM and SCM

|  | MCM (OFDM) | SCM (QAM or VSB) |
|---|---|---|
| Modulation and processing | Multicarrier flexibly modulated freq.-domain processing (robust to time-impulses but vulnerable to freq.-tone interferences) | Single-carrier fixed-modulated time-domain processing (robust to tone interferences but sensitive to time impulses) |
| Equalization | Freq.-domain equalization (multiplication of coefficients) | Time-domain equalization (training and/or blind adaptation) |
| Synchronization | Pilot-aided technique (pilot carriers and pilot symbols) | Data-directed or pilot-aided methods (decision feedback or pilot and sync) |
| ISI immunity | Guard interval (time domain) | Pulse shaping (freq. domain) |
| Diversity | Time/frequency (interleaving) | Time (interleaving) |

**TABLE 5.29**  Bandwidth Efficiency ($R_b/B$) Equations for OFDM, QAM, and VSB

|  | $M$-QAM | $L$-VSB | OFDM/$M$-QAM |
|---|---|---|---|
| Ideal | $\log_2 M$ | $\log_2 L$ | $[N/(1 + N)]\log_2 M$, $N$ = # of carriers |
| Practical | $\log_2 M/(1 + \alpha)$ $\alpha$ = roll-off factor | $\log_2 L/(1 + \alpha)$ $\alpha$ = roll-off factor | $[N/(1 + N)][1 - N_g/N]\log_2 M$ $N_g$ = cyclic extension length |

insertion in OFDM or by roll-off factor in QAM or VSB. Table 5.29 lists the equations for calculating the bandwidth efficiencies for the three modulation methods: $M$-QAM, $L$-VSB, and OFDM/$M$-QAM (i.e., each subchannel is $M$-QAM).

2. The difference in PAR between MCM and SCM is theoretically a function of the number of carriers, denoted $N$, given by the following:[45]

$$\Delta(\text{dB}) = 10 \cdot \log_{10}(N) \qquad (5.74)$$

For $N = 1000$, for example, the difference is 30 dB. However, this can rarely occur, because the input data are well-scrambled so that the chances that the maximal value is attained are extremely low. Practically, the PAR of an OFDM signal can be estimated based on its Gaussian amplitude distribution. For the purpose of comparison, Table 5.30 lists the PARs

**TABLE 5.30**  PAR Distributions for MCM and SCM Signals

| Cumulative distribution | OFDM | 8-VSB | 64-QAM |
|---|---|---|---|
| 99.00% | 8.30 dB | 5.0 dB | 5.0 dB |
| 99.90% | 10.4 dB | 6.2 dB | 6.0 dB |
| 99.99% | 11.8 dB | 6.9 dB | 7.0 dB |

of OFDM/64-QAM, 8-VSB, and SCM/64-QAM, respectively.[45] The table indicates that the OFDM signal generally has 3 to 4 dB higher PAR than the corresponding SCM signal.

***Performance Comparisons of MCM and SCM.***   Because of the lack of head-to-head comparative testing between MCM and SCM in various channel impairments, the following performance comparisons will be based on theoretical analysis, simulation, and qualitative assessments.

1. Under AWGN, both OFDM and QAM or VSB have comparable BER performance. However, since an OFDM signal has a Gaussian distribution (whereas QAM or VSB usually has an equal probable signal constellation) there might be, in theory, a few tenths of a dB SNR advantage for OFDM.[37] This is due to the fact that theoretically, to achieve the channel capacity bound requires that the signal points be Gaussian-distributed.[37]

2. Under impulse noise, OFDM is inherently more robust than QAM or VSB due to its long symbol interval and a multiplex of many carriers. To illustrate, Fig. 5.67 shows the impulse noise performances between OFDM and QAM (Fig. 5.67*a*) and between multicarrier offset QAM and single-carrier VSB (Fig. 5.67*b*). In the figures the Middleton class-A impulse noise model[17] is used in part (*a*), and Hirosaki's model[39] is used in part (*b*) (where SNR = 28 dB and 0 dBr = 0.39 $V_{peak}$/75W). The results indicate that MCM has about a 10-dB advantage over SCM in the case of impulse noise.

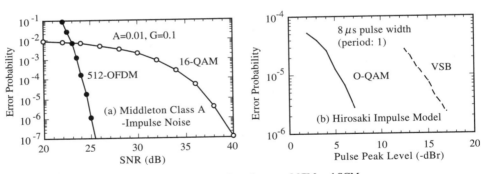

**FIGURE 5.67**   Impulse noise performance comparisons between MCM and SCM.

3. For the case of frequency-domain and cochannel interferences, OFDM without spectral shaping is more sensitive compared to SCM, because the interferences may destroy several subchannels and thereby the orthogonality of OFDM, whereas for SCM the interferences can be averaged out by the whole signal spectrum.

4. For the case of carrier offset and phase noise, OFDM has more stringent requirements for local oscillators and tuners than SCM due to the division of multiple carriers and the close spacing between carriers. As an example, Table 5.31 shows the SNR degradations at a BER of

**TABLE 5.31**  Carrier Offset Effect on OFDM and SCM

| Normalized freq. offset ($\Delta f T_s$) | | $5 \times 10^{-5}$ | $1 \times 10^{-4}$ | $2 \times 10^{-4}$ | $5 \times 10^{-4}$ | $1 \times 10^{-3}$ |
|---|---|---|---|---|---|---|
| SNR (dB) | OFDM/QPSK | 0.1 | 0.2 | 0.8 | $\gg 1$ | — |
| Degradation | SCM/QPSK | <0.01 | 0.01 | 0.02 | 0.05 | 0.14 |

$10^{-3}$ for a 256-carrier OFDM/QPSK and an SCM/QPSK.[15] It is clear that the OFDM cannot tolerate a frequency offset even at $\Delta f T_s = 10^{-3}$, whereas the SCM/QPSK gives less than 0.15-dB SNR degradation at the same offset.

5. In the case of linear distortions, studies have shown that OFDM suffers 1 to 2-dB deficiency compared to SCM using decision-feedback equalizers (e.g., in a cable environment) but may essentially achieve the similar performance if OFDM employs optimum subchannel constellation and power assignment scheme.[49]

6. For multipath and fading time-dispersive channels, OFDM typically employs a guard interval and/or frequency-domain equalization (complex multiplication). The SCM QAM or VSB, on the other hand, employs a time-domain adaptive equalization. It has been shown that an SCM using frequency-domain equalization significantly outperforms an uncoded OFDM and performs comparably to COFDM.[15] Figure 5.68 illustrates such a case, where a 1024-carrier OFDM is compared to an SCM with a 1024-tap frequency equalizer in the presence of amplitude-fading distortion, with and without coding.[15] The channel coding is a $K = 7$, 1/2-rate convolutional code. In the figure the dashed BER curves correspond to OFDM, and the solid curves correspond to SCM. The dotted BER line corresponds to COFDM with weighted coding instead of conventional coding.[15]

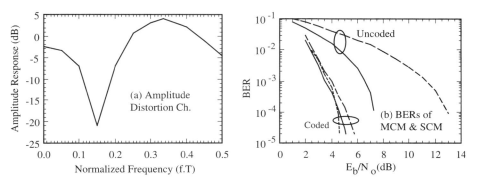

**FIGURE 5.68**  Comparison of MCM and SCM in strong fading. (From Sari et al,[15] ©1995 IEEE, with permission.)

7. For nonlinear distortions, since OFDM has a higher PAR than SCM, it thus requires higher-output back-off (OBO) than SCM in order to reduce the saturation effect. A recent study involving a traveling-wave tube amplifier,[15] for example, shows that a 512-carrier OFDM/QPSK has an optimum OBO of 4.3 dB at $10^{-3}$ BER, which gives a total 6.8-dB SNR degradation. The SCM/QPSK, on the other hand, has an optimum OBO of 0.4 dB and only 0.8 SNR degradation at BER $= 10^{-3}$.

Table 5.32 summarizes the performance comparisons for MCM and SCM.

**TABLE 5.32**   A Summary of Channel Performance of MCM and SCM

| Channel impairments/mod. | MCM (OFDM) | SCM (QAM or VSB) |
|---|---|---|
| AWGN channel | Equivalent | Equivalent |
| Impulse noise | Inherently immune | Sensitive; need coding/interleaving |
| Freq. interference (Cochannel interference) | Vulnerable; need spectral shaping or optimal loading | Inherently immune (improve by comb filtering) |
| Linear distortion (Multi-path, fading, etc.) | Guard interval, freq. EQ and/or coding | Time or freq. equalization and/or coding |
| Carrier offset/phase noise | Need accurate sync | Less stringent requirement |
| Nonlinear distortion | High OBO | Low OBO |

## 5.6   OTHER DIGITAL MODULATION TECHNIQUES[1–2,49–57]

In addition to QAM, VSB, or OFDM, other digital modulation methods have also found widespread usage, particularly in power-limited channels. The methods involve modulating either the phase or frequency of a carrier signal while keeping the amplitude of the signal constant. In other situations where it may be necessary to trade the bandwidth and power efficiencies for better security and interference immunity, the signal may be modulated by spreading its spectrum far beyond the minimum bandwidth required to transmit the information data, i.e., the spread spectrum technique.

### 5.6.1   *M*-ary Phase-Shift Keying (*M*-PSK) Modulation

In *M*-PSK the phase of the signal carrier changes in a discrete form, taking on one of *M* possible values, i.e., $\theta_k = 2k\pi/M$, $k = 0, 1, \ldots, M - 1$. For coherent *M*-PSK the modulation/demodulation steps are similar to that of *M*-QAM, except that the decision is made on phase only. For noncoherent PSK the receiver involves nonlinear devices such as envelope or square-law detector, or delayed correlator (for differential coding), in order to extract relevant carrier phases. As an example, Fig. 5.69 shows two demodulator structures: coherent binary PSK (BPSK) and differential PSK (DPSK).

**FIGURE 5.69**   Binary PSK demodulator structures.

The carrier synchronization of *M*-PSK can be effectively done using either the times-*n* method or the decision feedback method.[2,14] For symbol timing recovery the early-gate method is especially appropriate.[2]

***Signal Representation and Characteristics.***   The general expression for an *M*-PSK signal waveform, during each symbol interval $T_s$, can be written as follows:[50]

$$s_k(t) = \sqrt{\frac{2E_s}{T_s}} \cos\left(2\pi f_c t + \frac{2\pi k}{M}\right), \quad k = 0, 1, \ldots, M - 1 \tag{5.75}$$

where $E_s$ is the symbol energy. Note that Eq. (5.75) may also be expressed in in-phase quadrature form and can therefore be represented with two-dimensional constellation. It is easy to see that the constellation of $M$-PSK is a circle of radius $\sqrt{E_s}$, on which the $M$ signal points are equally spaced. The constellation mapping can be done using Gray coding or differential coding.

For coherent $M$-PSK the optimum receiver assumes a pair of correlators or matched filters with reference signals in phase quadrature, and a phase discriminator (decision device), which computes the phase estimate: $\hat{\theta} = \tan^{-1}\{x_Q/x_I\}$, $x_Q$ and $x_I$ being the two correlator outputs, and then selects from the set $\{s_k(t), k = 0, 1, \ldots, M - 1\}$ a signal whose phase is closest to the estimate $\hat{\theta}$.

The power spectrum of $M$-PSK, similar to $M$-QAM, is of the following form:[50]

$$S_B(f) = 2E_s \frac{\sin^2 \pi T_s f}{(\pi T_s f)^2} \tag{5.76}$$

assuming the baseband pulse is rectangular. Considering that the signal bandwidth of $M$-PSK is defined as $B = 1/T_s$ (i.e., the boundary of the spectral main lobe) the bandwidth efficiency of $M$-PSK[2] is $R_b/B = \log_2 M$, which is equivalent to $M$-QAM.

***Modulation Formats.*** There are several modulation formats for $M$-PSK, particularly for $M \le 4$. The formats are QPSK, offset (or staggered) QPSK (OQPSK), $\pi/4$-QPSK, and DQPSK. Table 5.33 summarizes the modulation characteristics of these formats.[51]

**TABLE 5.33** Modulation Properties of Various 4-PSK Formats

| 4-PSK formats | QPSK | OQPSK | $\pi/4$-QPSK | DQPSK |
|---|---|---|---|---|
| Phase values | $0$–$3\pi/2$ or $\pi/4$–$7\pi/4$ | $0$–$3\pi/2$ or $\pi/4$–$7\pi/4$ | $\pi/4$, $3\pi/4$, $-3\pi/4$, $-\pi/4$ | Differential (input + previous) |
| Max. phase change | 2-bit duration $\pm 180$ degrees | 1-bit duration $\pm 90$ degrees | 2-bit duration $\pm 135$ degrees | 2-bit duration |
| Bit mapping and modulation | Even I/Q bit map Coherent/2 BPSKs | I/Q offset by 1 bit Coherent/2 BPSKs | 2 QPSK shift by $\pi/4$ Coherent/incoherent | I/Q 1-symbol delay Noncoherent |
| Usage | Standard | Power efficient | (O)QPSK Trade-Off | Phase Ambiguity |

***Channel Impairment Performance.*** In AWGN the BER of $M$-PSK can be obtained using the decision region $-\pi/M \le \theta \le \pi/M$. Table 5.34 lists the BER equations for various $M$-PSK schemes.[50–51] In the table, $\gamma = E_b/N_o$ is the SNR per bit, and the approximations are valid for large $\gamma$. It is seen that the $M$-DPSK (both coherent and noncoherent) has higher BER than coherent $M$-PSK, or equivalently requires higher SNR to achieve the same BER. For example, the DQPSK needs about 2.3 dB more in SNR to achieve the same BER as QPSK, while the coherent DQPSK gives twice the BER as the QPSK but has negligible SNR loss.

**TABLE 5.34** BER Expressions for Various $M$-PSK Systems

| Modulation | BER | Modulation | BER |
|---|---|---|---|
| BPSK | $= 1/2 \operatorname{erfc}\{\sqrt{\gamma}\}$ | Coh. DQPSK | $\approx \operatorname{erfc}\{\sqrt{\gamma}\}$ |
| Coh. DPSK | $\approx \operatorname{erfc}\{\sqrt{\gamma}\}$ | DQPSK | $\approx 1/2 \operatorname{erfc}\{\sqrt{4\gamma}\sin(\pi/8)\}$ |
| DPSK | $\approx 1/2 \exp(-\gamma)$ | $M$-PSK | $\approx \operatorname{erfc}\{\sqrt{\gamma \log_2 M}\sin(\pi/M)\}/\log_2 M$ |
| QPSK/OQPSK | $\approx 1/2 \operatorname{erfc}\{\sqrt{\gamma}\}$ | $M$-ary DPSK | $\approx \operatorname{erfc}\{\sqrt{2\gamma \log_2 M}\sin(\pi/2M)\}/\log_2 M$ |

For multipath and (Rayleigh) fading channels, Table 5.35 lists the BER equations for binary PSK schemes.[51] In the table, $\gamma_{av} = E_b\bar{\alpha}^2/N_\alpha$ is the average SNR per bit ($\alpha$ being the fading attenuation factor). It is readily shown from the equations that much higher SNR would be required to maintain the same BER as the AWGN case. Further, for frequency-selective fading, irreducible BER floor can be observed for $M$-PSK due to time-varying spread.[51]

**TABLE 5.35** BER Equations of Binary PSK in Rayleigh Fading

| Modulation | BER |
|---|---|
| BPSK | $= 1/2\{1 - \sqrt{\gamma_{av}/(1 + \gamma_{av})}\} \approx 1/4\gamma_{av}$ |
| DPSK | $= 1/2\{1 - \sqrt{\gamma_{av}/(2 + \gamma_{av})}\} \approx 1/2\gamma_{av}$ |

*Applications.* One of the primary applications for $M$-PSK is in satellite systems, because of its inherent power efficiency advantage, where QPSK and 8-PSK are commonly used.[49,52] In DBS and in Eureka-256 satellite HDTV systems,[49,52] for example, QPSK and 2/3-rate trellis-coded 8-PSK, all concatenated with a RS (255,239) code, are used, providing error-free transmission.

In cable, BPSK has been utilized in traditional addressable converters for return and low data rate polled systems.[53] Recently, QPSK-based cable modems are being developed for potential consumer access to the Internet and for two-way cable services such as video-on-demand.[54]

In wireless applications QPSK and its other formats have been selected in several mobile and cellular radio and telephone standards, which are listed in Table 5.36.[51]

**TABLE 5.36** PSK Schemes Used in Various Wireless Standards

| | Wireless Standard | | | |
|---|---|---|---|---|
| | PHS | PACS/WACS | IS-54 | PDC |
| **Modulation** | $\pi/4$ DQPSK | $\pi/4$ QPSK/QPSK | $\pi/4$ DQPSK | $\pi/4$ DQPSK |
| **Data Rate** | 384 Kbps | 384 Kbps/500 Kbps | 48.6 Kbps | 42 Kbps |

### 5.6.2 *M*-ary Frequency-Shift Keying (*M*-FSK) Modulation

$M$-FSK modulation involves switching the frequency of a constant-amplitude carrier signal among $M$ values according to the $M$ possible symbols. In binary FSK (BFSK), for example, two frequencies are used to represent binary digits 1 and 0. $M$-FSK can be modulated coherently or noncoherently. For coherent $M$-FSK (CMFSK) the modulation/demodulation steps are similar to those of $M$-QAM, except that the decision is made on frequency only. For noncoherent $M$-FSK (NCMFSK) envelope or square-law nonlinear detectors are used in the receiver. For CMFSK the carrier and symbol timing recovery can be done in a similar way to $M$-PSK. For NCMFSK, carrier synchronization is not a concern.

*Signal Representation and Characteristics.* The general expression for an $M$-FSK signal waveform, during each symbol interval $T_s$, can be written as follows:[50]

$$s_k(t) = \sqrt{\frac{2E_s}{T_s}} \cos\left[\frac{\pi(n_c + k)t}{T_s}\right], \quad k = 0, 1, \ldots, M \tag{5.77}$$

where the carrier frequency $f_c$ is equal to $n_c/2T_s$ for some fixed integer $n_c$. Note that since the individual frequencies are separated by $1/2T_s$ hertz, the signals $\{s_k, k = 0, 1, \ldots, M\}$ are orthogonal:

$$\int_0^{T_s} s_i(t)s_j(t)\,dt = 0, \qquad i \neq j \tag{5.78}$$

Hence, $M$-FSK is also referred to as orthogonal modulation.

The spectral analysis of $M$-FSK is in general complicated. However, for BFSK the power spectral density can be obtained as follows,[50]

$$S_B(f) = \frac{E_b}{2T_b}\left[\delta\left(f - \frac{1}{2T_b}\right) + \delta\left(f + \frac{1}{2T_b}\right)\right] + \frac{8E_b\cos^2(\pi T_b f)}{\pi^2(4T_b^2 f^2 - 1)^2} \tag{5.79}$$

where $T_b$ is the bit duration. It can be seen that the spectrum of BFSK contains two discrete frequency components, which are useful to aid receiver synchronization with the transmitter.

The bandwidth efficiencies for CMFSK and NCMFSK signals, respectively, can be obtained as follows, based on the corresponding channel bandwidth definitions,[51]

$$R_b/B = 2\log_2 M/(3 + M) \qquad \text{(for CMFSK)} \tag{5.80a}$$

$$R_b/B = 2\log_2 M/M \qquad \text{(for NCMFSK)} \tag{5.80b}$$

which shows that the bandwidth efficiency of $M$-FSK decreases with increasing $M$. Hence, $M$-FSK is bandwidth-inefficient. However, since the signals are orthogonal in $M$-FSK, the power efficiency increases with $M$. Further, $M$-FSK can be amplified using nonlinear amplifiers without any performance degradation. Table 5.37 provides a listing of bandwidth efficiencies of both CMFSK and NCMFSK signals for various $M$ values.[50–51]

**TABLE 5.37**  Bandwidth Efficiency of CMFSK and NCMFSK

| $M$ | 2 | 4 | 8 | 16 | 32 | 64 |
|---|---|---|---|---|---|---|
| CMFSK, $R_b/B$ | 0.4 | 0.57 | 0.55 | 0.42 | 0.29 | 0.18 |
| NCMFSK, $R_b/B$ | 1 | 1 | 0.75 | 0.50 | 0.31 | 0.19 |

***Channel Impairment Performance.***    For AWGN the BER of $M$-FSK can be bounded by the following:[50]

$$P_{eb} \leq (M/4)\text{erfc}\{\sqrt{\gamma\log_2 M/2}\} \qquad \text{(CMFSK)} \tag{5.81a}$$

$$P_{eb} \leq (M/4)\exp(-\gamma\log_2 M/2) \qquad \text{(NCMFSK)} \tag{5.81b}$$

where the equality holds for $M = 2$ (BFSK). It is readily seen that as $M$ increases, the SNR needed to achieve a desired BER decreases; that is, $M$-FSK is power-efficient. Table 5.38, for example, lists the SNRs of $M$-FSK to achieve a BER of $10^{-6}$ with increasing $M$.

**TABLE 5.38**  Required SNRs of $M$-FSK at BER $= 10^{-6}$

| | $M$ | | | | | |
|---|---|---|---|---|---|---|
| | 2 | 4 | 8 | 16 | 32 | 64 |
| CMFSK, $\gamma$ (dB) | 13.5 | 10.9 | 9.42 | 8.42 | 7.66 | 7.08 |
| NCMFSK, $\gamma$ (dB) | 14.2 | 11.52 | 10.0 | 8.98 | 8.2 | 7.6 |

In multipath and (Rayleigh) fading channels, the BER of BFSK is given by the following:[51]

$$P_{eb} = \{1 - \sqrt{\gamma_{av}/(2 + \gamma_{av})}\}/2 \approx 1/2\gamma_{av} \qquad \text{(Coherent BFSK)} \qquad (5.82a)$$

$$P_{eb} = 1/(2 + \gamma_{av}) \approx 1/\gamma_{av} \qquad \text{(Noncoherent BFSK)} \qquad (5.82b)$$

which are again significantly higher than the BER in the AWGN channel.

*Applications.*    FSK is popular in communication systems where simplicity and economy are more important than bandwidth efficiency. For example, BFSK was used in voice-grade telephone modems,[50] providing asynchronous low-data-rate (300–1800 b/s) transmission. In these modems, either full-duplex or half-duplex, two frequencies (e.g., 2.2 kHz and 1.2 kHz), represent "space" and "mark," respectively. In cable, FSK has been utilized in traditional addressable converters for return-channel and low-data-rate polled systems.[53]

Perhaps the most common application of FSK is for frequency-hopping spread spectrum systems, in which NCMFSK is employed as the data modulation. The description of this system will be given later.

### 5.6.3    Continuous-Phase Modulation (CPM)

CPM refers to a broad class of frequency modulation techniques in which the carrier phase varies continuously. Compared to the linear, memoryless modulation such as PSK, CPM is a nonlinear modulation with memory. The demodulation of CPM typically uses the Viterbi algorithm, based on its trellis properties.

The synchronization of CPM requires three levels:[1] (1) carrier phase recovery, (2) symbol or baud timing, and (3) superbaud timing, modulo $K$ (cycle period of phase modulation indices). One technique that has been used to acquire such timing signals is the times-$n$ method.

*Signal Representation and Characteristics.*    The general form for a CPM signal is given by the following,[1]

$$s_{CPM}(t) = \sqrt{2E_s/T_s}\cos[2\pi f_c t + 2\pi\int_{-\infty}^{t}\sum_{i=-\infty}^{\infty}h_i a_i g(\tau - iT_s)\,d\tau] \qquad (5.83)$$

where $a_i$ equals $\pm 1, \pm 3, \ldots, \pm(M-1)$ for any $i$, $\{h_i\}$ is a set of phase-modulation indices, and $g(t)$ is a frequency pulse-shape function that is zero for $t < 0$ and $t > LT_s$ ($L$ is an integer) and nonzero otherwise. Properties of CPM include the following:[1–2]

1. If $h_i = h$ for all $i$, and $g(t)$ is rectangular with amplitude $1/2T_s$ and $L = 1$, then CPM reduces to a special modulation case known as *continuous-phase FSK* (CPFSK).

2. In general, $h_{i+K} = h_i$, $K$ being the cyclic period. Thus, CPM is also called "multi-$h$ CPM."

3. If $L = 1$ a full-response CPM results; if $L > 1$ a partial-response CPM results. Hence, a CPM signal can be generated by choosing different pulse shapes $g(t)$ and by varying the modulation indices $h_i$ and the alphabet size $M$.

The signal-space representation of CPM may best be described in terms of phase trajectory or trellis diagram.[1–2] Figure 5.70 shows the example of a binary CPM for $h_i = h = 1/2$.

*Minimum Shift Keying (MSK).*    MSK is a special form of binary CPM or CPFSK with modulation index $h = 1/2$ and rectangular frequency pulse shape.[2] Thus, the waveform of the MSK signal can be obtained from Eq. (5.82) as follows:

$$s(t) = \sqrt{2E_b/T_b}\cos[2\pi(f_c + a_i/4T_b)t - i\pi a_i/2 + \pi\sum_{k=-\infty}^{i-1}a_k/2], \quad iT_b \leq t \leq (i+1)T_b$$
$$(5.84)$$

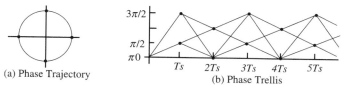

(a) Phase Trajectory

(b) Phase Trellis

**FIGURE 5.70** Signal-space diagram for binary CPM (CPFSK).

where $a_i = \pm 1$. From Eq. (5.84), it is seen that the MSK signal has one of two possible frequencies $f_{1,2} = f_c \pm 1/4T_b$ during each symbol interval. The frequency separation is $f_1 - f_2 = 1/2T_b$; this is the minimum frequency spacing for the two sinusoids to be coherently orthogonal,[2] hence the name MSK. Since this frequency spacing is half of that of the conventional BFSK (i.e., $1/T_b$), MSK is also often referred to as *fast FSK*.[1] Alternatively, Eq. (5.84) may also be written as[2]

$$s(\tau) = A_b a_{2i} \cos[\pi(T - 2i\tau b)/2\tau b] \cos 2\pi f_c \tau - A_b a_{2i+1} \sin[\pi(T - 2i\tau b - \tau b)2\tau b] \sin 2\pi f_c T \tag{5.85}$$

where $A_b = \sqrt{(2E_b/T_b)}$. This representation of MSK is equivalent to that of the OQPSK, in which the pulse shape is one-half cycle of a sinusoid. Note that the bit transitions on the sine/cosine carrier components are offset by $T_b$ seconds, and the phase varies linearly at each bit transition.

The constellation of the MSK signal is therefore also two-dimensional, with four message points $(\pm \sqrt{E_b}, \pm \sqrt{E_b})$ representing four possible phase states $(0, \pi/2, \pi, 3\pi/2)$, which correspond to the transmission of binary digit zero or one, as illustrated in Table 5.39.[50]

**TABLE 5.39** MSK Constellation Characteristics and Mapping

| Constellation points | Phase states | Transmitted binary digit |
|---|---|---|
| $\pm \sqrt{E_b}, -\sqrt{E_b}$ | $0(\pi), \pi/2$ | 1(0) |
| $\pm \sqrt{E_b}, \sqrt{E_b}$ | $0(\pi), 3\pi/2$ | 0(1) |

The modulator/demodulator structure of MSK based on Eq. (5.84) is shown in Fig. 5.71. Notice that in the receiver the integration in the quadrature channel is delayed by $T_b$ seconds to account for the offset. The power spectrum of MSK can be obtained by the Fourier transform of the half-cycle sinusoidal shape pulse:[50]

$$S_B(f) = (32E_b/\pi^2)(16T_b^2 f^2 - 1)^{-2} \cos^2 2\pi T_b f \tag{5.86}$$

Using the first null as bandwidth, the spectral efficiency of MSK is 1.33 b/Hz/s, which is lower than that of QPSK (2 b/Hz/s). However, it can be seen that MSK has much lower sidelobes than QPSK and therefore yields lower out-of-band power emissions.

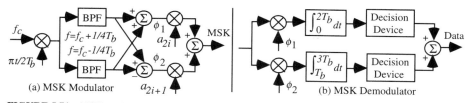

(a) MSK Modulator

(b) MSK Demodulator

**FIGURE 5.71** MSK modulator/demodulator scheme.

***Gaussian MSK (GMSK).*** GMSK is a derivative of MSK that uses Gaussian frequency pulse shaping. A GMSK signal can be generated simply by passing the data stream through a low-pass premodulation filter and frequency modulator (FM). For demodulation GMSK can be coherently detected as MSK or noncoherently detected using standard FM discriminators. The carrier recovery of GMSK can be accomplished using the times-$n$ (e.g., times-4) method.[51] The pulse-shape filter transfer function of GMSK is given by the following:[51]

$$H_G(f) = \exp[-2\ln 2(f/B_{3dB})^2] \tag{5.87}$$

where $B_{3dB}$ is the 3-dB bandwidth of the filter. In practice, the GMSK signal may be defined by the product of the filter bandwidth and baseband symbol duration, i.e., $B_{3dB}T_s$. It can be shown that as $B_{3dB}T_s$ decreases, the main lobe of the GMSK spectrum narrows and the sidelobe levels fall off rapidly compared to MSK ($B_{3dB}T_s = \infty$).[51] For example, with $B_{3dB}T_s = 0.5$, the second sidelobe level of GMSK is 30 dB below main lobe, whereas for MSK the second sidelobe level is 20 dB below main lobe. Hence, GMSK preserves the power efficiency of MSK (constant envelope) but introduces more bandwidth efficiency than MSK.

However, a trade-off exists. Notice that the impulse response of the Gaussian filter [Eq. (5.87)] has a duration greater than $T_s$. Thus, ISI will result when data passes through the filter. Further, the ISI increases as the product $B_{3dB}T_s$ decreases. Hence, whereas a small $B_{3dB}T_s$ increases the bandwidth efficiency and reduces spectral sidelobes, the induced ISI will degrade system performance. Studies have shown that the range $B_{3dB}T_s = 0.25-0.5$ gives a good compromise between spectral compactness and performance degradation.[55]

***Channel Impairment Performance.*** For AWGN the general BER expression of CPM may be obtained based on a union bound approach. For large SNR the BER can be approximated by the following:[1,51]

$$P_{eb} = C_o\text{erfc}\{D_{min,N}/2\sqrt{N_o}\} \tag{5.88}$$

where $C_o$ is a positive constant and $D_{min,N}$ is the minimum distance between two pairs of data sequences containing $N$ symbols. For the case of GMSK and MSK, then, we have the following,[51]

$$P_{eb} = 1/2\text{erfc}\{\sqrt{(\alpha\gamma)}\} \tag{5.89}$$

where $\alpha$ is a constant related to the product $B_{3dB}T_s$. For example, if $\alpha = 1$, Eq. (5.89) is the BER of MSK. If $\alpha = 0.68$, Eq. (5.89) becomes the BER of GMSK with $B_{3dB}T_s = 0.25$.

For Rayleigh fading the average BER of (coherent) GMSK can be derived as follows:[51]

$$P_{eb} = \{1 - \sqrt{[\delta\gamma_{av}/(1 + \delta\gamma_{av})]}\}/2 \approx 1/4\delta\gamma_{av} \tag{5.90}$$

where $\delta = 0.68$ for $B_{3dB}T_s = 0.25$, and $\delta = 0.85$ for $B_{3dB}T_s = \infty$. In frequency selective fading, irreducible BER floor for MSK may be induced due to time-varying Doppler shift. For GMSK, however, the penalty can be reduced to minimum with the choice of the $B_{3dB}T_s$ product.[51]

***Applications.*** CPM schemes are attractive in hostile environments because they exhibit excellent power efficiency (constant envelope) and bandwidth efficiency characteristics. Particularly, MSK and GMSK have found widespread applications in mobile and cellular radio communication systems and have been adopted in several wireless standards. Table 5.40 lists the CPM schemes used in several digital cellular, cordless, and personal communication systems (PCS).[51] In the table, GFSK (Gaussian-filtered BFSK) is equivalent to GMSK. In addition, the U.S. cellular digital packet data (CDPD) and RAM mobile systems also employ GMSK,[51] providing a channel data rate of 8–19.2 kb/s.

**TABLE 5.40**   GMSK Schemes in Different Wireless Standards

| | Wireless standard | | | |
| --- | --- | --- | --- | --- |
| | CT2/CT2+ | DECT | GSM | DCS 1800 |
| Modulation | GFSK | GFSK | GMSK ($B_{3dB}T_s = 0.3$) | GMSK |
| Data rate | 72 kb/s | 1152 kb/s | 270.833 kb/s | 270.833 kb/s |

### 5.6.4  Spread Spectrum Modulation Techniques

There are two basic features that distinguish a spread spectrum modulation (SSM) from a conventional modulation:[1–2] (1) transmission bandwidth far beyond the minimum required bandwidth, and (2) coding to spread, independent of information data, and despread, in a synchronized fashion, the signal band. Figure 5.72 shows the basic block diagram of SSM. The spreading is performed prior to transmission by multiplying the data-bearing signal, typically with a pseudonoise (PN) sequence or code. The PN code provides white noise–like (pseudo) randomness properties and yet has a periodic correlation property to aid data recovery. The despreading at the receiver is accomplished coherently by correlation techniques using the same PN code.

**FIGURE 5.72**   Basic block diagram of SSM.

***SSM Characteristics and Parameters.***   There are many reasons for spreading the spectrum. At the heart of SSM is perhaps the protection against interfering or jamming signals. As depicted in Fig. 5.73, it is clear that after the despread, while the signal returns to its original form, the interference gets spread in bandwidth and the in-band interference power is very much reduced at the baseband output.[55] Other benefits of SSM include low probability of intercept and high security, resistance and constructive utilization of multipath fading, multiuser random access capability, and high-resolution ranging and time-delay measurements.[1,55]

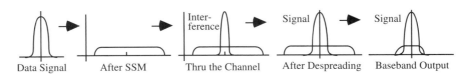

**FIGURE 5.73**   Basic concept of SSM.

Several parameters are often used to describe an SSM system. The first one is the chip rate. In SSM, to widen the spectrum, each data bit is effectively "chopped" by a PN code into a number of small time increments, called *chips*. Thus, the chip rate is the inverse of the chip time, which specifies the transmitted data rate after the spreading. The second parameter is the processing gain, which is defined as the spreading ratio,[55] i.e., chip rate/data rate. This is effectively the number of chips per bit. Alternatively, the processing gain is also the ratio of the spread bandwidth to the baseband data bandwidth. The third figure of merit, derived from the processing gain, is the jamming-to-signal margin (JSM), defined as the ratio of the processing gain to the required SNR.[55] Thus, the higher the processing gain of an SSM system, the larger the JSM that can be tolerated.

***Direct Sequence Modulation (DSM).*** DSM is one of the most popular SSM techniques. A DSM system, as shown in Fig. 5.72, is formed by directly multiplying the data-bearing pulses with a PN sequence of pulses $p(t)$, each having a chip time $T_c$. The form of $p(t)$ can be written as follows:[55]

$$p(t) = \sum_{n=-\infty}^{\infty} p_n g(t - nT_c) \tag{5.91}$$

where $p_n$ denotes the PN code, and $g(t)$ is the basic pulse shape, which is rectangular. The resulting DSM signal can then be written as follows:[55]

$$s(t) = \sqrt{2E_s/T_s}\,d(t)p(t)\cos(2\pi f_c t + \varphi) \tag{5.92}$$

where $d(t)$ is the data-bearing signal of symbol duration $T_s$. Because the signal is of the BPSK format before the spreading, this DSM is also commonly referred to as *DS/BPSK*. The demodulation of the DS/BPSK can be done by means of code synchronization and coherent or differentially coherent detection. The code synchronization requires the alignment of the local PN waveform to within one chip time with that of the received DSM signal;[55] this can be accomplished via searching and tracking for a high degree of correlation between two PN codes.

Assume that the received signal contains interference (ignoring noise), i.e., $y(t) = s(t) + i(t)$. Then, after despreading, the signal at the input of the BPSK coherent detector is given by the following:

$$r(t) = p(t)y(t) = \sqrt{(2E_s/T_s)}\,d(t)\cos(2\pi f_c t + \varphi) + p(t)i(t) \tag{5.93}$$

where we have used $p^2(t) = 1$. The modulated nature of $p(t)$ forces the interference $i(t)$ to widen its spectrum. It can thus be shown that at the detector output the interference power is reduced by a factor $T_c/T_s = 1/G_p$, which is the inverse of the processing gain.

Other types of DSM formats have also been explored. In particular, for DS/QPSK and DS/MSK, the spreading modulation is achieved simultaneously on two carriers using two PN sequences which are in phase quadrature.[1] The primary reason for using quadrature modulation is its low probability of detection and low sensitivity to some types of jamming.[1]

***Frequency-Hopping Modulation (FHM).*** FHM is another popular SSM technique, which can be formed by nonlinearly modulating a train of pulses with a PN sequence generated by pseudo-random frequency shifts. Thus, in FHM, the carrier in effect changes (hops) randomly from one frequency to another, resulting in a "sequential" spreading of the signal spectrum. The general form of the PN signal $p(t)$ for this case is given by the following:

$$p(t) = \sum_{n=-\infty}^{\infty} \exp\{j(2\pi f_n + \varphi_n)\}g(t - nT_h) \tag{5.94}$$

where $T_h$, the duration of the pulse $g(t)$, is called the hop time, and $\{\varphi_n\}$ is a sequence of random phases associated with the generation of the hops. A common modulation format for FHM is $M$-FSK, collectively called FH/MFSK, which has a complex form $\exp\{j[2\pi(f_c + d(t))t]\}$, where $d(t)$ is an $M$-level digital waveform representing the information frequency modulation at a rate $1/T_s$ symbols per second. Thus, the FH/MFSK signal may be written as[55]

$$s(t) = \sqrt{(2E_s/T_s)}p(t)\cos\{2\pi[f_c + d(t)]t\} \tag{5.95}$$

In reality, as shown in Fig. 5.74 for an FH/MFSK modem, the chip signal $p(t)$ is not generated in the transmitter. Rather, the signal is generated by mixing the $M$-FSK output with that of a

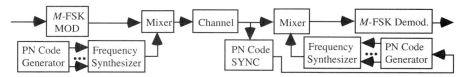

**FIGURE 5.74**   An FH/MFSK modem.

frequency synthesizer driven directly by the PN code generator. In particular, successive (not necessarily disjoint) $k$-chip segments of a PN sequence drive the synthesizer, which hops the carrier over $2^k$ distinct frequencies. On a single hop the signal bandwidth is the same as that of $M$-FSK with $M = 2^k$. However, when averaged over $2^k$ hops, the signal bandwidth then equals the spread bandwidth, denoted as $B_{ss}$. Hence, the processing gain for FHM is $G_p = B_{ss}/B$, where $B$ is the single-hop bandwidth. Indeed, the FHM bandwidths on the order of several GHz are currently attainable, which is an order of magnitude larger than that achievable with DSM.[55]

The received FHM signal is dehopped (mixed) to its original $M$-FSK form by the local frequency synthesizer synchronous with that of the transmitter, whereas the interference, if present, will be hopped onto a wide-band spectrum and its effect is therefore diminished.

Based on the hop rate, the FHM may be classified as slow-frequency hopping (SFH) and fast-frequency hopping (FFH) systems. SFH occurs if there are multiple symbols per hop, i.e., the hop time $T_h = NT_s$ ($N$ an integer). Note that in FHM, it is customary to refer a "chip" as the shortest duration of an individual FH/MFSK tone.[55] The chip rate $R_c$, for FH/MFSK, is then defined as the highest system clock rate, i.e., $R_c = \max(R_h, R_s)$. For slow FH/MFSK, a chip then equals an $M$-FSK symbol, thus, $R_c = R_s = R_b/\log_2 M \geq R_h$. Hence, at each hop, the spacing between the SFH tones is usually an integer multiple of $R_s$, to ensure no cross-talk or spillover in adjacent tones.[55] Assuming a minimum spacing (i.e., $R_s$) the entire spread band is then partitioned into a total of $N_t = B_{ss}/R_s$ equally spaced FH tones. One popular arrangement is to group these $N_t$ tones into $N_b = N_t/M = 2^k$ contiguous, nonoverlapping bands, each with bandwidth $MR_s$ and center frequency representing a hop carrier assigned to a unique $k$-tuple PN generator.[55] The processing gain for this SFH system thus equals $G_p = N_b = 2^k$.

For the case of FFH there is more than one hop per data symbol, i.e., $T_s = NT_h$. Hence, in FFH each hop is a chip, i.e., $R_c = R_h \geq R_s$. The spacing of FFH tones is then equal to the hop rate, and the spread spectrum is partitioned into a total of $N_t = B_{ss}/R_h$ equally spaced FH tones, each of which is assigned to a unique $k$-tuple PN generator.[55] The detection of an FFH signal can be operated in one of two modes:[50] (1) for each symbol, a decision is made on each frequency-hop chip received, and a simple majority vote rule can be used to estimate the dehopped FFH signal based on all $\log_2 M$ chip decisions; and (2) for each symbol, likelihood functions are computed as functions of the total signal received over all $\log_2 M$ chips, and the largest one is chosen. A receiver based on the second detection mode is optimum in the sense that it minimizes the average SER for a given $E_b/N_o$ SNR.[50]

***Other Types of SSM Techniques.***    One less popular SSM is the time-hopping modulation (THM), in which a time interval, selected to be much larger than the reciprocal of the data rate, is segmented into a large number of time slots.[2,55] The information symbols are then transmitted in pseudorandomly selected time slots. In this case the spreading is accomplished by time division. For instance, suppose that a time interval $T_f$ is subdivided into 1000 time slots of width $T_f/1000$ each. Then, if a BPSK having a data rate $R_b$ is used for THM, the output bit rate is $1000R_b$, and the required bandwidth is $B_{ss} = 1000R_b$. Now, denote the pulse width of time slot as $T_f/M_T$ ($M_T$ the time increment). Then, the PN signal for a THM system can be written as follows:[55]

$$p(t) = \sum_{n=-\infty}^{\infty} g[t - (n + a_n/M_T)T_f] \tag{5.96}$$

where $a_n$ denotes the pseudorandom position (one of the $M_T$ uniformly spaced locations). For THM, delay modulation is often used, in which the modulated signal takes the following form:[55]

$$s(t) = p[t - d(t)] \cos[j(2\pi f_c + \varphi_T)] \tag{5.97}$$

where $d(t)$ is usually a PSK. At the receiver the dehopping of the time slots removes the delays introduced by the time-hopping signal $p(t)$ and restores the original information signal, and again spreads the interfering signals, if they are present.

Finally, hybrid SSM techniques can be formed to achieve higher processing gain and better performance against interferer(s) than can any of the SSM methods acting alone. One common hybrid SSM is the DS/FH system, in which a PN sequence is used in combination with frequency hopping.[1-2] Here, the signal transmitted on a single hop consists of a DSM signal, which is demodulated coherently; however, the received signals from different hops are combined noncoherently.[2] Because coherent detection is performed within a hop, this hybrid SSM has an advantage over the pure FHM, albeit the increased complexity.

***Channel Impairment Performance.***   Note that SSM has no effect on AWGN, whose spectrum is much wider than the spread bandwidth. Thus, the SSM performance in AWGN is the same as the nonspread case and will not be given here. In what follows, we discuss the performances of SSM in three different cases: jammers, multipath, and multiaccess interference.

One main SSM function is against jammers. Conversely, a "smart" jammer may defeat the function of an SSM system. The jammers in practical use can be classified as noise and tone types. Noise-type jammers possess band-limited Gaussian noise nature with high average power, whereas tone-type jammers consist of sinusoids whose frequencies lie inside the spread band. Table 5.41 summarizes the BER equations for binary DSM and FHM systems under AWGN and various jammer conditions.[1,55] In the table, $J$ is the jammer power, $P_s$ is the signal power, $\theta$ is a random phase associated with a single-tone jammer with uniform distribution in $[0, 2\pi]$, and $G_p = B_{ss}/R_b$ is the processing gain. The JSM is given by $(J/P_s) = G_p - E_b/N_J$ ($N_J = J/B_{ss}$). From the table, it can be seen that (1) the BER reduces to the nonspread case as $G_p P_s/J \to \infty$; (2) a worst-case (maximum) BER exists, for an optimum $\rho$, for partial-band or pulsed jammers; i.e., for DSM,[1] the maximum BER $P_{eb,\max} = 0.37(G_p P_s/J)^{-1}$ for $G_p P_s/J > 2$, and for FHM,[1] $P_{eb,\max} = 0.083(G_p P_s/J)^{-1}$ for $G_p P_s/J > 1/2$; and (3) for multiple-tone jammers, the effect on DSM is the same as that of partial-band jammer noise, whereas a single-tone jammer is not effective for FHM. Finally, it should be noted that although jammers are effective against an uncoded SSM system, use of FEC in SSM can mitigate most of the jammer problems.[1-2]

Another function of SSM is to provide multiple (say, $K$) user access capability. Here, each user has an independent PN code or a hop pattern. In this case the performance of SSM may be assessed against $K - 1$ interferers. For this case the BERs of DS/BPSK (or QPSK) and FH/BFSK are given, respectively, by the following:[51]

$$P_{eb} = \tfrac{1}{2}\operatorname{erfc}\{\sqrt{[N_o/E_b + 2(K - 1)/(3N_c)]}\} \quad \text{(DS/B(Q)PSK)} \tag{5.98}$$

$$P_{eb} = \tfrac{1}{2}\exp\left(-\frac{E_b}{N_o}\right)\left[1 - \frac{1}{N_h}\left(1 + \frac{1}{N_b}\right)\right]^{K-1} + \tfrac{1}{2}\left\{1 - \left[1 - \frac{1}{N_h}\left(1 + \frac{1}{N_b}\right)\right]^{K-1}\right\} \quad \text{(FH/BFSK)}$$

$$\tag{5.99}$$

where $N_c$ is the number of chips per symbol, $N_h$ is the number of hopping slots, and $N_b$ is the number of bits per hop. Note that Eq. (5.99) is valid when users hop their frequencies asynchronously. It is readily shown that if $E_b/N_o = \infty$, i.e., when thermal noise is not a factor, both Eqs. (5.98) and (5.99) approach an irreducible BER floor due to multiple access interference. This is the so-called "near-far problem."[51] This problem is particularly harmful to DSM, because a close-in user may dominate the received signal energy at the receiver, making the Gaussian assumption invalid. For FHM, however, the problem is less critical, since it is unlikely that all users will utilize the same frequency simultaneously.

**TABLE 5.41**  Summary of SSM BER Performance in Jammer Environments

| | Jammer Type | | | |
|---|---|---|---|---|
| | Barrage noise | Partial-band noise | Pulsed noise | Tone jammers |
| Characteristics | Broad-band, equal to spread bandwidth | A fraction of spread band, $0 \le \rho \le 1$ | "On" a fraction of time, $0 \le \rho \le 1$ | Single or multiple sinusoidal tones |
| DS/BPSK DS/QPSK DS/MSK BER | $\dfrac{1}{2}\operatorname{erfc}\sqrt{\dfrac{1}{(N_o/E_b)+(J/P_s/G_p)}}$ | Same as that of barrage jammer | $\dfrac{\rho}{2}\operatorname{erfc}\sqrt{\dfrac{1}{(N_o/E_b)+(J/P_s/\rho G_p)}}$ $+\dfrac{1-\rho}{2}\operatorname{erfc}\sqrt{\dfrac{E_b}{2N_o}}$ | *Single tone:* $\dfrac{1}{4\pi}\displaystyle\int_0^{2\pi}\operatorname{erfc}\left(-\sqrt{\dfrac{S}{N_{\text{tot}}}}\right)d\theta$ $\dfrac{S}{N_{\text{tot}}}=\dfrac{1}{\dfrac{N_o}{E_b}+\dfrac{4J/P_s}{G_p}\cos^2\theta}$ |
| Slow FH/BFSK BER | $\dfrac{1}{2}\exp\left(\dfrac{-1}{(N_o/E_b)+(J/P_s/G_p)}\right)$ | $\dfrac{\rho}{2}\exp\left(\dfrac{-1}{(N_o/E_b)+(J/P_s/\rho G_p)}\right)$ $+\dfrac{1-\rho}{2}\exp\left(\dfrac{-E_b}{2N_o}\right)$ | Same as that of partial-band jammer | *Multitone:* $0.5\qquad G_p P_s/J < 2$ $\dfrac{J/P_s}{G_p}\qquad \dfrac{2}{G_p}\le\dfrac{P_s}{J}\le 1$ $0\qquad 1 < P_s/J$ |

The third function of SSM is multipath suppression. This is achieved either by frequency or time diversity.[51] For example, in DSM, if the multipath signals are delayed by more than one chip duration, they appear like uncorrelated noise at the receiver. In FHM the carrier hopping rate can be made fast enough relative to the differential delay between the direct and indirect paths, so that all (or most) of the multipath energy will fall in frequency slots that are orthogonal to that of the desired signal, and thereby the effect of multipath is diminished.[50] However, a more effective way is to coherently combine the time-delayed versions of the original signal to boost the SNR at the receiver. A receiver that does this function is called the *RAKE* receiver, which gets its name from the way it "rakes" in all the incoming signals to form an "equalized" signal.[51,55] A RAKE receiver is composed of a bank of correlators and a tapped-delay line with tap spacing equal to chip time. As the received signal passes through the delay line, each correlator attached to each tap detects a strongest multipath component. The correlator outputs are then weighted and combined for bit decisions. In effect, RAKE provides a form of diversity prior to bit decisions, so as to enhance signal in fading and improve SSM reception.

***Applications.***    SSM is a well-established technology having roots in military communications. Two key military applications of SSM are for antijam (AJ) and low-probability of detection (LPD).[4] The former is used in a hostile environment to attenuate or avoid jammers; by contrast, the latter is used in a secure situation to hide the desired signal from detection by others to avoid jam or exploitation in any manner. An example of a military SSM system is the Joint Tactical Information Distribution System (JTIDS) developed for AJ communications and location for troops in combat situations.[1] The system uses a hybrid DS/FH/TH spread spectrum to provide highly efficient signaling for the anticipated future battle ground.

In commercial and consumer sectors, SSM techniques have been applied to various wired and wireless communication systems to mitigate severe interference and multipath fading problems and/or provide multiaccess and signal capture capabilities. In particular, since the FCC's release of the ISM bands (900, 2400, 5700 MHz) for commercial usage,[55] the commercial application of SSM has increased dramatically. Table 5.42 lists some of today's SSM applications. In addition, SSM techniques have also been adopted in several communication standards, as listed in Table 5.43.[51,55–57]

**TABLE 5.42**    A List of SSM Applications

| Applications | Usage ex | SSM Used |
|---|---|---|
| Cable modem | Internet, 2-way | DS/B(Q)PSK |
| Power line | Home security | DS/BPSK |
| Satellite | Access, GPS | DS/B(Q)PSK |
| WLAN | PCMCIA | DS, FH |
| Radio | Access, remote | DS, FH |
| Cellular | Access, phone | DS, FH |

**TABLE 5.43**    A List of Communication Standards that Have Adopted SSM

| Standard | Application | RF band | Ch. BW | Modulation | PN sequence | Ch. data rate |
|---|---|---|---|---|---|---|
| IEEE 802.11 | Wireless LAN | 2.4–2.483 GHz | — | DS/D(Q)PSK | 11-bit barker | 1-2 Mb/s |
| EIA/IS-60 | CEBus/home | 902–908 MHz | 2.1 MHz | DS/BPSK | 2520 chips | 10 kb/s |
| TIA/IS-95 | Cellular/mobile | 869–894 MHz | 1.25 MHz | DS/QPSK | 15-bit short | 1.23 Mb/s |
|  | Phone, PCS | 824–849 MHz |  | DS/OQPSK | 42-bit long |  |
| GSM | Cellular/mobile | 935–960 MHz | 200 kHz | Slow | 217 hops/s | 270.8 kb/s |
|  | Phone, PCS | 890–915 MHz |  | FH/GMSK |  |  |

***Code-Division Multiple Access (CDMA).***    An important application of spread spectrum is the CDMA system for multiaccess communications. In conventional multiaccess systems, users access a communication channel via either frequency assignment (i.e., frequency-division multiple access [FDMA]) or time-slot assignment (i.e., time-division multiple access [TDMA]). In CDMA, however, all users may access the channel using the same frequency at the same time; here, each user is given a unique code that performs DSM or FHM. The code assignment is such that each code is approximately orthogonal (i.e., with low cross-correlation) with all the other codes. The receiver then performs a time-correlation operation to detect only the specific desired code. All other code signals will appear as noise due to decorrelation. Compared to a traditional FDMA or TDMA system, a CDMA system offers several attractive features:[50–51]

1. *Soft capacity limit.*    Performance degrades gradually as the number of users is increased. Thus, it is relatively easy to add on or remove users without disrupting the system.
2. *Asynchronous.*    External synchronization is not necessarily needed; that is, the transmission times of a user's data symbol do not have to coincide with those of the other users.
3. *Interference rejection.*    Multipath and fading are effectively combatted by spreading. In practice, CDMA may tolerate an interference that is 18 dB larger than the signal.[57]

CDMA has now been widely used in satellite and terrestrial wireless systems. In satellite application, DS-CDMA is promising for low-data-rate systems and low earth-orbiting satellite systems.[55] The most widespread use of CDMA today is perhaps for mobile, cellular, and PCS communications, where a number of CDMA-based standards are under development.[51] The fundamental CDMA standard is the IS-95, which specifies a common air interface for cellular phones as well as mobile and base stations at 824–849 MHz (reverse link) and at 869–894 MHz (forward link).[51,57] In IS-95 the maximum user data rate is 9600 b/s. User data are spread in three DSM stages to achieve a total transmission rate of 1.2288 Mb/s, as illustrated in Fig. 5.75 for forward- and reverse-link modulation processes. But, as seen in Fig. 5.75, the spreading steps are different for the two links due to different requirements.[51] In addition, the forward link (base station to mobile) employs baseband QPSK and a pilot signal for reference timing, whereas the reverse link uses OQPSK for noncoherent detection and an asynchronous transmission.

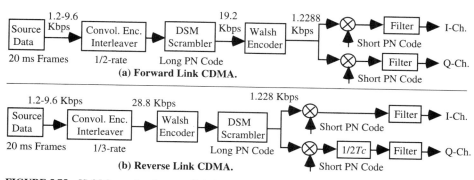

**FIGURE 5.75**    IS-95 forward and reverse link CDMA processing blocks.

## GLOSSARY

**ACATS**    FCC Advisory Committee on Advanced Television Service.

**ADSL**    Asymmetrical digital subscriber line. A twisted-pair–based distribution system providing transmission rates up to 6.3 Mb/s.

**ADTV** Advanced digital television. An HDTV system proposed by the Advanced Television Research Consortium, consisting (in 1992) of David Sarnoff Research Center, North American Philips, Thomson Consumer Electronics, NBC, and Compression Labs, Inc.

**AFC** Automatic frequency control.

**ASK** Amplitude shift keying. A digital amplitude modulation method.

**ATV** Advanced television; the term is often used interchangeably with HDTV.

**AWGN** Additive white Gaussian noise.

**BER** Bit error rate. A measure of the probability of bit error in a communication system.

**CCDC** Channel Compatible DigiCipher. An HDTV system proposed by the American Television Alliance, consisting of General Instruments and Massachusetts Institute of Technology.

**CCITT** International Telegraph and Telephone Consultative Committee. A standardization body for telecommunications.

**CDMA** Code-division multiple access, based on spread-spectrum code assignments.

**CEBus** Consumer electronics bus. A multimedia distribution system in the home.

**CMA** Constant modulus algorithm for updating tap-weight coefficients in blind equalization.

**CMFSK** Coherent multilevel frequency shift keying.

**CPM** Continuous-phase modulation.

**CSO** Composite second order. A clustering of second-order frequency beats at 1.25 MHz above the visual carriers in cable systems.

**CTB** Composite triple beat. A clustering of third-order frequency beats around the visual carriers in cable systems.

**CT2** Cordless telephone–2. A second generation cordless telephone system introduced in Great Britain in 1989.

**DAB** Digital audio broadcasting.

**DAVIC** Digital Audio-Visual Council. A nonprofit association whose purpose is to identify, select, augment, develop, and obtain the endorsement of formal standards bodies of specifications of interfaces, protocols, and architectures of digital audiovisual applications and services.

**DBS** Direct Broadcast Satellite, providing video programming services to subscribers.

**DCS 1800** Digital Communication System-1800, which is an extension of the Global System for Mobile Communications standard in the 1800-MHz band for personal communication networks.

**DDCR** Decision-directed carrier recovery. A carrier recovery scheme that uses the decisions feedback from the receiver bit or symbol decision detection circuit.

**DECT** Digital European Cordless Telephone. A cordless telephone standard developed by the European Telecommunications Standards Institute in 1992.

**DFE** Decision feedback equalizer. A nonlinear equalizer consisting of a feed-forward equalization filter and a feedback equalization filter.

**DMC** Discrete memoryless channel, which is specified by the transition probabilities of input and output symbols.

**DMT** Discrete multitone. A multicarrier modulation scheme.

**DSC-HDTV** Digital spectrum–compatible HDTV. An HDTV system proposed by Zenith Electronics and AT&T in 1992.

**DSM** Direct sequence modulation. A spread spectrum modulation scheme that directly modulates the information-bearing data with a spreading code sequence.

**DTTB**   Digital television terrestrial broadcasting.

**DVB**   Digital video broadcasting.

**EIA**   Electronic Industries Association. A standard-setting body for consumer electronics.

**FDMA**   Frequency-division multiple access, based on frequency assignments.

**FEC**   Forward error correction.

**FHM**   Frequency-hopping modulation. A spread-spectrum modulation scheme that is used to nonlinearly modulate a train of pulses with a sequence of pseudorandom frequency shifts.

**FIR**   Finite impulse response, commonly referring to an implementation structure of a filter.

**FPLL**   Frequency- and phase-locked loop.

**FSE**   Fractionally spaced equalizer, whose tap spacing is a fraction of a data symbol duration.

**FSK**   Frequency shift keying. A digital frequency-modulation scheme.

**GA**   Grand Alliance. A U.S. HDTV consortium consisting of General Instruments, Zenith Electronics, AT&T, David Sarnoff Research Center, Thomson Consumer Electronics, North Philips Electronics, and Massachusetts Institute of Technology.

**GMSK**   Gaussian minimum shift keying. A minimum shift keying with Gaussian frequency pulse shaping.

**GPS**   Global positioning system. A satellite ranging system based on direct sequence spread spectrum for mobile users to determine their location.

**GSM**   Global System for Mobile Communications. A second-generation cellular system standard set by the European Technical Standards Institute 1990.

**HD-DIVINE**   A Scandinavian digital terrestrial HDTV project proposed in 1991.

**HDSL**   High-speed digital subscriber line, providing T1 rate services.

**HDTV**   High-definition television. An advanced digital television providing high resolution.

**HFC**   Hybrid fiber and coaxial network.

**ICI**   Intercarrier interference, associated with multicarrier modulation systems.

**IEEE**   Institute of Electrical and Electronics Engineers.

**IM**   Intermodulation, referring to distortions generated by nonlinear devices in a system.

**IS-54**   Interim Standard-54. A dual mode standard for U.S. digital cellular system and Advance Mobile Phone Service based on time-division multiple access, set by EIA/TIA in 1990.

**IS-95**   Interim Standard-95. An air interface standard for U.S. code-division multiple access and Advance Mobile Phone Service dual-mode operation for mobile and cellular users.

**ISDN**   Integrated Services Digital Network, providing voice, data, and video services with rates up to 64 kb/s.

**ISI**   Intersymbol interference. The additive contribution (interference) to a received sample from transmitted symbols other than the symbol to be transmitted.

**ISM bands**   Industrial, scientific, and medical frequency bands 902–928 Mhz, 2.4–2.4835 GHz, and 5.725–5.85 GHz.

**ITU**   International Telecommunication Union. An intergovernmental organization responsible for setting global standards.

**LE**   Linear equalizer. A linear filter that is used to reduce intersymbol interference.

**LMS**   Least mean square. An algorithm for adaptively adjusting the tap coefficients of an equalizer based on the use of (noise-corrupted) estimates of the gradients.

**MAP**   Maximum *a posteriori*. A detection hypothesis that maximizes the *a posteriori* probability.

**MCM**   Multicarrier modulation. A parallel digital modulation scheme that simultaneously modulates a number of carriers.

**MMDS**   Multichannel multipoint distribution service. A microwave local distribution system providing multiple television channel programs (up to 33 channels) to subscribers.

**MMSE**   Minimum mean-square error.

**MSK**   Minimum shift keying. A special case of continuous phase-FSK wherein the peak frequency deviation is equal to half the bit rate, which is the minimum frequency spacing for two frequency-shift keying signals.

**NCMFSK**   Noncoherent multilevel frequency shift keying.

**NTSC**   National Television System Committee. The term also refers to the current U.S. color television system, which was specified and developed in 1950–1953.

**OBO**   Output back-off. The output power of a power amplifier in a modulation system that is needed to back off from the saturation point of the amplifier to avoid nonlinear distortions.

**OFDM**   Orthogonal frequency division modulation or multiplexing. A multicarrier modulation scheme that involves multiplexing a number of low-speed modulated carriers for transmission.

**OQPSK**   Offset quadrature phase-shift keying, where one quadrature arm is offset by one bit duration relative to the other arm.

**PACS**   Personal Access Communication System. A third-generation personal communications system originally developed and proposed by Bellcore in 1992; it later became a radio interface standard for U.S. Personal Communication Service in the 1920–1930-MHz band.

**PAM**   Pulse amplitude modulation. A digital amplitude-modulation scheme.

**PAR**   Peak-to-average power ratio (usually expressed in decibels [dB]) that is used to determine the robustness of a modulated signal against nonlinear distortion.

**PCMCIA**   Personal Computer Memory Card Industry Association. A nonprofit trade association founded in 1989 to define a standard memory card interface for computers.

**PCS**   Personal communication systems or services.

**PDC**   Pacific Digital Cellular. A cellular standard that was developed in 1991 to provide for needed capacity in congested cellular bands in Japan.

**PFD**   Phase and frequency detector used for decision-directed carrier recovery.

**PHS**   Personal Handy Phone System, a Japanese air interface standard set by the Research and Development Center for Radio Systems; its network interface was specified by the Telecommunications Technical Committee of Japan in 1993.

**PSK**   Phase-shift keying. A digital phase-modulation scheme.

**QAM**   Quadrature-amplitude modulation. A combined phase- and amplitude-modulation scheme that is used to modulate the carrier signal in phase quadrature.

**QPSK**   Quadrature or quaternary phase-shift keying. A digital phase-modulation scheme that is used to modulate the carrier signal in phase quadrature.

**RF**   Radio frequency that is above the audio and below infrared frequencies. The frequency range is 10 kHz to 10 GHz, allocated for radio use.

**SCM**   Single-carrier modulation, in which information data is modulated on one carrier for transmission.

**SER**   Symbol error rate. A measure of probability of symbol errors in a communication system.

**SFN**   Single-frequency network. A network with transmitters operating on the same frequency.

**SNR**   Signal-to-noise ratio, usually expressed in decibels (dB).

**SONET**   Synchronous Optical Network, which specifies data transmission over optical fiber with rates ranging from 51.84 Mb/s to 9.95 Gb/s.

**SSM**   Spread-spectrum modulation. A digital modulation scheme that spreads the system bandwidth far greater than the minimum required bandwidth.

**TCM**   Trellis-code modulation. A combined coding and modulation technique.

**TDMA**   Time-division multiple access. A time-slot–based multiuser access scheme.

**THM**   Time-hopping modulation. A spread-spectrum technique using pseudorandom time-hop patterns as the spreading sequence code; it is analogous to pulse-position modulation.

**TIA**   Telecommunication Industry Association. A trade organization that provides services such as government relations, market activities, educational programs, and standards-setting activities.

**TOV**   Threshold of visibility. A bit-error-rate HDTV threshold of $3 \times 10^{-6}$, at which value the impairment effect first becomes visible in the picture.

**TR**   Timing recovery for symbol synchronization.

**TTIB**   Transparent tone in band. A pilot-aided carrier recovery technique.

**VSB**   Vestigial sideband. The transmitted portion of one sideband; the sideband is suppressed in a transmitter having a gradual cut-off in the neighborhood of the carrier frequency. Meanwhile, the other sideband is transmitted without suppression.

**WACS**   Wireless Access Communication System, developed to provide wireless connections for local exchange carriers.

**WLAN**   Wireless local-area network.

## REFERENCES

1. R. E. Ziemer and Peterson, R. L., *Digital Communications and Spread Spectrum,* Macmillan, New York, 1985.

2. J. G. Proakis, *Digital Communications,* 2nd ed., McGraw-Hill, New York, 1989.

3. T. Noguchi et al., "Modulation techniques for microwave digital radio," *IEEE Communications Magazine,* vol. 24, no. 10, Oct. 1986, pp. 21–30.

4. Y. Wu et al., "Evaluation of channel coding, modulation, and interference of digital ATV terrestrial transmission systems," *IEEE Transactions on Broadcasting,* vol. 40, June 1994, pp. 75–81.

5. R. E. Ziemer, "Character error probabilities for M-ary signaling in impulsive noise environments," *IEEE Transactions on Communications,* COM-15, 1967, pp. 32–44.

6. O. Shimbo, *Transmission Analysis in Communication Systems,* vols. 1 and 2, Computer Science Press, Rockville, MD, 1988.

7. W. T. Webb and Hanzo, L., *Modern Quadrature Amplitude Modulation,* Pentech Press, London, and IEEE Press, Piscataway, N.J., 1994.

8. H. Stewart, "16-QAM modems in satellites," *Communications System Design,* July 1995, pp. 36–40.

9. D. A. Bryan, "QAM for terrestrial and cable transmission," *IEEE Transactions on Consumer Electronics,* vol. 41, no. 3, Aug. 1995, pp. 383–391.

10. T. S. Rzeszewski (ed.), *Digital Video,* IEEE Press, New York, 1995.

11. M. Sablatash, "Transmission of all-digital advanced television: State of the art and future directions," *IEEE Transactions on Broadcasting,* vol. 40, June 1994, pp. 102–121.

12. K. Ramchandran et al., "Multiresolution broadcast for digital HDTV using joint source/channel coding," *IEEE Journal of Selected Areas in Communications,* vol. 11, Jan. 1993, pp. 6–23.

13. C. W. Rhodes, "Measuring peak and average power of digitally modulated advanced television systems," *IEEE Transactions on Broadcasting,* vol. 38, no. 4, Dec. 1992, pp. 197–201.

14. A. Leclert and Vandamme, P., "Universal carrier recovery loop for QASK and PSK signal sets," *IEEE Transactions on Communications,* Jan. 1983, pp. 130–136.

15. H. Sari et al., "Transmission techniques for digital terrestrial TV broadcasting," *IEEE Communications Magazine,* vol. 33, no. 2, Feb. 1995, pp. 100–109.

16. D. D. Falconer, "Jointly adaptive equalization and carrier recovery in two-dimensional signal communication systems," *Bell System Technical Journal,* vol. 55, Mar. 1976, pp. 317–334.

17. Y.-S. Kim, "Performance of high level QAM in the presence of impulsive noise and co-channel interference in multipath fading environment," *IEEE Transactions on Broadcasting,* vol. 36, 1990, pp. 170–174.

18. J. B. Waltrich, "The impact of microreflections on digital transmission over cable and associated media," *NCTA Technical Papers,* 1992, pp. 321–336. Also in *Digital Video,* edited by T. S. Rzeszewski, IEEE Press, New York, 1995.

19. Q. Shi, "Asymptotic clipping noise distribution and its impact on M-ary QAM transmission of optical fiber," *IEEE Transactions on Communications,* COM-43, 1985, pp. 2077–2084.

20. K. Laudel, "Performance results of a low-cost alternative equalizer for 64/256-QAM demodulation in a CATV receiver," *NCTA Technical Papers,* May 1995, pp. 227–240.

21. CableLabs, *Results and Analysis of the Cable Portion of the ATV Modem Tests,* CableLabs Report, Feb. 1994.

22. Special issues on HDSL, *IEEE Journal Selected Areas of Communications,* vol. 9, no. 6, and on copper-wire access technologies, *IEEE Journal Selected Areas of Communications,* vol. 13, no. 9, Dec. 1995.

23. D. Minoli, *Video Dial Tone Technology,* McGraw-Hill, New York, 1995.

24. F. E. Froehlich and Kent, A. (eds.), *Encyclopedia of Telecommunications,* vol. 6, Marcel Dekker, New York, 1993.

25. L. Goldberg, "IC opens 500-channel frontier to cable systems," *Electronics Design,* Nov. 1994.

26. "European Telecommunication Standard on Digital Broadcasting Systems for Television, Sound, and Data Services: Cable Systems," Draft Document, Valbonne, France, 1994.

27. "DAVIC 1.0 Specification on Lower Layer Protocols and Physical Interface," Rev. 4.1, 1995.

28. "Grand Alliance HDTV System Specification-Chapter IV: VSB Transmission System Description," Draft Document, Feb. 1994.

29. G. Sgrignoli et al., "VSB modulation used for terrestrial and cable broadcasts," *IEEE Transactions on Consumer Electronics,* vol. 41, Aug. 1995, pp. 367–382.

30. "Grand Alliance Record of Test Report," ATTC, Alexandria, Va., Oct. 1995.

31. K. J. Kerpez, "A comparison of QAM and VSB for hybrid fiber/coax digital transmission," *IEEE Transactions on Broadcasting,* vol. 41, Mar. 1995, pp. 9–16.

32. "Special Report on ATV System Recommendation by FCC Advisory Committee on Advanced Television Service," *IEEE Transactions on Broadcasting,* vol. 39, Mar. 1993.

33. L. W. Lockwood, "VSB and QAM," *CED,* Dec. 1995, pp. 16–23.

34. G. Sgrignoli and Hrishnamurthy, G., "VSB Digital Transmission over MMDS Channels," Zenith Report for the Wireless Cable Digital Alliance, 1995.

35. H. Samueli et al., and Lee, R., "QAM vs. VSB," *CED,* Dec. 1994.

36. J. A. C. Bingham, "Multicarrier modulation for data transmission: An idea whose time has come," *IEEE Communications Magazine,* vol. 28, no. 5, May 1990, pp. 5–14.

37. Y. Wu and Zou, W. Y., "COFDM: An overview," *IEEE Transactions on Broadcasting,* vol. 41, no. 1, Mar. 1995, pp. 1–8.

38. B. Le Floch et al., "Coded orthogonal frequency division multiplex," *Proceedings IEEE,* vol. 83, no. 6, June 1995.

39. B. Hirosaki et al., "Advanced groupband data modem using orthogonally multiplexed QAM technique," *IEEE Transactions on Communications,* vol. COM-34, no. 6, June 1986, pp. 587–592.

40. B. Le Floch et al., "Digital sound broadcasting to mobile receivers," *IEEE Transactions on Consumer Electronics,* vol. 35, Aug. 1989, pp. 493–503.

41. P. J. Tourtier et al., "Multicarrier modem for digital HDTV terrestrial broadcasting," *Signal Processing: Image Communications,* vol. 5, Dec. 1993, pp. 379–403.

42. T. De Couasnon et al., "Results of the first digital terrestrial television broadcasting field-tests in Germany," *IEEE Transactions on Consumer Electronics,* vol. 39, Aug. 1993, pp. 668–674.

43. J. Ahn and Lee, H. S., "Frequency domain equalization of OFDM signals over frequency nonselective Rayleigh fading channels," Electronic Letters, vol. 29, Aug. 1993, pp. 1476–1477.

44. T. Pollet and Moeneclaey, M., "BER sensitivity of OFDM systems to carrier frequency offset and Wiener phase noise," *IEEE Transactions on Communications,* vol. 43, Feb. 1995, pp. 191–193.

45. Y. Wu et al., "OFDM for digital television terrestrial distribution over channels with multipath and non-linear distortion," *Proceedings International Conference HDTV'94,* Turin, Italy, Oct. 1994.

46. J. A. C. Bingham and Jacobsen, K., "CATV reverse channels using synchronized DMT," *IEEE Standards Committee 802.14/95-001-003,* Feb. 1995, and also "A proposal for an SDMT PHY layer," *IEEE Standards Committee 802.14/95-138,* Nov. 1995.

47. R. Citta and Lee, R., "Practical implementation of a 43 MBit/sec (8 Bit/HZ) digital modem for cable television," *1993 NCTA Technical Papers,* May 1993, pp. 271–279.

48. J. Henderson and James, B., "Status Report on ATV Transmission System Recommendation and Test Results for Transmission Expert Group and Specialist Group," Feb. 1994.

49. R. De Gaudebzi and Luise, M. (eds.), *Audio and Video Digital Broadcasting Systems and Techniques,* Elsevier Science, 1994.

50. S. Haykin, *Digital Communications,* Wiley, New York, 1988.

51. T. S. Rappaport, *Wireless Communications, Principles and Practice,* IEEE Press, 1996.

52. B. Schweber, "Direct satellite broadcast," *Electronic Design News,* Dec. 21, 1995, pp. 53–58.

53. D. T. Gall, "Digital modulation on coaxial/fiber hybrid systems," *Communications Technology,* Jan. 1995.

54. P. Kagan, *Cable Modems: A Strategic Analysis,* Paul Kagan Associates, Calif., 1995.

55. J. D. Gibson (ed.), *The Mobile Communications Handbook,* CRC Press, Boca Raton, Fla., 1996.

56. "EIA Home Automation Systems (CEBus) IS-60.3, Part 5—RF Physical Layer and Medium Specifications," Global Engineering Documents, Irvine, Calif., June 1993.

57. P. Whipple, "The CDMA standard," *Applied Microwave & Wireless,* Spring 1994, pp. 24–37.

## *ABOUT THE AUTHOR*

Qun Shi received the B.S. degree in radio engineering from Xidian University (Northwest Telecommunication Engineering Institute), Xian, China, in 1982, and the M.S. and Ph.D. degrees in electrical engineering from the University of Pennsylvania, Philadelphia, in 1986 and 1991, respectively.

He was a research fellow in the department of electrical engineering of the University of Pennsylvania from 1985 to 1990, working in the area of spatial spectrum estimation and array signal processing. He joined the Communication Systems Technology Laboratory of Panasonic Technologies, Inc., in 1990, where he is now a senior member of the technical staff, involved in the development and design of MPEG-2 systems and digital communications modems and systems for fiber optic, cable, and wireless communications. His current research interests include optical fiber communications, analog/digital communications, wireless communications, digital signal processing, and video compression.

Dr. Shi is a senior member of IEEE and a member of the Society of Telecommunications Engineers (SCTE).

# CHAPTER 6
# FUZZY LOGIC

**Timothy J. Ross**
*University of New Mexico*

## 6.1 INTRODUCTION

> Complexity in the world generally arises from uncertainty in the form of ambiguity. Complexity and ambiguity are features of problems that have been addressed by humans since man could think; they are ubiquitous features that imbue most social, technical, and economic problems faced by the human race. Why is it then that computers, which have been designed by humans after all, are not capable of addressing complex and ambiguous issues? How can humans reason about real systems, when the complete description of a real system often requires more detailed data than a human could ever hope to recognize simultaneously and assimilate with understanding? It is because humans have the capacity to reason approximately, a capability that computers currently do not have. In reasoning about such systems humans simply approximate behavior, thereby maintaining only a generic understanding about the problem. Fortunately, for our ability to understand complex systems, this generality and ambiguity is sufficient for human comprehension.

This quote from Dr. Zadeh's *principle of incompatibility* suggests that complexity and ambiguity (imprecision) are correlated: "The closer one looks at a real-world problem, the fuzzier becomes its solution."[1]

As we learn more and more about a system, its complexity decreases and our understanding increases. As complexity decreases, the precision afforded by computational methods becomes more useful in modeling the system. For systems with little complexity, hence little uncertainty, closed-form mathematical expressions provide precise descriptions of the systems. For systems that are a little more complex, but for which significant data exists, model-free methods (such as artificial neural networks) provide a powerful and robust means to reduce some uncertainty through learning, based on patterns in the available data. Finally, for the most complex systems where little numerical data exists and where only ambiguous or imprecise information may be available, fuzzy reasoning provides a way to understand system behavior by allowing us to interpolate approximately between observed input and output situations. The imprecision in fuzzy models is therefore generally quite high. Fuzzy systems can implement crisp inputs and outputs and, in this case, produce a nonlinear functional mapping just as do algorithms. All of the models are mathematical abstractions of the real physical world. The point is, however, to match the model type with the character of the uncertainty exhibited in the problem. In situations where precision is apparent, for example, fuzzy systems are not as efficient as more precise algorithms in providing us with the best understanding of the problem. On the other hand, fuzzy systems can focus on modeling problems with imprecise or ambiguous information.

After Dr. Zadeh's seminal paper on fuzzy sets,[2] many theoretical developments in fuzzy logic took place in the United States, Europe, and Japan. From the mid 1970s to the present, however, Japanese researchers have been a primary force in advancing the practical implementation of the theory; they have done an excellent job of commercializing this technology and now have a few thousand patents in the area. Much of the success of the new products associated with the fuzzy technology is due to fuzzy logic, and some is also due to the advanced sensors used in these products.

Fuzzy logic affects many disciplines[3] (see Chapter 28). In videography, for instance, Fisher, Sanyo, and others make "fuzzy logic" camcorders, which offer "fuzzy focusing" and "image stabilization." Mitsubishi manufactures a fuzzy air conditioner that allows temperature changes according to human comfort indexes. Matsushita builds a fuzzy washing machine that combines smart sensors with fuzzy logic. The sensors detect the color, kind, and quantity of grit in clothes, and a fuzzy microprocessor selects the most appropriate of 600 available combinations of water temperature, detergent amount, and wash and spin cycle times. The Japanese city of Sendai has a 16-station subway system that is controlled by a fuzzy computer. The ride is so smooth riders do not need to hold straps, and the controller makes 70 percent fewer judgment errors in acceleration and braking than human operators. Nissan introduced a fuzzy automatic transmission and a fuzzy antiskid braking system in one of its recent luxury cars. Even Tokyo's stock market has had at least one stock-trading portfolio based on fuzzy logic that outperformed the Nikkei Exchange average. In Japan there are fuzzy golf diagnostic systems, fuzzy toasters, fuzzy rice cookers, fuzzy vacuum cleaners, and many other industrial fuzzy control processes. In fact, the number of fuzzy consumer products and fuzzy applications involving new patents is increasing so rapidly that, in order to stay competitive, many U.S. companies, over the last five years, have been launching their own internal fuzzy projects.

Fuzzy set theory provides a means for representing uncertainties. Historically, probability theory has been the primary tool for representing uncertainty in mathematical models. Because of all this, all uncertainty was assumed to follow the characteristics of random uncertainty. A random process is one where the outcomes of any particular realization of the process are strictly a matter of chance, wherein a prediction of a sequence of events is not possible. What is possible for random processes is a precise description of the *statistics* of the long-run averages of the process. However, not all uncertainty is random. There are, of course, forms of uncertainty that are nonrandom and hence not suited to treatment or modeling by probability theory. In fact, it could be argued that the overwhelming amount of uncertainty associated with complex systems and issues, which humans address on a daily basis, is nonrandom in nature. Fuzzy set theory is a marvelous tool for modeling the kind of uncertainty associated with vagueness, with imprecision, and/or with a lack of information regarding a particular element of the problem at hand.

### 6.1.1 Fuzzy Sets and Membership

Making decisions about processes that contain nonrandom uncertainty, such as the uncertainty in natural language, has been shown to be less than perfect. The idea proposed by Lotfi Zadeh suggested that *set membership* is the key to decision-making when faced with uncertainty. As Dr. Zadeh stated in his seminal paper of 1965,[2]

> The notion of a fuzzy set provides a convenient point of departure for the construction of a conceptual framework which parallels in many respects the framework used in the case of ordinary sets, but is more general than the latter and, potentially, may prove to have a much wider scope of applicability, particularly in the fields of pattern classification and information processing. Essentially, such a framework provides a natural way of dealing with problems in which the source of imprecision is the absence of sharply defined criteria of class membership rather than the presence of random variables.

Suppose we are interested in the height of people. We can easily assess whether someone is over 6 feet tall. In a binary sense, they either are or are not, based on the accuracy, or imprecision, of our measuring device. For example, if "tall" is a set defined as heights equal to or greater than 6 feet, a computer would not recognize an individual of height 5 feet, 11.999 inches as being a member of the set "tall." But how do we assess the uncertainty in the following question: Is the person *nearly* 6 feet tall? The uncertainty in this case is due to the vagueness or ambiguity of the adverb *nearly*. A 5-foot 11-inch person could clearly be a member of the set of people "nearly 6 feet" tall. In the first situation the uncertainty of whether a person, whose height is unknown, is 6 feet tall or not is binary; he either is or is not, and we can produce a probability assessment of that prospect based on height data from many people. But the uncertainty of whether a person is nearly 6 feet tall is nonrandom. The degree to which the person approaches a height of 6 feet is fuzzy. In reality, "tallness" is a matter of degree and is relative. Among peoples of the TuTsi in Rwanda and Burundi in Africa a height for a male of 6 feet is considered short. So, 6 feet can be tall in one context and short in another. In the real (fuzzy) world, the set of tall people can overlap with the set of not-tall people—an impossibility in the world of binary logic.

This notion of set membership, then, is central to the representation of objects within a universe to sets defined on the universe. Classical sets contain objects that satisfy precise properties of membership; fuzzy sets contain objects that satisfy imprecise properties of membership, i.e., membership of an object in a fuzzy set can be approximate. For example, the set of heights *from 5 to 7 feet* is crisp; the set of heights in the region *around 6 feet* is fuzzy. To elaborate, suppose we have an exhaustive collection of individual elements (singletons), $x$, that make up a universe of information (discourse), $X$. Further, various combinations of these individual elements make up sets, say $A$, on the universe. For crisp sets an element $x$ in the universe $X$ is either a member of some crisp set $A$ or it is not. This binary issue of membership can be represented mathematically with the indicator function,

$$\chi_A(x) = \begin{cases} 1, & x \in A \\ 0, & x \notin A \end{cases} \tag{6.1}$$

where the symbol $\chi_A(x)$ gives the indication of an unambiguous membership of element $x$ in set $A$, and the symbols $\in$ and $\notin$ denote "contained in," and "not contained in," respectively. For our example of the universe of heights of people, suppose set $A$ is the crisp set of all people with $5.0 \le A \le 7.0$ feet, shown in Fig. 6.1a. A particular individual, $x_1$, has a height of 6.0 feet. The membership of this individual in crisp set $A$ is equal to 1, or full membership given symbolically as $\chi_A(x_1) = 1$. Another individual, say $x_2$, has a height of 4.99 feet. The membership of this individual in set $A$ is equal to 0, or no membership, so $\chi_A(x_2) = 0$, also seen in Fig. 6.1a. In these cases the membership in a set was binary: either an element is a member of a set or it isn't.

Zadeh extended the notion of binary membership to accommodate various "degrees of membership" on the real continuous interval [0, 1], where the endpoints of 0 and 1 conform to no membership and full membership, respectively, just as the indicator function does for crisp sets,

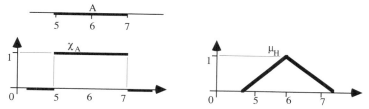

**FIGURE 6.1**    Height membership functions for (*a*) a crisp set *A*, and (*b*) a fuzzy set *H*.

but where the infinite number of values in between the endpoints can represent various degrees of membership for an element $x$ in some set on the universe. The sets on the universe $X$ that can accommodate "degrees of membership" were termed by Zadeh as "fuzzy sets." Continuing further with the example on heights, consider a set $H$ consisting of heights *near 6 feet*. Since the property *near 6-feet* is fuzzy, there is not a unique membership function for $H$. Rather, the analyst must decide what the membership function, denoted $\mu_H$, should look like. Plausible properties of this function might be normality [$\mu_H(6) = 1$], monotonicity (the closer $H$ is to 6, the closer $\mu_H$ is to 1), and symmetry (numbers equidistant from 6 should have the same value of $\mu_H$).[4] Such a membership function is illustrated in Fig. 6.1*b*. A key difference between crisp and fuzzy sets is their membership function; a crisp set has a unique membership function, whereas a fuzzy set can have an infinite number of membership functions to represent it. For fuzzy sets the uniqueness is sacrificed, but flexibility is gained because the membership function can be adjusted to maximize the utility for a particular application.

James Bezdek provided one of the most lucid comparisons between crisp and fuzzy sets.[4] Crisp sets of real objects are equivalent to, and isomorphically described by, a unique membership function, such as $\chi_A$ in Fig. 6.1*a*. However, there is no set-theoretic equivalent of "real objects" corresponding to $\chi_A$. Fuzzy sets are *always* functions, which map a universe of objects, (say, $X$) onto the unit interval [0, 1]; that is, the fuzzy set $H$ is the function $\mu_H$ that carries $X$ into [0, 1]. Hence, *every* function that maps $X$ onto [0, 1] is a fuzzy set. Although this is true in a formal mathematical sense, many functions that qualify on the basis of this definition cannot be suitable fuzzy sets. But they *become* fuzzy sets when, and only when, they match some intuitively plausible semantic description of imprecise properties of the objects in $X$.

The membership function embodies the mathematical representation of membership in a set, and the notation used throughout this text for a fuzzy set is a set symbol with a "tilde" underscore, say $\underset{\sim}{A}$, and where the functional mapping is given by the following,

$$\mu_{\underset{\sim}{A}}(x) \in [0, 1] \tag{6.2}$$

where the symbol $\mu_{\underset{\sim}{A}}(x)$ gives the degree of membership of element $x$ in fuzzy set $\underset{\sim}{A}$. Therefore, $\mu_{\underset{\sim}{A}}(x)$ is a value on the unit interval that measures the degree to which element $x$ belongs to fuzzy set $\underset{\sim}{A}$; equivalently, $\mu_{\underset{\sim}{A}}(x) = \text{Degree}(x \in \underset{\sim}{A})$.

To summarize, there is a clear distinction between fuzziness and randomness. Fuzziness describes the ambiguity of an event, whereas randomness describes the uncertainty in the occurrence of the event. The event will occur or not occur; but is the description of the event unambiguous enough to measure its occurrence or nonoccurrence?

***Fuzzy Sets.*** In classical, or crisp, sets, the transition for an element in the universe between membership and nonmembership in a given set is abrupt and well-defined (i.e., "crisp"). For an element in a universe that contains fuzzy sets, this transition can be gradual. This transition between various degrees of membership can be thought of as conforming to the fact that the boundaries of the fuzzy sets are vague and ambiguous. Hence, membership in this set of an element from the universe is measured by a function that attempts to describe vagueness and ambiguity.

A fuzzy set, then, is a set containing elements that have various degrees of membership in the set. This idea is contrasted with classical, or crisp, sets because members of a crisp set would not be members unless their membership was full or complete in that set (i.e., their membership is assigned a value of 1). Elements in a fuzzy set, because their membership can be a value other than complete, can also be members of other fuzzy sets on the same universe.

Elements of a fuzzy set are mapped to a universe of "membership values" using a function-theoretic form. As shown in Eq. (6.2), fuzzy sets are denoted in this text by a set symbol with a tilde understrike; so, for example, $\underset{\sim}{A}$ would be the "fuzzy set" $A$. This function maps elements of a fuzzy set $\underset{\sim}{A}$ to a real-numbered value on the interval 0 to 1. If an element in the universe (say, $x$) is a member of fuzzy set $\underset{\sim}{A}$, then this mapping is given by Eq. (6.2), or $\mu_{\underset{\sim}{A}}(x) \in [0, 1]$.

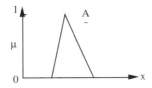

**FIGURE 6.2**   Membership function for fuzzy set $\underset{\sim}{A}$.

This mapping is shown in Fig. 6.2 for a fuzzy set.

A notation convention for fuzzy sets when the universe of discourse, $X$, is discrete and finite, is given as follows for a fuzzy set $\underset{\sim}{A}$:

$$\underset{\sim}{A} = \frac{\mu_{\underset{\sim}{A}}(x_1)}{x_1} + \frac{\mu_{\underset{\sim}{A}}(x_2)}{x_2} + \cdots = \sum_i \frac{\mu_{\underset{\sim}{A}}(x_i)}{x_i} \quad (6.3)$$

In the notation the horizontal bar is not a quotient, but rather a delimiter. In addition, the numerator in each individual expression is the membership value in set $\underset{\sim}{A}$ associated with the element of the universe indicated in the denominator of each expression. The summation symbol is not for algebraic summation, but rather denotes the collection or aggregation of each element; hence the "+" signs are not the algebraic "add," but rather are a function-theoretic union.

*Fuzzy Set Operations.*   Define three fuzzy sets $\underset{\sim}{A}$, $\underset{\sim}{B}$, $\underset{\sim}{C}$ on the universe $X$. For a given element $x$ of the universe, the following function-theoretic operations for the set-theoretic operations of union, intersection, and complement are defined for $\underset{\sim}{A}$, $\underset{\sim}{B}$, $\underset{\sim}{C}$ on $X$.

Union:
$$\mu_{\underset{\sim}{A}\cup\underset{\sim}{B}}(x) = \mu_{\underset{\sim}{A}}(x) \vee \mu_{\underset{\sim}{B}}(x) \quad (6.4)$$

Intersection:
$$\mu_{\underset{\sim}{A}\cap\underset{\sim}{B}}(x) = \mu_{\underset{\sim}{A}}(x) \wedge \mu_{\underset{\sim}{B}}(x) \quad (6.5)$$

Complement:
$$\mu_{\overline{\underset{\sim}{A}}}(x) = 1 - \mu_{\underset{\sim}{A}}(x) \quad (6.6)$$

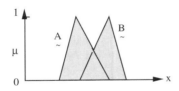

**FIGURE 6.3**   Union of fuzzy sets $\underset{\sim}{A}$ and $\underset{\sim}{B}$.

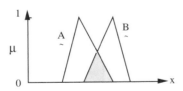

**FIGURE 6.4**   Intersection of fuzzy sets $\underset{\sim}{A}$ and $\underset{\sim}{B}$.

**FIGURE 6.5**   Complement of fuzzy set $\underset{\sim}{A}$.

Venn diagrams for these operations, extended to consider fuzzy sets, are shown in Figs. 6.3 through 6.5.

Any fuzzy set $\underset{\sim}{A}$ defined on a universe $X$ is a subset of that universe. Also, by definition, just as with classical sets, the membership value of any element $x$ in the null set $\emptyset$ is 0 and the membership value of any element $x$ in the whole set $X$ is 1. Note that the null set and the whole set are not fuzzy sets in this context (no tilde understrike). The appropriate notation for these ideas is as follows:

$$\underset{\sim}{A} \subseteq X \rightarrow \mu_{\underset{\sim}{A}}(x) \leq \mu_x(x) \quad (6.7a)$$

$$\mu_\emptyset(x) = 0 \quad \text{for all } x \in X, \quad (6.7b)$$

$$\mu_x(x) = 1 \quad \text{for all } x \in X, \quad (6.7c)$$

The collection of all fuzzy sets and fuzzy subsets on $X$ is denoted as the fuzzy power set $P(X)$. It should be obvious, based on the fact that all fuzzy sets can overlap, that the cardinality of the fuzzy power set is infinite; that is, $P(x) \rightarrow n_{p(x)} = \infty$.

*DeMorgan's laws* for classical sets also hold for fuzzy sets, as denoted by the following expressions:

$$\left(\underset{\sim}{A} \cap \underset{\sim}{B}\right) = \overline{\underset{\sim}{A}} \cup \overline{\underset{\sim}{B}} \tag{6.8a}$$

$$\left(\overline{\underset{\sim}{A} \cup \underset{\sim}{B}}\right) = \overline{\underset{\sim}{A}} \cap \overline{\underset{\sim}{B}} \tag{6.8b}$$

As enumerated before, all other operations on classical sets also hold for fuzzy sets, except for the excluded middle laws. These two laws do not hold for fuzzy sets because of the fact that since fuzzy sets can overlap, a set and its complement can also overlap. The *excluded middle laws*, extended for fuzzy sets, are expressed by the following:

$$\underset{\sim}{A} \cup \overline{\underset{\sim}{A}} \neq X \tag{6.9a}$$

$$\underset{\sim}{A} \cap \overline{\underset{\sim}{A}} \neq \varnothing \tag{6.9b}$$

Extended Venn diagrams comparing the excluded middle laws for classical (crisp) sets and fuzzy set are shown in Figs. 6.6 and 6.7, respectively.

## 6.2 FUZZY RELATIONS

Fuzzy relations are developed by allowing the relationship between elements of two or more sets to take on an infinite number of degrees of relationship between the extremes of "completely related" and "not related," which are the only degrees of relationship possible in crisp relations. In this sense, fuzzy relations are to crisp relations as fuzzy sets are to crisp sets; crisp sets and relations are more constrained realizations of fuzzy sets and relations.

Fuzzy relations map elements of one universe (say, $X$) to those of another universe (say, $Y$) through the Cartesian product of the two universes. However, the "strength" of the relation between ordered pairs of the two universes is not measured with the characteristic function (as in the case of crisp relations), but rather with a membership function expressing various "degrees" of strength of the relation on the unit interval $[0, 1]$. Hence a fuzzy relation $\underset{\sim}{R}$ is a mapping from the Cartesian space $X \times Y$ to the interval $[0, 1]$, where the strength of the mapping is expressed by the membership function of the relation for ordered pairs from the two universes, or $u_{\underset{\sim}{R}}(x, y)$.

### 6.2.1 Operations on Fuzzy Relations

Let $\underset{\sim}{R}$ and $\underset{\sim}{S}$ be fuzzy relations on the Cartesian space $X \times Y$. Then the following operations apply for the membership values for various set operations:

Union: $\qquad\qquad \mu_{\underset{\sim}{R} \cup \underset{\sim}{S}}(x, y) = \max[\mu_{\underset{\sim}{R}}(x, y), \mu_{\underset{\sim}{S}}(x, y)] \tag{6.10}$

Intersection: $\qquad\quad \mu_{\underset{\sim}{R} \cap \underset{\sim}{S}}(x, y) = \min[\mu_{\underset{\sim}{R}}(x, y), \mu_{\underset{\sim}{S}}(x, y)] \tag{6.11}$

Complement: $\qquad\quad\; \mu_{\overline{\underset{\sim}{R}}}(x, y) = 1 - \mu_{\underset{\sim}{R}}(x, y) \tag{6.12}$

Containment: $\qquad\quad \underset{\sim}{R} \subset \underset{\sim}{S} \Rightarrow \mu_{\underset{\sim}{R}}(x, y) \leq \mu_{\underset{\sim}{S}}(x, y) \tag{6.13}$

### 6.2.2 Fuzzy Cartesian Product and Composition

Because fuzzy relations in general are fuzzy sets, we can define the Cartesian product to be a relation between two or more fuzzy sets. Let $\underset{\sim}{A}$ be a fuzzy set on universe $X$ and $\underset{\sim}{B}$ be a fuzzy set on

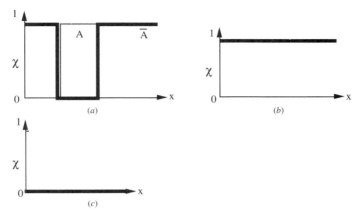

**FIGURE 6.6** Excluded middle laws for crisp sets: (*a*) crisp set $A$ and its complement; (*b*) crisp $A \cup \overline{A}$ (law of excluded middle); and (*c*) crisp $A \cap \overline{A} = \emptyset$ (law of contradiction).

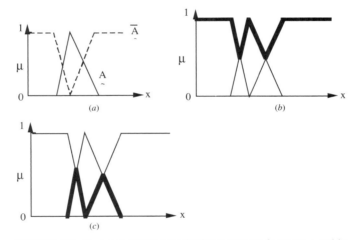

**FIGURE 6.7** Excluded middle laws for fuzzy sets: (*a*) fuzzy set $\underset{\sim}{A}$ and its complement; (*b*) fuzzy $\underset{\sim}{A} \cup \overline{\underset{\sim}{A}} \neq X$ (law of excluded middle); and (*c*) fuzzy $\underset{\sim}{A} \cap \overline{\underset{\sim}{A}} = \emptyset$ (law of contradiction).

universe $Y$; then the Cartesian product between fuzzy sets $\underset{\sim}{A}$ and $\underset{\sim}{B}$ will result in a fuzzy relation $\underset{\sim}{R}$, which is contained within the full Cartesian product space, or

$$\underset{\sim}{A} \times \underset{\sim}{B} = \underset{\sim}{R} \subset X \times Y \tag{6.14}$$

where the fuzzy relation $\underset{\sim}{R}$ has membership function,

$$\mu_{\underset{\sim}{R}}(x, y) = \mu_{\underset{\sim}{A} \times \underset{\sim}{B}}(x, y) = \min[\mu_{\underset{\sim}{A}}(x), \mu_{\underset{\sim}{B}}(y)] \tag{6.15}$$

The Cartesian product defined by $\underset{\sim}{A} \times \underset{\sim}{B} = \underset{\sim}{R}$, Eq. (6.14), is implemented in the same fashion as the cross-product of two vectors. Again, the Cartesian product is not the same operation as the arithmetic product. In the case of two-dimensional relations ($r = 2$), the former employs the idea of pairing of elements among sets, whereas the latter uses actual arithmetic products between elements of sets. More can be found on fuzzy arithmetic in Ross.[3] Each of the fuzzy sets could be thought of as a vector of membership values; each value is associated with a particular element in each set. For example, for a fuzzy set (vector) $\underset{\sim}{A}$ that has four elements (hence column vector of size $4 \times 1$) and for a fuzzy set (vector) $\underset{\sim}{B}$ that has five elements (hence a row vector size of $1 \times 5$), the resulting fuzzy relation $\underset{\sim}{R}$ will be represented by a matrix of size $4 \times 5$; i.e., $\underset{\sim}{R}$ will have four rows and five columns.

Fuzzy composition can be defined just as it is for crisp (binary) relations. Suppose $\underset{\sim}{R}$ is a fuzzy relation on the Cartesian space $X \times Y$; $\underset{\sim}{S}$ is a fuzzy relation on $Y \times Z$; and $\underset{\sim}{T}$ is a fuzzy relation on $X \times Z$. Then fuzzy max-min composition is defined in terms of the set-theoretic notation and membership function-theoretic notation as follows,

$$\text{Let } \underset{\sim}{T} = \underset{\sim}{R} \cdot \underset{\sim}{S}$$

$$\mu_{\underset{\sim}{T}}(x, z) = \bigvee_{y \in Y} \left( \mu_{\underset{\sim}{R}}(x, y) \wedge \mu_{\underset{\sim}{S}}(y, z) \right)$$

(6.16a)

and fuzzy max-product composition is defined in terms of the membership function-theoretic notation as follows:

$$\mu_{\underset{\sim}{T}}(x, z) = \bigvee_{y \in Y} \left( \mu_{\underset{\sim}{R}}(x, y) \cdot \mu_{\underset{\sim}{S}}(y, z) \right)$$

(6.16b)

It should be pointed out that neither crisp nor fuzzy compositions have inverses in general; that is,

$$\underset{\sim}{R} \cdot \underset{\sim}{S} \neq \underset{\sim}{S} \cdot \underset{\sim}{R}$$

(6.17)

This result, Eq. (6.17), is general for any matrix operation—fuzzy or otherwise—that must satisfy consistency between the cardinal counts of elements in respective universes. Even for the case of square matrices, the composition inverse [represented by Eq. (6.17)] is not guaranteed.

## 6.3   FUZZY AND CLASSICAL LOGIC

### 6.3.1   Classical Logic

In classical predicate logic a simple proposition, $P$, is a linguistic, or declarative, statement contained within a universe of elements (say, $X$) that can be identified as being a collection of elements in $X$ that are strictly true or strictly false. Hence, a proposition $P$ is a collection of elements (that is, a set) where the truth values for all elements in the set are either all true or all false. The veracity (truth) of an element in the proposition $P$ can be assigned a binary truth value, called $T(P)$, just as an element in a universe is assigned a binary quantity to measure its membership in a particular set. For binary (Boolean) predicate logic, $T(P)$ is assigned a value of 1 (truth) or 0 (false). If $U$ is the universe of all propositions, then $T$ is a mapping of the elements, $u$, in these propositions (sets) to the binary quantities (0, 1):

$$T: u \in U \rightarrow \{0, 1\}$$

(6.18)

All elements $u$ in the universe $U$ that are true for proposition $P$ constitute the *truth set* of $P$. Those elements $u$ in the universe $U$ that are false for proposition $P$ constitute the *falsity set* of $P$.

In logic we need to postulate the boundary conditions of truth values just as we do for sets; that is, in function-theoretic terms we need to define the truth value of a universe of discourse. For a universe $Y$ and the null set $\varnothing$, we define the following truth values: $T(Y) = 1$ and $T(\varnothing) = 0$.

Now let $P$ and $Q$ be two simple propositions on the same universe of discourse that can be combined using the following five logical connectives,

**1.** Disjunction ($\vee$)
**2.** Conjunction ($\wedge$)
**3.** Negation ($-$)
**4.** Implication ($\rightarrow$)
**5.** Equivalence ($\leftrightarrow$ or $\Leftrightarrow$)

to form logical expressions involving the two simple propositions. These connectives can be used to form new propositions from simple propositions.

The disjunction connective, the logical "or," is the term used to represent what is commonly referred to as the "inclusive or." The natural language term "or" and the logical "or" are different. The natural language "or" differs from the logical "or" in that the former implies exclusion (denoted in the literature as the *exclusive-or;* see this chapter for more details). For example, "soup or salad" on a restaurant menu implies choosing one or the other option, but not both. The inclusive or is the most often employed in logic; the inclusive or ("logical or" as used here) implies that a compound proposition is true if either of the simple propositions is true or they are both true.

The equivalence connective arises from dual implication, that is for some propositions $P$ and $Q$, if $P \rightarrow Q$ and $Q \rightarrow P$, then $P \leftrightarrow Q$.

Now define sets $A$ and $B$ from universe $X$ (universe $X$ is isomorphic with universe $U$), where these sets might represent linguistic ideas or thoughts. A *propositional calculus* (sometimes called the *algebra of propositions*) will exist for the case where proposition $P$ measures the truth of the statement that an element $x$ from the universe $X$ is contained in set $A$ and the truth of the statement that this element $x$ is contained in set $B$, or more conventionally,

$P$: truth that $x \in A$

$Q$: truth that $x \in B$

where truth is measured in terms of the truth value; i.e., if $x \in A$, $T(P) = 1$; otherwise $T(P) = 0$. If $x \in B$, $T(Q) = 1$; otherwise $T(Q) = 0$, or using the characteristic function to represent truth (1) and falsity (0), the following notation results:

$$\chi_A(x) = \begin{cases} 1, & x \in A \\ 0, & x \notin A \end{cases} \tag{6.19}$$

A notion of *mutual exclusivity* arises in this calculus. For the situation involving two propositions $P$ and $Q$, where, $T(P) \cap T(Q) = \varnothing$, we have that the truth of $P$ always implies the falsity of $Q$ and vice versa; hence $P$ and $Q$ are mutually exclusive propositions.

The five logical connectives defined above can be used to create compound propositions, where a compound proposition is defined as a logical proposition formed by logically connecting two or more simple propositions. Just as we are interested in the truth of a simple proposition, predicate logic also involves the assessment of the truth of compound propositions. For the case of two simple propositions, the resulting compound propositions are defined below in terms of their binary truth values.

Given a proposition $P$: $x \in A$, $\overline{P}$: $x \notin A$, we have the following for the logical connectives:

Disjunction                    $P \vee Q \Rightarrow x \in A$ or $B$

Hence $T(P \vee Q) = \max[T(P), T(Q)]$ (6.20a)

Conjunction                    $P \wedge Q \Rightarrow x \in A$ and $B$

Hence $T(P \wedge Q) = \min[T(P), T(Q)]$ (6.20b)

Negation          If $T(P) = 1$, then $T(\overline{P}) = 0$; if $T(P) = 0$, then $T(\overline{P}) = 1$          (6.20c)

Implication       $P \rightarrow Q \Rightarrow x \in A, B$

                 Hence $T(P \rightarrow Q) = T(\overline{P} \cup Q)$                      (6.20d)

Equivalence      $P \leftrightarrow Q \Rightarrow x \in A, B$

$$\text{Hence } T(P \leftrightarrow Q) = \begin{cases} 1, & \text{for } T(P) = T(Q) \\ 0, & \text{for } T(P) \neq T(Q) \end{cases} \tag{6.20e}$$

The logical connective "implication," i.e., $P \rightarrow Q$ ($P$ implies $Q$) presented here is also known as the *classical implication*, to distinguish it from an alternative form offered in the 1930s by Lukasiewicz, a Polish mathematician who was first credited with exploring logics other than Aristotelian (classical or binary) logic,[3] and from several other forms (see the end of this chapter). In this implication the proposition $P$ is also referred to as the *hypothesis* or the *antecedent*, and the proposition $Q$ is also referred to as the *conclusion* or the *consequent*. The compound proposition $P \rightarrow Q$ is true in all cases except where a true antecedent $P$ appears with a false consequent.

Hence, the classical form of the implication is true for all propositions of $P$ and $Q$ except for those propositions that are in both the truth set of $P$ and the false set of $Q$, i.e.,

$$T(P \rightarrow Q) = \overline{\left( T(P) \cap T(\overline{Q}) \right)} \tag{6.21a}$$

This classical form of the implication operation requires some explanation. For proposition $P$ defined on set $A$ and a proposition $Q$ defined on set $B$, the implication "$P$ implies $Q$" is equivalent to taking the union of elements in the complement of set $A$ with the elements in the set $B$ [this result can also be derived by using DeMorgan's laws on Eq. (6.21)]. That is, the logical implication is analogous to the set-theoretic form,

$$P \rightarrow Q \equiv \overline{A} \cup B \text{ is true} \equiv \text{either "not in } A\text{" or "in B"}$$

$$\text{So that } T(P \rightarrow Q) = T(\overline{P} \vee Q) = \max[T(\overline{P}), T(Q)] \tag{6.21b}$$

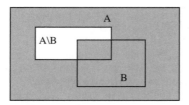

**FIGURE 6.8** Graphical analog of the classical implication operation. The shaded area is the set: $A\backslash B = A \cup B = \overline{(A \cap B)}$. If $x$ is in $A$ and $x$ is not in $B$ then $A \rightarrow B$ fails $\equiv A\backslash B$ (difference).

This is linguistically equivalent to the statement, "$P \rightarrow Q$ is true" when either "not $A$" or "$B$" is true (logical or). Graphically, this implication and the analogous set operation are represented by the Venn diagram in Fig. 6.8. As noted in the diagram, the region represented by the difference $A\backslash B$ is the set region where the implication $P \rightarrow Q$ is false (the implication "fails"). The shaded region in Fig. 6.8 represents the collection of elements in the universe where the implication is true.

Now, with two propositions ($P$ and $Q$) each being able to take on one of two truth values (true or false, 1 or 0), there will be a total of $2^2 = 4$ propositional situations. These situations are illustrated, along with the appropriate truth values, for the propositions $P$ and $Q$ and the various logical connectives between them in the truth table (Table 6.1). The values in the last five columns of Table 6.1 are calculated using the expressions in Eqs. (6.20) and (6.21). In Table 6.1 "$T$" or "1" denotes true and "$F$" or "0" denotes false.

**TABLE 6.1**    Truth Table for Various Compound Propositions

| $P$ | $Q$ | $\bar{P}$ | $P \vee Q$ | $P \wedge Q$ | $P \rightarrow Q$ | $P \leftrightarrow Q$ |
|-----|-----|-----------|------------|--------------|-------------------|-----------------------|
| T(1) | T(1) | F(0) | T(1) | T(1) | T(1) | T(1) |
| T(1) | F(0) | F(0) | T(1) | F(0) | F(0) | F(0) |
| F(0) | T(1) | T(1) | T(1) | F(0) | T(1) | F(0) |
| F(0) | F(0) | T(1) | F(0) | F(0) | T(1) | T(1) |

Suppose the implication operation involves two different universes of discourse; $P$ is a proposition described by set $A$, which is defined on universe $X$, and $Q$ is a proposition described by set $B$, which is defined on universe $Y$. Then the implication $P \rightarrow Q$ can be represented in set-theoretic terms by the relation $R$, where $R$ is defined by

$$R = (A \times B) \cup (\bar{A} \times Y) \equiv \text{IF } A, \text{ THEN } B$$

$$\text{If } x \in A \qquad \text{where } x \in X \text{ and } A \subset X \tag{6.22}$$

$$\text{Then } y \in B \quad \text{where } y \in Y \text{ and } B \subset Y$$

This implication, Eq. (6.22), is also equivalent to the linguistic rule form: IF $A$, THEN $B$. The graphic shown below in Fig. 6.9 represents the space of the Cartesian product $X \times Y$, showing typical sets $A$ and $B$, and superposed on this space is the set-theoretic equivalent of the implication. That is,

$$P \rightarrow Q: \text{If } x \in A, \text{ then } y \in B, \quad \text{or} \quad P \rightarrow Q \equiv \bar{A} \cup B$$

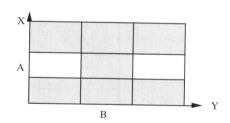

**FIGURE 6.9**    The Cartesian space showing the implication IF $A$, THEN $B$.

The shaded regions of the compound Venn diagram in Fig. 6.9 represent the truth domain of the implication, IF $A$, THEN $B(P \rightarrow Q)$. Another compound proposition in linguistic rule form is the expression,

IF $A$, THEN $B$, ELSE $C$

Linguistically, this compound proposition could be expressed as follows:

IF $A$, THEN $B$, or

IF $A$, THEN $C$

In predicate logic this rule has the form

$$(P \rightarrow Q) \vee (\bar{P} \rightarrow S)$$

where

$$P: x \in A, \quad A \subset X$$
$$Q: y \in B, \quad B \subset Y \tag{6.23}$$
$$S: y \in C, \quad C \subset Y$$

The set-theoretic equivalent of this compound proposition is given by the following:

IF $A$ THEN $B$ ELSE $C \equiv (A \times B) \cup (\bar{A} \times C) = R = \text{Relation on } X \times Y$    (6.24)

The graphic shown in Fig. 6.10 illustrates the shaded region representing the truth domain for this compound proposition for the particular case where $B \cap C = \varnothing$.

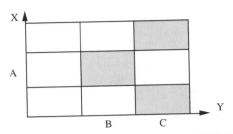

**FIGURE 6.10**   Truth domain for IF *A*, THEN *B*, ELSE *C*.

***Deductive Inferences.***   The *modus ponens* deduction is used as a tool for inferencing in rule-based systems. A typical IF-THEN rule is used to determine whether an antecedent (cause or action) implies a consequent (effect or reaction). Suppose we have a rule of the form, IF *A*, THEN *B*, where *A* is a set defined on universe *X* and *B* is a set defined on universe *Y*. As discussed before, this rule can be translated into a relation between sets *A* and *B*, that is, recalling Eq. (6.19), $R = (A \times B) \cup (\overline{A} \times Y)$. Now suppose a new antecedent (say, *A′*) is known. Can we use *modus ponens* deduction to infer a new consequent[3] (say, *B′*), resulting from the new antecedent, that is, in rule form IF *A′*, THEN *B′*? The answer, of course, is yes, through the use of the composition operation. Because "*A* implies *B*" is defined on the Cartesian space $X \times Y$, *B′* can be found through the following set-theoretic formulation, again from Eq. (7.22),

$$B' = A' \cdot R = A' \cdot [(A \times B) \cup (\overline{A} \times Y)] \tag{6.25}$$

where the symbol · denotes the composition operation. *Modus ponens* deduction can also be used for the compound rule, IF *A*, THEN *B*, ELSE *C*, where this compound rule is equivalent to the relation defined by the following:

$$R = (A \times B) \cup (\overline{A} \times C) \tag{6.26}$$

For this compound rule, if we define another antecedent *A′*, the following possibilities exist, depending on: (1) whether *A′* is fully contained in the original antecedent *A*, (2) whether *A′* is contained only in the complement of *A*, or (3) whether *A′* and *A* overlap to some extent, described as follows:

$$\text{If } A' \subset A, \text{ then } y = B$$
$$\text{If } A' \subset \overline{A}, \text{ then } y = C$$
$$\text{If } A' \cap A \neq \varnothing, A' \cap \overline{A} \neq \varnothing, \text{ then } y = B \cup C$$

The rule, IF *A*, THEN *B* (proposition *P* is defined on set *A* in universe *X*, and proposition *Q* is defined on set *B* in universe *Y*)—i.e., $P \rightarrow Q \Rightarrow R = (A \times B) \cup (\overline{A} \times Y)$—is then defined in function-theoretic terms as follows,

$$\chi_R(x, y) = \max\{[\chi_A(x) \wedge \chi_B(y)], [(1 - \chi_A(x)) \wedge 1]\} \tag{6.27}$$

where $\chi(\cdot)$ is the characteristic function as defined before.

The compound rule IF *A*, THEN *B*, ELSE *C* can also be defined in terms of a matrix relation as $R = (A \times B) \cup (A \times C) \Rightarrow (P \rightarrow Q) \vee (\overline{P} \rightarrow S)$, as given by Eqs. (6.23) and (6.24), where the membership function is determined as follows:

$$\chi_R(x, y) = \max\{[\chi_A(x) \wedge \chi_B(y)], [(1 - \chi_A(x)) \wedge \chi_C(y)]\} \tag{6.28}$$

### 6.3.2   Fuzzy Logic

The restriction of classical logic to a two-valued logic has created many interesting paradoxes over the ages. For example, does the "liar from Crete" lie when he claims, "All Cretians are liars"? If he is telling the truth, his statement is false. But if his statement is false, he is not telling the truth. The statement can't be both true and false.

The only way for the "liar from Crete" paradox to be solved is if the statement is both true and false, simultaneously. This can be shown using set notation.[5] Let $S$ be the proposition that the barber shave himself and $\overline{S}$ (not-$S$) that he does not. Then since $S \rightarrow \overline{S}$ ($S$ implies not-$S$), and $\overline{S} \rightarrow S$, the two propositions are logically equivalent: $S \leftrightarrow \overline{S}$. Equivalent propositions have the same truth value; hence,

$$T(S) = T(\overline{S}) = 1 - T(S)$$

which yields the expression

$$T(S) = 1/2$$

As seen, paradoxes reduce to half-truths (or half-falsities) mathematically. In classical binary (bivalued) logic, however, such conditions are not allowed; i.e., only $T(S) = 1$ or $0$ is valid.

There is a more subtle form of paradox that can also be addressed by a multivalued logic. Consider the classical *sorities* (a heap of syllogisms) paradoxes, for example, the case of a liter-full glass of water. Often this example is called the "optimist's conclusion" (is the glass half-full or half-empty when the volume is at 500 milliliters?). Is the liter-full glass still full if we remove one milliliter of water? Is the glass still full if we remove two milliliters of water, three, four, or 100 milliliters? If we continue to answer no, then eventually we will have removed all the water, and an empty glass will still be characterized as full! At what point did the liter-full glass of water become empty? Perhaps at 500 milliliters full? Unfortunately, no single milliliter of liquid provides for a transition between full and empty. This transition is gradual, so that as each milliliter of water is removed, the truth value of the glass being full gradually diminishes from a value of 1 at 1000 milliliters to 0 at 0 milliliters. Hence, many problems require a multivalued logic other than the classic binary logic that is so prevalent today.

A fuzzy logic proposition, $\underset{\sim}{P}$, is a statement involving some concept without clearly defined boundaries. Linguistic statements that tend to express subjective ideas and that can be interpreted slightly differently by various individuals typically involve fuzzy propositions. Most natural language is fuzzy, in that it involves vague and imprecise terms. Statements describing a person's height or weight, or assessments of people's preferences about colors or menus can be used as examples of fuzzy propositions. The truth value assigned to $P$ can be any value on the interval $[0, 1]$. The assignment of the truth value to a proposition is actually a mapping from the interval $[0, 1]$ to the universe $U$ of truth values $T$:

$$T: u \in U \rightarrow \{0, 1\} \tag{6.29}$$

As in classical binary logic, we assign a logical proposition to a set in the universe of discourse. Fuzzy propositions are assigned to fuzzy sets. Suppose proposition $P$ is assigned to fuzzy set $\underset{\sim}{A}$. Then the truth value of a proposition, denoted $T(\underset{\sim}{P})$, is given by

$$T(\underset{\sim}{P}) = \mu_{\underset{\sim}{A}}(x), \quad 0 \le \mu_{\underset{\sim}{A}} \le 1 \tag{6.30}$$

Equation (6.30) indicates that the degree of truth for the proposition $\underset{\sim}{P}: x \in \underset{\sim}{A}$ is equal to the membership grade of $x$ in the fuzzy set $\underset{\sim}{A}$.

The logical connectives of negation, disjunction, conjunction, and implication are also defined for fuzzy logic. These connectives are given in Eqs. (6.31a) through (6.31d) for two simple propositions: proposition $\underset{\sim}{P}$, defined on fuzzy set $\underset{\sim}{A}$, and proposition $\underset{\sim}{Q}$, defined on fuzzy set $\underset{\sim}{B}$.

Negation

$$T(\overline{\underset{\sim}{P}}) = 1 - T(\underset{\sim}{P}) \tag{6.31a}$$

Disjunction

$$\underset{\sim}{P} \vee \underset{\sim}{Q} \Rightarrow x \text{ is } \underset{\sim}{A} \text{ or } \underset{\sim}{B}$$

$$T(\underset{\sim}{P} \vee \underset{\sim}{Q}) = \max[T(\underset{\sim}{P}), T(\underset{\sim}{Q})] \tag{6.31b}$$

Conjunction

$$P \wedge Q \Rightarrow x \text{ is } A \text{ and } B$$

$$T(P \wedge Q) = \min[T(P), T(Q)] \tag{6.31c}$$

Implication

$$P \rightarrow Q \Rightarrow x \text{ is } A, \text{ then } x \text{ is } B$$

$$T(P \rightarrow Q) = T(\overline{P} \vee Q) = \max[T(\overline{P}), T(Q)] \tag{6.31d}$$

As before in binary logic, the implication connective can be modeled in rule-based form; $P \rightarrow Q$ is (IF $x$ is $A$, THEN $y$ is $B$) and it is equivalent to the following fuzzy relation, $R$, just as it is in classical logic, $R = (A \times B) \cup (\overline{A} \times Y)$. Recall Eq. (6.22), whose membership function is expressed by the following formula:

$$\mu_R(x, y) = \max\{[\mu_A(x) \wedge \mu_B(y)], [1 - \mu_A(x)]\} \tag{6.32}$$

When the logical conditional implication is of the compound form—IF $x$ is $A$, THEN $y$ is $B$, ELSE $y$ is $C$—then the equivalent fuzzy relation, $R$, is expressed as $R = (A \times B) \cup (\overline{A} \times C)$ in a form just as Eq. (6.5), whose membership function is expressed by the following formula:

$$\mu_R(x, y) = \max[\{\mu_A(x) \wedge \mu_B(y)\}, \{[1 - \mu_A(x)] \wedge \mu_C(y)\}] \tag{6.33}$$

***Approximate Reasoning.***    The ultimate goal of fuzzy logic is to form the theoretical foundation for reasoning about imprecise propositions; such reasoning is referred to as *approximate reasoning*.[6,7] Approximate reasoning is analogous to predicate logic for reasoning with precise propositions, and hence is an extension of classical propositional calculus that deals with partial truths.

Suppose we have a rule-based format to represent fuzzy information. These rules are expressed in conventional antecedent-consequent form, such as the following:

*Rule 1*    IF $x$ is $A$, THEN $y$ is $B$, where $A$ and $B$ represent fuzzy propositions (sets).

Now suppose we introduce a new antecedent (say, $A'$), and we consider the rule that follows.

*Rule 2*    IF $x$ is $A'$, THEN $y$ is $B'$.

From information derived from rule 1, is it possible to derive the consequent in rule 2, $B'$? The answer is yes, and the procedure is *fuzzy composition*. The consequent $B'$ can be found from the composition operation, $B' = A' \cdot R$, which is analogous to Eq. (6.23).

## 6.4    IF-THEN RULE-BASED SYSTEMS

In the field of artificial intelligence (machine intelligence) there are various ways to represent knowledge. Perhaps the most common way to represent human knowlege is to form it into natural language expressions of the type,

IF *premise (antecedent),* THEN *conclusion (consequent)*

This expression is commonly referred to as the IF-THEN "rule-based" form. It typically expresses an inference such that if a fact (premise, hypothesis, antecedent) is known, then we can infer, or derive, another fact called a *conclusion* (consequent). This form of knowledge representation, characterized as "shallow knowledge," is quite appropriate in the context of linguistics because it expresses human empirical and heuristic knowledge in our own language of communication. It does not, however, capture the "deeper" forms of knowledge usually associated with intuition, structure, function, and behavior of the objects around us—simply because these latter forms of knowledge are not readily reduced to linguistic phrases or representations. The rule-based system is distinguished from classical expert systems in the sense that the rules composing a rule-based system might derive from sources other than human experts and, in this context, are distinguished from expert systems. This chapter confines itself to the broader category of fuzzy rule-based systems (of which expert systems could be seen as a subset) because of their prevalence and popularity in the literature, because of their preponderant use in the engineering practice, and because the rule-based form makes use of linguistic variables as its antecedents and consequents. As illustrated earlier in this chapter, these linguistic variables can be naturally represented by fuzzy sets and logical connectives of these sets.

### 6.4.1  Canonical Rule Forms

In general, there exist three general forms for any linguistic variable.[8] These forms are assignment statements, conditional statements, and unconditional statements. Examples of each of these are given as follows:

*Assignment statements:*

$x = $ large

Banana's color $=$ yellow

$x \approx s$

$x$ is not large and not very small

*Conditional statements:*

IF the tomato is red THEN the tomato is ripe

IF $x$ is very hot THEN stop

IF $x$ is large THEN $y$ is small ELSE $y$ is not small

*Unconditional Statements:*

Go to 9

Stop

Divide by $x$

Turn the pressure higher

The assignment statements "restrict" the value of a variable to a specific quantity. The unconditional statements may be thought of as conditional "restrictions" with their IF clause condition being the universe of discourse of the input conditions, which is always true. An unconditional restriction such as "output is low" could be written as follows:

IF any conditions THEN output is low, or

IF anything THEN low.

Hence, the rule base under consideration could be described as using a collection of conditional restrictive statements. These statements may also be modeled as fuzzy conditional statements, such as IF condition $C^1$ THEN restriction $R^1$. The unconditional restrictions might be in the following form:

$R^1$: The output is $B^1$

AND

$R^2$ : The output is $B^2$

AND

etc.

where $B^1$, $B^2$, ... are fuzzy consequents. Table 6.2 is the rule-based system composed of a set of conditional rules. Hence, the canonical rule set may be put in the form shown therein.

**TABLE 6.2.** The Canonical Form for a Fuzzy Rule-Based System

| Rule 1 | IF condition $C^1$, THEN restriction $R^1$ |
|--------|--------------------------------------------|
| Rule 2 | IF condition $C^2$, THEN restriction $R^2$ |
| $\vdots$ | $\vdots$ |
| Rule $r$ | IF condition $C^r$, THEN restriction $R^r$ |

In general, the unconditional as well as conditional statements place some restrictions on the consequent of the rule-base process based on certain immediate or past conditions. These restrictions are usually manifested in terms of vague natural language words that can be modeled using fuzzy mathematics.

Consider the problem of the control of an industrial furnace with some restrictive statements:

If the temperature is hot, then the pressure is rather high.

If the temperature is cold, then the pressure is very low.

If the temperature is warm, then the pressure is medium and not high. (Etc.)

The vague term "rather high" in the first statement places a fuzzy restriction on the pressure, based on a fuzzy "hot" temperature condition in the antecedent.

In summary, the fuzzy level of understanding and describing a complex system basically is put in the form of a set of restrictions on the output based on certain conditions of the input. Restrictions are generally modeled by fuzzy sets and relations. These restriction statements, unconditional as well as conditional, are usually connected by linguistic connectives such as "and", "or", or "else". The restrictions $R^1$, $R^2$, ..., $R^r$ apply to the output actions, or consequents of the rules.

### 6.4.2 Aggregation of Fuzzy Rules

Most rule-based systems involve more than one rule. The process of obtaining the overall consequent (conclusion) from the individual consequents contributed by each rule in the rule base is known as *aggregation of rules*. In determining an aggregation strategy, two simple extreme cases exist:[8]

**1.** *Conjunctive system of rules.* In the case of a system of rules that have to be jointly satisfied, the rules are connected by "and" connectives. In this case the aggregated output (consequent), $y$, is found by the fuzzy intersection of all individual rule consequents, $y^i$, where $i = 1, 2, \ldots, r$, as follows,

$$y = y^1 \text{ and } y^2 \text{ and} \ldots \text{and } y^r \tag{6.34}$$

or

$$y = y^1 \cap y^2 \cap \cdots \cap y^r \tag{6.35}$$

which is defined by the membership function

$$\mu_y(y) = \min(\mu_{y^1}(y), \mu_{y^2}(y), \ldots, \mu_{y^r}(y)) \quad \text{for } y \in Y \tag{6.36}$$

**2.** *Disjunctive system of rules.* For the case of a disjunctive system of rules where the satisfaction of at least one rule is required, the rules are connected by the "or" connectives. In this case the aggregated output is found by the fuzzy union of all individual rule contributions, as follows,

$$y = y^1 \text{ or } y^2 \text{ or} \ldots \text{or } y^r$$

or

$$y = y^1 \cup y^2 \cup \cdots \cup y^r \tag{6.37}$$

which is defined by the membership function

$$\mu_y(y) = \max(\mu_{y^1}(y), \mu_{y^2}(y), \ldots, \mu_{y^r}(y)) \quad \text{for } y \in Y \tag{6.38}$$

### 6.4.3 Graphical Techniques of Inference

Fuzzy relations use mathematical procedures to conduct inferencing of IF-THEN rules [see Eqs. (6.32) and (6.33)]. These procedures can be implemented on a computer for processing speed. Sometimes, however, it is useful to be able to conduct the inference computation manually with a few rules to check computer programs or to verify the inference operations. Conducting the matrix operations illustrated in Eqs. (6.32) and (6.33) for a few rule sets can quickly become quite onerous. Graphical methods have been proposed that emulate the inference process and make manual computations involving a few simple rules straightforward.[3] To illustrate this idea, we consider a simple two-rule system where each rule comprises two antecedents and one consequent. This is analogous to a two-input and single-output fuzzy system. The graphical procedures illustrated here can be easily extended and will hold for fuzzy rule bases (or fuzzy systems) with any number of antecedents (inputs) and consequents (outputs). A fuzzy system with two non-interactive inputs $x_1$ and $x_2$ (antecedents) and a single output $y$ (consequent) is described by a collection of $r$ linguistic IF-THEN propositions,

$$\text{IF } x_1 \text{ is } \underset{\sim}{A}_1^k \text{ and } x_2 \text{ is } \underset{\sim}{A}_2^k \text{ THEN } y^k \text{ is } \underset{\sim}{B}^k \quad \text{for } k = 1, 2, \ldots, r \tag{6.39}$$

where $\underset{\sim}{A}_1^k$ and $\underset{\sim}{A}_2^k$ are the fuzzy sets representing the $k$th-antecedent pairs, and $\underset{\sim}{B}^k$ are the fuzzy sets representing the $k$th consequent.

In the following material, we consider two cases of two-input systems: (1) the inputs to the system are represented by fuzzy sets and we use a max-min inference method, and (2) the inputs to the system are represented by fuzzy sets and we use a max-product inference method.

***Case 1.*** Input $(i)$ and input $(j)$ are fuzzy variables described by fuzzy membership functions. The rule-based system is described by Eq. (6.39), so for a set of disjunctive rules, where $k = 1, 2, \ldots, r$, the aggregated output using a Mamdani implication[3] will be given by the following:

$$\mu_{\underset{\sim}{B}^k}(y) = \max_k [\min\{\max[\mu_{\underset{\sim}{A}_1^k}(x) \wedge \mu(x_1)], \max[\mu_{\underset{\sim}{A}_2^k}(x) \wedge \mu(x_2)]\}] \tag{6.40}$$

Equation (6.40) has a very simple graphical interpretation, which is illustrated in Fig. 6.11. In this figure the fuzzy inputs are represented by triangular membership functions [input $(i)$ and input $(j)$ in the figure]. The intersection of these inputs and the stored membership functions for the antecedents ($A_{11}$, $A_{12}$ for the first rule, and $A_{21}$, $A_{22}$ for the second rule) results in triangles. The maximum value of each of these intersection triangles results in a membership value, the minimum of which is propagated for each rule (because of the "and" connective between antecedents of each rule [Eq. (6.39)]. Figure 6.11 shows the aggregated consequent resulting from a disjunctive set of rules (the outer envelope of the individual truncated-consequents) and a defuzzified value, $Y^*$, resulting from a centroidal defuzzification method.[3]

***Case 2.***    Input ($i$) and input ($j$) are fuzzy variables described by fuzzy membership functions, and the inference method is a max-product method.[3] The resulting expression for this inference for the $r$ disjunctive rules would be as follows:

$$\mu_{\underline{B}^k}(y) = \max_k[\max[\mu_{\underline{A}_1^k}(x) \wedge \mu(x_1)] \cdot \max[\mu_{\underline{A}_2^k}(x) \wedge \mu(x_2)]] \tag{6.41}$$

Equation (6.41) also has a simple graphical interpretation, which is illustrated in Fig. 6.12. In this figure the fuzzy inputs are represented by triangular membership functions [input ($i$) and input ($j$) in the figure]. As before, the intersection of these inputs and the stored membership functions for the antecedents ($A_{11}, A_{12}$ for the first rule, and $A_{21}, A_{22}$ for the second rule) results in other triangles. The maximum value of each of these intersection triangles results in a membership value, the minimum of which is propagated for each rule (because of the "and" connective between the antecedents of each rule [Eq. (6.39)]). Figure 6.12 shows the aggregated consequent resulting from a disjunctive set of rules (the outer envelope of the individual scaled-consequents) and a defuzzified value, $Y^*$, resulting from a centroidal defuzzification method.

## 6.5  *NUMERICAL SIMULATION*

Most physical processes in the real world are nonlinear. It is our abstraction of the real world that leads us to the use of linear systems in modeling these processes. The linear systems are simple, understandable, and, in many situations, provide acceptable simulations of the actual processes that we observe. Unfortunately, it is only the simplest of systems that we are able to model with linear system theory and only a very small fraction of the nonlinear systems that have verifiable solutions. The bulk of the physical processes that we must address are, unfortunately, too complex to reduce to algorithmic form—linear or nonlinear. Most observable processes have only a small amount of information available with which to develop an algorithmic understanding. The vast majority of information we have on most processes tends to be nonnumeric and nonalgorithmic. Most of the information is fuzzy and linguistic in form.

If a process can be described algorithmically, we are able to describe the solution set for a given input set. If the process is not reducible to algorithmic form, perhaps the input-output features of the system are at least observable or measurable. This section deals with systems that cannot be simulated with conventional crisp or algorithmic approaches but which can be simulated because of the presence of other information—observed or linguistic—using fuzzy nonlinear simulation methods.

We wish to use fuzzy rule-based systems as suitable representations of simple and complex physical systems. For this purpose, a fuzzy rule-based system consists of a set of rules that represents the engineer's understanding of the behavior of the system, a set of input data observed going into the system, and a set of output data coming from the system. The input and output data can be numerical, or they can be nonnumeric observations. Figure 6.13 shows a general static physical system, which could be a simple mapping from the input-space to the output-space, an industrial control system, a system identification problem, a pattern recognition process, or a decision-making process.

We have shown that a fuzzy relation can also represent a logical inference. The fuzzy implication, IF $\underline{A}$ THEN $\underline{B}$, is known as the *generalized modus ponens* form of inference. There are numerous techniques for obtaining a fuzzy relation $\underline{R}$ that will represent this inference in the form of a fuzzy relational equation.

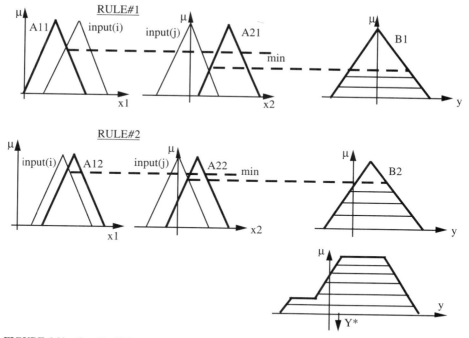

**FIGURE 6.11**    Graphical Mamdani (max-min) implication method with fuzzy inputs.

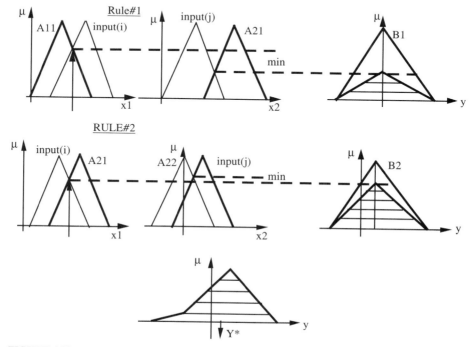

**FIGURE 6.12**    Graphical correlation-product inference using fuzzy inputs.

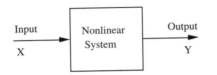

**FIGURE 6.13**   A general static physical system with observed inputs and outputs.

Suppose our knowledge concerning a certain nonlinear process is not algorithmic, like the algorithm $y = x^2$, but rather is in some other more complex form. This more complex form could be data observations of measured inputs and measured outputs. Relations can be developed from this data, which is analogous to a look-up table. Alternatively, the complex form of the knowledge of a nonlinear process could also be described with some linguistic rules of the form IF $\underset{\sim}{A}$ THEN $\underset{\sim}{B}$. For example, suppose we are monitoring a thermodynamic process involving an input heat of a certain temperature and an output variable, pressure. We observe that when we use heat of a "low" temperature, we get out of the process a "low" pressure; when we input heat of a "moderate" temperature, we see a "high" pressure in the system; when we place "high-temperature" heat into the thermodynamics of the system, the output pressure reaches an "extremely high" value; and so on. This process is shown in Figure 6.14, where the inputs are now *not* points in the input universe (heat) and the output universe (pressure), but *patches* of the variables in each universe. These patches represent the fuzziness in describing the variables linguistically. Obviously, the mapping (or relation) describing this relationship between heat and pressure is fuzzy. That is, patches from the input space map, or relate, to patches in the output space, and the relations $R_1$, $R_2$, and $R_3$ in Fig 6.14 represent the fuzziness in this mapping. In general, all the patches, including those representing the relations, overlap because of the ambiguity in their definitions.

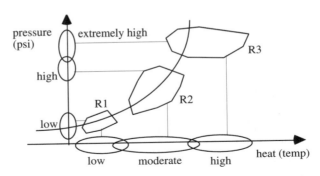

**FIGURE 6.14**   A fuzzy nonlinear relation matching patches in the input space to patches in the output space.

Each of the patches in the input space shown in Fig. 6.14 could represent a fuzzy set (say, $\underset{\sim}{A}$) defined on the input variable (say, $x$), and each of the patches in the output space could be represented by a fuzzy set (say, $\underset{\sim}{B}$), defined on the output variable (say, $y$). Each of the patches lying on the general nonlinear function path could by represented by a fuzzy relation (say, $\underset{\sim}{R}^k$, where $k = 1, 2, \ldots, r$, which represents $r$ possible linguistic relationships between input and output). Suppose we have a situation where a fuzzy input (say, $x$) results in a series of fuzzy outputs (say, $y^k$), depending on which fuzzy relation $\underset{\sim}{R}^k$ is used to determine the mapping. Each of these relationships, as listed in Table 6.3, could be described by what is called a *fuzzy relational equation*, where $y^k$ is the output of the system contributed by the $k$th rule and has membership function given by $\mu_{y^k}(y)$.

Both $x$ and $y^k$ ($k = 1, 2, \ldots, r$) can be written as single-variable fuzzy relations, of dimensions $1 \times n$ and $1 \times m$, respectively. The unary relations, in this case, are actually similarity

**TABLE   6.3**   System of Fuzzy Relational Equations

| $\underset{\sim}{R}^1$ | $y^1 = x \cdot \underset{\sim}{R}^1$ |
|---|---|
| $\underset{\sim}{R}^2$ | $y^2 = x \cdot \underset{\sim}{R}^2$ |
| $\vdots$ | $\vdots$ |
| $\underset{\sim}{R}^r$ | $y^r = x \cdot \underset{\sim}{R}^r$ |

relations between the elements of the fuzzy set and a most typical or prototype element, usually with membership value equal to unity.

## 6.5.1  Partitioning

The fuzzy relation $\underset{\sim}{R}$ is very context-dependent; hence it has local properties with respect to the Cartesian space of the input and output universes. This fuzzy relation results from the Cartesian product of the fuzzy sets representing the inputs and outputs of the fuzzy nonlinear system. However, before the relation $\underset{\sim}{R}$ can be determined, a more fundamental concern is the question of how to *partition* the input and output spaces (universes of discourse) into meaningful fuzzy sets. There are three frequently used methods for the definition of the membership functions (partitioning) of the input fuzzy sets (antecedents) as well as the output fuzzy sets (consequents). These partitioning methods are (1) prototype categorization, (2) degree of similarity, and (3) similarity-as-distance.

All three of these methods employ a fuzzy system where the knowledge regarding the functional relation of the system is determined through observation or is extracted from interviews with a human. The knowledge can be put in the form of linguistic IF-THEN rules such as "if the input is medium-negative then the output is positive-low," or "if the input is high the output is medium," and so on. In this way the human observer is simply giving the functional relation between pairs of prototype points "medium-negative" and "high" in the input space and the prototype points "positive-low" and "medium" in the output space. A crisp binary relation between prototype input points and prototype output points is all we are able to extract from the observer. The exact numerical values of these prototype variables and the corresponding membership functions that these prototype variables represent is unclear to us. These values are highly subjective and intuitive for the human, even though they may not know the exact values. Some intuitive methods used by humans for assigning membership functions to linguistic values in their propositions about a system have been described,[8] and one of the most prevalent situations is provided here.

There are situations where the ranges of the input and output variables are known some way. In this situation the interval from the lower limit to the higher limit is divided into $n$ equal partitions and the midpoint of each partition is taken to represent the prototype point for that partition, i.e., a membership equal to unity. For the two extreme partitions the minimum and maximum endpoints are assumed to be the corresponding prototype points. It is further assumed that the value of membership function corresponding to a typical prototype point is equal to zero at all other prototype points, and the value of any point in a particular fuzzy set is proportional to the distance of that point from the prototype point in that set. Based on this assumption, $n$ triangular-shaped membership functions will be defined on the interval for each of the input and output variables. Figure 6.15 is an illustration of this procedure for some input variable range defined on the interval $[X_{\min}, X_{\max}]$. Some intuitive linguistic values, such as negative-big or positive-medium, might be given to the fuzzy sets $P1$

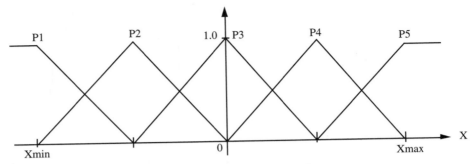

**FIGURE 6.15**   Fuzzy partitions for equally spaced prototype points.[8]

(Partition-1), $P2$ (Partition-2), and so on for $n$ partitions. Once the membership functions are determined for the input and output variables, the fuzzy system relational matrices (i.e., $\underset{\sim}{R}$'s) are found.

### 6.5.2   Fuzzy Associative Memories (FAMs)

Consider a fuzzy system with $n$ inputs and a single output. Also assume that each input universe of discourse (i.e., $X_1, X_2, \ldots, X_n$) is partitioned into $k$ fuzzy partitions. Based on the canonical fuzzy model given in Table 6.3 for a nonlinear system, the total number of possible rules governing this system is given by the following,

$$l = k^n \tag{6.42a}$$

$$l = (k + 1)^n \tag{6.42b}$$

where $l$ is the maximum possible number of canonical rules. Equation (6.42b) is to be used if the partition "anything" is to be used; otherwise, Eq. (6.42a) determines the number of possible rules. The actual number of rules, $r$, necessary to describe a fuzzy system is much less than $l$; i.e., $r \ll l$. This is due to the interpolative-reasoning capability of the fuzzy model and the fact that the fuzzy membership functions for each partition overlap. If each of the $n$ noninteractive inputs is partitioned into a different number of fuzzy partitions (say $X_1$ is partitioned into $k_1$ partitions and $X_2$ is partitioned into $k_2$ partitions and so forth) then the maximum number of rules is given by the following:

$$l = k_1 k_2 k_3 \cdots k_n \tag{6.43}$$

For a small number of inputs (e.g., $n = 1$, $n = 2$, or $n = 3$) there exists a compact form of representing a fuzzy rule-based system. This form is illustrated for $n = 2$ in Fig. 6.16. In the figure there are seven partitions for input $A$ ($A1$ to $A7$), there are five partitions for input $B$ ($B1$ to $B5$), and there are four partitions for the output variable, $C$ ($C1$ to $C4$). This compact graphical form is called a *fuzzy associative memory table,* or FAM table. As can be seen from the FAM table, the rule-based system actually represents a general nonlinear mapping from the input space of the fuzzy system to the output space of the fuzzy system. In this mapping the patches of the input space are being applied to the patches in the output space. Each rule, or equivalently, each fuzzy relation from input to the output is actually representing a fuzzy point of data that characterizes the nonlinear mapping from input to output.

In the FAM table shown in Fig. 6.16 we see that the maximum number of rules for this situation using Eq. (6.43) is $l = k_1 k_2 = 7(5) = 35$, but, as seen in the figure, the actual number of rules is only $r = 21$.

| Input B \ Input A | A1 | A2 | A3 | A4 | A5 | A6 | A7 |
|---|---|---|---|---|---|---|---|
| B1 | C1 |  | C4 | C4 |  | C3 | C3 |
| B2 |  | C1 |  |  |  | C2 |  |
| B3 | C4 |  | C1 |  |  | C2 | C2 |
| B4 | C3 | C3 |  | C1 |  | C1 | C2 |
| B5 | C3 |  | C4 | C4 | C1 |  | C3 |

**FIGURE 6.16**    FAM table for a two-input, single-output fuzzy rule-based system.

### 6.5.3  Example Simulation Using Fuzzy Rule-Based Systems

For the nonlinear function $y = 10 \sin x_1$ we will develop a fuzzy rule-based system using four simple fuzzy rules to approximate the output $y$. The universe of discourse for the input variable $x_1$ will be the interval $[-180°, 180°]$ in degrees, and the universe of discourse for the output variable $y$ is the interval $[-10, 10]$.

First, we will partition the input space $x_1$ into five simple partitions on the interval $[-180°, 180°]$, and we will partition the output space $y$ on the interval $[-10, 10]$ into three membership functions, as shown in Figs. 6.17a and 6.17b, respectively.

In Figs. 6.17a and 6.17b the acronyms NB, NS, Z, PS and PB refer to the linguistic variables "negative-big," "negative-small," "zero," "positive-small," and "positive-big," respectively.

Second, we develop four simple rules, which are listed in Table 6.4, which we think emulate the system dynamics (in this case the system is the nonlinear equation $y = 10 \sin x_1$—we are observing the harmonics of this system), and which make use of the linguistic variables shown in Fig. 6.17. The FAM table for these rules is given in Table 6.5.

The FAM table of Table 6.5 is one-dimensional because there is only one input variable, $x_1$. As seen in Table 6.5 all rules listed in Table 6.4 are accommodated. The four rules expressed are not all expressed in canonical form (some have disjunctive antecedents), but if they were transformed into canonical form, they would represent the five rules provided n the FAM table in Table 6.5.

In developing an approximate solution for the output $y$ we select a few input points and conduct a graphical inference method similar to that illustrated earlier. We will use the centroid method for defuzzification. Let us choose four crisp singletons as the input: $x_1 = \{-135°, -45°, 45°, 135°\}$. For input $x_1 = -135°$, rules 3 and 4 are fired, as shown in Figs. 6.18c and 6.18d (see also Table 6.4). For input $x_1 = -45°$, rules 1, 3, and 4 are fired. Figures 6.18a and 6.18b show the graphical inference for input $x_1 = -45°$, which fires rule 1, and for $x_1 = 45°$, which fires rule 2.

For input $x_1 = -45°$, rules 3 and 4 also get fired, and we get results similar to those shown in Fig. 6.18c and 6.18d after defuzzification:

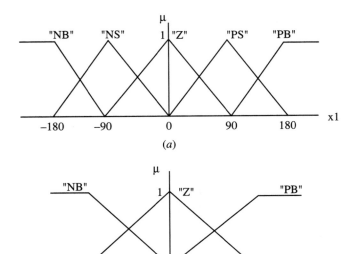

**FIGURE 6.17**   Fuzzy membership functions for the (*a*) input and (*b*) output spaces, where (*a*) has five partitions for the input variable, $x_1$, and (*b*) has three partitions for the output variable, *y*.

**TABLE 6.4**   Four Simple Rules for $y = 10 \sin x_1$

| 1 | IF $x_1$ is "Z" or "PB," THEN *y* is "Z" |
|---|---|
| 2 | IF $x_1$ is "PS," THEN *y* is "PB" |
| 3 | IF $x_1$ is "Z" or "NB," THEN *y* is "Z" |
| 4 | IF $x_1$ is "NS," THEN *y* is "NB" |

**TABLE 6.5**   FAM for the Four Simple Rules

| $x_1$ | NB | NS | Z | PS | PB |
|---|---|---|---|---|---|
| *y* | Z | NB | Z | PB | Z |

Rule 3: $y = 0$

Rule 4: $y = -7$

For $x_1 = 45°$, rules 1, 2, and 3 are fired (see Fig. 6.18*d* for rule 2) and we get the following results for rules 1 and 3 after defuzzification:

Rule 1: $y = 0$

Rule 3: $y = 0$

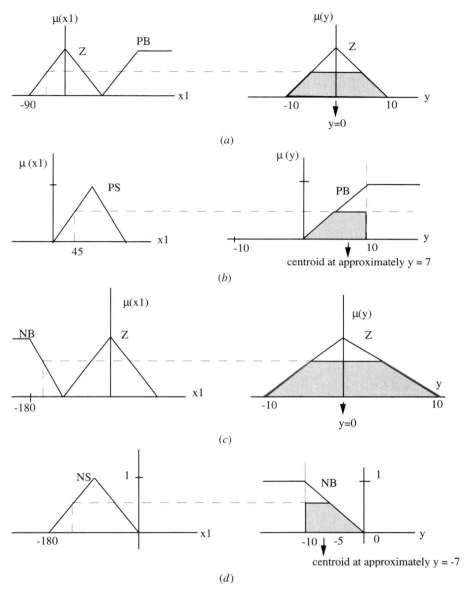

**FIGURE 6.18**  Graphical inference method showing membership propagation and defuzzification: (*a*) input $x_1 = -45°$ fires rule 1, (*b*) input $x_1 = 45°$ fires rule 2, (*c*) input $x_1 = -135°$ fires rule 3, (*d*) input $x_1 = 135°$ fires rule 4.

For $x_1 = 135°$, rules 1 and 2 are fired and we get, after defuzzification, results that are similar to those shown in Fig. 6.18*d*:

Rule 1: $y = 0$

Rule 2: $y = 7$

When we combine the results, we get an aggregated result summarized in Table 6.6 and shown graphically in Fig. 6.19. The $y$ values in each column of Table 6.6 are the defuzzified results from various rules firing for each of the inputs, $x_i$. When we aggregate the rules using the union operator (disjunctive rules), this has the effect of taking the maximum value for $y$ in each of the columns in Table 6.5. The plot in Fig. 6.19 represents the maximum $y$ for each of the $x_i$, and it represents a fairly accurate portrayal of the true solution. More rules would result in a closer *fit* to the true sine curve.

**TABLE 6.6**   Defuzzified Results for Simulation of $y = 10 \sin x_1$

| $x_1$ | $-135°$ | $-45°$ | $45°$ | $135°$ |
|-------|---------|--------|-------|--------|
| $y$   | 0       | 0      | 0     | 0      |
|       | $-7$    | 0      | 0     | 7      |
|       |         | $-7$   | 7     |        |

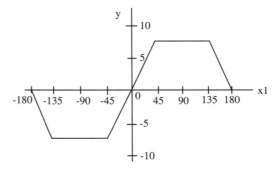

**FIGURE 6.19**   Simulation of nonlinear system $y = 10 \sin x_1$ using 4-rule fuzzy rule-base.

## 6.6  FUZZY CONTROL

A control system is a group of physical components arranged in such a manner as to alter, to regulate, or to command, through a *control action,* another physical system so that it exhibits certain desired characteristics or behavior. Control systems are typically of two types: open-loop control systems, in which the control action is independent of the physical system output, and closed-loop control systems (also known as *feedback control systems*), in which the control action is dependent on the physical system output. Examples of open-loop control systems are a toaster, where the amount of heat is set by a human, and an automatic washing machine, where the controls for water temperature, spin-cycle time, etc., are preset by the human. In both these cases the control actions are not a function of the output of the toaster or the washing machine. Examples of feedback control are a room temperature thermostat, where the room temperature is sensed and a heating or cooling unit is activated when a certain threshold temperature is reached, and an autopilot mechanism, where automatic course corrections to a plane are made when certain preset headings or altitudes are sensed by the instruments in the plane cockpit.

In order to control any physical variable, we must know the variable by measuring it. The system for measurement of the controlled signal is the *sensor.* The physical system under control is called a *plant.* In a closed-loop control system, certain forcing signals of the system (called *inputs*) are determined by the responses of the system (called *outputs*). To obtain a satisfactory response and characteristics for the closed-loop control system, it is necessary to connect an additional system into the loop. This system is known as a *compensator* or a *controller.* The general form of a closed-loop control system is illustrated in Fig. 6.20.[9]

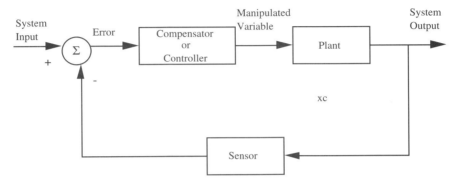

**FIGURE 6.20**   A closed-loop control system.

Control systems are sometimes divided into two classes. If the object of the control system is to maintain a physical variable at some constant value in the presence of disturbances, it is called a *regulatory type* of control or a *regulator.* The room temperature control and autopilot are examples of regulatory types of controllers. The second class of control systems is the tracking type of controllers. In this scheme of control, a physical variable is required to follow or track some desired time function. An example of this type of system is an automatic aircraft-landing system, in which the aircraft follows a ramp to the desired touchdown point.

The control problem is stated as follows.[9] The output or the response of the physical system under control (i.e., the plant) is adjusted as required by the *error signal.* The error signal is the difference between the actual response of the plant, as measured by the sensor system, and the desired response, as specified by a *reference input.* In the following section we derive different forms of common mathematical models describing a closed-loop control system.

### 6.6.1   Control System Design Problem

The general problem of feedback control system design is defined as obtaining a generally non-linear vector-valued function $\underset{\sim}{\mathbf{h}}(\cdot)$ defined as follows:[8]

$$\underset{\sim}{\mathbf{u}}(t) = \underset{\sim}{\mathbf{h}}[t, \underset{\sim}{\mathbf{x}}(t), \underset{\sim}{\mathbf{r}}(t)] \tag{6.44}$$

where $\mathbf{u}(t)$ is the input to the plant or process, $\underset{\sim}{\mathbf{r}}(t)$ is the reference input, and $\underset{\sim}{\mathbf{x}}(t)$ is the state vector. The feedback control law $\underset{\sim}{\mathbf{h}}$ is supposed to stabilize the feedback control system and result in a satisfactory performance.

In the case of a time-invariant system with a regulatory type of controller, where the reference input is a constant set-point, the vast majority of controllers are based on the general models given in Eqs. (6.45) or (6.46), i.e., either a full state-feedback or an output feedback, as shown in the following:

$$\underline{u}(t) = \underline{h}[\underline{x}(t)] \tag{6.45}$$

$$\underline{u}(t) = \underline{h}\left[y(t), \dot{y}, \int y \, dt\right] \tag{6.46}$$

In the case of a simple single-input and single-output system and a regulatory type of controller, the function $\underline{h}$ takes one of the following forms:

$$\underline{u}(t) = K_p \cdot e(t) \tag{6.47}$$

for a proportional or P controller;

$$\underline{u}(t) = K_p \cdot e(t) + K_I \cdot \int e(t) \, dt \tag{6.48}$$

for a proportional plus integral or PI controller;

$$\underline{u}(t) = K_p \cdot e(t) + K_D \cdot \dot{e}(t) \tag{6.49}$$

for a proportional plus derivative or PD controller;

$$\underline{u}(t) = K_p \cdot e(t) + K_I \cdot \int e(t) \, dt + K_D \cdot \dot{e}(t) \tag{6.50}$$

for a proportional plus derivative plus integral or PID controller, where $e(t)$, $\dot{e}(t)$, and $\int e(t) \, dt$ are the output error, error derivative, and error integral, respectively; and

$$\underline{u}(t) = -[k_1 \cdot x_1(t) + k_2 \cdot x_2(t) + \cdots + k_n \cdot x_n(t)] \tag{6.51}$$

for a full state-feedback controller.

The problem of control system design is defined as obtaining the generally nonlinear function $\underline{h}(\cdot)$ in the case of nonlinear systems, coefficients $K_P$, $K_I$, and $K_D$ in the case of an output-feedback, and coefficients $k_1, k_2, \ldots, k_n$ in the case of a full state-feedback control policy for linear systems. The function $\underline{h}(\cdot)$ in Eqs. (6.45) and (6.46) describes a general nonlinear surface that is known as a control or decision surface.

***Control (Decision) Surface.*** The concept of a control surface or decision surface is central in fuzzy control systems methodology.[8] In the following we define this very important concept. The function $\underline{h}$ as defined in Eqs. (6.44) through (6.46) is, in general, defining a nonlinear hypersurface in an $n$-dimensional space. For the case of linear systems with output feedback or state feedback it is a hyperplane in an $n$-dimensional space. This surface is known as the *control* or *decision surface*. The control surface describes the dynamics of the controller and is generally a time-varying nonlinear surface. Because of unmodeled dynamics present in the design of any controller, techniques should exist for adaptively tuning and modifying the control surface shape.

Fuzzy logic rule-based expert systems use a collection of fuzzy conditional statements derived from an expert knowledge base to approximate and construct the control surface.[10–12] This paradigm of control system design is based on interpolative and approximate reasoning. Fuzzy rule-based controllers, or system identifiers, are generally model-free paradigms. Fuzzy logic rule-based expert systems are universal nonlinear function approximators, and any nonlinear function (e.g., control surface) of $n$ independent variables and one dependent variable can be approximated to any desired precision.

Alternatively, artificial neural networks are based on analogical learning and try to learn the nonlinear decision surface through adaptive and converging techniques on the basis of numerical data available from input-output measurements on the system variables and some performance criteria.

***Control System Design Stages.***    In order to obtain the control surface for a nonlinear and time-varying real-world complex dynamic system, a number of simplifying steps are used in modeling a controller for the system. The seven basic steps in designing a controller for a complex physical system are as follows:[3]

1. Large-scale systems are decentralized and decomposed into a collection of decoupled sub-systems.
2. The temporal variations of plant dynamics are assumed to be "slowly varying."
3. The nonlinear plant dynamics are locally linearized about a set of operating points (see rotation assumptions in the example in Section 6.6.2).
4. A set of state variables, control variables, or output features are made available.
5. A simple P, PD, PID (output feedback), or state-feedback controller is designed for each decoupled system. The controllers are of regulatory type and are fast enough to perform satisfactorily under tracking control situations. Optimal controllers might also prove useful.
6. The first five steps just mentioned introduce uncertainties. There are also uncertainties due to external environment. The controller design should be made as close as possible to the optimal one based on the control engineer's best available knowledge—in the form of input-output numerical observations data and analytic, linguistic, and intuitive information regarding the plant dynamics and external world.
7. A supervisory control system—either automatic or with a human expert operator—forms an additional feedback control loop to tune and adjust the controller's parameters in order to compensate for the effects of uncertainties and variations due to unmodeled dynamics.

***Assumptions in a Fuzzy Control System Design.***    In addition to the seven simplifying steps just mentioned, a number of assumptions are implicit in a fuzzy control system design. Six basic assumptions are commonly made whenever a fuzzy logic–based control policy is selected. These assumptions are as follows:[3]

1. The plant is observable and controllable: state, input, and output variables are usually available for observation and measurement or computation.
2. There exists a body of knowledge in the form of a set of expert production linguistic rules, engineering common sense, intuition, a set of input/output measurements data, or an analytic model that can be fuzzified and the rules extracted.
3. A solution exists.
4. The control engineer is looking for a good enough solution and not necessarily the optimum one.
5. We desire to design a controller to the best of our available knowledge and within an acceptable precision range.
6. The problems of stability and optimality are still open problems in fuzzy controller design.

***Simple Fuzzy Logic Controllers.***    First-generation (nonadaptive) simple fuzzy logic controllers are generally depicted by the block diagram shown in Fig. 6.21.

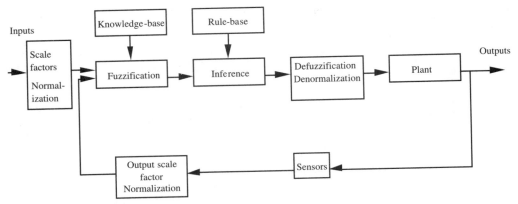

**FIGURE 6.21**  A simple fuzzy-logic control-system block diagram.

The knowledge-base module in Fig. 6.21 contains the knowledge regarding all the input and output fuzzy partitions. It will include the term-set and the corresponding membership functions defining the input variables to the fuzzy rule-based system and the output variables, i.e., control actions to the plant under control.

The steps in designing a simple fuzzy logic control system are as follows:

1. Identify the variables (inputs, states, and outputs) of the plant.

2. Partition the universe of discourse or the interval spanned by each variable into a number of fuzzy subsets each labeled by a linguistic label (subsets include all the elements in the universe).

3. Assign or determine a membership function for each fuzzy subset.

4. Assign the fuzzy relationships between the input, or state, fuzzy subsets on the one hand and the output fuzzy subsets on the other hand (rule-set).

5. Choose appropriate scaling factors for the input and output variables in order to normalize the variables to the [0, 1] or [−1, 1] interval.

6. Fuzzify the inputs to the controller.

7. Use fuzzy approximate reasoning to infer the output contributed from each rule.

8. Aggregate the fuzzy outputs recommended by each rule.

9. Apply defuzzification to find a crisp output.

These steps will now be applied to an example problem in the fuzzy control of an inverted pendulum.[3]

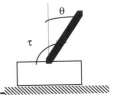

**FIGURE 6.22**  Inverted pendulum-control problem.

### 6.6.2  Classical Fuzzy Control Problem: Inverted Pendulum

Figure 6.22 shows the classic inverted-pendulum system, which has been an interesting case in control theory for many years. We want to design and analyze a fuzzy controller for the simplified version of the inverted-pendulum system shown in Fig. 6.22. The differential equation describing the system is given as follows,[13,14]

$$-ml^2 d^2\theta/[dt^2 + (mlg)\sin(\theta)] = \tau = u(t) \tag{6.52}$$

where $m$ is the mass of the pole located at the tip point of the pendulum, $l$ is the length of the pendulum, $\theta$ is the deviation angle from vertical in the clockwise direction, $\tau = u(t)$ is the torque applied to the pole in the counterclockwise direction [$u(t)$ is the control action], and $g$ is the gravitational acceleration constant.

Assuming $x_1 = \theta$ and $x_2 = d\theta/dt$ to be the state variables, the state-space representation for the nonlinear system defined by Eq. (6.52) is given by

$$dx_1/dt = x_2$$

$$dx_2/dt = (g/l)\sin(x_1) - (l/ml^2)u(t)$$

It is known that for very small rotations, $\theta$, we have $\sin(\theta) = \theta$, where $\theta$ is measured in radians. This relation is used to linearize the nonlinear state-space equations, which yields the following:

$$dx_1/dt = x_2$$

$$dx_2/dt = (g/l)x_1 - (l/ml^2)u(t)$$

If $x_1$ is measured in degrees and $x_2$ is measured in degrees per second, by choosing $l = g$ and $m = 180/(\pi \cdot g^2)$, the linearized and discrete-time state-space equations can be represented as follows:

$$x_1(k + 1) = x_1(k) + x_2(k)$$

$$x_2(k + 1) = x_1(k) + x_2(k) - u(k)$$

For this problem we assume the universe of discourse for the two variables to be $-2° \le x_1 \le 2°$ and $-5$ dps $\le x_2 \le 5$ dps (dps = degrees per second).

*Step 1.*    First we want to construct three membership functions for $x_1$ on its universe, i.e., for the values positive (P), zero (Z) and negative (N), as shown in Fig. 6.23. We then construct three membership functions for $x_2$ on its universe, i.e., for the values positive (P), zero (Z), and negative (N), as shown in Fig. 6.24.

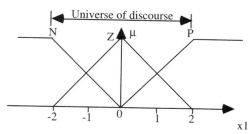

**FIGURE 6.23**    Input $x_1$ partitioned.

*Step 2.*    To partition the control space (output), we will construct five membership functions for $u(k)$ on its universe, which is $-24 \le u(k) \le 24$, as shown in Fig. 6.25 (note the figure has seven partitions, but we will use only five for this problem).

*Step 3.*    We now construct nine rules (even though we may not need this many) in a $3 \times 3$ FAM table, Table 6.7, for this system, which will involve $q$ and $\theta$ in order to stabilize the inverted-pendulum system. The entries in this table are the control actions, $u(k)$.

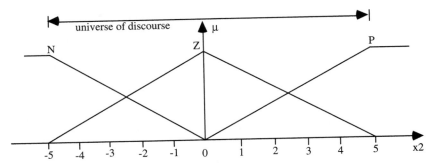

**FIGURE 6.24**    Input $x_2$ partitioned.

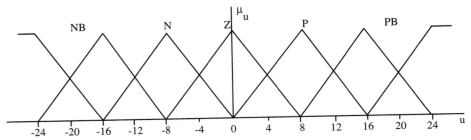

**FIGURE 6.25**    Output, $u(k)$, partitioned into seven partitions (only five used).

**TABLE 6.7**    FAM Table

| | $x_2$ | | |
|---|---|---|---|
| $x_1$ | $P$ | $Z$ | $N$ |
| $P$ | PB | $P$ | $Z$ |
| $Z$ | $P$ | $Z$ | $N$ |
| $N$ | $Z$ | $N$ | NB |

*Step 4.*    Using the rules expressed in Table 6.7 we will now conduct a simulation of this control problem. In conducting the simulation we will use the graphical method to conduct the fuzzy operations. To start the simulation we will use the following crisp initial conditions:

$$x_1(0) = +1° \text{ and } x_2(0) = -4 \text{ dps}$$

Then we will conduct 4 cycles of simulation using the matrix difference equations, given earlier, for the discrete steps $0 \le k \le 3$. Each simulation cycle will result in membership functions for the two input variables. The FAM table will produce a membership function for the control action $u(k)$. We will defuzzify the membership function for the control action using the centroid method and then use the recursive difference equations to solve for new values of $x_1$ and $x_2$. Each simu-

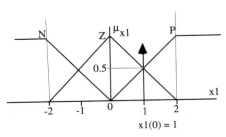

**FIGURE 6.26**    Initial condition for $x_1$.

lation cycle after $k = 0$ will begin with the previous values of $x_1$ and $x_2$ as the input conditions to the next cycle of the recursive difference equations. Figures 6.26 and 6.27 show the initial conditions for $x_1$ and $x_2$, respectively.

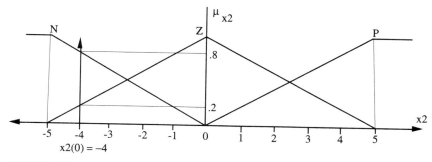

FIGURE 6.27   Initial condition for $x_2$.

From the FAM table (Table 6.7) we have the following:

If $(x_1 = P)$ and $(x_2 = Z)$ then $(u = P)$          min$(.5, .2) = .2(P)$

If $(x_1 = P)$ and $(x_2 = N)$ then $(u = Z)$          min$(.5, .8) = .5(Z)$

If $(x_1 = Z)$ and $(x_2 = Z)$ then $(u = Z)$          min$(.5, .2) = .2(Z)$

If $(x_1 = Z)$ and $(x_2 = N)$ then $(u = N)$          min$(.5, .8) = .5(N)$

The union of the truncated fuzzy consequents for the control variable $u$ are shown in Fig. 6.28 or, in final form with the defuzzified control value, in Fig. 6.29.

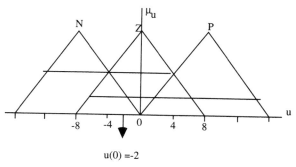

$u(0) = -2$

FIGURE 6.28   Union of fuzzy consequents fired by rules.

We have completed the first cycle of the simulation. Now we take the value of the defuzzified control variable (i.e., $u = 1$) and, using the system equations, find the initial conditions for the next iteration.

$$x_1(1) = x_1(0) + x_2(0) = 1 - 4 = -3$$

$$x_1(1) = x_1(0) + x_2(0) - u(0) = 1 - 4 - (-2) = -1$$

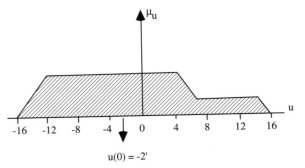

**FIGURE 6.29** Union of fuzzy consequents and defuzzifed value.

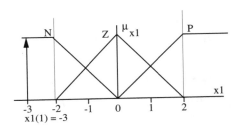

**FIGURE 6.30** Initial condition for second cycle for $x_1$.

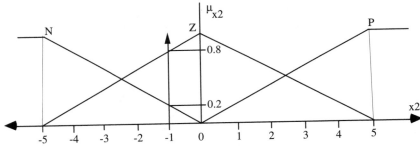

**FIGURE 6.31** Initial condition for second cycle for $x_2$.

From this we get the initial conditions for the second cycle as follows,

$$x_1(1) = -3$$
$$x_2(1) = -1$$

which are shown graphically in Figs. 6.30 and 6.31.

From the FAM table (Table 6.7) we have the following:

If $(x_1 = N)$ and $(x_2 = N)$ then $(u = NB)$ $\qquad$ min(1, .2) = .2(NB)

If $(x_1 = N)$ and $(x_2 = Z)$ then $(u = N)$ $\qquad$ min(1, .8) = .8(N)

The union of the fuzzy consequents and the resulting defuzzified output are shown in Fig. 6.32, where the defuzzified value is $u = -9.6$.

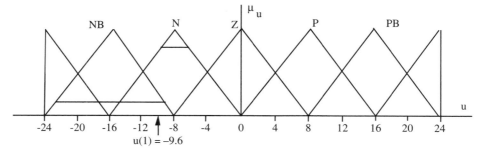

**FIGURE 6.32**   Truncated consequents and defuzzified output for second cycle.

As before, we now use $u = -9.6$ to find initial conditions for the third-cycle iteration.

$$x_1(2) = x_1(1) + x_2(1) = -3 - 1 = -4$$

$$x_2(2) = x_1(1) + x_2(1) - u(1) = -3 - 1 - (-9.6) = +5.6$$

And we get initial conditions,

$$x_1(2) = -4$$

$$x_2(2) = 5.6$$

which are shown graphically in Figs. 6.33 and 6.34, for $x_1$ and $x_2$, respectively.

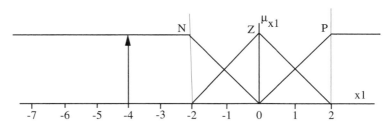

**FIGURE 6.33**   Initial condition for third cycle for $x_1$.

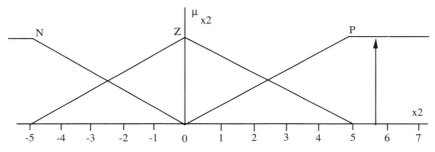

**FIGURE 6.34**   Initial condition for third cycle for $x_2$.

From the FAM table (Table 6.7) we have the following,

If $(x_1 = N)$ and $(x_2 = P)$ then $(u = Z)$          $\min(1, 1) = 1(Z)$

and the resulting consequent and defuzzified control variable, $u$, is shown in Fig. 6.35.

$$x_1(3) = x_1(2) + x_2(2) = -4 + 5.6 = 1.6$$

$$x_2(3) = x_1(2) + x_2(2) - u(2) = -4 + 5.6 - (0.0) = 1.6$$

And we get initial conditions,

$$x_1(3) = 1.6$$

$$x_2(3) = 1.6$$

which are shown graphically in Figs. 6.36 and 6.37, for $x_1$ and $x_2$, respectively.

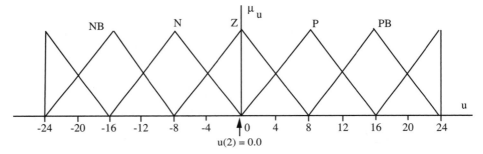

**FIGURE 6.35** Defuzzified output for third cycle of simulation.

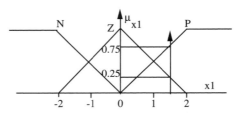

**FIGURE 6.36** Initial condition for next iteration for $x_1$.

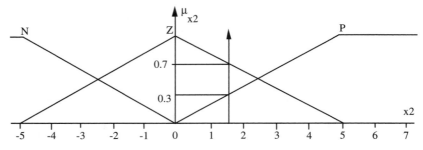

**FIGURE 6.37** Initial condition for next iteration for $x_2$.

From Table 6.7 we get the following:

| | |
|---|---|
| If $(x_1 = Z)$ and $(x_2 = P)$ then $(u = P)$ | $\min(.25, .3) = .3(P)$ |
| If $(x_1 = Z)$ and $(x_2 = Z)$ then $(u = Z)$ | $\min(.25, .7) = .25(Z)$ |
| If $(x_1 = P)$ and $(x_2 = P)$ then $(u = PB)$ | $\min(.75, .3) = .3(PB)$ |
| If $(x_1 = P)$ and $(x_2 = Z)$ then $(u = P)$ | $\min(.75, .7) = .7(P)$ |

which are shown graphically in Fig. 6.38 for $u$, where the defuzzified value is $u = 5.28$.

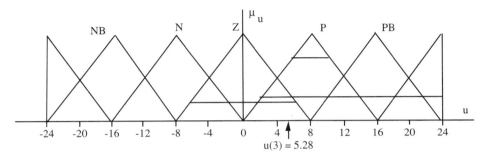

**FIGURE 6.38**   Defuzzified output for next iteration.

As before, we will use $u = 5.28$ to find initial conditions for the next iteration,

$$x_1(4) = x_1(3) + x_2(3) = 1.6 + 1.6 = 3.2$$
$$x_2(4) = x_1(3) + x_2(3) - u(3) = 1.6 + 1.6 - (5.28) = -2.08$$

and we get the following initial conditions:

$$x_1(5) = 3.2$$
$$x_2(5) = -2.08$$

and $u(4) = 1.12$; we conclude the simulation at this point.

*Step 5.*   We now plot the four simulation cycle results for $x_1$, $x_2$, and $u(k)$ in Fig. 6.39. In these figures the length and direction of the arrow are proportional to the angular velocity and direction of the pendulum, respectively.

## 6.7 SUMMARY

This chapter has presented the basics of fuzzy sets, membership, fuzzy relations, fuzzy logic, fuzzy rule-based systems, fuzzy simulation, and fuzzy control. The utility of fuzzy logic transcends all engineering disciplines, and all scientific fields as well. Perhaps the greatest current interest in fuzzy logic in the commercial marketplace is the use of fuzzy logic in the development of new "smart" consumer products. These products employ the elements of fuzzy control in their designs.

Beyond the material provided in this chapter is a new era in fuzzy logic, where this new technology will be combined with the newest digital forms of fuzzy logic controllers. New generations of fuzzy logic controllers are based on the integration of conventional digital and fuzzy controllers. Digital fuzzy clustering techniques can be used to extract the linguistic IF-THEN from numerical data. In general, the trend is toward the compilation and fusion of different

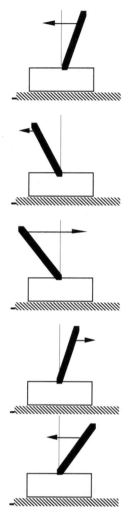

**FIGURE 6.39** Inverted-pendulum simulation results for four iteration cycles. From top to bottom: initial conditions, $x_1(0) = 1°$, $x_2(0) = 14$ dps; iteration 1, $x_1(1) = -3$, $x_2(1) = -1$; iteration 2, $x_1(2) = -4$, $x_2(2) = 5.6$; iteration 3, $x_1(3) = 1.6$, $x_2(3) = 1.6$; iteration 4, $x_1(3) = 3.2$, $x_2(3) = -2.08$.

forms of knowledge representation for the best possible identification and control of ill-defined complex systems. The two new paradigms, artificial neural networks and fuzzy systems, try to understand a real-world system starting from the very fundamental sources of knowledge—i.e., patient and careful observations, digital measurements, experience, and intuitive reasoning and judgments—rather than starting from a preconceived theory or mathematical model. Advanced fuzzy controllers will use adaptation capabilities to tune the membership functions' vertices or supports or to add or delete rules to optimize the performance and compensate for the effects of any internal or external perturbations. New fuzzy learning systems will try to learn the membership functions or the rules, most certainly in digital form. Finally, principles of genetic algorithms will be used to find the best string representing an optimal class of digital input or output membership functions in digital form.

Major sections of this chapter were taken from *Fuzzy Logic with Engineering Applications,* McGraw-Hill, 1995, and permission has been granted for publication here.

## *GLOSSARY*

**Antecedent**   The hypothesis of a rule; the IF portion.

**Approximate reasoning**   The various inferences that deal with vague linguistic propositions.

**Cardinality**   The total possible number of sets in a universe.

**Cartesian product**   The intersection of two sets.

**Complement**   The obverse of a set; the region of a universe outside a particular set.

**Conjunction**   The logical intersection of two linguistic propositions (*A* **and** *B*).

**Consequent**   The conclusion of a rule; the THEN portion.

**Containment**   One set fully contained within another.

**Control surface**   The graphical result of a set of fuzzy rules plotted in the input and output space.

**Crisp sets**   Sets with precisely described boundaries.

**Deductive inference**   A logical method to deduce conclusions from a body of facts.

**Defuzzification (defuzzify)**   To reduce a fuzzy (ambiguous) quantity to a crisp (unambiguous) one.

**Disjunction**   The logical union of two linguistic propositions (*A* **or** *B*).

**Equivalence**   The relation holding between two propositions that are logically identical.

**Feedback control system**   A closed-loop system where the control actions are a part of the input.

**Fuzzification (fuzzify)**   To convert a crisp (unambiguous) quantity to a fuzzy (ambiguous) one.

**Fuzzy associative memory**   A mapping that relates fuzzy inputs to fuzzy outputs.

**Fuzzy composition**   A mathematical operation for relating elements from different domains.

**Fuzzy sets**   Sets with ambiguously defined boundaries.

**Implication**   A logical premise that deduces a conclusion from a hypothesis.

**Intersection**   The common (shared) region of overlapping sets.

**Inverted pendulum**   An upside-down pendulum; hence an unstable one-dimensional system.

**Membership**   The degree of *belongingness* of an element from a universe to a set on this universe.

**Mutual exclusivity**   Two or more sets whose boundaries do not intersect or overlap.

**Negation**   The logical opposite of a proposition; negation of "yes" is "no."

**Partitioning**   The procedure used to subdivide a domain or universe into two or more fuzzy sets.

**Power set**   The total collection of subsets possible on a universe of sets.

**Predicate logic**   A classical form of logic based on the use of predicate terms to define propositions.

**Propositional calculus**   A mathematical foundation for describing predicate logic.

**Union**   The entire region defined by two or more sets, whether or not they overlap.

## REFERENCES

1. L. Zadeh, "Outline of a new approach to the analysis of complex systems and decision processes," *IEEE Transactions on Systems, Man, and Cybernetics,* SMC-3, 1973, pp. 28–44.

2. L. Zadeh, "Fuzzy sets," *Information and Control,* no. 8, pp. 1965, 338–353.

3. T. Ross, *Fuzzy Logic with Engineering Applications,* McGraw-Hill, New York, 1995.

4. J. Bezdek, "Editorial: Fuzzy models—What are they, and why?" *IEEE Transactions on Fuzzy Systems,* vol. 1, 1993, pp. 1–5.

5. B. Kosko, *Neural Networks and Fuzzy Systems,* Prentice Hall, Englewood Cliffs, N.J., 1992.

6. L. Zadeh, "The concept of a linguistic variable and its application to approximate reasoning—Part 3," *Information Science,* no. 9, 1976, pp. 43–80.

7. L. Zadeh, "A theory of approximate reasoning," in *Machine Intelligence,* J. Hayes, Michie, D., and Mikulich, L. (eds.), Halstead Press, New York, 1979, pp. 149–194.

8. N. Vadiee, "Fuzzy rule-based expert systems—I," in *Fuzzy Logic and Control: Software and Hardware Applications,* M. Jamshidi, Vadiee, N., and Ross, T. (eds.), Prentice Hall, Englewood Cliffs, N.J., 1993, pp. 51–85.

9. C. L. Phillips and Harbor, R. D., *Feedback Control Systems,* Prentice Hall, Englewood Cliffs, N.J., 1988, pp. 2–4.

10. E. H. Mamdani and Gaines, R. R. (eds.), *Fuzzy Reasoning and Its Applications,* Academic Press, London, 1981.

11. J. B. Kiszka, Gupta, M. M., and Nikfrouk, P. N., "Some properties of expert control systems," in *Approximate Reasoning in Expert Systems,* M. M. Gupta, Kendal, A., Bandler, W., and Kiszka, J. B. (eds.), Elsevier Science B.V. (North-Holland), 1985, pp. 283–306.

12. M. Sugeno (ed.), *Industrial Application of Fuzzy Control,* North-Holland, New York, 1985.

13. T. Kailath, *Linear Systems,* Prentice Hall, Englewood Cliffs, N.J., 1980, pp. 209–211.

14. J. J. Craig, *Introduction to Robotics—Mechanics and Control,* Addison-Wesley, 1986, pp. 173–176.

## ABOUT THE AUTHOR

Timothy J. Ross is Professor and Regents' Lecturer, Department of Civil Engineering, University of New Mexico. He is a structural engineer with 22 years experience in the fields of computational mechanics, hazard survivability, fuzzy logic, stochastic processes, risk assessment, and expert systems. He has been or is presently the principal investigator on numerous research projects for NASA, the Department of Energy, the National Science Foundation, and others. Ross was formerly president and chairman of the IntelliSys Corp., an Albuquerque, N.M., R&D firm. Prior to that he was a senior research engineer at the Air Force Weapons Laboratory, now the Phillips Laboratory, and a vulnerability engineer for the Defense Intelligence Agency. He received a B.S. in civil engineering from Washington State University in 1971, an M.S. in structural engineering from Rice University in 1973, and a Ph.D. from Stanford University in 1983. He is a fellow in the American Society of Civil Engineers.

# CHAPTER 7
# CHANNEL CODES FOR DIGITAL MAGNETIC AND OPTICAL RECORDS

**Kees A. Schouhamer Immink**
*Philips Research Laboratories*

## 7.1 INTRODUCTION

A code is a set of rules for replacing a given source sequence with another sequence, which is recorded or transmitted. The aim of this transformation is to improve both the reliability and efficiency of the channel. The reliability is commonly expressed in terms of the probability of receiving the wrong information—that is, information that differs from what was originally transmitted. The quantity efficiency is related to the amount of information that can be stored given the physical limitations of the recorder.

The coding problem is partitioned into two main categories: source and channel coding. Source coding is, roughly speaking, a technique to reduce the source symbol rate by removing the redundancy in the source signal (see Chapter 4). In recorder systems, channel encoding is commonly accomplished in two successive steps: error-correction code and recording (or modulation) code. The output generated by the recording code is stored on the storage medium in the form of binary physical quantities, for example, pits and lands or (for magnetic media) positive and negative magnetizations. Schematically, the elements of the coding steps in a digital recorder are similar to those of a "point-to-point" communication link. During readout the decoder constitutes the input sequences as accurately as possible. Error-correction codes make it possible for the receiver to detect and/or correct the majority of the errors that may occur in the received message. There are many different families of error-correcting codes. Of major importance for recoding applications is the family of Reed-Solomon codes (see Chapter 4).

The arrangement, called *recording code,* on which this chapter will concentrate, converts the input stream to a signal suitable for the specific physical recorder requirements. For example, in optical recording, information is recorded in the form of pits and the absence of pits, called *lands.* If source 1s would be written as pits and source 0s as lands, then a long sequence of source 0s would mean that, as pits are not written, the track would be absent for some undefined time. This may pose problems with the tracking during reading, and track loss could be the result. This serious difficulty can be circumvented by using a recording code, which transforms the source sequence into a sequence where such long strings of 0s cannot occur.

### 7.1.1   Recording Codes

Many kinds of recording codes are found in the recording area. In this chapter three types of recording codes will be discussed: run-length–limited sequences, deconstrained codes, and generation of pilot-tracking tones.

*Run-length–Limited Sequences.*   Recording codes based on run-length–limited (RLL) sequences have found widespread application in optical and magnetic disc recording practice. RLLs are characterized by two parameters, $(d + 1)$ and $(k + 1)$, which represent the minimum and maximum run length of the sequence, respectively. The maximum run-length constraint guarantees a clock pulse within some specified time, which is needed for the clock regeneration at the receiver. The minimum run-length constraint is dictated to increase the storage capacity.

*Deconstrained Codes.*   Another noteworthy family of codes found in recording systems is the group of the spectral null code—codes that have vanishing spectral density at specific frequencies. Among these, codes with a spectral null at the zero frequency have predominated in applications. These are referred to as DC-free or DC-balanced codes.

*Generation of Pilot-Tracking Tones.*   Codes can also be used to tailor the signal to other spectral properties. To illustrate this idea, a discussion follows on how recording codes have been created to produce low-frequency pilot-tracking tones that can be exploited for servo tracking purposes. A list of recording codes used in consumer electronics storage products is shown in Table 7.1. A comprehensive discussion of recording codes can be found in the literature.[2,3]

**TABLE 7.1**   Survey of Recording Codes and Their Application Areas

| Device | Code | Type |
|---|---|---|
| Compact Disc | EFM | RLL, DC-free |
| DVD | EFMPlus | RLL, DC-free |
| R-DAT | 8–10 | DC-free |
| Floppy and hard disk | (2,7) or (1.7) | RLL |
| DCC | ETM | DC-free |
| Scoopman[1] | LDM-2 | RLL, DC-free |
| DVC | 24 → 25 | DC-free with pilot tones |

## 7.2   RUN-LENGTH–LIMITED SEQUENCES

Codes based on RLL sequences are the state-of-the-art cornerstone of current disc recorders whether their nature is magnetic or optical. The length of time (usually expressed in channel bits) between consecutive transitions is known as the *run length*. As stated previously, RLL sequences are characterized by two parameters, $(d + 1)$ and $(k + 1)$, which stipulate the minimum and maximum run length, respectively, that may occur in the sequence. The parameter $d$ controls the highest transition frequency and thus has a bearing on intersymbol interference when the sequence is transmitted over a bandwidth-limited channel. In the transmission of binary data it is generally desirable that the received signal be self-synchronizing or self-clocking. Timing is commonly recovered with a phase-locked loop, which adjusts the phase of the detection instant according to observed transitions of the received waveform. The maximum run-length parameter $k$ ensures adequate frequency of transitions for synchronization of the read clock. The grounds on which $d$ and $k$ are chosen, in turn, depend on various factors such as the channel response, the desired data rate (or information density), and the jitter and noise characteristics. The rate $1/2(d = 2, k = 7)$ or rate $2/3(d = 1, k = 7)$ codes are applied in rigid or floppy disk drives, and the EFM code (rate $= 8/17, d = 2, k = 10$) is employed in the Compact Disc and its derivatives: CD-ROM, CD-I, and MiniDisc.

In order to describe RLL codes it is convenient to introduce another constrained sequence, which is closely related to an RLL sequence. A $dk$-limited binary sequence [or simply "$(dk)$ sequence"] satisfies simultaneously the following two conditions:

**1.** $d$ constraint.  Two logical 1s are separated by a run of consecutive 0s of length at least $d$.

**2.** $k$ constraint.  Any run of consecutive 0s is of length at most $k$.

In general, a $(dk)$ sequence is not employed in optical or magnetic recording without a simple coding step. A $(dk)$ sequence is converted into an RLL channel sequence in the following way. Let the channel signals be represented by a polar sequence $\{y_i\}$, $y_i \in \{-1, 1\}$. The channel signals represent the positive or negative magnetization of the recording medium, or pits or lands when dealing with optical recording. The logical 1s in the $(dk)$ sequence indicate the positions of a transition $1 \rightarrow -1$ or $-1 \rightarrow 1$ of the corresponding RLL sequence. The $(dk)$ sequence,

$$0\ 1\ 0\ 0\ 0\ 1\ 0\ 0\ 1\ 0\ 0\ 0\ 1\ 1\ 0\ 1\ldots$$

would be converted to the RLL channel sequence:

$$1\ -1\ -1\ -1\ -1\ 1\ 1\ 1\ -1\ -1\ -1\ -1\ 1\ 1\ -1\ -1\ 1\ 1\ldots$$

The mapping of the waveform by this coding step is known as *precoding,* a confusing term because it is in fact a "postcoding" process. Waveforms that are transmitted without such an intermediate coding step are referred to as non–return-to-zero (NRZ). Coding techniques using the NRZ format are generally accepted in digital optical and magnetic recording practice. It can readily be verified that the minimum and maximum distances between consecutive transitions of the RLL sequence derived from a $(dk)$ sequence is $d + 1$ and $k + 1$ symbols, respectively. In other words, the RLL sequence has the virtue that at least $d + 1$ and at most $k + 1$ consecutive like symbols occur. Table 7.2 gives some parameters of RLL codes that have found practical application.

**TABLE 7.2** Various Codes with Run-Length Parameters $d$ and $k$

| $d$ | $k$ | $R$ | Name | Type |
|---|---|---|---|---|
| 0 | 1 | 1/2 | Biphase | DC-free |
| 1 | 3 | 1/2 | MFM, Miller | |
| 2 | 7 | 1/2 | (2,7) | |
| 1 | 7 | 2/3 | (1,7) | |
| 1 | 4 | 1/2 | LDM-2 | DC-free |
| 2 | 10 | 8/17 | EFM | DC-free |

## 7.2.1  Number of $d$ Sequences

This section addresses the problem of counting the number of sequences of a certain length that comply with $dk$ constraints. For the sake of clerical convenience, let us start with the enumeration of $d$ sequences. Let $N_d(n)$ denote the number of distinct $(d)$ sequences of length $n$ and then define as follows:

$$N_d(n) = 0, \quad n < 0$$
$$N_d(0) = 1$$

$$(7.1)$$

The number of $(d)$ sequences of length $n > 0$ is found with the following recursive relations:[4]

$$N_d(n) = n + 1, \qquad\qquad 1 \leq n \leq d + 1$$

$$N_d(n) = N_d(n - 1) + N_d(n - d - 1), \qquad n > d + 1 \tag{7.2}$$

Table 7.3 lists the number of distinct $(d)$ sequences as a function of the sequence length $n$, with the minimum run length $d$ as a parameter. The number of $(dk)$ sequences of length $n$ can be computed in a similar fashion. Given Table 7.3, it is elementary to design RLL codes.

**TABLE 7.3**  The Number of Distinct $(d)$ Sequences as a Function of the Sequence: Length $n$ and the Minimum Run Length $d$ as a Parameter

| $d =$ | $n = 4$ | 5 | 6 | 7 | 8 | 9 | 10 | 11 | 12 | 13 | 14 |
|---|---|---|---|---|---|---|---|---|---|---|---|
| 5 | 8 | 13 | 21 | 34 | 55 | 89 | 144 | 233 | 377 | 610 | 987 |
| 4 | 6 | 9 | 13 | 19 | 28 | 41 | 60 | 88 | 129 | 189 | 277 |
| 4 | 5 | 7 | 10 | 14 | 19 | 26 | 36 | 50 | 69 | 95 | 131 |
| 4 | 5 | 6 | 8 | 11 | 15 | 20 | 26 | 34 | 45 | 60 | 80 |
| 4 | 5 | 6 | 7 | 9 | 12 | 16 | 21 | 27 | 34 | 43 | 55 |

### 7.2.2  Design Example of a Block Code

To clarify how a code can be designed, a simple illustrative case is a $(1, \infty)$ block code, where source blocks of length $m = 3$ are translated onto codewords of length $n = 5$.

In Table 7.4 the left column tabulates the eight possible source words along with their decimal representation. The codewords, tabulated in the right column, can be freely cascaded without violating the $d = 1$ constraint because the first symbol of the codewords has been selected to be a 0. There are exactly eight codewords of length $n - 1 = 4$ that meet the specified $d = 1$ constraint (see Table 7.3). It is straightforward to generalize the preceding implementation example to encoders that generate sequences with an arbitrary value of the minimum run length. To that end, choose some appropriate codeword length $n$. Set the first $d$ symbols of each codeword to 0. The number of codewords that meet the given run-length condition is $N_d(n - d)$, which can be computed with Eq. (7.2) or by looking it up in Table 7.3. A maximum run-length constraint can be incorporated in the code rules in a straightforward manner. For instance, in the $(d = 1)$ code previously described, the first codeword symbol is preset to 0. If, however, the last symbol of the preceding codeword and the second symbol of the actual codeword to be conveyed are both 0, then the first codeword symbol can be set to 1 without violating the $d = 1$ channel constraint.

**TABLE 7.4**  Simple Fixed-Length $(d = 1)$ Block Code

| Source | | Output |
|---|---|---|
| 0 | 000 | 00000 |
| 1 | 001 | 00001 |
| 2 | 010 | 00010 |
| 3 | 011 | 00100 |
| 4 | 100 | 00101 |
| 5 | 101 | 01000 |
| 6 | 110 | 01001 |
| 7 | 111 | 01010 |

This extra rule—the merging rule—which governs the selection of the first symbol, can be implemented quite smoothly with some extra hardware. The EFM (eight-to-fourteen modulation) code to be discussed in the next section is a generalization of the previous simple block-code example. The next sections will outline implementations of codes that have been applied in digital consumer electronics.

### 7.2.3  Rate-1/2, (2,7) RLL Code

In the previous example of an RLL block code, the source words and codewords have a unique one-to-one relationship. The rate-1/2 ($d = 2$, $k = 7$) code, which has been used in almost all hard disk drives, is completely different. Table 7.5 shows the code table of this code. The encoding of the incoming data is accomplished by dividing the source sequence into 2-, 3-, and 4-bit partitions to match the entries in the code table and then mapping them into the corresponding channel representations.

**TABLE  7.5**  Variable-Length  Synchronous (2,7) Code

| Data | | Code |
|------|------|------|
| 10 | ← → | 1000 |
| 11 | ← → | 0100 |
| 011 | ← → | 000100 |
| 010 | ← → | 001000 |
| 000 | ← → | 100100 |
| 0011 | ← → | 00100100 |
| 0010 | ← → | 00001000 |

The codebook is to be used as follows. For example, let the source sequence be 010111010. After the appropriate parsing, one obtains the following:

$$in : 010\ 11\ 10\ 10 \ldots$$

which, using Table 7.5, is transformed into the corresponding output sequence:

$$001000\ 0100\ 1000\ 1000 \ldots$$

Decoding of the received message can be achieved with an 8-stage shift register. The incoming message is shifted into the register every two-channel clock cycle. The contents of the register are decoded with simple logic. Channel bit errors propagate to at maximum of four decoded data bits. The hardware requirements of this code are extremely attractive.

### 7.2.4  Rate-2/3, (1,7) Code

Jacoby and Kost described a rate-2/3, (1,7) code with full-word look-ahead used in a particular magnetic disk file.[5] To understand the algorithm of the 2/3-rate look-ahead code, begin with the basic encoding table presented in Table 7.6. The 2/3-rate code is quite similar to a fixed-length block code, where data words of two bits are converted into codewords of three bits. The basic encoding table lists this conversion for the four basic source words. Encoding is done by taking

**TABLE 7.6**   Basic Coding Table (1,7) Code

| Data | Code |
|------|------|
| 00 | 101 |
| 01 | 100 |
| 10 | 001 |
| 11 | 010 |

one source word at a time and always looking ahead to the next source word. After conversion of the source symbols to code symbols—and provided there is no violation of the $d$ constraint at the codeword boundaries—the first codeword (the first three bits) will be made final.

There is always the possibility that the last word, up to the point reached in the encoding process, may change when one looks ahead to the next word. When the $d$ constraint is violated, there are four combinations of codewords that may lead to this, and then substitutions are required in order to eliminate successive 1s. The process of substitutions in these four combinations is revealed in Table 7.7.

**TABLE 7.7**   Substituting Coding Table (1,7) Code

| Data | Code |
|------|------|
| 00.00 | 101.000 |
| 00.01 | 100.000 |
| 10.00 | 001.000 |
| 10.01 | 010.000 |

The encoding function can be expressed in the form of Boolean equations. In decoding the codewords, those cases in which substitution is made during encoding can be treated without ambiguity, because all three bits of the succeeding codeword are simultaneously 0.

## 7.2.5   EFM Code

EFM, used in the Compact Disc, is both a DC-free and RLL code. There are two reasons why EFM suppresses the low-frequency components. In the first place, the servosystems for track following and focusing are controlled by low-frequency signals, so that low-frequency components of the information signal could interfere with the servosystems. In the Compact Disc system the frequency range from 20 kHz to 1.5 MHz is used for information transmission, whereas the servosystems operate on signals in the range 0 to 20 kHz. The second reason is that low-frequency disturbances in the retrieved signal resulting from fingerprints or dirt on the disc can be filtered out without distorting the data signal itself. The (high-pass) filter should be chosen to pass the signal and to reject the low-frequency noise. It cannot simultaneously do both completely.

Under EFM rules the data bits are translated eight at a time into 14 channel bits, with minimum run-length parameter $d = 2$ and a maximum run-length parameter $k = 10$ channel bits (this means at least 2 and at most 10 successive 0s between successive 1s). Part of the EFM coding table is presented in Table 7.8, which shows the decimal representation of the 8-bit source word (left column) and its 14-bit channel representation.

It should be appreciated that the codewords are described in NRZI (non–return-to-zero inverted) notation, which means that a 1 represents a transition of either positive or negative

**TABLE 7.8**   Part of the EFM Coding Table

| Data | Code |
| --- | --- |
| 100 | 01000100100010 |
| 101 | 00000000100010 |
| 102 | 01000000100010 |
| 103 | 00100100100010 |
| 104 | 01001001000010 |
| 105 | 10000001000010 |
| 106 | 10010001000010 |
| 107 | 10001001000010 |
| 108 | 01000001000010 |
| 109 | 00000001000010 |
| 110 | 00010001000010 |
| 111 | 00100001000010 |
| 112 | 10010010000010 |
| 113 | 00100000100010 |
| 114 | 01000010000010 |
| 115 | 00000010000010 |
| 116 | 00010001000010 |
| 117 | 00100001000010 |
| 118 | 01001000000010 |
| 119 | 00001001001000 |
| 120 | 10010000000010 |
| 121 | 10001000000010 |
| 122 | 01000000000010 |
| 123 | 00001000000010 |

polarity, and a 0 represents the absence of a transition. As a result, the lengths of the pits and lands recorded lie between 3 and 11 unity lengths. At least two bits, called *merging bits,* are required to ensure that the run-length conditions continue to be satisfied when the codewords are cascaded. If the run length is in danger of becoming too short, 0s are chosen for the merging bits. If the run length is too long, a 1 is chosen for one of the merging bits. By so doing, a large measure of freedom is retained in the choice of the merging bits. This freedom is used for minimizing the low-frequency content of the signal. In itself, two merging bits would be sufficient for continuing to satisfy the run-length conditions. A third merging bit is necessary, however, to give sufficient freedom for effective suppression of low-frequency content, even though it entails a loss of 6 percent of the information density on the disc.

The merging bits contain no information, and they are removed from the bit stream in the demodulator. Figure 7.1 illustrates, finally, how the merging bits are determined. The measure of the low-frequency content is the running digital sum (RDS). It is the difference between the totals of pit and land lengths accumulated from the beginning of the disc. At the top, two 8-bit data words are shown. From the $d = 2$ rule, the first of the merging bits in this case must be a 0. This position is marked "$x$." In the two following positions, the choice is free. These are marked "$m$." The three possible choices "xmm" = "000," "010," and "001" would give rise to the patterns of pits as illustrated and to the indicated waveforms of the RDS, on the assumption that the RDS was equal to 0 at the beginning. The system now opts for the merging combination that makes the RDS at the end of the second codeword as close to zero as possible, i.e., "000" in this case. If the initial value had been −3, the merging combination "001" would have been chosen. The power spectral density (PSD) function of conventional EFM has been obtained by computer simulation. Results are plotted in Fig. 7.2. Both axes are normalized for fixed user bit rate $f_b$.

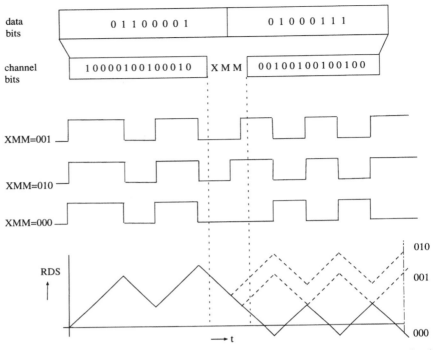

**FIGURE 7.1**  Strategy for minimizing the running digital sum (RDS). Eight user bits are translated into 14 channel bits. The 14 bits are merged by means of three merging bits in such a way that the run-length conditions continue to be satisfied. The proviso that there should be at least two 0s between 1s requires a 0 at the first merging bit position. In this case, therefore, there are three alternatives for the merging bits: "000," "010," and "001." The encoder chooses the alternative that gives the lowest absolute value of the RDS at the end of a new codeword, i.e., "000" in this case.

## 7.2.6  EFMPlus

An extension of the Compact Disc family, the Digital Versatile Disc (DVD), is a proposal for a new optical recording medium with a storage capacity seven times higher than the conventional Compact Disc (see Chapter 12). The major part of the capacity increase is achieved by the use of optics, shorter laser wavelength, and a larger numerical aperture that reduces the spot diameter by a factor of 1.5. The track formed by the recorded pits and lands as well as the track pitch can be reduced by the same factor. The storage capacity is further increased by a complete redesign of the logical format of the disc, including a more powerful error-correction and recording code, EFMPlus.[6] The rate of EFMPlus is 8:16 and it is therefore capable of recording 6 percent more user information than is possible with EFM, whose rate is 8:17.

The principle of operation of the encoder can be represented by a finite-state sequential machine with an 8-bit input, a 16-bit output, and four states, which are functions of the (discrete) time. The states are said to be connected by edges and the edges, in turn, are labeled with tags called "words." A word in this context is a 16-bit sequence that obeys the prescribed ($d = 2, k = 10$) constraints. Each of the four states is characterized by the type of words that enter, or leave, the given state. The states and word sets are characterized as follows:

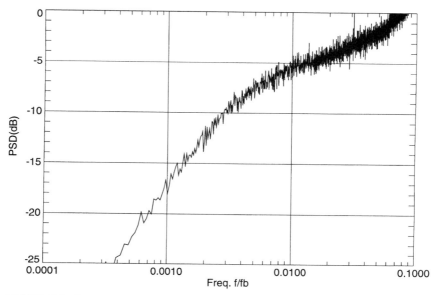

**FIGURE 7.2**   Spectrum of conventional EFM.

- Words entering state 1 end with $\{0, 1\}$ trailing 0
- Words entering state 2 and 3 end with $\{2, \ldots, 5\}$ trailing 0s
- Words entering state 4 end with $\{6, \ldots, 9\}$ trailing 0s

The words leaving the states are chosen in such a way that the concatenation of words entering a state and those leaving that state obey the $(d = 2, k = 10)$ channel constraints. Words emerging from states 2 and 3 comply with the above run-length constraints, but they also comply with other conditions. Words leaving state 2 have been selected such that the first (msb) bit, $x_1$, and the thirteenth bit, $x_{13}$, are both equal to 0. In a similar fashion, words leaving state 3 are characterized by the fact that the 2-tuple $x_1 x_{13}$ does not equal 00. The attributes of the four states just defined guarantee that any walk through the graph, stepping from state to state, produces a $(d = 2, k = 10)$-constrained sequence by reading the words tagged to the edges that connect the states.

Given these definitions, it is a simple matter to compute the number of edges (or words) that leave each of the four states and find that from each of the states at least 351 words are leaving. An encoder is constructed by assigning a source word to each of the 351 edges that leave each state. Excess edges are removed. Also, edges are removed that could generate the unique synchronization pattern. As a result, each edge in the graph now has two labels—namely, a 16-bit word and a source word number from 0 to 344. Given the source word and the encoder state, the encoder will transmit the word "tagged" to the same edge as the source word at hand. The encoder requires accommodation for only 256 source words. The surplus, $344 - 256 = 88$, words will be used for controlling the low-frequency power.

The encoder is defined in terms of three sets (the inputs, the outputs, and the states) and two logical functions (the output function and the next-state function). The specific codeword, denoted by $\underset{\sim}{x}_t$, transmitted by the encoder at instant $t$ is, of course, a function of the source word $\underset{\sim}{b}_t$ that enters the encoder, but depends further on the particular state, $s_t$, of the encoder. Similarly, the "next" state at instant $t+1$ is a function of $\underline{x}_t$ and $s_t$. The output function $h(\cdot)$ and the next-state function $g(\cdot)$ can be succinctly written as follows:

$$\underline{x}_t = h(\underline{b}_t, s_t)$$

$$s_{t+1} = g(\underline{b}_t, s_t)$$

Both the output function $h(\cdot)$ and the next-state function $g(\cdot)$ are described by four lists with 351 entries. A part of the output function and the next-state function is listed in Table 7.9.

**TABLE 7.9**   Part of the EFMPlus Coding Table

| $i$ | $h(i, 1),\ g(i, 1)$ | $h(i, 2),\ g(i, 2)$ | $h(i, 3),\ g(i, 3)$ | $h(i, 4),\ g(i, 4)$ |
|---|---|---|---|---|
| 0 | 0010000000001001, 1 | 0100000100100000, 2 | 0010000000001001, 1 | 0100000100100000, 2 |
| 1 | 0010000000010010, 1 | 0010000000010010, 1 | 1000000100100000, 3 | 1000000100100000, 3 |
| 2 | 0010000100100000, 2 | 0010000100100000, 2 | 1000000000010010, 1 | 1000000000010010, 1 |
| 3 | 0010000001001000, 2 | 0100010010000000, 4 | 0010000001001000, 2 | 0100010010000000, 4 |
| 4 | 0010000010010000, 2 | 0010000010010000, 2 | 1000000100100000, 2 | 1000000100100000, 2 |
| 5 | 0010000000100100, 2 | 0010000000100100, 2 | 1001001000000000, 4 | 1001001000000000, 4 |
| 6 | 0010000000100100, 3 | 0010000000100100, 3 | 1000100100000000, 4 | 1000100100000000, 4 |
| 7 | 0010000001001000, 3 | 0100000000010010, 1 | 0010000001001000, 3 | 0100000000010010, 1 |
| 8 | 0010000010010000, 3 | 0010000010010000, 3 | 1000010010000000, 4 | 1000010010000000, 4 |

Table 7.9 has an entry column that describes the source (input) word $i$ by an integer between 0 and 255. The table also shows $h(i, s)$ the 16-bit output to a particular input $i$ when the encoder is in one of the four states $s$. The words are written in NRZI notation. The third, fifth, seventh, and ninth columns show the next-state function $g(i, s)$. Let the encoder graph be initialized at state 1 and let the source sequence be 8, 3, 4. The response to input 8, while in state 1, equals $h(8, 1) =$ (see Table 7.9) "0010000010010000." The new state becomes $g(8, 1) = 3$. As a result, the response to input 3, while in state 3, is "0010000001001000." In the next clock cycle the encoder state becomes $g(3, 3) = 2$. From state 2 with the input equal to 4, one finds from the table that the corresponding output is $h(4, 2) = $ "0010000010010000."

Under EFM rules, it suffices to observe 14 of the 17 channel bits that constitute an EFM codeword. In contrast, decoding of the new code is done by a logic array that translates 16 channel bits of the current codeword plus the first bit, $x_1$, and the thirteenth bit, $x_{13}$, of the upcoming codeword into 8 bits.

The encoder defined earlier can freely accommodate 344 source words. Considering that accommodation for only 256 source words is needed, the surplus 88 words can be used for minimizing the power at low frequencies, in short, the DC-control. The suppression of low-frequency components or DC-control is done in the same vein as in the EFM code, namely, by controlling the running digital sum (RDS). The surplus words are used as an alternative channel representation of the source words $0, \ldots, 87$. The full encoder is described by two tables called *main* and *substitute tables*, respectively (Fig. 7.3). The main table describes an encoder table of 256 inputs. The substitute table shows a similar table of 88 words, which act as alternative representations of the source words $0, \ldots, 87$ of the main table. The source words $0, \ldots, 87$ can thus be represented by the designated entries of the main table or alternatively by the entries of the substitute table. The power spectral density of the new code has been computed by a computer program, which has simulated the encoder algorithm. Results are plotted in Fig. 7.4. Note that the PSD is only 1–2 dB worse than that of conventional EFM.

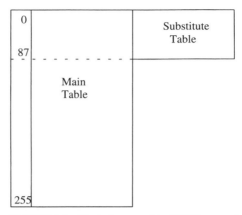

**FIGURE 7.3**    Block diagram of the EFMPlus encoder.

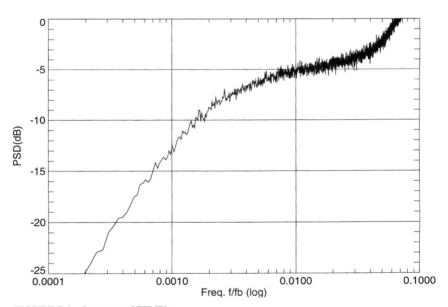

**FIGURE 7.4**    Spectrum of EFMPlus.

## 7.3   DC-FREE CODES

Binary sequences with spectral nulls at zero frequency have found widespread application in optical and magnetic recording systems. *DC-balanced codes,* as they are also called, have a long history, and their application is certainly not confined to recording practice. Since the early days of digital communication over cable, DC-balanced codes have been employed to counter the effects of low-frequency cut-off due to coupling components, isolating transformers, etc.

Suppression of the low-frequency components is achieved by restricting the unbalance of the transmitted positive and negative pulses. In optical recording, DC-balanced codes are employed to circumvent or reduce interaction between the data written on the disc and the servosystems that follow the track.

### 7.3.1   Properties of DC-Balanced Sequences

The running digital sum (RDS) of a sequence plays a significant role in the analysis and synthesis of codes whose spectrum vanishes at the low-frequency end. Let

$$\{x_i\} = \{\ldots, x_{-1}, x_0, \ldots, x_i, \ldots,\} x_i \in \{-1, 1\}$$

be in binary sequence. The (running) digital sum $z_i$ is defined as follows:

$$z_i = \sum_{j=-\infty}^{i} x_j = z_{i-1} + x_i \tag{7.3}$$

Figure 7.5 portrays the various signals. For example, the RDS is supposed to stay within the prescribed limits, and as a result the RDS may take at maximum seven values. An essential point (which is not difficult to prove) is that if $z_i$ is bounded, the spectral density vanishes at DC. All codes designed to suppress the low frequencies do so by guaranteeing that the RDS stays within a limited range.

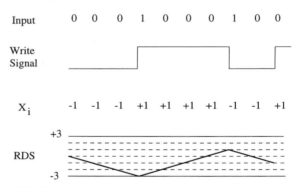

**FIGURE 7.5**   Running digital sum (RDS) versus time. Input symbols (NRZI) are translated into the (NRZ) write signal and the channel bits $x_i$. In this example the RDS assumes at most seven values.

### 7.3.2   Code Implementations

Practical coding schemes devised to achieve suppression of low-frequency components are mostly constituted by block codes. The source digits are grouped in source words of $m$ digits. The source words are translated using a conversion table, known as *codebook,* into blocks of $n$ digits. The essential principle of operation of a channel encoder that translates arbitrary source data into a DC-free channel sequence is remarkably simple. Essentially, there are three approaches that have actually been used for DC-balanced code design: zero-disparity code, low-disparity code, and polarity bit code.

The disparity of a codeword is defined as the excess of the number of 1s over the number of 0s in the codeword. Thus the codewords 000110 and 100111 have disparity $-2$ and $+2$, respectively. Of special interest are zero-disparity codewords, which contain equal numbers of 1s and 0s. The obvious method for constructing DC-balanced codes is to employ codewords that contain an equal number of 1s and 0s or, stated alternatively, to employ zero-disparity codewords that have a one-to-one correspondence with the source words.

In low-disparity codes, the translations source word to codeword are not one-to-one. The source words operate with two alternative translations, which are of equal or opposite disparity. During transmission, the choice of a specific translation is made in such a way that the accumulated disparity, or the running digital sum, of the encoded sequence, after transmission of the new codeword, is as close to zero as possible.

The third coding method is attractive because no look-up tables are required for encoding and decoding. The $(n - 1)$ source symbols are supplemented by one symbol called the *polarity bit.* The encoder has the option to transmit the $n$-bit words without modification or to invert all symbols. Again, as in the low-disparity code, the choice of a specific translation is made in such a way that the accumulated disparity is as close to zero as possible.

## 7.3.3  Zero-Disparity Coding Schemes

Probably the most obvious basic method to generate DC-free sequences (and certainly the easiest to describe) is constituted by zero-disparity codewords. In this scheme each source word is uniquely represented by a codeword that contains equally many 1s and 0s. Clearly, zero-disparity codewords are possible only if the code-word length $n$ is even. The number of zero-disparity codewords, $N_0$, of binary symbols ($n$ even) is given by the following binomial coefficient:

$$N_0 = \binom{n}{n/2}$$

Table 7.10 shows the number of zero-disparity codewords as a function of the codeword length $n$. Table 7.10 also presents the code rate, $R_0$, given by

$$R_0 = \frac{1}{n} \log_2 N_0$$

The zero-disparity codewords can be concatenated without a merging rule. That is, the sequence is encoded without information about the history and, obviously, there is a fixed relationship between codewords and source words.

**TABLE  7.10**  Number of Zero-Disparity Codewords and Code Rate Versus Codeword Length $n$

| $n$ | $N_0$ | $R_0$ | $N_1$ | $R_1$ |
|---|---|---|---|---|
| 2 | 2 | 0.500 | 3 | 0.792 |
| 4 | 6 | 0.646 | 10 | 0.830 |
| 6 | 20 | 0.720 | 35 | 0.855 |
| 8 | 70 | 0.766 | 126 | 0.872 |
| 10 | 252 | 0.798 | 462 | 0.885 |
| 12 | 924 | 0.821 | 1716 | 0.895 |
| 14 | 3432 | 0.839 | 6435 | 0.904 |

In the most elementary case, $n = 2$, there are two codewords—namely, $(1, -1)$ and its inverse, $(-1, 1)$. This block code (called *biphase* or *Manchester code*) is a popular encoding function in low-end magnetic disc drives. The reasons for the popularity are self-clocking capability and relatively simple encoding and decoding circuits.

The spectra of zero-disparity codes are shown in Fig. 7.6. It can be seen that the spectra indeed vanish and are also small in the neighborhood of the zero frequency. Note that in the frequency range where the power spectral density function is small, the spectral notch width becomes larger when the codeword length $n$ and (see Table 7.10) rate of the code becomes smaller. For a wider frequency range of suppressed components, one has to pay more in terms of redundancy of the sequence. This reflects what we intuitively expect and illustrates very well the old cliché, "there is no such thing as a free lunch."

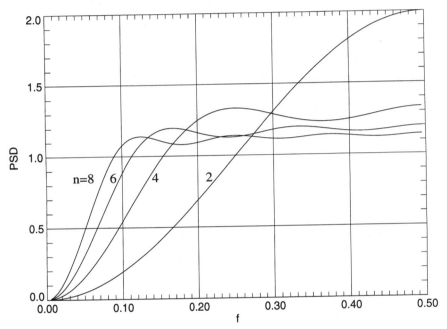

**FIGURE 7.6** Power density function of zero-disparity codes versus frequency with the codeword length $n$ as a parameter.

### 7.3.4 Low-Disparity Coding Schemes

Besides the set of zero-disparity codewords, sets of codewords with nonzero disparity are used in low-disparity codes. Source words can be represented by two alternative channel representations of opposite disparity. The encoder opts for a particular channel representation with the aim of minimizing the absolute value of the running digital sum after transmission of the new codeword. A simple example of a rate-3/4 code is shown in Table 7.11.

The left column shows the list of the eight 3-bit source words. The word transmitted is taken from the column denoted "State 1" or "State 2." Let's say that the encoder is in state 1. Then, for example, the source word 111 is represented by 1101. On the other hand, if the encoder

**TABLE 7.11** Simple Low-Disparity DC-Free Code

| Source | State 1 | State 2 |
|--------|---------|---------|
| 000 | 0011 | 0011 |
| 001 | 0101 | 0101 |
| 010 | 1001 | 1001 |
| 011 | 0110 | 0110 |
| 100 | 1010 | 1010 |
| 101 | 1100 | 1100 |
| 110 | 1110 | 0001 |
| 111 | 1101 | 0010 |

would have occupied state 2, 0010 would have been transmitted. If a zero-disparity codeword is transmitted (source words 000 to 101), then the encoder remains in the same state. If, however, a low-disparity codeword is transmitted, then the encoder changes state. It is easily verified that, as codewords of zero or alternating disparity are transmitted, the running digital sum of the sequence will be limited. Clearly, the number of source words, $N_1$, that can be catered by low-disparity codes is given by the following ($n$ even):

$$N_1 = \binom{n}{n/2} + \binom{n}{n/2 + 1}$$

Table 7.10 shows $N_1$ and the code rate

$$R_1 = \frac{1}{n} \log_2 N_1$$

as a function of the codeword length $n$.

*8b10b Code.* In digital recording on magnetic tape it has been found that a rate-8/10, DC-balanced channel code has attractive features both in terms of system penalty and hardware realization. Most of the implementations in use are block codes that translate 1 byte (8 bits) into 10 channel symbols. Clearly, a zero-disparity block code is impossible because the number of available 10-bit zero-disparity codewords, 252, is smaller than the required 256. A two-state encoder offers the freedom of at maximum $252 + 210 = 462$ codewords. Because only 256 codewords are required, this evidently offers a large variety of choices. The codebook can be tailored to particular needs, such as minimum DC-content and/or ease of implementation. It is therefore not surprising that numerous variants have been described in the literature—patent literature in particular.

Table 7.12 shows the main parameters of selected DC-constrained 8b10b codes documented in the literature. A code of great practical interest is the Fukuda2 code, employed as the recording code in the DAT digital audio tape recorder.[8] The 8b10b code used in the DAT recorder is designed to function well in the presence of cross-talk from neighboring tracks, allow the use of the rotary transformer, and have a small ratio of maximum-to-minimum run length in order to ease overwrite erasure. Essentially, the code operates on the low-disparity principle. The encoder has two states, and the codebook contains 153 zero-disparity codewords and 103 codewords of disparity $\pm 2$. The hardware of the encoder and decoder has been reduced by computer optimizing of the relationship between the 8-bit source words and the 10-bit codewords. This has been done so that the conversion can be performed in a small programmed logic array.[8] The maximum run length produced by the code is 4. A variant of this 8b10b code is also reported by Fukuda et al.

**TABLE 7.12**   Main Parameters of $R = 8/10$
DC-Constrained Codes

| Reference | $N$ | $T_{max}$ |
|---|---|---|
| Shirota[7] | 7 | 6 |
| Fukuda1[8] | 7 | 5 |
| Fukuda2[8] | 7 | 4 |
| Immink[9] | 6 | 5 |

The DCC recorder is also equipped with a low-disparity rate-8/10 DC-free code.[9] The code was designed in such a way that the channel sequence can take at most six digital sum values. The maximum run length is five channel symbols.

### 7.3.5   Polarity Bit Principle

The polarity bit code is a DC-balanced code where the transmitter appends one extra bit to $(n-1)$ source symbols. The extra bit is initially set to 1. If the accumulated disparity at the start of the transmission of a codeword and the disparity of the new $n$-bit codeword have the same sign, then all symbols in the $n$-bit word (including the polarity bit) are inverted before transmission; otherwise, the $n$-bit word is left intact. The accumulated disparity is tallied by a reversible counter connected to the output line. The receiving translator is simpler, since it merely observes a possible inversion of a received word by inspecting the sign of the polarity bit and, depending on whether this is positive or negative, reads the remaining bits directly or complemented. The polarity bit encoding principle has the virtue of a very simple implementation, since there is no need for look-up tables. The polarity bit code is, at this moment, the only feasible option for constructing high-rate DC-free codes.

## 7.4   PILOT-TRACKING TONES

Dynamic track-following mechanisms are at present used in consumer-type video or digital audio tape recorders that aim to alleviate the mechanical inaccuracies of the recorder. In the digital videocassette (DVC) recorder (see Chapter 20), the servo position information is recorded as low-frequency components called *pilot-tracking tones*.[10] The principle of operation is essentially as follows. On even tracks the pilot tone has a frequency $f_1$ and on odd tracks the pilot tone frequency is $f_2$. Servo position information is developed from the read-back signals by subtracting the amplitude of the $f_2$ component from that of the $f_1$ component. These two components can be separated with band filters because they differ in frequency. As the head moves off-track in one direction, the amplitude of one component decreases and the amplitude of the other increases. On the basis of the difference error signal and the control technique employed, this then instructs the control mechanism. In other words, by observing the difference between the pilot tone amplitudes of adjacent tracks, one can tell whether movement is off-track to the right or to the left. Experiments have shown that adding analog pilot tones to the write current will result in serious interference between the digital data and the tones. Furthermore, the precise amplitude of an analog pilot tone depends on the characteristics of the head-tape combination, which means careful adjustment of the write current is necessary for each of the heads. This fact precludes the technique, for example, of simply adding a sinusoidal waveform to the binary data. The only option available is to embed the tones in the recorded binary sequence.

### 7.4.1 Description of the 24 → 25 Channel Code

The main parameters of the 24 → 25 code can be summarized as follows:

- Rate = 24/25.
- Minimum run length = 1.
- Maximum run length = 9. This maximum run-length feature is not guaranteed, but the probability of occurrence of run lengths larger than 9 is smaller than $5 \times 10^{-5}$.
- Perfectly DC-free, $-3$ dB at $f \leq f_b/500$.
- Provisions for tracking (channel bit rate 21 Mb/s):

  Pilot tone $f_1 = f_b/90$ (230 kHz), SNR $\geq 20$ dB in 300 Hz.

  Pilot tone $f_2 = f_b/60$ (350 kHz), SNR $\geq 20$ dB in 300 Hz.

  Notches around the pilot tones $f_1$ and $f_2$ stabilize the amplitude.

  Notch depth $\geq 5$ dB (resolution bandwidth 300 Hz). Amplitude stability within 0.3 dB peak-to-peak. Notches at frequencies $f_1$ and $f_2$ improve the tracking SNR by reducing the in-track code noise.

- No coding/decoding tables are needed.

### 7.4.2 Tracking Format

The recorder system uses two pairs of heads that are diametrically positioned with respect to one another on the drum. The tracking format requires three types of tracks with different frequency spectra. The track types—F0, F1, and F2—are listed in Table 7.13. The footprint for the proposed scanner configuration is shown in Table 7.14.

**TABLE 7.13**  Definition of Types of Heads

| Type | Notches | Pilot |
|------|---------|-------|
| F0 | DC, $f_1$, $f_2$ | No pilot tone |
| F1 | DC, $f_1$, $f_2$ | $f_1$ |
| F2 | DC, $f_1$, $f_2$ | $f_2$ |

**TABLE 7.14**  Configuration of Heads and Spectra

| | Tracks | | | | | |
|---|---|---|---|---|---|---|
| | F0 | F1 | F0 | F2 | F0 | F1 |
| Head number | 0 | 1 | 2 | 3 | 0 | 1 |

It can be seen in Table 7.14 that each head pair has one head of type F0 (no pilot tone), whereas the second head is alternately of type F1 or F2. During the reading of the F0-type tracks, the cross-talk is measured from the neighboring tracks, which are of type F1 or F2. Servo position information is developed by subtracting the amplitude of the $f_1$ component from the $f_2$ component. These two components can be separated with band filters since they differ in frequency. As the head moves off-track in one direction, the amplitude of one component decreases, whereas

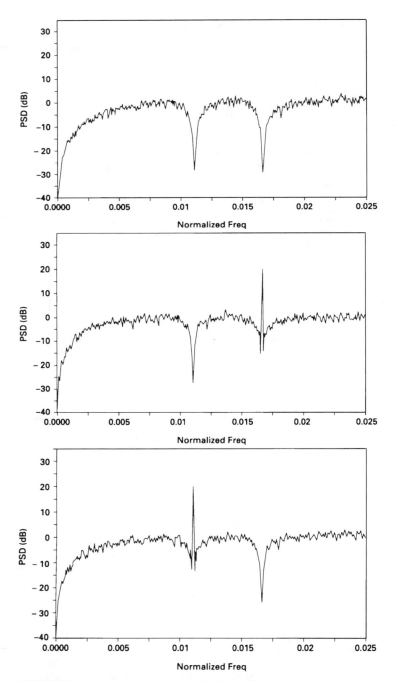

**FIGURE 7.7**    Spectra of the track types F0, F1, and F2.

the amplitude of the other increases. The difference error signal is used for steering the actuator on which the head pairs are mounted. The spectral notches are required to cancel any interference between the written data and the servo detection system.

The $24 \rightarrow 25$ code operates similarly to the polarity bit code. A polarity bit is stuffed in each sequence of 24 successive input data bits to create the freedom to shape the spectrum of the encoded bit stream. By setting this control bit to 0 or 1, the polarity of 25 bits at the output of the coder can be chosen. The sender opts for that value of the control bit that minimizes the accumulated signal power of the transmitted codewords at three predefined frequencies: DC, $f_1$, and $f_2$. Furthermore, with subtraction of the desired pilot tone before "measurement" of the power at the pilot tone frequencies, the pilot tone will appear at the output of the precoder. Should the maximum run length on the channel become larger than the predefined maximum (i.e., nine) the spectral optimization is overruled. The resulting spectra are shown in Fig. 7.7.

## GLOSSARY

**Code**    Set of rules for transforming a sequence of digital symbols into another one with some specified properties. Examples are source code, error correcting code, and recording code.

**DC-free code**    Sequences generated under the rules of a DC-free code show a spectral null at the zero frequency.

**Disparity**    The difference between the numbers of 0s and 1s in a binary codeword. For example, the disparity of "11011" is $4 - 1 = 3$.

**Eight-to-fourteen modulation (EFM)**    An example of a run-length–limited code used in the Compact Disc.

**Merging bits**    These are required to ensure that the run-length conditions continue to be satisfied when codewords are cascaded.

**Pilot-tracking tones**    These are used in the Digital Videocassette (DVC) Recorder for deriving the servo position information.

**Recording code**    A recording code converts the input stream to a signal suitable for the specific physical recorder requirements.

**Run length**    The length of time usually expressed in channel bits between consecutive transitions.

**Run-length–limited code**    The run lengths of the generated sequence lie between specified minimum and maximum values.

## REFERENCES

1. T. Himeno, Tanaka, M., Katoku, T., Matsumoto, K., Tamura, M., and Min-Jae, H., "High-density magnetic tape recording by a nontracking method," *Electronics and Communications in Japan, Part 2,* vol. 76, no. 5, 1993, pp. 83–93.

2. K. A. S. Immink, *Coding Techniques for Digital Recorders,* Prentice Hall International (U.K.) Ltd., Englewood Cliffs, N.J., 1991.

3. J. Watkinson, *The Art of Digital Audio,* Focal Press, London, 1988.

4. D. T. Tang, and Bahl, L. R., "Block codes for a class of constrained noiseless channels," *Information and Control,* vol. 17, 1970, pp. 436–461.

5. G. V. Jacoby, and Kost, R., "Binary two-thirds rate code with full word look-ahead," *IEEE Transactions on Magnetics,* vol. MAG-20, no. 5, Sept. 1984, pp. 709–714. See also M. Cohn, Jacoby, G. V., and Bates C.A., III, U.S. Patent 4,337,458, June 1982.

6. K. A. S. Immink, "EFMPlus: The coding format of the multimedia compact disc," *IEEE Transactions on Consumer Electronics,* vol. CE-41, August 1995, pp. 491–497.

7. N. Shirota, "Method and apparatus for reducing dc components in a digital information system," U.S. Patent 4,387,364, 7 June 1983.

8. S. Fukuda, Kojima, Y., Shimpuku, Y., and Odaka, K., "8/10 modulation codes for digital magnetic recording," *IEEE Transactions on Magnetics,* vol. MAG-22, no. 5, September 1986, pp. 1194–1197.

9. K. A. S. Immink, "Construction of binary DC-constrained codes," *Philips Journal of Research*, vol. 40, 1985, pp. 22–39.

10. J. A. H. Kahlman, and Immink, K. A. S., "Channel code with embedded pilot tracking tones for DVCR," *IEEE Transactions on Consumer Electronics,* vol. CE-41, no. 1, February 1995, pp. 180–185.

## ABOUT THE AUTHOR

Kees A. Schouhamer Immink received the M.S. and Ph.D. degrees from the Eindhoven University of Technology. He joined the Philips Research Laboratories in Eindhoven, the Netherlands, in 1968, where he currently holds the position of research fellow. He is also guest professor at the Institute of Experimental Mathematics at Essen University, Germany.

Dr. Immink started his career in consumer recording technology in 1974 when he joined the research group "optics," where he and his colleagues conducted pioneering experiments with optical video disc recording. His main contributions were in the fields of tracking servosystems and electronic signal processing. He participated, in 1979, in the discussions between Sony and Philips that led to the worldwide standard for the Compact Disc. He developed the recording format EFM, which was adopted as the standard for the Compact Disc and has been extensively employed in a variety of digital audio players and home-storage products such as CD-ROM, CD-I, and MiniDisc. He contributed to the design and development of a variety of consumer digital audio and video recorders such as the Compact Disc Video, DAT, Digital Compact Cassette (DCC), and, very recently, the DVD.

Dr. Immink has written numerous articles, coauthored three books, and has been granted more than 30 U.S.-issued patents. He is a fellow of the Audio Engineering Society (AES), the Institution of Electrical Engineers (IEE), and the Institute of Electrical and Electronics Engineers (IEEE). For his part in the digital audio and video revolution he was awarded the AES Silver Medal, the IEE J. J. Thomson Medal, The Society of Motion Picture and Television Engineers (SMPTE) Alexander M. Poniatoff Gold Medal for Technical Excellence, and the IEEE Masaru Ibuka Consumer Electronics Award. Dr. Immink is a member of the Royal Netherlands Academy of Arts and Sciences.

# KEY ENABLING STANDARDS

# CHAPTER 8
# MPEG STANDARDS

## SECTION 1
## DIGITAL VIDEO CODING STANDARDS

**Thomas Sikora**
*Heinrich-Hertz Institute, Berlin, Germany*

## 8.1 INTRODUCTION

Modern image and video compression techniques today offer the possibility to store or transmit the vast amount of data necessary to represent digital images and video in an efficient and robust way. New audiovisual applications in the fields of communications, multimedia, and broadcasting became possible based on digital video-coding technology. As manifold as applications for image coding are today, so too are the different approaches and algorithms and so too were the first hardware implementations and even systems in the commercial field, such as private teleconferencing systems.[1,2] However, with the advances in VLSI technology it became possible to open more application fields to a larger number of users and therefore the necessity for video-coding standards arose. Commercially, international standardization of video communication systems and protocols aims to serve two important purposes: interoperability and economy of scale. Interworking between video communication equipment from different vendors is a desirable feature for users and equipment manufacturers alike. It increases the attractiveness for buying and using video communication equipment because it enables large-scale international video data exchange via storage media or via communication networks. An increased demand can lead to economy of scale—the mass production of VLSI systems and devices—which in turn makes video equipment more affordable for a wide field of applications and users.

From the beginning of the 1980s on, a number of international video and audio standardization activities started within CCITT, followed by CCIR and ISO/IEC.[3] The Moving Picture Experts Group (MPEG) was established in 1988 in the framework of the Joint ISO/IEC Technical Committee (JTC 1) on Information Technology with the mandate to develop standards for coded representation of moving pictures, associated audio, and their combination when used for storage and retrieval on digital storage media with a bit rate up to about 1.5 Mb/s. The standard was called MPEG-1 and was issued in 1992. The scope of the group was later extended to provide appropriate MPEG-2 video and associated audio compression algorithms for a wide range of audiovisual applications at substantially higher bit rates not successfully covered or envisaged by the MPEG-1 standard. Specifically, MPEG-2 was given the charter to provide video quality not lower than NTSC/PAL and up to CCIR 601 quality with bit rates targeted between 2 and 10 Mb/s. Emerging applications—such as digital cable TV distribution, networked database services via ATM, digital VTR applications, and satellite and terrestrial digital broadcasting distribution—were seen to benefit from the increased quality expected to result from the emerging MPEG-2 standard. The MPEG-2 standard was released in 1994. Table 8.1 summarizes the primary applications and quality requirements targeted by the MPEG-1 and MPEG-2 video standards together with examples of typical video input parameters and compression ratios achieved.

**TABLE 8.1** Typical MPEG-1 and MPEG-2 Coding Parameters

|  | MPEG-1 | MPEG-2 |
| --- | --- | --- |
| Standardized | 1992 | 1994 |
| Main application | Digital video on CD-ROM | Digital TV (and HDTV) |
| Spatial resolution | CIF format (1/4 TV) | TV (4 × TV) |
|  | ≈ 288 × 360 pixels | ≈ 576 × 720 pixels |
|  |  | (1152 × 1440 pixels) |
| Temporal resolution | 25–30 frames/s | 50–60 fields/s |
|  |  | (100–120 fields/s) |
| Bit rate | 1.5 Mb/s | ≈ 4 Mb/s |
|  |  | (≈ 20 Mb/s) |
| Quality | Comparable to VHS | Comparable to NTSC/PAL for TV |
| Compression ratio over PCM | ≈ 20–30 | ≈ 30–40 |
|  |  | (≈ 30–40) |

The MPEG-1 and MPEG-2 video compression techniques developed and standardized by the MPEG group have developed into important and successful video-coding standards worldwide, with an increasing number of MPEG-1 and MPEG-2 VLSI chip sets and products becoming available on the market. One key factor for the success is the generic structure of the MPEG standards, supporting a wide range of applications and applications-specific parameters.[3–4] To support the wide range of applications profiles a diversity of input parameters—including flexible picture size and frame rate—can be specified by the user. Another important factor is the fact that the MPEG group only standardized the decoder structures and the bitstream formats. This allows a large degree of freedom for manufacturers to optimize the coding efficiency (or, in other words, the video quality at a given bit rate) by developing innovative encoder algorithms even after the standards were finalized.

The purpose of Section 1 is to provide an overview of the MPEG-1 and MPEG-2 video-coding algorithms and standards and their role in video communications. Section 8.2 reviews the basic concepts and techniques that are relevant in the context of the MPEG video compression standards. In Sections 8.3 and 8.4 the MPEG-1 and MPEG-2 video-coding algorithms are outlined in more detail. Furthermore, the specific properties of the standards related to their applications are presented. Section 8.5 discusses the performance of the standards and their success in the marketplace.

## 8.2 FUNDAMENTALS OF MPEG VIDEO COMPRESSION ALGORITHMS

Generally speaking, video sequences contain a significant amount of statistical and subjective redundancy within and between frames. The ultimate goal of video source coding is the bit-rate reduction for storage and transmission by exploring both statistical and subjective redundancies and to encode a "minimum set" of information using entropy-coding techniques. This usually results in a compression of the coded video data compared to the original source data. The performance of video compression techniques depends on the amount of redundancy contained in the image data as well as on the actual compression techniques used for coding (see Chapter 3). With practical coding schemes, a balance between coding performance (high compression with sufficient quality) and implementation complexity is targeted. For the development of the MPEG compression algorithms the consideration of the capabilities of "state-of-the-art" (VLSI) technology foreseen for the life cycle of the standards was most important.

Depending on the applications requirements, one may envisage "lossless" and "lossy" coding of the video data. The aim of lossless coding is to reduce image or video data for storage and transmission while retaining the quality of the original images—the decoded image quality is required to be identical to the image quality prior to encoding. In contrast the aim of lossy coding techniques—and this is relevant to the applications envisioned by MPEG-1 and MPEG-2 video standards—is to meet a given target bit rate for storage and transmission. Important applications comprise transmission of video over communications channels with constrained or low bandwidth and the efficient storage of video. In these applications high video compression is achieved by degrading the video quality: the decoded image "objective" quality is reduced compared to the quality of the original images prior to encoding (i.e., taking the mean-squared error between both the original and reconstructed images as an objective image quality criterion). The smaller the target bit rate of the channel is, the higher the necessary compression of the video data and usually the more coding artifacts become visible. The ultimate aim of lossy coding techniques is to optimize image quality for a given target bit rate subject to objective or subjective optimization criteria. It should be noted that the degree of image degradation (both the objective degradation and the amount of visible artifacts) depends on the complexity of the image or video scene as much as on the sophistication of the compression technique: for simple textures in images and low video activity a good image reconstruction with no visible artifacts may be achieved even with simple compression techniques.

### 8.2.1 The MPEG Video Coder Source Model

The MPEG digital video-coding techniques are statistical in nature. Video sequences usually contain statistical redundancies in both temporal and spatial directions. The basic statistical property upon which MPEG compression techniques rely is inter-pel correlation, including the assumption of simple correlated translatory motion between consecutive frames. Thus, it is assumed that the magnitude of a particular image pel can be predicted from nearby pels within the same frame (using intraframe coding techniques) or from pels of a nearby frame (using interframe techniques). Intuitively, it is clear that in some circumstances (i.e., during scene changes of a video sequence), the temporal correlation between pels in nearby frames is small or even vanishes. The video scene then assembles a collection of uncorrelated still images. In this case intraframe coding techniques are appropriate to explore spatial correlation to achieve efficient data compression. The MPEG compression algorithms employ discrete cosine transform (DCT) coding techniques on image blocks of $8 \times 8$ pels to efficiently explore spatial correlations between nearby pels within the same image. However, if the correlation between pels in nearby frames is high (i.e., in cases where two consecutive frames have similar or identical content), it is desirable to use interframe DPCM coding techniques employing temporal prediction (motion-compensated prediction between frames). In MPEG video-coding schemes, an adaptive combination of temporal motion-compensated prediction and transform coding of the remaining spatial information is used to achieve high data compression (hybrid DPCM/DCT coding of video).

Figure 8.1 depicts an example of intraframe pel-to-pel correlation properties of images, here modeled using a rather simple, but nevertheless valuable statistical model. The simple model assumption already inherits basic correlation properties of many "typical" images upon which the MPEG algorithms rely—namely, the high correlation between adjacent pixels and the monotonical decay of correlation with increased distance between pels. This model assumption will be used later to demonstrate some of the properties of transform domain coding.

### 8.2.2 Subsampling and Interpolation

Almost all video-coding techniques described here make extensive use of subsampling and quantization prior to encoding. The basic concept of subsampling is to reduce the dimension of the input video (horizontal dimension and/or vertical dimension) and thus the number of pels to be coded prior to the encoding process. It is worth noting that for some applications video is also subsampled in the temporal direction to reduce frame rate prior to coding. At the receiver the decoded

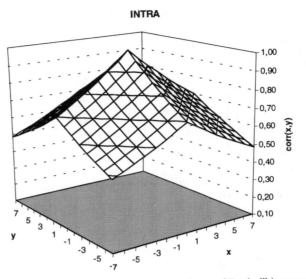

**FIGURE 8.1**   Spatial interelement correlation of "typical" images as calculated using a AR(1) Gauss-Markov image model with high pel-pel correlation. Variables $x$ and $y$ describe the distance between pels in horizontal and vertical image dimensions, respectively.

images are interpolated for display. This technique may be considered one of the most elementary compression techniques that also makes use of specific physiological characteristics of the human eye and thus removes subjective redundancy contained in the video data—i.e., the human eye is more sensitive to changes in brightness than to chromaticity changes. Therefore the MPEG coding schemes first divide the images into $YUV$ components (one luminance and two chrominance components). Next the chrominance components are subsampled relative to the luminance component with a $Y{:}U{:}V$ ratio specific to particular applications (i.e., with the MPEG-2 standard, a ratio of 4:1:1 or 4:2:2 is used).

### 8.2.3  Motion-Compensated Prediction

Motion-compensated prediction is a powerful tool to reduce temporal redundancies between frames and is used extensively in MPEG-1 and MPEG-2 video-coding standards as a prediction technique for temporal DPCM coding. The concept of motion compensation is based on the estimation of motion between video frames; i.e., if all elements in a video scene are approximately spatially displaced, the motion between frames can be described by a limited number of motion parameters (i.e., by motion vectors for the translational motion of pels). In this simple example the best prediction of an actual pel is given by a motion-compensated prediction pel from a previously coded frame. Usually, both prediction error and motion vectors are transmitted to the receiver. However, encoding motion information with each coded image pel is generally neither desirable nor necessary. Because the spatial correlation between motion vectors is often high, it is sometimes assumed that one motion vector is representative for the motion of a block of adjacent pels. To this aim, images are usually separated into disjoint blocks of pels (i.e., $16 \times 16$ pels in MPEG-1 and MPEG-2 standards), and only one motion vector is estimated, coded, and transmitted for each of these blocks (Fig. 8.2).

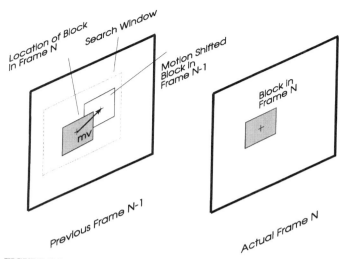

**FIGURE 8.2**   Block-matching approach for motion compensation. One motion vector (MV) is estimated for each block in the actual frame $N$ to be coded. The motion vector points to a reference block of the same size in a previously coded frame $N - 1$. The motion-compensated prediction error is calculated by subtracting each pel in a block with its motion-shifted counterpart in the reference block of the previous frame.

In the MPEG compression algorithms the motion-compensated prediction techniques are used for reducing temporal redundancies between frames, and only the prediction error images—the difference between original images and motion-compensated prediction images—are encoded. In general the correlation between pels in the motion-compensated interframe error images to be coded is reduced compared to the correlation properties of intraframes in Fig. 8.1 due to the prediction based on the previous coded frame.

### 8.2.4   Transform Domain Coding

Transform coding has been studied extensively during the last two decades and has become a very popular compression method for still-image coding and video coding. The purpose of transform coding is to decorrelate the intra- or interframe error image content and to encode transform coefficients rather than the original pels of the images. To this aim, the input images are split into disjoint blocks of pels $b$ (i.e., of size $N \times N$ pels). The transformation can be represented as a matrix operation using a $N \times N$ transform matrix $A$ to obtain the $N \times N$ transform coefficients $c$ based on a linear, separable, and unitary forward transformation:

$$c = AbA^{\mathrm{T}}$$

Here, $A^{\mathrm{T}}$ denotes the transpose of the transformation matrix $A$. Note that the transformation is reversible, because the original $N \times N$ block of pels $b$ can be reconstructed using a linear and separable *inverse* transformation:[*]

$$b = A^{\mathrm{T}}cA$$

---

[*]For a unitary transform the inverse matrix $A^{-1}$ is identical with the transposed matrix $A^{\mathrm{T}}$, that is, $A^{-1} = A^{\mathrm{T}}$.

Among many possible alternatives, the discrete cosine transform (DCT), applied to smaller image blocks of usually 8 × 8 pels, has become the most successful transform for still-image and video coding.[5] In fact, DCT-based implementations are used in most image- and video-coding standards due to their high decorrelation performance and the availability of fast DCT algorithms suitable for real-time implementations. VLSI implementations that operate at rates suitable for a broad range of video applications are commercially available today.

A major objective of transform coding is to make as many transform coefficients as possible small enough so that they are insignificant (in terms of statistical and subjective measures) and need not be coded for transmission. At the same time it is desirable to minimize statistical dependencies between coefficients with the aim to reduce the amount of bits needed to encode the remaining coefficients. Figure 8.3 depicts the variance (energy) of an 8 × 8 block of intraframe DCT coefficients based on the simple statistical model assumption already discussed in Fig. 8.1. Here, the variance for each coefficient represents the variability of the coefficient as averaged over a large number of frames. Coefficients with small variances are less significant for the reconstruction of the image blocks than coefficients with large variances. As may be seen in Fig. 8.3, on average only a small number of DCT coefficients need to be transmitted to the receiver to obtain a valuable approximate reconstruction of the image blocks. Moreover, the most significant DCT coefficients are concentrated around the upper-left corner (low DCT coefficients) and the significance of the coefficients decays with increased distance. This implies that higher DCT coefficients are less important for reconstruction than lower coefficients. Also employing motion-compensated prediction, the transformation using the DCT usually results in a compact representation of the temporal DPCM signal in the DCT domain—which essentially inherits a statistical coherency similar to that of the signal in the DCT domain for the intraframe signals in Fig. 8.3 (although

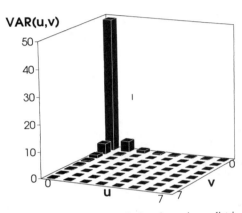

**FIGURE 8.3**   The figure depicts the variance distribution of DCT coefficients "typically" calculated as average over a large number of image blocks. The variance of the DCT coefficients was calculated based on the statistical model used in Fig. 8.1. The variables $u$ and $v$ describe the horizontal and vertical image transform domain variables within the 8 × 8 block. Most of the total variance is concentrated around the dc DCT coefficient ($u = 0, v = 0$).

with reduced energy). This is the reason why MPEG algorithms employ DCT coding also for interframe compression successfully.[3]

The DCT is closely related to the discrete Fourier transform (DFT), and it is of some importance to realize that the DCT coefficients can be given a frequency interpretation close to the DFT. Thus low DCT coefficients relate to low spatial frequencies within image blocks and high DCT coefficients to higher frequencies. This property is used in MPEG coding schemes to remove subjective redundancies contained in the image data based on human visual systems criteria. Because the human viewer is more sensitive to reconstruction errors related to low spatial frequencies than to high frequencies, a frequency-adaptive weighting (quantization) of the coefficients according to the human visual perception (perceptual quantization) is often employed to improve the visual quality of the decoded images for a given bit rate.

The combination of the two techniques just described—temporal motion-compensated prediction and transform domain coding—can be seen as the key elements of the MPEG coding standards. A third characteristic element of the MPEG algorithms is that these two techniques are processed on small image blocks (of typically 16 × 16 pels for motion compensation and 8 × 8 pels for DCT coding). For this reason the MPEG coding algorithms are usually referred to as *hybrid block-based DPCM/DCT algorithms*.

## 8.3  MPEG-1: A GENERIC STANDARD FOR CODING OF MOVING PICTURES AND ASSOCIATED AUDIO FOR DIGITAL STORAGE MEDIA AT UP TO ABOUT 1.5 MEGABITS PER SECOND

The video compression technique developed by MPEG-1 covers many applications from interactive systems on CD-ROM to the delivery of video over telecommunications networks. The MPEG-1 video-coding standard is thought to be generic. To support the wide range of applications profiles, a diversity of input parameters including flexible picture size and frame rate can be specified by the user. MPEG has recommended a constraint parameter set: every MPEG-1 compatible decoder must be able to support at least video source parameters up to TV size, including a minimum number of 720 pixels per line, a minimum number of 576 lines per picture, a minimum frame rate of 30 frames per second, and a minimum bit rate of 1.86 Mb/s. The standard video input consists of a noninterlaced video picture format. It should be noted that the application of MPEG-1 is by no means limited to this constrained parameter set.

The MPEG-1 video algorithm has been developed with respect to the JPEG and H.261 activities. It was sought to retain a large degree of commonality with the CCITT H.261 standard so that implementations supporting both standards would be plausible. However, MPEG-1 was primarily targeted for multimedia CD-ROM applications, requiring additional functionality supported by both encoder and decoder. Important features provided by MPEG-1 include frame-based random access of video, fast-forward/fast-reverse (FF/FR) searches through compressed bitstreams, reverse playback of video, and editability of the compressed bitstream.

### 8.3.1  The Basic MPEG-1 Interframe Coding Scheme

The basic MPEG-1 (as well as the MPEG-2) video compression technique is based on a macroblock structure, motion compensation, and the conditional replenishment of macroblocks. As outlined in Fig. 8.4a, the MPEG-1 coding algorithm encodes the first frame in a video sequence in intraframe coding mode (I-picture). This means that each subsequent frame is coded using interframe prediction (P-pictures); only data from the nearest previously coded I- or P-frame is used for prediction. The MPEG-1 algorithm processes the frames of a block-based video sequence.

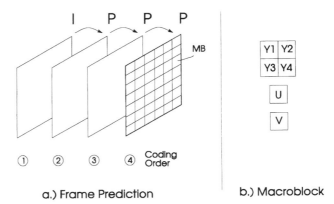

a.) Frame Prediction    b.) Macroblock

**FIGURE 8.4**  (*a*) Illustration of I-pictures (I) and P-pictures (P) in a video sequence. P-pictures are coded using motion-compensated prediction based on the nearest previous frame. Each frame is divided into disjoint macroblocks (MB). (*b*) With each macroblock, information related to four luminance blocks ($Y1$, $Y2$, $Y3$, $Y4$) and two chrominance blocks ($U$, $V$) is coded. Each block contains $8 \times 8$ pels.

This means that each color input frame in a video sequence is partitioned into nonoverlapping macroblocks as depicted in Fig. 8.4$b$. Each macroblock contains blocks of data from both luminance and co-sited chrominance bands—four luminance blocks ($Y_1$, $Y_2$, $Y_3$, $Y_4$) and two chrominance blocks ($U$, $V$), each with size $8 \times 8$ pels. Thus the sampling ratio between $Y$:$U$:$V$ luminance and chrominance pels is 4:1:1.

The block diagram of the basic hybrid DPCM/DCT MPEG-1 encoder and decoder structure is depicted in Fig. 8.5. The first frame in a video sequence (I-picture) is encoded in intra mode without reference to any past or future frames. At the encoder the DCT is applied to each $8 \times 8$ luminance and chrominance block and, after output of the DCT, each of the 64 DCT coefficients is uniformly quantized ($Q$). The quantizer stepsize (sz) used to quantize the DCT coefficients within a macroblock is transmitted to the receiver. After quantization, the lowest DCT coefficient (DC coefficient) is treated differently from the remaining coefficients (AC coefficients). The DC coefficient corresponds to the average intensity of the component block and is encoded using a differential DC prediction method.[*] The nonzero quantizer values of the remaining DCT coefficients and their locations are then zigzag scanned and run-length entropy-coded using variable-length code (VLC) tables.

## HYBRID DCT/DPCM CODING SCHEME

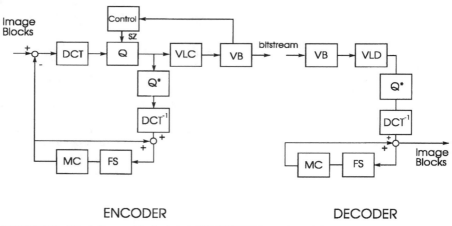

ENCODER                              DECODER

**FIGURE 8.5**   Block diagram of a basic hybrid DCT/DPCM encoder and decoder structure.

The concept of zigzag scanning of the coefficients is outlined in Fig. 8.6. The scanning of the quantized DCT domain two-dimensional signal followed by variable-length code-word assignment for the coefficients serves as a mapping of the two-dimensional image signal into a one-dimensional bitstream. The nonzero AC coefficient quantizer values (length, ●) are detected along the scan line as well as the distance (run) between two consecutive nonzero coefficients. Each consecutive (run, length) pair is encoded by transmitting only one VLC codeword. The purpose of zigzag scanning is to trace the low-frequency DCT coefficients (containing most energy) before tracing the high-frequency coefficients.[†]

---

[*]Because there is usually strong correlation between the DC values of adjacent $8 \times 8$ blocks, the quantized DC coefficient is encoded as the difference between the DC value of the previous block and the actual DC value.

[†]The location of each nonzero coefficient along the zigzag scan is encoded relative to the location of the previous coded coefficient. The zigzag scan philosophy attempts to trace the nonzero coefficients according to their likelihood of appearance to achieve an efficient entropy coding. With reference to Fig. 8.3, the DCT coefficients most likely to appear are concentrated around the DC coefficient with decreasing importance. For many images the coefficients are traced efficiently using the zigzag scan.

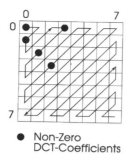

● Non-Zero
   DCT-Coefficients

**FIGURE 8.6**   Zigzag scanning of the quantized DCT coefficients in an $8 \times 8$ block. Only the nonzero quantized DCT coefficients are encoded. The possible locations of nonzero DCT coefficients are indicated in the figure. The zigzag scan attempts to trace the DCT coefficients according to their significance. With reference to Fig. 8.3, the lowest DCT coefficient (0,0) contains most of the energy within the blocks and the energy is concentrated around the lower DCT coefficients.

The decoder performs the reverse operations, first extracting and decoding (VLD) the variable-length coded words from the bitstream to obtain locations and quantizer values of the nonzero DCT coefficients for each block. With the reconstruction ($Q^*$) of all nonzero DCT coefficients belonging to one block and subsequent inverse DCT ($DCT^{-1}$) the quantized block pixel values are obtained. By processing the entire bitstream all image blocks are decoded and reconstructed.

For coding P-pictures, the previously I- or P-picture frame $N-1$ is stored in a frame store (FS) in both encoder and decoder. Motion compensation (MC) is performed on a macroblock basis—only one motion vector is estimated between frame $N$ and frame $N-1$ for a particular macroblock to be encoded. These motion vectors are coded and transmitted to the receiver. The motion-compensated prediction error is calculated by subtracting each pel in a macroblock from its motion-shifted counterpart in the previous frame. An $8 \times 8$ DCT is then applied to each of the $8 \times 8$ blocks contained in the macroblock, followed by quantization ($Q$) of the DCT coefficients with subsequent run-length coding and entropy coding (VLC). A video buffer (VB) is needed to ensure that a constant target bit rate output is produced by the encoder. The quantization stepsize (sz) can be adjusted for each macroblock in a frame to achieve a given target bit rate and to avoid buffer overflow and underflow.

The decoder uses the reverse process to reproduce a macroblock of frame $N$ at the receiver. After decoding the variable-length words (VLD) contained in the video decoder buffer (VB) the pixel values of the prediction error are reconstructed ($Q^*$ operations and $DCT^{-1}$ operations). The motion-compensated pixels from the previous frame $N-1$ contained in the frame store (FS) are added to the prediction error to recover the particular macroblock of frame $N$.

The advantage of coding video using the motion-compensated prediction from the previously reconstructed frame $N-1$ in an MPEG coder is illustrated in Fig. 8.7$a$–8.7$d$ for a typical test sequence. Figure 8.7$a$ depicts a frame at time instance $N$ to be coded and Fig. 8.7$b$ the reconstructed frame at instance $N-1$, which is stored in the frame store (FS) at both encoder and decoder. The block motion vectors (MV, see also Fig. 8.2) depicted in Fig. 8.7$b$ are estimated by the encoder motion estimation procedure and provide a prediction of the translatory motion displacement of each macroblock in frame $N$ with reference to frame $N-1$. Figure 8.7$b$ depicts the pure frame difference signal (frame $N$ to frame $N-1$) that is obtained if no motion-compensated prediction is used in the coding process—thus all motion vectors are assumed to be zero. Figure 8.7$d$ depicts the motion-compensated frame difference signal when the motion vectors in Fig. 8.7$b$ are used for prediction. It is apparent that the residual signal to be coded is greatly reduced using motion compensation if compared to pure frame difference coding in Fig. 8.7$c$.

### 8.3.2   Conditional Replenishment

An essential feature supported by the MPEG-1 coding algorithm is the possibility to update macroblock information at the decoder only if needed—if the content of the macroblock has changed in comparison to the content of the same macroblock in the previous frame (conditional macroblock replenishment). The key for efficient coding of video sequences at lower bit rates is the selection of appropriate prediction modes to achieve conditional replenishment. The MPEG standard is distinguished mainly between three different macroblock coding types (MB types):

**FIGURE 8.7a**   Frame at time instance $N$ to be coded.

**FIGURE 8.7b**   Frame at instance $N - 1$ used for prediction of the content in frame $N$ (note that the motion vectors depicted in the image are not part of the reconstructed image stored at the encoder and decoder).

**FIGURE 8.7c**    Prediction error image obtained without using motion compensation. All motion vectors are assumed to be zero.

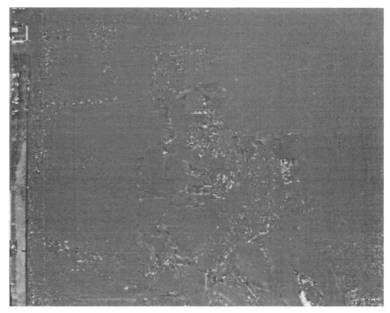

**FIGURE 8.7d**    Prediction error image to be coded if motion compensation prediction is employed.

*Skipped MB.* Prediction from previous frame with zero motion vector. No information about the macroblock is coded or transmitted to the receiver.

*Inter MB.* Motion-compensated prediction from the previous frame is used. The MB type, the MB address and, if required, the motion vector, the DCT coefficients, and quantization stepsize are transmitted.

*Intra MB.* No prediction is used from the previous frame (intraframe prediction only). Only the MB type, the MB address, and the DCT coefficients and quantization stepsize are transmitted to the receiver.

### 8.3.3    Specific Storage Media Functionalities

For accessing video from storage media the MPEG-1 video compression algorithm was designed to support important functionalities such as random access and fast-forward (FF) and fast-reverse (FR) playback functionalities. To incorporate the requirements for storage media and to further explore the significant advantages of motion compensation and motion interpolation, the concept of B-pictures (bidirectional predicted/bidirectional interpolated pictures) was introduced by MPEG-1. This concept is depicted in Fig. 8.8 for a group of consecutive pictures in a video sequence. Three types of pictures are considered. Intrapictures (I-pictures) are coded without reference to other pictures contained in the video sequence, as already introduced in Fig. 8.4. I-pictures allow access points for random access and FF/FR functionality in the bitstream but achieve only low compression. Interframe predicted pictures (P-pictures) are coded with reference to the nearest previously coded I-picture or P-picture, usually incorporating motion compensation to increase coding efficiency. Because P-pictures are usually used as reference for prediction for future or past frames they provide no suitable access points for random access functionality or editability. Bidirectional predicted/interpolated pictures (B-pictures) require both past and future frames as references. To achieve high compression, motion compensation can be employed based on the nearest past and future P-pictures or I-pictures. B-pictures themselves are never used as references.

The user can arrange the picture types in a video sequence with a high degree of flexibility to suit diverse applications requirements. As a general rule, a video sequence coded using

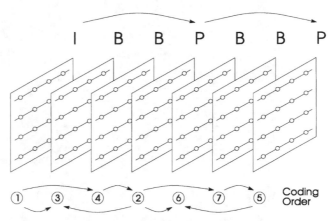

**FIGURE 8.8**  I-pictures (I), P-pictures (P), and B-pictures (B) used in a MPEG-1 video sequence. B-pictures can be coded using motion-compensated prediction based on the two nearest already coded frames (either I-picture or P-picture). The arrangement of the picture-coding types within the video sequence is flexible to suit the needs of diverse applications. The direction for prediction is indicated in the figure.

I-pictures only (I I I I I I ...) allows the highest degree of random access, FF/FR, and editability but achieves only low compression. A sequence coded with a regular I-picture update and no B-pictures (i.e., I P P P P P P I P P P P ...) achieves moderate compression and a certain degree of random access and FF/FR functionality. Incorporation of all three pictures types [i.e., as depicted in Fig. 8.8 (I B B P B B P B B I B B P ...)] may achieve high compression, reasonable random access, and FF/FR functionality but also increases the coding delay significantly. This delay may not be tolerable for applications such as videotelephony or videoconferencing.

### 8.3.4  Rate Control

An important feature supported by the MPEG-1 encoding algorithms is the possibility to tailor the bit rate (and thus the quality of the reconstructed video) to specific applications require-ments by adjusting the quantizer stepsize (sz) in Fig. 8.5 for quantizing the DCT coefficients. Coarse quantization of the DCT coefficients enables the storage or transmission of video with high compression ratios but, depending on the level of quantization, may result in significant coding artifacts. The MPEG-1 standard allows the encoder to select different quantizer values for each coded macroblock; this enables a high degree of flexibility to allocate bits in images where needed to improve image quality. Furthermore, it allows the generation of both constant and variable bit rates for storage or real-time transmission of the compressed video.

Compressed video information is inherently variable in nature. This is caused by the generally variable content of successive video frames. To store or transmit video at constant bit rate, it is therefore necessary to buffer the variable bitstream generated in the encoder in a video buffer (VB) as depicted in Fig. 8.5. The input into the encoder VB is variable over time and the output is a constant bitstream. At the decoder the VB input bitstream is constant and the output used for decoding is variable. MPEG encoders and decoders implement buffers of the same size to avoid reconstruction errors.

A rate control algorithm at the encoder adjusts the quantizer stepsize depending on the video content and activity to ensure that the video buffers will never overflow—while at the same time targeting to keep the buffers as full as possible to maximize image quality. In theory, over-flow of buffers can always be avoided by using a large enough video buffer. However, besides the possibly undesirable costs for the implementation of large buffers, there may be additional disadvantages for applications requiring low delay between encoder and decoder, such as for the real-time transmission of conversational video. If the encoder bitstream is smoothed using a video buffer to generate a constant bit rate output, a delay is introduced between the encoding process and the time the video can be reconstructed at the decoder. Usually, the larger the buffer is, the larger the delay introduced is.

MPEG has defined a minimum video buffer size that needs to be supported by all decoder implementations. This value is identical to the maximum value of the VB size that an encoder can use to generate a bitstream. However, to reduce delay or encoder complexity, it is possible to choose a virtual buffer size value at the encoder smaller than the minimum VB size that needs to be supported by the decoder. This virtual buffer size value is transmitted to the decoder before sending the video bitstream.

The rate control algorithm used to compress video is not part of the MPEG-1 standard and it is thus left to the implementers to develop efficient strategies. It is worth emphasizing that the efficiency of the rate control algorithms selected by manufacturers to compress video at a given bit rate heavily affects the visible quality of the video reconstructed at the decoder.

### 8.3.5  Coding of Interlaced Video Sources

The standard video input format for MPEG-1 is noninterlaced. However, coding of interlaced color television with both 525 and 625 lines at 29.97 and 25 frames per second, respectively, is an important application for the MPEG-1 standard. A suggestion for coding Rec.601 digital color television signals has been made by MPEG-1 based on the conversion of the interlaced source to a progressive intermediate format. In essence, only one horizontally subsampled field of each interlaced video input frame is encoded, i.e., the subsampled top field. At the receiver the even field is predicted from the decoded and horizontally interpolated odd field for display. The necessary preprocessing steps required prior to encoding and the postprocessing required after decoding are described in detail in the Informative Annex of the MPEG-1 International Standard.[6]

## 8.4  MPEG-2 STANDARD FOR GENERIC CODING OF MOVING PICTURES AND ASSOCIATED AUDIO

Worldwide, MPEG-1 is developing into an important and successful video-coding standard with an increasing number of products becoming available on the market. A key factor for this success is the generic structure of the standard supporting a broad range of applications and applications-specific parameters. However, MPEG continued its standardization efforts in 1991 with a second phase (MPEG-2) to provide a video-coding solution for applications not originally covered or envisaged by the MPEG-1 standard. Specifically, MPEG-2 was given the charter to provide video quality not lower than NTSC/PAL and up to CCIR 601 quality. Emerging applications—such as digital cable TV distribution, networked database services via ATM, digital VTR applications, and satellite and terrestrial digital broadcasting distribution—were seen to benefit from the increased quality expected to result from the new MPEG-2 standardization phase. Work was carried out in collaboration with the ITU-T SG 15 Experts Group for ATM Video Coding and in 1994 the MPEG-2 Draft International Standard (which is identical to the ITU-T H.262 recommendation) was released.[2] The specification of the standard is intended to be generic—hence the standard aims to facilitate the bitstream interchange among different applications, transmission, and storage media.

Basically, MPEG-2 can be seen as a superset of the MPEG-1 coding standard and was designed to be backward-compatible to MPEG-1—i.e., every MPEG-2–compatible decoder can decode a valid MPEG-1 bitstream. Many video-coding algorithms were integrated into a single syntax to meet the diverse applications requirements. New coding features were added by MPEG-2 to achieve sufficient functionality and quality; thus, prediction modes were developed to support efficient coding of interlaced video. In addition, scalable video-coding extensions were introduced to provide additional functionality, such as embedded coding of digital TV and HDTV, and graceful quality degradation in the presence of transmission errors.

However, implementation of the full syntax may not be practical for most applications. MPEG-2 has introduced the concepts of *profiles* and *levels* to stipulate conformance between equipment not supporting the full implementation. Profiles and levels provide means for defining subsets of the syntax and thus the decoder capabilities required to decode a particular bitstream. This concept is illustrated in Tables 8.2 and 8.3.

As a general rule, each profile defines a new set of algorithms added as a superset to the algorithms in the profile that follows. A level specifies the range of the parameters that are supported by the implementation (i.e., image size, frame rate, and bit rates). The MPEG-2 core algorithm at MAIN profile features nonscalable coding of both progressive and interlaced video sources. It is expected that most MPEG-2 implementations will at least conform to the MAIN profile at the MAIN level that supports nonscalable coding of digital video with approximately digital TV parameters—a maximum sample density of 720 samples per line and 576 lines per frame, a maximum frame rate of 30 frames per second, and a maximum bit rate of 15 Mb/s.

### 8.4.1  MPEG-2 Nonscalable Coding Modes

The MPEG-2 algorithm defined in the MAIN profile is a straightforward extension of the MPEG-1 coding scheme to accommodate coding of interlaced video while retaining the full range of

**TABLE 8.2** Upper Bound of Parameters at Each Level of a Profile

| Level | Parameters |
|---|---|
| High | 1920 samples/line<br>1152 lines/frame<br>60 frames/s<br>80 Mb/s |
| High 1440 | 1440 samples/line<br>1152 lines/frame<br>60 frames/s<br>60 Mb/s |
| Main | 720 samples/line<br>576 lines/frame<br>30 frames/s<br>15 Mb/s |
| Low | 352 samples/line<br>288 lines/frame<br>30 frames/s<br>4 Mb/s |

**TABLE 8.3** Algorithms and Functionalities Supported with Each Profile

| Profile | Algorithms |
|---|---|
| High | Supports all functionality provided by the spatial scalable profile plus the provision to support<br>• 3 layers with the SNR and spatial scalable coding modes<br>• 4:2:2 *YUV* representation for improved quality requirements |
| Spatial scalable | Supports all functionality provided by the SNR scalable profile plus an algorithm for<br>• Spatial scalable coding (2 layers allowed)<br>• 4:0:0 *YUV* representation |
| SNR scalable | Supports all functionality provided by the MAIN profile plus an algorithm for<br>• SNR scalable coding (2 layers allowed)<br>• 4:2:0 *YUV* representation |
| Main | Nonscalable coding algorithm supporting functionality for<br>• Coding-interlaced video<br>• Random access<br>• B-picture prediction modes<br>• 4:2:0 *YUV* representation |
| Simple | Includes all functionality provided by the MAIN profile but<br>• Does not support B-picture prediction modes |

functionality provided by MPEG-1. Identical to the MPEG-1 standard, the MPEG-2 coding algorithm is based on the general hybrid DCT/DPCM coding scheme outlined in Fig. 8.5, incorporating a macroblock structure, motion compensation, and coding modes for conditional replenishment of macroblocks. The concept of I-pictures, P-pictures, and B-pictures as introduced in Fig. 8.8 is fully retained in MPEG-2 to achieve efficient motion prediction and to assist random access functionality. Notice that the algorithm defined with the MPEG-2 SIMPLE profile is basically identical to the one in the MAIN profile, except that no B-picture prediction modes are allowed at the encoder. Thus the additional implementation complexity and the additional frame stores necessary for the decoding of B-pictures are not required for MPEG-2 decoders only conforming to the SIMPLE profile.

*Field and Frame Pictures.*  MPEG-2 has introduced the concept of frame pictures and field pictures along with particular frame prediction and field prediction modes to accommodate coding of progressive and interlaced video. For interlaced sequences it is assumed that the coder input consists of a series of odd (top) and even (bottom) fields that are separated in time by a field period. Two fields of a frame may be coded separately (field pictures, see Fig. 8.9). In this case each field is separated into adjacent nonoverlapping macroblocks and the DCT is applied on a field basis. Alternatively, two fields may be coded together as a frame (frame pictures), similar to conventional coding of progressive video sequences. Here, consecutive lines of top and bottom fields are simply merged to form a frame. Notice that both frame pictures and field pictures can be used in a single video sequence.

*Field and Frame Prediction.*  New motion-compensated field prediction modes were introduced by MPEG-2 to efficiently encode field pictures and frame pictures. An example of this new concept is illustrated and simplified in Fig. 8.9 for an interlaced video sequence—here assumed to contain only three field pictures and no B-pictures. In field prediction, predictions are made independently for each field by using data from one or more previously decoded fields; i.e., for a top field a prediction may be obtained from either a previously decoded top field

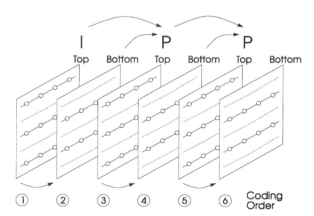

**FIGURE 8.9**  The concept of field pictures and an example of possible field prediction. The top fields and the bottom fields are coded separately. However, each bottom field is coded using motion-compensated interfield prediction based on the previously coded top field. The top fields are coded using motion-compensated interfield prediction based on either the previously coded top field or the previously coded bottom field. This concept can be extended to incorporate B-pictures.

(using motion-compensated prediction) or from the previously decoded bottom field belonging to the same picture. Generally the interfield prediction from the decoded field in the same picture is preferred if no motion occurs between fields. An indication of which reference field is used for prediction is transmitted with the bitstream. Within a field picture, all predictions are field predictions.

Frame prediction forms a prediction for a frame picture based on one or more previously decoded frames. In a frame picture either field or frame predictions may be used and the particular prediction mode preferred can be selected on a macroblock-by-macroblock basis. It must be understood, however, that the fields and frames from which predictions are made may have themselves been decoded as either field or frame pictures.

MPEG-2 has introduced new motion compensation modes to efficiently explore temporal redundancies between fields, namely, the *dual prime prediction* and the motion compensation based on $16 \times 8$ blocks. A discussion of these methods is, however, beyond the scope of this chapter.

***Chrominance Formats.***   MPEG-2 has specified additional $Y:U:V$ luminance and chrominance subsampling ratio formats to assist and foster applications with highest video quality requirements. Next to the 4:2:0 format already supported by MPEG-1, the specification of MPEG-2 is extended to 4:2:2 formats suitable for studio video-coding applications.

### 8.4.2   MPEG-2 Scalable Coding Extensions

The scalability tools standardized by MPEG-2 support applications beyond those addressed by the basic MAIN profile coding algorithm. The intention of scalable coding is to provide interoperability between different services and to flexibly support receivers with different display capabilities. Receivers either not capable or not willing to reconstruct the full-resolution video can decode subsets of the layered bitstream to display video at lower spatial or temporal resolution or with lower quality. Another important purpose of scalable coding is to provide a layered video bitstream that is amenable for prioritized transmission. The main challenge here is to reliably deliver video signals in the presence of channel errors, such as cell loss in ATM-based transmission networks or cochannel interference in terrestrial digital broadcasting.

Flexibly supporting multiple resolutions is of particular interest for interworking between HDTV and standard definition television (SDTV), in which case it is important for the HDTV receiver to be compatible with the SDTV product. Compatibility can be achieved by means of scalable coding of the HDTV source and the wasteful transmission of two independent bitstreams to the HDTV and SDTV receivers can be avoided. Other important applications for scalable coding include video database browsing and multiresolution playback of video in multimedia environments.

Figure 8.10 depicts the general philosophy of a multiscale video-coding scheme. Here, two layers are provided—each layer supporting video at a different scale; i.e., a multiresolution representation can be achieved by downscaling the input video signal into a lower-resolution video (downsampling spatially or temporally). The downscaled version is encoded into a base-layer bitstream with reduced bit rate. The upscaled reconstructed base-layer video (upsampled spatially or temporally) is used as a prediction for the coding of the original input video signal. The prediction error is encoded into an enhancement-layer bitstream. If a receiver is either not capable or not willing to display the full-quality video, a downscaled video signal can be reconstructed by only decoding the base-layer bitstream. It is important to notice, however, that the display of the video at highest resolution with reduced quality is also possible by only decoding the lower–bit rate base layer. Thus scalable coding can be used to encode video with a suitable bit rate allocated to each layer in order to meet specific bandwidth requirements of transmission channels or storage media. Browsing through video databases and transmission of video over heterogeneous networks are applications expected to benefit from this functionality.

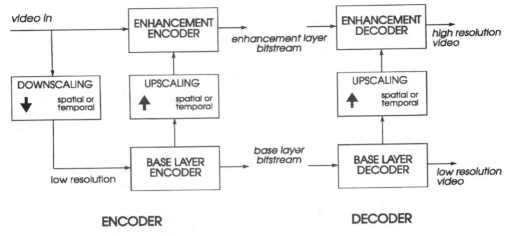

**FIGURE 8.10**   Scalable coding of video.

During the MPEG-2 standardization phase it was found impossible to develop one generic scalable coding scheme capable of suiting all of the diverse applications requirements envisaged. Although some applications are constricted to low implementation complexity, others call for very high coding efficiency. As a consequence, MPEG-2 has standardized three scalable coding schemes: SNR (quality) scalability, spatial scalability, and temporal scalability—each of them targeted to assist applications with particular requirements. The scalability tools provide algorithmic extensions to the nonscalable scheme defined in the MAIN profile. It is possible to combine different scalability tools into a hybrid coding scheme; i.e., interoperability between services with different spatial resolutions and frame rates can be supported by means of combining the spatial scalability and the temporal scalability tool into a hybrid layered-coding scheme. Interoperability between HDTV and SDTV services can be provided along with a certain resilience to channel errors by combining the spatial scalability extensions with the SNR scalability tool.[7] The MPEG-2 syntax supports up to three different scalable layers.

*Spatial Scalability.*   This tool has been developed to support displays with different spatial resolutions at the receiver: lower–spatial resolution video can be reconstructed from the base layer. This functionality is useful for many applications including embedded coding for HDTV/TV systems, allowing a migration from a digital TV service to higher–spatial resolution HDTV services.[8,9] The algorithm is based on a classical pyramidal approach for progressive image coding.[10,11] Spatial scalability can flexibly support a wide range of spatial resolutions but adds considerable implementation complexity to the MAIN profile coding scheme.

*SNR Scalability.*   This tool has been developed primarily to provide graceful degradation (quality scalability) of the video quality in prioritized transmission media. If the base layer can be protected from transmission errors, a version of the video with gracefully reduced quality can be obtained by decoding the base layer signal only. The algorithm used to achieve graceful degradation is based on a frequency (DCT-domain) scalability technique. Both layers in Fig. 8.10 encode the video signal at the same spatial resolution. A detailed outline of a possible implementation of a SNR scalability encoder and decoder is depicted in Fig. 8.11. The method is implemented as a simple and straightforward extension to the MAIN profile MPEG-2 coder and achieves excellent coding efficiency.

At the base layer the DCT coefficients are coarsely quantized and transmitted to achieve moderate image quality at reduced bit rate. The enhancement layer encodes and transmits the difference between the nonquantized DCT coefficients and the quantized coefficients from the

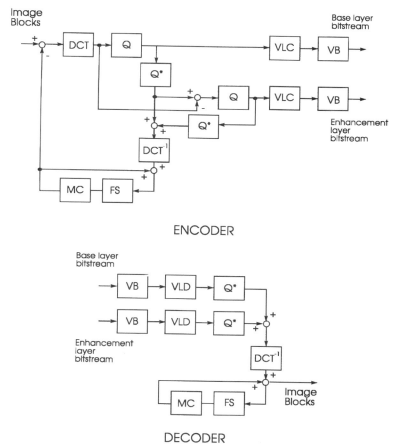

ENCODER

DECODER

**FIGURE 8.11** (*Top*) A possible implementation of a two-layer encoder for SNR-scalable coding of video. (*Bottom*) Decoder.

base layer with finer quantization stepsize. At the decoder the highest-quality video signal is reconstructed by decoding both the lower- and the higher-layer bitstreams.

It is also possible to use this method to obtain video with lower spatial resolution at the receiver. If the decoder selects the lowest $N \times N$ DCT coefficients from the base-layer bitstream, nonstandard inverse DCTs of size $N \times N$ can be used to reconstruct the video at reduced spatial resolution.[12,13] However, depending on the encoder and decoder implementations the lowest-layer downscaled video may be subject to drift.[14]

***Temporal Scalability.*** This tool was developed with an aim similar to spatial scalability—steroscopic video can be supported with a layered bitstream suitable for receivers with stereoscopic display capabilities. Layering is achieved by providing a prediction of one of the images of the stereoscopic video (i.e., left view) in the enhancement layer based on coded images from the opposite view transmitted in the base layer.

***Data Partitioning.*** This is intended to assist with error concealment in the presence of transmission or channel errors in ATM, terrestrial broadcast, or magnetic recording environments.

Because the tool can be used entirely as a postprocessing and preprocessing tool to any single-layer coding scheme, it has not been formally standardized with MPEG-2 but is referenced in the Informative Annex of the MPEG-2 DIS document.[8] The algorithm is similar to the SNR scalability tool, based on the separation of DCT coefficients and is implemented with very low complexity compared to the other scalable coding schemes. To provide error protection, the coded DCT coefficients in the bitstream are simply separated and transmitted in two layers with different error likelihood.

## 8.5 CONCLUSION

International standardization in image coding has made a remarkable evolution from a committee-driven process dominated by telecoms and broadcasters to a market-driven process incorporating industries, telecoms, network operators, satellite operators, broadcasters, and research institutes. With this evolution the actual work of the standardization bodies has also changed considerably—from discussion circles of national delegations to international collaborative R&D activities. The standardization process has become significantly more efficient and faster; the reason is that standardization has to follow the accelerated speed of technology development because standards are otherwise in danger of obsolescence before they are agreed upon by the standardization bodies.

It has to be understood that video-coding standards have to rely on compromises between what is theoretically possible and what is technologically feasible. Standards can only be successful in the marketplace if the cost/performance ratio is well balanced. This is specifically true in the field of image and video coding, where a large variety of innovative coding algorithms exist but may be too complex for implementation with state-of-the-art VLSI technology.

In this respect the MPEG-1 standard provides efficient compression for a large variety of multimedia terminals with the additional flexibility provided for random access of video from storage media and supporting a diversity of image source formats. A number of MPEG-1 encoder and decoder chip sets from different vendors are currently available on the market. Encoder and decoder PC boards have been developed using MPEG-1 chip sets. A number of commercial products use the MPEG-1 coding algorithm for interactive CD applications, such as the CD-I product.

The MPEG-2 standard is becoming more and more successful because there is a strong commitment from industries, cable and satellite operators, and broadcasters to use this standard. Digital TV broadcasting, pay TV, pay-per-view, video-on-demand, interactive TV, and many other future video services are the applications envisaged. Many MPEG-2 MAIN profile at MAIN level decoder prototype chips are already developed. The worldwide acceptance of MPEG-2 in consumer electronics will lead to large production scales, making MPEG-2 decoder equipment cheap and therefore also attractive for related areas such as video communications and storage and multimedia applications in general.

## REFERENCES

1. W. Chen and Hein, D., "Motion compensated DXC system," *Proceedings of the 1986 Picture Coding Symposium,* Tokyo, vol. 2–4, April 1986, pp. 76–77.

2. B. R. Halhed, "Videoconferencing codecs: Navigating the maze," *Business Communication Review,* vol. 21, no. 1, 1991, pp. 35–40.

3. R. Schafer and Sikora, T., "Digital video coding standards and their role in video communications," *Proceedings of the IEEE,* vol. 83, 1995, pp. 907–923.

4. T. Sikora, "The MPEG-1 and MPEG-2 digital video coding standards," *IEEE Signal Processing Magazine,* to be published.

5. N. Ahmed, Natrajan, T., and Rao, K. R., "Discrete cosine transform," *IEEE Transactions on Computers,* vol. C-23, no. 1, Dec. 1984, pp. 90–93.

6. ISO/IEC 11172-2, "Information technology—Coding of moving pictures and associated audio for digital storage media at up to about 1.5 Mb/s—Video," Geneva, 1993.

7. J. De Lameilieure and Schafer, R., "MPEG-2 image coding for digital TV," *Fernseh und Kino Technik,* Jahrgang, Germany, vol. 48, March 1994, pp. 99–107 (in German).

8. ISO/IEC JTC1/SC29/WG11 N0702 Rev., "Information technology—Generic coding of moving pictures and associated audio: Recommendation H.262," Draft International Standard, Paris, Mar. 25, 1994.

9. J. De Lemeilierue and Schamel, G., "Hierarchical coding of TV/HDTV within the German HDTVT project," *Proceedings International Workshop on HDTV'93,* Ottawa, Canada, Oct. 1993, pp. 8A1.1–8A1.8.

10. A. Puri and Wong, A., "Spatial domain resolution scalable video coding," *Proceedings SPIE Visual Communication and Image Processing,* Boston, Mass., Nov. 1993.

11. P. J. Burt and Adelson, E., "The Laplacian pyramid as a compact image code," *IEEE Transactions on Communications,* vol. COM-31, 1983, pp. 532–540.

12. C. Gonzales and Viscito, E., "Flexibly scalable digital video coding," *Signal Processing: Image Communication,* vol. 5, no. 1–2, Feb. 1993.

13. T. Sikora, Tan, T. K., and Ngan, K. N., "A performance comparison of frequency domain pyramid scalable coding schemes," *Proceedings Picture Coding Symposium,* Lausanne, Switzerland, Mar. 1993, pp. 16.1–16.2.

14. A. W. Johnson, Sikora, T., Tan, T. K., and Ngan, K. N., "Filters for drift reduction in frequency scalable video coding schemes," *Electronic Letters,* vol. 30, no. 6, 1994, pp. 471–472.

## *ABOUT THE AUTHOR*

Thomas Sikora, as chairman of the MPEG (Moving Picture Experts Group) video group, is responsible for the development and standardization of the video compression algorithms for the MPEG-2 and MPEG-4 video-coding standards.

Sikora received the Dipl.-Ing. degree and the Dr.-Ing. degree in electrical engineering from Bremen University, Germany, in 1985 and 1989, respectively. From 1985 to 1989 he was employed with the Free University Berlin, conducting research on digital signal-processing algorithms. In 1990 he joined Siemens Ltd. and Monash University, Melbourne, Australia, as a project leader responsible for video compression research activities in the Australian Universal Broadband Video Codec consortium. Since 1994 he has been directing the research activities on very-low-bit-rate video coding in the image-processing department of the Heinrich-Hertz Institute in Berlin, Germany.

Sikora has been extensively involved in several European research groups for a number of years and has been appointed chairman of the European COST 211*ter* video compression research group. He has published more than 50 papers relating to image processing and image coding.

# CHAPTER 8
# MPEG STANDARDS

## SECTION 2
## DIGITAL AUDIO CODING STANDARDS

**Peter Noll**
*Technische Universität Berlin, Germany*

## 8.6   MPEG STANDARDIZATION ACTIVITIES

Advances in digital technology have made it possible to use digital audio and digital video in an increasing number of applications. In 1988, in response to a growing need for a common format for coding of audiovisual signals, the International Organization for Standardization (ISO/IEC) established a Working Group ISO/IEC/JTC SC29/WG 11, the Moving Pictures Experts Group (MPEG), with the mandate to develop standards for the coding of moving pictures and associated audio information—for digital storage media at up to about 1.5 Mb/s. It soon became clear that such standards should be of high interest for a wide range of communications-based and storage-based applications.

MPEG's initial effort was the MPEG Phase 1 (MPEG-1) coding standard IS 11172, supporting bit rates around 1.2 Mb/s for video (with video quality comparable to that of today's analog videocassette recorders) and around 256 kb/s for two-channel audio (with audio quality comparable to that of today's Compact Discs). The coding standard IS 11172 consists of five parts: video, audio, systems, compliance testing, and software simulation.[1] Descriptions covering various parts of the audio standard have been published.[2-4] The MPEG-1 audio-coding algorithm has been the result of a collaborative effort of the ASPEC group (AT&T, CNET, FhG/Erlangen University, TCE) and the MUSICAM group (CCETT, IRT, Philips, Matsushita). In the standardization process of the audio-coding algorithms there was a close cooperation between MPEG and the European EUREKA 147 project, which was the platform for developing a digital audio-broadcasting system (see Section 8.10.2).

In 1990 MPEG started the second phase of its work (MPEG-2) to develop extensions to MPEG-1 that would allow for greater input-format flexibility and higher bit rates for high-quality video (including high-definition TV) in bit rate ranges from 3 to 15 Mb/s and that would provide new audio features, including low–bit rate digital audio at lower sampling frequencies and various forms of multichannel audio.[5]

The current MPEG-4 activity addresses a broader field of applications, including interactive services, mobile communications, database access, etc. MPEG-4/Audio will merge the whole range of audio from high fidelity audio coding and speech coding down to synthetic speech and synthetic audio. In order to fulfill these requirements, MPEG-4/Audio will consist of a family of tools and algorithms with a high degree of configurability.

The MPEG activities were originally aimed at coding of audiovisual information for digital storage media, such as magneto-optical disks (MOD), digital audio tape (DAT), read-only and interactive CD, etc. Presently, the MPEG standard is about to become a universal standard in many application areas with totally different requirements in the fields of consumer electronics, professional audio processing, telecommunications, and broadcasting.[6] For example, it is possible for MPEG-1/Audio to provide a subjective quality that is equivalent to Compact Disc (CD) quality (16-bit PCM) at stereo rates between 384 kb/s (layer I) and 128 kb/s (layer III) for many types of music. A number of applications of MPEG-1 and MPEG-2/Audio will be described in Section 8.10.

### 8.6.1 Audio Bandwidths and Bit Rates

Typical audio signal classes are telephone speech, wideband speech, and wideband audio, all of which differ in bandwidth, dynamic range, and in listener expectation of offered quality. In addition, wideband audio usually will be presented in a stereophonic or multichannel audio format.

The conventional digital format for wideband audio signals is PCM of the Compact Disc (CD); it is today's de facto standard of digital audio representation and has made digital audio popular. On a CD, audio signals are stored with a 44.1-kHz sampling rate and a 16-bit amplitude resolution. The resulting stereo net bit rate is $2 \times 44.1 \times 16 \times 1000 \approx 1.41$ Mb/s. However, the CD needs a significant overhead for a run-length–limited line code, which maps 8 information bits into 14 bits, for synchronization and for error correction, resulting in a 49-bit representation of each 16-bit audio sample. Hence the gross stereo bit rate is $1.41 \times 49/16 \approx 4.32$ Mb/s. The digital audio tape (DAT) format has a PCM net bit rate identical to that of the CD (at 44.1-kHz sampling rate), but a smaller overhead bit rate. Table 8.4 lists these parameters and those of two more recent storage systems, Philips Digital Compact Cassette (DCC)[7] and Sony's 64-mm optical or magneto-optical MiniDisc (MD)[8] (see also Section 8.10).

**TABLE 8.4** Audio Bit Rates for Various Storage Devices*

| Storage device | Format | Audio rate | Overhead | Total bit rate |
|---|---|---|---|---|
| Compact Disc (CD) | PCM | 1.41 Mb/s | 2.91 Mb/s | 4.32 Mb/s |
| Digital audio tape (DAT) | PCM | 1.41 Mb/s | 1.67 Mb/s | 3.08 Mb/s |
| Digital Compact Cassette (DCC) | PASC | 384 kb/s | 384 kb/s | 768 kb/s |
| MiniDisc (MD) | ATRAC | 292 kb/s | 718 kb/s | 1.01 Mb/s |

\* Stereophonic signals, sampled at 44.1 kHz.
DAT and DCC systems also support sampling rates of 32 and 48 kHz.

### 8.6.2 Bit Rate Reductions

Although high-bit-rate channels and networks become more easily accessible, low-bit-rate coding of audio signals has retained its importance. The main motivations for low-bit-rate coding are the need to minimize transmission costs or to provide cost-efficient storage, the demand to transmit over channels of limited capacity such as mobile radio channels, and the need to support variable-rate coding in packet-oriented networks. We have seen rapid progress in bit-rate compression of speech and audio signals. Linear prediction, subband coding, transform coding, as well as various forms of vector quantization and entropy-coding techniques have been used to design efficient coding algorithms that can achieve substantially more compression than was thought possible only a few years ago. Speech and audio coding are similar in that quality is, in both cases, based on the properties of human auditory perception. On the other hand, speech can be coded very efficiently because a speech production model is available, whereas nothing similar exists for audio signals. Table 8.5 lists typical bit rates and algorithms for coding speech (bandwidth approximately 3 kHz), wideband speech (bandwidth approximately 7 kHz), and wideband audio (bandwidth approximately 20 kHz).

A wide range of speech- and audio-coding algorithms are available;[9–14] however, most speech coders are not suitable for coding audio signals because of limited bandwidths and/or other constraints. In the following we shall only cover MPEG/Audio coding, a class of wideband (high-fidelity) audio representations including multichannel audio.

**TABLE 8.5**    Important Audio and Speech Coding Standards and Products

| Type of coder | Standard or product | Mono bit rates in kb/s | Region |
|---|---|---|---|
| Audio coding (20 kHz; stereo)* | | | |
| Subband | MPEG/Audio, layers I and II | Wide range | World |
| Subband/ATC | MPEG/Audio, layer III | Wide range | World |
| Subband/ATC | Sony MiniDisc ATRAC | 146 | World |
| ATC | AT&T PAC | Wide range | U.S. |
| ATC | Dolby AC-2 | 128, 192 | U.S. |
| Wideband speech coding (7 kHz) | | | |
| Subband/ADPCM | ITU-T G.722 | 48, 56, 64 | World |
| Not defined yet | Not defined yet | 16, 24 | World |
| Speech coding (3 kHz) for telecommunications | | | |
| PCM | ITU-T G.711 | 64 | World |
| ADPCM† | ITU-T G.726 | 16, 24, 32, 40 | World |
| Embedded ADPCM | ITU-T G.727 | 16, 24, 32, 40 | World |
| Low-delay CELP | ITU-T G.728 | 16 | World |
| CELP | ITU-T G.729 | 8 | World |
| Multipulse, CLEP | ITU-T G.723 | 5.3,6.3 | World |
| Not defined yet | Not defined yet | 4 | World |
| Speech coding (3 kHz) for mobile radio | | | |
| RPE-LTP | GSM | 13 | World |
| QCELP | IS 96 | ≤ 8.5‡ | U.S. |
| VSELP | IS 54 | 7.95 | U.S. |
| VSELP | GSM half rate | 5.6 | Europe |
| VSELP | JDC | 6.7 | Japan |
| PSI-CELP | JDC half rate | 3.45 | Japan |
| Speech coding (3 kHz) for secure voice | | | |
| CELP | FS 1016 | 4.8 | U.S.DoD |
| MELP | FS 1017 | 2.4 | U.S.DoD |

\* MPEG also offers multichannel versions. The corresponding coding schemes of AT&T and Dolby are MPAC and AC-3, respectively.
† ITU-T G.726 is the successor of ITU-T G.721 and G.723.
‡ Variable rate coding.

The example of MPEG audio coding and other algorithms shows that significant reductions in bit rates can be obtained only by perception-based forms of adaptive noise shaping, to be covered next.

## 8.7   PERCEPTUAL AUDIO CODING

In recent audio-coding algorithms, perceptual coding has played an important role. Compression is achieved by transforming the input signal into the frequency domain and by removing information that is perceptually irrelevant.[15] This *lossy* compression is achieved by exploiting the well-known effects of auditory masking. In the process of hearing, the inner ear performs short-term critical band analyses where frequency-to-place transformations occur along the basilar membrane. Simultaneous masking is a frequency domain phenomenon in which a low-level signal (the maskee) can be made inaudible (masked) by a simultaneously occuring stronger signal (the masker), if masker and maskee are close enough to each other in frequency.[16] Early results on the tone-masking-noise problem showed that there is a limited band of frequencies centered at each tone in which the just-noticeable noise energy remains constant, even though the noise bandwidth and/or in-band noise shape are changed. This band of frequencies is called the *critical band* and corresponds to a physical measurement in the cochlea. The auditory system can roughly be described as a bandpass filterbank, consisting of strongly overlapping bandpass filters with critical band bandwidths in the order of 50 to 100 Hz for signals below 500 Hz and up to 5000 Hz for signals at high frequencies. Twenty-five critical bands covering frequencies up to 20 kHz have to be taken into account. Figure 8.12 compares the linear frequency scale with the critical band scale (the index of critical bands is often given in Bark). Here, the linear scale is given in multiples of 750-Hz-wide subbands; 24 subbands cover the frequency range from 0 to 24 kHz. This partition describes the MPEG/Audio filterbank (see Section 8.8.3). Figure 8.12 shows that critical bands have very small bandwidths at low frequencies. On the other hand, at high frequencies critical band bandwidths cover a number of subbands.

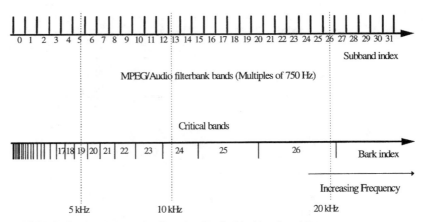

**FIGURE 8.12**    Comparison of uniform-band and critical-band partitions.

Simultaneous masking is greatest in the critical band in which the masker is located, and it is effective to a lesser degree in neighboring bands. A masking threshold can be measured below which the low-level signal will not be audible. This masked signal can consist of low-level signal contributions, quantization noise, aliasing distortion, or transmission errors. The masking threshold (in the context of source coding, also known as *threshold of just-noticeable distortion* [JND][15]) varies with time. It depends on the sound pressure level (SPL), the frequency of the masker, and on characteristics of masker and maskee. Take the example of the masking threshold for the narrow-band masker shown at the right side of Fig. 8.13: around 1 kHz the two maskees will be masked as long as their individual sound pressure levels are below the masking threshold. Note that the slope of the masking threshold is steeper toward lower frequencies; i.e., higher frequencies are more easily masked. The distance between masker and masking threshold is smaller in noise-masking-tone experiments than in tone-masking-noise experiments; i.e., noise is a better masker than a tone is. A tone-masking-noise example is given in the left part of Fig. 8.13. Noise-masking-tone experiments have shown that the level of the maskee (the tone) must be smaller by 3 to 6 dB, whereas in tone-masking-noise experiments the level of the maskee must be around

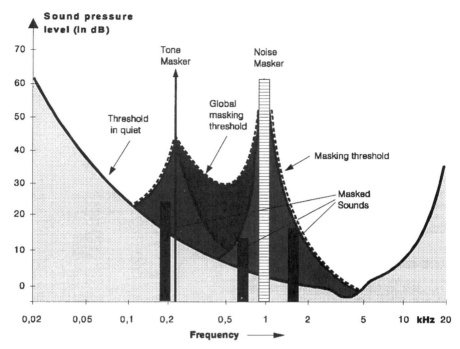

**FIGURE 8.13**    Threshold in quiet and masking thresholds. Acoustical events in the shadowed areas will not be audible.

$(14.5 + B)$ dB lower in level than that of the masker; $B$ is the Bark index of the given critical band.

Without a masker, a signal is inaudible if its sound pressure level is below the threshold in quiet, which depends on frequency and covers a dynamic range of more than 60 dB as shown in the lower curve of Fig. 8.13.

The distance between the level of the masker and the masking threshold is called *signal-to-mask ratio* (SMR). Assume an $m$-bit quantization of an audio signal. Within a critical band the quantization noise will not be audible as long as the quantizer signal-to-noise ratio, $SNR(m)$, is higher than the maximum SMR value within that given critical band. Noise and signal contributions outside the particular critical band will also be masked, although to a lesser degree, if their SPL is below the masking threshold. The perceivable distortion in a given subband is measured by the noise-to-mask ratio, $NMR(m) = SMR - SNR(m)$ (in dB). Within a critical band, coding noise will not be audible as long as $NMR(m)$ is negative.

If the source signal consists of many simultaneous maskers, a global masking threshold can be computed that describes the threshold of just-noticeable distortions as a function of frequency. This global masking threshold has been sketched in Fig. 8.13 as a dotted line for an example with two maskers—a tone at the left and a small-band noise at the right.

In addition to simultaneous masking, the time domain phenomenon of temporal masking plays an important role in human auditory perception. It may occur when two sounds appear within a small interval of time. Depending on the individual sound pressure levels, the stronger sound may mask the weaker one. Temporal masking can help to mask pre-echoes caused by the spreading of a sudden large quantization error over the actual coding block. The duration within which premasking applies (a weak signal occurs before a strong signal and is masked) is only in the order of a few milliseconds, whereas that of postmasking (a weak signal occurs after a strong signal and is masked) is in the order of 50 to 200 ms. Both pre- and postmasking are being exploited in current audio-coding algorithms.

In digital coding of audio signals at low bit rates, the error criterion has to be in favor of an output signal that is useful to the human receiver rather than favoring an output signal that follows and preserves the input waveform. Therefore, the dependence of human auditory perception on frequency and the accompanying perceptual tolerance of errors must influence encoder designs; noise-shaping techniques can shift coding noise to frequency bands where it is masked. The noise shifting must be dynamically adapted to the actual short-term input spectrum in accordance with the signal-to-mask ratio, which can be done in different ways. Frequency domain coders are of particular interest because they offer a direct method for noise shaping. Subband- or transform-based frequency mappings of the audio signals (or combinations thereof) are typically used in the encoding process (see Section 8.8). Figure 8.14 depicts the structure of a perception-based encoder that exploits auditory masking. The quantization of the individual frequency band signals is controlled by the signal-to-mask ratio (SMR) versus frequency curve, from which the needed amplitude resolution (and hence the bit allocation and rate) in each frequency band is derived. The SMR is typically determined from a high resolution, say, a 1024-point FFT-based spectral analysis of the audio block to be coded.

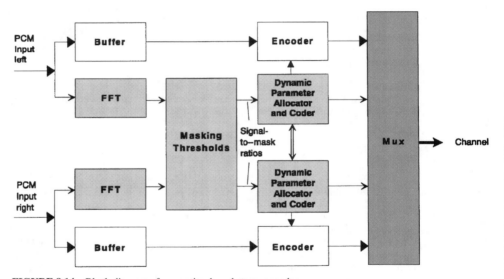

**FIGURE 8.14**    Block diagram of perception-based stereo encoder.

If the necessary bit rate for a complete masking of distortion is available, the coding scheme will be perceptually transparent; i.e., the decoded signal is then subjectively indistinguishable from the source signal. The 16-bit sample PCM of the Compact Disc often serves as a reference for audio. In practical designs we cannot go to the limits of just-noticeable distortion, because postprocessing of the acoustic signal by the end-user and multiple encoding/decoding processes in transmission links have to be considered. Moreover, our current knowledge about auditory masking is very limited. Generalizations of masking results, derived for simple and stationary maskers and for limited bandwidths, may be appropriate for many source signals but may fail for others. Therefore, as an additional requirement, we need a sufficient safety margin in practical designs of such perception-based coders.

## 8.8    *MPEG-1/AUDIO CODING*

The MPEG-1 audio-coding standard is the first international standard for the compression of digital audio signals. It can be applied to audiovisual or to audio-only data streams. The standard specifies the syntax of the coded bitstream, defines the decoding process, and provides compliance tests. This allows for different encoding algorithms, provided that the bitstream can be decoded by an MPEG-compliant decoding. The standard combines features of MUSICAM and ASPEC coding algorithms.[17,18] Main steps of development toward the MPEG-1/Audio standard have been described.[2,19]

### 8.8.1    Layers, Operating Modes, Sampling Rates, and Bit Rates

The MPEG-1/Audio standard consists of three layers (I, II, and III) of increasing complexity, delay, and subjective performance. From a hardware and software standpoint, the higher layers incorporate the main building blocks of the lower layers. MPEG layer I and II coders have very similar structures.

A standard full MPEG-1/Audio decoder is able to decode bitstreams of all three layers. The standard supports also MPEG-1/Audio layer X decoders (X = I, II, or III). Usually a layer II decoder will be able to decode bitstreams of layers I and II; a layer III decoder will be able to decode bitstreams of all three layers. MPEG-1 and MPEG-2/Audio support the sampling rates and maximum bit rates of Table 8.6. In addition, other bit rates can be chosen in the free format mode.

**TABLE 8.6**    MPEG/Audio Sampling Rates and Maximum Stereo or Multichannel Bit Rates (at a Sampling Frequency of 48 kHz)

| Layer | Sampling rates | Maximum bit rates |
|-------|----------------|-------------------|
| MPEG-1/Audio | | |
| I | 32, 44.1, and 48 kHz | 448 kb/s |
| II | 32, 44.1, and 48 kHz | 384 kb/s |
| III | 32, 44.1, and 48 kHz | 320 kb/s |
| MPEG-2/Audio: extension of MPEG-1 | | |
| I | 16, 22.05, and 24 kHz | 256 kb/s |
| II | 16, 22.05, and 24 kHz | 160 kb/s |
| III | 16, 22.05, and 24 kHz | 160 kb/s |
| MPEG-2: multichannel audio | | |
| I | 32, 44.1, and 48 kHz | 1130 kb/s |
| II | 32, 44.1, and 48 kHz | 1066 kb/s |
| III | 32, 44.1, and 48 kHz | 1002 kb/s |

MPEG/Audio has four modes: mono, stereo, dual with two separate channels (useful for bilingual programs), and joint stereo. In the optional joint stereo mode interchannel dependencies and interchannel auditory masking are exploited to reduce the overall bit rate by using a technique called *intensity stereo* (see Section 8.8.2).

## 8.8.2 Basic Structure

The basic structure of the MPEG/Audio coders follows that of perception-based coders (see Fig. 8.14). Frequency domain coders with dynamic allocations of bits (and hence of quantization noise contributions) to subbands or transform coefficients offer an easy and accurate way to control the quantization noise in order to make full use of auditory masking.[2,10,14,15,17,18,20] Redundancy (the nonflat short-term spectral characteristics of the source signal) and irrelevancy (signals below the psychoacoustical thresholds) are exploited to reduce the transmitted data rate with respect to PCM. This is achieved (1) by mapping the source spectrum into frequency bands to generate near-uncorrelated spectral components, and (2) by quantizing these separately. Two coding categories exist: subband coding (SBC) and transform coding (TC). The differentiation between these two categories is mainly due to historical reasons. Both use an analysis filterbank in the encoder to decompose the input signal into subsampled spectral components. The spectral components are called *subband samples* if the filterbank has low frequency resolution; otherwise, they are called *spectral lines* or *transform coefficients*.

In MPEG/Audio the audio signal is converted into spectral components via an analysis filterbank; layers I and II make use of a subband filterbank, layer III employs a hybrid filterbank—a cascade of the subband filterbank of Layers I and II and a linear transform. Each spectral component is then quantized and coded with the goal to keep the quantization noise below the masking threshold. The number of bits for each subband and a scale factor are determined on a block-by-block basis. The number of quantizer bits is obtained from a dynamic bit allocation algorithm (layers I and II) that is controlled by a *psychoacoustic model*. The subband code words, the scale factor, and the bit allocation information are multiplexed into one bitstream along with a header and optional ancillary data. In the decoder the synthesis filterbank reconstructs blocks of audio output samples from the demultiplexed bitstream.

## 8.8.3 Frequency Mapping in Layers I and II

Layer I and II coders employ subband coding; the source signal is fed into an analysis filterbank consisting of $M = 32$ bandpass filters, which are equally spaced and contiguous in frequency so that the set of subband signals can be recombined additively to produce the original signal or a close version thereof (Fig. 8.15). At a 48-kHz sampling rate each band has a width of $24{,}000/32 = 750$ Hz; hence, at low frequencies, a single subband covers a number of adjacent critical bands (see Fig. 8.12). The subband signals are resampled (critically decimated) at twice the nominal bandwidth of the bandpass filters, i.e., at a rate of 1500 Hz. This decimation results in an aggregate number of subband samples that equals that in the source signal. A polyphase filter structure is used for the frequency mapping; its filters have 512 coefficients, which implies impulse responses of 5.33-ms length at a sampling rate of 48 kHz. Polyphase structures are computationally very efficient because an FFT can be used in the filtering process, and they are of moderate complexity and low delay. On the negative side, the filters are equally spaced, and therefore the frequency bands do not correspond well to the critical band partition, as mentioned before. The impulse response of subband $k$, $h_{\text{sub}(k)}(n)$, is obtained by multiplication of the impulse response of a single prototype lowpass filter, $h(n)$, by a modulating function of the form $\cos[(2k + 1)\pi n/M + \varphi(k)]$, with $M = 32$, $k = 0, 1, \ldots, 31$, and $n = 0, 1, \ldots, 511$, which shifts the low-pass response to the appropriate subband frequency range. The prototype lowpass filter has a 3-dB bandwidth of $750/2 = 375$ Hz, and the center frequencies are at odd multiples thereof (all values at a 48-kHz sampling rate). Details about the coefficients of the prototype filter and the phase shifts are given in the ISO/MPEG standard.[1] Details about an efficient implementation of the filterbank can be found in the literature,[21,22] and, again, in the standardization documents.

In the receiver the sampling rate of each subband is increased to that of the source signal by filling in the appropriate number of zero samples. Interpolated subband signals appear at the bandpass outputs of the synthesis filterbank. The sampling processes may introduce aliasing distortion due to the overlapping nature of the subbands. If perfect filters (such as two-band

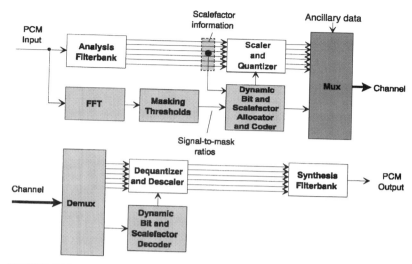

**FIGURE 8.15**    Structure of MPEG-1/Audio encoder and decoder, layers I and II.

quadrature mirror filters or polyphase filters) are applied, aliasing terms will cancel and the sum of the bandpass outputs equals the source signal in the absence of quantization.[21,23,24] With quantization, aliasing components will not cancel ideally. However, the errors will be inaudible in MPEG/Audio coding if a sufficient number of bits is used.

### 8.8.4    Frequency Mapping in Layer III

Layer III coding employs a hybrid filterbank—i.e., a cascade of a filterbank (with its short delays) and a linear discrete transform that splits each subband sequence further in frequency content to achieve a higher frequency resolution. In transform coding, a block of input samples is linearly transformed via a discrete transform into a set of near-uncorrelated transform coefficients. These coefficients are then quantized and transmitted in digital form to the decoder. In the decoder an inverse transform maps the signal back into the time domain. In the absence of quantization errors the synthesis yields exact reconstruction. Typical transforms are the discrete cosine transform (DCT), calculated via an FFT, and modified versions thereof.

The MPEG-1/Audio layer III filterbank of Fig. 8.16 is a cascade of the subband mapping of layer I and II coders and a 6-point or 18-point *modified DCT* (MDCT) block transform with 50-percent overlap.[24] Without quantization, MDCTs are free from block boundary effects between successive blocks, they have a higher transform-coding gain than the DCT, and their basis functions correspond to better bandpass responses. In the presence of quantization, block boundary effects are deemphasized due to the doubling of the filter impulse responses resulting from the overlap.

The MDCT windows contain 12 or 36 subband samples, respectively. Because the first stage of the frequency mapping generates 32 subbands, the maximum number of frequency components is $32 \times 18 = 576$, each representing a bandwidth of only $24,000/576 = 41.67$ Hz, less than or in the range of the bandwidths of the critical bands at low frequencies (see Fig. 8.12). The 18-point block transform provides better frequency resolution and is normally applied, whereas the 6-point block transform provides better time resolution and is applied in case of expected pre-echoes. The appearance of pre-echoes is similar to copying effects on analog tapes. Consider the case in which a silent period is followed by a percussive sound, such as from castanets

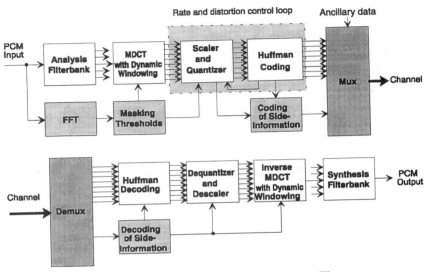

**FIGURE 8.16**   Structure of MPEG-1/Audio encoder and decoder, layer III.

or triangles, within the same coding block. Such an onset ("attack") will cause comparably large instantaneous quantization errors. In block transforms the inverse transform in the decoding process will distribute such errors over the block; similarly, in subband coding, the decoder bandpass filters will spread such errors. In both mappings pre-echoes can become distinctively audible, especially at low bit rates with comparably high error contributions. Pre-echoes can be masked by the time domain effect of premasking if the time spread is of short length (in the order of a few milliseconds). Therefore, they can be reduced or avoided by using blocks of short lengths. However, a larger percentage of the total bit rate is typically required for the transmission of side information if the blocks are shorter. A solution to this problem is to switch between block sizes of different lengths as proposed by Edler (window switching).[25] The small blocks are only used to control pre-echo artifacts during nonstationary periods of the signal; otherwise, the coder switches back to long blocks. It is clear that the block size selection has to be based on an analysis of the characteristics of the actual audio coding block. In principle, a pre-echo is assumed when an instantaneous demand for a high number of bits occurs. Depending on the nature of potential pre-echoes, all or a smaller number of transforms are switched. Two special MDCT windows, a start window and a stop window, are needed in case of transitions between short and long blocks and vice versa to maintain the time domain alias cancellation feature of the MDCT.[22,25]

### 8.8.5   Dynamic Bit Allocation in Layer I and Layer II Coders

Frequency domain coding significantly gains in performance if the number of bits assigned to each of the quantizers of the transform coefficients is adapted to the short-term spectrum of the audio-coding block on a block-by-block basis. In the mid-1970s, Zelinski and Noll proposed a DCT mapping and a dynamic bit allocation algorithm based on the DCT-based short-term spectral envelope. It was demonstrated that significant SNR-based and subjective improvements could be reached with that form of adaptive transform coding (ATC).[20,26]

In MPEG/Audio encoding the number of quantizer levels for each spectral component is obtained from a dynamic bit allocation rule that is controlled by a psychoacoustic model. The bit allocation algorithm selects one uniform midtread quantizer out of a set of available quantizers such that both the bit rate requirement and the masking requirement are met. The iterative procedure minimizes the noise-to-mask ratio (NMR) in each subband (see Section 8.7). It starts with the number of bits for the samples and scale factors set to zero. In each iteration step the quantizer signal-to-noise ratio, $SNR(m)$, is increased for the one subband quantizer producing the largest value of noise-to-mask ratio at the quantizer output. (The increase is obtained by allocating one more bit). For that purpose, the noise-to-mask ratio $NMR(m) = SMR - SNR(m)$ is calculated as the difference (in dB) between the actual quantization noise level and the minimum global masking threshold. The standard provides tables with estimates for the quantizer signal-to-noise ratio $SNR(m)$ for a given $m$.

Block companding is used in the quantization process; i.e., blocks of decimated samples are formed and divided by a scale factor such that the sample of largest magnitude is unity. In layer I blocks of 12 decimated and scaled samples are formed in each subband (and for the left and right channel) and there is one bit allocation for each block. At a 48-kHz sampling rate, 12 subband samples correspond to 8 ms of audio. There are 32 blocks, each with 12 decimated samples. These $32 \times 12 = 384$ subband samples represent 384 audio samples.

In layer I, information about the block-wise bit allocation is sent as 4-bit values with a total number of $2 \times 32 \times 4 = 256$ bits for a stereophonic signal. If, in a given subband, the signal is below the hearing threshold, a value of zero is sent to indicate that this subband is empty. The scale factors for all subbands (except those that are empty) are quantized in a 6-bit format; the total bit rate for the scale factors is $2 \times 32 \times 6 = 376$ bits.

In layer II in each subband a 36-sample superblock is formed of three consecutive blocks of 12 decimated samples corresponding to 24 ms of audio at a 48-kHz sampling rate. There is one bit allocation for each 36-sample superblock. The 32 superblocks, each with 36 decimated samples, together represent $32 \times 36 = 1152$ audio samples. A redundancy reduction technique is used for the transmission of the scale factors. Depending on the significance of the changes between the three consecutive scale factors, one, two, or all three scale factors are transmitted, together with a 2-bit scale factor select information. Compared with layer I, the bit rate for the scale factors is reduced by around 50 percent.[2] Through this reduction the layer II coder achieves a better performance. Additionally, a finer quantization is provided.

The scaled and quantized spectral subband components are transmitted to the receiver together with scale factor, scale factor select (layer II), and bit allocation information. Quantization with block companding provides a very large dynamic range of more than 120 dB. In layer I, 14 midtread quantizers are available with $2^n - 1$ steps ($n = 2, 3, \ldots, 15$). Layer II allows for a maximum of 15 bits, but only in the low-frequency region. In the mid- and high-frequency region the maxima are 7 and 3 bits, respectively. Layer II includes, unlike layer I, quantizers with 5 and 9 steps. In order to reduce the bit rate, the code words of three successive subband samples resulting from quantizing with 3-, 5-, and 9-step quantizers are assigned one common codeword. The savings in bit rate is about 40 percent.[2]

Figure 8.17 shows an example of the statistics of the assigned number of quantizer bits in all subbands. Shown is the layer II distribution of bit assignments for a high-quality speech signal about 30 seconds in length. Note, for example, that quantizers with 10 or more bits of resolution are only employed in the first few subbands (each subband corresponds to a 750-Hz bandwidth), and that subbands of high index are quantized with not more than five bits per sample. In particular, note that no bits have been assigned to subbands 25 and above (frequencies above 18 kHz).

### 8.8.6  Dynamic Bit Allocation in Layer III Coders

Layer III of the MPEG-1/Audio coding standard introduces many new features, in particular a switched hybrid filterbank. In addition it employs an analysis-by-synthesis approach, an advanced pre-echo control, and nonuniform quantization of the MDCT output samples. Huffman coding, based on 32 predefined code tables, and additional run-length coding are applied to represent the quantizer indices in an efficient way. The encoder maps the variable-word-length

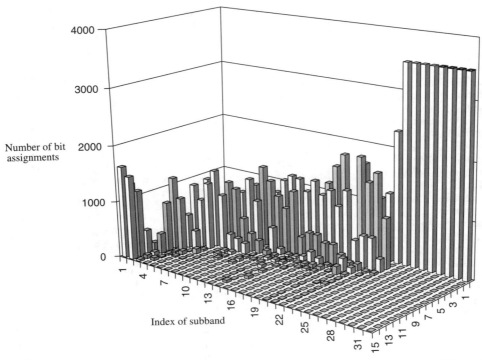

**FIGURE 8.17**   Distribution of bit assignments vs. subband index. The distribution is based on about 30 seconds of MPEG-1/Audio layer II coded speech.

code words of the Huffman code tables into a constant bit rate by monitoring the state of a bit reservoir. The bit reservoir ensures that the decoder buffer neither underflows nor overflows when the bitstream is presented to the decoder at a constant rate.[1,2] In passing, we note that layer III is the only layer that provides decoder support for variable bit rate coding.

In order to keep the quantization noise in all critical bands below the global masking threshold (noise allocation), an iterative analysis-by-synthesis method is employed, whereby the process of scaling, quantization, and coding of spectral data is carried out within two nested iteration loops.[1,2] The decoding follows that of the encoding process.

## 8.8.7 Psychoacoustic Models

We have already mentioned that the adaptive bit allocation algorithm is controlled by a psychoacoustic model. This model computes signal-to-mask ratios (SMR) taking into account the short-term spectrum of the audio block to be coded and knowledge about noise masking. The model is only needed in the encoder, which makes the decoder less complex; this asymmetry is a desirable feature for audio playback and audio-broadcasting applications.

The normative part of the ISO/MPEG standard describes the decoder and the meaning of the encoded bitstream,[1] but the encoder is not standardized—thus leaving room for an evolutionary improvement of the encoder. In particular, different psychoacoustic models can be used, ranging from very simple (or none at all) to very complex ones based on quality and implementability requirements. Information about the short-term spectrum can be derived in various ways, for example, as an accurate estimate from an FFT-based spectral analysis of the audio input samples. If the frequency resolution of the subband- or transform-based frequency mapping is high enough, the SMR can be derived directly from the subband samples or transform coefficients without running a FFT-based spectral analysis in parallel.[7,20] Encoders can also be optimized for a certain application. All these encoders can be used with complete compatibility with all existing MPEG-1/Audio decoders.

The informative part of the standard gives two examples of FFT-based models;[1] see also the literature.[2,4,5,22] Both models identify, in different ways, tonal and nontonal spectral components and use the corresponding results of tone-masks-noise and noise-masks-tone experiments in the calculation of the global masking thresholds. Details are given in the standard; experimental results for both psychoacoustic models are described in the literature.[22] A 512-point FFT is proposed for layer I, and a 1024-point FFT for layers II and III. In both models the audio input samples are Hann-weighted. Model 1, which may be used for Layers I and II, computes for each masker its individual masking threshold, taking into account its frequency position, power, and tonality information. The global masking threshold is obtained as the sum of all individual masking thresholds and the absolute masking threshold. The SMR is then the ratio of the maximum signal level within a given subband and the minimum value of the global masking threshold in that given subband.

Model 2, which may be used for all layers, is more complex. Tonality is assumed when a simple prediction indicates a high prediction gain, and the masking thresholds are calculated in the cochlea domain—i.e., properties of the inner ear are taken into account in more detail. Finally, in case of potential pre-echoes, the global masking threshold is adjusted appropriately.

### 8.8.8  Stereo Coding

The left and right channels of an audio stereo pair can be coded separately, but there is a significant advantage in coding these channels jointly because they are generally correlated. The perceptual coder of Fig. 8.14 is able to exploit correlations and irrelevancy between the left and right channel. Stereo coding can be used to increase the audio quality at low bit rates or to reduce the bit rates. Similarly, improvements are obtained from a joint coding of all channels of a multichannel signal (see Section 8.9).

It is known that, above 2 kHz and within each critical band, the human auditory system bases its perception of stereo imaging more on the temporal envelope of the audio signal than on its temporal fine structure. Therefore, the MPEG/Audio compression algorithm supports a stereo-coding mode called *intensity stereo coding,* which reduces the total bit rate without violating the spatial integrity of the stereophonic signal.

In intensity stereo mode the encoder codes some upper-frequency subband outputs with a single sum signal $L + R$ (or some linear combination thereof) instead of sending independent left ($L$) and right ($R$) subband signals. The decoder reconstructs the left and right channels based only on the single $L + R$ signal and on independent left and right channel scale factors. Hence the spectral shape of the left and right outputs is the same within each intensity-coded subband but the magnitudes are different.[27] This optional joint stereo mode will only be effective if the required bit rate exceeds the available bit rate, and it will only be applied to subbands corresponding to frequencies of around 2 kHz and above. Layer III has an additional option: The mono/stereo (M/S) mode encodes the left and right channel signals as middle ($L + R$) and side ($L - R$) channels. In addition, the M/S mode can be combined with the intensity mode.

Joint coding of an audio stereo pair needs a very careful design; otherwise, a number of undesirable artifacts can occur. One example is binaural unmasking: the $L$ and $R$ coders may mask the corresponding $L$ and $R$ distortions completely, whereas the binaural signal may have significant unmasked noise. Joint coding and unmasking will be mentioned again in Section 8.9 in the context of multichannel audio coding.

### 8.8.9   Decoding

The decoding is straightforward: the subband sequences are reconstructed on the basis of blocks of subband samples taking into account the decoded scale factor and bit allocation information. If a subband has no bits allocated to it, the samples in the subband are set to zero. Each time the subband samples of all 32 subbands have been calculated, they are applied to the synthesis filterbank, and 32 consecutive 16-bit PCM format audio samples are calculated. If available, as in bidirectional communications or in recorder systems, the encoder (analysis) filterbank can be used in a reverse mode in the decoding process.

### 8.8.10   Frame and Multiplex Structure

MPEG uses a flexible ATM-like packet protocol with headers/descriptors for multiplexing audio and video bitstreams in one stream with the necessary information to keep the streams synchronized when decoding. Figure 8.18 shows the frame structure of MPEG-1/Audio coded signals (layer I and layer II). Each frame has a header; its first part contains 12 synchronization bits, 20 bit system information, and an optional 16-bit cyclic redundancy check code (CRC) to check the header for errors. Its second part contains side information about the bit allocation and the scale factors (and, in layer II, scale factor select information). As main information, a frame carries a total of $32 \times 12$ subband samples (corresponding to 384 PCM audio input samples—equivalent to 8 ms at a sampling rate of 48 kHz) in layer I, and a total of $32 \times 36$ subband samples in layer II (corresponding to 1152 PCM audio input samples—equivalent to 24 ms at a sampling rate of 48 kHz). Note that the frames are autonomous: each frame contains all information necessary for decoding. Therefore, each frame can be decoded independently from previous frames; it defines an entry point for audio storage and audio-editing applications. The lengths of the frames are not fixed, due to (1) the length of the main information field, which depends on bit rate and sampling frequency; (2) the side information field, which varies in layer II; and (3) the ancillary data field, the length of which is not specified. In MPEG-2 (multichannel coding) this ancillary field is used partly to

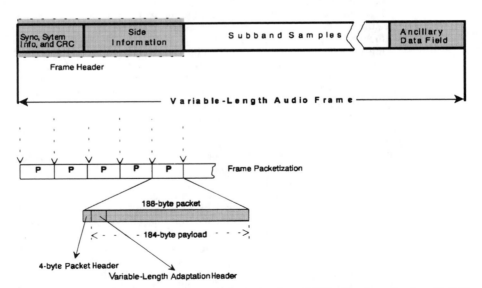

**FIGURE 8.18**   MPEG-1 frame structure and packetization. Layer I: 384 subband samples; layer II: 1152 subband samples.

transmit multichannel extension signals. In the European DAB application this field can carry program-associated data (PAD), text-based information, and/or still images and video sequences.

A packet structure is used for multiplexing audio, video, and ancillary data bitstreams in one stream. The variable-length MPEG frames are broken down into packets. The packet structure uses 188-byte packets, consisting of a 4-byte header followed by 184 bytes of payload (see Fig. 8.18). The header includes a sync byte, a 13-bit field called the *packet identifier* to inform the decoder about the type of data, and additional information. For example, a 1-bit payload unit start indicator indicates if the payload starts with a frame header. No predetermined mix of audio, video, and ancillary data bitstreams is required, and the mix may change dynamically. Hence, servives can be provided in a very flexible way. If additional header information is required, for example, for periodic synchronization of audio and video timing, a variable-length adaptation header can be used as part of the 184-byte payload field. Although the lengths of the frames are not fixed, the interval between frame headers is held constant (within a byte) by making use of padding bytes.

### 8.8.11   Subjective Quality

It is possible for MPEG-1/Audio to provide a subjective quality that is equivalent to Compact Disc (CD) quality (16-bit PCM) at stereo rates between 384 kb/s (layer I) and 128 kb/s (layer III) for many types of music. Because of its high dynamic range, MPEG-1/Audio has potential to exceed the quality of a CD.[6,28]

The MPEG standardization process included extensive subjective tests and objective evaluations of parameters such as complexity and overall delay. The MPEG listening tests were carried out under very similar and carefully defined conditions with around 60 experienced listeners. Approximately 10 test sequences were used, and the sessions were performed in stereo with both loudspeakers and headphones. In order to detect even small impairments, the 5-point ITU-R impairment scale was used in all experiments.[29,30] Critical test items were chosen in the tests to evaluate the coders by their worst-case (not average) performance. The MPEG-1/Audio coding standard leaves room for encoder-based improvements by using better psychoacoustic models because such models are not normative elements of the standard. Indeed, subjective results of more recent MPEG-1 listening tests show that many improvements have been achieved in the course of the last years.

### 8.8.12   Lower Sampling Rates

Lower sampling rates (16, 22.05, and 24 kHz) have been defined in MPEG-2 for better audio quality at bit rates at, or below, 64 kb/s per channel;[5] see Table 8.6. The corresponding maximum audio bandwidths are 7.5, 10.3, and 11.25 kHz. Possible applications include the transmission of medium-quality audio and wideband speech. A reduction in sampling frequency by a factor of 2 doubles the frequency resolution of the filterbanks and transforms and therefore improves the performance of the coders—in particular those with a low-frequency resolution (layers I and II). The syntax, semantics, and coding techniques of MPEG-1 are maintained except for a small number of parameters.

## 8.9   MPEG-2/AUDIO MULTICHANNEL CODING

A logical further step in digital audio is the definition of a multichannel audio representation system to create a convincing, lifelike soundfield for audio-only applications and for audio-visual systems, including video conferencing, videophony, multimedia services, and electronic cinema. Multichannel systems can also provide multilingual channels and additional channels for the visually impaired (a verbal description of the visual scene) and for the hearing impaired (dialogue with enhanced intelligibility). Recommended loudspeaker configurations, referred to as *3/2-stereo,*—with a left and a right channel (L and R), an additional center channel C, and two side/rear surround channels (LS and RS) augmenting the L and R channels—offer an improved realism of auditory ambience with a stable frontal sound image and a large listening area. Multichannel digital audio systems support *p/q* presentations with *p* front and *q* back channels and also provide the possibilities of transmitting two independent stereophonic programs, and/or of a number of commentary or multilingual channels. Typical combinations of channels are given in Table 8.7.

**TABLE 8.7**    Multichannel Configurations

| | | |
|---|---|---|
| 1 channel | 1/0-configuration | center (mono) |
| 2 channels | 2/0-configuration | left, right (stereophonic) |
| 3 channels | 3/0-configuration | left, right, center |
| 4 channels | 3/1-configuration | left, right, center, mono-surround |
| 5 channels | 3/2-configuration | left, right, center, surround left, surround right |

Recommendations of ITU-R, EBU, and SMPTE 775 provide a set of downward-mixing equations if the number of loudspeakers is to be reduced (downward compatibility). An additional *low frequency* enhancement (LFE or subwoofer) channel is particularly useful for HDTV applications. It can be added, optionally, to any of the configurations. A 3/2-configuration with five high-quality full-range channels plus a subwoofer channel is often called a *5.1 system.* The LFE channel extends the low-frequency content between 15 Hz and 120 Hz in terms of both frequency and level. One or more loudspeakers can be positioned freely in the listening room to reproduce this LFE signal. (The film industry uses a similar system for their digital sound systems).

Redundancies and irrelevancy (such as interchannel dependencies and interchannel masking effects, respectively) have to be exploited to obtain small overall bit rate. In addition, stereophonic-irrelevant components of the multichannel signal, which do not contribute to the localization of sound sources, may be identified and reproduced in a monophonic format to further reduce bit rates. State-of-the-art multichannel coding algorithms make use of such effects. It should be noted that such joint coding needs a careful design; otherwise, it may produce artifacts such as unmasking of noise.

The second phase of MPEG is called MPEG-2. In its audio part it includes two multichannel audio coding standards, one of which is forward- and backward-compatible with MPEG-1/Audio.[5,31-32] Forward compatibility means that an MPEG-2 multichannel decoder is able to properly decode MPEG-1 mono or stereophonic signals. Backward compatibility (BC) means that existing MPEG-1 stereo decoders, which only handle two-channel audio, are able to reproduce a meaningful basic 2/0 stereo signal from an MPEG-2 multichannel bitstream so as to serve the need of users with simple mono or stereo equipment. *Non–backward-compatible* (NBC) multichannel coders will not be able to feed a meaningful bitstream into an MPEG-1 stereo decoder. On the other hand, NBC codecs have more freedom in producing a high-quality reproduction of audio signals.

## 8.9.1    Subjective Quality

Early subjective tests, independently run at German Telekom and BBC (UK) under the umbrella of the MPEG-2 standardization process, have shown a satisfactory average performance of non–backward-compatible (NBC) and of backward-compatible (BC) coders. The tests were carried out with experienced listeners and critical test items at low bit rates (320 and 384 kb/s). However, all codecs showed deviations from transparency for some of the test items.[27,33]

Recently, extensive formal subjective tests[34] have been carried out to compare two MPEG-2 NBC coders, operating, respectively, at 256 and 320 kb/s, and a backward-compatible MPEG-2 layer II coder,* operating at 640 kb/s. All coders showed a very good performance, with a slight advantage of the 320 kb/s MPEG-2 NBC coder compared with the 640 kb/s MPEG-2 layer II BC coder. The performances of those coders are indistinguishable from the original (if the EBU definition of "indistinguishable quality" is applied).[35]

### 8.9.2  Backward-Compatible (BC) MPEG-2/Audio Coding

Backward compatibility is achievable by the use of compatibility matrices. The MPEG-1 signals $L$ and $R$ are replaced by the matrixed signals $L0 = L + 0.707C + 0.707LS$ and $R0 = R + 0.707C + 0.707RS$. Other parameter choices are possible.

The signals $L0$ and $R0$ are transmitted in MPEG-1 format in transmission channels $T1$ and $T2$. Channels $T3$, $T4$, and $T5$ together form the multichannel extension signal (Fig. 8.19). They have to be chosen such that the decoder can recompute the complete 3/2-stereo multichannel signal. Interchannel redundancies and masking effects are taken into account to find the best choice. A simple example is $T3 = C$, $T4 = LS$, and $T5 = RS$. In MPEG-2 the matrixing can be done in a very flexible and even time-dependent way.

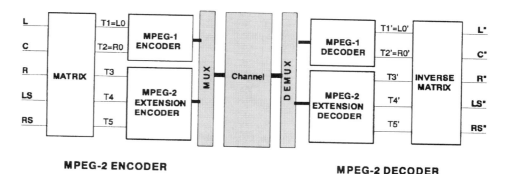

**FIGURE 8.19**  Backward-compatible MPEG-2 multichannel audio bitstreams.

The basic frame format is identical to the MPEG-1 bitstream. Backward compatibility is achieved by transmitting the channels $L0$ and $R0$ in the subband-sample section of the MPEG-1 audio frame and by transmitting all multichannel extension signals, $T3$, $T4$, and $T5$, in the first part of its ancillary data field, which is ignored by MPEG-1 decoders (Fig. 8.19). The length of the ancillary data field is not specified in the standard. If the decoder is of type MPEG-1, it uses the 2/0-format front left and right down-mix signals, $L0'$ and $R0'$, directly. If the decoder is of type MPEG-2, it recomputes the complete 3/2-stereo multichannel signal with its components $L''$, $R''$, $C''$, $LS''$, and $RS''$ via *dematrixing* of $L0'$, $R0'$, $T3'$, $T4'$, and $T5'$.

With matrixing, unmasking of quantization noise may appear.[36] It may be caused in the dematrixing process when sum and difference signals are formed. In certain situations a sum or difference signal component can disappear in a specific channel. Because this component was supposed to mask the quantization noise in that channel, that noise may become audible. Note

---

*A 1995 version of this latter coder was used, so its test results do not reflect any subsequent enhancements.

that the masking signal will still be present then in the multichannel representation but it will appear on a different loudspeaker. Measures against unmasking effects have been described in the literature.[37]

MPEG-1 decoders have a bit-rate limitation (384 kb/s in layer II). In order to overcome this limitation, the MPEG-2 standard allows for a second bitstream (extension bitstream) to provide compatible multichannel audio at higher rates.

We finally note that if bit rates are not of high concern, a simulcast transmission may be employed, in which an MPEG-1 bitstream is multiplexed with a complete MPEG-2 bitstream in order to support backward compatibility without matrixing techniques.

With backward compatibility it is possible to introduce multichannel audio at any time in a smooth way without making existing two-channel stereo decoders obsolete. Additionally, the program provider and the consumer can switch from two-channel to multichannel at any time. An important example is the European Digital Audio Broadcast system, which will require MPEG-1 stereo decoders in the first generation but may offer MPEG-2 multichannel audio at a later point.

### 8.9.3 Advanced Audio Coding (AAC)

A second standard within MPEG-2 supports applications that do not request compatibility with the existing MPEG-1 stereo format. Therefore, matrixing and dematrixing are not necessary and the corresponding potential artifacts disappear. The AAC multichannel coding mode will have sampling rates, audio bandwidths, and channel configurations of MPEG-2/Audio. It is based on the evaluation and definition of six basic modules, each to be selected from a number of proposals. The modules are the optional preprocessing module, time to frequency mapping module, psychoacoustics model, joint-coding module, quantization and coding modules, and bitstream formatter.[38]

The MPEG-2 AAC standard combines the efficiency of high-resolution filter banks, prediction techniques, and Huffman coding.[39] To serve different needs, the standard offers three profiles: the main profile, the low-complexity profile, and the sampling-rate-scalable profile. For example, in its main profile the filter bank is a modified discrete cosine transform of blocklength 2048 or 256, and it allows for a frequency resolution of 23.43 Hz and a time resolution of 2.6 ms (both at a sampling rate of 48 kHz). In the case of the long blocklength, the window shape can be varied dynamically as a function of the signal, a temporal noise shaping tool is offered to control the time dependence of the quantization noise, time-domain predictions with second-order backward-adaptive linear predictors reduce the bit rate for coding subsequent subband samples in a given subband, and iterative non-uniform quantization and noiseless coding are applied.

The low-complexity profile does not employ temporal noise shaping and time-domain prediction, whereas in the sampling-rate-scalable profile a preprocessing module is added that allows for sampling rates of 6, 12, 18, and 24 kHz.

A detailed description of the MPEG-2 AAC multichannel standard can be found in the literature.[39] The standard offers high quality at lowest possible bit rates between 320 and 384 kb/s for five channels; it will find many applications, both for consumer and professional use.

MPEG-2 AAC will become an international standard in July 1997 as an extension to MPEG-2 (ISO/MPEG 13818-7).

### 8.9.4 MPEG-4/Audio Coding

Activities within MPEG-4 aim at proposals for a broad field of applications including multimedia. MPEG-4 will offer higher compression rates, and it will merge the whole range of audio from high-fidelity audio coding and speech coding down to synthetic speech and synthetic audio. It will also offer new functionalities such as time scale changes, pitch control, edibility, database access, and scalability, which enables the extraction of a subset from the transmitted bitstream sufficient to generate audio signals with lower bandwidth or lower quality, depending on channel capacity or decoder complexity.

To represent, integrate, and exchange pieces of audio-visual information, MPEG-4 offers standard tools that can be combined to satisfy specific user requirements.[40] A number of such configurations may be standardized. A syntactic description will be used to convey to a decoder the choice of tools made by the encoder. This description can also be used to describe new algorithms and download their configuration to the decoding processor for execution. The current tool set supports audio and speech compression at monophonic bit rates ranging from 2 to 64 kb/s. Three core coders are used: a parametric coding scheme for low-bit-rate speech coding, an analysis-by synthesis coding scheme for medium bit rates (6 to 16 kb/s), and a subband/transform-based coding scheme for higher bit rates. These three coding schemes have been integrated into a so-called verification model that describes the operations both of encoders and decoders and is used to carry out simulations and optimizations. In the end, the verification model will be the embodiment of the standard.[40] MPEG-4 will become an international standard in November 1998.

## 8.10   APPLICATION FIELDS

Consumer electronics, professional audio, telecommunications, broadcasting, and multimedia will gain from MPEG/Audio compression technologies. A number of MPEG-based consumer products have recently reached the audio market. A few typical application fields are described in the following. It is important to realize that MPEG standards do not specify the operations of the encoder but specify only the syntax of the encoded bitstream and the decoding process. Thus, they provide flexibility in the specifications so that they can be optimized for different applications. In particular, psychoacoustic modeling can be done differently; hence, subjective quality of MPEG/Audio compression will depend on the actual encoder algorithm.

### 8.10.1   Digital Storage

The Philips Digital Compact Cassette (DCC) employs layer I of the MPEG-1/Audio coder running at a 384-kb/s stereo rate, the algorithm is called *PASC* (Precision Audio Subband Coding).[7]

Proposed digital surround systems for the consumer market are currently based on a 5.1-channel configuration. In movie theaters a 7.1-channel configuration is becoming popular because of an improved front-back stability of the stereo image and an improved impression of spaciousness. The two additional loudspeakers are positioned at the left and the right side of the center loudspeaker. It is expected that consumer demand will vary between 2.0-channel, 5.1-channel, and 7.1-channel reproduction. A scalable configuration can be based on the MPEG-1 and MPEG-2 standards by down-mixing the 7-channel signal into a 5-channel signal and subsequently down-mixing the latter one into a 2-channel signal.[41] The 2-channel signal, three contributions from the 5-channel signal, and two contributions from the 7-channel signal can then be transmitted or stored. The decoder uses the 2-channel signal directly, or it employs matrixing to reconstruct 5- or 7-channel signals. This scalable configuration may be applied in the Digital Versatile Disc (DVD).

Solid-state audio playback systems (e.g., for announcements), with the compressed data stored on chip-based memory cards or smart cards will find many applications. A recent example is NEC's prototype silicon audio player, which uses a one-chip MPEG-1/Audio layer II decoder and offers 24 minutes of stereo at its recommended stereo bit rate of 192 kb/s.[42] Future applications will include various interactive multimedia and interchange media services based on memory cards.

### 8.10.2  Digital Audio Broadcasting

In Europe a project group named Eureka 147, set up in 1987, has worked out a digital audio broadcasting (DAB) system (see Chapter 14) able to cope with the problems of digital broadcasting.[43,44] ITU-R recommended the MPEG-1/Audio coding standard for DAB after it made extensive subjective tests. Layer II of this standard will be used at rates between 32 and 192 kHz per mono channel. The layer III version is recommended for commentary links* at a low rate. The sampling rates will be 48 kHz or 24 kHz, and part of the ancillary data field will be used for program-associated data (PAD information). Typical examples are dynamic range control, display of program titles, and text with graphic features. The European DAB system will have a significant bit rate overhead for error correction based on punctured convolutional codes. Source-adapted channel coding (i.e., an unequal error protection (UEP)) that is in accordance with the sensitivity of individual bits or a group of bits to channel errors will be applied.[45] The DAB specification provides means to meet different protection requirements, ranging from mobile radio to cable distribution. In the case of UEP, code rates ranging from $R = 1/4$ (300 percent overhead) to $R = 3/4$ (25 percent overhead)† will be provided. Additionally, error concealment techniques will be applied to provide a graceful degradation in case of severe errors. Concealment techniques will also be evoked if the cyclic redundancy check (CRC) code in the audio frame header detects errors.

In the United States a standard has not yet been decided upon.[44] Simulcasting analog and digital versions of the same audio program in the FM terrestrial band (88–108 MHz) is an important issue (whereas the European solution is based on new channels). One of the candidates in the EIA-NRSC‡ contest for a DAB standard is the MPEG-based DAB system of the Eureka 147 group.

As examples of satellite-based digital broadcasting, we mention the Hughes DirecTV satellite subscription television system and ADR (Astra Digital Radio), both of which make use of MPEG-1 layer II. As a further example, the Eutelsat SaRa system will be based on layer III coding.

### 8.10.3  ISDN Transmission

Narrowband Integrated Services Digital Networks (ISDN) customers have physical access to a number of relatively low-cost dial-up digital telecommunications channels. For the transmission of audio signals the basic-rate interface is of interest; it consists of one or two 64-kb/s $B$ channels and one 16-kb/s $D$ channel (which supports signaling but can also carry user information). Other configurations are possible, including $p \times 64$ kb/s ($p = 1, 2, 3, \dots$) services. These rates indicate that ISDN offers useful channels for a practical distribution of stereophonic and multichannel audio signals. Because ISDN is a bidirectional service, it also provides upstream paths for future on-demand and interactive audiovisual just-in-time audio services. One example of transmission of commentary grade material is an MPEG-2 coding at a lower sampling rate and a bit rate of 48 kb/s with an additional transmission of ancillary data at 16 kb/s.

### 8.10.4  Audiovisual Communications and Multichannel Audio

Advanced digital TV systems will provide HDTV delivery to the public by terrestrial broadcasting and a variety of alternate media and other full-motion high-resolution video and high-quality

---

*A commentary link must be capable of delivering speech of exellent quality to the listener; for musical program material a reduced level of performance can be accepted.

† A code rate is the ratio of source (audio) data rate and gross data rate after channel coding.

‡ Electronic Industries Association–National Radio Systems Committee.

multichannel surround audio. The U.S. Grand Alliance HDTV system and the European Digital Video Broadcast system both make use of the MPEG-2 video compression system and of the MPEG-2 transport layer. The systems differ in the way the audio signal is compressed: the Grand Alliance system will use Dolby's AC-3 transform coding technique,[41,42] whereas the European system will use the MPEG-2/Audio algorithm.

Many films are already in 5.1 multichannel format, although at least four different and incompatible formats for digital multichannel audio exist.

## 8.11   CONCLUSION

The ISO/MPEG-1/Audio coding algorithm, with its three layers, has been widely accepted as international standard. MPEG/Audio coders are controlled by psychoacoustic models which may be improved, thus leaving room for an evolutionary improvement of codecs. A better understanding of binaural perception and of stereo presentation will lead to new proposals.

Digital multichannel audio improves stereophonic images and will be of importance both for audio-only and multimedia applications. MPEG-2/Audio offers a backward-compatible coding scheme that is of high relevance in order to introduce multichannel audio at any time in a smooth way without making existing MPEG-1/Audio two-channel stereo decoders obsolete. Ongoing research will result in enhanced multichannel representations by making better use of interchannel correlations and interchannel masking effects to bring the bit rates further down. The non–backward-compatible MPEG Advanced Audio Coding scheme with its new modular approach and its flexibility may become a very successful standard in the future.

### 8.11.1   WWW Information about MPEG/Audio

Web information about MPEG is available at different addresses. The official MPEG home page has the address http://drogo.cselt.stet.it/mpeg/. There one can find crash courses in MPEG and ISO, overviews of current activities, MPEG requirements, work plans, and information about documents and standards. Links lead to collections of frequently asked questions. One example is http://www.vol.it/MPEG/ which includes a list of MPEG, multimedia, and digital video related products. Other examples are http://www.bok.net/~tristan/MPEG/audio.html#audio-softwares with detailed information about MPEG/Audio resources, software, and audio test bitstreams.

## GLOSSARY

**AAC**     Advanced Audio Coding (MPEG-2)

**ADPCM**     Adaptive differential pulse code modulation

**ATC**     Adaptive transform coding

**ATM**     Asynchronous time multiplex

**ATRAC**     Adaptive transform acoustic coding (Sony)

**BBC**     British Broadcasting Corporation

**BC**     Backward compatibility

**CCETT**     Centre commun d'ètudes de télécommunications et de télédiffusion

**CD**     Compact Disc

**CELP**     Code-excited linear predictive coding

**CNET**   Centre national d'ètudes de télécommunications

**CRC**   Cyclic redundancy check

**DAB**   Digital audio broadcasting

**DAT**   Digital audio tape

**DCC**   Digital Compact Disc

**DCME**   Digital circuit multiplex equipment

**DVD**   Digital versatile disc

**EBU**   European Broadcast Union

**EIA-NRSC**   Electronic Industries Association—National Radio Systems Committee

**FFT**   Fast Fourier transform

**FhG**   Fraunhofer-Gesellschaft

**GSM**   Group spèciale mobile

**HDTV**   High-definition television

**IRT**   Institut für Rundfunktechnik, München (FRG)

**IS**   International Standard

**ISDN**   Integrated services digital network

**ISO**   International Organization of Standardization

**ITU-T**   International Telecommunications Union–Telecommunications Standardization Sector

**ITU-R**   International Telecommunications Union–Radio Standardization Sector

**JDC**   Japanese digital cellular

**LFE**   Low-frequency enhancement

**MDCT**   Modified discrete cosine transform

**MELP**   Mixed-excitation linear predictive coding

**MOD**   Magneto-optical disk

**MPAC**   Multichannel perceptual audio coding (AT&T)

**MPEG**   Moving Pictures Expert Group

**M/S mode**   Mono/stereo mode

**NBC**   Non-backward compatibility

**NMR**   Noise-to-mask ratio

**PAC**   Perceptual audio coding (AT&T)

**PAD**   Program-associated data

**PASC**   Precision audio subband coding (Philips)

**PCM**   Pulse code modulation

**PSI-CELP**   Pitch-synchronous innovation code excited linear predictive coding

**QCELP**   Qualcomm code excited linear predictive coding

**RPE-LTP**   Regular pulse excited linear predictive coding with long-term prediction

**SMR**   Signal-to-mask ratio

**TCE**   Thomson Consumer Electronics

**TIA**   Telecommunications Industries Association

**UEP**   Unequal error protection

**VSELP**   Vector-sum excited linear predictive coding
**WWW**   World Wide Web

## REFERENCES

1. ISO/IEC JTCl/SC29, "Information Technology—Coding of Moving Pictures and Associated Audio for Digital Storage Media at up to about 1.5 Mbit/s—IS 11172 (Part 3, Audio)," 1992.

2. K. Brandenburg and Stoll, G., "The ISO/MPEG-audio codec: A generic standard for coding of high quality digital audio," *Journal of the Audio Engineering Society,* vol. 42, no. 10, Oct. 1994, pp. 780–792.

3. R. G. van der Waal; Brandenburg, K., and Stoll, G., "Current and future standardization of high-quality digital audio coding in MPEG," *Proceedings IEEE ASSP Workshop on Applications of Signal Processing to Audio and Acoustics,* New Paltz, N.Y., 1993.

4. P. Noll and Pan, D., "ISO/MPEG audio coding," *International Journal of High Speed Electronics and Systems,* 1997 (submitted).

5. ISO/IEC JTCl/SC29, "Information Technology—Generic Coding of Moving Pictures and Associated Audio Information—IS 13818 (Part 3, Audio)," 1994.

6. L. M. van de Kerkhof and Cugnini, A. G., "The ISO/MPEG audio coding standard," *Widescreen Review,* 1994.

7. A. Hoogendorn, "Digital compact cassette," *Proceedings of the IEEE,* vol. 82, no. 10, Oct. 1994, pp. 1479–1489.

8. K. Tsutsui, Suzuki, H., Shimoyoshi, O., Sonohara, M., Akagari, K., and Heddle, R. M., "Adaptive transform acoustic coding for MiniDisc," *93rd Audio Engineering Society Convention,* San Francisco, Calif., 1992, preprint 3456.

9. P. Noll, "Wideband speech and audio coding," *IEEE Communications Magazine,* vol. 31, no. 11, 1993, pp. 34–44.

10. P. Noll, "Digital audio coding for visual communications," *Proceedings of the IEEE,* vol. 83, no. 6, June 1995.

11. A. Gersho, "Advances in speech and audio compression," *Proceedings of the IEEE,* vol. 82, no. 6, 1994, pp. 900–918.

12. A. S. Spanias, "Speech coding: A tutorial review," *Proceedings of the IEEE,* vol. 82, no. 10, Oct. 1994, pp. 1541–1582.

13. N. S. Jayant, Johnston, J. D., and Shoham, Y., "Coding of wideband speech," *Speech Communication,* vol. 11, 1992, pp. 127–138.

14. N. S. Jayant and Noll, P., *Digital Coding of Waveforms: Principles and Applications to Speech and Video,* Prentice Hall, Englewood Cliffs, N.J., 1984.

15. N. S. Jayant, Johnston, J. D., and Safranek, R., "Signal compression based on models of human perception," *Proceedings of the IEEE,* vol. 81, no. 10, 1993, pp. 1385–1422.

16. E. Zwicker and Taste, H., *Psychoacoustics,* Springer, Berlin, 1990.

17. Y. F. Dehery, Stoll, G., and Kerkhof, L. v.d., "MUSICAM source coding for digital sound," *17th International Television Symposium,* Montreux, Switzerland, June 1991, pp. 612–617.

18. K. Brandenburg, Herre, J., Johnston, J. D., Mahieux, Y., and Schroeder, E. F., "ASPEC: Adaptive Spectral Perceptual Entropy Coding of high quality music signals," *90th Audio Engineering Society Convention,* Paris, 1991, preprint 3011.

19. H. G. Musmann, "The ISO audio coding standard," *Proceedings IEEE Globecom,* Dec. 1990.

20. R. Zelinski and Noll, P., "Adaptive transform coding of speech signals," *IEEE Transactions on Acoustics, Speech and Signal Processing,* vol. ASSP-25, Aug. 1977, pp. 299–309.

21. J. H. Rothweiler, "Polyphase quadrature filters, a new subband coding technique," *Proceedings International Conference ICASSP'83,* 1983, pp. 1280–1283.

22. D. Pan, "A tutorial on MPEG/Audio compression," *IEEE Transactions on Multimedia,* 1995.

23. D. Esteban and Galand, C., "Application of quadrature mirror filters to split band voice coding schemes," *Proceedings International Conference on Acoustics, Speech, and Signal Processing (ICASSP '87),* 1987, pp. 191–195.

24. J. Princen and Bradley, A., "Analysis/synthesis filterbank design based on time domain aliasing cancellation," *IEEE Transactions on Acoustics Speech, and Signal Processing,* vol. ASSP-34, 1986, pp. 1153–1161.

25. B. Edler, "Coding of audio signals with overlapping block transform and adaptive window functions," (in German), *Frequenz,* vol. 43, 1989, pp. 252–256.

26. R. Zelinski and Noll, P., "Approaches to adaptive transform speech coding at low bit rates," *IEEE Transactions on Acoustics, Speech and Signal Processing,* vol. ASSP-27, no. 1, Aug. 1979, pp. 89–95.

27. D. Meares and Kirby, D., "Brief Subjective Listening Tests on MPEG-2 Backwards Compatible Multichannel Audio Codecs," ISO/IEC JTCl/SC29/WG 11: Aug. 1994.

28. R. G. van der Waal, Oomen, A. W. J., and Griffiths, F. A., "Performance comparison of CD, noise-shaped CD and DCC," in *Proceedings 96th Audio Engineering Society Convention,* Amsterdam, 1994, preprint 3845.

29. T. Ryden, Grewin, C., and Bergman, S., "The SR Report on the MPEG Audio Subjective Listening Tests in Stockholm April/May 1991," ISO/IEC JTCl/SC29/WG 11: Doc.-No. MPEG 91/010, May 1991.

30. H. Fuchs, "Report on the MPEG/Audio Subjective Listening Tests in Hannover," ISO/IEC JTCl/SC29/WG 11: Doc.-No. MPEG 91/331, Nov. 1991.

31. B. Grill et al., "Improved MPEG-2 audio multi-channel encoding," *96th Audio Engineering Society Convention,* Amsterdam, 1994, preprint 3865.

32. G. Stoll et al, "Extension of ISO/MPEG-Audio layer II to multi-channel coding: The future standard for broadcasting, telecommunication, and multimedia application," *94th Audio Engineering Society Convention,* Berlin, 1993, preprint 3550.

33. F. Feige and Kirby, D., "Report on the MPEG/Audio Multichannel Formal Subjective Listening Tests," ISO/IEC JTCl/SC29/WG 11: Doc.-No. 0685, March 1994.

34. ISO/IEC/JTC1/SC29, "Report on the Formal Subjective Listening Tests of MPEG-2 NBC Multichannel Audio Coding," Document N1371, Oct. 1996.

35. ITU-R Document TG 10-2/3, Oct. 1991.

36. W. R. T. ten Kate et al., "Matrixing of bit rate reduced audio signals," *Proceedings International Conference on Acoustics, Speech, and Signal Processing* (ICASSP '92), 1992, vol. 2, pp. II-205–II-208.

37. W. R. T. ten Kate, "Compatibility matrixing of multi-channel bit-rate-reduced audio signals," *96th Audio Engineering Society Convention,* Amsterdam, 1994, preprint 3792.

38. K. Brandenburg and Bosi, M., "Overview of MPEG-audio: Current and future standards for low bit-rate audio coding," *99th Audio Engineering Society Convention,* New York, 1995, preprint 4130.

39. W.M. Bosi et al., ISO/IEC MPEG-2 Advanced Audio Coding," 101th Audio Engineering Society Convention, Los Angeles, 1996, preprint 4382.

40. ISO/IEC/JTC1/SC29, "Description of MPEG-4," Document N1410,Oct. 1996.

41. W. R. T. ten Kaze, Akagiri, K., van de Kerkhof, L. M., and Kohut, M. J., "Scalability in MPEG Audio Compression. From Stereo via 5.1-Channel Surround Sound to 7.1-Channel Augmented Sound Fields," 100th Audio Engineering Society Convention, Copenhagen, 1996, preprint 4196.

42. A. Sugiyama et al., "A New Implementation of the Silicon Audio Player Based on an MPEG/Audio Decoder LSI," Technical Report DSP94-99 (1994-12) of the IEICE, 1994, pp. 39–45.

43. ETSI, "European Telecommunication Standard," Draft prETS 300 401, Jan. 1994.

44. R. K. Jurgen, "Broadcasting with Digital Audio," *IEEE Spectrum,* March 1996, pp. 52–59.

45. C. Weck, "The error protection of DAB," *Audio Engineering Society Conference: DAB—The Future of Radio,* London, May 1995.

46. C. Todd et al., "AC-3: Flexible perceptual coding for audio transmission and storage," *96th Audio Engineering Society Convention,* Amsterdam, 1994, preprint 3796.

47. R. Hopkins, "Choosing an American digital HDTV terrestrial broadcasting system," *Proceedings of the IEEE,* vol. 82, no. 4, 1994, pp. 554–563.

## *ABOUT THE AUTHOR*

Peter Noll is professor of telecommunications at the Technical University of Berlin, Germany, and is also director of the Institute of Telecommunications there. His previous positions included professor of electrical engineering and statistical communication theory at the University of Bremen, Germany, and research on coding of medium-band speech signals at the Heinrich-Hertiz-Institut für Nachrichtentechnik, Berlin.

Noll is a member of the ISO/IEC Working Group 11 (Moving Pictures Experts Group) and has acted as chairman (ISO/IEC Rapporteur) of its Subgroup of Audio from 1991 to 1995. This activity led to standards for stereophonic (MPEG-1) and multichannel (MPEG-2) audio coding.

Noll has authored or coauthored numerous journal papers and technical reports and is coauthor of the book *Digital Coding of Waveforms: Principles and Applications to Speech and Video.* In 1982 he was elected to the grade of IEEE Fellow "for contributions to adaptive quantization and coding of speech signals." He is a member of the Berlin Brandenburg Academy of Science, the former Prussian Academy of Science.

# CHAPTER 8
# MPEG STANDARDS

## SECTION 3
## THE MPEG SYSTEMS LAYER STANDARD[1-5]

**Qun Shi**
*Panasonic Technologies, Inc.*

## 8.12 INTRODUCTION

The ISO/IEC 11172-1 (MPEG-1) and 13818-1 (MPEG-2) specifications are the system layer portions of the MPEG international standards that specify the combination of one or more audio and video elementary streams, and/or data streams, into single or multiple system streams, which are suitable for storage and transmission.[1,2] The MPEG-1 system layer is designated mainly for "error-free" storage medium such as CD-ROM. The MPEG-2 system layer is designated for both the storage media such as video server and "error-prone" transmission such as terrestrial broadcast, cable, and satellite systems. The specifications, especially for MPEG-2, provide salient features and capabilities for video and audio synchronization, stream multiplex, packet and stream identification, error resilience, buffer management, random access and program insertion, private data and conditional access, as well as interoperability with other networks such as those using asynchronous transfer mode (ATM).

The 11172-1 and 13818-1 documents provide detailed syntactical and semantic rules for the MPEG system standard.[1,2] This section attempts to provide a concise description of the standard in an easy-to-understand manner and address the main features of the standard and applications.

The MPEG standard, developed between 1990 and 1994, intends to provide digitally compressed full-motion quality video for applications ranging from low-end entertainment video to high-end high-definition television (HDTV). Specifically, MPEG-1 provides a data rate of 2.0 Mb/s or less, whereas MPEG-2 provides the data rates from 3 Mb/s up to 60 Mb/s. In MPEG a multiplexed bitstream consists of two layers: a compression layer (inner layer) and a system layer (outer layer). The compression layer defines the coded MPEG audio and video data streams. The system layer defines the functions necessary for combining one or more compressed data streams in a system. Specifically, the system layer supports four basic functions: (1) multiplexing of multiple bitstreams, (2) synchronous display of multiple coded bitstreams, (3) buffer management and control, and (4) time recovery and identification. In addition, the system layer also supports random access, program insertion, and conditional access operations.

For MPEG-1 the system coding is specified in a stream form, which consists of a sequence of packs combining one or more elementary stream packets with a common time base. For MPEG-2 the system coding is specified in two forms: program stream (PS) and transport stream (TS). The PS, which is analogous to the MPEG-1 system stream, combines one or more streams of packetized elementary stream (PES) packets having a common time base and is designed for use in relatively error-free environments. On the other hand, the TS is used to combine one or more programs made up of PES-coded data with one or more independent time bases, into a single stream, and is designated for use in error-prone environments, such as transmission or storage in lossy or noisy media.[2] The PS and TS are logically constructed from a common PES layer, and the two may be converted from each other using the same PES layer information. Figure 8.20 illustrates the MPEG system layer multiplex operation and relationships.

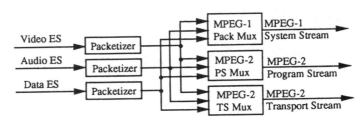

**FIGURE 8.20** MPEG system layer overview.

Note that the MPEG system specifications do not define the architecture or implementation of encoder or decoder. However, the bitstream properties imposed by the standard impose functional and performance requirements on encoder and decoders. For example, the encoders must meet minimum clock tolerance requirements for buffer management.[1,2] Notwithstanding this and other requirements, a considerable degree of freedom exists in the design and implementation of encoders and decoders.

## 8.13 MPEG-1 SYSTEM LAYER DESCRIPTION

The hierarchical structure of the MPEG-1 system layer is shown in Fig. 8.21, in which three sublayers of coding syntax are defined. These are the ISO 11172 stream sublayer, the pack sublayer, and the packet sublayer:

- The ISO 11172 stream sublayer consists of a sequence of packs followed by an end code.
- The pack sublayer includes important system-specific information fields for multiplex-wide operations, such as start code, system clock reference (SCR), mux rate, and (optional) system header, as well as the packet sublayer data.

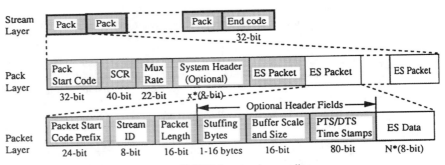

**FIGURE 8.21** Hierarchical structure of MPEG-1 system layer coding.

- The packet sublayer comprises packets that are made up of elementary video and audio streams. The packet fields include, in addition to elementary stream (ES) data, packet start code, stream identifier (stream_id), packet length, decoding and presentation time stamps (DTS and PTS), buffer sizes, etc. These fields are designed for stream-specific operations involving demultiplexing and synchronizing playback of multiple ESs.

### 8.13.1 Stream Multiplexing

The construction of the multiplexed bitstream from different ESs is facilitated in both pack and packet sublayers. In a packet, the start code, consisting of a 24-bit prefix and an 8-bit stream_id, is used to identify the ES. Note that a packet never contains data from more than one ES and byte ordering is preserved. Thus, after removing the packet headers, packet data from all packets with a common stream_id are concatenated to recover the original ES. The ESs can be MPEG audio, MPEG video, and private or padding data. As a limit, up to 32 MPEG audio and 16 MPEG video streams may be multiplexed simultaneously.

The multiplex construction in the encoder is constrained in such a way as to ensure the system target decoder (STD) buffers do not overflow or underflow. In general, short packets require less STD buffering but more system coding overhead than large packets.

### 8.13.2 Audio and Video Synchronization

The end-to-end synchronization of audio and video streams is effectively accomplished by means of time stamps, i.e., the presentation time stamps (PTS) and decoding time stamps (DTS), which are encoded in the packet header portion. The synchronization occurs when encoders record time stamps at capture time, when the time stamps propagate with the associated coded data to decoders, and when decoders use those time stamps to begin display. Figure 8.22 illustrates the process of the end-to-end synchronization using prototypical encoders and decoders. The time stamps are in units of 90 kHz and encoded into a 33-bit value. For DTS and PTS the values encoded can be respectively expressed as follows,[1]

$$PTS = NINT\{system\_clock\_Frequency * [tp_n(k)]\}\%2^{33} \tag{8.1}$$

$$DTS = NINT\{system\_clock\_Frequency * [td_n(k)]\}\%2^{33} \tag{8.2}$$

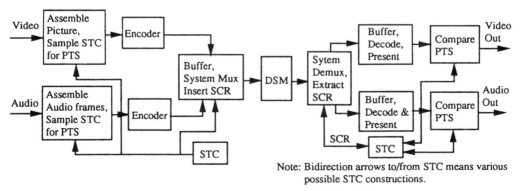

FIGURE 8.22 End-to-end synchronization of MPEG-1 prototype encoder and decoder. (From ISO/IEC 11172-1:1993, ©ISO/IEC, reproduced with permission of the International Organization for Standardization, ISO.)

where $tp_n(k)$ is the presentation time of presentation unit (PU) $P_n(k)$ and $td_n(k)$ is the decoding time of access unit (AU) $A_n(k)$. The PU $P_n(k)$ corresponds to the first AU that commences in the packet, where the AU refers to a picture start code of a video stream or a sync word of an audio frame. Thus, the DTS and PTS indicate the intended times of decoding and presentation in the decoder, respectively. There are several constraints on the coding of the time stamps:

• The interval of PTS insertion in the packet should be less than 700 ms.
• For each ES the PTS is encoded in the packet in which the first access unit of that ES starts.
• A DTS appears in the packet if and only if a PTS exists and differs in value.

### 8.13.3  System Clock Recovery

There is also a time stamp field in the pack layer representing the system clock reference (SCR) and the common time master for the multiplexed bitstream. The SCR represents the intended time of arrival of the system clock field at the input of the STD. It is also a 33-bit number in units of 90 kHz and encoded by

$$SCR = NINT\{system\_clock\_Frequency * [tm(i)]\}\%2^{33} \qquad (8.3)$$

where $tm(i)$ is the input arrival time expressed as follows,[1]

$$tm(i) = \frac{SCR(i')}{system\_clock\_frequency} + \frac{i - i'}{(mux\_rate * 50)} \qquad (8.4)$$

where $i$ is the index of any byte in the pack, $i'$ is the index of the last byte of the SCR field, and mux_rate is the multiplexed bit rate encoded in the pack.

At the system decoder a single time-master exists for decoder synchronization and for avoiding buffer overflow and underflow. All other synchronized entities slave their timing to the master clock. If a decoder attempts to have more than one simultaneous time-master it may experience problems with buffer management or synchronization. The system decoder can choose a video or audio ES decoder or a data source (e.g., a digital storage medium) as the time-master. Typically, as shown in Fig. 8.22, the SCR fields are extracted to recover the system time-clock (STC) as time-master, and the PTS/DTS are then extracted to compare to the current values of STC for decoding and presentation.

### 8.13.4  Buffer Management

The management of (decoder) buffer underflow and overflow is based on the STD model, which specifies the times at which each data byte enters and leaves the buffer of each ES decoder in terms of a common system time clock. ES decoder buffers can be guaranteed not to overflow or underflow during decoding as long as the data stream conforms to the specification, and the complete decoding system is synchronized in terms of SCR, PTS/DTS. The STD model is illustrated in Fig. 8.23. The STD buffer size is specified in the packet header field (13-bit). Specifically, for the audio stream the actual buffer size $BS_n$ is equal to the buffer size value encoded in the packet header times 128 and is bounded by 4096 bytes if the multiplexed stream is a constraint system parameter stream (CSPS). For the video stream the buffer size $BS_n$ is equal to the encoded buffer size value times 1024 and is bounded by $46 \times 1024$ bytes if a CSPS applies. Note that the STD assumes ideal decoding and presentation, i.e., no delays; the ESs buffered are decoded instantaneously by decoders and displayed at their presentation times. In practice, however, additional buffer size must be needed to account for processing delays in the decoder as well as multiplexing operations.

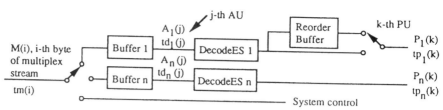

**FIGURE 8.23**    System target decoder for MPEG-1 system stream.

### 8.13.5    Time Identification

The absolute time of presentation of the material contained in the coded data stream is indicated in the PTS fields, which are defined as modulo $2^{33}$ values of the 90-kHz STC. If required, the PTS fields can be converted into a time-code format (e.g., SMPTE). The use of 90-kHz STC is based on the divisibility of 90 kHz by the nominal video picture rates of 24 Hz, 25 Hz, 29.97 Hz, and 30 Hz. The use of 33-bit encoding allows the elapsed times of about 26 hours to have distinct coded values.

## 8.14    MPEG-2 PACKETIZED ELEMENTARY STREAM (PES) LAYER DESCRIPTION

The PES layer (analogous to the packet layer of MPEG-1 system stream) is the inner layer portion of the MPEG-2 multiplexed stream upon which the transport or program streams are logically constructed and provides stream-specific operations. The PES layer supports the following functions:

- A common base of conversion between transport and program streams.
- Time stamps for video and audio synchronization and associated timing.
- Stream identifications for stream multiplexing and demultiplexing.
- Other services such as scrambling, VCR functions, private data, etc.

MPEG-2 requires that the video and audio ES data must be PES-packetized before inserting into TS or PS. The PES packets are variable in length and relatively long but upper-bounded by 64 kbyte. For video ES the packet length can be set to zero, indicating the nonfixed length. Normally, a PES packet consists of a video AU or an audio AU.

### 8.14.1    PES Packet Structure and Definition

The packet hierarchy structure of PES is shown in Fig. 8.24. In specific, a PES packet has a 9-byte fixed header, which includes packet start code (prefix plus stream_id), packet length, control flags for optional header fields and length for optional fields. The optional header fields include PTS/DTS fields, trick mode (VCR functions), CRC, and private data. Some of the primary functions of the fields are listed in Table 8.8. Note that the PES stream is not defined for interchange and interoperability; it has to work with either TS or PS for system-level control. Again, up to 16 video and 32 audio ESs can be used.

In the case of inserting a PES packet into TS packets, PES headers must be aligned with TS packet headers in order to facilitate the decoding of AUs. A single PES packet may span many TS packets; in such a case the PES packet chunks must appear in consecutive TS packets of the same packet identifier (PID) value. Note, however, that these TS packets may be freely interleaved with other TS packets of different PIDs that carry different ESs.

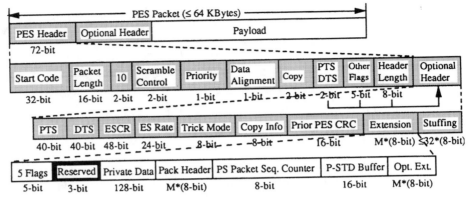

**FIGURE 8.24** PES packet structure.

**TABLE 8.8** Basic Functions of PES Fields

| PES Fields | Basic Functions |
|---|---|
| Packet start code (prefix + stream_id) | Packet synchronization/identification |
| Random access indicator, data alignment indicator | Payload identification |
| PTS/DTS flag, PTS/DTS fields | Presentation synchronization |
| Packet/header lengths, payload, stuffing | Data transfer |
| Scramble control | Payload scramble |
| Extension/private flags, private data field | Private information transfer |
| Trick mode | DSM VCR functionality |
| CRC flag, previous PES CRC | Network maintenance |

## 8.15 MPEG-2 TRANSPORT STREAM LAYER DESCRIPTION

The MPEG-2 transport stream (TS) is tailored for transmission of MPEG-coded data and other data in error-prone environments and includes features for error resiliency and packet loss detection. A TS may contain one or more programs; here, each program is composed of one or several ESs having a common time base. Different transport streams can also be combined into a single TS, i.e., a system TS. The encapsulation of ES data (AUs) into TS packets is illustrated in Fig. 8.25, where the ES data is first packetized via PES and then inserted into short, fixed-length TS packets. The TS is designed in such a way that certain operations on the TS are possible

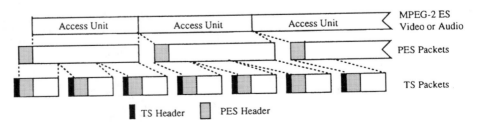

**FIGURE 8.25** Encapsulation MPEG-2 ES data into TS packets.

with minimum effort:

* Demultiplex and retrieve the ES data from one program within the TS.
* Remultiplex the TS with one or more programs into a TS with a single program.
* Extract the TS packets from different transport streams and produce as output another TS.
* Demultiplex a TS into one program and convert it to a PS containing the same program.
* Convert a PS into a TS to carry it over a lossy medium, and then recover a valid PS.

### 8.15.1  TS Packet Structure and Definition

A TS packet is of 188-byte fixed length and has a packet structure as shown in Fig. 8.26. Specifically, a TS packet consists of a 4-byte header containing a sync byte (47 hex) and a 3-byte prefix, including a 13-bit PID and system indicators; an optional adaptation header containing program reference clock (PCR); control flags and fields; private data and stuffing bytes, if any; and a payload section containing the PES or system data. Some of the major functions of the fields are given in Table 8.9. In particular, the 13-bit PID identifies—via the program-specific information (PSI) tables to be discussed later—the contents of the data contained in the TS packet. Transport packets of one PID value carry data of one and only one ES. As indicated in Fig 8.26 and

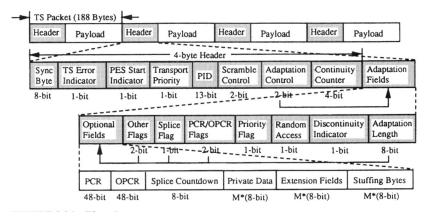

**FIGURE 8.26**  TS packet structure.

**TABLE 8.9**  Basic Functions of TS Header Fields

| TS header fields | Basic functions |
| --- | --- |
| Sync byte | Packet synchronization |
| PID | Packet identification, demux, sequence integrity |
| Continuity counter, error indicator | Packet sequence integrity, error resilience |
| PES start indicator, adaptation control | Payload synchronization |
| Discontinuity indicator, PCR | Decoder time-base recovery, error resilience |
| Transport scramble control | Payload scrambling |
| Private data flag and data bytes | Private data transfer |
| Data bytes, stuffing bytes | Data transfer |

Table 8.9, the TS packet structure supports time base (clock) recovery, random access, program insertion, and error resilience.

### 8.15.2 Time-Base Recovery and Synchronization

The MPEG-2 system layer embodies an end-to-end constant delay timing model in which all digitized pictures and audio samples that enter the encoder are presented exactly once each—at the output of the decoder.[2] As such, the sample rates (i.e., the video picture rate and the audio frame rate) are precisely the same at the decoder as they are at the encoder. The delay is the sum of encoding, encoder buffering, multiplexing, transmission or storage, demultiplexing, decoder buffering, decoding, and presentation. The system coding contains timing information (time stamps) that is used to ensure end-to-end constant delay. The time stamps are PCR and PTS/DTS.

As shown in Fig. 8.27, the transport decoder typically employs a local system clock whose frequency and phase match those of the encoder. The clock, if allowed to free-run, would produce an incorrect time base for program delivery, because the frequency of the local clock is highly unlikely to be identical to that of the original encoder. This leads to the primary function of the PCR coding in the TS adaptation field. PCR values, sent periodically in the TS, are used to correct the clock value. PCR is a 42-bit field coded in two parts: the first part, PCR base, is a 33-bit value in units of 90 kHz, and the second part is a 9-bit PCR extension field in units of 27 MHz (system clock frequency). The value encoded in the PCR field specifies the byte arrival time $t(i)$, where $i$ refers to the byte containing the last bit of the PCR base field:[2]

$$\text{PCR base}(i) = [(\text{System clock frequency} * t(i))\ \text{DIV}\ 300]\%2^{33} \tag{8.5}$$

$$\text{PCR extension}(i) = [(\text{System clock frequency} * t(i))\ \text{DIV}\ 1]\%300 \tag{8.6}$$

$$\text{PCR}(i) = \text{PCR base}(i) * 300 + \text{PCR extension}(i) \tag{8.7}$$

Usually, the first PCR initializes the counter in the clock generation, and as the clock runs, subsequent PCR values are compared to the clock values for fine adjustment. Because a PCR represents the correct time base for the program when it arrives, the difference between it and the local clock value can be used to drive the instantaneous frequency of a VCO and either slow down or speed up the clock, as appropriate. Further, the constraint interval between successive PCRs (less than or equal to 100 ms) ensures the clock recovery circuit to be stable. The output of the VCO is then a 27-MHz oscillator signal and is used as the system clock frequency for the decoder. The clock may also be used to reconstruct the video picture rate or audio frame rate.

The audio and video synchronization is achieved through the PTS field frequently inserted in the PES header. Similar to MPEG-1, the PTS is a 33-bit value in units of 90 kHz (the 27-MHz

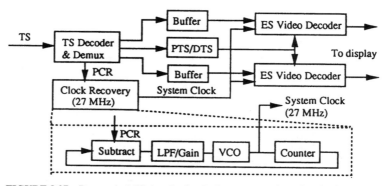

**FIGURE 8.27** Prototypical TS decoder for clock recovery and synchronization.

system clock frequency divided by 300), indicating the time that the PU (which results from decoding the AU associated with the PTS) should be presented to the user. The audio PTS and video PTS fields are both samples from a common time clock (in a program). As shown in Fig. 8.27, the extracted PTS/DTS values are compared to the reconstructed local clock values for fine presentation time adjustments. Moreover, the discontinuity indicator in TS and the requirement on new PTS after discontinuity ensures the resynchronization of video and audio as may be happening in channel changing and/or random access.

### 8.15.3  System Multiplex Hierarchy and Program Specific Information (PSI)

In overview, the TS embodies two levels of multiplexing mechanism. First, a program TS is formed by multiplexing (interleaving) one or more individual TS packets of ES data (with or without PES) sharing a common time base and a control TS that describes the program. Each individual ES TS and the control TS are identified by the unique PIDs in the corresponding TS header. Secondly, different TS programs with different time bases are combined asynchronously, forming a system TS, under the control of a system-level map TS (its PID = 0) that describes the identities of the programs. The mechanism of TS multiplexing is further illustrated in Fig. 8.28. At the TS decoder, the demultiplexing performs the inverse functions. Thus, MPEG-2 adopts a hierarchical view of systems: (a) a network may have one or more transport streams, (b) a TS may contain one or more programs, and (c) a program (TS) may contain one or more ES transport streams. Each layer of the hierarchy encapsulates the ones underneath it. Hence, for example, a program number has significance only within a TS. However, different program numbers can be duplicated in different TS streams in the same network. Similarly, the ES PIDs in one program are meaningful only in that program. Note that *PID collision* may occur when mapping packets from one TS to another. In such a case the PIDs must be reassigned accordingly.

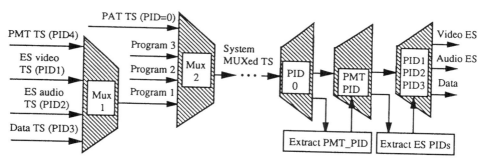

**FIGURE 8.28**   TS multiplexing/demultiplexing mechanism layer hierarchy.

The control bitstreams that coordinate the TS multiplexing are collectively called the *program-specific information (PSI) tables.* MPEG-2 has four recognized PSI tables, in which the contents of three of them are specified. They are listed in Table 8.10 along with their functions. Figure 8.29 shows the syntax structure of the PSI tables. The PSI tables can have up to 255 sections (except the program map section), and each section has a maximum of 1 kbyte. The PSI tables are directly inserted in TS packets without any PES encoding, and the corresponding PSI-bearing TS packets are freely intermixed with those carrying ES data. For supporting random access and frequent channel changing, PSI-bearing TS packets may need to occur at a 10- to 20-Hz rate.[2] The PSI tables also contain descriptors that provide descriptions of programs and ES information to aid decoding and presentation. Further, private sections are given for carrying user-defined private information.

**TABLE 8.10** PSI Tables and Functions

| PSI table name | Structure type | PID | Description and functionality |
|---|---|---|---|
| Program association table (PAT) | MPEG-2 specified | 0 | Associates program number with PMT PID; points to network table (program 0) |
| Program map table (PMT) | MPEG-2 specified | User given | Maps program number to program element (TS) PIDs, PCR-PID, and descriptors |
| Network information table (NIT) | Private | User given | Specifies physical network parameters, e.g., FDM frequencies, transponder number, etc. |
| Conditional access table (CAT) | MPEG-2 specified | 1 | Associates one or more (private) EMM streams and other CA data with PIDs |

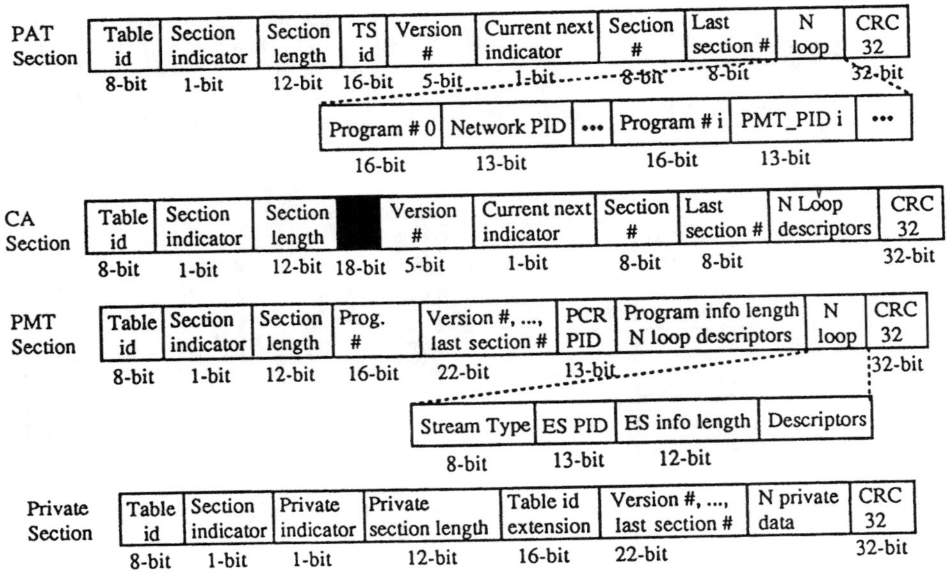

**FIGURE 8.29** PSI table section syntax.

## 8.15.4 Buffer Management

The buffer management of the MPEG-2 transport layer is defined based on the transport stream system target decoder (T-STD) shown in Fig. 8.30. The T-STD describes a hypothetical decoder model specifying the buffer sizes at various decoding stages and byte-removal rates out of each buffer. Tables 8.11 and 8.12 show a list of the buffer sizes and data removal rates for video, audio, and system data, respectively. For video, only the main level, main profile case is given.

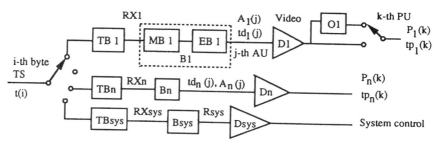

**FIGURE 8.30**    MPEG-2 TS system target decoder model. (From ISO/IEC 13818-1:1996, ©ISO/IEC, reproduced with the permission of the International Organization for Standardization, ISO.)

**TABLE 8.11**    T-STD Transport and Main Buffer Sizes

|  | Video (main level, main profile) | Audio | System data |
|---|---|---|---|
| TS buffer | 512 bytes | 512 bytes | 512 bytes |
| Main buffer | BSmux + BSoh + VBVmax | BSmux + BSoh + BSdec | 1536 bytes |
|  | BSmux = 0.004 (s) * Rmax | = 3584 bytes |  |
|  | BSoh (PES overhead) | BSnux = 736 bytes |  |
|  | = (1/750) * Rmax | BSdec + BSoh ≤ 2848 bytes |  |

**TABLE 8.12**    T-STD Transport and Main Buffer Byte Removal Rates

| Remove rate | Video (main level, main profile) | Audio | System data |
|---|---|---|---|
| TS buffer | RXn = 1.2 * Rmax | 2 Mb/s | 1Mb/s |
| Main buffer | RBXn = Rmax | — | Rsys = max{80 kb/s * TS_rate * 8/500} |

Additionally, it is required that the TS buffers (TBn or TBsys) be empty once every second to avoid overflow. The main buffers shall not overflow or underflow, except when the low delay flag in video is set.

### 8.15.5    Error Resilience

The transport layer provides error resiliency mechanisms that are appropriate and tailored to the transmission data. This is illustrated in Table 8.13.

### 8.15.6    Other Features

The other important features supported by the TS layer may be local program insertion and conditional access (for encryption). The local program insertion or commercial insertion is facilitated via the splicing flag and splice countdown field in the TS adaptation header. The two fields define a splice point (when the flag is set and the countdown is zero), at which the "old" TS can be discarded and the "new" TS inserted. The splicing is carried out on a TS packet boundary. Generally, the splicing and program insertion steps include (1) searching for splice points in TS; (2) extracting relevant TS packets; (3) modifying PMT of the corresponding

**TABLE 8.13**  Error Resilience Avenues in MPEG-2 TS Layer

| Error resiliency | Relation to TS structure | Functionality and usage |
|---|---|---|
| TS error indicator | 1-bit in TS header | Indicate any uncorrectable error(s) associated with the TS packet. |
| Priority indicator | 1-bit in TS header | Indicate relative priority of TS packet and give preferential error protection for high-priority TS. |
| Continuity counter | 4-bit in TS header | Detect packet loss in a sequence and discontinuity. |
| CRCs | 16-bit CRC in PES<br>32-bit CRC in PSI | Provide network maintenance (PES) and error detection (PSI). |

program; and (4) modifying or inserting, if necessary, PCR and PTS in the first packet of the new stream to prevent buffer overflow or underflow. A splice can also be seamless or non-seamless. A seamless splice is such that no decoding discontinuity occurs after splicing—which means no presentation discontinuity before and after the insertion. A nonseamless splice obviously results in decoding discontinuity after splicing, producing noncontinuous PU presentations (i.e., a dead time between display of two consecutive pictures or audio frames). In this case the decoders may be designed to freeze the pictures or mute the audio frames to overcome the discontinuity.

The conditional access in the TS is supported in two features. First, there is a two-bit scrambling control field that signals the decoder (1) whether the TS packet was scrambled or not, and (2) which scrambling key (odd or even) is used if scrambled. Secondly, "private" data can be encoded in a TS header or a CA table, which provide encrypted scrambling key(s) and further scrambling information such as entitlement control message (ECM) and entitlement management message (EMM), which specify scrambling control parameters and authorization levels. At the TS decoder the information and parameters are then used for decryption and descrambling using the smart card based on the Digital Encryption Standard (DES).

## 8.16  MPEG-2 PROGRAM STREAM LAYER DESCRIPTION

Similar to the MPEG-1 system stream, MPEG-2 PS is also designed specifically for use in storage media and allows one or more ESs to be combined into a single stream. Data from each ES is multiplexed and encoded along with information (e.g., time stamps) that allows ESs to be displayed in synchronism.

### 8.16.1  PS Packet Structure and Definition

Similar to that shown in Fig. 8.21, a PS consists of a sequence of packs, which is further comprised of PES packets. The pack sublayer is designated for multiplex-wide operations involving data parsing and retrieval, clock recovery, and buffer management, whereas the PES sublayer is designated for stream-specific operations that involve stream demultiplex and synchronizing playback of multiple ESs (e.g., video and audio). The PS begins with a system header (within pack) that may be repeated. The system header carries a summary of the system parameters defined in the stream, such as video and audio stream boundaries, PS STD buffer bounds, etc. The PS packets are variable in length.

## 8.16.2  Stream Multiplexing

The multiplexing in the PS is an intermix of PES packets, each identified with a unique stream_id. Thus, one PS contains one program of ESs only, with a common time base. Similar to TS, there is a program stream map providing a definition of the ESs in the PS and their relationship to each other. The map is sent as a PES packet with a special stream_id.

## 8.16.3  Clock Recovery and Synchronization

The system clock used in the PS is called the *system clock reference* (SCR), which is 27 MHz. The SCR is encoded in the pack immediately after the pack start code and is divided into two parts: a 33-bit base field and a 9-bit extension field. To aid the clock recovery at the decoder, it is required that the SCR should be sent in an interval not exceeding 700 ms. For video and audio synchronization the PTS/DTS fields are again used. The requirements for PTS/DTS are the same as those used in TS.

## 8.16.4  Buffer Management

The buffer sizes specified in the PS STD are given by the PS STD buffer size parameter encoded in the PS pack syntax:

$$\text{BSn} = \text{PS-STD\_buffer\_size} * 1024, \text{ if STD\_buffer\_scale} = 1 \tag{8.8}$$

$$\text{BSn} = \text{PS-STD\_buffer\_size} * 128, \text{ if STD\_buffer\_scale} = 0 \tag{8.9}$$

Two constraints are imposed as follows so that buffers BS1 to BSn will be neither underflow nor overflow:

**1.** $0 \leq F_n(t) \leq \text{BSn}$ for all $t$ and $n$, where $F_n(t)$ is the instantaneous fullness of STD buffer Bn.

**2.** The delay caused by the STD buffering shall be less than or equal to 1 second except for still-picture video (in which case the delay is less than or equal to 1 minute).

However, in certain cases, buffer underflow is allowed, such as in the case when the low_delay_ flag is set to "1" in the video sequence header.[2]

## 8.16.5  Conversion to TS

As noted, it is possible and reasonable to convert between PS and TS by means of PES. This is made possible by the syntax structures of PS and TS. PES packets may, with some constraints, be mapped directly from the payload of one multiplexed bitstream into the payload of another multiplexed bitstream. It is possible to identify the correct order of PES packets in a PS to assist with this. Certain other information necessary for conversion (e.g., the relationship between ESs) is available in tables and headers in both streams. Such information must be available and correct in any stream before and after the conversion. Moreover, time stamps may also need to be modified to allow smooth conversion.

## 8.17  MPEG SYSTEM LAYER APPLICATIONS

As mentioned, the MPEG system layer supports the applicability of compressed video, audio, and data in broad areas from digital storage to terrestrial digital television broadcast. Described as follows are some of the application areas.

### 8.17.1 CD-ROM

As noted, the MPEG-1 system layer is designed for CD-ROM usage, providing data rates up to 1.5 Mb/s. Table 8.14 lists some typical system layer parameters as used in an example given in annex A of the ISO 11172-1 specification.

**TABLE 8.14** MPEG-1 System Layer Parameters Used in CD-ROM

| MPEG-1 system parameters | CD-ROM application |
|---|---|
| Pack length (one pack per sector) | 2324 bytes (1/75 s) |
| Total bit rate | $8 * 2324 * 75 = 1.3944$ Mb/s |
| Pack payload | 1 audio, or 1 video, 1 padding packet |
| Packet length | 2250 bytes |
| Interleaving and multiplexing | 6 video sectors per audio sector; repeats every 3 s (225 sectors) |
| STD buffer size | 46 kbyte (constraint system) |

### 8.17.2 Video Server and Video-on-Demand

In the digital storage medium (DSM), MPEG-2 PS is used when the application bitstream rates are higher than 1.5 Mb/s. The applications include video server, video-on-demand, interactive video services, and video network. The interface protocol for controlling the bitstream transfer between user and DSM is done by the DSM command and control (CC) protocol. Figure 8.31 illustrates an application system using MPEG-2 PS and DSM CC. Specifically, the DSM CC provides the means to do the following:

- Select an ISO/IEC 13818-1 system stream (e.g., PS) on which to perform the succeeding operations, such as creation of a new stream.
- Retrieve and play (including forward, rewind, and pause) the bitstream.
- Cause storage of a valid bitstream for a specified duration or strop.

The DSM CC may be packetized into PES and sent via PS or TS.

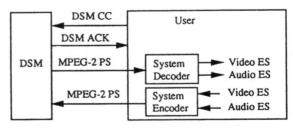

**FIGURE 8.31** MPEG-2 PS used in DSM with DSM CC.

### 8.17.3 Grand Alliance Transport System for HDTV

The U.S. HDTV system developed by the Grand Alliance (GA) gives a direct application of MPEG-2 TS for digital terrestrial broadcast.[3] The GA's transport system uses a subset of MPEG-2 TS and imposes more stringent constraints on the TS bitstream in order to aid fast data acquisition and synchronization to support channel hopping and/or random access. One of the constraints is GA's program paradigms developed for PSI transmission and acquisition. Table 8.15 lists the imposed constraints for program paradigms and Table 8.16 gives an example specifically related to PID assignments.

**TABLE 8.15**  GA's Program Paradigms and Constraints

| | |
|---|---|
| (TV) program number range: | 1 to 255 |
| TS carrying PSI: | No adaptation field |
| (Program) Base_PID: | Program number * 0010 hex |
| PAT (section 0) rate: | 100 ms maximum spacing |
| PMT rate: | 400 ms maximum spacing |

**TABLE 8.16**  Example of PID Assignments for a Program in GA

| Program number: 52 (0×0034h) | | base_PID: 0×0340h (h = hex) | |
|---|---|---|---|
| Name | PID definition | Example PID | Description |
| PMT_PID | base_PID + 0 × 0000h | 0×0340h | PID for PMT of the program |
| Video_PID | base_PID + 0 × 0001h | 0×0341h | PID for video TS for the program |
| PCR_PID | base_PID + 0 × 0001h | 0×0341h | PCR is carried in the video TS |
| Audio(A)_PID | base_PID + 0 × 0004h | 0×0344h | PID for primary audio TS |
| Audio(B)_PID | base_PID + 0 × 0005h | 0×0345h | PID for secondary audio TS |
| Data_PID | base_PID + 0 × 000Ah | 0×034Ah | PID for data TS for the program |

### 8.17.4  Digital Video Broadcast (DVB) Transport

The DVB standard,[4] developed under the auspices of the European Broadcasting Union and the European Telecommunications Standards Institute, employs a subset of MPEG for its transport layer and imposes additional constraints. For example, the PCR-occurring frequency interval in DVB is 40 ms, instead of 100 ms specified by MPEG-2. Moreover, for service applications, DVB expands the PSI tables to include six new service tables. The definition of the new service information is given in Table 8.17. The syntax structure of these new table sections conform to that of PSI table sections. The minimum time interval between the PSI sections is 25 ms. The new service descriptors include information relating to bouquet, delivery system, NVOD reference, time-shifted event, etc.

### 8.17.5  Transport Through ATM Network

MPEG-2 and ATM are two key technologies needed for the deployment of broad-band video distribution networks for video services in both residential and business applications. There are several ways of carrying MPEG-2 TS packets over ATM with an appropriate adaptation layer (AAL). To ensure the satisfactory performance of carrying MPEG-2 video signals over ATM, several key technical issues must be addressed:

• Selection of a suitable AAL on which MPEG-2 TS can be transported.

• Error control by means of error concealment and/or error detection/correction.

• Timing recovery in the presence of cell delay variation (CDV).

Bellcore's GR-2901-CORE document specifies that constant-bit-rate coded MPEG-2 video streams be transported as MPEG-2 TS over ATM using AAL5 with a null convergence sublayer.[5]

**TABLE 8.17**  New Program Service Tables Defined by DVB

| New PSI | PID and Table_id | Service definition |
|---|---|---|
| Network information table (NIT) | PID = 0x10h<br>Table_id = 0x40(1)h | Network name, descriptors for satellite/cable/terrestrial delivery systems (freq. modulation). |
| Bouquet association table (BAT) | PID = 0x11h<br>Table_id = 0x4Ah | Information regarding bouquets (list of services, etc.), under the admin. of bouquet provider. |
| Service description table (SDT) | PID = 0x11h<br>Table_id = 0x42(6)h | Service descriptors (service provider, service name), under admin. of broadcaster/service provider. |
| Event information table (EIT) | PID = 0x12h<br>Table_id = 0x50–0x5F | Event descriptors (start time, duration, title . . . ); content descriptors (event category). |
| Time and date table (TDT) | PID = 0x10h<br>Table_id = 0x70h | Information relating to time and date, under the admin. of broadcaster/service provider. |
| Running status table (RST) | PID = 0x13h<br>Table_id = 0x71h | Status of an event (running, not running, pausing, etc.), event update. |

The requirement specifies two cases of the mapping:

- Under normal (default) conditions, two TS packets are inserted into one AAL5 protocol data unit (PDU) in eight ATM cells; here, the first TS has no PCR and the second TS may or may not contain PCR field.
- When the first TS carries a PCR, the packet is encapsulated as an AAL5-PDU of one TS packet, appended with stuffing bytes into five ATM cells and forwarded immediately, without waiting for the second TS to arrive.

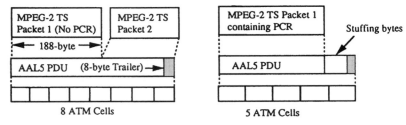

**FIGURE 8.32**  The mapping of MPEG-2 TS into ATM AAL5.

Figure 8.32 illustrates the mapping. The adoption of a null convergence sublayer requires that the MPEG-2 decoder handle timing recovery and error concealment. Such mechanisms are provided in the systems and video layers of the MPEG-2 bitstream. This scheme is simple and efficient: it provides minimum overhead for transporting MPEG bitstreams but relies on the MPEG layer to do timing recovery and error control. An alternative scheme is to use AAL1, because it provides a constant delay across the network, facilitating timing recovery at the application systems layer via synchronous residual time stamps and limited error recovery capabilities.[5]

## *GLOSSARY*

**Access unit**    A coded representation of a presentation unit. For audio an access unit is the coded representation of an audio frame; for video an access unit is the coded representation of a picture.

**Bit rate**    The rate at which the compressed bitstream is delivered from the channel to the input of a decoder.

**Byte aligned**    A bit in a coded bitstream is byte-aligned if its position is a multiple of 8 bits from the first bit in the stream.

**CAT (conditional access table)**    Provides the association between one or more conditional access systems, their entitlement management message streams, and any special parameters associated with them.

**CBR (constant bit rate)**    An operation in which the bit rate is constant from start to finish of the compressed bitstream.

**CRC (cyclic redundancy check)**    Verifies the correctness of data.

**CSPS (constraint system parameters stream)**    An MPEG-2 program stream that conforms to the bounds specified in the MPEG-2 system layer specification.

**DSM (digital storage medium)**    A digital storage or transmission device or system.

**DSM CC (digital storage medium command and control)**    A protocol intended to provide the basic control functions and operations specific to managing an MPEG-2 system bitstream on digital storage media.

**DTS (decoding time stamp)**    A field that may be present in a PES packet header that indicates the time that an access unit is decoded in the system target decoder.

**ECM (entitlement control message)**    Private conditional access information that specifies control words and possibly other, typically stream-specific, scrambling, and/or control parameters.

**EMM (entitlement management message)**    Private conditional access information that specifies authorization levels or the services of specific decoders.

**ES (elementary stream)**    A generic term for one of the coded video, coded audio, and other coded bit data in PES packets.

**Packet**    A packet consists of a header followed by a number of contiguous bytes from an elementary data stream.

**PAT (program association table)**    Provides the association between a program number and the PID value of the transport stream packets, which carry the program definition.

**Payload**    A generic term referring to the bytes that follow the header bytes in a packet.

**PCR (program clock reference)**    A time stamp in the transport stream from which decoder timing is derived.

**PES (packet elementary stream)**    Consists of PES packets, all of whose payloads consist of data from a single elementary stream, and all of which have the same stream identifier.

**PID (packet identifier)**    A unique integer value used to associate elementary streams of a program in a single or multiple program MPEG-2 transport stream.

**PMT (program map table)**    Provides the mappings between program numbers and the elementary streams that make up the programs.

**Program**   A collection of elementary streams with a common time base.

**PS (program stream)**   Consists of contiguous PES packets that comprise one or more video or audio elementary streams.

**PSI (program-specific information)**   Consists of normative data necessary for the demultiplexing of transport streams and the successful regeneration of programs.

**PTS (presentation time stamp)**   A field that may be present in a PES packet header that indicates the time that a presentation unit is presented in the system target decoder.

**PU (presentation unit)**   A decoded audio frame or a decoded picture.

**Random access**   The process of beginning to read and decode the coded bitstream at an arbitrary point.

**SCR (system clock reference)**   A time stamp in the program stream from which decoder timing is derived.

**Scrambling**   The alteration of the characteristics of a video, audio, or coded data stream in order to prevent unauthorized reception of the information in a clear form.

**Splicing**   The concatenation, performed on the system level, of two different elementary streams. The resulting system stream conforms totally to the MPEG-2 system standard. The process may result in discontinuities in time base and decoding.

**Start code**   A 32-bit code word embedded in the coded bitstream that is unique. The start codes are used for several purposes, including identifying some of the layers in the coding syntax, and they consist of a 24-bit prefix (0x000001) and an 8-bit stream ID.

**STD (system target decoder)**   A hypothetical reference model of a decoding process used to describe the semantics of the MPEG-multiplexed bitstream.

**Stream ID (stream identifier)**   An 8-bit unique word used to define a video, an audio, or other PES packets.

**Time stamp**   Indicates the time of an event such as the arrival of a byte or the presentation of a presentation unit.

**TS (transport stream)**   Contiguous packets, each of which is 188-bytes in length, consisting of one or more PES packets.

**VBR (variable bit rate)**   An operation in which bit rate varies with time during the decoding of a compressed bitstream.

## REFERENCES

1. ISO/IEC 11172-1, *International Standard: Coding of Moving Pictures and Associated Audio for Digital Storage Media at up to about 1.5 Mbit/s—Part 1: Systems,* 1st ed., ANSI, New York, Aug. 1993.
2. ISO/IEC 13818-1, *International Standard: Generic Coding of Moving Pictures and Associated Audio: Systems,* Geneva, Switzerland, Nov. 1994.
3. Draft Grand Alliance Transport System Specification, Doc. T3/S8-027, June 1994.
4. "Digital Broadcasting Systems for Television, Sound and Data Services; Framing Structure, Channel Coding and Modulation Cable System," European Telecommunication Standard, June 1994.
5. "Video Transport over Asynchronous Transfer Mode (ATM)," Generic Requirements GR-2901-CORE, May 1995.

## ABOUT THE AUTHOR

See "About the Author" in Chapter 5.

# CHAPTER 9
# THE ISO JPEG AND JBIG STILL IMAGE–CODING STANDARDS

## SECTION 1
## THE JPEG STANDARD

**Jechang Jeong**
*Department of Electronic Communication Engineering*
*Hanyang University*

## 9.1 INTRODUCTION

When one attempts to store or transmit digital images, the biggest difficulty encountered is the huge amount of image data. For instance, an image of $480 \times 640$ (resolution of the IBM-PC VGA display) with 8 bits/pixel occupies approximately 2.5 Mb of storage. Therefore, image compression is needed, for which JPEG (Joint Photographic Experts Group) is the emerging standard.[1,2,3]

JPEG signifies either the international still-image compression standard jointly prepared by ISO-IEC (International Standards Organization–International Electrotechnical Commission) and ITU-T (International Telecommunication Union–Telecommunication Sector), or the committee of ISO-IEC that is in charge of the standard.

More specifically in the context of a standard, JPEG is the standard for digital compression and coding of continuous-tone still images—monochrome or color. Note that JPEG excludes bilevel images as the input image, because bilevel images require different processing for efficient compression. The standard for bilevel images was separately prepared and is called *JBIG* (Joint Bilevel Image Experts Group; IS 11544). Section 2 of this chapter describes the JBIG standard.

In the context of a committee, JPEG stands for ISO-IEC/ JTC1/ SC29/WG1 (JTC = Joint Technical Committee; SC = subcommittee; WG = working group). Note that, under the same subcommittee, there are other relevant working groups, called *MPEG* (Moving Picture Experts Group; WG11) and *MHEG* (Multimedia Hypermedia Experts Group; WG12). The CCITT(International Telegraph and Telephone Consultative Committee; renamed ITU-T in 1993) SG VIII collaborated closely with the JPEG committee for establishment of the standard; for example, they provided JPEG with coding requirements for still-image compression applications.

The JPEG standardization activity began in 1982. After a decade of joint efforts (including 12 proposals submitted for evaluation in 1987), the main part of the standards was finally approved in 1992 as IS 10918 and T.81 by ISO-IEC and ITU-T, respectively.

## 9.2 THE STRUCTURE OF JPEG

The JPEG standard is a typical hybrid coding technique in that many compression tools—including orthogonal transform, predictive coding, quantization, run-length coding, and entropy coding—are combined to achieve good compression performance. The standard specifies the format of the input/output images and the algorithm of the compression/decompression process. It provides algorithms for both lossless and lossy compression. Unlike H.261 or MPEG, the JPEG standard allows the image format to be rather flexible. There are four different compression modes in JPEG.

- The baseline system (sequential DCT-based lossy mode) is the basic mode and is the minimum configuration supported by all JPEG encoders/decoders. This mode will be described in greater detail in this chapter.
- The extended DCT-based lossy mode improves the baseline system and provides more functionalities and further compression.
- The lossless mode provides perfect reconstruction of the original image at the cost of lowered compression ratio.
- The hierarchical mode enables compression and decompression of the image at different resolutions. The compression algorithm used here is one of the previous three modes.

## 9.3   BASELINE SYSTEM: SEQUENTIAL DCT-BASED LOSSY MODE

The baseline system is the fundamental lossy compression mode in JPEG. Every JPEG decoder should at least be able to decode the baseline mode. Simplified block diagrams of the encoder and the decoder of this mode are shown in Fig. 9.1 and Fig. 9.2, respectively.

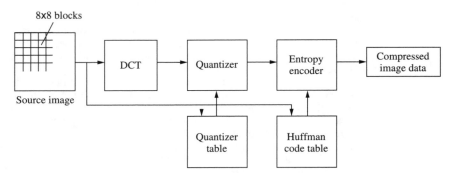

**FIGURE 9.1**   Simplified block diagram of a JPEG baseline encoder.

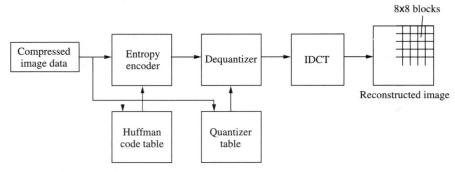

**FIGURE 9.2**   Simplified block diagram of a JPEG baseline decoder.

### 9.3.1   Source Image Format

In this mode the input image is represented by 8 bits (0–255) per pixel. There can be several such image planes. A common choice of the number of image planes is one (monochrome) or three (color). Each plane of the original image is partitioned into $8 \times 8$ pixel units, called *blocks*.

### 9.3.2   DCT (Discrete Cosine Transform)

Each block goes through an orthogonal transform, called *DCT,* for energy compaction and decorrelation. The DCT was invented by Ahmed, Natarajan, and Rao in 1974.[4] Since then there have been some variations in the definition of the DCT, but here we shall use the one that is adopted in the international standards of H.261, JPEG, and MPEG.

$$F(u, v) = \frac{2c(u)c(v)}{n} \sum_{j=0}^{n-1} \sum_{k=0}^{n-1} f(j, k) \cos\left[\frac{(2j + 1)u\pi}{2n}\right] \cos\left[\frac{(2k + 1)v\pi}{2n}\right] \tag{9.1}$$

where

$$c(w) = \begin{cases} 1/\sqrt{2}, & \text{for } w = 0 \\ 1, & \text{for } w = 1, 2, \ldots, n - 1 \end{cases}$$

Among many well-known orthogonal transforms, the DCT provides nearly optimal performance in image compression in terms of energy compaction and decorrelation. Ideally, the KLT (Karhunen-Loeve transform, or Hotelling transform) is the optimum orthogonal transform for compression purpose, but its kernel functions are signal-dependent because they are derived from eigenanalysis of the covariance matrix of the stationary signal.[5] Theoretically, the DCT kernel functions are very close to those of the KLT of the first-order Markov sequence having high correlation ( > 0.95) between adjacent samples. Note that typical images are well approximated by 2-D first-order Markov sequences with a very high correlation coefficient.

At the decoder, the 2-D inverse DCT (IDCT) is carried out to produce the $8 \times 8$ spatial domain pixels from the dequantized $8 \times 8$ DCT coefficients.

$$f(j, k) = \frac{2}{n} \sum_{u=0}^{n-1} \sum_{v=0}^{n-1} c(u)c(v)F(u, v) \cos\left[\frac{2(j + 1)u\pi}{2n}\right] \cos\left[\frac{2(k + 1)v\pi}{2n}\right] \tag{9.2}$$

A number of efficient algorithms and architectures for fast computation of DCT/IDCT have been reported.[6]

### 9.3.3   Quantization

After the $8 \times 8$ pixel block is transformed into the frequency domain through the 2-D DCT, each of the 64 real-valued DCT coefficients of the block is quantized to be represented by an integer. The quantization is a lossy process, and, as a reward of loss, a higher compression ratio is generally obtained.

The human visual system (HVS) is incorporated in the quantization process. That is, the human eyes are less sensitive to the quantization noise at higher frequencies, and the quantization at higher frequencies is usually coarser for balanced human perception. For this purpose, two quantization tables with 64 entries each are often used—one for luminance, the other for chrominance. There are no default quantization tables in the JPEG standard. The tables are instead provided by the application. Thereby, the quantization of the 64 DCT coefficients can be controlled separately, and some frequency components can be treated more significantly, depending on the application.

Figure 9.3 is a typical quantization matrix for luminance, which is obtained empirically and included in the JPEG standard as an informative part. The table entries are 8-bit integers and are determined according to the HVS, the sampling structure of the image, and the desired compression ratio (or desired image quality). The same tables are used at the encoder (quantizer) and at the decoder (dequantizer).

| | | | | | | | |
|---|---|---|---|---|---|---|---|
| 16 | 11 | 10 | 16 | 24 | 40 | 51 | 61 |
| 12 | 12 | 14 | 19 | 26 | 58 | 60 | 55 |
| 14 | 13 | 16 | 24 | 40 | 57 | 69 | 56 |
| 14 | 17 | 22 | 29 | 51 | 87 | 80 | 62 |
| 18 | 22 | 37 | 56 | 68 | 109 | 103 | 77 |
| 24 | 35 | 55 | 64 | 81 | 104 | 113 | 92 |
| 49 | 64 | 78 | 87 | 103 | 121 | 120 | 101 |
| 72 | 92 | 95 | 98 | 112 | 100 | 103 | 99 |

**FIGURE 9.3**    A typical 8 × 8 quantization table for luminance.

The DCT coefficients are quantized as follows:

$$S(u, v) = \text{Nearest Integer} \left[ \frac{F(u, v)}{Q(u, v)} \right] \tag{9.3}$$

where $\{F(u, v)\}$ are DCT coefficients and $\{Q(u, v)\}$ are the quantization matrix elements. At the decoder the inverse quantization is carried out as follows:

$$\hat{F}(u, v) = S(u, v)Q(u, v) \tag{9.4}$$

### 9.3.4  Entropy Coding

The quantized DCT coefficients are integer values. Because of the energy-compaction (into low frequencies) property of the DCT and the increasing-with-frequency property of the quantization tables, many of the high-frequency coefficients are reduced to zero. Utilizing this statistical characteristic, further data compression is possible through entropy coding. Unlike the quantization process, this step is a lossless (reversible) process.

The dc and ac coefficients are handled separately. The dc coefficients carry a significant amount of the image information and have high correlation between adjacent blocks. As shown in Fig. 9.4, the difference in the dc values between the current and the previous blocks are taken first, then it is classified into several categories, and finally it is Huffman-coded[7] using the dc Huffman table.

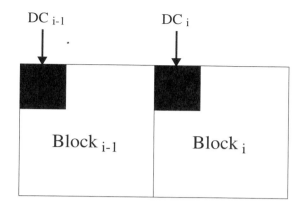

$$DIFF = DC_i - DC_{i-1}$$

**FIGURE 9.4**   Differential PCM of dc coefficients for efficient entropy coding.

Taking into account that most nonzero values are concentrated in the low-frequency region, the ac coefficients are run length–coded first using the zigzag scan as shown in Fig. 9.5. It results in a sequence of symbols of (RUN, LEVEL), where RUN is the number of successive zeros and LEVEL is the nonzero coefficient following the successive zeros. When the last LEVEL of the block is reached, the corresponding (RUN, LEVEL) is followed by the EOB (end-of-block) mark. As is the case with dc, the (RUN, LEVEL) symbol is classified into several categories depending on the value of LEVEL, and then it is Huffman-coded using the ac Huffman table.

| 0 | 1 | 5 | 6 | 14 | 15 | 27 | 28 |
|---|---|---|---|----|----|----|----|
| 2 | 4 | 7 | 13 | 16 | 26 | 29 | 42 |
| 3 | 8 | 12 | 17 | 25 | 30 | 41 | 43 |
| 9 | 11 | 18 | 24 | 31 | 40 | 44 | 53 |
| 10 | 19 | 23 | 32 | 39 | 45 | 52 | 54 |
| 20 | 22 | 33 | 38 | 46 | 51 | 55 | 60 |
| 21 | 34 | 37 | 47 | 50 | 56 | 59 | 61 |
| 35 | 36 | 48 | 49 | 57 | 58 | 62 | 63 |

**FIGURE 9.5**   Zigzag scanning order for $8 \times 8$ DCT coefficients.

## *9.4*   *OTHER MODES*

### 9.4.1   Extended DCT-Based Lossy Mode

This mode is an extension to the baseline DCT-based lossy mode for further functionalities and data reduction. First, the sample precision of input images includes 12 bits/sample (professional application) as well as 8 bits/sample (consumer application) used in the baseline mode. Second, the progressive coding is included in addition to the sequential coding adopted in the baseline mode. The progressive coding presents the image roughly at first, and then the image is gradually refined. There are two methods in the JPEG progressive mode:

- Spectral selection method: Low-frequency components of the quantized DCT coefficients of each block are passed to the entropy coder first, and then higher-frequency components are passed successively.
- Successive approximation method: The most significant bits of all quantized DCT coefficients of each block are passed to the entropy coder first, and then less significant bits are passed successively.

Third, the entropy coding adopts the arithmetic coding[8] in addition to the Huffman coding used in the baseline mode. The compression efficiency of the arithmetic coder is slightly better (approximately 10 percent at maximum) than that of the Huffman coder. The JPEG arithmetic coder is a descendent of the well-known Q-coder and is called the *QM-coder*. It needs no code table because it is automatically adapted to the source statistics. Nevertheless, the arithmetic coder is more sophisticated and, unlike the Huffman coder (which has no patent involved), it involves patents filed by IBM, etc.[1] The complexity and the patents of the arithmetic coder seem to be the main reason that most JPEG applications simply use the Huffman coding rather than the arithmetic coding. From the discussions so far, it can be seen that there are 12 combinations in the extended system—that is, the combination of 8 or 12 bits/sample; sequential or progressive spectral or progressive successive; and Huffman or arithmetic.

### 9.4.2   Lossless Mode

This mode is a reversible compression technique with a lower compression ratio. It is used for applications such as medical images, where no loss is permitted. Unlike the DCT-based compression, the lossless mode uses pixel-by-pixel operations rather than block-by-block operations. As shown in Fig. 9.6, it consists of a predictor for DPCM and an entropy coder for the error signal. The current pixel $x$ to be coded is predicted from three adjacent previous pixels using one of eight predictors listed in Fig. 9.7. The index of the selected predictor and the residual error are entropy-coded, where either the Huffman coder or the arithmetic coder is used.

### 9.4.3   Hierarchical Mode

This mode enables compression of an image at different resolutions. The original image is first decimated by two, both horizontally and vertically, with an appropriate antialiasing filter. The reduced image is then interpolated by two in both directions, and is subtracted from the original image to yield the error image. This process is repeated until the desired level of image hierarchy is obtained; see Fig. 9.8.

The hierarchical mode uses either the lossless compression mode or the DCT-based lossy mode for image compression. This mode requires increased (33.3 percent for the infinite-level case) image storage capacity, and the resultant coded bitstream is considerably increased as well. But at the decoder, part of the total bitstream can be decoded to yield the desired smaller image (or full-size image with lower resolution if interpolation is used after decoding). This feature of spatial scalability is also adopted later by MPEG-2.

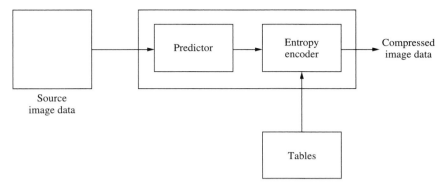

**FIGURE 9.6**   JPEG lossless mode block diagram.

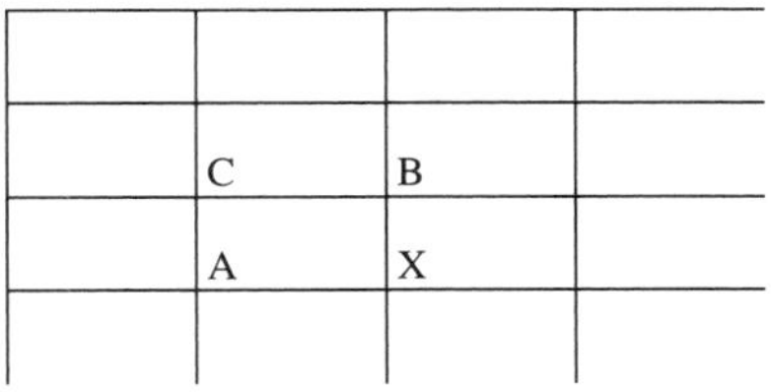

|   |   |   |   |
|---|---|---|---|
|   |   |   |   |
|   | C | B |   |
|   | A | X |   |
|   |   |   |   |

| Selection Value | Prediction |
|:---:|:---:|
| 0 | No Prediction |
| 1 | $\hat{X} = A$ |
| 2 | $\hat{X} = B$ |
| 3 | $\hat{X} = C$ |
| 4 | $\hat{X} = A + B - C$ |
| 5 | $\hat{X} = A + (B - C)/2$ |
| 6 | $\hat{X} = B + (A - C)/2$ |
| 7 | $\hat{X} = (A + B)/2$ |

**FIGURE 9.7**   Predictors for lossless coding.

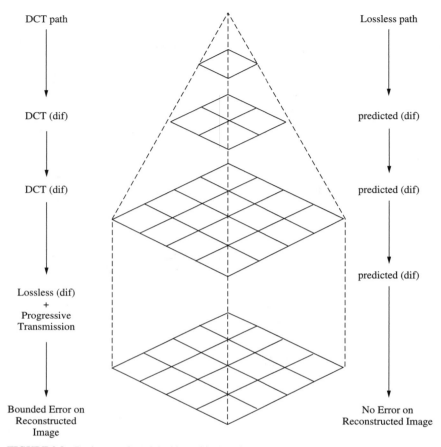

DCT path

DCT (dif)

DCT (dif)

Lossless (dif)
+
Progressive
Transmission

Bounded Error on
Reconstructed
Image

Lossless path

predicted (dif)

predicted (dif)

predicted (dif)

No Error on
Reconstructed Image

**FIGURE 9.8**    Basic operation of the hierarchical mode.

## 9.5  CONCLUSION

In JPEG lossy modes we note the following characteristics regarding the image quality and the compression ratio:[2]

*0.25 to 0.50 bits/pixel*   Moderate quality; sufficient for some applications.
*0.50 to 0.75 bits/pixel*   Good quality; sufficient for many applications.
*0.75 to 1.50 bits/pixel*   Excellent quality; sufficient for most applications.
*1.50 to 2.00 bits/pixel*   Not distinguishable from the original in most cases; sufficient for almost all applications—even for highest-quality requirements.

In JPEG lossless mode the typical compression ratio is about 2:1 due to simplicity of the algorithm.

JPEG can be used for various applications, including image storage in computers and multimedia devices, digital still cameras, color facsimile, color printers, color scanners, DTP (desktop publishing), medical imaging, and video phones. In addition, JPEG may also be applied to moving picture compression, in which each frame can be independently compressed based on JPEG. This scheme, called *MJPEG* (motion JPEG), is less efficient than the MPEG standard because, unlike MPEG, JPEG does not exploit frame-by-frame correlation resident in moving pictures. Nevertheless, MJPEG is still an attractive solution because it is much more cost-effective than MPEG and provides frame-by-frame editability.

## REFERENCES

1. International Standards Organization: "Information technology—Digital compression and coding of continuous-tone still images," International Standard ISO-IEC IS 10918, ISO-IEC JTC 1, 1993.

2. W. B. Pennebaker and Mitchell, J. L., *JPEG Still Image Data Compression,* Van Nostrand Reinhold, New York, 1993.

3. G. K. Wallace, "The JPEG still picture compression standard," *Communications of the ACM,* vol. 34, no. 4, Apr. 1991, pp.30–44.

4. N. Ahmed, Natarajan, T., and Rao, K. R., "Discrete cosine transform," *IEEE Transactions on Computers,* vol. 23, Jan. 1974, pp. 90–93.

5. N. S. Jayant and Noll, P., *Digital Coding of Waveforms,* Prentice Hall, Englewood Cliffs, N.J., 1984.

6. B. G. Lee, "A new algorithm to compute the discrete cosine transform," *IEEE Transactions on Acoustics, Speech, and Signal Processing,* vol. ASSP-32, no. 6, Dec. 1984, pp. 1243–1245.

7. D. A. Huffman, "A method for the construction of minimum redundancy codes," *Proceedings of the IRE,* vol. 40, Sept. 1952, pp. 1098–1101.

8. G. Langdon, "An introduction to arithmetic coding," *IBM Journal of Research and Development,* vol. 28, Mar. 1984, pp. 135–149.

## ABOUT THE AUTHOR

Jechang Jeong is a professor of image processing, multimedia, and digital communications in the Department of Electronic Communication Engineering at Hanyang University in Seoul, Korea. He received a B.S. from Seoul National University, an M.S. from Kaist, and a Ph.D. from the University of Michigan in Ann Arbor. Prior to joining the staff at Hanyang University, he worked as a researcher for the Korean Broadcasting System, as a research and technical assistant and a research fellow at the University of Michigan, and as a senior researcher at Samsung Electronics. He has written over 40 technical papers and holds more than 30 patents in image processing, signal processing, multimedia, and digital communications.

# CHAPTER 9
# THE ISO JPEG AND JBIG STILL IMAGE–CODING STANDARDS

## SECTION 2
## THE JBIG STANDARD

**Whoi-Yul Kim**
*Department of Electronic Engineering*
*Hanyang University*

## 9.6  INTRODUCTION

JBIG, the Joint Bilevel Image Experts Group, was formed in 1988. In the following discussion, a bilevel image will be one that is black and white, or any other form of two-color image. The JBIG standard was preceded by facsimile standards known as *CCITT Group 3* and *CCITT Group 4,* which are important in facsimile and document-storage applications that deal with black-and-white documents. Originally, the goals of JPEG (Joint Photographic Experts Group) included compression of bilevel images. Because it was not possible to produce an algorithm that worked well on both bilevel and continuous-tone images, it was decided to separate JBIG from JPEG and to establish a standard for the progressive encoding of bilevel images. The standard was prepared by ISO-IEC/JTC/SC2/WG9 and CCITT/SGV III.

The goals of JBIG were similar to those of JPEG, with the major applications of facsimile images and the accessing of large image databases. To support such applications, the technique also needed to be capable of providing both progressive and sequential image buildup. This would make it adaptable to various types of display and printing devices with a wide range of image resolutions and varying image quality. Here, the term "progressive" means that the resolution and quality of the image improves gradually, with a low-resolution image rendered first and then a higher resolution image as more data become available.

At a series of JBIG meetings, five major algorithms were selected.[1] These are progressive transmission of binary images by hierarchical coding, progressive encoding of facsimile image using edge decomposition, progressive encoding of predicted signal according to classified patterns, progressive adaptive bilevel image compression, and progressive coding scheme using block reduction. Similar to JPEG, JBIG was elevated to ISO working group status in 1990. Several introductory papers have been published on the JBIG standard.[1–3] The reader is encouraged to refer to that standard for more detail.[4]

## 9.7   *GENERAL OVERVIEW*

Although JBIG was formed to concentrate on bilevel images only, it can handle a general class of bilevel image by decomposing an image plane into a set of bit planes. Gray scale or color images can also be compressed by bit-plane decomposition. The method is also progressive. A progressive encoding system first transmits a compressed image prepared by a reduced-resolution version of the image and then enhances it as needed by transmitting additional information. Encoding is done from high resolution to low resolution, and decoding is done in reverse order. A progressive system provides two distinctive benefits. First, the image can be adaptively displayed or printed on devices with different resolution capabilities—for example, on fax machines or laser printers. Second, when creating thumbnail or iconlike small representations, the system can browse a large image database. That is, when a number of computing systems are connected by network to share a large image database, the overall system performance depends heavily upon the data traffic of the network. The effective compression ratio goes up much higher because the user can interrupt the transmission of the rest of the image if it is not needed.

The progressive bilevel coding technique consists of the following procedure:

1. Repeatedly reduce the resolution of a bilevel image. The reduced image has one-half the number of pixels per line and one-half the number of lines per image. As the reduction continues, the image size may reduce to a single pixel, but may not necessarily do so. The lowest or the coarse resolution image is called the *base layer.*

2. The image is transmitted from the lowest resolution to the highest ones. First the base layer image is transmitted after being encoded by arithmetic encoding. Next, the immediately higher resolution image is sent.

3. Instead of transmitting the next level directly after the arithmetic encoding, however, a prediction is made based on the previously transmitted pixel. If the prediction is possible (both transmitter and receiver follow the same rule in prediction), the predicted pixel is not transmitted.

4. Compatibility between progressive and sequential coding is achieved by dividing an image into several stripes. In the progressive mode the whole image is divided into a set of stripes and processed with the procedure previously described. The entire algorithm consists of several procedures: image reduction, typical reduction, deterministic prediction, and binary arithmetic coding.

## 9.8   *DATA ORGANIZATION*

Figure 9.9 shows the graphical illustration of the overall description of the encoded data stream, "Bilevel Image Entity" (BIE). The data stream is transmitted from the leftmost field to the right. The BIE consists of a bilevel image header (BIH) and bilevel image data (BID), whose lengths vary with data. Basically, BIH contains the physical description as well as the methods employed in compressing the image to transmit. The BIH is further decomposed into the subfields shown in the second block. The third block shows that the Order byte controls the method or transmission order of stripes, where the basic unit of each field is 1/8 byte. The last byte, Options, controls several methods of prediction techniques.

The minimum unit for $I_D$ in Fig. 9.9 will be roughly 10 to 25 dots per inch (dpi). The resolution of the image may increase from 12.5 dpi to 25, 50, 100, and 200 by repeatedly doubling the resolution. The image may not be legible at low resolution, but still is very useful for page layout and other applications, such as icon generation for quick recognition in browsing a large-image database on a computer terminal. The specification for progressive transmission capability is not restricted to the number $D$, which is the resolution doubling. If the progressive capability does

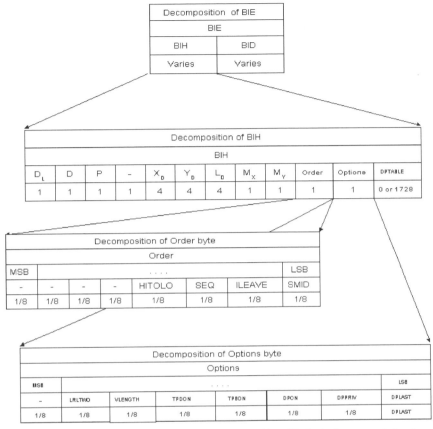

**FIGURE 9.9** Bilevel image entity and header decomposition. ($D_L$ is the lowest-resolution layer starting from zero. $D$ is the number of doublings in resolution (the number of differential layers or the number of reduced resolution images), that is, $2^D$ = image size. $X_D$, $Y_D$ are the horizontal and vertical dimensions in pixels, respectively. $I_D$ is the highest-resolution image or image to encode consisting of $I_D = X_D + Y_D$ pixels. $R_D$ is the sampling resolution of the image in $I_D$ in dpi, i.e., 200 dpi, 400 dpi, etc. (typically 300 or 600 dpi for laser printers). $P$ is the number of bit planes (1 for bilevel image). The dash (—) signifies that the item is unused. $L_0$ is the number of stripes in the lowest resolution.)

not exist, for example in hard-copy fax, it can be set to zero. This specification can also be used for the gray-scale image by coding bit planes independently, as if they were bilevel images. Of course, there is no limit to the number of bit planes, but the gray-coding method of gray-scale images yields superior results.

## 9.9 STRIPES AND DATA ORDERING

Sequential coding requires no buffer memory, but progressive coding does need memory for a page buffer because the lower layer is used in coding the higher layer. Coding in the progressive-compatible sequential mode is said to be "compatible" with coding in the progressive mode as long as the data streams at the encoder and decoder carry exactly the same information. In order to achieve this compatibility, an image is broken, before compression, into small parts called *stripes,* or horizontal bands. All the rest of the different resolution layers have the same number of stripes. Obviously, a stripe has a vertical size much smaller than that of the entire page. In a typical example, let $L_0$ be the number of 8-mm lines per stripe. The number of stripes in an image of letter size will be about 35 (35 mm $\times$ 8 = 280 mm = 28 cm). Therefore, progressive-compatible sequential coding does require a stripe buffer that is much smaller than a page buffer, as well as additional memory for adaptive entropy coding of each resolution layer and bit plane (if necessary).

The illustration in Fig. 9.10 shows a hierarchical representation of the image with three layers and three stripes per layer. More specifically, one bit-plane image of resolution 100 dpi is divided into three stripes: 6, 7, and 8. The lower-resolution layers of 50 dpi and 25 dpi also have the same number of stripes: 3, 4, 5 and 0, 1, 2, respectively. Stripe 0 corresponds to the lower resolution version of stripe 3, and stripe 3 to stripe 6, respectively. There are several ways to send these stripes. In Fig. 9.9, the HITOLO bit controls the resolution order, and SEQ controls the progressive versus sequential mode. When the HITOLO bit is set, the image is sent from the high-resolution to the low-resolution one. When SEQ is on, the image is coded, sent, and displayed sequentially; i.e., the image grows in raster order at the receiving end. For example, as illustrated in Table 9.1, when both the HITOLO and SEQ bits are off, the stripes of the image are sent sequentially from the lowest resolution to the highest in order, i.e., 0 1 2, 3 4 5, and 6 7 8. When HITOLO is off and SEQ is on, the stripe that corresponds to the same region in each layer is sent from the lowest resolution layer to the higher ones, i.e., 0 3 6, 1 4 7, and 2 5 8. Tables 9.1 and 9.2 show the orderings of the stripes in transmission for 1-bit and 2-bit images, respectively, in detail. When an image consists of two bit planes (4-level image), there will be 12 stripes (Fig. 9.11). When the bit plane is more than one, there are two more bits: one ILEAVE for controlling the interleaving state of multiple bit planes, and the other SMID for indexing over stripes. When ILEAVE is on, each pair of bit planes at the same resolution is processed first. That is, assuming that no other bit is set, the lowest-resolution bit-plane pairs are processed by the bit planes first, as shown in Fig. 9.11, and the stripes are processed in the following order: 00, 01, 02, 03, 04, and 05 in the lowest resolution, and so on. On the other hand, when SMID is on, the stripes are processed earlier than the bit plane in each resolution layer; i.e., the stripes belonging to the same region of the bit planes are processed first over the bit planes by the following order: 00, 03, 01, 04, 02, and 05. Therefore, there may be 16 states that can be controlled by 4 bits, but only 12 cases are considered.

As will be explained later, the processed or compressed data for each stripe is independent of stripe ordering. All that changes as HITOLO, SEQ, ILEAVE, and SMID vary is the order in which the data are concatenated onto a data stream. For simplicity, the remainder of this discussion will assume only one bit plane and the subscript $p$ will be dropped from $C_{s,d,p}$.

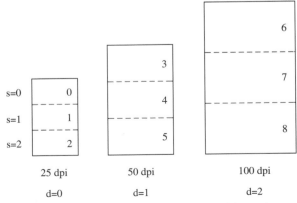

**FIGURE 9.10**    One-bit plane with three layers and three stripes.

**TABLE 9.1**  Possible Bi-level Data Orderings

| HITOLO | SEQ | Example order | | |
|--------|-----|------|------|------|
| 0 | 0 | 0,1,2 | 3,4,5 | 6,7,8 |
| 0 | 1 | 0,3,6 | 1,4,7 | 2,5,8 |
| 1 | 0 | 6,7,8 | 3,4,5 | 0,1,2 |
| 1 | 1 | 6,3,0 | 7,4,1 | 8,5,2 |

**TABLE 9.2**  Possible Multiplane Data Orders

| HITOLO | SEQ | ILEAVE | SMID | Example Order |
|--------|-----|--------|------|---------------|
| 0 | 0 | 0 | 0 | (00,01,02 06,07 08 12, 13, 14) (03, 04, 05 09, 10, 11 15, 16, 17) |
| 0 | 0 | 1 | 0 | (00, 01, 02 03, 04, 05) (06, 07, 08 09, 10, 11) (12, 13, 14 15, 16, 17) |
| 0 | 0 | 1 | 1 | (00, 03 01, 04 02, 05) (06, 09 07, 10 08, 11) (12, 15 13, 16 14, 17) |
| 0 | 1 | 0 | 0 | (00, 06, 12 03, 09, 15) (01, 17, 13 04, 10, 16) (02, 18, 14 05, 11, 17) |
| 0 | 1 | 0 | 1 | (00, 06, 12 01, 07, 13 02, 08, 14) (16, 09, 15 04, 10, 16 05, 11, 17) |
| 0 | 1 | 1 | 0 | (00, 03 16, 09 12, 15) (01, 04 07, 10 13, 16) (02, 05 08, 11 14, 17) |
| 1 | 0 | 0 | 0 | (12, 13, 14 06, 07, 08 00, 01, 02) (15, 16, 17 09, 10, 11 03, 04, 05) |
| 1 | 0 | 1 | 0 | (12, 13, 14 15, 16, 17) (06, 07, 08 09, 10, 11) (00, 01, 02 03, 04, 05) |
| 1 | 0 | 1 | 1 | (12, 15, 13, 16 14, 17) (06, 09 07, 10 08, 11) (00, 03 01, 04 02, 05) |
| 1 | 1 | 0 | 0 | (12, 06, 00 15, 09, 03) (13, 07, 01 16, 10, 04) (14, 08, 02 17, 11, 05) |
| 1 | 1 | 0 | 1 | (12, 06, 00 13, 07, 01 14, 08, 02) (15, 09, 03 16, 10, 14 17, 11, 05) |
| 1 | 1 | 1 | 0 | (12, 15 06, 09 00, 03) (12, 16 07, 10 01, 04) (14, 17 08, 11 02, 05) |

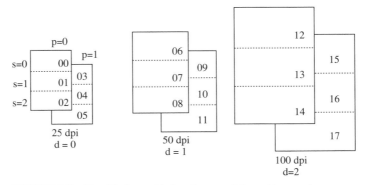

**FIGURE 9.11**  Two-bit plane with three layers and three stripes per layer.

## 9.10  *ENCODER AND DECODER FUNCTIONAL BLOCKS*

Figure 9.12 illustrates the conceptual block diagram for reducing the image by half until it reaches the lowest resolution layer (base layer). All $C_{s,d}$ values are transmitted after the encoding process to the receiver from the lowest to the highest order of $s$ and $d$. The differences between the layers are encoded by blocks, the functions of which are identical. Hence, a description of the operation is needed at only one layer. The input to the block is the high-resolution layer and the

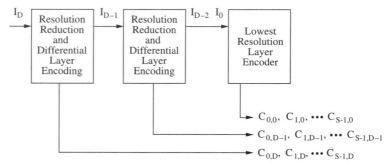

**FIGURE 9.12** Composition of encoder.

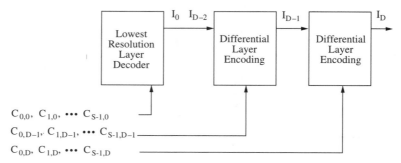

**FIGURE 9.13** Composition of decoder.

output is the lower-resolution layer. The decoding process is analogous to the encoding process, in reverse order, as shown in Fig. 9.13.

## 9.11 RESOLUTION REDUCTION

The resolution reduction block creates by subsampling a low-resolution image with half as many rows and columns as the original image. As illustrated in Fig. 9.14, the original image in the top layer is divided into $2 \times 2$ blocks of pixels, and each of these $2 \times 2$ pixels maps onto one pixel in a reduced resolution image in the lower layer. The two-dimensional representation of this situation is depicted in Fig. 9.15 with the same annotation of each pixel. This representation scheme will be used for the rest of this discussion because of its simplicity. The subsampling results in a poor-quality image that contains thin lines and dithered image or text. To avoid these problems, JBIG recommends a special algorithm that preserves as much detail as possible in the lower-resolution image. The method is based upon the context of the image and depends on the pixel values of the high-resolution layer, as well as those in the lower-resolution layer that are already determined. At the boundary of an image (for example, at the very beginning and end of an image, such as the top and bottom lines of the image), the previous and next line will be assumed to be the same as the top and last line, respectively. Similarly, the leftmost and the rightmost columns will also follow the same rule, called the *edge rule*.

For example, Fig. 9.16a shows the annotation for the pixels that participate to determine the pixel value. The pixels in the circles with "?" are determined by the pixels at both the high- and

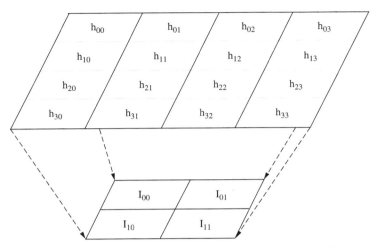

**FIGURE 9.14**  High- and low-resolution pixels in three-dimensional graphic.

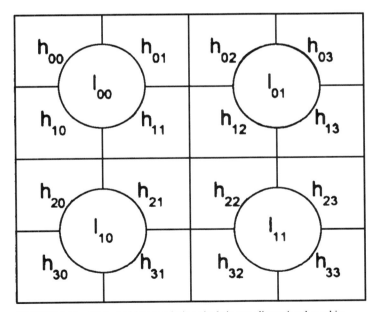

**FIGURE 9.15**  High- and low-resolution pixels in two-dimensional graphic.

low-resolution images. If, for example, the pixels 11, 10, and 9 are already determined, and the pixel "?" is to be determined, the following equation is used to determine the pixel "?":

$$4h_{22} + 2(h_{23} + h_{33}) + h_{33} + (h_{11} - l_{00}) + 2(h_{21} - l_{10}) + (h_{31} - l_{10}) + 2(h_{12} - l_{01}) + (h_{13} - l_{01})$$
$$= 4h_{22} + 2(h_{12} + h_{21} + h_{23} + h_{33}) + (h_{11} + h_{13} + h_{31} + h_{33}) - 3(l_{10} + l_{01}) - l_{00}$$

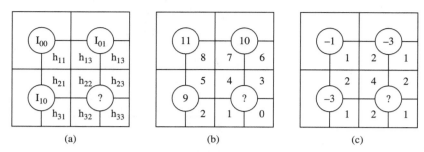

**FIGURE 9.16**    Pixels used to determine the "?" marked pixel value at low resolution. Annotation for pixels used to determine (*a*) the pixel value, (*b*) bit order of the pixels for indexing, and (*c*) pixel weightings.

Instead of evaluating the equation for every single pixel in the low-resolution image, a pre-computed lookup table is used with a default resolution reduction algorithm. The index for the lookup table is determined by the following example. In Fig. 9.17, "0" indicates the zeroth-order bit (the lowest-order bit or least significant bit) and 11 the highest-order bit. For example, when the pixels denoted as 0, 1, 3, and 4 are the color black, then, intuitively, the pixel with "?" should also be black.

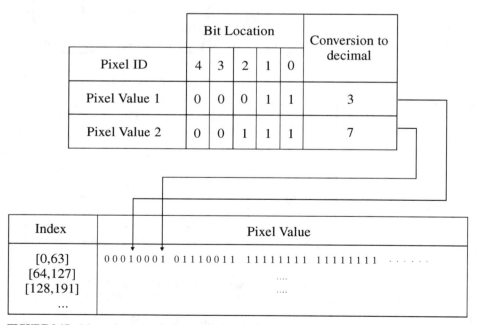

**FIGURE 9.17**    Map to determine the low-resolution pixel.

The index table that coincides with this situation is the lower table of Fig. 9.17. Here, the binary representation of "0 0 0 1 1" becomes 3. In the lookup table, the first seven pixel values are "0 0 0 1 0 0 0 1." Therefore, the fourth (it is 3 because the index always starts from 0) pixel value becomes 1. Similarly, when the eighth (or 7, starting from 0) value is 1, the binary representation of 7 becomes "1 1 1," which means that the pixel values corresponding to the pixels from 0 to 2 become all ones. At the edge of an image, the previous edge rules apply; i.e., the pixel above the first line of the image is considered the same as the first line. Similarly, the left pixels or the first pixels of all lines are considered the same. The algorithm is identical for all different resolution layers and all bit planes (if an image is more than binary-level).

## 9.12  PREDICTION

When a differential layer is being encoded (Fig. 9.18) or decoded, much of the compression is achieved by use of a predictor template to predict new values from the pixels in the template. When the prediction is correct, the predicted pixel value need not be encoded or decoded. In doing this, two kinds of prediction are employed in JBIG: typical prediction (TP) and deterministic prediction (DP).

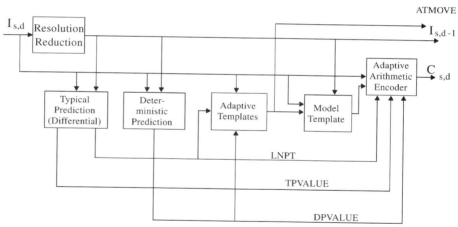

**FIGURE 9.18**   Differential-layer encoder.

### 9.12.1  Typical Prediction

The typical prediction (TP) block is used to speed up the processing, but it also provides some coding gain. Typical prediction means the correct prediction is almost always made. The transmitter looks ahead for pixels and reports TP errors, but depends on the probability of the image. Differential-layer TP looks for regions of solid color when the current pixel in the high-resolution layer is in the region. No further processing is necessary and thus it is possible to avoid coding over 95 percent of the pixels.

### 9.12.2 Differential-Layer Typical Prediction

Differential-layer TP is controlled by the TPDON (typical prediction of differential layer ON) bit in the option fields of BIH. To process the pixels across the boundary of the image or stripe, the same edge rule is applied. As illustrated in Fig. 9.19, where only part of the interested pixels are shown, the pixel "a" is not typical if all eight neighboring pixels (from "f" to "m" in the lower-resolution layer in the circle) are the same colors. But more than one pixel in four in the high-resolution image is not the same. If it is not typical, the virtual pixel for the virtual pixel LNTP is inserted and encoded by the arithmetic coder.

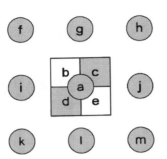

**FIGURE 9.19** A nontypical line in the differential layer in typical prediction.

Although most of the lines are typical in the differential layer in TP, only a small portion of the image is not "typical" in the lowest-resolution layer. Unlike the case with the differential layer previously discussed, if the current line differs from the previous line at any pixel, it is not typical in the lowest-resolution layer. At the top of the image, it becomes not typical because it differs from the background. If not typical, a pseudopixel is again inserted in the coded image at the virtual location prior to the beginning of the line. Coding the changes of LNTP is more efficient than coding LNTP itself. The SLNTP (same LNTP) is also encoded by the arithmetic coder.

### 9.12.3 Deterministic Prediction

The purpose of deterministic prediction (DP) is to provide coding gain, typically about 7 percent. Deterministic prediction implies that the predicted value is always correct. This prediction is much more precise than that of TP and is determined by looking up a rule in a table indexed by the state of the predictor pixels. DP is a table-driven algorithm. In order to obtain the deterministic prediction, the values of particular surrounding pixels in the low-resolution image and causal high-resolution image are used to index a table to check for determinicity. When the pixel is to be coded, the higher-resolution image is inferable from the previously processed pixels, i.e., in the previous line or pixel of the current line (causally related). On the lower-resolution image, the current pixel is said to be deterministically predictable. If so, the DP block flags any such pixels and inhibits their coding by the arithmetic coder. That is, the pixels that can be predicted deterministically are not flagged and not encoded by the arithmetic encoder. The DP table is dependent on the particular resolution reduction method used. Therefore, the provision is made to use the same rule for resolution reduction. If a private resolution reduction algorithm is used, it needs to be sent to the decoder; otherwise, a default algorithm is used.

### 9.12.4 Default DP Table

In the decoding process the low-resolution image is read first. In order to predict the current pixel value, both the low-resolution layer and the previous pixels that are read or predicted in the current layer are used. The phase value is determined by the relative location of the pixel to decode with respect to the pixel that corresponds to it in the lower-resolution image. Depending on the phase value, the number of pixels varies as shown in Table 9.3. The DP table (equivalent to the index table) is constructed in the same way as in the resolution reduction process, except that Fig. 9.20 denotes the order of bits in the table. For example, when the pixel 8 is to be decoded, the

**TABLE 9.3**  DP Pixels for Each Spatial Phase

| Phase | Target pixel | Reference pixels | Number of hits with default resolution reduction |
|-------|-------------|------------------|-----------------------------------|
| 0 | 8 | 0,1,2,3,4,5,6,7 | 20 |
| 1 | 9 | 0,1,2,3,4,5,6,7,8 | 108 |
| 2 | 11 | 0,1,2,3,4,5,6,7,8,9,10 | 526 |
| 3 | 12 | 0,1,2,3,4,5,6,7,8,9,10,11 | 1044 |

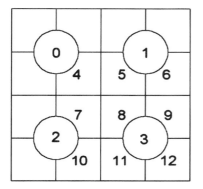

**FIGURE 9.20**    Labeling of pixels used by DP.

phase with respect to the pixel 3 is 0, and zero to seven pixels are used for indexing according to the table. The total number of indices amounts to 20 when the default resolution reduction algorithm is used. The entries in the table give 0, 1, or 2. A "2" indicates that it is not possible to make a deterministic prediction.

### 9.12.5   Model Template

The model template provides the arithmetic coder with an integer called the *context*. The binary representation of the integer consists of the colors (binary levels) of the pixels adjacent to the pixel to predict. The set includes three types of information. The first consists of the neighboring pixels in the high-resolution layer. The second consists of the pixels in the lower-resolution layer that are already determined (causal). The third consists of the pixels for the spatial phase, which denotes the orientation of the high-resolution pixel with respect to the corresponding low-resolution image of the pixel to code. The neighboring pixels are confined by the template. Two kinds of templates are used depending on the layer types—one for the lowest-resolution layer and one for differential layers. The lowest-resolution layer has two types: a 3-line model template and a 2-line model template (Fig. 9.21).

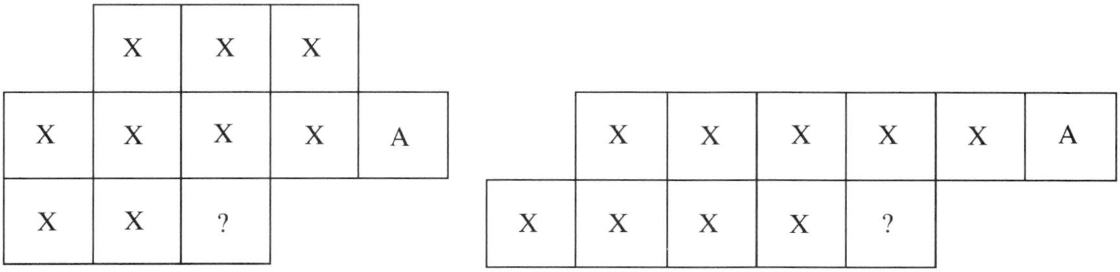

(a) Three line model template

(b) Two line model template

**FIGURE 9.21**    Model templates.

### 9.12.6 Adaptive Template

The purpose of the adaptive-context template is to take advantage of horizontal periodicity, which often occurs in half-tone and dithered images. The adaptive template achieves about 80 percent of the compression. In general, the 3-line model template shown in Fig. 9.21*a* yields a higher compression ratio (about 5 percent), whereas the template in Fig. 9.21*b* has slightly higher-speed performance. All templates for both the lowest-resolution layers and for differential layers are adaptable, depending upon the context of the pixels in the template. The adaptive template searches for periodicity in the image. When it finds periodicity, the template changes so that the pixel preceding the current pixel by this periodicity is incorporated into the template. The template consists of the pixels marked X and A, as shown in Fig. 9.22. The pixel denoted by "?" corresponds to the pixel to be coded and is not part of the template. The one marked A is a special pixel and can adaptively move around outside of the template, so it is called the *adaptive* or *AT pixel.* Once the location of A is determined to yield the best performance for the rest of the pixels in a stripe, it does not change its location within a stripe or an image. Although infrequent, when such a change is made, a control sequence ATMOVE is added to the output stream. Hence, decoders need not do any processing to search for the correct setting for AT.

The arithmetic encoder notes TP and DP to determine if it is necessary to code the current pixel. Similar to the resolution reduction module, the pixel values in the template constitute a context.

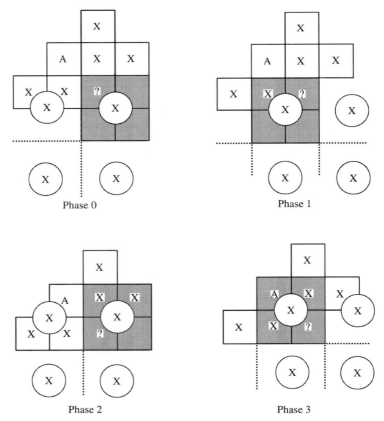

**FIGURE 9.22** Model templates for differential-layer coding.

### 9.12.7  Differential Layer Encoder

In the differential layers, another type of template is used containing lower and higher differential layers. In Fig. 9.22, each template has five X's and one A in the higher layer and four X's in the lower layer, to constitute a 10-bit context. The other two bits are used to represent phase information. Therefore, 12 bits in total are required to generate 4096 possible contexts. These contexts are used to identify the arithmetic coder adapter to be used.

### 9.12.8  Lowest-Resolution Layer Encoder

The encoder used for the lowest-resolution layer is conceptually simpler than the ones used in the differential layer encoders because no resolution reduction and deterministic prediction are needed. Furthermore, the TP, AT, and MT blocks are different from the ones in the differential layers. In this lowest-resolution layer, TP is mainly for speed processing. On typical images with text and graphics, about 40 percent of pixels are eliminated from the processing. Because there are 10 pixels in these templates, contexts associated with the templates can take on different values in the index table. This context is used to identify which arithmetic coder is to be used for encoding the pixel.

## 9.13  MISCELLANEOUS NOTES

JBIG recommends several methods for testing arithmetic and full algorithms. Also suggested are minimum support for free parameters, several indexing tables for resolution reduction and DP tables, and a method of generating an artificial image of size $1960 \times 1951$. These methods are provided for the purpose of testing the integrity of the proposed system, sharing both hardware and software, and exchanging decodable image data. The design of the probability estimation table for a binary arithmetic encoder is also included. A list of patents can be found in the standard, including the addresses for the companies to which the patents are assigned.

## 9.14  PERFORMANCE

The proposed system is five times as effective as the previously developed MMR encoding scheme, especially on computer-generated images of printed characters. On a gray scale image rendered by half-toning or dithering, the compression ratio is two to thirty times as great. Since JPEG handles continuous-tone images with some loss but can also handle images without any loss, this approach can be used for JPEG's lossless mode as an alternative. Experimental results have shown that the compression ratio is up to six bits per pixel higher than that of JPEG. For six to eight bits per pixel, the ratio is quite similar. In another experiment with 13 different types of images, JBIG outperformed for all images over G4 MMR.[1] The highest compression was achieved when two layers were used. That is, the compression ratio was highest when the number of layers was two—not one or more than two. Depending upon the image type used for testing, JBIG performance of the progressive mode is similar to the one in the sequential mode, and the sequential mode is often better. For data browsing, however, the progressive mode is more appropriate.

## REFERENCES

1. S. J. Urban, "Review of standards for electronic imaging for facsimile systems," *Journal of Electronic Imaging,* vol. 1, no. 1, Jan. 1992, pp. 5–21.

2.  W. B. Pennebaker and Mitchell, J. L., *JPEG Still Image Data Compression,* Van Nostrand Reinhold, New York, 1993.

3.  R. B. Arps, Truong, T. K., Lu, D. J., Pasco, R. C., and Friedman, T. D., "A multipurpose VLSI chip for adaptive data compression of bilevel images," *IBM Journal of Research and Development,* vol. 32, no. 6, Nov. 1988, pp. 774–794.

4.  International Telecommunication Union, "Information technology—Coded representation of picture and audio information—Progressive bilevel image compression," ITU-T Recommendation T.82, Mar. 1993.

## ABOUT THE AUTHOR

Whoi-Yul Kim is an assistant professor in the Department of Electronic Engineering at Hanyang University, Seoul, Korea. He received the B.S. degree in electronic engineering from Hanyang University in 1980, the M.S. in electrical engineering from Pennsylvania State University, University Park, in 1983, and the Ph.D. from Purdue University in 1989. His previous work experience was as a research assistant at Pennsylvania State University and Purdue University and as an assistant professor in the School of Engineering and Computer Science at the University of Texas, Dallas. He has published 14 technical papers in computer vision and image processing.

## ACKNOWLEDGMENT

This chapter is based on ITU Recommendation T.82 (03/93) and Corrigendum 1 (03/95)–Information Technology with the prior authorization of the International Telecommunication Union as copyright holder. The ITU material can be obtained from International Telecommunication Union, General Secretariat–Sales and Marketing Service, Place des Nations, CH-1211 Geneva 20 (Switzerland).

# CHAPTER 10
# ITU-T RECOMMENDATION H.261 VIDEO CODER-DECODER

## Jae Jeong Hwang
*Department of Radiocommunication Engineering*
*Kunsan National University*

## Beom Ryeol Lee
*Electronics and Telecommunication Research Institute*

## 10.1  INTRODUCTION

### 10.1.1  ITU-T H.320 Series

Video coding is one of the key technologies in this multimedia era, and its international standardization is essential for efficient interchange of audiovisual information. Since the late 1980s there has been much demand for visual telephony through historical telephone lines with their narrow bandwidths. The International Telecommunications Union (ITU-T)—formerly the International Telegraph and Telephone Consultative Committee or CCITT—H-series standard[1-3] was established in November 1989 and is applicable to videophone or videoconferencing. H.261, entitled "Video codec for audiovisual services at $p \times 64$ kb/s," is a video coding standard formulated by the CCITT SG15 Specialists Group on Coding for Visual Telephony. H.261 specifies a real-time encoding-decoding system with a delay time less than 150 ms. There are five related recommendations in the H.320 series developed by CCITT SG15, as follows:

- H.221   Frame structure for a 64 to 1920 kb/s channel in audiovisual teleservices.
- H.230   Frame-synchronous control and indication signals for audiovisual systems.
- H.242   System for establishing communication between audiovisual terminals using digital channels up to 2 Mb/s.
- H.261   Video coding/decoding for audiovisual services at $p \times 64$ kb/s (in the range of 1 to 30).
- H.320   Narrow-band visual telephone systems and terminal equipment (audio coding is indirectly specified).

### 10.1.2  Hybrid Coding

For hybrid coding the coding algorithm is similar to that in the typical still-image coder-decoder (codec), the JPEG (see Chapter 9), in which a discrete cosine transform (DCT) is used as a main data compression tool in intraframe, and the transform coefficients are coded by variable length coding (VLC)—e.g., a Huffman coder. But the primary difference between the two standards is the use of motion-compensated temporal prediction adopted in H.261. The motion-compensated temporal prediction errors are mapped into the DCT domain, followed by quantization and VLC. Therefore, the coding algorithm employs a hybrid of motion-compensated interframe coding and intraframe coding. The former takes advantage of the strong correlation among video frames to reduce the temporal redundancy. The latter is used to remove spatial redundancy in single-frame images. Intraframe coding is invoked when the interframe prediction is not efficient, and the input signal is passed directly to the DCT process. Generally speaking, spatial and temporal redundancy reduction are considered together in order to achieve the highest possible compression ratio.[4]

In the predictive mode the motion-compensated predictive error, defined as the difference between the current block and the best-match block, is coded. The encoder may transmit a motion vector (MV) to the decoder. Otherwise, the current block is coded directly. If the best-match block data are not quantitatively close to the current block, the motion-compensated predictive error will be large. Coding the current block without MC is more advantageous.

Using an $8 \times 8$ DCT for removing intraframe correlation, zigzag order for scanning the transform coefficients, and run-length coding for zero-valued coefficients after quantization are common modules in all three typical standards: JPEG, MPEG (Moving Picture Experts Group; see Chapter 8), and H.261. In H.261, motion estimation is applied to a video sequence to improve the prediction between successive frames, but it is not as complex as in MPEG, which is designed for a number of interactive applications. This is because application fields of H.261 are limited to videophone and videoconference applications, which have relatively slow-moving objects. For the purpose of simplex or duplex communication (different from JPEG) the transmission rate has to be controlled in the range of $p \times 64$ kb/s, where $p$ is up to 30. Naturally, data-compressed image transmission is more sensitive to channel errors. Error resilience, including synchronization and concealment techniques, are required in the transmission coder shown in Fig. 10.1.

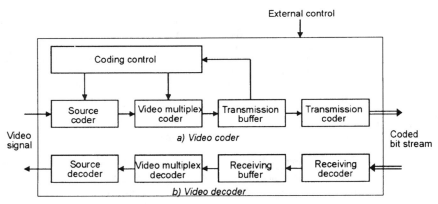

**FIGURE 10.1** Simplified diagram of the H.261 Codec.[3]

### 10.1.3 Nonspecified Parameters

The standard does not specify all parameters or exact components for each function. The following operations are not subject to H.261 and are left to the designer:

- The conversion algorithm to video data format (common intermediate format, or CIF) from any video source (National Television System Committee [NTSC], Phase-alternating Line [PAL], Séquentiel couleur avec mémoire [SECAM], ITU-R 601, etc.), and vice versa.

- Motion estimation in the encoder (one motion vector per macroblock may be transmitted).
- The arithmetic process for computing FDCT (mismatch error criteria are defined).
- Processing mode decisions such as intra or inter, motion compensation or no motion compensation (MC/No MC), loop filter or not, coded or not.
- How to design a loop filter in the encoder.
- All pre- and post-processing.
- How to decode BCH (511,493) error-correction code.

The following sections describe video data format as a compromise based on the different television formats. The video compression algorithm based on 2-D DCT in the macroblock unit is discussed, as are motion estimation and compensation techniques based on integer picture element accuracy. Finally, we address transmission environment and new directions for improving codec performance.

## 10.2  SOURCE FORMAT

### 10.2.1  Color Space Conversion

Television displays generate colors by mixing lights of the additive primaries: red, green, and blue. The color space obtained through combining the three colors can be determined by drawing a triangle on a special color chart with each of the base colors as an endpoint. The classic color chart used in early television specification was established by the Commission International de L'Eclairage (CIE). Using the CIE chart as a guideline, the National Television System Committee (NTSC) defined the transmission of signals in a luminance and chrominance format, namely $YIQ$, representing the luminance ($Y$), in-phase chrominance ($I$), and quadrature chrominance ($Q$) coordinates, respectively. In Europe the PAL format and the SECAM format were later established, based on $YUV$ color space. The digital equivalent of $YUV$ is $YC_bC_r$, where the $C_b$ corresponds to the analog $U$ component and the $C_r$ corresponds to the analog $V$ component, though with different scaling factors.[5]

$YC_bC_r$ are obtained from digital gamma-corrected $RGB$ signals as follows:

$$Y = 0.299R + 0.587G + 0.114B \tag{10.1}$$

$$C_b = -0.169R - 0.331G + 0.500B$$

$$C_r = 0.500R - 0.419G - 0.081B$$

The color-difference signals are given by:

$$(B - Y) = -0.299R - 0.587G + 0.886B \tag{10.2}$$

$$(R - Y) = 0.701R - 0.587G - 0.114B$$

where the values for $(B - Y)$ have a range of $\pm 0.886$ and for $(R - Y)$ a range of $\pm 0.701$, whereas those for $Y$ have a range of 0 to 1. To restore the signal excursion of the color-difference signals to unity ($-0.5$ to $+0.5$), $(B - Y)$ is multiplied by a factor 0.564 (0.5 divided by 0.886) and $(R - Y)$ is multiplied by a factor 0.713 (0.5 divided by 0.701). Thus $C_b$ and $C_r$ are the renormalized blue and red color-difference signals, respectively.

Given that the luminance signal is to occupy 220 levels (16 to 235), the luminance signal has to be scaled to obtain the decimal value, $\bar{Y}$. Similarly, the color difference signals are to occupy 224 levels, and the zero level is to be level 128. The decimal values for the three components are expressed as follows:

$$\overline{Y} = 219Y + 16 \tag{10.3}$$

$$\overline{C}_b = 224[0.564(B - Y)] + 128 = 126(B - Y) + 128$$

$$\overline{C}_r = 224[0.713(R - Y)] + 128 = 160(R - Y) + 128$$

where the corresponding level number after quantization is the nearest integer. This is useful for most of the applications where 8-bit binary encoding is adopted. The inverse conversion (after subtracting 128, if the values of $C_b$ and $C_r$ are level shifted) is given by the following:

$$R = Y + 1.402C_r \tag{10.4}$$

$$G = Y - 0.344C_b - 0.714C_r$$

$$B = Y + 1.772C_b$$

There are several $YC_bC_r$ sampling formats, such as 4:4:4, 4:2:2, and 4:1:1 (4:2:0). The sampling format 4:2:2 implies that the sampling rates of $C_b$ and $C_r$ are one-half that of the $Y$ component, and so on. H.261 uses the 4:2:0 format—four luminance samples and two chrominance samples co-sited in the center of the luminance samples, as shown in Fig. 10.2.

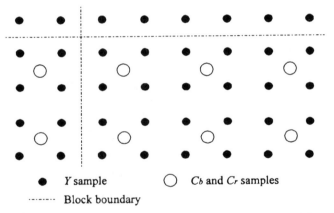

●   $Y$ sample      ○   $C_b$ and $C_r$ samples

------ Block boundary

**FIGURE 10.2**   Positioning of luminance and chrominance pels (4:2:0) in H.261.

The $YC_bC_r$ format concentrates most of the image information in the luminance and less in the chrominance. The result is that the $YC_bC_r$ elements are less correlated and can therefore be coded separately without much loss in efficiency. Another advantage comes from reducing the transmission rates of the $C_b$ and $C_r$ chrominance components, because the human visual system is less sensitive to the chrominance change than to the luminance change.

## 10.2.2   Common Intermediate Format

To accommodate both 525-line (NTSC) and 625-line (PAL/SECAM) systems, H.261 can accept the video signal in two formats: One is CIF, which consists of 360 picture elements (pels) by

288 lines for luminance and 180 pels by 144 lines for chrominance in an orthogonal arrangement, as shown in Fig. 10.3. For videophone applications where low bit rates are required, another format is allowed: quarter-CIF (QCIF), which has 180 pels by 144 lines for the luminance and 90 pels by 72 lines for chrominance signals. The frames are noninterlaced (progressive) and the temporal resolution can be 30, 15, 10, or 7.5 pictures per second. The frame rate is 29.97 frames per second for the NTSC-compatible systems.[6]

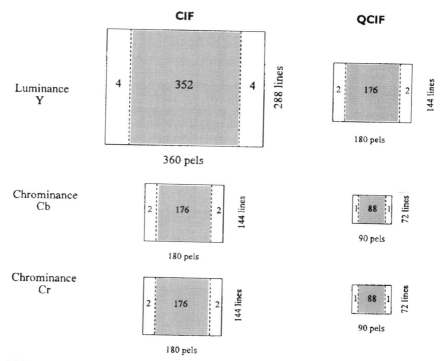

**FIGURE 10.3**   Formats and resolutions for CIF/QCIF picture.

For the luminance component, 360 pels per line are adopted, just 2:1 decimation from 720 active pels per line in both the TV systems (ITU-R 601). Two parameters are adopted based on sharing the horizontal resolution burdens: one parameter, 288 active lines, is influenced by the 625/50 (PAL/SECAM) system, which has 576 active lines (twice the CIF resolution); the other parameter is frame frequency 29.97, which is based on the 525/60 (NTSC) system. It may be emphasized that for both CIF and QCIF configurations, the frames are noninterlaced (progressive). The processing of images is block-based and the basic block is 8 × 8 pels. As the motion estimation is based on 16 × 16 luminance, $Y$, blocks, vertical stripes of 4 picture elements on either side of the CIF picture, are cropped out, resulting in 352 pels per line and 288 lines per picture resolution. This is called the *significant pel area* (SPA). Henceforth, for CIF and QCIF the term "picture" will be used rather than "frame" to stress that it is noninterlaced. The picture resolution of the SPA is now integer-divisible by the 16 × 16 block. Corresponding SPA for the QCIF and the chrominance components of the CIF are shown in Fig. 10.3.

## *10.3   VIDEO DATA STRUCTURE*

### 10.3.1   Hierarchical Data Structure

H.261 defines a consistent hierarchical structure so that the decoder may decode the received bitstream without any ambiguity. Hierarchy denotes several layers in the overall decoding process. Another reason for hierarchical structure is that the loss of a whole frame can be avoided by the number of hierarchical synchronizations. Certain regions that have uncorrectable errors can be duplicated or interpolated. A hierarchical structure with four layers of video data is shown in Fig. 10.4.

The four layers and corresponding luminance ($Y$) pels in each layer are distinguished by the following:

- Picture layer: $352 \times 288$ pels (1584 basic blocks).
- Group of block (GOB) layer: $176 \times 48$ pels (132 basic blocks).
- Macroblock (MB) layer: $16 \times 16$ pels (4 basic blocks).
- Block layer: $8 \times 8$ pels (basic block).

Each layer consists of header information and data for the following layer. The header includes several kinds of information including sync, numbering, quantizer, motion vector, extensibility, etc.

### 10.3.2   Picture Layer

A CIF picture consists of 12 GOBs, as shown in Fig. 10.5. Data for each picture layer consist of a picture header followed by data for GOBs. The picture header includes a 20-bit picture start code and other information, such as video format (CIF or QCIF), temporal reference, etc., as follows:

- Picture start code (PSC).   (20-bit, 0000 0000 0000 0001 0000).
- Temporal reference (TR), (5-bit FLC).   Indicates the source picture number, including the number of nontransmitted pictures for the decoder's use (how to use it is outside the scope of H.261). It is not for informing the picture rate.
- Picture type (PTYPE), (6-bit FLC).   This informs the split-screen indicator on/off (bit 1), document camera indicator on/off (bit 2), freeze picture release on/off (bit 3), and source format CIF ("1") or QCIF ("0") (bit 4).
- Picture extra insertion (PEI) bit.   Signals the presence of the following optional data field.
- Spare bits (PSPARE).   For future use (0, 8, 16, . . . , bits).

### 10.3.3   Group of Block Layer

A GOB consists of 33 megabytes in a CIF picture. Start codes (PSC in picture layer and GOB start code [GBSC] in GOB layer) are used to synchronize a decoder under transmission error conditions. The GOB layer consists of a GOB header followed by data for a 16-bit GOB start code and other information, such as position of the GOB, and quantizer information for the GOB until overridden by any subsequent MB quantizer information, etc., as follows:

- GBSC.   (16-bit, 0000 0000 0000 0001).
- Group number (GN) representing 12 GOBs (4-bit FLC).

PICTURE LAYER

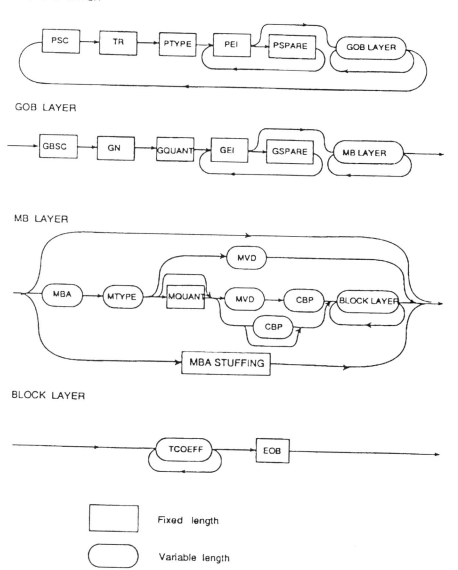

GOB LAYER

MB LAYER

BLOCK LAYER

**FIGURE 10.4**  Syntax diagram for hierarchical video multiplexing. Each layer is composed of possible headers and corresponding low-level layers.[3]

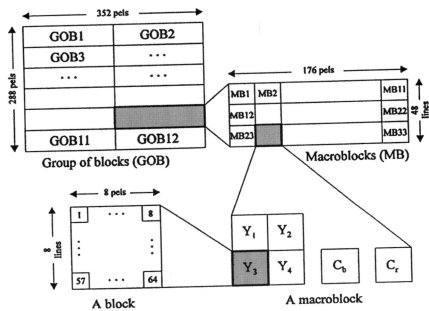

**FIGURE 10.5**    Hierarchical block structure in a CIF picture.

- Group quantizer information (GQUANT), 5-bit FLC.    This indicates one of the 31 quantizers to be used in a GOB until overridden by any subsequent MQUANT information.
- GEI and GSPARE.    Similar to PEI and PSPARE, respectively.

### 10.3.4  Macroblock Layer

A macroblock comprises four luminance blocks and two chrominance blocks in the order shown in Fig. 10.5. The MB layer consists of an MB header followed by data for the six blocks. The MB header includes a VLC for the MB address. It is followed by a VLC for MB type, indicating whether it is intraframe or interframe, with or without motion compensation and/or loop filter. Depending on a particular MB type, various combinations of video side information may follow.

- MB address (MBA), indicating its position within a GOB ( up to 11-bit VLC). MBA is the difference between the absolute address of the macroblock and the last transmitted macroblock in a GOB. It is only included in transmitted macroblocks. MBA for the first transmitted macroblock in a GOB is the absolute address. If the bitstream buffer is underflow, the 11-bit MBA stuffing code (0000 0001 111) is transmitted.
- MB-type information (MTYPE), VLC in Table 10.1, indicating which data elements are present in the macroblock.
- MB quantizer information (MTYPE), VLC in Table 10.1, to replace the quantizer defined by GQUANT.
- Motion vector data (MVD), up to 11 bits and 32 VLCs, means the differential motion vector (horizontal and vertical MV components) from the motion vector of the preceding MB.
- Coded block pattern (CBP), up to 9 bits and 63 VLCs.

**TABLE 10.1** VLC Table for Macroblock Type (MTYPE)[3]

| | MQUANT | MVD | CBP | TCOEFF | VLC |
|---|:---:|:---:|:---:|:---:|---|
| Intra | • | | | • | 0001 |
| Intra | • | | | • | 0000 001 |
| Inter | | | • | • | 1 |
| Inter | • | | • | • | 0000 1 |
| Inter + MC | | • | | | 0000 0000 1 |
| Inter + MC | | • | • | • | 0000 0001 |
| Inter + MC | • | • | • | • | 0000 0000 01 |
| Inter + MC + LF | | • | | | 001 |
| Inter + MC + LF | | • | • | • | 01 |
| Inter + MC + LF | • | • | • | • | 0000 01 |

"•" means that the item is present in the MB.
TCOEFF is the transform coefficient.
It is possible to apply the loop filter (LF) in a non-motion-compensated MB by declaring it as MC + LF, but with a zero MV.

### 10.3.5 Block Layer

The block layer contains code words for the DCT coefficients of a block followed by a fixed-length end-of-block (EOB) code to indicate the end of a block coding. The coefficients are coded using a two-dimensional (2-D) VLC. Not every block in an MB need be transmitted.

- Transform coefficient (TCOEFF). Zigzag scanning and 8-bit FLC or up to 66 different 13-bit VLCs.

- EOB (2-bit, 10) indicating that no nonzero coefficients follow in the block.

## 10.4 SOURCE CODING TECHNIQUES

### 10.4.1 Discrete Cosine Transform

If the video energy of the image of low spatial frequency is slowly varying, then a transform can be used to concentrate the energy into very few coefficients. The transform method chosen by H.261 is the two-dimensional $8 \times 8$ discrete cosine transform (DCT), a transform studied extensively for image compression. The energy is preserved in the transform domain and the signal can be recovered completely by the inverse transform, because it is an orthogonal transform, which implies that the mapping is unique and reversible. Implementing the inverse transform (IDCT) is essentially the same as implementing the forward transform. In fact, the same software and hardware (VLSI) chip designed for the forward transform can be used with minor modification for implementing the inverse transform. Taking advantage of the recursive relations, the same chip can implement different sizes of transforms, such as $(4 \times 4)$, $(4 \times 8)$, $(8 \times 8)$, and $(16 \times 16)$, at video rates.

DCT is by no means an optimal transform. The Karhunen-Loeve transform (KLT), although statistically optimal, has limited applications—the main drawbacks being that it is not a fixed transform and does not have a fast algorithm. DCT, on the other hand, comes very close to the KLT (especially for the first-order Markov signals) in performance. Added to this are the fast algorithms and recursive structures that weight heavily in its favor. One major disadvantage of the DCT is the block artifact that dominates at very low bit rates. This feature is a characteristic of all block transforms.[4]

DCT is a separable transform, as is IDCT. The 2-D DCT can be obtained by performing a one-dimensional DCT on the columns (rows) and a one-dimensional DCT on the rows (columns). An explicit formula for the two-dimensional $8 \times 8$ DCT can be written in terms of the pel values, $f(i, j)$, and the frequency domain transform coefficients, $F(u, v)$.

$$F(u, v) = \left(\frac{2}{8}\right) c_u c_v \sum_{i=0}^{7} \sum_{j=0}^{7} f(i, j) \cos\left[\frac{(2i + 1)u\pi}{16}\right] \cos\left[\frac{(2j + 1)v\pi}{16}\right]$$

$$c_x = \begin{cases} 1/\sqrt{2}, & x = 0 \\ 1, & \text{otherwise} \end{cases} \tag{10.5}$$

The transformed output from the two-dimensional DCT will be ordered so that the mean value, the DC coefficient, is in the upper left corner and the higher-frequency coefficients progress by distance from the DC coefficient. The higher vertical frequencies are represented by higher row numbers and the higher horizontal frequencies are represented by higher column numbers.

The inverse transform is written as follows:

$$f(i, j) = \left(\frac{2}{8}\right) \sum_{u=0}^{7} \sum_{v=0}^{7} c_u c_v F(u, v) \cos\left[\frac{(2i + 1)u\pi}{16}\right] \cos\left[\frac{(2j + 1)v\pi}{16}\right] \tag{10.6}$$

Because the DCT is unitary, the maximum value of each $8 \times 8$ DCT coefficient is limited to a factor of eight times the original values; i.e., an eight-bit input value can be represented by an 11-bit transformed value.

### 10.4.2   DCT/IDCT Mismatch Problem

A block diagram of the H.261 source encoder is shown in Fig. 10.6. It is a hybrid of the DCT and DPCM schemes with motion estimation. The basic reason for inverse transform is that prediction (motion compensation) is performed with the previous reconstructed image. Therefore, forward and inverse transforms coexist in the prediction loop. If the two IDCTs in the encoder and decoder—one for the DPCM loop and the other for reconstruction of the transmitted image—are implemented with different accuracies, the difference between the two IDCT outputs will accumulate, leading to degraded reconstructed images. This is called *DCT/IDCT mismatch error.*[7] A DCT process that comprises irrational number operations requires full-precision arithmetic. Because of the round-off operations and the accumulative nature, any slightly different IDCT outputs cause substantial distortions. The accumulated mismatch error can be removed by a forced updating process, giving a new intra coding mode. In H.261 a macroblock should be updated at least once for every 132 times it is transmitted.

### 10.4.3   Motion Estimation and Compensation

Because most frames in an image sequence look very similar excepting shifts due to movement, such as a pan of the video camera across a scene, the same block need not be coded twice if the displacement vector from the previous image is sent. Motion estimation (ME) can improve the prediction accuracy between adjacent frames/pictures. This technique falls into two categories: pel-by-pel ME, called *pel recursive algorithms* (PRA), and block-by-block ME, called *block-matching algorithms* (BMA).[4]

PRAs have been used rarely because they are inherently complex and the ME algorithms sometimes run into convergence problems. In BMA, motion of a block of pels (say, $M \times N$)

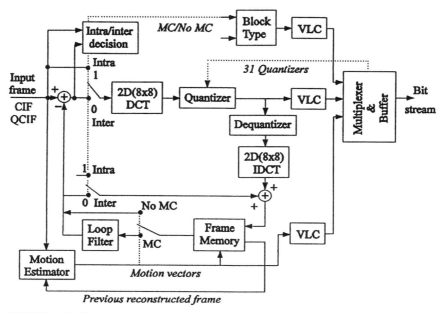

**FIGURE 10.6**    Block diagram of the H.261 source encoder.

within a picture interval is estimated. The range of the motion vector is constrained by the search window. BMA also ignores rotational motion and assumes all pels within the $(M \times N)$ block have the same uniform motion. Under these limitations the goal is to find the *best match* (or least distortion) between the $(M \times N)$ block in the present picture and a corresponding block in the previous picture within a search window, say, of size $(M + 2m_2) \times (N + 2n_1)$, shown in Fig. 10.7. Hence the MV range is $\pm n_1$ pels and $\pm m_2$ lines. In case of fast motion or scene change, ME may not be effective. Also, the pels in the $(M \times N)$ block can conceivably be moving in different directions. When the BMA is no longer useful in the ME process, one can switch to intraframe coding—i.e., no prediction, just 2-D transform of the $(M \times N)$ block. This adaptive mode requires overhead because the decoder has to track the exact mode of operation. ME of a small block size is much more meaningful compared to ME of a large block size. The penalty, of course, is the increased number of bits to represent the large number of motion vectors.

Brute, or full, search implies that for every pel and line displacement within the search window, the cost function has to be computed and compared so as to find the location corresponding to the optimal cost function. The distortion between the block in the present frame and the displaced block in the previous frame can be defined by mean squared error (MSE), mean absolute error (MAE), cross-correlation function, etc. In H.261 the MV range is $\pm 15$ pels or lines per picture, and the ME compares a $16 \times 16$ macroblock in the luminance throughout a small search window of the previous frame. How to design the ME is optional in the encoder. Brute search for integer pel/line MV resolution therefore requires computation and comparison of $(2m_2 + 1) \times (2n_1 + 1) = (31 \times 31)$ cost functions. Extensive investigations have shown that the MAE performs as well as the MSE in MC prediction. The industry has therefore designed and developed the hardware based on the MAE as the cost function given by the following:

$$M(i, j) = \frac{1}{MN} \sum_{m=1}^{M} \sum_{n=1}^{N} \left| X_{m,n} - X_{m+i,n+j}^{R} \right| \qquad |i| \le m_2, |j| \le n_1 \qquad (10.7)$$

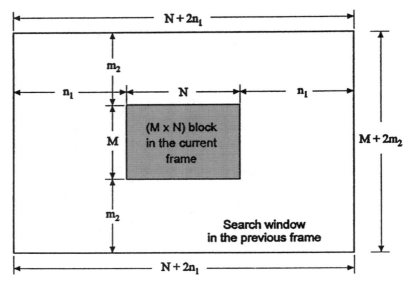

**FIGURE 10.7**    Motion estimation of a $(M \times N)$ block in the previous frame within a $(M + 2m_2) \times (N + 2n_1)$ search window. MV range is $\pm n_1$ pels and $\pm m_2$ lines per frame interval.

The displacement with the smallest absolute macroblock difference, determined by the sum of the absolute values of the pel-to-pel difference throughout the block, is considered the motion-compensation vector for that particular macroblock. The chrominance motion vector is just the luminance motion vector divided in half.

### 10.4.4  Quantizer Design

Two types of quantizers are applied for quantizing DCT coefficients in an H.261 encoder/decoder: a uniform quantizer and a nearly uniform quantizer. The intra DC coefficient is uniformly quantized with a step size of 8 and no dead-zone. Each of the other 31 quantizers for intra AC and inter DC/AC coefficients are quantized by a nearly uniform quantizer with a central dead-zone around zero, as shown in Fig. 10.8. The input between $-T$ and $+T$ ($T$ is a threshold) is quantized to level zero. Except for the dead zone the step size $Q$ is uniform. The step size is an even integer in the range of 2 to 62, which represents 31 quantizers. Because many AC coefficients have near zero levels, the midtread quantizer with zero as one of the output values is used.

The quantization step size represents the distance between possible values of the quantized signal. By varying the step size, the amount of information to describe a particular pel or block of pels can be changed. Larger step sizes result in less information being required, but accuracy is reduced in the representation. Smaller step sizes result in better quality at the expense of increased information to be transmitted.

GQUANT and MQUANT are the related headers in the hierarchical structure. GQUANT has initial information of quantizer step size. The step size is the same for all coefficients within a macroblock but can be changed for each macroblock that has MQUANT information. GQUANT may be replaced by MQUANT in the macroblock layer and the MQUANT by another possible MQUANT.

In the H.261 encoder the input image has 8-bit gray levels. Dynamic range of the levels is changed by the encoding techniques (DPCM, quantizer, and DCT). The sample precisions and dynamic ranges are summarized in Table 10.2.

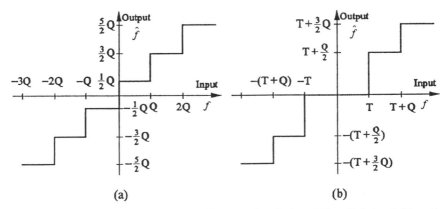

**FIGURE 10.8**  Quantizers in H.261: *(a)* uniform quantizer for intra DC coefficient and *(b)* nearly uniform midtread quantizer for intra AC and inter DC/AC coefficients with a dead zone $2T$. The step size $Q$ can be adaptively changed from MB to MB in increments of 2 from 2 to 62.

**TABLE 10.2**  Dynamic Range of Data at Each Coding Stage

|                  | Precision (bits/pel) | Dynamic range   |
|------------------|----------------------|-----------------|
| Input            | 8                    | 0 to 255        |
| DPCM             | 9                    | −255 to 255     |
| FDCT             | 12                   | −2047 to 2047   |
| Quantized output | 8                    | −127 to 127     |

### 10.4.5  Buffer Rate Control

In the H.261 coder the length of the coded bitstream is dependent on the image properties, complexity, motion, and scene changes. The easy way to control the output bit rate is to change the quantizer step sizes. Reference Model 8 inspects buffer status after coding 11 macroblocks.[8] If the buffer status is full, there is no room for accepting residual information in the frame. One should increase step size by a feedback control signal, as shown in Fig. 10.6. This is called the liquid level control model.[9] If the buffer status is zero, the encoder can use the fine quantizer that requires more bits. The rate control algorithm is left totally to the designer, but the encoder must control its output bitstream to comply with the requirements of the hypothetical reference decoder (HRD). The number of bits for the $(N + 1)$th coded picture $d_{N+1}$ must satisfy the following equation:

$$d_{N+1} > b_N + \int_{t_N}^{t_{N+1}} R(t)\,dt - 4R_{\max}/29.97 \qquad (10.8)$$

where $b_N$ is buffer occupancy just after the time $t_N$. $R(t)$ is the instantaneous video bit rate, and $R_{\max}$ is the maximum bit rate. Note that if one complete coded picture is in the buffer, then all the data for the earliest picture are removed from the buffer.

### 10.4.6  Data Reconstruction

Decision levels for the transform coefficients are not defined in H.261. However, reconstruction levels $\hat{F}_{uv}$ for the quantized levels $Q_l$ are obtained as follows:

- In the case where $Q_n$ is odd:

  If $Q_l > 0$, then

  $$\hat{F}_{uv} = Q_n \times (2 \times Q_l + 1)$$

  If $Q_l < 0$, then

  $$\hat{F}_{uv} = Q_n \times (2 \times Q_l - 1)$$

- In the case where $Q_n$ is even:

  If $Q_l > 0$, then

  $$\hat{F}_{uv} = Q_n \times (2 \times Q_l + 1) - 1$$

  If $Q_l < 0$, then

  $$\hat{F}_{uv} = Q_n \times (2 \times Q_l - 1) + 1$$

- In the case where $Q_l = 0$:

  $$\hat{F}_{uv} = 0$$

where $Q_n$ and $Q_l$ are quantizer numbers from 1 through 31 and quantized levels from $-127$ through 127, respectively. The above equations yield the reconstruction levels as shown in Table 10.3. Quantizer numbers from 1 to 31 are transmitted by either GQUANT in the GOB layer or MQUANT in the MB layer. All reconstruction levels are symmetrical except for the clipped $2047/-2048$. Note that the dead zone around the zero level becomes larger as $Q_n$ increases.

The transform coefficients are quantized in the range of $-127$ to 127 for AC and 1 to 254 for intra DC (level 255 forbidden) and sequentially scanned by zigzag order. The intra DC coefficient is uniformly quantized with a step size of 8 with no dead zone. The resulting values

**TABLE 10.3**   Reconstruction Levels for All Coefficients Other Than Intra DC[3]

| $Q_1$ | | | | | | | | | |
|---|---|---|---|---|---|---|---|---|---|
| | | | | | | $Q_n$ | | | |
| | 1 | 2 | 3 | 4 | ... | 9 | ... | 30 | 31 |
| $-127$ | $-255$ | $-509$ | $-765$ | $-1019$ | ... | $-2048$ | ... | $-2048$ | $-2048$ |
| $-126$ | $-253$ | $-505$ | $-759$ | $-1011$ | ... | $-2048$ | ... | $-2048$ | $-2048$ |
| ⋮ | ⋮ | ⋮ | ⋮ | ⋮ | ... | ⋮ | ... | ⋮ | ⋮ |
| $-3$ | $-7$ | $-13$ | $-21$ | $-27$ | ... | $-63$ | ... | $-209$ | $-217$ |
| $-2$ | $-5$ | $-9$ | $-15$ | $-19$ | ... | $-45$ | ... | $-149$ | $-155$ |
| $-1$ | $-3$ | $-5$ | $-9$ | $-11$ | ... | $-27$ | ... | $-89$ | $-93$ |
| 0 | 0 | 0 | 0 | 0 | ... | 0 | ... | 0 | 0 |
| 1 | 3 | 5 | 9 | 11 | ... | 27 | ... | 89 | 93 |
| 2 | 5 | 9 | 15 | 19 | ... | 45 | ... | 149 | 155 |
| 3 | 7 | 13 | 21 | 27 | ... | 63 | ... | 209 | 217 |
| ⋮ | ⋮ | ⋮ | ⋮ | ⋮ | ... | ⋮ | ... | ⋮ | ⋮ |
| 56 | 113 | 225 | 339 | 451 | ... | 1017 | ... | 2047 | 2047 |
| 57 | 115 | 229 | 345 | 459 | ... | 1035 | ... | 2047 | 2047 |
| ⋮ | ⋮ | ⋮ | ⋮ | ⋮ | ... | ⋮ | ... | ⋮ | ⋮ |
| 126 | 253 | 505 | 759 | 1011 | ... | 2047 | ... | 2047 | 2047 |
| 127 | 255 | 509 | 765 | 1019 | ... | 2047 | ... | 2047 | 2047 |

**TABLE 10.4**  Reconstruction Levels for Intra DC Coefficient[3]

| FLC | Decimal level | $\hat{F}_{uv}$ of intra DC |
|---|---|---|
| 0000 0001 | 1 | 8 |
| 0000 0010 | 2 | 16 |
| ⋮ | ⋮ | ⋮ |
| 0111 1111 | 127 | 1016 |
| 1111 1111 | 255 | 1024 |
| 1000 0001 | 129 | 1032 |
| ⋮ | ⋮ | ⋮ |
| 1111 1110 | 254 | 2032 |

are represented with an 8-bit FLC as shown in Table 10.4. All AC and inter DC coefficients are coded by VLC with "RUN" and "LEVEL" ($Q_l$). H.261 defines levels in the VLC table smaller than those in MPEG because it mainly treats slow-moving video.

The remaining combinations of (RUN, $Q_l$) are encoded with a 20-bit word, consisting of 6 bits ESCAPE, 6 bits RUN, and 8 bits $Q_l$. Quantizer level zero is not necessary to denote a code in both cases, because it only increases "RUN." The other 8-bit FLCs are 2's complement–ordered from −127 to 127.

The H.261 algorithm has been applied to a tennis sequence as an example. Figure 10.9 shows the PSNR at 64, 128, and 256 kb/s and 5 frame/second transmission rate. During a scene change at about the 90th frame, the prediction error is so large that the coding efficiency becomes worse. Figure 10.10 shows the reconstructed 100th frames at 64 kb/s and 128 kb/s. The blocking artifact of the block transform is objectionable in Fig. 10.10 (top) because the human visual system is sensitive to regular patterns such as the staircase effect. Note that the picture quality depends on contents of the sequence.

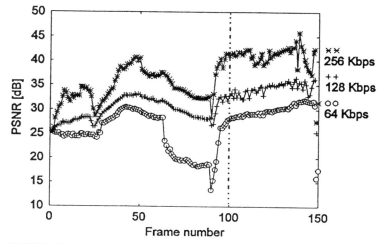

**FIGURE 10.9**  PSNR for "tennis" sequence.

**FIGURE 10.10**   Reconstructed frame 100 of CIF sequence "tennis" at (*top*) 64 kb/s and (*bottom*) 128 kb/s.

## 10.5   PROCESSING MODE SELECTION

### 10.5.1   Intra/Inter Mode Decision

The four mode decisions (intra or inter, MC or no MC, loop-filtering or not, and coded or not coded) have to be included in the header MTYPE (Table 10.1). Intra or inter mode may be decided, for example, on the basis of the energies of the luminance prediction error (Fig. 10.6). If the energy is abnormally large, it implies scene change or fast motion when interpicture prediction even with motion estimation can be ineffective. In this case, full accuracy information of the DC coefficient is needed for the next inter prediction. If the energy is too small, however, it implies that there is no significant change between consecutive pictures and the macroblock need not be transmitted. In inter mode, the input value of 2-D DCT in Fig. 10.6 is the displaced block difference (DBD) when MC is used, and it is the block difference (BD) when no MC is used.

Figure 10.11 shows the two decision areas for intra or inter mode, which was simulated during the standardization.[10] MSE is based on a $16 \times 16Y$ block (macroblock). The "VAR input" is variance of input macroblock. Generally, from Fig. 10.11, the intra area has smaller variance and greater MSE. This curve is empirically optimized. The inter mode includes the boundary line. Although the algorithm is not mandatory, many manufacturers usually employ this approach.

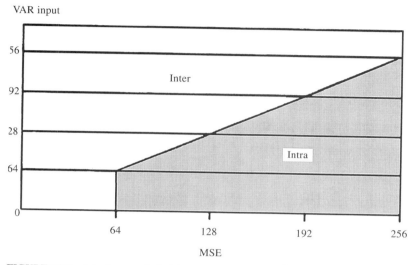

**FIGURE 10.11**    Intra/inter mode decision curve as an example.[8]

## 10.5.2   MC/No MC Mode Decision

The motion compensation is performed on the macroblock with a maximum displacement of $\pm15$ pels/picture allowed in each dimension. The previous frame is stored in the frame memory so that each macroblock can be motion compensated. The motion compensated block is the difference (error) between the best-match block and the current to-be-coded block. If the energy of the motion compensated block with a zero displacement is roughly less than the energy of the motion compensated block with best-match displacement, then zero displacement motion compensation is used; otherwise, motion vector compensation is used. The decoder finds the best-matching macroblock from the previous frame by using the MV and replaces the current MB in this mode.

Although the motion compensation is optional in the encoder, motion estimation and compensation result in an efficient inter mode compression. Thus most of H.261-based codecs employ this technique with frame memory. The basic requisites are as follows:

1. Both horizontal and vertical components of the motion vector must have integer values not exceeding $\pm15$ pels/picture.
2. The ME is based on the $16 \times 16$ luminance block in the macroblock.
3. The MV for the $8 \times 8$ color difference blocks is derived by halving the MV of luminance and then truncating to integer value. For example, if the MV is $(-5, -6)$, then the MV for chrominance is $(-2, -3)$.

The criteria to determine MC/no MC mode can be decided by the relationship of BD and DBD, as shown in Fig. 10.12, which is experimentally optimized. Note that no MC includes the

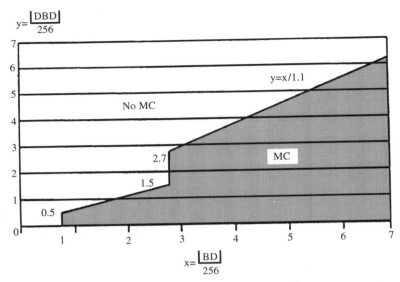

**FIGURE 10.12**    MC/no MC mode decision curve as an example.[8]

boundary line. Denoting $F_n(j, k)$ and $F_{n-1}(j, k)$ as the pel intensities of the current frame $n$ and previous frame $n - 1$, respectively (row $j$ and column $k$), the block difference, BD, is defined as follows:

$$BD = \sum_{MB} |F_n(j, k) - F_{n-1}(j, k)| \tag{10.9}$$

where $\sum_{MB}$ implies summation over the $16 \times 16Y$ block. Similarly, the displaced block difference DBD is defined as follows:

$$DBD = \sum_{MB} |F_n(j, k) - F_{n-1}(j + MV_x, k + MV_y)| \tag{10.10}$$

where $MV_x$ and $MV_y$ are the horizontal and vertical components of the MV of the $16 \times 16$ block.

### 10.5.3  Loop Filter On/Off Decision

Loop filtering (optional) is normally activated to remove the blocking artifacts associated with the motion compensation. The $3 \times 3$ (1 2 1) loop filter processes the data in each $8 \times 8$ block without overlap between the blocks. Loop filtering can be tied with the MC operation; i.e., if the motion vector is nonzero, the filter is on. No separate overhead for the filter on/off is then needed. However, the MC/no MC switch does not always control the loop filter shown in Fig. 10.6. Even in MC mode the LF may not be activated as shown in Table 10.1. A loop filter is optional at the encoder but is mandatory for the decoder implementation.

### 10.5.4  Coded/Not Coded Mode Decision

Any transformed block may be coded or not coded, indicated by CBP (Table 10.1). For example, it is not necessary to code and transmit the macroblock that has the same contents as the previous one. In intra mode all blocks have to be coded and CBP information is not required. In case of no nonzero transform coefficients (TCOEFF) or only motion vector data (MVD), CBP also has no meaning. In Table 10.1, CBP is not transmitted in intra mode and MVD-only mode. This implies that the macroblock has to be coded and transmitted in the two modes. If the CBP signal is defined, only the selected blocks in a macroblock are coded, and the remaining blocks are reproduced from the previous picture.

Figure 10.13 shows codeword lengths of the possible 63 patterns for the four luminance and two color blocks. Note that the most probable (the shortest length) pattern is four luminance

| | | 63 Patterns codework length | | | |
|---|---|---|---|---|---|
| | | $C_b$ $C_r$ | ▢■ | ■▢ | ■■ |
| $Y_0$ $Y_1$ / $Y_2$ $Y_3$ | (0) | - | 5 | 5 | 6 |
| | (4) | 4 | 7 | 7 | 8 |
| | (8) | 4 | 7 | 7 | 8 |
| | (16) | 4 | 7 | 7 | 8 |
| | (32) | 4 | 7 | 7 | 8 |
| | (12) | 5 | 8 | 8 | 8 |
| | (48) | 5 | 8 | 8 | 8 |
| | (20) | 5 | 8 | 8 | 8 |
| | (40) | 5 | 8 | 8 | 8 |
| | (24) | 6 | 8 | 8 | 9 |
| | (36) | 6 | 8 | 8 | 9 |
| | (28) | 5 | 8 | 8 | 9 |
| | (44) | 5 | 8 | 8 | 9 |
| | (52) | 5 | 8 | 8 | 9 |
| | (56) | 5 | 8 | 8 | 9 |
| | (60) | 3 | 5 | 5 | 6 |

■ Coded block ('1')    ▢ Non-Coded block ('0')

**FIGURE 10.13**   Coded block pattern and code-word length for macroblock transmission.[3]

block coding (code length 3). Figures in parentheses are decimal numbers, $D_n$, derived by the following,

$$D_n = 32Y_0 + 16Y_1 + 8Y_2 + 4Y_3 + 2C_b + C_r \qquad (10.11)$$

which implies 63 patterns when the coded block is represented by 1. Note that all zero (i.e., nontransmission) of a macroblock is not defined here, as this can be indicated by macroblock addressing (MBA). The MPEG-1 video-coding algorithm also uses this technique.

## 10.6    TRANSMISSION ENVIRONMENT

### 10.6.1    H.320 Videophone

For the purpose of transmitting the bitstream as per H.261 for videophone applications, the transmission coder includes the other necessary information, i.e., the audio, data, and control signals. The audio signal is processed by separate standards or by proprietary algorithms. ITU-T G.722 deals with ADPCM in the broad band (50 Hz to 7.0 kHz),[11] and G.711 uses A-law or $\mu$-law in the narrow band (300 Hz to 3.4 kHz).[12]

The videophone transmission (see Chapter 27) can be established on two commercial networks: public switched telephone network (PSTN, non-ISDN) and integrated services digital network (ISDN). However, interworking problems should be solved between any two different videotelephone systems. The transmission channel of PSTN is a 3.1 kHz analog channel with modem, whereas an ISDN B-channel is a 64-kb/s digital channel. From this basic difference, some kind of interworking functions—including audio transcoding, end-to-end signaling, and mode switching—are needed between the two videotelephone systems. With the advent of new transmission media, the H.261 has to interwork with them.[13]

The videophones on an ISDN channel, sometimes called *H.320 videophones,* include several recommendations (H.221, H.242, H.230, G.711, G.722, I.400, etc.) for system aspects of

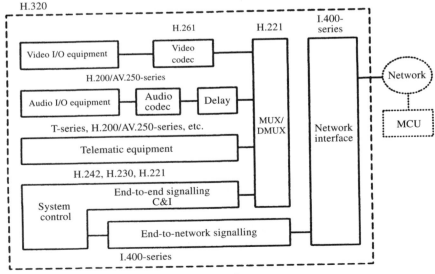

**FIGURE 10.14**    A generic visual telephone system.[3]

audiovisual terminals and systems. MUX/DMUX is used to multiplex/demultiplex coded video, coded audio, and data signals corresponding to bit rates of $p \times 64$ kb/s with a frame structure specified in H.221. The network interface complies with the Recommendation I.400 series, typically $D_0$ (16 kb/s), $D_1$ (64 kb/s), B (64 kb/s), $H_0$ (384 kb/s), etc. System control is used to control the network access, establishing communication by end-to-end signaling.[14]

A schematic diagram of H.320, which covers the technical requirements for narrow-band telephone services defined in H.200/AV.120 series recommendations (channel rates not exceeding 1920 kb/s), is shown in Fig. 10.14. *Video I/O equipment* includes cameras, monitors, and video-processing units (split-screen and other capabilities). *Audio I/O equipment* includes microphones/loudspeakers and audio-processing units (echo cancellation and other capabilities). *Telematic equipment* includes visual aids such as electronic blackboards, still-picture transceivers, etc. *Delay* in the audio path compensates for video codec delay to maintain lip synchronization.

## 10.7   NEW DIRECTIONS

### 10.7.1   Asymmetric Digital Subscriber Line

Although the H.261 standard was designed for two-way videoconferencing, some additional applications may emerge. For example, educational applications are envisioned in which a lecture is encoded in real time and transmitted to remote classrooms and to students at home. At the transmission rate of 1.5 Mb/s, it could provide sufficient picture quality for coverage of the lecturer, graphics from databases, and visuals from an overhead projector.

Another application can be the ADSL (asymmetric digital subscriber line).[15] It adopts a data-over-voice approach and employs frequency-division duplex (FDD) for the two directions of digital data transmission. The high-rate data (digital video) and the low-rate control signal are transmitted over pair-wire lines or fiber/coaxial networks. An H.261 decoder can be employed for the service module at the subscriber side of the ADSL network. ADSL service provides distribution of video, multimedia, and interactive programs to a customer's premises. Initial application of the ADSL will provide video on demand (VOD), whereas further interests can be in interactive educational programming, news and sports services, fax store and forward, and multimedia-based e-mail.

### 10.7.2   ISO/IEC MPEG-4

ISO/IEC MPEG-4 is a response to the expected need to represent audiovisual information for a variety of applications where either the channel has a low bit rate or the storage medium has a limited capacity, but long audiovisual sequences are to be stored. In addition to traditional waveform-based coding methods, more advanced techniques such as object-based coding, model-based coding, etc., are being considered. The standard is expected to reach international standard (IS) stage in November 1998. MPEG-4 was initially aimed at very-low-bit-rate coding (less than 64 kb/s), and its activities during 1995 have resulted in extended applications, tools, algorithms, and profiles. The first verification model (VM) for the MPEG-4 simulation was released in January 1996. The experts group on very-low-bit-rate visual telephony (LBC group) of ITU-T SG15 is finalizing a set of near-term recommendations for PSTN video telephony, named H.263. The transmission rates are typically less than 64 kb/s. The reference model is named the test model for near-term (TMN),[16] providing encoder reference. The video-coding algorithm of the LBC group was developed in close collaboration with MPEG-4, whereas network-related matters (error protection, etc.) and specific requirements for the videophone application are primarily the responsibility of ITU-T SG15.

Contrary to conventional coding methods (which efficiently represent 2-D waveforms of image signals) object-based or model-based coding represents image signals using structural image models, which in some sense take into account the 3-D properties of the scene. A major advantage of these new coding methods is that they describe image content in a structural way. An object-based coder analyzes input images and extracts three parameters: motion, shape, and color information.[17] The analysis is based on source models of moving 2-D or 3-D objects, and object-dependent parameter coding is used. The compression ratio can be increased when the decoder synthesizes the image using these parameters. A model-based approach also relies on object information.[18] The objects are segmented from the background. Initial segmentation is obtained if the image is divided into changed and unchanged regions. Moving objects are detected and modeled as significant information to be transmitted. The procedures might be hierarchical ordering. The decoder receives the output image by means of the hierarchical model parameters. However, to obtain such a model description for a real scene, criteria and algorithms are necessary for performing the segmentation, analysis, modeling, and synthesis. Often, such a whole 3-D modeling is difficult to develop. Therefore, simple and stable algorithms are frequently used. In fact, the block-based (H.263-related) coders outperformed in the subjective evaluation and testing at the MPEG meeting in November 1995.

### 10.7.3   ITU-T H.263

The video-coding algorithm of Draft Recommendation H.2631[19] is an extended version of ITU-T H.261. It is based on motion-compensated hybrid DPCM/DCT coding with considerable improvements to fit bit rates less than 64 kb/s. Reference Model 8 was the last reference model used for simulating H.261, and TMN series is the test model for H.263. The main structure of TMN5 is the same as in RM8. Some important deviations from RMB are as follows:

- Inclusion of various formats such as sub-QCIF and 16CIF.
- Advanced prediction mode: half pel motion estimation, median-based MV prediction, 4 MVs per macroblock, and overlapped macroblock MC.
- Unrestricted MV mode: when MV points outside the picture area, use edge pels.
- Syntax-based arithmetic coding (SAC) mode is possible to change the given VLC tables.
- PB-frames mode (forward and bidirectional prediction) is similar to those in MPEG.
- Weighted quantizer matrix for B-blocks.
- No loop filter. No macroblock addressing.
- 1-bit coded or noncoded macroblock information in MB layer (separate coded block patterns for luminance [CBPY] and chrominance [MCBPC] components and for intra/inter mode).
- 2-bit differential quantizer information in MB layer and 5-bit quantizer information in picture layer and in GOB layer.
- 3-D VLC (last-run level) for coding the transform coefficients.
- VLC for B-blocks.

Application areas of H.263 are aimed at bit rates up to 64 kb/s. The coder may be used for bidirectional or unidirectional visual communications such as videophone, videoconferencing, mobile telephone, etc. Because of these very low transmission rates, no error handling is included in Draft Recommendation H.263.

## GLOSSARY

**ADPCM**   Adaptive differential pulse code modulation.

**ADSL**   Asymmetric digital subscriber line.

**BCH**   Bose-Chaudhuri-Hocquenghem.

**BD**   Block difference.

**BMA**   Block matching algorithm.

**bpp**   Bits per pixel (picture element or pel).

**CBP**   Coded block pattern.

**CBPY**   Coded block pattern for luminance.

**CCITT**   International Telephone and Telegraph Consultative Committee.

**CIF**   Common intermediate format.

**DBD**   Displaced block difference.

**DCT**   Discrete cosine transform.

**DPCM**   Differential pulse code modulation.

**EOB**   End of block.

**FDCT**   Forward DCT.

**FLC**   Fixed-length coding.

**GBSC**   GOB start code.

**GOB**   Group of blocks.

**HVS**   Human visual system.

**IDCT**   Inverse DCT.

**ISDN**   Integrated Services Digital Network.

**ITU-R**   ITU Radiocommunication Sector.

**ITU-T**   ITU Telecommunication Standardization Sector.

**JPEG**   Joint Photographic Experts Group.

**kb/s**   Kilobits per second.

**KLT**   Karhunen-Loeve transform.

**LF**   Loop filter.

**MAE**   Mean absolute error.

**MB**   Macroblock.

**Mbps**   Megabits per second.

**MBA**   Macroblock address.

**MC**   Motion compensation or model compliance.

**MCBPC**   Macroblock type and coded block pattern for chrominance.

**ME**   Motion estimation.

**MPEG**   Moving Picture Experts Group.

**MSE**   Mean square error.

**MV**   Motion vector.

**MVD**   Motion vector data.

**NTSC**   National Television System Committee.

**PAL**   Phase-alternating line.

**PRA**   Pel recursive algorithm.

**PSNR**   Peak signal-to-noise ratio.

**PSTN**   Public Switched Telephone Network.

**QCIF**   Quarter CIF.

**SECAM**   Séquentiel couleur avec mémoire.

**SPA** Significant pel area.

**TMN** Test model for near-term solution (H.263).

**VAR** Variance.

**VLC(D)** Variable length coding (decoding).

**VOD** Video on demand.

## REFERENCES

1. A. Tabatabai, Mills, M., and Liou, M. L., "A review of CCITT $p \times 24$ kbps video coding and related standards," *International Electronic Imaging Exposition and Conference,* Oct. 1990, pp. 58–61.

2. "Draft revision of recommendation H.261: Video codec for audiovisual services at $p \times 64$ kb/s," *Signal Processing: Image Communication,* vol. 2, Aug. 1990, pp. 221–239.

3. CCITT Study Group XV, "Recommendations of the H-Series," Report R37, Aug. 1990.

4. K. R. Rao and Hwang, J. J., *Techniques and Standards for Digital Image/Video/and Audio Coding,* Prentice Hall, Englewood Cliffs, N. J.,1996.

5. K. Jack, *Video Demystified,* Hightest Publications, 1993.

6. S. Nishimura et al., "NTSC-CIF mutual conversion processor," *SPIE, Visual Communications and Image Processing IV,* vol. 1199, Nov. 1989, pp. 885–894.

7. IEEE. *IEEE Standard Specification for the Implementations of 8 x 8 Inverse Discrete Cosine Transform,* IEEE Std. 1180-1990, Mar. 18, 1991.

8. CCITT SG15 WP/1/Q4 Specialist Group on Coding for Visual Telephony, *Description of Reference Model 8 (RM8),* Document 525, June 1989.

9. R. Plompen, *Motion Video Coding for Visual Telephony,* PTT Research Neher Laboratories, 1989.

10. CCITT SG15 WP/1/Q4 Specialist Group on Coding for Visual Telephony, *Description of Reference Model 6 (RM6),* Document 396, Oct. 1988.

11. ITU, *CCITT Recommendation G.722: 7 kHz Audiocoding within 64 kbits/s,* Geneva, Switzerland, 1986.

12. ITU, *CCITT Recommendation G.711: Pulse Code Modulation (PCM) of Voice Frequencies,* Geneva, Switzerland, 1986.

13. Y. Endo et al., "Development of CCITT standard video codec: Visuallink 5000," *NEC Research and Development,* vol. 32, Oct. 1991, pp. 30–38.

14. M. L. Liou, "Visual telephony as an ISDN application," *IEEE Communications Magazine,* vol. 28, Feb. 1990, pp. 30–38.

15. D. W. Lin, Chen, C., and Hsing, T. R., "Video on phone lines: Technology and applications," *Proceedings of the IEEE,* vol. 83, Feb. 1995, pp. 175–193.

16. ITU-T SG15 WP15/1, *Video Codec Test Model, TMN5,* Jan. 31, 1995.

17. M. Hoetter, "Object-oriented analysis-synthesis coding based on moving two-dimensional objects," *Signal Processing: Image Communication,* vol. 2, Dec. 1990, pp. 409–428.

18. M. Buck and Diehl, N., "Model-based image sequence coding," in *Motion Analysis and Image Sequence Processing,* M.I. Sezan and R. L. Lagendijk (eds.), Kluwer Academic 1993.

19. M. Carr et al., "Motion video coding in CCITT SG15—the video multiplex and transmission coding," *Globecom '88,* Hollywood, Fla., Dec. 1988, pp. 1005–1010.

## BIBLIOGRAPHY

Araki, T., et al, "The architecture of a vector digital signal processor for video coding," *ICASSP '92,* San Francisco, Calif., Mar. 1992, pp. 681–684.

Bloom, W., "Compensation for non-uniform illumination in videotelephone images," *IEEE Visual Signal Processing and Communications Workshop,* Melbourne, Australia, Sept. 21–22, 1993.

Carr, M., et al., "Motion video coding in CCITT SG15—the video multiplex and transmission coding," *Globecom '88,* Hollywood, Fla., Dec. 1988, pp. 992–996.

De Sequeira, M. M., and F. M. Pereira, "Global motion compensation and motion vector smoothing in an extended H.261 recommendation," *SPIE, Video Communications and PACS for Medical Applications,* Berlin, vol. 1977, Apr. 1993, pp. 226–237.

Diehl, N., "Object-oriented motion estimation and segmentation in image sequences," *Signal Processing: Image Communication,* vol. 3, Feb. 1991, pp. 23–56.

Dufaux, F., and M. Kunt, "Multigrid block matching motion estimation with an adaptive local mesh refinement," *SPIE, Visual Communications and Image Processing '92,* Boston, Mass., vol. 1818, Nov. 1992, pp. 97–109.

Ebrahimi, T., and F. Dufaux, "Efficient hybrid coding of video for low bitrate applications," *ICC '93,* Geneva, Switzerland, May 1993, pp. 522–526.

Eleftheriadis, A., and A. Jacquin, "Low bit rate model-assisted H.261 compatible coding of video," *IEEE International Conference on Image Processing,* Washington, D.C., Oct. 1995.

Fadzil, M. H. A., and T. J. Dennis, "Video subband VQ coding at 64 kb/s using short-kernel filter banks with an improved motion estimation technique," *Signal Processing: Image Communication,* vol. 3, Feb. 1991, pp. 3–22.

Fujiwara, H., et al., "An all-ASIC implementation of a low bit-rate video codec," *IEEE Trans. on Circuits and Systems for Video Technology,* vol. 2, June 1992, pp. 123–134.

Girod, B., "Motion compensation: Visual aspects, accuracy, and fundamental limits," in *Motion Analysis and Image Sequence Processing,* M. I. Sezan and R. L. Lagendijk (eds.), Kluwer Academic, 1993.

Giunta, G., T. R. Reed, and M. Kunt, "Image sequence coding using oriented edges," *Signal Processing: Image Communication,* vol. 2, Dec. 1990, pp. 429–440.

Guichard, J., et al., "Motion video coding in CCITT SG XV—Hardware trials," *Globecom '88,* Hollywood, Fla., Dec. 1988, pp. 37–42.

Herpel, C., D. Hepper, and D. Westerkamp, "Adaptation and improvement of CCITT Reference Model 8 video coding for digital storage media applications," *Signal Processing: Image Communication,* vol. 2, Aug. 1990, pp. 171–185.

Harborg, E., "A real-time wideband CELP coder for a videophone application," *ICASSP-94,* Adelaide, Australia, vol. 2, Apr. 1994, pp. 121–124.

Ibaraki, H., et al., "Design and evaluation of programmable video codec board," *ICSPAT '94,* Dallas, Tex., Oct. 1994, pp. 922–927.

Jain, P. C., W. Schlenk, and M. Riegel, "VLSI implementation of two–dimensional DCT processor in real-time video codec," *IEEE Trans. on Consumer Electronics,* vol. 38, Aug. 1992, pp. 537–544.

Lin, D. W., M. L. Liou, and K. N. Ngan, "Improvement of low bit rate video coding performance," *Proc. IEEE Workshop on Visual Signal Processing and Communications,* Hsinchu, Taiwan, ROC, June 1991, pp. 1–4.

Liou, M. L., "Overview of $p \times 64$ kb/s video coding standard," *Communications of the ACM,* vol. 34, Apr. 1991, pp. 60–63.

Loos, R. P., et al., "Hybrid coding with pre-buffering and pre-analysis in a software-based codec environment," *Signal Processing: Image Communication,* vol. 3, Feb. 1991, pp. 57–69.

Nicol, R. C., and N. Mukawa, "Motion video coding in CCITT SG XV—The coded picture format," *Globecom '88,* Hollywood, Fla., Dec. 1988, pp. 992–996.

Okubo, S., et al., "Hardware trials for verifying recommendation H.261 on $p \times 64$ kb/s video codec," *Signal Processing: Image Communication,* vol. 3, Feb. 1991, pp. 71–78.

Pang, K., H. G. Lim, and S. C. Hall, "A low complexity H.261-compatible software video decoder," *Australian Telecommunications, Networks, and Applications Conference (ATNAC '94),* Melbourne, Australia, Dec. 1994.

Pereira, F. M., D. Cortez, and P. Nunes, "Mobile videotelephone communications: the CCITT H.261 chances," *SPIE, Video Communication and PACS for Medical Applications,* Berlin, vol. 1977, Apr. 1993, pp. 168–179.

Plompen, R., "Motion video coding in CCITT SG15—The video source coding," *Globecom '88,* Hollywood, Fla., Dec. 1988, pp. 997–1004.

Plompen, R., et al., "Motion video coding: A universal coding approach," *SPIE/SPSE Symposium on Electronic Imaging Science and Technology,* Feb. 1990.

Ruetz, P.A., et al., "A high-performance full-motion video compression chip set," *IEEE Trans. on Circuits and Systems for Video Technology,* vol. 2, June 1992, pp. 111–122.

Sikora, T., T. K. Tan, and K. K. Pang, "A two layer pyramid image coding scheme for interworking of video services in ATM," *SPIE, Visual Communication and Image Processing '91,* Boston, Mass., vol. 1605, Nov. 1991, pp. 624–634.

Sun, M. T., et al., "Coding and interworking for videotelephony," *International Symposium on Circuits and Systems,* Chicago, June 1993, pp. 20–23.

Suwa, A., et al., "A video quality improvement technique for videophone/videoconferencing terminal," *IEEE Visual Signal Processing and Communications, Workshop,* Melbourne, Australia, Sept. 21–22, 1993.

Takishima, Y., M. Wada, and H. Murakami, "An analysis of optimum frame rate in low bit rate video coding," *SPIE, Visual Communication and Image Processing '91,* Boston, Mass., vol. 1605, Nov. 1991, pp. 635–645.

Tirso, A., and M. Luo, "Subband coding of videoconference sequence at 384 kbps," *International Conference on Signal Processing Applications and Techniques,* Santa Clara, Calif., Sept. 28-Oct. 1, 1993.

Westerink, P. H., J. Biemond, and F. Muller, "Subband coding of image sequences at low bit rates," *Signal Processing: Image Communication,* vol. 3, Feb. 1991, pp. 23–56.

Wondrow, M. A., et al., "Role of standard compressed video teleconference codecs in the transmission of medical image data," *IS &T/SPIE Symposium in Electronic Imaging* (abstract), San Jose, Calif., vol. 2188, Feb. 1994, pp. 243–244.

Wu, H., et al., "Real-time H.261 software based codec on the Power Mac," *SPIE/IS&T, Electronic Imaging: Science & Technology,* San Jose, Calif., vol. 2419, Feb. 1995, pp. 492–498.

Yu, G. S., and M. K. Liu, "Temporal and spatial interleaving of H.261 compression for lossy transmission," *SPIE Visual Communication and Image Processing,* Taipei, Taiwan, vol. 2501, May 1995, pp. 1478–1485.

Zhang, X., J. F. Arnold, and M. C. Cavenor, "A study of motion compensated adaptive quadtree schemes for coding videophone sequences on the broadband ISDN," *IEEE Visual Signal Processing and Communications Workshop,* Melbourne, Australia, Sept. 21–22, 1993, pp. 81–84.

## ABOUT THE AUTHORS

Jae Jeong Hwang is an associate professor of radiocommunication engineering at the Kunsan National University, Korea. He received his B.S., M.S., and Ph.D. degrees from the Chonbuk National University, Korea, in 1983, 1986, and 1992, respectively. He spent the academic year 1990–1991 as a visiting scientist at the Hannover and Wuppertal University in Germany and the academic year 1993–1994 as a visiting professor at the University of Texas at Arlington. He is the coauthor of *Techniques and Standards for Digital Image, Video, and Audio Coding* (Prentice Hall, 1996). His current research interests include very-low-bit-rate image and video coding, video data transmission over wireless channels, and ATM networks.

Beom Ryeol Lee received the B.S. and M.S. degrees from the Chonbuk National University, Korea, in 1987 and 1989, respectively. Since 1989, he has been with the Department of Multimedia Research, ETRI. His current research interests lie in video coding, mobile videophone, and real-time computing.

## ACKNOWLEDGMENT

The authors acknowledge that Figs. 10.1 and 10.4 and Tables 10.1, 10.3, and 10.4 have been extracted from ITU-T Recommendation H.261 (03/93) and reproduced with the prior authorization of the International Telecommunication Union (ITU) as copyright holder. The complete ITU publication from which the material was extracted can be obtained from the International Telecommunication Union, General Secretariat—Sales and Marketing Service, Place des Nations, CH-1211 Geneva 20, Switzerland. Telephone: 41 22 730 61 41 (English), 41 22 730 61 42 (French). Fax: 41 22 730 51 94. The authors also acknowledge that this work was supported in part by KOSEF, under Grant 951-0913-081-1.

# CHAPTER 11
# PRINCIPLES AND STANDARDS OF OPTICAL DISC SYSTEMS

**Boudewijn van Dijk and Jaap G. Nijboer**
*Philips Consumer Electronics*
*Optical and Magnetic Media Systems*

## 11.1   THE STORY OF A FAMILY

The Compact Disc (CD) is surely one of the major technological innovations of our era. Beginning as a pure, high-quality sound reproduction system, it rapidly developed into an entire family of systems, with applications extending across the entire landscape of data storage and distribution. This family did not arise by accident. Each member is a fully compatible development of its predecessor, and each has helped to pave the way to succeeding systems (Fig. 11.1).

The precursor of the CD was LaserVision, the LP-sized optical disc with analog video and audio. Originating as a linear playback system in 1978, it went interactive in 1981. By that time, digital optical recording technology was making great strides, leading to the launch of the Compact Disc–Digital Audio system (CD-DA) in 1982.

The first data storage and interchange system, CD-ROM, arrived in 1984. It was conceived as a computer peripheral. Three years later, in 1987, the stand-alone, interactive multimedia system CD-i made its appearance. A link between these two platforms, which in fact gives CD-ROM true multimedia functionality, is CD-ROM XA, born in 1989.

In 1990 the CD family was taken down a significant new path: recordability. CD-WO (write once) allowed users to make their own nonerasable optical disc recordings.

To achieve compatibility between CD-i and other playback systems, CD-i Bridge was established in 1991. This paved the way for Photo CD in the same year, Karaoke CD in the year following, and Video CD (embracing Karaoke CD) in 1993. All these are, in fact, applications of CD-i Bridge.

To complete the story to date, CD-i was extended with full-motion video in 1993, and 1995 saw the advent of the Enhanced Music CD (a mixed application of CD-Audio and CD-ROM XA on a multisession CD) and the rewritable Compact Disc, CD-RW.

The growth of the CD family has always been relatively simple and perfectly logical. It has to be for the technology to thrive in the way it has. Let us examine how it all happened and the current state of the relationships between the various members.

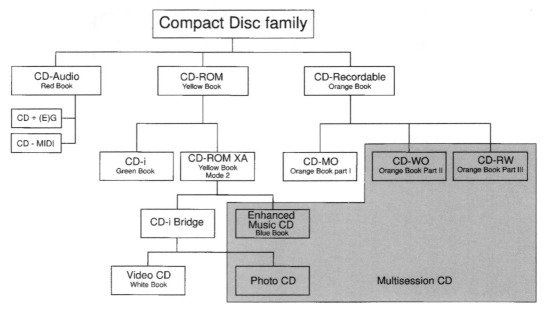

**FIGURE 11.1** CD family tree.

## 11.2 HOW THE CD FAMILY CAME TO BE

### 11.2.1 Optical Recording

The idea of optical storage was conceived in the late 1960s. Among the pioneers were Klaas Compaan and Piet Kramer of the Philips Research Laboratories. In a prerecorded optical disc, the recording consists of a spiral track of microscopic depressions (pits) in a highly reflective surface. With a pitch of about 1.6 μm, the track of an optical disc is 60 times denser than the old LP groove. On a standard 120-mm diameter, 1.2-mm-thick disc, it can be nearly 6 km long (Fig. 11.2).

The reflective surface is of aluminum sputtered on a transparent (polycarbonate) injection-molded disc. Another protective layer is deposited over the aluminum, and this is the surface on which the label is printed.

The recording is read by a laser beam, focused through the transparent disc onto the reflective layer. By a combination of multipath interference and beam diffraction, the pits attenuate the returning beam. The resulting intensity of the reflected light, which constitutes the recorded data, is detected by a photodiode array.

A very simplified explanation of the beam attenuation is as follows: multipath interference arises from the fact that the pit depth is about equal to a quarter wavelength of the laser light. Light reflected from the pit bottom is thus out of phase with light reflected from the surface, and the two tend to cancel out (Fig. 11.3). Furthermore, the width of the pit means that it acts as a diffraction slit, with multipath interference (orders) in several directions (Fig. 11.4). This phenomenon reduces the intensity of the zeroth order (central aperture) of the reflected beam by a minimum of 60 percent.

The detectable data density is limited by the maximum spatial resolution of the read-out spot, which is inversely proportional to the laser wavelength, and directly proportional to the numerical aperture (NA) of the focusing lens (NA $= n \sin \alpha$, where $n$ is the refractive index of the substrate and $\alpha$ is the angle between the optical axis and the outer edge of the focused beam).

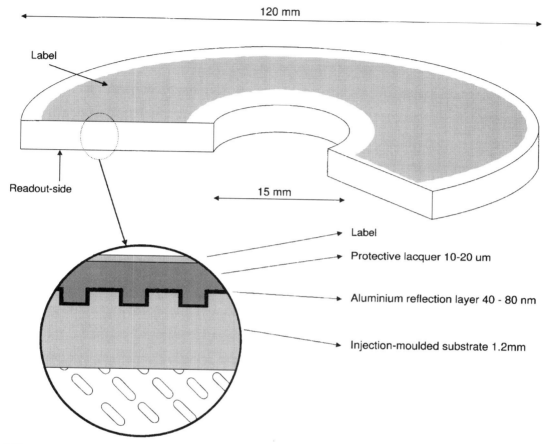

**FIGURE 11.2**    Main structure of a Compact Disc.

Compact Disc systems employ an aluminum gallium arsenide (AlGaAs) semiconductor laser at a wavelength of the order of 780 nm. With an NA of around 0.45, the (half-intensity) diameter of the spot is around 0.9 μm, somewhat wider than the 0.6-μm pit width.

An array of two or four photodiodes is necessary because, in addition to the data to be retrieved, differential signals have to be derived to control the tracking and focusing of the laser spot on the information track. The performance of the tracking servo is critical: as a consequence of eccentricity and turntable tolerances, the track can have a side-to-side swing of 300 μm, but the laser must stay within 0.4 μm of the track axis. Likewise, although the disc may rise and fall by as much as 1 mm, the depth of focus at the reflective layer is only about 2 μm.

## 11.2.2  Audio: CD-DA

By the end of the 1970s, analog sound reproduction was reaching its limits of performance. The dynamic range of even a new LP could be as low as 60 dB, with a channel separation of only 30 dB. These figures look distinctly disappointing today. Furthermore, the phonograph record had always been liable to wear, deterioration, and accidental damage.

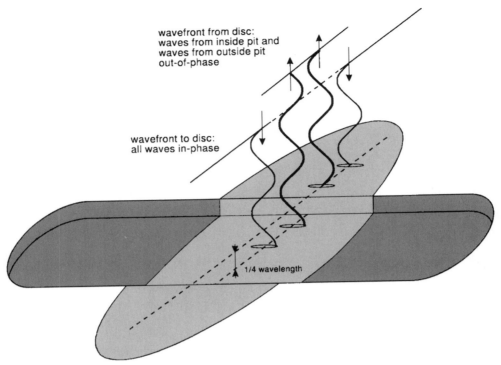

wavefront from disc:
waves from inside pit and
waves from outside pit
out-of-phase

wavefront to disc:
all waves in-phase

1/4 wavelength

**FIGURE 11.3** Light extinction due to phase differences.

Various efforts had already been made to record analog video and audio on optical discs. These led, in fact, to LaserVision. More importantly, they led to CD-DA. Optical recordings, buried inside the disc and read by laser beam, did not suffer from wear. They were well protected from damage. And, by using digital techniques, several new features became possible:

- The discs could be made to very high audio specifications.

- They could be made very easy and convenient to play, with fast access to individual tracks.

- The discs could be very compact (12-cm diameter for more than one hour's play). Consequently, and significantly, the disc players could be very compact too.

- Erroneous data could be corrected by means of an error correction system.

Common interests brought Philips and Sony together to work on this project. Following the successful demonstration of a laboratory prototype by Philips in 1979, the Compact Disc Digital Audio System Description was proposed in 1981. It rapidly gained industry acceptance and became a de facto world standard. Compact Disc–Digital Audio quickly caught the attention of consumers and professionals alike. A major reason for this was that, in establishing the system as a worldwide standard, Philips and Sony were determined, on full support of the licensees, to guarantee compatibility and to create confidence in market development.

***The Structure of CD-DA.*** CD is a digital storage and playback medium. As such, it is based on pits, encoded in a specific way. Although designed for audio, CD has proved flexible enough

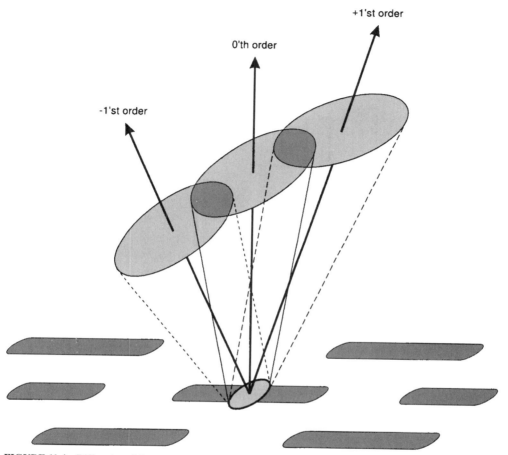

**FIGURE 11.4** Diffraction of the readout spot due to slit behavior of the track.

to adapt readily to the requirements of other types of information, while still maintaining a compatible data and disc structure.

The original requirement was to convert sound (analog audio) into a digital format suitable for registration on a disc, in such a fashion that this format could be restored (decoded) to analog form by a special player.

***Encoding: The Bitstream.*** The first stage of encoding is the generation of a digital bitstream directly representing the sound. The (time- and amplitude-continuous) analog signal is converted into discrete digital samples by sampling and quantization (Fig. 11.5). To allow for a maximum audio frequency of 20 kHz, a sampling frequency of 44.1 kHz was chosen. This is slightly over twice the required maximum audio frequency, and so satisfies Nyquist's sampling theorem.

The sampled amplitude is divided into equal parts by uniform quantization: one word of 16 bits each for both the left and right stereo channels, giving a total of 32 bits (= 4 bytes) per sample—and an inherent signal-to-noise ratio in excess of 96 dB. The resulting bitstream is pulse code–modulated (PCM) and has a net bit rate of $44.1 \times 10^3 \times 32 = 1.41$ Mb/s.

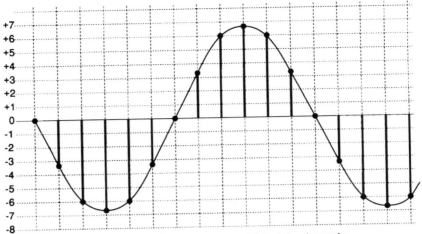

Sampling:  measuring the analog audio values at equidistant time intervals.

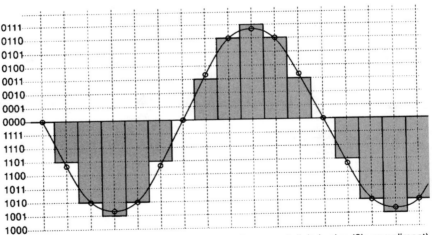

Quantizing:  rounding the sampled audio values to the nearest digital value (2's-compliment).

**FIGURE 11.5**   Sampling and quantization of the audio signal.

***Encoding: Error Detection and Correction.***   To make it possible to detect and correct errors during playback, parity bytes are added to the bitstream. This calls for a second level of structure to enable the parity pattern to be imposed. An audio frame containing six audio samples (24 bytes) was chosen. The error correction code adopted was the Cross-Interleaved Reed-Solomon Code (CIRC). Eight bytes (4 bytes each of CIRC cross-interleaving components C1 and C2) are added to each audio frame. Cross interleaving has the effect of shuffling the order of the data, which spreads out any burst error into smaller errors, enabling the error correction system to correct them.

***Encoding: The Subcode.*** For playback control and display functions, one more byte is allocated to each audio frame. This is the subcode byte. The subcode has its own channel, separate from the main (audio) channel, with its own structure and error detection/correction code.

***Encoding: Eight-to-Fourteen Modulation (EFM).*** In storing the CD signal on the disc, the run length of the modulated-channel bit sequences is limited to a minimum of 3 and a maximum of 11 successive bits having the same value (0 or 1). This is out of spatial frequency considerations, the requirement for self-clocking of the decoded signal, and a wish to optimize the data density.

To satisfy these conditions, EFM is imposed (Fig. 11.6). Each byte (8 bits) of audio, CIRC, and subcode data is transposed by a look-up table into 14 channel bits. To ensure that the proper pattern with run lengths of minimum 3 and maximum 11 channel bits is maintained between units, and to suppress noise in the servo-band frequencies ($< 20$ kHz), 3 merging bits are added to each 14-bit EFM unit. Finally, to allow recognition of individual frames, a sync pattern is added to each frame. In the EFM format a frame is thus composed of 33 EFM codes of 17 bits each, plus 27 synchronization bits. This unit of 588 channel bits is known as an EFM frame.

These overheads result in a channel bit rate of 4.32 Mb/s. With the CD disc geometry, this allows a theoretical maximum playing time of 79 minutes and 58 seconds.

$$T = \frac{\pi \cdot (R_{end}^2 - R_{start}^2)}{v \cdot p} \tag{11.1}$$

where $T$ is playing time, $R_{end}$ is the end radius of track, $R_{start}$ is the start radius of track, $v$ is the linear velocity, and $p$ is the track pitch. To allow for practical limitations and manufacturing tolerances, the maximum playing time is in fact limited to about 74 minutes.

***Encoding: The Subcode Frame.*** The eight bits in a subcode byte each form part of a separate information channel (subcode channels P through W). The P channel marks the lead-in, program, and lead-out areas of the disc, and the intervals between tracks. In the lead-in area the Q channel carries a table of contents (TOC), containing the starting address of each track in the program area, repeated several times. In the program area the Q channel carries track numbers, times, and in-track indices (see Fig. 11.7). In different modes it also registers the manufacturer's disc catalog number and ISRC code. All Q-channel data are protected by a cyclic redundancy check against errors.

Channels R through W are allocated for text and graphics purposes and are used in CD+G and CD-MIDI discs. The data are block structured, and error protection is by Reed-Solomon codes.

One complete cycle of subcode information extends over a block of 98 EFM frames and is known as a *subcode frame*. At 24 audio bytes per EFM frame, a subcode frame thus equates to $98 \times 24 = 2352$ bytes of actual audio data (Fig. 11.8).

The data rate is 75 subcode frames per second, allowing a time code in minutes, seconds, and frames. This is accurate enough to point to the start of a music track. In fact, a big advantage of the CD was quickly perceived to be the ease and convenience with which music tracks could be found and played. It is also usefully related to the 25 and 30 frames per second rates of video recordings.

***Recording.*** By means of a laser, the EFM signal is recorded onto a master disc as a continuous spiral track. To achieve maximum data density, the recording is made at a constant linear velocity (CLV). The recording, which begins at the inner radius of the disc, starts with a lead-in area containing P- and Q-channel subcode, including the table of contents (TOC). Then comes the program area, consisting of up to 99 tracks, followed by the lead-out area, both with P- and Q-channel subcode containing time and track information. From the master, stampers are made for use in the injection molding process, by which the familiar CD is replicated.

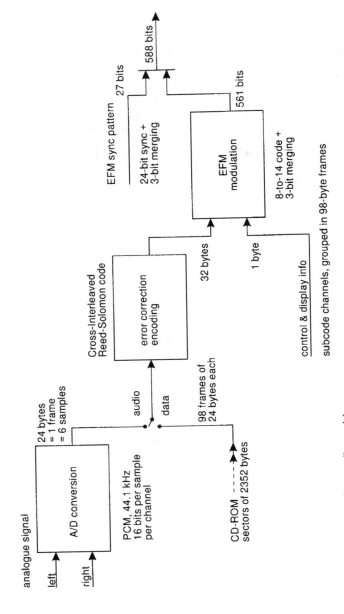

**FIGURE 11.6** CD-DA encoding model.

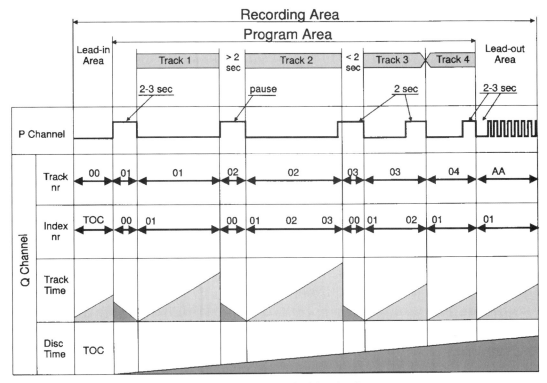

**FIGURE 11.7**  General track layout as given by channels P and Q of the subcode.

***Playback.***   In the CD player the disc again spins at the same constant linear velocity, which means that the turntable motor gradually slows down from about 540 rpm at the beginning, to as low as 200 rpm at the end, depending on the length of the recording.

***Decoding.***   The encoding format, specified in the CD-DA System Description, or Red Book, enables a CD player (Fig. 11.9) to read data from the disc using a laser pickup, and to derive the EFM signal from it.

For good audio quality the main channel data have to arrive at the output of the decoder at a fixed and very constant rate. The EFM data rate, however, is variable, because it is directly proportional to any CLV speed variation on the disc (e.g., due to eccentricity). The turntable motor control system, with its limited bandwidth, is not able to cancel these variations with sufficient accuracy. The demodulated EFM data are therefore forwarded to a first-in/first-out (FIFO) buffer, of which the input is clocked by the EFM bit clock, and the output is clocked by the X-tal system clock. The filling factor of the FIFO buffer is used to control the turntable motor speed and thus the average data input rate from the disc. As the FIFO buffer fills up, the motor is slowed down and vice versa. The data that flow into the error correction circuit are thus delayed, but arrive at a fixed (X-tal clock) rate at the output.

The subcode data and their clock signal are extracted directly from the EFM demodulator, which means that the rate varies proportionally to the CLV speed variations. Because of the variable time delay of the main channel decoding, which depends on the filling factor of the FIFO buffer, and the varying data rate at the output of the subcode processor, there is a variable

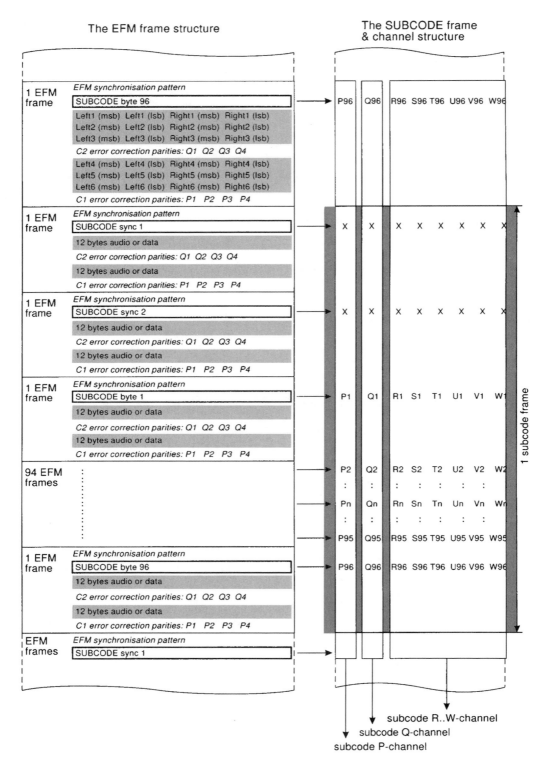

**FIGURE 11.8** EFM and subcode frame structures.

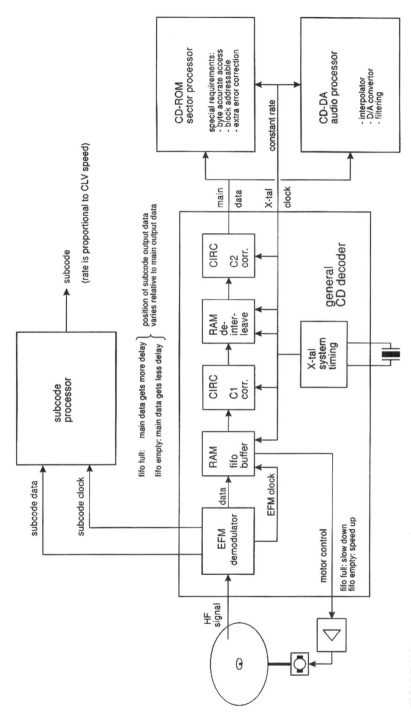

**FIGURE 11.9** General block diagram of a CD player.

time difference between the arrival of the main channel data and the arrival of the subcode data at their respective outputs. For the access and display functions in CD-Audio, these differences are small enough not to be disturbing. For CD-ROM however, as will be seen later, special requirements are needed to improve the access accuracy.

Errors are corrected as far as possible, and the data are de-interleaved. Remaining errors can then be smoothed out by an interpolation scheme. Finally, the reconstituted PCM signal is restored to analog form by a digital-to-analog converter. The effectiveness of this chain of digital signal processing is reflected in the success of CD-DA.

### 11.2.3   Computers: CD-ROM

When CD-DA arrived, personal computers were also in their infancy. In those days, magnetic memories were, by today's standards, small in capacity and costly. The industry was crying out for a bigger, cheaper storage and interchange medium—an alternative to magnetic storage media such as the floppy disk and the hard disk.

Meanwhile, on a CD, it was not sound underneath that polycarbonate shell; it was data. Here, in fact, was a durable, nonvolatile, low-cost storage medium that could hold hundreds of megabytes. Suitably formatted, the disc became a large-capacity, easily installable, removable, portable, mailable read-only memory for computers: CD-ROM. And it retained the essential backward compatibility with CD-DA; CD-ROM drives could play CD-DA discs or tracks. Today, an ever-growing number of CD-ROM drives are in use as computer peripherals.

***The Structure of CD-ROM.***   Sound waves are continuous in time and amplitude. They follow a pattern which, to some extent, is predictable. To reconstruct the pattern, CD-DA uses interpolation to smooth out any error burst too long for the CIRC error correction to handle. And for track finding, addressing only really needs to be accurate enough to find the gaps between the music tracks.

Computer data, on the other hand, is intrinsically random. In general, it has no pattern and cannot be predicted. All errors have to be corrected if the data retrieved are to be usable. Furthermore, fast and bit-accurate access to individual frames (sectors) is essential.

***Sectors.***   CD-ROM therefore uses a 2352-byte subcode-related frame of CD-DA as its basic unit, or sector (Fig. 11.10). To ensure data integrity and to facilitate addressing, further sync, a header, and the facility for extra error detection and correction codes are included, in addition to actual user data (Fig. 11.11).

Data such as audio, video, and graphics—derived from time and amplitude continuous information—do not need the extra error correction. For this type of data, CD-ROM provides

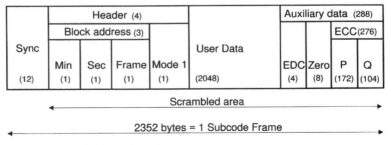

**FIGURE 11.10**   CD-ROM mode 2 sector structure.

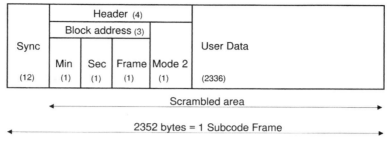

**FIGURE 11.11**   CD-ROM mode 2 sector structure.

a simpler mode 2 sector, which holds 15 percent more data than the regular mode 1 sector. With these provisos, CD-ROM uses the same basic format as CD-DA (Fig. 11.12).

*Access.*   Data are accessed in two stages: the subcode finds the region on the disc, and the required sector is then precisely identified from its address in minutes, seconds, and frames, contained in the sector header. The 4-byte header also includes a mode-indication byte.

As with CD-DA, the information is stored as one long spiral, as opposed to a number of concentric circles as, for example, on a floppy disc. The information is read at a relatively low constant linear velocity compared with a floppy disc's constant angular velocity. In skipping from one point to another on the disc, the angular velocity often needs adjusting before the data can be read. Access times are therefore slower than with magnetic media. However, players that run at double, quad, or even higher speeds are now available to shorten access times and to improve throughput rate.

*File Systems.*   The basic ability to access sectors is still not enough. Computer systems are accustomed to handling files, and different systems have always done so in different ways. In itself, the CD-ROM System Description (Yellow Book) is open and specifies no logical file and directory structure, as different computer systems have their own particular structures. However, to reach a practical degree of compatibility between systems, some form of common structure was deemed by the industry to be necessary. This was supplied by the High Sierra Group, a working group with representation from leading companies. The results of the High Sierra Group's deliberations were eventually expressed in ISO 9660. ISO 9660 specifies a file system that is used, among others, on IBM PC–compatible systems, and has been included in later CD standards. For Unix systems the ISO 9660–compatible Rock Ridge Interchange Protocol (RRIP) has been adopted for CD-ROM. ISO 9660 guarantees the usability of the same CD-ROM disc on different computer systems. For Macintosh computers the proprietary Hierarchical File System (HFS) is often used, whose discs cannot be used on other computer systems.

*ISO 9660.*   ISO 9660 defines a hierarchical file structure, as opposed to the linear organization of tracks. A directory hierarchy allows files to be organized in a more flexible way. As a starting point for access, ISO defines a disc label with a primary volume descriptor (PVD), with a predefined location on the disc. The PVD contains the location of the root directory and a path table that stores the addresses of all files. In this way any file can be accessed by a single seek operation. It is, of course, the static nature of the file system that makes such preprocessing possible.

ISO 9660 supports interleaving of files (Fig. 11.13). In contrast to a track, which is a contiguous sequence of data, a file consists of sectors that need not be stored contiguously. In general, sectors belonging to one file can be positioned on the disc in groups of equal sizes, with gaps of constant size between successive groups. These gaps are available for storage of other files.

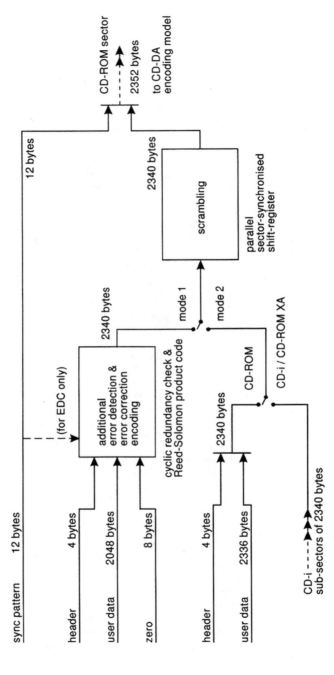

**FIGURE 11.12** CD-ROM encoding model.

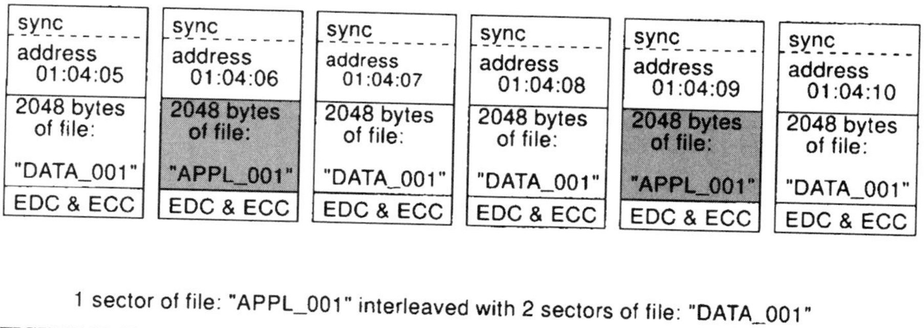

| sync | sync | sync | sync | sync | sync |
|---|---|---|---|---|---|
| address 01:04:05 | address 01:04:06 | address 01:04:07 | address 01:04:08 | address 01:04:09 | address 01:04:10 |
| 2048 bytes of file: | 2048 bytes of file: | 2048 bytes of file: | 2048 bytes of file: | 2048 bytes of file: | 2048 bytes of file: |
| "DATA_001" | "APPL_001" | "DATA_001" | "DATA_001" | "APPL_001" | "DATA_001" |
| EDC & ECC | EDC & ECC | EDC & ECC | EDC & ECC | EDC & ECC | EDC & ECC |

1 sector of file: "APPL_001" interleaved with 2 sectors of file: "DATA_001"

**FIGURE 11.13**    Example of interleaved files in mode 1 sectors.

Unlike other file systems in which files may be scattered over the disc in random fashion, this regular structure not only permits random access to any one sector in a file, but also supports the implementation of what are called *real-time files*. These files are characterized by constraints on the rate at which they can be processed and played back. The constant rate of readout of successive interleaved sectors from disc suffices to overcome such restraints. The ability to read a number of interleaved, real-time files simultaneously—allowing synchronized audio, video, and text processing—is used, for example, in CD-i.

Interleaving can also be employed to economize on disc space. If the processing rate of a real-time file is less than 75 sectors per second, it can be interleaved with other unrelated (real-time) files to fill up the disc space until a joint rate of 75 sectors per second is reached.

*CD-ROM Applications.*    CD-ROM was developed as a peripheral for a range of computer systems, each of which makes its own application arrangements. CD-ROM discs are particularly well suited to publishing large databases, offering a cost-effective alternative to on-line systems. They are also widely used for distribution of software such as technical manuals and works of reference.

In addition to the data track(s) a CD-ROM disc can also contain audio tracks. These discs are known as "mixed-mode" discs. Their application can be found, for instance, in games.

## 11.2.4    Multimedia: CD-Interactive (CD-i)

Not everyone feels comfortable with computers. But everyone could learn the value of a CD-ROM–like data carrier in supplying information—if only it could be made easy to use, and particularly if it could handle the forward-looking concept of multimedia.

Multimedia applications, with their intensive use of audio-visual material, are an interesting application area for CD. Whereas CD-DA works with 1.41 megabits of audio data per second, video is generally far more demanding with respect to storage requirements. For both audio and video, there is usually a trade-off between quantity and quality.

Thus arose CD-i. In this stand-alone concept (a development based on CD-ROM) a complete hardware and software system is specified. The disc contains the application as well as the user data. The player has a built-in computer. And the system—with its own file and directory structure and real-time operating system—can handle a broad range of specified levels of audio and video, as well as text and graphics. The way that the media are to be integrated is also defined, as well as a choice of pointing devices and keyboard user interfaces. At its introduction, CD-i offered the first true multimedia environment.

*The Structure of CD-i.*    An important issue in multimedia applications is real-time operation. The audio-video material has to be played without perceptible discontinuity and must be carefully synchronized. As with CD-ROM, CD-i has two different sector formats—one with additional ECC (error correction code) (form 1) for "error-sensitive" data (Fig. 11.14) and one without additional ECC (form 2) for "capacity demanding" applications such as audio and video (Fig. 11.15). These two formats use the basic CD-ROM mode 2 sector.

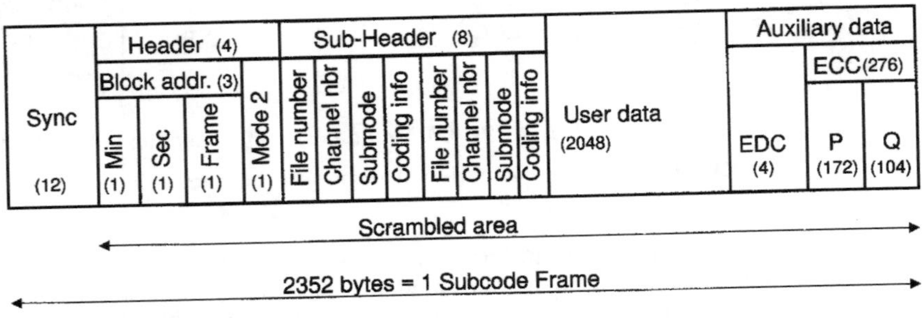

| Sync (12) | Header (4) | | | | | Sub-Header (8) | | | | | | | | User data (2048) | Auxiliary data | | |
|---|---|---|---|---|---|---|---|---|---|---|---|---|---|---|---|---|---|
| | Block addr. (3) | | | Mode 2 (1) | | File number | Channel nbr | Submode | Coding info | File number | Channel nbr | Submode | Coding info | | EDC (4) | ECC (276) | |
| | Min (1) | Sec (1) | Frame (1) | | | | | | | | | | | | | P (172) | Q (104) |

Scrambled area

2352 bytes = 1 Subcode Frame

**FIGURE 11.14**   CD-i form 1 sector structure.

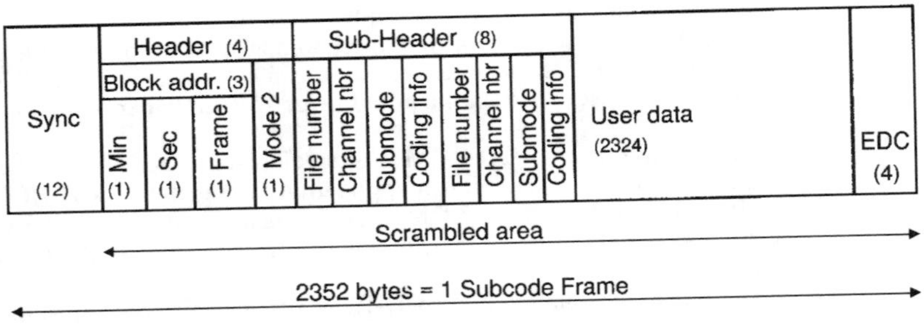

| Sync (12) | Header (4) | | | | | Sub-Header (8) | | | | | | | | User data (2324) | EDC (4) |
|---|---|---|---|---|---|---|---|---|---|---|---|---|---|---|---|
| | Block addr. (3) | | | Mode 2 (1) | | File number | Channel nbr | Submode | Coding info | File number | Channel nbr | Submode | Coding info | | |
| | Min (1) | Sec (1) | Frame (1) | | | | | | | | | | | | |

Scrambled area

2352 bytes = 1 Subcode Frame

**FIGURE 11.15**   CD-i form 2 sector structure.

At the access times involved, these requirements cannot be met by random access to the relevant material on the disc. Consequently, audio-video sequences are stored on the disc in such an interleaved way that they can be played back in a single read.

Synchronization of the various types of data can be achieved by interleaving files (each containing one type of data) or by interleaving the various types of data within one file. For the latter option, CD-i uses the concept of interleaved channels. Files containing such time-critical information are known as *real-time files*.

To process several streams of real-time information synchronously, a single real-time file may consist of channels, which can store various streams of synchronized real-time information. Channels are composed of sectors. The sectors of the channels within one file can be interleaved in arbitrary fashion. One channel, for example, could be used for video, and several other channels for the accompanying audio in various languages. Any combination of channels of one file can be chosen in real time during play.

In order to handle lower data rates, the CD-i System Description (Green Book) allows the scattering of files along the spiral track in a regular structure. In this way files can be read at the required speed while the laser scans the disc at a fixed speed. Scattered files can be interleaved to save disc space. See also the example in Fig. 11.16.

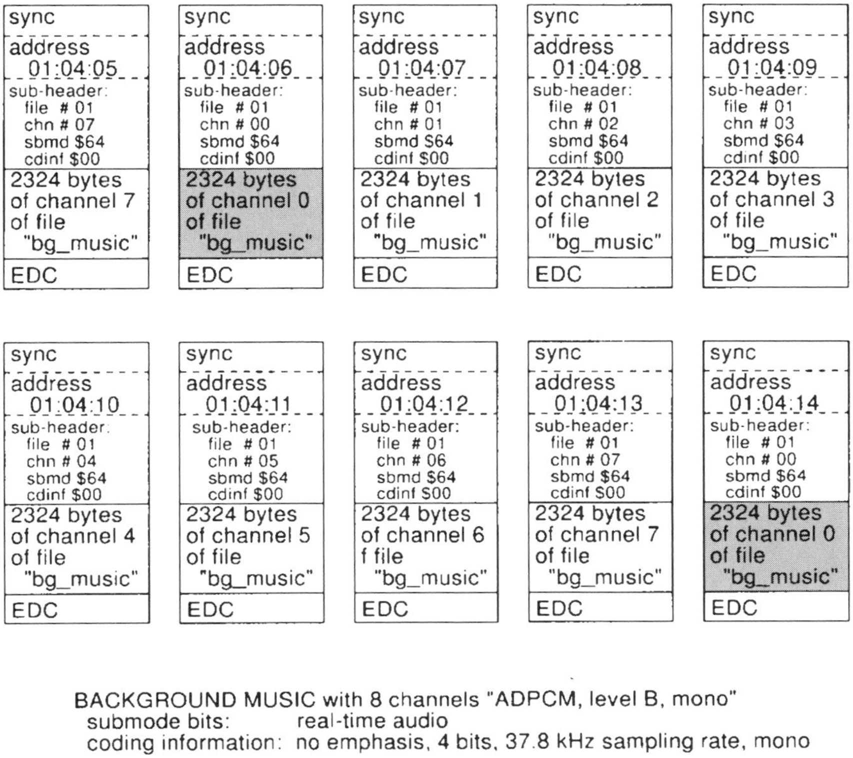

| sync | sync | sync | sync | sync |
|---|---|---|---|---|
| address 01:04:05 | address 01:04:06 | address 01:04:07 | address 01:04:08 | address 01:04:09 |
| sub-header: file # 01 chn # 07 sbmd $64 cdinf $00 | sub-header: file # 01 chn # 00 sbmd $64 cdinf $00 | sub-header: file # 01 chn # 01 sbmd $64 cdinf $00 | sub-header: file # 01 chn # 02 sbmd $64 cdinf $00 | sub-header: file # 01 chn # 03 sbmd $64 cdinf $00 |
| 2324 bytes of channel 7 of file "bg_music" | 2324 bytes of channel 0 of file "bg_music" | 2324 bytes of channel 1 of file "bg_music" | 2324 bytes of channel 2 of file "bg_music" | 2324 bytes of channel 3 of file "bg_music" |
| EDC | EDC | EDC | EDC | EDC |

| sync | sync | sync | sync | sync |
|---|---|---|---|---|
| address 01:04:10 | address 01:04:11 | address 01:04:12 | address 01:04:13 | address 01:04:14 |
| sub-header: file # 01 chn # 04 sbmd $64 cdinf $00 | sub-header: file # 01 chn # 05 sbmd $64 cdinf $00 | sub-header: file # 01 chn # 06 sbmd $64 cdinf $00 | sub-header: file # 01 chn # 07 sbmd $64 cdinf $00 | sub-header: file # 01 chn # 00 sbmd $64 cdinf $00 |
| 2324 bytes of channel 4 of file "bg_music" | 2324 bytes of channel 5 of file "bg_music" | 2324 bytes of channel 6 f file "bg_music" | 2324 bytes of channel 7 of file "bg_music" | 2324 bytes of channel 0 of file "bg_music" |
| EDC | EDC | EDC | EDC | EDC |

BACKGROUND MUSIC with 8 channels "ADPCM, level B, mono"
submode bits:      real-time audio
coding information:  no emphasis, 4 bits, 37.8 kHz sampling rate, mono

**FIGURE 11.16**   Example of CD-i sectors of file with interleaved channels.

Although compatible with the Red and Yellow Books, the Green Book lays down a disc format with five different sector types (audio, video, data, empty, and message). All of these sectors are basically CD-ROM mode 2, whereby the user data include a subheader field with extra addressing information relating to file number, real-time channel number, submode (or sector administration, including selection of the appropriate level of error correction and synchronization), and coding of the data type (Fig. 11.17).

*File Structure.*    Specifications are also laid down for a file and directory structure (which are based on ISO 9660 but are not fully compatible), the audio data, real-time video data, program-related data, and the OS-9–based real-time operating system, CD-RTOS.

*Full-Motion Video.*    An extension of the CD-i standard is full-motion video, coded to the MPEG-1 (Motion Picture Experts Group) standard as defined in ISO 11172 (see Chapter 8). Only CD-i players that conform to the extension can play full-motion video.

*Audio and Video Formats.*    The CD-i audio formats are the pulse code modulation (PCM) of CD-DA and the adaptive differential pulse code modulation (ADPCM) at three audio quality levels. For full-motion MPEG-1 video, the accompanying audio is also encoded according to the MPEG-1 standard, for which the sampling frequency is always 44.1 kHz, whereas the average number of bits per stereo sample ranges from 0.36 to 5.08 bits.

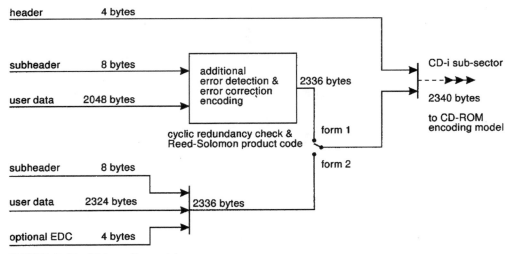

**FIGURE 11.17**    CD-i encoding model.

In the CD-i standard, ADPCM levels A, B, and C are recognized. Levels B and C are also used in CD-ROM XA. Level A uses 8 bits per sample; levels B and C use 4 bits per sample. The sampling frequencies used to measure the audio signal are 37.8 kHz for levels A and B and 18.9 kHz for level C. This brings the bandwidth (the maximum frequency to be reproduced) to 17 and 8.5 kHz, respectively (Table 11.1).

**TABLE 11.1**    CD-i Audio Quality Levels

| Format | Sampling frequency | Bits/sample | Bandwidth | Number of channels | Quality |
|---|---|---|---|---|---|
| CD-DA | 44.1 kHz | 16 | 20 kHz | 1 | Stereo |
| CD-i ADPCM | | | | | |
| Level A | 37.8 kHz | 8 | 17 kHz | 2 | Stereo |
| | | | | 4 | Mono |
| Level B | 37.8 kHz | 4 | 17 kHz | 4 | Stereo |
| | | | | 8 | Mono |
| Level C | 18.9 kHz | 4 | 8.5 kHz | 8 | Stereo |
| | | | | 16 | Mono |

The CD-i video formats are as follows: RGB (red, green, blue), DYUV (delta luminance/chrominance), DYUV+QHY (quantized high-resolution luminance), CLUT (color look-up table), and MPEG-1 for full-motion video. Except for the lowest CLUT with only a limited color range, all can be used to store natural, true-color images. The rate at which data can be read from the disc restricts the use of RGB, DYUV, or DYUV+QHY, for full-motion video, to a part of the screen only. For cartoonlike images and animations requiring just a few colors, CLUT is the usual choice. Such images can be compressed by run-length encoding of sequences of pixels of the same color, instead of encoding each individual pixel.

## 11.2.5  Applications

*Link 1—CD-ROM XA.*   If multimedia programs could be run on CD-i players, then why not on regular (personal) computers? By including a start-up program on a basically CD-ROM disc, CD-ROM XA (extended architecture) makes this possible, as long as the computer system itself supports the required video and audio decoding.

In order to create a certain level of compatibility for audio and video applications, several CD-i system options have been included in the CD-ROM XA System Description. These include, among others, audio compression (ADPCM levels B and C), multichannel audio, some video formats, and file interleaving. CD-ROM XA therefore also adopted the mode 2 form 1/2 sector format of CD-i.

*Link 2—CD-i Bridge.*   If a CD-i application program is added to a CD-ROM XA disc, that disc can be played on both computer systems and CD-i players. And if certain files containing information concerning disc contents and applications are also given fixed locations on the disc, it can also be played on players employing simple microcontrollers, and dedicated to particular applications. CD-i Bridge provides this opportunity. An open standard itself, it allows specific additional details to be defined for individual applications. Photo CD and Video CD are both products of this concept. As such, they are not truly systems in themselves, but CD-i Bridge applications.

*Link 3—Multisession CD.*   Originally defined for CD–Write Once, the multisession technique now also finds applications on prerecorded media. For a CD-WO disc to be readable by standard CD-ROM drives, it needs to have a lead-in area with the table of contents in the subcode. This TOC, however, can only be generated when the disc is finished. The process of finishing the disc and writing the lead-in and lead-out area is called *finalizing* the CD-WO disc, and it can also be performed on a partially written disc. To allow the unwritten part of the disc to be used later, the technique of *sessions* was introduced.

A session can be seen as a "complete disc" within a CD, with its own lead-in area, program area, and lead-out area. A CD can contain more than one session and is then called a *Multisession CD*. On such a disc, all lead-in areas, except the final one, contain a pointer to the next (possible) session. By using these pointers a Multisession-Compatible CD-ROM drive can access the full contents of the disc.

An example of Multisession CD is the recordable Photo CD, in which each roll of developed film is put into one session on the disc. The disc is filled up step by step as films are shot and developed, and all sessions can be read on CD-ROM and CD-i drives.

Another powerful feature of Multisession CDs is used in the Enhanced Music CD. Normal CD-Audio players can only access the first session on such a disc, which is a normal audio session. The second session, a CD-ROM XA session—with additional information that can be handled by CD-ROM drives in (personal) computers—is thus automatically hidden from CD-audio players.

*Video CD.*   Also based on CD-i Bridge, Video CD stores digital video by using the CD-i full-motion video extension. Consequently, the pictures are encoded in conformance with the MPEG-1 standard as defined in ISO 11172. By using the compression rate of MPEG-1, 74 minutes of full-motion, full-screen video (with accompanying audio) can be recorded on a 12-cm Video CD disc. Video CD is defined in the Video CD specification, also known as the White Book. Major applications of Video CD are linear movies, music video, and Karaoke.

Besides the ability to play on a variety of systems (including CD-i and Video CD players; and personal computer systems equipped with a CD-ROM drive and the appropriate hardware or software for MPEG decoding), the specification explicitly mentions the possibility of connecting a full-motion video adapter to the digital output of a CD-DA player. In such a case, however, the player would have to be able to accept non–CD-DA discs and to pass nonaudio information to its digital output. (Note: Video CD should not be confused with the earlier CD Video, which is a hybrid of CD-DA and LaserDisc).

***Photo CD.*** The CD-i Bridge–based system, jointly developed by the Eastman Kodak Company and Philips, stores high-resolution photographic images for display on a TV set or print-out on a hard-copy device. It provides a means of scanning photographic film, processing the images thereon, and recording them as a series of digitally coded images of different resolutions on a Photo CD disc. These discs may be either prerecorded or CD–Write Once (CD-WO) discs. In the latter case, images can be recorded on disc in a number of sequential sessions. Prerecorded discs may also contain more than one session.

The highest-resolution image contains $3072 \times 2048$ picture elements or pixels. However, images may be acceptable at lower resolutions, depending on the intended use. Each image is therefore stored at five resolutions, obtained by repetitively decreasing the highest-resolution image by a factor of two in both the horizontal and vertical directions. In increasing order of resolution, these are called "Base/16," "Base/4," "Base," "4Base," and "16Base" (Fig. 11.18). The first three of these are stored directly on the disc: for many video systems these are easy to access and may have all the necessary image quality. For 4Base resolution, only the difference between an interpolated Base image and the complete 4Base image is stored, using variable-length encoding. Likewise, an interpolated 4Base image is used to store a 16Base image. On this basis, a 12-cm Photo CD disc can contain more than 100 images.

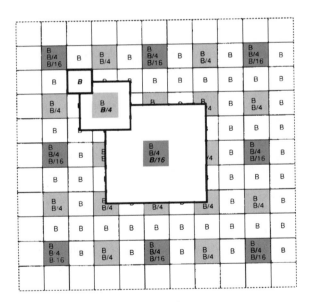

All fields constitute the Base picture,
all gray fields constitute the Base/4 picture,
the dark gray fields constitute the Base/16 picture.

**FIGURE 11.18** Resolutions on Photo CD.

Photo CDs are well suited for the photographic archives of both consumers and professionals. The standard also supports photo presentations with accompanying audio.

***CD–Background Music.*** A rather different CD-i–based system, CD–Background Music is a purely audio application using CD-i ADPCM audio coding with interleaving to realize a low

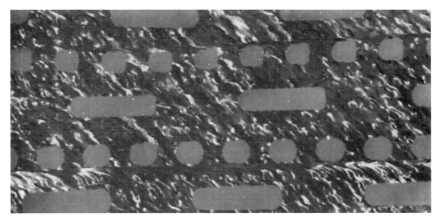

**FIGURE 11.19**    Pit structure of CD-RW disc.

bit rate (Fig. 11.16). Because of the ADPCM compression, only 1/8 of the CD data rate is needed. By applying eightfold interleaving, this data rate is reached with one channel, while playing the disc at nominal speed, without any waste of capacity. Depending on the quality level required, a single disc can contain more than 20 hours of music. CD–Background Music discs only play on dedicated players.

***Enhanced Music CD.***    The Enhanced Music CD, also referred to as *CD-Extra,* is the latest branch on the ever-growing CD family tree. The disc is a Multisession CD with a CD-Audio session and a CD-ROM XA session. The CD-Audio session is in itself a "complete Compact Disc" with a lead-in, program, and lead-out area. It will play on any CD-Audio player.

The CD-ROM XA session, a "second CD," contains data files related to the audio tracks in the first session, with information such as the title of the album, titles of the songs, lyrics, video, artwork, graphics, etc. The CD-ROM XA session can be used in computer systems equipped with a CD-ROM drive, or in dedicated Enhanced Music CD players.

The format, as laid down in the Enhanced Music CD specification, or Blue Book, enables full-featured interactive use of the disc on multimedia computer systems, although it still behaves exactly like a CD-Audio disc on CD-Audio players.

***Other Derived, Prerecorded Standards.***    Certain other read-only standards, derived in whole or in part from CD, although less widely used, are worth mentioning. CD-Video (CD-V) is specified as an addition to the Red Book. It can store up to 5 or 6 minutes of analog video in combination with CD-Audio (the same way as LaserDisc) together with 20 minutes of CD-Audio on a 12-cm disc. CD-Video is not to be confused with Video CD.

CD+G, CD+EG, and CD-MIDI are also based on CD-Audio. Through its subcode channels R through W, the CD-DA standard allows for encoding of music-related data. CD+EG defines the way in which simple text and (extended) graphics can be recorded for display on either a player display or a TV set. CD-MIDI similarly defines the way of recording data conforming to the musical instrument digital interface (see Chapter 24).

### 11.2.6    CD-Recordable Systems

All the systems so far described are prerecorded systems and lack the facility of recording enjoyed by tape systems. CD-Recordable systems address this deficiency.

***CD-MO.***   The first version of CD-Recordable, CD-MO, used a magneto-optical recording technology. MO technology is not compatible with the basic optical parameters, like reflectivity, of CD. As a result, it has not gained wide acceptance. It should be noted, however, that the MiniDisc system (see Chapter 23) is based on CD-MO–like technology.

***CD-WO.***   CD-Write Once (CD-WO) has the advantage of full compatibility with all prerecorded CD systems. It also introduces the additional concept of multisession (hybrid) discs. The key to CD-WO is an organic dye coating (the photoabsorption layer) applied over a substrate containing a wobbled tracking groove. The wobble frequency of the groove is FM-modulated with time-code information. The average wobble frequency is used to control the turntable motor speed, whereas the time-code information is used to position recordings on the disc.

The organic dye coating, in its turn, is covered with a reflective layer. The coating is initially transparent, and for recording it is heated by a laser beam. When the intensity of the laser spot passes a certain threshold, a bump appears in the layer. This is an irreversible process that drastically alters the optical characteristics. Using this technique, a pit (or rather, bump) pattern is written in the tracking groove by a relatively high-power laser in a dedicated recorder. The laser power required for recording is typically an order higher than the 0.5-mW laser power used for reading.

*Sessions.*   CD-WO sessions can be recorded at different times, and even on different recorders. Each session has its own lead-in, program, and lead-out area, and can itself be recorded in parts over any period of time. One session must be finalized before another can be begun, and every single session on a disc must conform to the CD-DA, CD-ROM, CD-ROM XA, or CD-i standard. The format of each session is indicated in the subcode in the lead-in area of that particular session. The first session of a CD-WO disc can be played by a regular player of the appropriate type, but a multisession player is needed to play all the sessions.

*CD-WO Applications.*   CD-WO offers speedy, convenient, large-capacity, long-term storage at a relatively low cost per megabyte. It can be used for do-it-yourself, one-off applications such as application prototype and mastering, data storage and retrieval, low-volume software distribution, in-house publishing, photo file archiving, and simple presentations.

***CD-ReWritable (CD-RW).***   The year 1995 saw the announcement of the CD-ReWritable (CD-RW) system description. CD-RW may be seen as a logical extension of the CD-Recordable series of systems based on the same pregroove structure as CD-WO with just another type of recording layer. CD-RW drives will be able to write, read, and rewrite CD-RW discs, as well as writing and reading CD-WO discs and reading all CD-ROM discs. Whereas CD-WO finds applications in small office environments, such as large-file exchange and archiving, CD-RW fulfills the role of back-up medium in business-critical applications and personal storage, as if it were a huge floppy disk. With only a minor modification in electronics, which all CD-ROM drive manufacturers can easily implement, future CD-ROM drives will also be able to read CD-RW discs.

*How CD-RW Recordings Are Made and Erased.*   An erasable CD technology must, of necessity, be in harmony with the existing CD systems. Philips therefore turned away from incompatible magneto-optics and persevered with phase-change technology, by itself quite challenging in compatibility terms. "Phase," in this context, refers to the physical aggregation state of the material used for the recording layer of the disc. There are two possible phases, with quite different optical properties. A low-reflectance domain of amorphous, or patternless, phase (equivalent to a CD pit) is produced when a laser heats the recording material rapidly above its melting point of 500 to 700 °C (Fig. 11.19). Cooling very quickly, the amorphous domain "freezes." On the other hand, if the recording material is heated to a lower temperature for a somewhat longer time, a higher-reflectance domain of polycrystalline phase is formed, equivalent to a CD land.

The life of CD-RW discs under this heat treatment is at present about 1000 write/erase cycles. Although this is considerably shorter than the life of MO discs, it can be expected to increase, and is in any case adequate for the foreseen applications.

*Readout Considerations.*   Phase-change tracks are read in the same way as regular CD tracks. The readout mechanism does no more than detect the transitions between low and high reflectivity and measure the length of the periods between those transitions. Although the reflectance is lower than for regular CDs, the relative proportions of light reflected from the amorphous and polycrystalline phases remains the same.

A question of gain therefore arises. The CD's reflectance specifications—70 percent minimum for lands, 28 percent maximum for pits—allowed the relatively insensitive photodiodes of the early 1980s to read the signal pattern reliably. But with these specifications, the development of a recording material is a practically impossible task. If 70 percent of the light is reflected, only 30 percent at most remains to change the phase. Today these levels are not necessary because modern photodiodes are able to detect much lower reflectance differences. All that is needed is the correct reflectance ratio and adequate amplification.

A realistic erasable system has to work on reflectivities of about one-third of the original CD figure. Once this constraint was accepted and written into the specification, phase-change CD-RW became a perfectly practical and reliable proposition. The new CD-R/RW drives (CD-WO drives with the CD-RW extension) will write and rewrite the CD-RW discs, as well as write CD-WO discs. And future CD-ROM drives, fitted with automatic gain control, will be able to read them as well.

The recording material for CD-RW consists of a layer of silver, indium, antimony, and tellurium. The polycrystalline phase reflects about 20 percent of the light; the amorphous phase only 8 percent. That meets the requirement for a minimum modulation of 60 percent and allows for writing by a laser of 10 to 15 mW.

CD-RW thus conforms to the original CD specification except in one small respect. The discs have equal dimensions and store the same quantity of data. The drives embody a small modification, but in other respects are identical. Once they are in quantity production, CD-R/RW drives need not be more expensive than their CD-WO predecessors.

*CD-RW Applications.* CD-RW is not intended to supersede CD-WO but rather to do the work that CD-WO cannot conveniently do, such as editing applications. All future Philips CD-R/RW drives will be designed to work with both types of disc.

### 11.2.7  Relationships in the CD Family

The way in which the members of the CD family are related is shown in three-dimensional form in Fig. 11.20. Each of the three different planes represents one of the CD media fields: Prerecorded, Write Once, and ReWritable. In each field the same basic CD systems occur: CD-DA (Red Book); CD-ROM (Yellow Book), including CD-ROM XA; and CD-i (Green Book). The interrelationships among these main systems is depicted by the standards which span them—Enhanced Music CD (Blue Book) and CD-i Bridge, with its present applications of Video CD (White Book) and Photo CD.

In addition, the three fields are all intersected by two planes of functional linkage: multisession and ISO 9660. The Multisession CD standard applies in general to systems in all three fields. ISO 9660 also applies to all three fields, but only with respect to the CD-ROM branch and, to a limited extent, to CD-i (as indicated by the broken line in Figure 11.20).

## 11.3  A ROAD MAP

Computers and consumer electronics have been converging for years now, a tendency accelerated by the development of the CD optical disc family, which opened the way to the recent explosive growth of multimedia systems. As the road map of Fig. 11.21 shows, digital video developed along a separate path from CD-DA, and is now in the market in the form of CD-i, Video CD, and MPEG-1 compression under ISO 9660.

As this convergence continues, the need is growing in the film industry for a new high-capacity video disc, and in the computer industry for an equally high-capacity ROM disc. Both sets of requirements are likely to be met by a new high-density disc, which is in essence a ROM

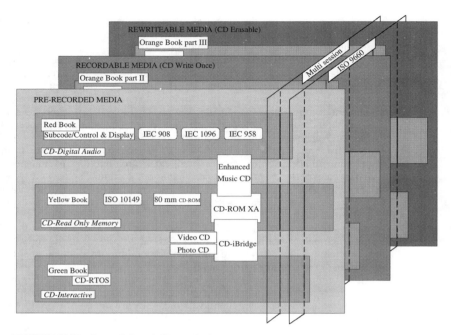

**FIGURE 11.20**  Interrelation of CD standards.

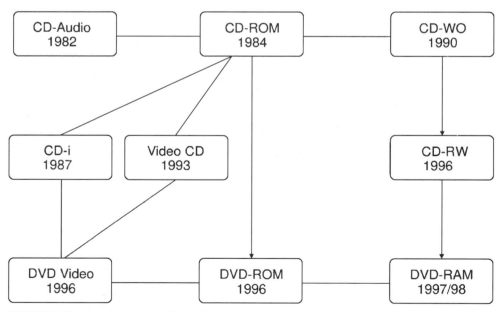

**FIGURE 11.21**  A road map toward future developments.

equally fitted for digital video and computer applications (see Chapter 12), as indicated at the bottom of Fig. 11.21.

For prerecorded media, the reduced pit and track dimensions and the improved format structure of high-density discs will mean greatly increased data capacity. This will lead to longer video playing times, allowing practically all films to be accommodated on single-disc albums, and an equal potential for far more sophisticated software programs.

The write-once and rewritable media will also benefit from the improved structure of the high-density format. For data retrieval, the recommended volume and file structure will be in accordance with the new universal disc format (UDF) standard (released by OSTA), leading to better compatibility with future recordable systems.

High-density rewritable CDs could also have the capability to record digital (MPEG-1 and MPEG-2) video, subject of course to satisfactory copying constraints. And there are other ways of making video recordings than copying—from camcorders, for instance. Future PCs, with high-density WO/RW drives, will be powerful tools for re-editing digital video material from a variety of sources. Fast and easy random access to recordings on disc (as opposed to tape), and virtually unlimited archival life, are solid benefits that will be felt by both consumers and professionals in the computer and video sectors.

The technologies developed under the auspices of the Compact Disc have played, and will continue to play, a major role in the phenomenon of multimedia consumer electronics.

## GLOSSARY

**Access time**    The time required by a CD-ROM drive to read the data requested from the CD-ROM disc and send it back to the computer.

**ADPCM**    Adaptive differential pulse code modulation is a method of compressed audio data storage in which it is not the value of the signal that is stored, but the difference from the previous sample (or measurement). This means that only 8 or 4 bits per sample are used, rather than the normal 16 bits. In the CD-i standards, the levels A, B, and C are recognized. Levels B and C are also used in CD-ROM XA. Level A uses 8 bits per sample and levels B and C use 4 bits per sample. The sampling frequencies used to measure the audio signal are 37.8 kHz for levels A and B, respectively, and 18.9 kHz for level C. This brings the bandwidth (the maximum frequency to be reproduced) to 17 and 8.5 kHz. For normal linear PCM the number of bits is 16, the sampling frequency is 44.1 kHz, and the audio bandwidth is 20 kHz. Using ADPCM, a maximum 16-fold reduction in storage requirement relative to linear PCM can be achieved (level C, mono).

**ASCII**    American Standard Code for Information Interchange: a coding scheme that represents characters in an 8-bit binary format. Almost every language uses the same coding for the first 128 symbols in this table. Different tables are in use for ASCII numbers 128 to 255.

**Backup**    A reserve copy of information so that, in case of emergency, the old situation can be restored.

**Bit**    The smallest unit of information within computer systems. Contraction of "binary" and "digit." A binary digit has the value 0 or 1.

**BER**    Bit error rate; expressed as the average ratio of the number of erroneous bits and the total number of processed bits. For CD-ROM the figure is typically about $10^{-15}$.

**BLER**    Block error rate; expressed as the number of blocks (= EFM frames) with at least one error, against the total number of blocks processed.

**Block**    In CD-ROM technology the data are stored on the CD-ROM in blocks, sometimes called *sectors* or *frames,* of 2048 bytes. Apart from the user-data, extra information is added (see *Mode 1/2, Form 1/2*).

**Byte**    A symbol or character consisting of 8 bits.

**CAV**    Constant angular velocity; a fixed number of disc revolutions per second.

**CD-DA** or **CD-Audio**    Compact Disc Digital Audio; the "normal" audio CD. The standard is defined by Philips and Sony in the Red Book.

**CD-i**    Compact Disc Interactive; a system for presenting information (text, images, and video) on a television screen. The standard is defined by Philips and Sony in the Green Book.

**CD-i Bridge disc**    A special type of CD-ROM XA disc with a CD-i application program. By using the CD-i Bridge disc concept, the discs can be played on a variety of players, such as CD-i, CD-ROM XA, etc. Examples are Photo CD and Video CD.

**CD-ROM**    Compact Disc–Read-Only Memory; a storage medium that can only be read, based on CD-Audio technology. Maximum capacity in mode 1: for 74 minutes, 681,984,000 bytes = 650 megabytes; for 63 minutes, 580,608,000 bytes = 553 megabytes, where 1 min = 60 s, 1 s = 75 frames, 1 frame = 2048 bytes. The standard is defined by Philips and Sony in the Yellow Book.

**CD-ROM XA**    "XA" stands for *extended architecture* and is a standard for CD-ROM in which a number of options from CD-i have been added. These include audio compression (ADPCM), multichannel audio (maximum 16), file interleaving, and image compression. A CD-ROM XA disc is a mode 2 disc in which the data are located in form 1 (2048 user bytes/sector) or form 2 (2324 user bytes/sector) sectors.

**CLV**    Constant linear velocity; a condition in which a uniform relative velocity is maintained between the disc and the pickup.

**Data compression**    A technique in which as much redundant information as possible is removed from the data; e.g., a repetitive sequence can be stored as the sequence value and the number of repetitions.

**Data conversion**    Method of transforming data from one electronic storage format into another format; e.g., converting a word processor text into ordinary ASCII code.

**Directory**    A file that contains information (name and location) about the files on a disk. Used with almost every storage medium (floppy, hard disk, CD-ROM).

**ECC**    Error correction code; redundant information used to correct erroneous bytes.

**ECMA**    European Computer Manufacturers Association; ECMA has close working relations with European and international standardization bodies such as IEC and ISO.

**EDC**    Error detection code; a check-sum over the bytes in a sector or frame. This enables decoders to conclude immediately that an error has occurred during the reading of the information. With the use of ECC, errors of this kind can be corrected to a certain extent. EDCs are applied for instance in the CD subcode and in the CD-ROM data sectors.

**EFM**    Eight-to-fourteen modulation; a method of modulation of 8 bits into sequences of 14 bits. These groups of 14 bits are "linked together" by 3 merging bits.

**Floppy disk**    A removable magnetic medium in a synthetic jacket. Information can be stored on it using a read/write head. The information is rewritable.

**FMV**    Full-motion (full-screen) video.

**Form 1**    Used in CD-i and CD-ROM XA; division of a mode 2 sector into sync (12 bytes), header (4 bytes), subheader (8 bytes), user data (2048 bytes), EDC (4 bytes), and ECC (276 bytes). This layout is used for normal data files (including photo CD).

**Form 2**    Used in CD-i and CD-ROM XA; division of a mode 2 sector into sync (12 bytes), header (4 bytes), subheader (8 bytes), user data area (2324 bytes), and EDC (4 bytes) or reserved. This layout is used in files where error correction is less important and the highest possible data rate is needed, such as (compressed) audio or moving images.

**GB**    Gigabyte, or 1024 megabytes.

**Hard disk**   A permanent (nonremovable) storage medium for computer data, based on a rotating disc with a magnetically sensitive layer. Information can be written on this and read again, using a read/write head. The information stored is rewritable.

**HFS**   Hierarchical file system used by Apple. Used for floppy and hard disks and for CD-ROM.

**High Sierra**   The predecessor of the ISO 9660 standard; published by the CD-ROM Ad Hoc Advisory Committee, also known as the High Sierra Group, on 28 May 1986. Use of this "standard" is no longer recommended. ISO 9660 is preferred.

**IEC**   International Electrotechnical Commission. See also **ISO.**

**Index**   A separate list of words or keys, sorted alphabetically or numerically, along with a reference to their location in the text of a database.

**Injection molding**   The process in which polycarbonate is forced under pressure against the stamper to produce a CD.

**Interactive media**   Media in which the user is required/expected to take action to find information, or to be provided with information.

**Interface**   Point of contact or border surface between two systems. These can be items of equipment (e.g., SCSI interface between computer and CD-ROM player) or software modules (e.g., user interface).

**ISO**   International Standardization Organization. The ISO, along with the IEC, forms a system for worldwide standardization as a whole.

**IT**   Information technology.

**JPEG**   A compression algorithm defined by the Joint Photographic Experts Group for continuous-tone images (color pictures). Permitting loss of information nonessential for human observation, compression ratios of 10–30 : 1 can be achieved. "Lossless" compression cannot currently go beyond 8 : 1.

**Karaoke**   Japanese word meaning "empty orchestra." A form of entertainment where guests (e.g., in a karaoke bar) use a microphone and sing along to the musical backing.

**KB**   Kilobyte, or 1024 bytes.

**LaserVision**   An analog optical video disc system.

**Mastering**   The process in which a glass master is produced (needed to make stampers), which are in turn used for the replication of CDs. The glass master is coated with photosensitive lacquer, which is illuminated on a laser beam recorder. The data for mastering come from a premaster tape or from a CD-WO disc.

**MB**   Megabyte or 1024 KB.

**Mode 1**   A CD-ROM sector with sync (12 bytes), header (4 bytes), user data (2048 bytes), EDC (4 bytes), zero (8 bytes), and ECC (276 bytes).

**Mode 2**   A CD-ROM sector with sync (12 bytes), header (4 bytes), and user data (2336 bytes). For CD-i and CD-ROM XA, two specific formats are specified in mode 2: form 1 and form 2.

**MPC**   A definition for the multimedia PC, including the CD-ROM drive, as put forward by Microsoft. The MPC logo tells the user that an MPC application will work on an MPC computer.

**MPEG**   A norm for the compression of motion video, developed and defined by the Motion Picture Experts Group. This has become the ISO standard and is used in CD-i and Video-CD players. The algorithm used (discrete cosine transform), makes an extremely high rate of compression (200 : 1) possible.

**NA**   Numerical aperture; a combined measure for the diameter and the focal distance of an optical lens.

**OSTA** Optical Storage Technology Association.

**PCM** Pulse code modulation; digital encoding of the amplitude of a signal. The number of bits to express the amplitude value is dependent on the required signal-to-noise (S/N) ratio. For CD-Audio this is 16 bits, resulting in S/N > 96 dB. The frequency with which the amplitude has to be measured (the sample rate) is dependent on the required bandwidth. For CD-Audio the sampling rate is 44.1 kHz, resulting in a bandwidth of 20 kHz. PCM does not use compression.

**Premaster tape** The magnetic tape used in CD-Audio and CD-ROM replication companies to produce the CD master with which the CDs themselves are pressed.

**Retrieval** Term for calling up information in databases. The retrieval usually takes place on the basis of indexes.

**Replication** The production of identical copies of a Compact Disc from a stamper.

**Scrambling** Randomizing data by adding a predefined pseudorandom bitstream to the data stream. Scrambling is applied to prevent the occurrence of fixed, repetitive data patterns.

**Stamper** The metal form used to press a CD in the injection-molding process.

**Subcode** Information (track, time, text, graphical, or MIDI) stored together with audio on a CD. The subcode is spread across eight channels (P through W). P and Q contain the track and time information shown on the display of an audio CD player.

**TOC** Table of contents; a list in the subcode in the lead-in area of the disc, containing the starting addresses of the tracks in the program area.

**Transfer rate** The speed with which information can be transferred, usually expressed in terms of kilobytes per second. A standard CD-ROM drive is rated at 150 kB/s. A double speed player can deliver 300 kB/s.

**UDF** Universal disc format; a proposed file format as a follow-up for ISO 9660.

**Volume** The CD-ROM term for a CD-ROM disc. In the case of extremely large databases it is possible to define a "volume set," consisting of a number of CD-ROM discs that together form a whole.

## BIBLIOGRAPHY

CD-DA: Compact Disc–Digital Audio, specified in the System Description Compact Disc–Digital Audio (Red Book), N.V. Philips and Sony Corporation.

CD-ROM: Compact Disc–Read-Only Memory, specified in the System Description Compact Disc–Read-Only Memory (Yellow Book), N.V. Philips and Sony.

CD-i: Compact Disc–Interactive, specified in the CD-i Full Functional Specification (Green Book), N.V. Philips and Sony.

CD-ROM XA; Compact Disc–Read-Only Memory Extended Architecture, specified in the System Description CD-ROM XA, N.V. Philips and Sony.

CD-i Bridge: Compact Disc–interactive Bridge, specified in the CD-i Bridge Specification, N.V. Philips and Sony.

Video CD: Video Compact Disc, specified in the Video CD Specification, N.V. Philips, JVC, Sony, and Matsushita.

Multisession CD: Multisession Compact Disc, specified in the Multisession CD Specification, N.V. Philips and Sony.

Enhanced Music CD: Enhanced Music Compact Disc, specified in the Enhanced Music CD Specification, N.V. Philips, Sony, and Microsoft Corp.

CD-WO: Compact Disc–Write Once, specified in the System Description Recordable Compact Disc Systems, Part II: CD-WO (Orange Book Part II), N.V. Philips and Sony.

CD-RW: Compact Disc–ReWritable, specified in the System Description Recordable Compact Disc Systems, Part III: CD-RW (Orange Book Part III), N.V. Philips and Sony.

Photo CD: Photo Compact Disc, specified in the System Description Photo CD, N.V. Philips and Eastman Kodak Co.

MPEG-1 Standard; Information technology—coding of moving pictures and associated audio for digital storage media at up to about 1.5 Mb/s.

Part 1: Systems, ISO/IEC11172-1:1993(E)

Part 2: Video, ISO/IEC11172-2:1993(E)

Part 3: Audio, ISO/IEC11172-3:1993(E)

ISO 9660: Information processing—volume and file structure of CD-ROM for information interchange.

ISO 10149: Information technology—data interchange on read-only 120-mm optical data discs (CD-ROM).

IEC 908: Compact Disc–Digital Audio system.

IEC 958: Digital audio interface.

IEC 1096: Methods of measuring the characteristics of reproducing equipment of digital audio Compact Discs.

DAG 320, DAG 321, DAG 322: Specification of the 3/4-inch cassette type CD master tape, N.V. Philips and Sony.

## ABOUT THE AUTHORS

Boudewijn van Dijk, after his graduation in electronics, joined the Philips Research Laboratories in 1970. For a number of years he worked on broadcast radio and cable television systems. In 1979 he joined the Philips business unit Optical Disc Mastering (ODM). At ODM he was responsible for the development of the Compact Disc signal processing electronics used in CD mastering equipment. In 1988 he became product manager responsible for CD premastering and mastering systems. After the merger of ODM and OD & ME into a new independent company ODME (Optical Disc Manufacturing Equipment) in 1991, he also became responsible for the marketing issues of CD test equipment and the Automatic Mastering System. In August 1995 he joined Philips Consumer Electronics as manager of standards and licensing. Boudewijn is coauthor of several publications including *Corporated CD-ROM Publishing; High Density CD, An Industrial Approach,* and *Dataconversion, A Premastering Activity.*

Jaap Nijboer received his electrical engineering degree in electronics from the University of Technology, Delft, the Netherlands, in 1974. Immediately after finishing his studies, he joined the Philips Portable Radio and Cassette-Recorder Design Group in Eindhoven, working on predevelopment and IC design activities. From 1978 he was involved in Compact Disc system research and development in the audio division of the main industry group Consumer Electronics of Philips. In 1987 he joined the business unit ODM. After the merger of ODM with OD&ME, he was involved in all aspects of CD and LD premastering, mastering, replication, and quality control. Since June 1995 Nijboer has been with Philips Media System B.V. as standardization officer for CD-related systems and products.

## ACKNOWLEDGMENT

The authors would like to express their gratitude to Bert Gall, Henk Hoeve, Peter Horn, Hans Mons, and Gerard Vos for their encouragement and support.

# CHAPTER 12
# THE DIGITAL VERSATILE DISC (DVD)

**Boudewijn van Dijk**
*Philips Consumer Electronics*

**Johannes G. F. Kablau**
*Philips Key Modules*

**Jaap G. Nijboer**
*Philips Key Modules*

## 12.1 INTRODUCTION

In this section an overview is given of the considerations and arguments that have led to the new high-density optical medium, the Digital Versatile Disc (DVD).

### 12.1.1 From Compact Disc to DVD

In the last 15 years, the Compact Disc (CD) has become the distribution medium of choice for music, multimedia, movies, computer operating and application software, and data distribution. More than 500 million players and billions of discs are currently in use worldwide.

The needs of the entertainment and personal computer industries, which are served by different CD systems, have been steadily converging. At the same time, applications demand more and more data storage capacity. The motion picture industry decided that a new optical disc should be able to store full-length movies, with a picture quality comparable to that of broadcast pictures, in both regular and wide-screen formats, with digital or analog surround sound and subtitles in several languages. At the same time, multimedia computer applications strive for increased reality, with more full-screen, high-quality video, 3-D animation, and multilingual hi-fi audio. The resulting demand for storage capacity exceeds the capacity of existing CD-ROMs by as much as an order of magnitude, and it has to be measured in gigabytes.

These were the pressures that produced a new high-capacity, universally applicable optical disc standard: DVD, or Digital Versatile Disc.

Optical disc technology has moved on from the original CD parameters laid down in the late 1970s to such an extent that DVD discs have the potential to offer more than 25 times the capacity of CD-ROMs (using a double-sided, dual-layer disc). And the new DVD players can handle the much faster data rates demanded by full-motion video and multimedia applications.

Advances in laser design and optics, together with the experience gained in disc manufacture, allow a smaller and more compressed pit pattern on a tighter track pitch, as well as the introduction of both dual-layer and double-sided disc formats.

Great strides in silicon technology have dramatically reduced both the size and relative cost per unit of memory and microprocessor chips. Faster and lower-cost data formatting and decoding components, capable of handling the higher data rates, have been developed.

MPEG-2, an improvement over the MPEG-1 digital audio-video compression standard (see Chapter 8) that incorporates multichannel (surround) sound and is enhanced by a faster and dynamically adaptable data rate, is laid down in an ISO/IEC specification.

### 12.1.2  Industry Requirements

The DVD specifications are the result of wide-ranging industry cooperation coordinated by the DVD Advisory Committee. The Motion Picture Studio Advisory Group defined the needs of the movie industry, and a technical working group defined the requirements for PC-based multimedia applications.

***The Movie Industry Requirements.***   The basic requirement of the movie industry was that the capacity of a single disc should be sufficient to hold a full-length motion picture with picture quality comparable to broadcast quality. Broadcast quality is superior to the quality of current consumer video systems. Audio had to be compatible with existing surround-sound formats, including the 5.1 channel digital surround-sound systems. Multiple sound channels to store three to five dubbed languages and subtitles in four to six languages all on the same disc were also asked for. Other requirements included the ability to handle multiple screen aspect ratios and to provide for parental control, regional control, and copy protection.

***The PC Industry Requirements.***   The basic requirement of the PC industry was for a single interchange standard for entertainment and PC applications, with backward (read) compatibility with existing CDs, and forward compatibility with future rewritable and write-once discs.

A single file system was needed for all contents and ranges of media, while the resulting player and disc replication costs were to be kept low, comparable with current CD-ROM drives and discs. No disc caddies were to be necessary.

Last, but not least, reliable data storage and retrieval was essential, with high and extendible on-line capacity, and high performance for both linear and interactive data.

## 12.2  DVD SYSTEM OVERVIEW AND MAIN PARAMETERS

The functional specifications as based upon the requirements of both the movie and the PC industry are given in Table 12.1.

### 12.2.1  DVD Family

The Digital Versatile Disc consists of two halves, each having a thickness of 0.6 mm. With a disc made like this and utilizing the possibility of data storage in several layers, a data capacity of more than 17 gigabytes can be achieved. An overview of the prerecorded DVD family is given

**TABLE 12.1**  DVD Main Parameters

| Parameter | Units | Compact Disc | DVD Single-layer | DVD Dual-layer |
|---|---|---|---|---|
| Wavelength of laser diode | nm | 780 | 650 | |
| NA of objective lens | | 0.45 | 0.60 | |
| Reference scanning velocity | m/s | 1.2–1.4 | 3.49 | 3.84 |
| Channel bit length | μm | 0.28–0.32 | 0.13 | 0.15 |
| Minimum pit length | μm | 0.83–0.97 | 0.40 | 0.44 |
| Track pitch | μm | 1.6 | 0.74 | |
| Disc diameter | mm | 120 | 120 | |
| Disc thickness | mm | 1.2 | $2 \times 0.6$ | |
| Data area inner radius | mm | 25 | 24 | |
| Data area outer radius | mm | 37.5 / 58 | 38 / 58 | |
| Maximum disc tilt | | 0.6° | 0.4° | |
| Maximum substrate thickness error | μm | ±100 | ±30 | +40, −50 |
| Maximum radial run-out | μm | 140 | 100 | |
| User data capacity | gigabytes | 0.68 | 4.70 | 8.54 |
| User data bit rate (maximum) | Mb/s | 1.23 | 11.08 | |

in Fig. 12.1. The total family of DVDs consists of the following members:

- A single-sided, single-layer disc having a maximum storage capacity of 4.7 gigabytes for a 12-cm disc
- A single-sided, dual-layer disc with a maximum storage capacity of 8.5 gigabytes
- A double-sided, single-layer disc having a maximum storage capacity of 9.4 gigabytes
- A double-sided, dual-layer disc featuring a maximum storage capacity of more than 17 gigabytes

The single-sided discs are read from only one side, whereas the double-sided discs have to be read from both sides of the disc. Since the DVD may contain two information layers for each readout side, several methods for readout can be applied.

## 12.2.2  Dual-Layer, Parallel Track Path

Using the parallel track path (PTP) on a dual-layer disc, both layers have a lead-in and a lead-out area as part of the information area. For each layer, the lead-in area is located at the inner radius of the disc and the lead-out area is located at the outer radius of the disc. This layout structure is comparable with the layout of the CD. The data are read, as with a CD, from the inner radius of the disc to the outer radius, for both information layers. The PTP method can be used for providing artist profiles and other background information in one track along with the movie in the other track, or quick-responding game structures achieved by means of quick access from layer to layer.

## 12.2.3  Dual-Layer, Opposite Track Path

For long-playing time movies or very high picture quality, the storage capacity of one information layer might not be sufficient. Here we are confronted with the need for a seamless continuation

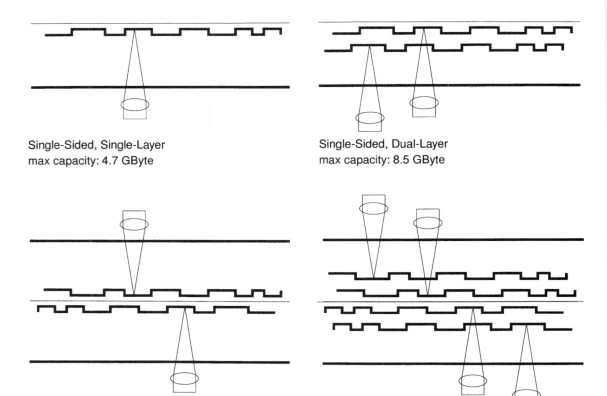

Single-Sided, Single-Layer
max capacity: 4.7 GByte

Single-Sided, Dual-Layer
max capacity: 8.5 GByte

Double-Sided, Single-Layer
max capacity: 9.4 GByte

Double-Sided, Dual-Layer
max capacity: 17 GByte

**FIGURE 12.1**    The DVD prerecorded disc family.

of the playback from one layer to the other. The method used for this is the so-called opposite track path (OTP) method, which is completely different from the PTP method.

The first information layer starts with a lead-in area at the inner radius of the disc and ends with a so-called middle area at the outer radius. The second information layer starts with a middle area at the outer radius and ends with a lead-out area at the inner radius of the disc. The reading of the data stored on the disc will start at the inner radius of the first information layer and proceed until the middle area of this layer is reached. Then a switch over to the middle area in the second information layer is made in order to continue the reading of the data from the outer radius up to the lead-out area at the inner radius of the second information layer. The several readout methods are given in Fig. 12.2.

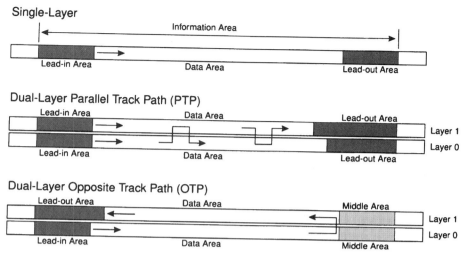

**FIGURE 12.2**    Readout possibilities.

## 12.3    *THE PHYSICAL FORMAT*

### 12.3.1    Capacity and Gain Factors Relative to CD

*Capacity.*    As with a CD, a DVD may be 120 or 80 mm in diameter. It may be single- or double-sided, and each side may carry either one or two layers of data. In the single-layer, single-sided format, the 120-mm disc has a basic capacity of 4.7 gigabytes, compared with the 680 megabytes of the equivalent CD. This represents a gain factor of 6.9. Similarly, the capacity of the 8-cm disc is 1.46 gigabytes, compared with the 195 megabytes of the CD.

By also defining dual layers on a single-sided disc, and single or dual layers on a double-sided version, the DVD specification allows for a potential maximum capacity of 17.1 gigabytes.

*Gain Factors.*    The bar chart of disc capacity gain, given in Fig. 12.3, shows how the overall capacity gain from CD to single-sided DVD can be expressed as a number of steps. Starting with a CD reference of 100, the use of a shorter-wavelength laser achieves a capacity gain of 1.44 times (proportional to $1/\lambda^2$). Using a pickup lens with a higher numerical aperture (NA) adds a further factor of 1.78 (proportional to $NA^2$). More severe disc production margins contribute a factor of 1.35 because of radial margins and a factor of 1.31 because of tangential margins. These factors together result in a total physical density increase of 4.54 times (see Table 12.2).

More efficient channel modulation (which is a modified version, known as EFM+, of the CD eight-to-fourteen modulation scheme, EFM, and which has the subcode deleted), error correction (RSPC), and sector formatting (with greatly reduced sector overhead) achieve gains of 1.12, 1.16, and 1.14, respectively, resulting in an additional data density increase of 1.49 times (see Table 12.2).

Together with a slightly larger (1.02 times) program area, this gives the single-sided, single-layer (SL) DVD disc a capacity gain of 6.9 times that of CD. The single-sided, dual-layer (DL) format adds an additional 1.82 increase in capacity, at 12.6 times that of CD. The double-sided

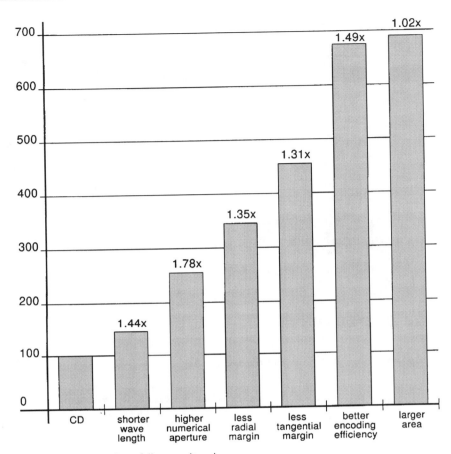

**FIGURE 12.3**   Bar chart of disc capacity gain.

versions give a doubling of the capacity, resulting in 13.8 times (SL) or over 25 times (DL) the capacity of CD.

Table 12.2 shows that the bit size and track pitch of DVD are both less than half that of CD, giving a density increase of more than 4.5 times. At the same time, encoding efficiency in sector formatting, error correction, and channel modulation is up by a factor of nearly 1.5, and the size of the program or information area is slightly larger. As mentioned previously, the total gain in user data capacity is by a factor of 6.9 for the SL disc.

On average, the DVD disc runs nearly three times faster than the CD disc, resulting in an increase of the user data transfer bit rate by a factor of nine.

The combination of greater storage capacity and faster data transfer produces operational parameters that are in full accordance with, and in many cases exceed, the industry requirements.

### 12.3.2   Density, Margins, and System Robustness

***Density.***   The physical (channel bit) density of the DVD format is 4.54 times that of CD. Most of this density increase is due to better optical resolution, caused by the smaller size of the laser

**TABLE 12.2** Main Parameters of DVD Compared with Those of CD-ROM

| Parameter | | CD-ROM mode 1, 74 min | DVD-ROM (single layer) | Gain factor |
|---|---|---|---|---|
| Spot diameter | $w = \lambda/(2 \cdot NA)$ | 0.87 μm | 0.54 μm | 1.60* |
| Track pitch | $p$ | 1.6 μm | 0.74 μm | 2.16 |
| Relative radial density | $p/w$ | 1.85 | 1.37 | 1.35† |
| Channel bit length | $c$ | 0.28 μm | 0.133 μm | 2.10 |
| Minimum pit length (3T) | $3c$ | 0.84 μm | 0.40 μm | 2.10 |
| Relative tangential density | $3c/w$ | 0.97 | 0.74 | 1.31† |
| Density | $d = p \cdot c$ | 0.45 μm²/ch. bit | 0.1 μm²/ ch. bit | 4.54‡ |
| Linear velocity | $v$ | 1.21 m/s | 3.49 m/s | 2.88 |
| Optical cut-off frequency | $f_{co} = v/w =$ | | | |
| | $= v \cdot 2 \cdot NA/\lambda$ | 1.4 MHz | 6.44 MHz | 4.6 |
| Channel bit rate | $f = v/c$ | 4.3218 MHz | 26.16 MHz | 6.05 |
| Highest EFM frequency | $f_{13} = f/6$ | 0.72 MHz | 4.36 MHz | 6.05 |
| Relative bandwidth | $f_{13}/f_{co}$ | 0.52 | 0.68 | 1.31§ |
| Encoding efficiency | $e$ (see Table 12.3) | 28.4% | 42.3% | 1.49 |
| User data bit rate | $f \cdot e$ | 1.2288 Mb/s | 11.08 Mb/s | 9.0 |
| Size of program area on disc | $a = \pi \cdot (R_{max}^2 - R_{min}^2)$ | 8600 mm² | 8760 mm² | 1.02 |
| User data capacity | $(a/d) \cdot e$ | 682 megabytes | 4700 megabytes | 6.9 |

*The smaller spot size increases the optical resolution, without sacrificing system margins.
†Both the higher relative radial density and higher relative tangential density reduce system margins.
‡Total increase of physical density is composed of
  (gain in spot area) · (gain in relative radial density) · (gain in relative tangential density) $= (1.6)^2 \cdot 1.35 \cdot 1.31 \approx 4.5$
§Gain in relative bandwidth ≡ gain in relative tangential density

spot focused on the information layer. The area of the laser spot is proportional to $(\lambda/NA)^2$, where $\lambda$ is the laser wavelength and NA is the numerical aperture of the focusing lens. For DVD, $\lambda$ is reduced to 650 nm, while NA goes up to 0.6. Together they account for an improvement of the optical resolution by a factor of 2.56. In this way, both the track pitch and the channel bit length are reduced by a factor of 1.6, giving a density increase of 2.56 without sacrificing margins.

*Margins.* The remaining 1.77-times gain in physical density had to be met by reducing the margins in the DVD system, as compared with CD. The relative track pitch (track pitch/spot diameter) is reduced by an additional factor of 1.35, and the relative channel bit length (channel bit length/spot diameter) is reduced by an additional factor of 1.31. Tighter production tolerances are therefore required for discs and drives. Years of experience in CD manufacture, however, have put these tolerances well within the capabilities of today's production facilities.

Because of the higher tangential density, the EFM bandwidth is relatively closer to the optical cut-off frequency than in the case of the CD. Therefore the amplitudes of the highest EFM frequencies are smaller compared with the "eye pattern" of CD. Figure 12.4 shows pit patterns and eye patterns (both equalized) of CD and DVD.

*System Robustness.* As explained in the previous paragraph, the physical density increase of DVD relative to CD plays an important role in the ultimately achieved capacity gain. With system margins considerably reduced, the robustness of the system has to be carefully considered. Here the main considerations are given that have led to the 4.5 physical density increase.

Physical density is essentially a trade-off between the obtained high capacity on the one hand and the reduced manufacturing tolerances and system robustness on the other hand. The key factor is that to read out the smaller pit structure on the disc, the laser diode wavelength has to be reduced or the numerical aperture of the objective lens has to be enlarged. The problem

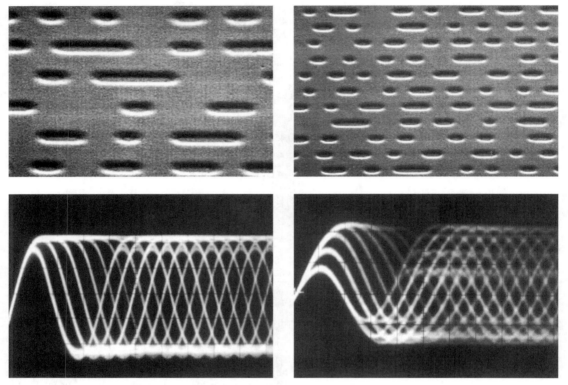

**FIGURE 12.4**    Pits and eye pattern of CD (*left*) and pits and eye pattern of DVD (*right*).

with reducing λ is the technical availability of laser diodes. The problem with enlarging the NA is the increased sensitivity to optical aberrations. Optical aberrations are caused by the following tolerances in the system:

- Wavefront error caused by defocus: proportional to $(\text{NA})^2/\lambda$
- Coma caused by disc tilt: proportional to $(t \cdot \text{NA}^3)/\lambda$ (with substrate thickness $t$)
- Spherical aberration caused by substrate thickness error: proportional to $(\text{NA}^4)/\lambda$.

λ can be fixed at 650 nm on the basis of the availability of the laser diode, and NA is set to 0.60, at which value the objective lens still can be manufactured reasonably well. The contribution from coma is essentially halved by selecting substrate thickness $t = 0.6$ mm, half the value for CD. By bonding two 0.6-mm substrates together, a technique well known from LaserDisc, a rigid disc structure is obtained that is mechanically compatible with that of conventional CD.

Given values for λ, NA, and $t$, the above tolerances as well as the tracking tolerance determine the level of optical aberrations that can be expected in day-to-day reality. The real question, however, is when (at what aberration level) is the system performance adversely affected. Here the channel bit density plays a crucial role: at higher density, less aberration can be tolerated. In other words, the robustness of the system becomes worse. To select the physical density at which the system is still sufficiently robust, the following procedure is useful. A so-called cocktail of tolerances, representing a *workable worst-case condition* for the combination of disc and drive, is defined. In this cocktail all system tolerances are used, not all to their full extent but in a realistic mix. Then values of the channel bit length and track pitch are selected at which the data can still be reliably decoded even in the presence of the cocktail defined above.

To judge the quality of the detected RF signal, the *data-to-clock jitter* is considered. This quantity measures the variance of the detected pit edge relative to the regenerated data clock and can thus be related to the probability of misdetection. Given the capabilities of the RSPC error correction system, one can use a data-to-clock jitter that is less than 15 percent of the channel bit as a criterion for reliable data detection.

The following cocktail of tolerances is considered realistic:

- 0.5-degree tilt (of disc and drive) in the radial direction
- 0.3-degree tilt (of disc and drive) in the tangential direction
- 0.4-$\mu$m defocus
- 30-$\mu$m substrate thickness error
- 50-nm tracking error

Since the development of LaserDisc, the diffraction theory associated with the readout of optical discs has been well known, and with today's computing power extensive simulations can be carried out. Figure 12.5 shows the results of simulated RF signals both without and in the presence of the cocktail of tolerances. From these eye patterns, it can be well imagined that in the presence of this cocktail, with fixed $t$, $\lambda$, and NA, and with increasing density, the resulting signal will become more and more critical.

In Fig. 12.6 the data-to-clock jitter is calculated as a function of the physical (or areal) density increase relative to CD, both in an aberration-free situation and in the presence of the cocktail defined previously. Here one can see that in an aberration-free situation, a physical density up to 6.2 times that of the CD would be possible with the chosen values for $t$, $\lambda$, and NA. However, such a high density would not lead to a robust system. When the cocktail of tolerances is considered, one has to restrain the physical density to 4.5 times that of the CD in order to guarantee reliable data detection and system robustness.

## 12.4   DUAL-LAYER DISCS

### 12.4.1   Double-Sided Discs

With DVD consisting of two bonded substrates, it is straightforward to have an information layer on the other side of the disc as well, creating a double-sided disc. Although this obviously doubles the total capacity, the merits for the user are quite limited. To access the other information layer, the disc has to be flipped over unless the drive is equipped with special facilities (a pickup on both sides of the disc or one that slides around it), which inevitably raise the hardware prices. Looking at a double-sided disc from the artwork point of view, the greater area of an unused side offers far more attractive options for a label compared with the tiny area available on the double-sided disc. So, for enhanced on-line, single-sided capacity, the dual layer disc has been introduced.

### 12.4.2   Outline

On a dual-layer disc two information layers are separated by a thin space layer (Fig. 12.7). The first layer (layer 0) is partially transmissive, allowing the second layer (layer 1) to be read out through layer 0. If the spacer thickness is much larger than the focal depth of the imaging optics, both layers can be read out from the same side, and individual layers can be accessed by simply stepping the focus actuator.

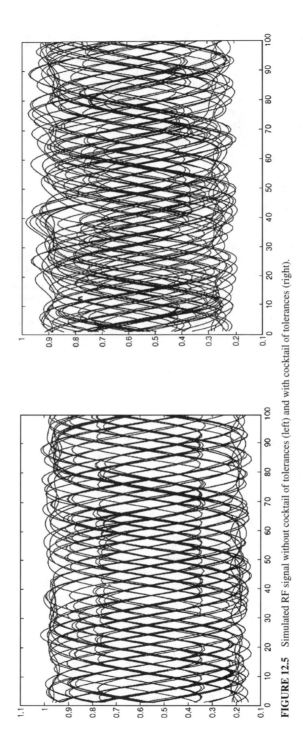

**FIGURE 12.5**  Simulated RF signal without cocktail of tolerances (left) and with cocktail of tolerances (right).

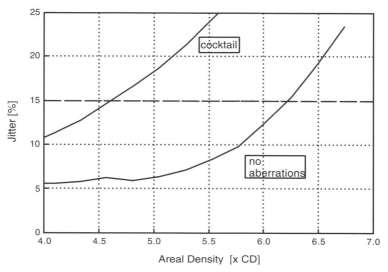

**FIGURE 12.6**    Calculated data-to-clock jitter as a function of the density.

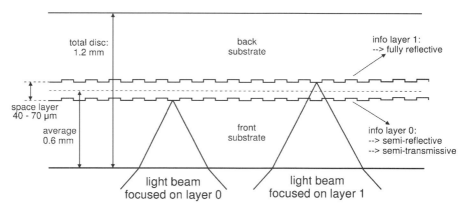

**FIGURE 12.7**    Outline of dual-layer disc.

### 12.4.3   Density

Reading both layers of a dual-layer disc with a single objective lens (corrected for only one thickness) inevitably leads to a thickness error of at least half the space layer thickness. This means that even when a perfect disc is read by an ideal drive, a significant amount of spherical aberration is present in the system. To achieve the same robustness as in the case of reading a single-layer disc, the tangential density is reduced by 10 percent. So the total physical density increase relative to a CD is reduced from a factor of 4.5 to a factor of 4.1. The total capacity per layer is reduced by the same proportion, from 4.7 gigabytes to 4.27 gigabytes, which yields 8.54 gigabytes of total disc capacity.

### 12.4.4 Design Considerations

One has to consider the separation of the two focus S-curves that are generated by the two information layers of the dual-layer disc. Obviously, well separated S-curves are a necessary condition for a reliable focus-jump mechanism. This is achieved by sufficient thickness of the space layer.

A second phenomenon that has to be dealt with is stray light on the photodetector. The light reflected by the other layer illuminates a large area that covers the photodetector, thus producing unwanted DC offsets in the electric signals. One has to realize that because of the highly coherent laser light, interference may also affect the readout signal. Given the NA of the system, these phenomena also result in a minimum value for the space layer thickness.

Finally, and most important, the total spherical aberration that can be allowed in the system determines the upper value of the space layer thickness specification. These considerations have led to a 40–70 μm space layer thickness range, which on one hand enables reliable detection and focus-jump capabilities and on the other hand enables manufacture of the dual-layer disc.

### 12.4.5 Manufacturing Method

The most straightforward method to make a dual-layer disc is by replicating the two information layers on each substrate of the disc. One is covered by the semireflecting (dielectric) layer, the other with a metal reflector. Then the two substrates are bonded together back to back by an optically transparent bonding method (e.g., ultraviolet bonding). Obviously this method can only be used to make single-sided dual-layer discs (8.5 gigabytes).

Another method is the 2P method. Here, layer 0 is again replicated in the substrate and covered with a dielectric layer. Layer 1, however, is replicated in a thin layer of 2P lacquer deposited on top of layer 0 and finally metalized. This method, first proposed by 3M, enables the production of double-sided dual layer discs featuring an impressive total capacity of 17.1 gigabytes.

The most critical parameters in dual-layer disc manufacturing are the homogeneity of the dielectric layer and the homegeneity of the space layer. Again, it has been the progressing state of manufacturing technology that has made possible a sophisticated product like the dual-layer disc.

## 12.5 TRACKING METHOD

In CD, several radial tracking schemes are utilized, of which the push-pull method is explicitly described in the Red Book system description. Likewise, several tracking schemes are possible for DVD, of which only differential phase detection (DPD) is specified and extensively described in the system description. In this section the basic principles and advantages of DPD are described.

### 12.5.1 Principle of Operation

The pit pattern of the disc can be seen as a grating structure in both radial and tangential directions, with diffracted orders in these directions as a result (Fig. 12.8). The main reflection from the disc forms the zeroth order (0, 0). Besides the first orders in the radial direction, $(+1, 0)$ and

$(-1, 0)$, and the first orders in the tangential direction, $(0, +1)$ and $(0, -1)$, there are also first orders in the diagonal directions: $(+1, +1)$, $(-1, -1)$, and $(+1, -1)$, $(-1, +1)$.

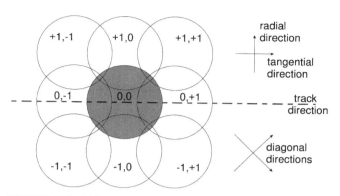

**FIGURE 12.8**   Diffraction pattern due to a two-dimensional grating.

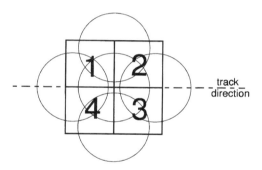

**FIGURE 12.9**   Orthogonal diffraction pattern as used for a CD.

Just like most conventional CD players, a DVD player is equipped with a four-quadrant photodetector (Fig. 12.9) on which an image of the spot on the disc is formed, including part of the first orders. It is the overlap of the first orders with the main (zeroth-order) reflection that generates the operational signals. The RF signal is generated by the tangential first orders, whereas the track crossing signals are generated by the radially diffracted orders.

The RF signal itself is obtained by taking the sum of all quadrant segments $(1 + 2 + 3 + 4)$, whereas the radial push-pull signal is obtained by subtracting the radial detector halves $(1 + 2) - (3 + 4)$.

The DPD tracking scheme operates on the basis of the first diagonally diffracted orders (see Fig. 12.10)—hence a mixture of both radial and tangential detections. The diagonal quadrant segments are summed by pair, $(1 + 3)$ and $(2 + 4)$, and the resulting diagonal signals (both containing RF signals) are compared in phase: $\phi(1 + 3) - \phi(2 + 4)$. If the scanning spot is on one side of the track, one diagonal signal will be slightly ahead of the other. On the other side

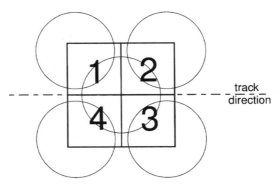

**FIGURE 12.10**    Diagonal diffraction pattern as used for DVD.

of the track, the situation will be reversed. Hence the output of a phase comparator circuit will yield a sawtooth-like signal (Fig. 12.11) which is perfect for a tracking servo designed to track the fine pit pattern on a disc.

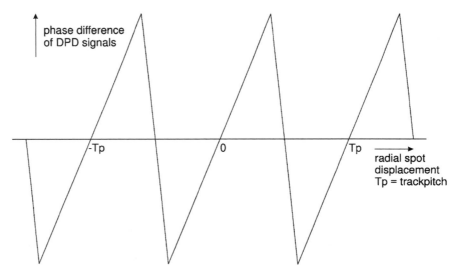

**FIGURE 12.11**    Phase difference signal as a function of the detracking.

### 12.5.2   Diagonal Gratings Model

Although the mathematics of the DPD tracking method is quite complex and therefore beyond the scope of this text, the following simple model is useful to comprehend the basics of DPD tracking. Consider a spot scanning a pit pattern on the disc (Fig. 12.12). When the spot is entering or leaving the pit, another orthogonal grating at 45 degrees to the pit pattern can be envisioned apart from the above-mentioned gratings in the radial and tangential directions. This diagonal

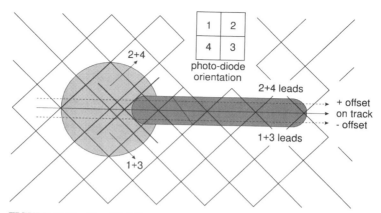

**FIGURE 12.12**    Simplified diagonal grating model.

grating, caused by the natural rounding of the pits, will create diffracted orders on the diagonal detector quadrants. When the spot is entering the pit on one side of the track, one grating is seen earlier than the other. Hence the diagonal detector signal associated with it is ahead of the other. When the spot is entering on the other side of the track, the situation is reversed.

### 12.5.3    Merits of the DPD Method

To enable push-pull detection on a CD, a slight deviation had to be introduced from the pit depth that is optimal for the RF signal detection. In the DVD system, however, considering the tighter margins as compared with those of the CD system, it was felt that no deviation from the optimum RF signal could be afforded. Moreover, the wider range of allowed pit depth would improve the yield in disc manufacturing.

As mentioned in Section 12.4.4, stray light is a potential problem in reading dual-layer discs. It will generate unwanted DC offsets on detector signals such as the push-pull and three-spot tracking error signals. DPD tracking has a significant advantage here because, unlike push-pull and three-beam tracking, it is basically a high-frequency method in which DC offsets have no influence.

## 12.6    INFORMATION AREA FORMAT

### 12.6.1    Information Area

The information area is that portion of the disc's surface containing recorded (or to be recorded) information. It consists of three parts: the lead-in area, the data area, and the lead-out area. In the case of the OTP disc, the lead-out area of layer 0 and the lead-in area of layer 1 are together known as the middle area (Fig. 12.2).

### 12.6.2    Sector Format

As on a CD-ROM, the information recorded on a DVD is formatted into sectors. A sector is the smallest addressable part of the information track that can be accessed independently. Depending on the stage of the signal processing, a sector (or group of sectors) is called a data sector, an ECC (Error Correction Code) block, a recording sector, or a physical sector.

A *data sector* is 2064 bytes long and consists of 2048 bytes of main data, 12 bytes of identification data (ID), and 4 bytes of error-detection code (EDC). As shown in Fig. 12.13, this is several hundred bytes shorter than the CD-ROM sector format.

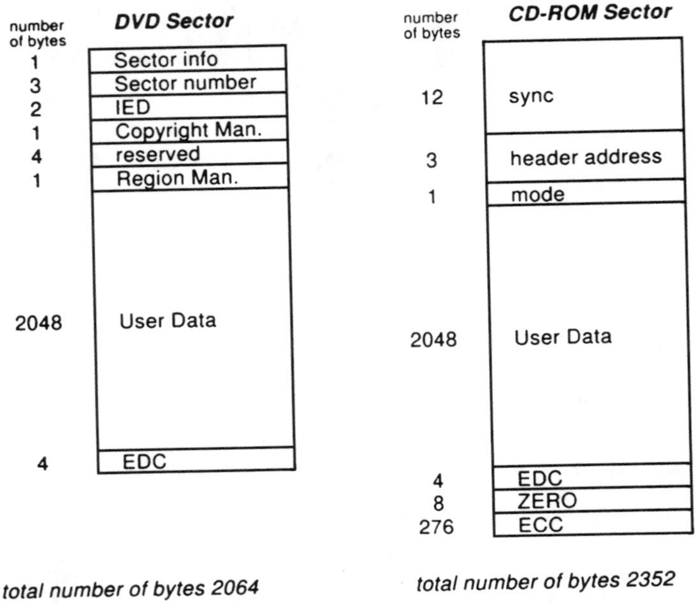

| number of bytes | **DVD Sector** | | number of bytes | **CD-ROM Sector** |
|---|---|---|---|---|
| 1 | Sector info | | | |
| 3 | Sector number | | 12 | sync |
| 2 | IED | | | |
| 1 | Copyright Man. | | | |
| 4 | reserved | | 3 | header address |
| 1 | Region Man. | | 1 | mode |
| 2048 | User Data | | 2048 | User Data |
| 4 | EDC | | 4 | EDC |
| | | | 8 | ZERO |
| | | | 276 | ECC |

*total number of bytes 2064*        *total number of bytes 2352*

**FIGURE 12.13**   DVD data sector, as compared with CD-ROM.

After scrambling the main data in the data sectors, Reed-Solomon error-correction coding information is added to each group of 16 data sectors to form an *ECC block* with supplemental inner-code parity (PI) and outer-code parity (PO) bytes.

The *recording sectors* are formed by interleaving the PO rows in the ECC block and dividing such a block again into 16 sectors.

Finally, EFM$^+$ channel modulation creates the *physical sector,* which is the actual format recorded on the disc.

**Data Sector.**   The 12 bytes of identification data (ID), which take the place of the sync and header found in CD-ROM data sector formatting, contain sector information and a unique sector number. The sector information defines such parameters as sector format type, tracking method, reflectivity, area type, data type, and layer number. The sector information and sector number are protected by an additional error detection code (IED). Furthermore, the "header" contains copyright management information and region management information. The sync as used in CD-ROM is not needed because DVD uses an advanced synchronization method at the EFM level. The extra ECC layer as used in CD-ROM is superfluous because the normal error correction incorporated in DVD is much more powerful than the CIRC in CD and is able to give the high reliability needed for data applications. The EDC is maintained, because it gives a very easy and powerful error-detection capability at the sector level.

***Error Correction (ECC Block).***    Thanks to the development of new ECC decoding algorithms, DVD error correction coding, using a Reed-Solomon product code, can be applied across a larger amount of data, with better correction capabilities. This reduces error correction redundancy to approximately 13 percent, or half that of CD.

After EDC calculation over the data sectors, scrambling data are added to the 2048 bytes of main data in the data sector. Then the error-correction code is applied over 16 data sectors, or one ECC block.

After the data bytes are put in a matrix of 192 rows by 172 columns, 16 bytes of PO-parity are added to each column (Fig. 12.14). Next, 10 bytes of PI-parity are added to each of the 208 rows (192 data + 16 formed by the PO-bytes) to form a Reed-Solomon product code with 208 rows and 182 columns (172 data + 10 formed by the PI-bytes). This code can correct at least five byte errors in each row and at least eight byte errors in each column. By applying several alternating calculations over rows and columns, much larger error patterns can be solved.

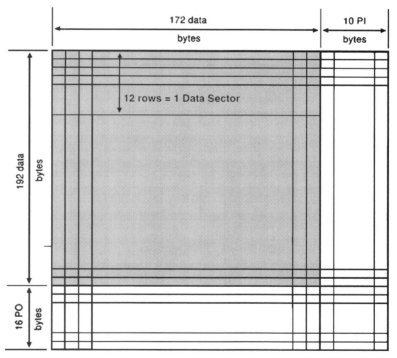

**FIGURE 12.14**    ECC block: product code RS(208,192,17) × RS(182,172,11).

***Recording Sector.***    Finally, the PO-rows are interleaved with the data rows in a regular order (12 data, 1 PO) and each interleaved ECC block is divided into 16 recording sectors (Fig. 12.15). In this way each recording sector contains the original data from 1 data sector plus 12·10 PI-bytes plus 1 PO-row of 182 bytes: together, 2366 bytes.

***Physical Sector.***    A physical sector is formed by splitting each row of a recording sector into two parts of 91 bytes each (Fig. 12.16). These bytes are converted to EFM[+] and preceded by a special EFM[+] sync pattern of 32 channel bits. By using eight different alternating sync patterns,

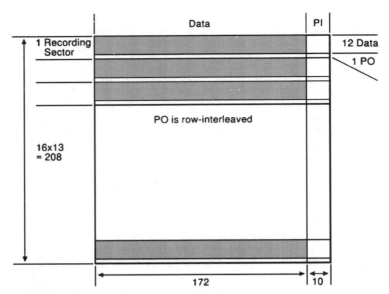

**FIGURE 12.15** Recording sector.

| 32 | 1456 channel bits | 32 | 1456 channel bits |
|---|---|---|---|
| SY0 | 91 bytes | SY5 | 91 bytes |
| SY1 | 91 bytes | SY5 | 91 bytes |
| SY2 | 91 bytes | SY5 | 91 bytes |
| SY3 | 91 bytes | SY5 | 91 bytes |
| SY4 | 91 bytes | SY5 | 91 bytes |
| SY1 | 91 bytes | SY6 | 91 bytes |
| SY2 | 91 bytes | SY6 | 91 bytes |
| SY3 | 91 bytes | SY6 | 91 bytes |
| SY4 | 91 bytes | SY6 | 91 bytes |
| SY1 | 91 bytes | SY7 | 91 bytes |
| SY2 | 91 bytes | SY7 | 91 bytes |
| SY3 | 91 bytes | SY7 | 91 bytes |
| SY4 | 91 bytes | SY7 | 91 bytes |
| Sync Frame | | Sync Frame | |

13 rows

**FIGURE 12.16** Physical sector consisting of 26 sync frames.

decoders can easily synchronize (or resynchronize after clock loss) their timing. A physical sector has 38,688 channel bits, which is equivalent to 2418 bytes before modulation. A sync frame has 1488 channel bits.

### 12.6.3 Channel Modulation

DVD uses a modified version of the CD eight-to-fourteen channel modulation scheme (EFM), which is known as EFM$^+$. This scheme converts eight-bit data directly into 16 channel bits.

The 16-bit channel codewords are chosen from the EFM$^+$ conversion table. There are 256 possible data symbol values and four possible run-length conditions, or states. For each data-symbol/state combination, the conversion table contains a 16-bit EFM$^+$ codeword and the value of the run-length state for the next data symbol to be modulated. Special design of the conversion tables not only allows unique mapping (from one EFM$^+$ codeword to one data symbol) during demodulation, but also limits error propagation.

The 16-bit codeword represents a small but serviceable saving of data capacity as compared with the 14-bit codeword and three merging bits of EFM used in CD. This enhanced version of EFM channel modulation is both more effective and more efficient than its predecessor, without compromising the performance of servo and data extraction systems.

All EFM$^+$ codewords are serialized into EFM$^+$ sync frames consisting of 91 converted data symbols (half a row from a recording sector) and a sync word of 32 channel bits. There are eight different sync words, all consisting of a run length of 14 channel bits (which does not appear in normal EFM$^+$) plus a specific channel bit pattern that identifies the sync word. The sync words identify different positions in the physical sector.

SYO: Start of the first sync frame of a physical sector

SY1–SY4: Start of an odd sync frame

SY5–SY7: Start of an even sync frame

***Low-Frequency Suppression.*** Suppression of the low-frequency components is very important to ensure stable tracking and accurate RF signal detection. For this reason, the running digital sum value (DSV) of the modulated data is maintained close to zero by using a substitute conversion table in addition to the main EFM$^+$ conversion table.

Taking into account the rules for a minimum run length of 3 channel bits and a maximum run length of 11 channel bits, there are 351 source words allowed by the run-length encoding. For historical reasons, 7 of these are discarded, leaving 344 available codewords. Of these, 256 are needed for the main table, leaving 88 for the substitute table. The substitute table, with different DSVs, is applied to the 88 lowest data symbols. Codewords are chosen from one or the other table as necessary to obtain the lowest absolute value for the running DSV.

The higher rate of EFM$^+$ encoding ($8 \longrightarrow 16$ instead of $8 \longrightarrow 14 + 3$) results in only a small increase, of about $+4$ dB, in the power in the low-frequency part of the data spectrum. Because of the higher channel bit rate (26.16 instead of 4.3218 MHz), the power within the same servo bandwidth is actually decreased by about $-12$ dB. For 2 kHz, a typical servo bandwidth, the net reduction is about $-8$ dB (2.5:1).

## 12.7  RELATIVE CODING EFFICIENCIES: DVD VERSUS CD

The relative encoding efficiencies of DVD and CD are summarized in Table 12.3.

## 12.8  LEAD-IN AREA, LEAD-OUT AREA, AND MIDDLE AREA

Except for some blocks at the end, the lead-in area is filled with ECC blocks with their main data set to 00h. The lead-in area ends with 2 ECC blocks containing reference code data, 30 ECC blocks with 00h data, 192 ECC blocks containing control data (physical format information, disc manufacturing information, and contents provider information), and another 32 ECC blocks with 00h data. Next to the lead-in area starts the data area. The reference code contains a specific fixed data pattern and can be used, for instance, for the setting of adaptive equalizer circuits. The physical format information concerns the book type and version, disc size and minimum readout rate, disc structure, recording density, data area allocation, BCA descriptor, and a number of reserved bytes. The disc manufacturing information is text or code data written by the disc manufacturer in free format. The contents provider information is text or code written by the

**TABLE 12.3** Data Encoding Efficiency: DVD versus CD

| Item | CD format | DVD format |
|---|---|---|
| | Data sector | |
| User data | 2048 bytes | 2048 bytes |
| Sync and header/ identification data | 16 bytes | 12 bytes |
| EDC + ECC/EDC | 288 bytes | 4 bytes |
| Total | 2352 bytes | 2064 bytes |
| Efficiency | 87% | 99% |
| | Error correction | |
| CIRC/RSPC | 24 data + 8 parity | 2064 data + 302 parity |
| Efficiency | 75% | 87% |
| | Modulation | |
| EFM/EFM$^+$ per frame | 32 · 8 data bits 588 EFM bits | 91 · 8 data bits 1488 EFM$^+$ bits |
| Efficiency | 44% | 49% |
| **Total efficiency** | **28.4%** | **42.4%** |

contents provider, again in free format. In the lead-out area and the middle area, all main data are set to 00h.

## 12.9 (RE)WRITABLE DVD

To develop applications for DVD, dedicated authoring tools are used. Given the size of the applications, there is a need for increasing storage capacity. Apart from the need for testing the application before going on to costly mastering and replication, the archiving of video data as well as the storage of large databases and computer programs call for the need of one-off DVDs. Therefore, as has been the case for CD, the logical next steps for DVD are the recordable DVD (DVD-R) and the rewritable DVD (DVD-RAM). DVD-R will allow for write-once recording, whereas DVD-RAM will be rewritable. At this time, the storage capacity for DVD-R is 3.9 gigabytes per side, and for DVD-RAM, 2.6 gigabytes per side can be stored.

## 12.10 COMPATIBILITY CONSIDERATIONS

When a new system such as DVD is introduced, the user requires compatibility. It may be divided into several components:

- The new drive is required to accept media of an older, already existing system. This is referred to as backward compatibility.
- When media of the new system also work with older, existing drives, the system is said to have full compatibility. This has been the key factor in the success of the CD-Recordable, where recorded discs play back on conventional CD drives.

- When the new drive is prepared to accept media yet to be introduced, it is said to have forward compatibility.

Because of the CD's dominant position in the market, users may have a considerable amount of capital invested in prerecorded CD software (data or music). That is why, in the case of DVD, backward compatibility with the CD family (or at least with CD-DA) is an absolute requirement. In the following paragraphs, an overview is given of the technical problems as well as the possible solutions associated with this requirement.

### 12.10.1  Spherical Aberration

First one needs to identify the problems in establishing backward compatibility with CD. The DVD decoder LSI, responsible for converting channel bits to user data, must be designed in such a way that the CD data streams can be handled as well. To find out where the problem lies, one has to compare the DVD's optical/mechanical specifications with those of the CD. Clearly, the mechanical specifications, such as outer diameter, disc thickness, and center hole, are identical. The pit pattern on a CD has, as extensively explained in the previous paragraphs, a density lower by a factor of 4.5. This represents no problem in itself. The DVD optical pickup is just "over-specified" for reading the relatively large CD pits. The main problem lies in the substrate thickness difference.

From the laws of optics, we know that when a light beam is focused through a medium other than a vacuum, spherical aberration proportional to the thickness of the medium will emerge in the wavefront. This aberration will keep the beam from being focused into a sharp spot unless measures are taken to compensate. In the case of CD and DVD, the laser light is focused through a polycarbonate substrate onto the information layer, and a considerable amount of spherical aberration is generated. In CD players the objective lenses are designed in such a way that the spherical aberration associated with a 1.2-mm substrate thickness is exactly compensated. In a DVD drive one has to find a way to overcome the spherical aberration caused by both 0.6-mm (DVD) and 1.2-mm (CD) substrates.

### 12.10.2  Backward Compatibility Solutions

Solutions proposed and used in DVD drives that overcome the problem of reading discs with two substrate thicknesses vary from simply adding CD optics to highly integrated methods. A brief overview is given in this section.

***Double Optics.***    A completely duplicated pickup for both CD and DVD readout was proposed by Sony. Clearly, low risk and short lead times are advantages of this approach, but the cost may be a little high. Important for readout of CD-R is the fact that the original 780-nm laser wavelength for CD readout is used.

A straightforward solution, first proposed by Mitsubishi Electric Corporation, is the so-called twin lens actuator (Fig. 12.17). Two objective lenses are employed, each compensated for its specific substrate thickness, contained in a single actuator that is capable of swinging each lens into position when required. The dynamic behavior of this actuator has to be designed carefully.

***Double-Focus Lens.***    A more integrated method was proposed by Matsushita Electric Inc. A double-focus objective lens was obtained by adding a Holographic Optical Element (HOE; see Fig. 12.18). The hologram, covering the center part of the lens, splits the light into zeroth and first orders. While the zeroth order is transmitted for DVD readout, the first order is diffracted onto another focal point, compensating for the appropriate amount of spherical aberration at the same time. Once the hologram can be replicated cheaply, this is potentially a low-cost solution.

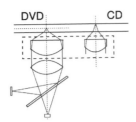

**FIGURE 12.17**    Twin lens actuator.

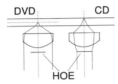

**FIGURE 12.18**    Hologram-type double-focus lens.

Because of the poor light efficiency, however, it is not suitable for (re)writable applications.

***Aperture Reduction and Spatial Filtering.***    To come to a really simple and hence cheap design capable of reading both CD and DVD, several vendors have proposed reducing the NA at the CD readout. Doing so will not eliminate the spherical aberration, but because of the aberration's strong (fourth power) relation with NA, it will reduce it to acceptable proportions while maintaining an acceptable spot size. As an aperture switch, a liquid crystal shutter has the reliability advantage over a mechanical shutter.

More sophisticated methods make use of the fact that when CD readout is attempted with a DVD lens, the spherical aberration is mainly concentrated at the outer edges of the pupil. So the low-aberration part can be selected by spatial filtering. For example, by adding a nontransmissive ring to the objective lens (rather than reducing the aperture) the low-aberration center part of the pupil is separated from the heavily distorted outer part, effectively removing the disturbing light rays on the detector.

### 12.10.3  Forward Compatibility

DVD drives featuring forward compatibility with (re)writable DVD will also be required on the market.

***DVD-R.***    As DVD is designed for easy compatibility, drive designers will have little trouble in making it forward compatible with DVD-R. Care has to be taken, however, that the slightly larger DVD-R track pitch (0.8 μm) is taken into consideration.

***DVD-RAM.***    When forward compatibility with DVD-RAM is considered, the following aspects will have to be tackled:

* Reflectivities might be very low, even lower than in dual-layer discs.
* DPD tracking may not be possible on phase-change-based media, even when they have been recorded.
* Randomly written discs will have to be handled.
* Embossed header information will have to be skipped.
* Single spiral land/groove format requires synchronous push-pull switching.

This will create a tremendous challenge for drive designers. But as history has shown, the customer requirement will be the key factor that determines the final product.

## *ABOUT THE AUTHORS*

For biographical information on Boudewijn van Dijk and Jaap Nijboer, see "About the Authors," Chapter 11.

Hans Kablau received his physical engineering degree from Delft University of Technology, the Netherlands in 1985 and joined Philips Electronics in 1986. He joined the Consumer Electronics division's "system evaluation" group where he worked on CD standards in general and on CD-Recordable in particular. From 1994, Hans Kablau has been active in setting the physical specifications for Multi-Media CD and later, from within the DVD consortium working groups, for DVD.

## ACKNOWLEDGMENTS

The authors would like to express their gratitude to Jan Bakx, Jacques Heemskek, Peter Horn, and Gerard Vos for their encouragement and support.

# KEY DIGITAL DELIVERY/RECEPTION SYSTEMS

# CHAPTER 13
# HDTV BROADCASTING
# AND RECEPTION

**Robert Hopkins**
*Advanced Television Systems Committee*
*Washington, D.C.*

## 13.1  INTRODUCTION

The Advisory Committee on Advanced Television Service (Advisory Committee) was formed by the U.S. Federal Communications Commission (FCC) in 1987 to "advise the FCC on the facts and circumstances regarding advanced television systems" for terrestrial broadcasting. The Advisory Committee objective also stated that the Advisory Committee should recommend a technical standard in the event the FCC decides that adoption of some form of advanced broadcast television is in the public interest. The Advisory Committee is organized into three subcommittees—one for planning, one for systems analysis and testing, and one for implementation. Further information on the objectives and organization of the Advisory Committee may be found in the literature.[1,2]

From 1987 to 1991 many technical system proposals were made to the Advisory Committee. These proposals were analyzed by technical experts. Tests were planned. Only five proposals survived the rigorous process. Then in mid-1990, the first digital high-definition television (HDTV) system was proposed to the Advisory Committee. Within seven months, three other digital HDTV systems were proposed. Tests on five HDTV systems (four digital, one analog) were conducted from September 1991 through October 1992. The results and conclusions were analyzed by the Special Panel of the Advisory Committee in February 1993 and are available,[3,4] as is a summary of the conclusions.[5,6] In short, the Special Panel found that there are major advantages in the performance of digital HDTV systems, that no further consideration should be given to analog-based systems, and that all of the systems produced good HDTV pictures in a 6-MHz channel, but that none of the systems was ready to be selected as the standard without implementing improvements. The Advisory Committee adopted the Special Panel report and encouraged the proponents of the four digital systems to combine their efforts into a "Grand Alliance." The Advisory Committee also authorized its Technical Subgroup to monitor ongoing developments.

This chapter is an updated version of an article, "Digital terrestrial HDTV for North America: The Grand Alliance HDTV System," that appeared in the *IEEE Transactions on Consumer Electronics*, Aug. 1994, pp. 185–198, ©1994 IEEE and reprinted with permission.

Within three months, in May 1993, the proponents of the four digital systems agreed to combine their efforts. The resulting organization was called the "Digital HDTV Grand Alliance." The members of the Grand Alliance are AT&T, David Sarnoff Research Center, General Instrument Corporation, Massachusetts Institute of Technology, Philips Electronics North America Corporation, Thomson Consumer Electronics, and Zenith Electronics Corporation.

In June 1993 the Grand Alliance submitted a preliminary technical proposal to the Technical Subgroup (video formats of 720 active lines and 960 active lines, video compression using MPEG-2 simple profile (no B-frames) with non-MPEG-2 enhancements, and MPEG-2 transport stream). Some subsystems were not specified by the Grand Alliance (audio compression and modulation), but were proposed to be the winner of subsystem tests to be conducted by the Grand Alliance. The Technical Subgroup began a review of the proposal within individual expert groups of the Technical Subgroup. The expert groups agreed with some portions of the proposal and made various suggestions on possible changes to other portions. The Grand Alliance, with assistance from the Audio Expert Group, performed tests on three different multichannel audio compression systems in July 1993.

In a meeting of the Technical Subgroup in October 1993, the Grand Alliance reported that their experiments showed that noncompatible enhancements to MPEG-2 did not produce a sufficient gain in picture quality to offset the loss of MPEG compatibility, that higher video compression performance could be obtained using B-frames, and that the AC-3* audio compression system exhibited the best overall technical performance in their tests. The Grand Alliance also reported that they had decided to replace the 960-active-line video format with a 1080-active-line video format. The Technical Subgroup approved the proposed subsystems.

The Grand Alliance, with assistance from the Transmission Expert Group, performed tests on 32 QAM (quadrature amplitude modulation) and 8-VSB (vestigial sideband) subsystems in January 1994. Both subsystems were tested also for high data rate cable transmission (256 QAM and 16-VSB). In February 1994 the Grand Alliance reported that the VSB system exhibited the best overall technical performance and proposed that the modulation subsystem be VSB. The Technical Subgroup approved the proposal. The VSB subsystem was subsequently subjected to field tests in Charlotte, North Carolina. Measurements were made at almost 200 sites during a three-month period.

This completed the selection of all subsystems of the Grand Alliance HDTV System. The Grand Alliance was authorized to construct a prototype for testing by the Advisory Committee. Laboratory tests began late in 1994. Early in 1995 the Advisory Committee decided to add standard definition television (SDTV) video formats to the proposed system. Characteristics of the resulting ATV system are shown in Table 13.1. (Note that the technical description of the Grand Alliance system in the subsequent sections of this chapter will discuss only the HDTV

**TABLE 13.1**   ATV System Characteristics

| | |
|---|---|
| Video formats | 1920 (H)×1080 (V), 16:9 aspect ratio, interlaced scan at 60 Hz, progressive scan at 30 Hz and 24 Hz |
| | 1280 (H) × 720 (V), 16:9 aspect ratio, progressive scan at 60 Hz, 30 Hz, and 24 Hz |
| | 704 (H) × 480 (V), 16:9 and 4:3 aspect ratios, interlaced scan at 60 Hz, progressive scan at 60 Hz, 30 Hz, and 24 Hz |
| | 640 (H) × 480 (V), 4:3 aspect ratio, interlaced scan at 60 Hz, progressive scan at 60 Hz, 30 Hz, and 24 Hz |
| | Vertical rates also at 59.94 Hz, 29.97 Hz, and 23.98 Hz |
| Video compression | MPEG-2 (main profile) |
| Audio compression | AC-3 |
| Transport | MPEG-2 transport stream |
| Modulation | 8-VSB |

---

*The AC-3 audio compression system was developed by Dolby Laboratories.

video formats proposed by the Grand Alliance.) During the summer of 1995 field tests were conducted. Also in 1995 the Advanced Television Systems Committee (ATSC) completed the documentation of the ATV system as the *ATSC Digital Television Standard.* In late 1995 the Advisory Committee recommended adoption of this ATV system to the FCC as the terrestrial HDTV broadcasting standard for the United States.

## 13.2   TECHNICAL OVERVIEW OF THE GRAND ALLIANCE HDTV SYSTEM

The Technical Subgroup approved specifications of the Grand Alliance HDTV System.[7] The information contained in the technical description that follows is taken from those specifications.

A simplified diagram of the Grand Alliance HDTV System encoder is shown in Fig. 13.1. The input video conforms to SMPTE proposed standards for the 1920 × 1080 system[8] or the 1280 × 720 system.[9] The input may contain either 1080 active lines or 720 active lines—the choice is left to the user. In either case, the number of horizontal picture elements (1920 or 1280) results in square pixels because the aspect ratio is 16:9. With 1080 active lines, the vertical rate can be 60 (or 59.94) fields per second with interlaced scan. With 720 active lines, the vertical rate can be 60 (or 59.94) frames per second with progressive scan. If the video input is from scanned film, the encoder will detect the frame rate (30, 29.97, 24, or 23.98 Hz) and convert the 60-Hz video to progressive scan video at the film frame rate.* Although the Grand Alliance prototype will not be designed to directly accept inputs at the 30- or 24-Hz frame rate, this would be possible in Grand Alliance encoders in the future. Anticipating this possibility, SMPTE plans to document the 1080 and 720 proposed standards also at picture rates of 30 and 24 Hz.

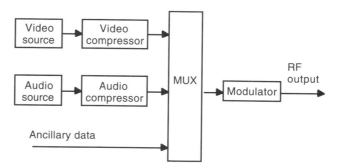

**FIGURE 13.1**   Grand Alliance encoder.

Video compression is accomplished in accordance with the MPEG-2 video standard[10] at the main profile/high level. The video encoder output is packetized in variable-length packets of data called packetized elementary stream (PES) packets. The video compression is explained in Section 13.3.

Audio compression is accomplished using the AC-3 system.[11,12] A standard for AC-3 is being documented by ATSC.[13] The audio encoder output also is packetized in PES packets. The audio compression is explained in Section 13.4.

---

*Throughout the remainder of this article, vertical rates of 60, 30, or 24 will be used. It should be understood that, in each case, the vertical rate also can be 59.94, 29.97, or 23.98 (1000/1001 times 60, 30, and 24). The capability to use either set of numbers allows eventual phase-out of the NTSC-based vertical rates.

The video and audio PES packets, along with any ancillary data (which could be in the form of PES packets), are presented to the multiplexer. The output of the multiplexer is a stream of fixed-length 188-byte MPEG-2 transport stream packets. Both the PES packets and the transport packets are formed in accordance with the MPEG-2 systems standard.[14] The multiplex and transport are explained in Section 13.5. The MPEG-2 transport stream packets are presented to the modulator, where the data are encoded for the channel and a modulated carrier is generated. The channel coding and modulation are explained in Section 13.6.

A summary of the specifications of the Grand Alliance HDTV system is given in Tables 13.2 through 13.5. Table 13.2 lists video specifications, Table 13.3 lists audio specifications, Table 13.4 lists transport specifications, and Table 13.5 lists transmission specifications.

**TABLE 13.2**   Video Specifications

| Video parameter | Format 1 | Format 2 |
|---|---|---|
| Active pixels | 1280 (H) × 720 (V) | 1920 (H) × 1080 (V) |
| Total samples | 1600 (H) × 787.5 (V) | 2200 (H) × 1125 (V) |
| Frame rate | 60 Hz progressive; 30 Hz progressive; 24 Hz progressive | 60 Hz interlaced; 30 Hz progressive; 24 Hz progressive |
| Chrominance sampling | 4:2:0 | |
| Aspect ratio | 16:9 | |
| Data rate | Selected fixed-rate (10–45 Mb/s)/variable | |
| Colorimetry | SMPTE 240M | |
| Picture-coding types | Intra coded (I); predictive coded (P); bidirectionally predictive coded (B) | |
| Video refresh | I-picture; progressive | |
| Picture structure | Frame | Frame; Field (60 Hz only) |
| Coefficient scan pattern | Zigzag | Zigzag; alternate zigzag |
| DCT modes | Frame | Frame; field (60 Hz only) |
| Motion compensation modes | Frame | Frame; field (60 Hz only); dual prime (60 Hz only) |
| P-frame motion vector range | Horizontal: Unlimited by syntax Vertical: −128, +127.5 | |
| B-frame motion vector range (forward and backward) | Horizontal: Unlimited by syntax Vertical: −128, +127.5 | |
| Motion vector precision | 1/2 pixel | |
| DC coefficient precision | 8 bits; 9 bits; 10 bits | |
| Rate control | Modified TM5 with forward analyzer | |
| Film mode processing | Automated 3:2 pulldown detection and coding | |
| Maximum VBV buffer size | 8 Mb | |
| Intra/inter quantization | Downloadable matrices (scene-dependent) | |
| VLC coding | Separate intra and inter run length; amplitude codebooks | |
| Error concealment | Motion-compensated frame holding (slice level)hfill | |

**TABLE 13.3**   Audio Specifications

| Audio parameter | |
|---|---|
| Number of channels | 5.1 |
| Audio bandwidth | 10–20 kHz |
| Sampling frequency | 48 kHz |
| Dynamic range | 100 dB |
| Compressed data rate | 384 kb/s |

**TABLE 13.4**  Transport Specifications

| Transport parameter | |
|---|---|
| Multiplex technique | MPEG-2 systems layer |
| Packet size | 188 bytes |
| Packet header | 4 bytes including sync |
| Number of services | |
|     Conditional access | Payload scrambled on service basis |
|     Error handling | 4-bit continuity counter |
|     Prioritization | 1 bit/packet |
| System multiplex | Multiple program capability described in PSI stream |

**TABLE 13.5**  Transmission Specifications

| Transmission parameter | Terrestrial mode | High-data-rate cable mode |
|---|---|---|
| Channel bandwidth | 6 MHz | 6 MHz |
| Excess bandwidth | 11.5% | 11.5% |
| Symbol rate | 10.76 Msymbols/s | 10.76 Msymbols/s |
| Bits per symbol | 3 | 4 |
| Trellis FEC | 2/3 rate | None |
| Reed-Solomon FEC | $(207,187)\ T = 10$ | $(207,187)\ T = 10$ |
| Segment length | 832 symbols | 832 symbols |
| Segment sync | 4 symbols per segment | 4 symbols per segment |
| Frame sync | 1 per 313 segments | 1 per 313 segments |
| Payload data rate | 19.3 Mb/s | 38.6 Mb/s |
| NTSC cochannel rejection | NTSC rejection filter in receiver | N/A |
| Pilot power contribution | 0.3 dB | 0.3 dB |
| C/N threshold | 14.9 dB | 28.3 dB |

## 13.3  *VIDEO COMPRESSION*

The bit rate required for an RGB HDTV studio signal with 1080 active lines, 1920 samples per active line, 8 bits per sample, and 30 pictures per second is $3 \times 1080 \times 1920 \times 8 \times 30 \simeq 1.5$ Gb/s with no bit rate reduction. To broadcast such a signal in a 6-MHz channel, with a service area comparable to the NTSC service area, requires that the data rate be compressed to something less than 20 Mb/s, a factor of 75. Techniques that can be used to accomplish this compression are source-adaptive processing, reduction of temporal redundancy, reduction of spatial redundancy, exploitation of the human visual system, and increased coding efficiency.

### 13.3.1.  Video Encoder

Source-adaptive processing is applied to the RGB components which, to a human observer, are highly correlated with each other. The RGB signal is changed to luminance and chrominance components to take advantage of this correlation. Furthermore, the human visual system is more sensitive to high frequencies in the luminance component than to high frequencies in the chrominance components. To take advantage of these characteristics, the chrominance components are low-pass filtered, and subsampled by a factor of two both horizontally and vertically.

Figure 13.2 is a diagram showing the essential elements in video compression. Temporal re-dundancy is reduced using the following process. In the motion estimator an input video frame, called a *new picture,* is compared with a previously transmitted picture held in the picture mem-ory. Macroblocks (an area 16 picture elements wide and 16 picture elements high) of the previous picture are examined to determine if a close match can be found in the new picture. When a close match is found, a motion vector is produced describing the direction and distance the macro-block moved. A predicted picture is generated by the combination of all the close matches, as shown in Fig 13.3. Finally, the new picture is compared with the predicted picture on a picture element–by–picture element basis to produce a difference picture.

The process of reducing spatial redundancy is begun by performing a discrete cosine trans-form (DCT) on the difference picture using $8 \times 8$ blocks. The first value in the DCT matrix (top

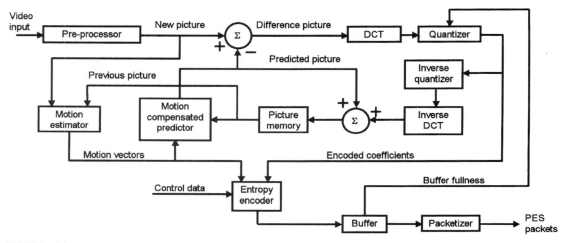

**FIGURE 13.2**   Video encoder.

Blocks of previous picture used to predict new picture.

Previous picture after using motion vectors to adjust block positions.

**FIGURE 13.3**   Predicted picture.

left corner) represents the DC value of the 64 picture elements of the $8 \times 8$ block. The other 63 values in the matrix represent the AC values of the DCT with higher horizontal and vertical frequencies as one moves to the bottom-right corner of the matrix. If there is little detail in the picture, these higher-frequency values become very small. The DCT values are presented to a quantizer which, in an irreversible manner, can "round off" the values. Quantization noise arises because of rounding off the coefficients. It is important that the round-off be done in a manner that maintains the highest possible picture quality. When quantizing the coefficients, the perceptual importance of the various coefficients can be exploited by allocating the bits to the perceptually more important areas. The quantizer coarseness is adaptive and is coarsest (has the fewest bits) when the quantization errors are expected to be least noticeable. The DCT coefficients are transmitted in a zigzag order, as shown in Fig. 13.4. When the picture is interlaced, the DCT coefficients are read in an alternate zigzag fashion. After rounding, the higher-frequency coefficients often have zero value. This leads to the frequent occurrence of several sequential zero-value coefficients.

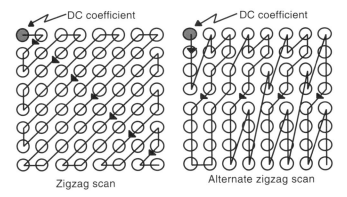

**FIGURE 13.4**   Scanning of DCT coefficients.

The quantizer output is presented to an entropy encoder, which increases the coding efficiency by assigning shorter codes to more frequently occurring sequences. An example of entropy encoding is found in the Morse code. The frequently occurring letter "e" is given the shortest one-symbol code, whereas the infrequently occurring letter "q" is given a longer four-symbol code. Another example is run-length coding, in which several sequential same-value coefficients can be represented with fewer bits by encoding the value of the coefficients and the number of times the coefficient is repeated rather than encoding the value of each and every repeated coefficient. This is especially useful when the higher-frequency DCT coefficients have zero value. Run-length coding is used in the Grand Alliance system. Huffman coding, also used in the Grand Alliance system, is one of the most common entropy encoding schemes.

The entropy encoder bitstream is placed in a buffer at a variable input rate, but it is taken from the buffer at a constant output rate. This is done to match the capacity of the transmission channel and to protect the decoder rate buffer from overflow or underflow. If the encoder buffer approaches maximum fullness, the quantizer is signaled to decrease the precision of coefficients to reduce the instantaneous bit rate. If the encoder buffer approaches minimum fullness, the quantizer is allowed to increase the precision of coefficients. The output of the buffer is packetized as a stream of PES packets. Because the transmitted picture is required also at the encoder for the motion-compensated prediction loop, the quantizer output is presented to the inverse quantizer, then to the inverse DCT, summed with the predicted picture, and then placed in the picture memory.

In the description thus far, it has been assumed that the picture used to predict the new picture was, in fact, the previous picture from the video source. An advantage may be gained in some cases by predicting the new picture from a future picture, or from both a past and a future picture. For example, after a video switch, a future frame is a better predictor of the current frame than is a past frame. In the MPEG standard, three types of frames are defined. An I-frame is a picture that is transmitted as a new picture, not as a difference picture. A P-frame is a picture that is predicted from a previous P- or I-frame. A B-frame is a picture that is predicted from both a past P- or I-frame and a future P- or I-frame. This is illustrated in Fig 13.5. Inclusion of B-frames requires an additional frame of storage in the decoder. Before the information describing the B-frame can be transmitted, the information for both anchor frames must be transmitted and stored. As a result, the transmission order is different from the display order.

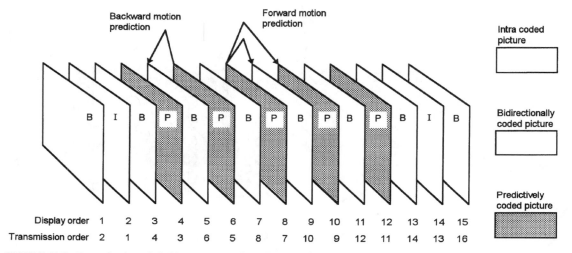

**FIGURE 13.5**    Example of a coded video sequence using I-frames, P-frames, and B-frames.

Because the two fields of an interlaced picture represent two different points in time, they can vary significantly when there is a lot of motion. In such a case it may be preferable to make the motion-compensated prediction based on fields rather than frames. This choice is facilitated by allowing both prediction modes. Another prediction mode, dual prime, is also supported. Dual prime is available only for interlaced video material and only when B-frames are not in use. It allows motion vectors determined in one field to be used in the other field.

With a motion-compensated prediction loop, refreshing the received image is necessary whenever the receiver is first turned on or tuned to another channel, after a loss of signal, and when major transmission errors occur. In each case the picture in the receiver memory will be different from the picture in the memory at the encoder. Because the transmitter cannot know when the pictures are different, it is necessary to periodically transmit the new picture, rather than the difference picture. Otherwise, errors will propagate in the receiver. Two refresh methods are allowed: I-frame refresh and progressive refresh. With I-frame refresh an entire frame is transmitted at a periodic rate. This is accomplished by transmitting the DCT coefficients of the new picture in place of the DCT coefficients of the difference picture. With progressive refresh the DCT coefficients of a group of blocks (macroblock) of the new picture are transmitted at a periodic rate in place of the DCT coefficients of the same group of blocks of the difference picture.

### 13.3.2 Video Decoder

The video decoder is shown in Fig. 13.6. Following depacketizing of the PES packets, the encoded coefficients and motion vectors are held in a buffer until they are needed to decode the next picture. The entropy decoder performs the inverse function of the entropy encoder. The encoded coefficients, after inverse quantization and inverse DCT, are added to the predicted picture to produce the new picture. The predicted picture is obtained by using the received motion vectors to move portions of the previously transmitted picture.

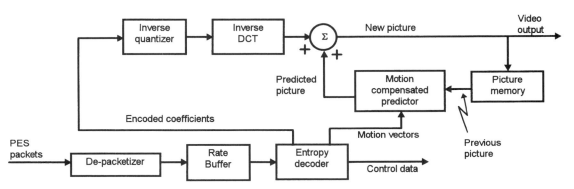

**FIGURE 13.6** Video decoder.

## 13.4  *AUDIO COMPRESSION*

The Grand Alliance audio system uses AC-3 technology. The main audio service can range from a simple monophonic service, through stereo, up to a six-channel surround-sound service (left, center, right, left surround, right surround, and subwoofer). The sixth channel conveys only low-frequency (subwoofer) information and is often referred to as 0.1 channel, for a total of 5.1 channels. Several services, in addition to the main audio service, can be provided. Examples are services for the hearing or visually impaired, dynamic range control, and multiple languages.

When the audio service is a multichannel service and mono or stereo outputs are required in the receiver, the down-mix is done in the decoder. The down-mix may be done in the frequency domain, reducing the complexity of mono and stereo receivers. The program originator can indicate in the bitstream which down-mix coefficients are appropriate for a given program.

The audio sampling rate is 48 kHz. With six channels and 18 bits per sample, the total bit rate before compression is $48,000 \times 6 \times 18 \cong 5$ Mb/s. The compressed data rate is 384 kb/s for the 5.1 channel service, representing a compression factor of 13.

### 13.4.1  AC-3 Encoder

Because of the frequency-masking properties of human hearing, a frequency domain representation of audio is used in the bit rate compression. As shown in the diagram of the AC-3 encoder in Fig 13.7, the audio input channels are transformed from the time domain to the frequency domain using the time domain–aliasing cancellation (TDAC) transform. The block size is 512 points. Each input time-point is represented in two transforms. The 512-point transform is done every 256 points, providing a time resolution of 5.3 ms at a 48-kHz sampling rate. The frequency

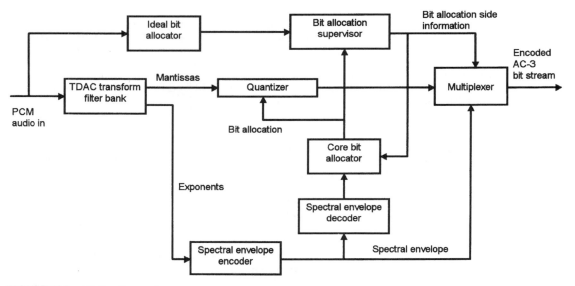

**FIGURE 13.7**   AC-3 audio encoder.

resolution is 93 Hz and is uniform across the spectrum. During transients, the encoder switches to a 256-point transform, giving a time resolution of 2.7 ms.

The output of the TDAC transform is a set of frequency coefficients for each channel. Each transform coefficient is encoded into an exponent and a mantissa. The exponent provides a wide dynamic range. The mantissa is encoded with limited precision, resulting in quantizing noise. The exponents of each channel are encoded into a representation of the overall signal spectrum, referred to as the *spectral envelope*. The time and frequency resolution of each spectral envelope is signal-dependent. The frequency resolution varies from 93 Hz to 750 Hz, depending on the signal. The time resolution varies from 5.3 ms to 32 ms. The algorithm that determines the time and frequency resolution of the spectral envelope is in the AC-3 encoder only, and thus may be improved in the future without affecting decoders in the field. The AC-3 encoder decodes the spectral envelope to make use of the identical information that will be available in the receiver. The decoded version is used as a reference in quantizing the transform coefficients and in determining the bit allocation.

Allocation of bits to the various frequency components of the audio signals is a critical part of the encoder design. AC-3 makes use of hybrid forward/backward adaptive bit allocation. With forward bit allocation, the encoder calculates the bit allocation and explicitly encodes the bit allocation into the bitstream. This method allows for the most accurate bit allocation because the encoder has full knowledge of the input signal. Also, the psychoacoustic model is resident only in the encoder and may be improved without affecting decoders in the field. With backward bit allocation, the bit allocation is calculated from the encoded data without explicit information from the encoder. This method is more efficient because all of the bits are available for encoding audio. Disadvantages of backward bit allocation are that the bit allocation must be computed from information in the bitstream that is not fully accurate, and that the psychoacoustic model cannot be updated because it is included in the decoder. The AC-3 encoder, with a hybrid forward/backward adaptive bit allocation, has a relatively simple backward adaptive core bit allocation routine, which runs in both the encoder and the decoder. The decoder psychoacoustic model can be adjusted by sending some parameters of the model forward in the bitstream. The encoder can compare the results of the bit allocation based on the core routine to an ideal allocation. If a better match can be made, the encoder can cause the core bit allocation in both the encoder and decoder to change. When it is not possible to approach the ideal allocation by varying parameters, the encoder can send bit allocation information directly. The multiple channels are allocated bits from a common bit pool.

### 13.4.2.  AC-3 Decoder

The AC-3 decoder, shown in Fig. 13.8, performs the inverse functions of the encoder. The input serial data are demultiplexed, producing the quantized mantissas, spectral envelope, and bit allocation side information. The spectral envelopes are decoded and the bit allocation is computed. After inverse quantization of the mantissas, they are combined with the exponents to form the frequency coefficients. The frequency coefficients are inverse-transformed to reproduce the original PCM audio signals.

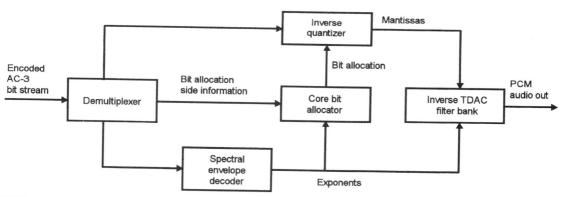

**FIGURE 13.8**    AC-3 audio decoder.

## 13.5   TRANSPORT

The Grand Alliance HDTV system uses a constrained subset of the MPEG-2 transport stream syntax. MPEG-2 defines two alternative approaches—program streams and transport streams. Program streams are designed for use in relatively error-free environments. Transport streams are designed for use in environments where errors are likely, such as transmission in noisy media. Because the Grand Alliance system is designed for terrestrial broadcasting, an environment where errors are likely, transport streams are the proper choice. Both approaches, however, are described here to illustrate the differences.

Both program streams and transport streams provide syntax to synchronize the decoding of the video and audio information while ensuring that data buffers in the decoders do not overflow or underflow. Both streams include time-stamp information required for synchronizing the video and audio. Both stream definitions are packet-oriented multiplexes. Program streams use variable-length packets. Transport streams use fixed-length 188-byte packets.

Another type of packet is the packetized elementary stream (PES) packet. After compression, video data and audio data are packaged into separate PES packets. PES packets may be fixed-length or variable-length. PES packets contain the complete information required to reconstruct the video or the audio. A program consists of elementary streams with a common time base, for example, video PES packets, audio PES packets, and possibly ancillary data PES packets, along with a control data stream.

The program stream results from combining one or more streams of PES packets, with a common time base, into a single stream. The transport stream results from combining one or more programs (each program consisting of one or more streams of PES packets with a common time base), with one or more independent time bases, into a single stream.

The three different types of packets discussed here—program stream packets, transport stream packets, and PES packets—are illustrated in Fig. 13.9. A system-level multiplex of two different programs is illustrated in Fig. 13.10. Each transport packet begins with a four-byte header. The contents of the packet and the nature of the data are identified by the packet header. The remaining 184 bytes are the payload. Individual PES packets, including the PES headers, are transmitted as the payload. The beginning of each PES packet is aligned with the beginning of the payload of a transport packet—stuffing bytes are used to fill partially full transport

Program Stream packets are designed for relatively error-free environments. Transport Stream packets are designed for environments where errors are likely. The Grand Alliance HDTV System uses Transport Stream packets.

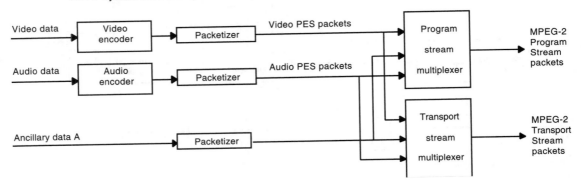

**FIGURE 13.9**  MPEG-2 packets.

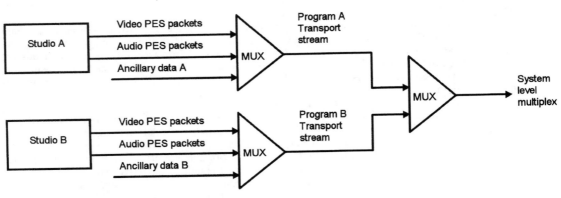

**FIGURE 13.10**  System-level multiplex.

packets. This means that every transport packet contains only one type of data—video, audio, or ancillary. The four-byte transport header also provides the functions of packet synchronization, error handling, and conditional access.

For conditional access, audio, video, and ancillary data can be scrambled independently. Information in the transport header of the individual packets indicates whether the payload in that packet is scrambled. The transport header is always transmitted in the clear. In the Grand Alliance system, scrambling is implemented only within transport packets—not within PES packets.

Sometimes additional header information is required. This is provided by the adaptation header, a variable-length field placed in the payload of the transport packet. Its presence is flagged in the transport header. Functions of this layer include synchronization (audio and video program timing), support for random entry into the compressed bitstream (tuning to a new channel), and support for local program insertion (inserting local programming into a network program).

The transport stream provides easy interoperability with asynchronous transfer mode (ATM) transmission. ATM cells consist of a five-byte header and a 48-byte payload. The ATM header is used primarily for networking purposes. There are various ways that the transport packets can be mapped into ATM cells. The transport packet size was selected to ease this transfer. Note that one transport packet (188 bytes including header) can fit into four ATM cells ($4 \times 48 = 192$-byte payload).

## 13.6  MODULATION

The VSB transmission system provides two modes, one for terrestrial broadcasting (8-VSB) and one for high-data-rate cable transmission (16-VSB). Both modes make use of Reed-Solomon coding, segment sync, a pilot, and a training signal. The terrestrial mode adds trellis coding. The symbol rate for both modes is 10.76 million symbols per second. The terrestrial mode uses 3 bits/symbol. Because the cable environment is less severe, a higher data rate is transmitted by using 4 bits/symbol and no trellis overhead. The C/N threshold for the terrestrial mode is 14.9 dB. The C/N threshold for the high-data-rate cable mode is 28.3 dB. The terrestrial mode has a payload data rate of 19.3 Mb/s. The high-data-rate cable mode has a payload data rate of 38.6 Mb/s.

The Reed-Solomon code is a (207,187) $t = 10$ code (the data block size is 187 bytes with 20 parity bytes added for error correction) and can correct up to 10 byte errors per block. A 2/3-rate trellis code is used in the terrestrial mode (one input bit is encoded into two output bits, whereas the other input bit is not encoded).

Data are transmitted according to the data frame shown in Fig. 13.11. The data frame begins with a first data field sync segment followed by 312 data segments, then a second data field sync segment followed by another 312 data segments. Each segment consists of 4 symbols of segment sync followed by 828 symbols of data. The symbols during segment sync and data field sync carry only 1 bit/symbol in order to make packet- and clock-recovery rugged.

In the terrestrial mode, one segment corresponds to one MPEG-2 transport packet, as follows. The number of bits of data plus FEC per segment is 2484 (828 symbols times 3 bits/symbol). The MPEG-2 transport packet contains 188 bytes. Not transmitting the sync byte makes this 187. Because Reed-Solomon encoding adds 20 bytes for every 187 payload bytes, the total becomes 207 bytes. Because trellis coding adds one bit for every two input bits, this number must be increased by the ratio 3/2, making the total 2484 bits. Thus, one segment is 2484 bits and one MPEG-2 transport packet requires 2484 bits in transmission.

The symbols modulate a single carrier using suppressed-carrier modulation. Before transmission, most of the lower sideband is removed. The resulting spectrum is flat, except for the band edges. A small pilot, used in the receiver to achieve carrier lock, is added 310 kHz above the lower band edge.

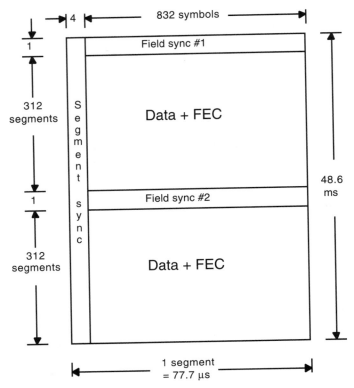

**FIGURE 13.11**   VSB data frame.

### 13.6.1   VSB Transmitter

A diagram of the VSB transmitter is shown in Fig. 13.12. The data randomizer performs an exclusive OR on the incoming data with a 16-bit maximum-length pseudorandom sequence (PRS) that is locked to the data frame. The data are randomized to ensure that random data are transmitted, even when the data are constant. Segment sync, data field sync, and Reed-Solomon parity bytes are not randomized. After randomizing, the signal is encoded using a (207,187)

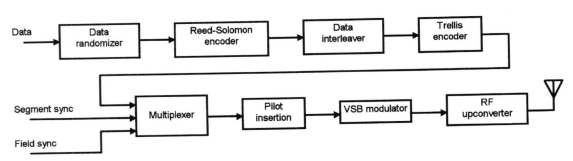

**FIGURE 13.12**   VSB transmitter.

$t = 10$ Reed-Solomon code. The interleaver, a 52–data segment (intersegment) diagonal byte interleaver, spreads data from one Reed-Solomon block over a longer time to give protection against burst errors.

The terrestrial transmission mode uses a 2/3-rate trellis code. The signaling waveform is a 3-bit one-dimensional constellation. To help protect the trellis decoder against short-burst interference (such as impulse noise or NTSC cochannel interference) 12-symbol code interleaving is employed in the transmitter. Twelve identical trellis encoders operate on interleaved data symbols. In the high-data-rate cable mode, there is only a simple mapper that converts data to multilevel symbols, as opposed to the trellis encoder/mapper used in the terrestrial mode.

Segment sync and field sync symbols are not Reed-Solomon–encoded, trellis-encoded, or interleaved. Field sync can serve five purposes. It can be used (1) as a means to determine the beginning of each data field, (2) as a training reference signal in the receiver, (3) in the receiver to determine whether the NTSC rejection filter should be used, (4) for system diagnostics, and (5) as a reset by the receiver phase tracker.

A small pilot, at the suppressed-carrier frequency, is added to the suppressed-carrier RF signal to allow robust carrier recovery in the receiver during extreme conditions. At the output of the multiplexer, the data signal takes the relative values of $\pm 1$, $\pm 3$, $\pm 5$, and $\pm 7$. To add the pilot, the relative value of 1.25 is added to every data and sync value. This has the effect of adding a small in-phase pilot to the baseband data signal in a digital manner, providing a highly stable and accurate pilot.

The baseband data signal is filtered with a complex filter to produce in-phase and quadrature components for orthogonal modulation. These two signals are converted to analog signals and then used to quadrature-modulate the IF carrier, creating a vestigial sideband IF signal by sideband cancellation.

The frequency of the RF up-converter oscillator in advanced television (ATV) terrestrial broadcasts is typically the same as the nominal NTSC carrier frequency and not an offset NTSC carrier frequency. ATV cochannel interference into NTSC is noiselike and does not change with offset. Even the pilot interference into NTSC is not significantly reduced with offset because it is so small and falls far down the Nyquist slope of NTSC receivers.

With ATV cochannel interference into ATV, carrier offset can prevent misconvergence of the receiver's adaptive equalizer. If the data field sync of the interfering signal occurs during the data field sync of the desired signal, the adaptive equalizer could misinterpret the interference as a ghost. A carrier offset equal to half the data segment frequency causes the interference to have no effect in the adaptive equalizer.

### 13.6.2  VSB Receiver

A diagram of the Grand Alliance prototype VSB receiver is shown in Fig. 13.13. After the signal has traversed the tuner, IF, and synchronous detector stages and the clocks and syncs have been recovered, the data are switched into an NTSC rejection filter if NTSC cochannel interference is detected. The NTSC comb filter is designed with seven nulls in the 6-MHz channel. The NTSC picture carrier falls near the second null; the NTSC color subcarrier falls at the sixth null; and the NTSC sound carrier falls near the seventh null. The filter is a 12-symbol feed-forward subtractive comb filter. Although the comb filter reduces NTSC cochannel interference, the data are also affected. Also, white noise performance is degraded by 3 dB. Therefore, if little or no NTSC interference is present, the comb filter is automatically switched out of the signal path. The NTSC comb filter is not required in the high-data-rate cable mode, because cochannel interference is not present on cable.

The equalizer/ghost canceler delivered for the Grand Alliance test in January 1994 used a least-mean-square (LMS) algorithm, adapting on the data field sync. By adapting on a known training signal, the circuit converges even in extreme conditions. After reaching convergence on the data field sync, the circuit is switched to equalize on the random data for high-speed tracking

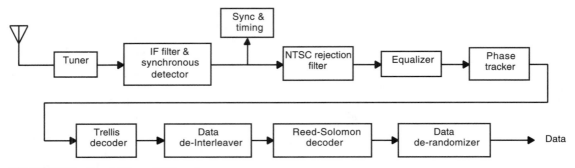

**FIGURE 13.13**    Grand Alliance VSB receiver.

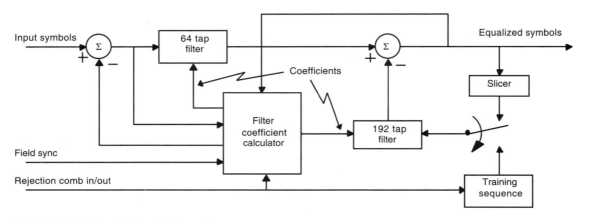

**FIGURE 13.14**    Grand Alliance VSB equalizer.

of moving ghosts such as airplane flutter. A diagram of the equalizer is shown in Fig. 13.14. The equalizer filter consists of two parts: a 64-tap feed-forward transversal filter, followed by a 192-tap decision-feedback section. Following the equalizer, the data symbols are used to detect and remove phase noise.

Because 12-symbol code interleaving is used in the trellis encoder, the receiver uses 12 trellis decoders in parallel. The trellis decoder has two modes, depending on whether the NTSC rejection filter is in use. When NTSC cochannel interference is detected, the NTSC rejection filter is switched into the signal path and a trellis decoder optimized for use in tandem with the comb filter is used. When NTSC interference is not detected, the NTSC rejection filter is switched out of the signal path and an optimal trellis decoder is used. In the high-data-rate cable mode, the trellis decoder is replaced by a slicer that translates the multilevel symbols into data.

The de-interleaver performs the inverse function of the transmitter interleaver. The (207,187) $t = 10$ Reed-Solomon decoder uses the 20 parity bytes to perform the byte-error correction on a segment-by-segment basis. The de-randomizer accepts the error-corrected data bytes from the Reed-Solomon decoder and applies to the data the same PRS code that was used at the transmitter.

## 13.7   CONCLUSIONS

The Grand Alliance HDTV system is the product of many people's efforts over many years. The visible effort began when the Advisory Committee on Advanced Television Service was formed in 1987. Not so visible at the outset was the effort of many engineers from several different organizations designing proposed systems. Those efforts really began to show in 1991, when testing of the proposed systems began.

The testing showed strong points and weak points in the original designs. One extremely strong point was the digital design that had been adopted in four of the five HDTV systems tested. After the testing and analyses were complete, the Grand Alliance was formed by the proponents of the digital systems. The Grand Alliance, working with the Advisory Committee, has designed a system that will satisfy the needs of North America. Subsystems have been selected based on technical excellence. The system has a great deal of flexibility to facilitate interoperability and is heavily based on international standards.

## GLOSSARY

**Interlaced Scanning**   A scanning process in which the distance from center to center of successively scanned lines is two or more times the nominal line width, and in which the adjacent lines belong to different fields.

**MPEG**   Moving Picture Experts Group.

**Packet**   A group of binary digits that is switched as a composite whole with data, control, and error elements arranged in a specified format.

**Pixel, Pel, Picture element**   The smallest area of a television picture capable of being de-lineated by an electric signal passed through the system or part of it.

**Progressive (sequential) scanning**   A rectilinear scanning process in which the distance from center to center of successively scanned lines is equal to the nominal line width.

**Quantize**   To subdivide the range of values of a variable into a finite number of nonover-lapping subranges or intervals, each of which is represented by an assigned value within the subrange.

**SMPTE**   Society of Motion Picture and Television Engineers.

**Vestigial sideband**   The transmitted portion of the sideband that has been largely suppressed by a transducer having a gradual cutoff in the neighborhood of the carrier frequency—the other sideband being transmitted without much suppression.

## REFERENCES

1. R. Hopkins and Davies, K. P., "Development of HDTV emission systems in North America," *IEEE Trans. on Broadcasting,* vol. 35, no. 3, Sept. 1989.

2. R. Hopkins and Davies, K. P., "HDTV emission systems approach in North America," *ITU Telecommunication Journal,* vol. 57, May 1990, pp. 330–336.

3. "ATV System Recommendation," *IEEE Trans. on Broadcasting,* vol. 39, no. 1, Mar. 1993, pp. 2–245.

4. "ATV System Recommendation," *Proceedings, 1993 NAB HDTV World Conference,* pp. 237–493.

5. R. Hopkins, "Progress on HDTV broadcasting standards in the United States," *Image Communication,* vol. 5, nos. 5–6, Dec. 1993, pp. 355–378.

6. R. Hopkins, "Choosing an American digital HDTV terrestrial broadcasting system," *Proceedings of the IEEE,* vol. 82, no. 4, Apr. 1994, pp. 554–563.

7. *Grand Alliance HDTV System Specification,* version 1.0, Apr. 14, 1994. Available from International Transcription Services, 2100 M Street NW, Suite 140, Washington, D.C. 20037. Telephone (202)857-3800.

8. SMPTE S17.394, "1920 × 1080 Scanning and Interface," proposed SMPTE standard for television.

9. SMPTE S17.392, "1280 × 720 Scanning and Interface," proposed SMPTE standard for television.

10. ISO/IEC DIS 13818-2, "MPEG-2 Video," draft international standard.

11. M. Davis, "The AC-3 multichannel coder," *AES 95th Convention,* Preprint 3774, Oct. 1993.

12. C. C. Todd, Davidson, G. A., Davis, M. F., Felder, L. D., Link, B. D., and Vernon, S., "AC-3: Flexible perceptual coding for audio transmission and storage," *AES 96th Convention,* Preprint 3796, Feb. 1994.

13. ATSC T3/S7-016, "Digital Audio Compression (AC-3)," draft ATSC standard.

14. ISO/IEC DIS 13818-1, "MPEG-2 Systems," draft international standard.

## BIBLIOGRAPHY

U.S. Advanced Television Systems Committee, *ATSC Digital Television Standard,* Washington, D.C., Sept. 1995.

U.S. Advanced Television Systems Committee, *Guide to the Use of the ATSC Digital Television Standard,* Washington, D.C., Oct. 1995.

U.S. Advanced Television Systems Committee, *Digital Audio Compression (AC-3),* Washington, D.C., Dec. 1995.

## ABOUT THE AUTHOR

Robert Hopkins received the B.S. degree in electrical engineering from Purdue University, West Lafayette, Indiana, and the M.S. and Ph.D. degrees from Rutgers University, New Brunswick, New Jersey. He is also a graduate of the Harvard Business School Program for Management Development. His current position is vice president and general manager of the Sony Pictures High Definition Center. From 1985 to 1996 he was the executive director of the United States Advanced Television Systems Committee (ATSC), a standards organization sponsored by more than 50 companies involved in HDTV. He was employed by RCA from 1964 to 1985 at the David Sarnoff Research Center, the Broadcast Systems Division, and as managing director of RCA Jersey Limited, Channel Islands, Great Britain. Dr. Hopkins is a fellow of SMPTE and a senior member of IEEE.

## ACKNOWLEDGMENT

The author wishes to thank several persons who reviewed for accuracy the original paper on which this chapter is based: Stan Baron of NBC, David Bryan of Philips Laboratories, Lynn Claudy of NAB, Carl Eilers of Zenith Electronics Corporation, James Gaspar of Panasonic Advanced Television Laboratory, John Henderson of Hitachi America, Robert Keeler of AT&T Bell Laboratories, Bernard Lechner, James McKinney of ATSC, Woo Paik and Robert Rast of General Instrument Corporation, Terrence Smith and Joel Zdepski of David Sarnoff Research Center, and Craig Todd of Dolby Laboratories.

# CHAPTER 14
# DIGITAL AUDIO BROADCASTING

## SECTION 1
## EUREKA 147—DIGITAL AUDIO BROADCASTING SYSTEM FOR EUROPE AND WORLDWIDE

**Franc Kozamernik**
*European Broadcasting Union*

## 14.1 INTRODUCTION

In Europe, studies on advanced digital audio broadcasting have been conducted since about 1985. Several possible technical approaches—including a simple single-carrier digital system, frequency hopping, spread spectrum, and code modulation—have been taken into close consideration. In 1987 the Eureka 147 Consortium was established and developed in close cooperation with the European Broadcasting Union (EBU), a novel transmission system based on the coded orthogonal frequency division multiplex (COFDM) approach and the advanced audio compression scheme called MASCAM (now MUSICAM). The Eureka 147 system ("Eureka" for short) was first presented to the general public on the occasion of the second session of the WARC-ORB Conference in 1988 in Geneva. It was agreed upon in its present configuration in 1991. Following further extensive demonstrations, field tests, and real-life scrutiny in several countries in Europe and worldwide, Eureka became an official European standard in December 1994,[1] following the usual public inquiry and voting procedure within the European Telecommunications Standards Institute (ETSI). In 1994 Eureka obtained worldwide recognition within the International Telecommunications Union (ITU) to become a recommended sound broadcasting system for both terrestrial and satellite delivery to all types of receivers, including mobile, portable, and, of course, fixed.[2-6] In November 1995 the Eureka system was officially selected as the Canadian system standard for digital radio broadcasting in the frequency band 1452–1492 MHz.

Eureka has all the technical, operational, and economical characteristics to become not only a brilliant piece of communication engineering, but also a universal delivery mechanism for all broadcasting applications—including audio, video, text, data, and multimedia signals, where mobile reception is a primary requirement. It is becoming a commercially successful product in the marketplace worldwide.

In July 1995 a new frequency allotment plan at VHF and in L-band was agreed upon for Europe. This plan provides sufficient frequencies for the start of terrestrial Eureka services and, at the same time, leaves the existing FM services in band II (i.e., 88–108 MHz) untouched, in the short term.

The world's first official Eureka services were inaugurated in the United Kingdom by BBC Radio, and in Sweden by the Swedish Broadcasting Corporation, on 27 September 1995. Several tens of pilot services are on the air in Germany. According to the current plans, more than 100 million people will be served with the Eureka signal in 1997. These extensive broadcasts should encourage electronics manufacturers to bring their consumer Eureka receiver products to the marketplace at an affordable price as soon as possible:

- Governments are facing the hard task of finding ways to share the finite radio spectrum between a mass of conflicting interests, and they welcome Eureka as a highly spectrum-efficient system.
- Broadcasters see the opportunity to offer more services of better quality and presentation for less production and emission cost.
- Manufacturers welcome the opportunity to sell large quantities of Eureka receivers and associated equipment.
- Network operators are keen to build the new distribution and transmitter networks that are required for Eureka terrestrial services.

Not least, the listener welcomes a new technology that offers more program choice and multipath- and interference-free CD-quality signal, as well as a very robust signal when listening in a vehicle or on a portable set.

## 14.2    OVERVIEW OF THE EUREKA SYSTEM

The Eureka system is designed to provide reliable, multiservice digital sound broadcasting for reception by mobile, portable, and fixed receivers, using a simple, nondirectional antenna. The system can be operated at any frequency between 30 MHz and 3 GHz (VHF and UHF) and is designed to provide excellent service reliability in a range of difficult receiving conditions, such as in densely populated urban and suburban areas. The system is intended for use on terrestrial transmitters and networks, and can be used on broadcasting satellites operating in L-band (or in S-band if required). Hybrid (i.e., satellite with complementary terrestrial) networks represent an interesting option, whereby a low-power satellite operating in L-band provides general coverage of a territory in such areas where the line-of-site situation is possible and a network of terrestrial transmitters (fed by the satellite and operating on the same frequency as the satellite) provides signals in obstructed areas. The same receiver is used to capture signals from the satellite and terrestrial transmitters.

In addition to supporting a wide range of sound-coding rates (and, hence, qualities), Eureka is also designed to have a flexible, general-purpose digital multiplex that can support a wide range of source and channel-coding options, including sound-program associated data and independent data/multimedia services.

The Eureka system is a rugged, yet highly spectrum- and power-efficient sound and data broadcasting system. It uses advanced digital techniques to remove redundancy and perceptually irrelevant information from the audio source signal; then, to provide strong error protection, it applies closely controlled redundancy to the signal to be transmitted. The transmitted information is spread in both the frequency and time domains so that the defects of channel distortions and fades may be eliminated from the recovered signal in the receiver—even when working in conditions of severe multipath propagation, whether stationary or mobile.

Efficient spectrum utilization is achieved by interleaving multiple program signals and, additionally, by a special feature of frequency reuse, which permits broadcasting networks to be extended, virtually without limit, by operating additional transmitters on the same radiated frequency. The latter feature is known as the Single-Frequency Network (SFN). This may also employ the gap-filling technique. In this case a gap filler transmitter receives and retransmits the signal on the same frequency without intervening demodulation. This provides coverage of shadowed areas that may arise within the overall coverage area provided by the main broadcast network transmitters.

Nevertheless, the relatively low cochannel protection ratio of the Eureka system also permits adjacent local coverage areas to be planned, on a continuously extending basis, with as few as four different frequency blocks.

### 14.2.1   Summary of the Major Features

Eureka provides a signal that carries a multiplex of several digital services simultaneously. The system bandwidth is about 1.5 MHz, providing a total transport bit rate capacity of just over 2.3 Mb/s in a complete "ensemble." Depending on the requirements of the broadcaster (transmitter coverage, reception quality), the amount of error protection provided is adjustable for each service independently, with a coding overhead ranging from about 33 percent to 300 percent (200 percent for sound). Accordingly, the available services bit rate ranges between about 1.7 Mb/s and 0.6 Mb/s.

The services may contain audio program data or other data services, and a data service may or may not be related to the audio program. The number and bit rate of each individual service is flexible, and receivers are generally able to decode several service components or services simultaneously.

The actual content of the flexible multiplex is described by the *Multiplex Configuration Information* (MCI). This is transported in a specific reserved part of the multiplex known as the Fast Information Channel (FIC), because it does not suffer the inherent delay of time-interleaving that is applied to the Main Service Channel (MSC). In addition, the FIC carries information on the services themselves and the links between the services. In particular, the following principal features have been specified:

- Audio bit rates from 384 kb/s down to 8 kb/s. This enables the multiplex to be configured to provide, typically, six high-quality stereo audio programs or up to, say, 40 mono programs with moderately rugged error protection. An example table of multiplex options for audio services is given in Table 14.1.

**TABLE 14.1**   Examples of Audio Service Capacities in a Eureka Ensemble

| Protection level | 3 | 4 |
|---|---|---|
| Mean code rate, $R_{\text{ave}}$ | ~0.5 | ~0.6 |
| Coded audio rate, kb/s | Number of audio services | |
| 64 | 18 | 20 |
| 128 | 9 | 10 |
| 192 | 6 | 7 |
| 224 | 5 | 6 |
| 256 | 4 | 5 |

- Program-associated data (PAD), embedded in the audio bitstream, for data that are directly linked to the audio program (e.g., dynamic range control, song lyrics, music/speech flag, etc.). The amount of PAD is adjustable (minimum 667 b/s), at the expense of capacity for the coded audio signal within the chosen audio bit rate.
- Data services, whereby each service can be a separately defined stream or can be divided further by means of a packet structure.

- Conditional access (CA), applicable to each individual audio or data service and to each individual packet of packet mode data. Specific subscriber management does not form part of the Eureka system specification;[1] however, Eureka provides CA transport and the actual signal scrambling mechanisms.

- Service information (SI) for (textual) information on the selected Eureka ensemble and selected program, and also complementary machine code for ease of operation of the receiver. Another important SI feature is to establish links between different services in the multiplex and links to other (related) services in another Eureka multiplex or even to existing FM/AM broadcasts.

## 14.3  OUTLINE OF SYSTEM IMPLEMENTATION

### 14.3.1  General Specifications

A conceptual block diagram of the Eureka system is shown in Figs. 14.1 and 14.2. Figure 14.1 shows a conceptual transmitter drive in which each service signal is coded individually at the source level and then error-protected and time-interleaved. Then it is multiplexed into the Main Service Channel (MSC) with other similarly processed service signals according to a predetermined but changeable services configuration. The multiplexer output is frequency interleaved and combined with multiplex control and service information, which travels in a Fast Information Channel (FIC) in order to avoid the delay of time-interleaving. Finally, very rugged synchronization symbols are added before applying orthogonal frequency division multiplexing (OFDM) and differential QPSK modulation onto a large number of carriers to form the Eureka signal.

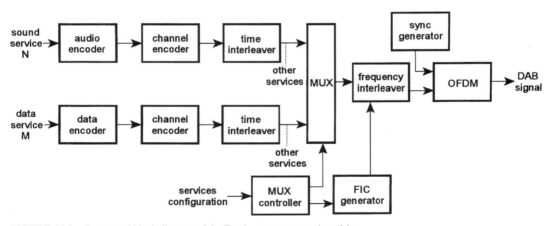

**FIGURE 14.1**   Conceptual block diagram of the Eureka system transmitter drive.

Figure 14.2 shows a conceptual receiver in which the received signal is selected, downconverted, and quadrature demodulated before applying it to an analog-to-digital converter pair. Thereafter, the receiver performs the transmitter operations of Fig. 14.1 in reverse order, having selected the desired Eureka ensemble and acquired synchronization. Thus selection is done in the analog tuner, which performs the tuning and filtering functions. The digitized output of the converter is first fed to the DFT (discrete Fourier transform) stage and differentially demodulated. This is followed by (1) time and frequency de-interleaving processes and (2) error correction

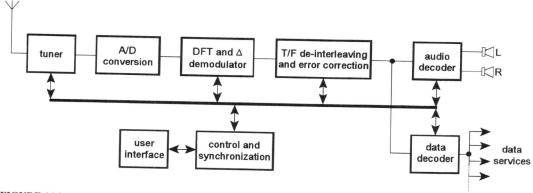

**FIGURE 14.2**   Conceptual block diagram of the Eureka system receiver.

to output the original coded services data. That data is further processed in an audio decoder (producing the left and right audio signals) or in a data decoder, as appropriate. The decoding of more than one service component from the same ensemble (e.g., an audio program in parallel with a data service) is practicable and provides interesting possibilities for new receiver features. The system controller is connected to the user interface and processes the user commands in accordance with the information contained in the FIC.

### 14.3.2   Audio Services

The audio source coding method is a perceptually based, low-bit-rate subband coding system for high-quality audio signals, standardized by ISO/IEC under the heading ISO/IEC 11172-3 (MPEG-1 audio), layer II.[4] The Eureka specification permits use of the flexibility of layer II except for the fact that only the standard studio sampling frequency of 48 kHz and a half-sampling rate of 24 kHz are used. Layer II is capable of processing mono, stereo, and dual-channel (such as a bilingual program), and various encoded bit-rate options are available (i.e., 8, 16, 24, 32, 48, 56, 64, 80, 96, 112, 128, 160, or 192 kb/s per monophonic channel). Extension to multichannel sound according to ISO/IEC 13818-3 is under consideration.[5] In stereophonic or dual-channel mode the encoder produces twice the bit rate of a mono channel. These options can be exploited by broadcasters depending on the intrinsic quality required and the number of sound programs to be broadcast. A stereophonic signal may be conveyed in the stereo mode or, particularly at lower bit-rates, in the joint stereo mode. This exploits the irrelevancy of the two channels of a stereophonic program to maximize the overall perceived audio quality.

Each audio service channel also contains a PAD (program-associated data) channel, having a variable capacity (minimum 0.667 kb/s), which can be used to convey information that is intimately linked to the sound program. This PAD channel is incorporated at the end of a Eureka audio frame compatible with the ISO standard and therefore cannot be subject to a different transmission delay. Typical examples are dynamic range control information, a dynamic label to convey program titles or lyrics, and speech/music indication. Additionally, text with graphics features, for example, may be conveyed in the PAD.

### 14.3.3   Data Services

In addition to the program-associated data that may be carried with the audio, general data may be conveyed as a separate service. This may be in the form of a continuous stream, segmented in 24 ms "logical frames," or arranged as packet data services. The resource allocated to a data service is arranged in multiples of 8 kb/s data rate, though individual packet data services may have much lower capacities and be bundled in a packet submultiplex. In general, the capacity available for independent data is necessarily limited by the capacity requirements of the audio program services making up the Eureka multiplex.

A Traffic Message Channel (TMC) is an example of a data service that can be carried in the FIC and also use the packet mode.

### 14.3.4  Service Information

The following elements of service information (SI) can be made available for display on a receiver:

- Basic program service label (i.e., the name of a program service)
- Time and date
- Dynamic program label (e.g., the program title, lyrics, names of artists)
- Program language
- Program type label (e.g., news, sports, classical music)

The following elements of service information can be used for control of a receiver:

- Cross-reference to the same service being transmitted in another Eureka signal or being simulcast by an AM or FM service.
- Transmitter identification information (e.g., for geographical selection of information). Transmitter network data can also be included, for example, for monitoring and control by the broadcasters.

### 14.3.5  System Organization and Service Control

In order that a receiver can gain access to any or all of the individual services with a minimum overall delay, precise information about the current and future content of the Main Service Multiplex (MUX) is set up and carried by the Fast Information Channel (FIC). This information is the Multiplex Configuration Information (MCI), which is machine-readable data. Data in the FIC are not time-interleaved, so the MCI does not suffer the delay inherent in the time-interleaving process applied to audio and general data services. However, these data are highly protected and repeated frequently to ensure their ruggedness. When the multiplex configuration is about to change, the new information, together with the timing of the change, is sent in advance within the MCI.

Essential items of SI, which concern the content of the MSC (i.e., for program selection), must also be carried in the FIC. More extensive text that is not required immediately on switching on a receiver (such as a list of all the day's programs) may be carried separately as a general data service.

The user of a receiver can select programs on the basis of the textual information carried in the SI by using the program service label, the program type label, or the language. The selection is then implemented in the receiver using the corresponding elements of the MCI. Provision is also made for the use of conditional access to services if desired.

If alternative sources of a chosen program service are available and an original digital service becomes untenable, then linking data carried in the SI (i.e., the "cross-reference") may be used to identify an alternative (e.g., an FM service) and switch to it. However, in such a case, the Eureka/FM receiver will switch back to the Eureka service as soon as reception is possible. This is a particularly important feature at the start of Eureka services because not all areas will be served from day one, and the ability to drop back to the same program on FM, where a simulcast is available, will help maintain service continuity.

### 14.3.6 Channel Coding and Time-Interleaving

The data representing each of the program services being broadcast (digital audio with some ancillary data, and maybe also general data) are subjected to energy dispersal scrambling, convolutional coding, and time-interleaving.

The convolutional encoding process involves adding redundancy to the service data using a code with a constraint length of 7. In the case of an audio signal, some source-encoded bits are given greater protection than others, following a preselected pattern known as the unequal error protection (UEP) profile. The average code rate—defined as the ratio between the number of source-encoded bits and the number of encoded bits after convolutional encoding—may take a value from 0.35 (the highest protection level) to 0.75 (the lowest protection level). Different average code rates can be applied to different audio sources, subject to the protection level required and the bit rate of the source-encoded data. For example, the protection level of audio services carried by cable networks may be lower than that of services transmitted in radio-frequency channels. General data services are convolutionally encoded using one of a selection of uniform rates, whereas data in the FIC are encoded at a constant one-third rate.

Time-interleaving improves the ruggedness of data transmission in a changing environment (e.g., reception by a moving receiver) and imposes a 384-ms transmission delay.

### 14.3.7 Main Service Multiplex

The encoded and interleaved data are fed to the Main Service Multiplexer, where, each 24 ms, the data are gathered in sequence into the multiplex frame. The combined bitstream output from the multiplexer is known as the Main Service Channel (MSC), which has a gross capacity of 2.3 Mb/s. Depending on the chosen convolutional code rate (which can be different from one application to another), this gives a net bit rate ranging from approximately 0.6 to 1.7 Mb/s, accommodated in a 1.5-MHz bandwidth Eureka signal. The Main Service Multiplexer is the point at which synchronized data from all of the program services using the multiplex are brought together.

### 14.3.8 Transmission Frame and Modes

The system provides three transmission mode options, which allow the use of a wide range of transmitting frequencies, up to 3 GHz for mobile reception. These transmission modes have been designed to cope with Doppler spread and delay spread, for mobile reception in the presence of multipath echoes.

Table 14.2 gives the temporal guard interval duration and nominal maximum transmitter separation and frequency range for mobile reception. The noise degradation at the highest frequency and in the most critical multipath condition, occurring infrequently in practice, is approximately 1 dB at 100 km/h and 4 dB at 200 km/h.

Table 14.2 shows that the use of higher frequencies imposes a greater limitation on the guard interval duration and hence on the maximum nondestructive echo delay. Mode I is most suitable for a terrestrial single-frequency network (SFN), because it allows the greatest transmitter

**TABLE 14.2**  Limiting Planning-Parameter Values for Each Transmission Mode

| System parameter | Transmission mode | | | |
|---|---|---|---|---|
| | I | II | III | IV |
| Guard interval duration | ~246 μs | ~62 μs | ~31 μs | ~124 μs |
| Nominal maximum transmitter separation for SFN | 96 km | 24 km | 12 km | 48 km |
| Nominal frequency range (for mobile reception) | ≤ 375 MHz | ≤ 1.5 GHz | ≤ 3 GHz | ≤ 1.5 GHz |

separations. Mode II is most suitable for local radio applications requiring one terrestrial transmitter, although it can also be used for a medium-scale SFN; in fact, larger transmitter spacings can be accommodated by inserting artificial delays at the transmitters and by using directive transmitting antennas. Mode III is most appropriate for cable, satellite, and complementary terrestrial transmission (since it is able to operate at all frequencies up to 3 GHz for mobile reception) and has the greatest tolerance of phase noise. Mode IV has been introduced recently in a compatible manner to allow for relatively large separation distances between the transmitters operating in SFN and in L-band. The parameters of mode IV lie between those of mode I and mode II.

In order to facilitate receiver synchronization, the transmitted signal is built up with a frame structure having a fixed sequence of symbols. Each transmission frame (see Fig. 14.3) begins with a null symbol for coarse synchronization (when no RF signal is transmitted), followed by a phase reference symbol for differential demodulation. The next symbols are reserved for the FIC, and the remaining symbols provide the MSC. The total frame duration $T_F$ is either 96 ms or 24 ms, depending on the transmission mode as given in Table 14.3. Each audio service within the MSC is allotted a fixed time slot in the frame.

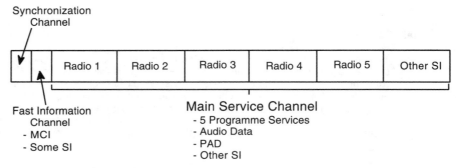

**FIGURE 14.3**    An example of a Eureka multiplex frame.

**TABLE 14.3**    Eureka Transmission Parameters for Each Transmission Mode

|  | Transmission mode | | | |
|---|---|---|---|---|
|  | I | II | III | IV |
| Usable frequency range | 300 MHz | 1.5 GHz | 3 GHz | 1.5 GHz |
| Number of carriers | 1536 | 384 | 192 | 768 |
| Carrier spacing | 1 kHz | 4 kHz | 8 kHz | 2 kHz |
| Total symbol duration | 1246 μs | 312 μs | 156 μs | 623 μs |
| Guard interval duration | 246 μs | 62 μs | 31 μs | 123 μs |
| Frame duration | 96 ms | 24 ms | 24 ms | 48 ms |
| Symbols per frame | 76 | 76 | 153 | 76 |
| Null Symbol | 1297 μs | 324.2 μs | 168 μs | 648.4 μs |

(Ensemble bandwidth = 1.536 MHz; sampling frequency = 2.048 MHz; number of FFT points = 1024)

### 14.3.9  Modulation with OFDM

In the presence of multipath propagation, some of the carriers are enhanced by constructive signals, whereas others suffer destructive interference (frequency selective fading). Therefore,

Eureka provides frequency interleaving by a rearrangement of the digital bitstream amongst the carriers, such that successive source samples are not affected by a selective fade. When the receiver is stationary, the diversity in the frequency domain is the prime means to ensure successful reception; the time diversity provided by time-interleaving provides further assistance to a mobile receiver. Consequently, multipath propagation is a form of diversity and is not considered to be a significant disadvantage for Eureka, in stark contrast to conventional FM or narrow-band digital systems where multipath propagation can completely destroy a service. Principal Eureka performance characteristics are given in Section 14.10.

## 14.4   PRINCIPAL ADVANTAGES OF EUREKA

### 14.4.1   CD Quality

Eureka has several advantages over conventional analog AM/FM broadcasting. The main benefit is that the high sound quality—normally indistinguishable from that of the CD—is effectively free from interference. However, Eureka also has a unique ability to serve the mobile audience, thus providing high-quality coverage wherever and whenever required.

### 14.4.2   Spectrum Efficiency

A further advantage of Eureka is that it is highly spectrum-efficient (Fig. 14.4). This means that it will be possible to increase the number of radio stations—initially by a factor of at least three when compared with FM—without congesting the radio waves. As more efficient audio-coding (compression) methods are introduced, it will be possible to carry even more radio programs with no degradation to existing services and without needing to modify existing receivers. A radio set of the future will thus make it possible to choose, for example, a favorite type of music station from among hundreds of music stations.

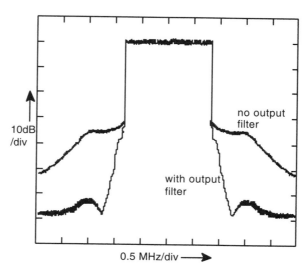

**FIGURE 14.4**   Example of a Eureka-transmitted signal spectrum (VHF band III).

### 14.4.3    Active Echoes

The Eureka system is able to use "passive echoes" so that they add in a constructive manner to the direct signals already received. The Eureka system is also able to constructively use "active echoes"—i.e., delayed signals generated by other cochannel transmitters. This leads to two important concepts: single-frequency networks (SFNs) and cochannel gap-fillers.

The SFN concept enables all transmitters covering a particular area with the same set of sound programs to operate on the same nominal radio-frequency channel, i.e., within the same frequency block. All SFN transmitters need to be synchronized, in terms of both frequency and time, and the transmitted bitstream must be identical. Although the signals emitted by the various transmitters are received with different time delays, the receiver recognizes this as a direct signal coming from the nearest transmitter, followed by active echoes coming from other transmitters in the SFN.

Gap-filling represents the second type of application that makes full use of the active echoes. A gap-filler acts rather like a mirror; it receives the signals from the main transmitter and retransmits them at low power on the same frequencies to provide coverage in an area where the main transmitter is not received satisfactorily. Although the listener receives signals from both the main transmitter and the gap-filler at slightly different times, the two sets of signals add together constructively to enhance the reception of the program. Gap-filling is useful both for terrestrial and satellite broadcasting systems.

As a result of these two concepts, Eureka eliminates the problem of having to retune car radios at frequent intervals. At present, long-distance drivers who are listening to an FM program are forced to retune as they move away from the area covered by one transmitter to that of another. With Eureka, however, a car radio does not need to be retuned because the desired station will be in the same frequency block everywhere within a national or regional service area.

### 14.4.4    Flexible Bit Rates

Eureka is a highly flexible and dynamically reconfigurable system. It can accommodate a range of bit rates between 8 and 384 kb/s, with a range of channel protection mechanisms.

Some broadcasters are interested in using low audio bit rates per audio channel—say, between 16 and 64 kb/s—in order to transmit more channels at slightly reduced quality, but sufficiently good for speech signals. With a bit rate of 32 kb/s per audio channel, the Eureka multiplex of 1.5 MHz can accommodate as many as 36 channels, with 1/2-channel protection level.

### 14.4.5    Eureka Transmission Modes

Technically, Eureka can be used at any frequency between 30 MHz and 3 GHz. This wide range of frequencies includes VHF bands I, II, and III; UHF bands IV and V; L-band (frequencies around 1.5 GHz); and S-band. Because the propagation conditions vary with frequency, four Eureka transmission modes are used (see Table 14.3). These modes are detected automatically by the receiver and are transparent to the user.

### 14.4.6    Data Services

Although audio has been its primary raison d'être, the Eureka 147 transmission system can also be used to carry a large variety of program-associated and independent data services. Many data services of the program-associated category will probably be transmitted from the outset and will be received by the first generation of Eureka consumer-type receivers. Later on, independent data services may also appear. These would be received by dedicated data receivers, including those incorporated in desktop and laptop computers. Two examples of this application are the electronic delivery of newspapers and the transmission of compressed video images, such as weather maps.

The Eureka system's immunity to multipath and other reception impairments will guarantee error-free data reception in the mobile environment. Hence, the Eureka 147 system is an ideal complement to the wired information highway distribution system now being established worldwide. Further data-transmission possibilities of Eureka are outlined in Section 14.7.

### 14.4.7   Future-Proofing

The Eureka is future-proof. Once the receiver has been purchased, it will not become obsolete as the digital technology develops, nor as new services and applications emerge.

In Europe, for example, Eureka delivery will commence via terrestrial networks. Nevertheless, the receivers designed for use with these terrestrial services should, in principle, also be able to receive future Eureka services delivered via satellite and cable. In other words, the Eureka 147 system will become a universal means to deliver sound programs and data, irrespective of the transmission medium used. (Studies are now being conducted on the use of the Eureka 147 system as a digital television delivery medium for mobile reception on small screens.)

## 14.5   *SPECTRUM ISSUES*

In Europe a planning meeting was convened in July 1995 under the auspices of the CEPT (European Conference of PTT Administrations). The aim of this meeting was to produce a special arrangement for the introduction of terrestrial transmissions of the Eureka system in the frequency bands 46–68 MHz, 174–240 MHz, and 1452–1467.5 MHz, as well as to prepare an associated frequency block allotment plan, taking into account the final requirements of the CEPT member countries.

The allotment plan drawn up at the meeting provides practically all member countries of the CEPT with two sets of frequency blocks, each of width 1.536 MHz. This is a vital prerequisite to the wide-scale launch of terrestrial Eureka services in Europe. Most of the CEPT countries have opted for frequency block allotments in VHF band III and in L-band.

In Europe, 85 frequency blocks can potentially be used for current and future Eureka services. The distribution of these frequency blocks is as follows:

- 12 blocks in VHF band I (47–68 MHz)
- 38 blocks in VHF band III (174–240 MHz)
- 23 blocks in L-band (1452–1492 MHz)
- 12 blocks in VHF band II (87–108 MHz)

Among these blocks, Eureka block numbers 1 to 59 have been considered by the CEPT planning meeting. Blocks 60 to 85 have been added to the plan by the EBU. Each frequency block carries a two- or three-character label, which is easier to remember than the center frequency of the block, and which is convenient for receiver manufacturers and consumers to use when initially programming their receivers.

The labeling system of the frequency blocks in VHF band I and band III is fully compatible with the existing VHF television channel numbers (i.e., channels 2 to 13). Each of these television channels can accommodate four Eureka blocks—six blocks in the case of channel 13. All of the frequencies listed in Table 14.3 comply with the 16-kHz raster as specified in the ETS 300 401 standard.

One of the important results of the meeting was a definition of the center frequency of each ensemble (i.e., frequency block). This information is very important for receiver manufacturers and may help substantially to simplify the receiver design; before the meeting, any frequency in the 16-kHz raster could be used as the center frequency, resulting in a very large number of possibilities. The number of defined center frequencies has now been reduced to match the total number of ensembles allocated in band III and in L-band (i.e., 61). These center frequencies are planned to be implemented in the first-generation Eureka receivers manufactured for the European market. The bases for planning of terrestrial services and the planning criteria are given in the literature.[7]

## 14.6   EUREKA—A SUITABLE SYSTEM FOR SATELLITE SOUND BROADCASTING

The majority view in Europe is that any sound-broadcasting satellite system (designed to cover large areas) will have the same modulation/coding system parameters as the ground-based system (designed to cover regional/national territories) so that the same receiver could be used. An essential requirement for any new satellite system is that it should be able to provide for mobile and portable reception in all types of propagation environments (rural, urban, etc.).

Although the Eureka system has been developed as a terrestrial system, there is no technical reason why it could not be used for satellite delivery as well. Many computer simulations have shown that this assumption may be true, but real experiments are needed to demonstrate that satellite delivery is both a technically viable and an economically attractive proposal.

Two such experiments have been conducted—one in Australia, the other in Mexico. The Australian test was carried out using the Optus B3 satellite at 1552 MHz. The trial in Mexico, carried out by the British Broadcasting Corporation (BBC), used the Solidaridad satellite. Both satellites were originally launched to provide mobile phone services; they were not specifically designed for multicarrier systems such as Eureka 147. Even so, the results showed that fixed and portable reception of Eureka signals via a satellite is technically feasible. Because of the low transmitting power of the test satellites, mobile reception was possible only under line-of-sight conditions.

In Munich a satellite simulation using a helicopter is being carried out jointly by the ESA, the IRT, and the BBC to determine the service-availability performance (i.e., percentage of coverage) for various elevation angles.

One of the outstanding issues to be clarified is whether or not it is possible to uplink from different feeder-link stations—programs that constitute the same multiplex. Studies are being undertaken within the Eureka 147 project to determine how on-air multiplexing at RF could be performed at the input to the satellite in a manner similar to the well-established time-division multiple-access (TDMA) technology used with FSS satellites. Preliminary results indicate that feeding the satellite from different uplink stations may not be a major technical problem. However, it will be necessary to coordinate between the different feeder-link stations—in terms of time synchronization (approximately only) and power control—but such coordination is quite normal in any TDMA and FDM system.

Because the Eureka is a multicarrier system, some output back-off at the satellite will be necessary to reduce the amount of intermodulation. Similar back-off will be necessary with the alternative FDM systems (such as the WorldSpace system), because multichannel operation already generates multicarriers. Thus, it is fair to assume that there will be no significant difference between the Eureka and the FDM systems in terms of the output back-off required at the satellite. In the Australian Eureka experiment the transponder operated satisfactorily with an output back-off of 2.2 dB.

International broadcasters are usually interested in large coverage areas. Therefore, a geostationary (GEO) satellite system could be satisfactory to cover low-latitude areas, such as most parts of Africa, Central and South America, India, Indonesia, etc. However, many regions of the world that wish to be covered are situated in the northern hemisphere (above 30–40 degrees latitude), including Europe, China, and Japan. For such areas the highly inclined elliptical orbit

(HEO) satellite concept—promoted currently by the Archimedes/MediaStar project of the European Space Agency—seems to be of interest, because it would enable greater penetration to mobile receivers in urban areas (due to the high elevation angle of the satellite and hence less shadowing of the signals). In practice, a combination of the HEO and GEO concepts may be an attractive solution. The Eureka addresses both GEO and HEO satellite solutions.

For international broadcasting all WARC-92 bands (i.e., bands located at 1.5, 2.3, and 2.6 GHz) should be considered. Preference should clearly be given to the 1.5-GHz band for technical and economic reasons (the best trade-off between the size of satellite transmit antenna and its transmit power). Preliminary studies have shown that, at 2.6 GHz, considerably larger transponder output power would be required (on the order of four times greater than that required at 1.5 GHz). Eureka transmission modes II, III, or IV are suitable for use at these frequencies.

## 14.7 EUREKA AS A MULTIMEDIA CARRIER

Within the Eureka 147 project, further developments are under way to study the use of the Eureka 147 system as a multimedia and data broadcasting system. This study is aimed at expanding the future use of the Eureka beyond the provision of excellent sound reception in adverse mobile and portable environments. In addition to the conventional audio services, the system is opening up many new opportunities to carry a number of nonaudio services, such as text, still pictures, moving images, etc.

The multiplex of the Eureka 147 system has been designed to carry a large number of digital services with a total bit rate of up to 1.7 Mb/s, organized in up to 64 stream- or packet-mode subchannels. Four different data transport mechanisms have been defined in the Eureka standard:

- Program-Associated Data (PAD)
- Fast Information Channel (FIC)
- Stream Mode (SM)
- Packet Mode (PM)

The choice of transport mechanism depends on the kind of data that it is necessary to transport. For example, the Program-Associated Data is suitable for services that bear a strong relationship to the audio signal. Because this data is taken from the audio frame, there is a trade-off between the intrinsic audio quality and the PAD data capacity.

The FIC channel was originally intended to carry information on the organization of the Eureka multiplex. Nevertheless, the FIC can carry a limited amount of additional information, such as paging and emergency warning messages. Dedicated (or special-purpose) receivers that decode only the FIC part of the multiplex may be significantly less complex than general-purpose Eureka receivers.

In Stream Mode a subchannel is assigned to a single data service, providing a fixed data rate (in multiples of 8 kb/s) with specific error correction. In Packet Mode a number of services may share the same subchannel. Packet headers contain a service address, which allows the receiver to restore the original data. The PM is a convenient way to carry asynchronous services (which use variable data rates).

Examples of Eureka data services currently being implemented are given as follows. These services may be presented either in the form of textual information (shown, for example, on a simple receiver display of, typically, between 8 and 128 characters), still pictures, or even video images.

- Program-associated services such as current song title, interpreter and performer, lyrics, news headlines, CD covers, etc.
- News, including events; traffic messages; weather, sport, stock market, travel, and tourist information.

- Traffic navigation by means of transmitted digitized road maps, combined with positional information provided via the GPS system.
- Advertisements and sales including sales catalogues, purchase offers, etc.
- Entertainment, including games and noncommercial bulletin boards.
- Closed user group services such as banking information, electronic newspapers, fax print-outs and remote teaching.

The Eureka standard offers two modes for text transmission: dynamic label and interactive text transmission (ITTS). The former mode is similar to the radio text feature of the Radio Data System (RDS) on the FM band. ITTS is a more sophisticated text transmission system. It allows for menu-driven operation and can also be used to transmit text at the rate a broadcaster prescribes. It can process several streams of textual information simultaneously to convey, for example, the same information in several languages or to transmit a program schedule at the same time as giving details of the program currently on-air.

Ideally, multimedia services should be fully interactive, in which case the consumer can communicate with the service provider's database. Because broadcasting services are one-way only, the return channel could be provided by GSM telephone (in the case of mobile Eureka receivers) or via a telephone line (in the case of a fixed receiver). Nevertheless, a semi-interactive mode is also possible. In this instance, information is down-loaded by the service provider to the user's data terminal and stored there as a database. All interactivity is then handled within the user's data terminal, but the database contents have to be updated regularly by the data service provider. The storage capacity of the user's terminal is a trade-off between the service transmission rate, the repetition rate, and the cost of the memory.

A key factor for the success of Eureka will be its ability to address each receiver individually. This will allow service providers to customize the "bouquet" of services provided to each user, and even to identify the user in an interactive transaction. This feature has some far-reaching implications, particularly for privately funded radio.

Studies are continuing on the suitable presentation of Eureka data services. Currently, data services specified in the ETSI standard have a text-based presentation. In order to improve the man-machine interface, Eureka will be enhanced to support a graphical user interface, such as Microsoft Windows. This will be of importance for screen-based services, which seem to be more relevant for stationary and portable receivers. For mobile receivers, synthesized speech-based interfaces are a better alternative, because they would be less distracting to drivers.

For the user's data terminal a unified transmission protocol would be very helpful, because no distinction between different transport mechanisms would be necessary. A software-based language for object-oriented page description is being developed to define a communication and a presentation layer. Such a unified protocol for the multimedia transport mechanism could be used not only with Eureka services, but also in other communications systems. Currently, within the Eureka 147 project, a standard receiver data interface is being specified to transfer the data carried within an ensemble—from the receiver to any external devices such as a PC, tape recorder, or conditional access decoder. A demonstration of both the audio and the multimedia usage of the Eureka 147 system was given in August 1995 at the IFA fair in Berlin.

## 14.8   RECEIVERS

In order to receive Eureka services, consumers will need to buy a new kind of receiver. The consumer Eureka receivers will also contain FM and AM circuits that will initially be analog. However, it will not be long before the AM and FM circuits in a Eureka receiver become digital. These all-digital AM/FM/Eureka receivers will be based on advanced computer technology, which will allow the down-loading of large quantities of information to program the radio set and its associated equipment (Digital Cassette Recorders, MiniDisc recorders, PCs, etc.). A receiver standard is being prepared by the EACEM (European Association of Consumer Electronics Manufacturers).[8] At the IFA-95 fair in Berlin, six manufacturers (Alpine, Bosch, Grundig, Kenwood, Philips, and Sony) displayed their current Eureka receivers. In fact, they look more like semiprofessional equipment; the Eureka part is in a separate box, mounted in the trunk with a link to an FM/RDS receiver in the dashboard. These first-generation car receivers are not generally available yet; they can only be purchased on special order and in limited quantities for evaluation purposes. The first mass-produced Eureka car radios are not expected until 1997; the first portable Eureka receivers are projected for 1999.

The industry has been carrying out a lot of research and development on further applications of the Eureka system, including data-only receivers, picture radios, advanced teletext full-bitstream video decoders, navigation systems, differential GPS, traffic information systems, Traffic Message Control (TMC), real-time packet-mode multiplexers/demultiplexers, fax, video-text, audio in conjunction with radiotext (dynamic labels), electronic newspaper publishing (including text and pictures in packet mode), and high-capacity storage using MiniDisc.

## 14.9   EUROPEAN STRATEGIES FOR THE INTRODUCTION OF EUREKA SERVICES

Currently, many pilot service trials and field tests at VHF and in L-band are being conducted all over Europe. In many countries there are plans to commence preoperational terrestrial services, primarily making use of the existing transmitter and distribution infrastructure. (The world's first official DAB services were inaugurated in the United Kingdom by BBC Radio and in Sweden by the Swedish Broadcasting Corporation on 27 September 1995.) Although the situation varies very much from country to country, it is clear that the critical mass has already been achieved and that the introduction of the Eureka in Europe is ensured. Many countries have developed their own strategy for the introduction of Eureka services. Many have established national forums (or "platforms") to coordinate efficiently among broadcasters, manufacturers, network operators, and others. In 1995 a European association, EuroDab Forum, was established to harmonize the implementation of Eureka services in Europe and elsewhere. The EuroDab Forum includes more than one hundred member organizations from all continents. Its principal objective is to prepare a concerted launch of Eureka services at Internationale Funkaustellung (IFA) 1997. It is planned that a population of 100 million or more will be able to receive Eureka signals in Europe and that first consumer receivers, both in the car and in the home, will become available at an affordable price.

## 14.10   PERFORMANCE OF THE EUREKA SYSTEM

Measurements have been carried out on both the MSC and the FIC. The data carried in the FIC are formatted in Fast Information Blocks (FIBs), and each FIB contains error detection bits, which allow a receiver to determine the integrity of a block; a receiver discards an FIB if it is found to contain one or more uncorrected errors. The performance of the FIC was assessed by measuring the block-error ratio of the FIBs, which, for a given number of received FIBs, is defined as the ratio between the number of discarded FIBs and the total number of FIBs received.

A receiver that makes full use of the standardized features of Eureka has a well-defined digital functionality. For example, the power of the error protection is dictated largely by the format of transmitted signal, and variations in the design of digital circuitry in receivers (e.g., Viterbi decoders) should have little influence. Therefore, any influence that an individual receiver has on the results of such measurements is likely to be caused by the performance of the analog circuitry. Particularly relevant are the following items: the noise figure of the front-end amplifier, the short-term stabilities of the local oscillators, and cross-talk or "self-interference" between the digital circuitry and the analog circuitry. These latter two items make the major contributions to what is generally referred to as the "implementation margin," that is, the departure from a hypothetical ideal receiver. The magnitude of the implementation margin is expected to be reduced as experience is gained in the design of receivers.

### 14.10.1   BER versus S/N for the MSC

A known pseudorandom binary sequence (PRBS) was input to the channel encoder and the Eureka signal was passed to the receiver through a simulated transmission channel. The BER was measured by comparing the sequence of bits output by the channel decoder in the receiver (following the OFDM demodulator) with the known PRBS. The experimental equipment that was used provides two options for transmission of general data, both of which were used: 64-kb/s data at rate 0.5 error protection and 24-kb/s data at rate 0.375 error protection. For the channel encoder, the error protection rate is defined as the ratio between the number of input bits and the number of output bits.

Two types of transmission channel were simulated using a fading-channel simulator, which can provide up to 12 independent paths:

- *Gaussian channel.*   Typical of static reception over a single line-of-sight propagation path (e.g., a satellite down-link). Within the channel bandwidth, white Gaussian noise is added to the signal to establish the S/N.

- *Simulated Rayleigh channel.*   Typical of mobile reception from a terrestrial transmitter in cases where a line-of-sight path makes less contribution to the total received signal power than the total power arising from multipath propagation. White Gaussian noise is added to the signal within the channel bandwidth to establish the average S/N, but the instantaneous S/N depends on instantaneous signal power, which is subject to fading.

For the second type, two different simulation "models" were used, representing propagation in rural and urban environments. Two different speeds of motion of the receiver were simulated— 130 km/h for the rural model and 15 km/h for the urban model.

For the rate 0.5 results presented here, the implications in terms of the quality of reproduced sound can be deduced by comparison with the results presented in the next section. Specifically for 224-kb/s stereo source coding and a Gaussian channel, the meaning of the BER is as follows:

$\leq 10^{-4}$ corresponds to unimpaired sound quality

$10^{-4}$ corresponds to the onset of impairment

$\geq 10^{-1}$ corresponds to failure

The results are presented in Figs. 14.5 through 14.9. All values of S/N are referred to the significant bandwidth of the Eureka signal, which is 1.537 MHz. Theoretical values (the results of software simulations) are included in Fig. 14.5 to demonstrate the typical magnitude of the implementation margin for the "third generation prototype" receiver design—about 1 dB.

### 14.10.2   Block-Error Ratio versus S/N for the FIC

Data carried in the FIC have fixed- and constant-rate 0.33 error protection, which is more powerful than would normally be used for sound program data. This is necessary because the FIC data are not subject to time-interleaving, in order to minimize the delay in configuring a receiver when it is initially switched on.

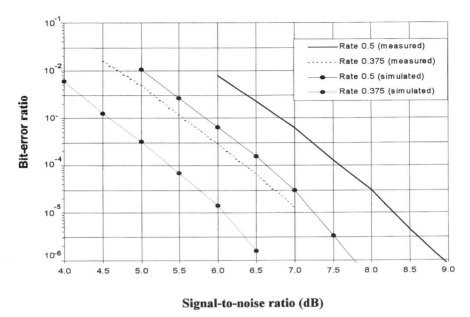

**FIGURE 14.5**    Bit-error ratio vs. signal-to-noise ratio for Eureka (transmission mode 1): Gaussian channel.

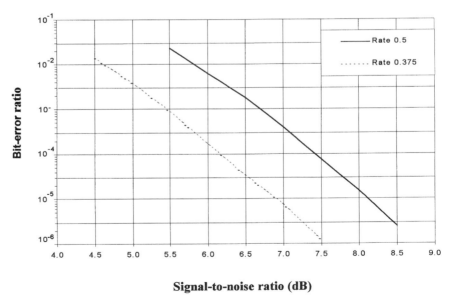

**FIGURE 14.6**    Bit-error ratio vs. signal-to-noise ratio for Eureka (transmission mode 2 or 3): Gaussian channel.

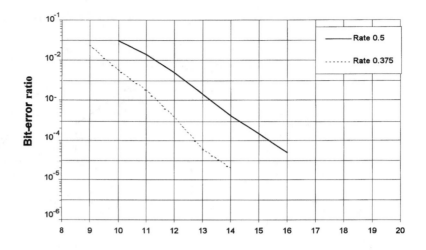

**FIGURE 14.7** Bit-error ratio vs. signal-to-noise ratio for Eureka (transmission mode 1, 226 MHz) simulated Rayleigh channel (rural environment, 130-km/h speed).

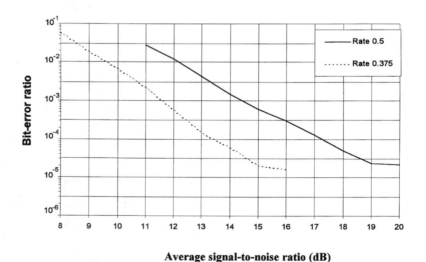

**FIGURE 14.8** Bit-error ratio vs. signal-to-noise ratio for Eureka (transmission mode 2, 1480 MHz): simulated Rayleigh channel (rural environment, 130-km/h speed).

The Eureka signal was passed to the receiver through a simulated transmission channel. The types of transmission channel that were simulated are the same as those described in 14.10.1, and the same comments about the expected accuracies apply in this case. The results are presented in Figs. 14.10 through Fig. 14.11. All values of S/N are referred to the 1.537-MHz significant bandwidth of the Eureka signal.

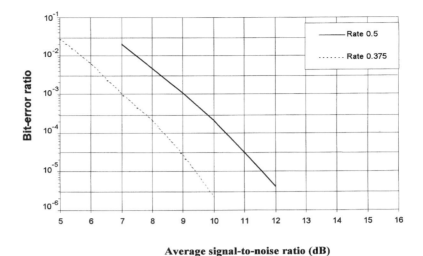

**FIGURE 14.9**    Bit-error ratio vs. signal-to-noise ratio for Eureka (transmission mode 3, 1480 MHz): simulated Rayleigh channel (urban environment, 15-km/h speed).

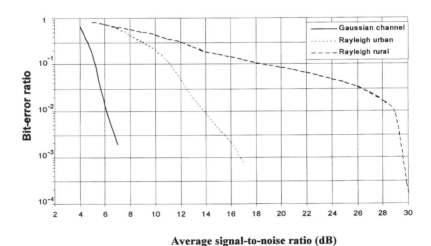

**FIGURE 14.10**    Block-error ratio vs. signal-to-noise ratio for Eureka (transmission mode 2, 226 MHz): Gaussian and simulated Rayleigh channels.

## 14.11  *SOUND QUALITY VERSUS RF SIGNAL-TO-NOISE RATIO*

A program of informal subjective tests was carried out by a panel of audio engineers to assess the sound quality provided by an experimental Eureka receiver in conditions where the radio-frequency S/N is sufficiently low for the sound quality to be impaired by disturbances resulting from errors.

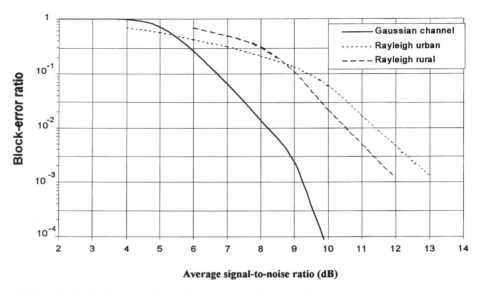

**FIGURE 14.11**    Block-error ratio vs. signal-to-noise ratio for Eureka (transmission mode 3, 226 MHz): Gaussian and simulated Rayleigh channels.

**TABLE 14.4**    Sound Quality vs. Signal-to-Noise Ratio for Eureka (Transmission Mode 1): Gaussian Channel

| Source coding | | Channel coding | Onset of impairment | Point of failure |
|---|---|---|---|---|
| Bit rate (kb/s) | Mode | Average rate | S/N (dB) | S/N (dB) |
| 256 | Stereo | 0.6 | 7.6 | 5.5 |
| 224 | Stereo | 0.6 | 8.3 | 5.9 |
| 224 | Stereo | 0.5 | 7.0 | 4.8 |
| 224 | Joint Stereo | 0.5 | 6.8 | 4.5 |
| 192 | Joint Stereo | 0.5 | 7.2 | 4.7 |
| 64 | Mono | 0.5 | 6.8 | 4.5 |

**TABLE 14.5**    Bit-Error Ratio vs. Signal-to-Noise Ratio for Eureka (Transmission Mode 1, 226 MHz) Simulated Rayleigh Channel (Urban Environment, 15-km/s speed)

| Source coding | | Channel coding | Onset of impairment | Point of failure |
|---|---|---|---|---|
| Bit rate (kb/s) | Mode | Average rate | S/N (dB) | S/N (dB) |
| 224 | Stereo | 0.6 | 18.0 | 11.0 |
| 224 | Stereo | 0.5 | 16.0 | 9.0 |

A selection of items of program material was chosen from the EBU SQAM Compact Disc. The selection included two particularly critical solo instruments—clarinet and glockenspiel—as well as male and female speech and an item of popular music.

The program material was fed into a laboratory Eureka transmission chain, commencing at the audio source-encoder and ending at the output of the receiver. The sound signals output by the receiver were reproduced in a listening room. The transmission path included equipment for establishing the S/N in a Gaussian channel and, using a fading channel simulator, in a Rayleigh channel. Two different simulation "models" were used in the case of a Rayleigh channel.

Listening tests were conducted in which the average S/N was reduced in 0.5-dB steps to establish, in sequence, the following two conditions:

- The onset of impairment, which is the point at which the effects of errors start to become noticeable. This was defined as the point where 3 or 4 error-related events could be heard in a period of about 30 seconds.

- The point of failure, which is the point at which a listener would probably stop listening to the program because it became unintelligible or because it no longer provided the enjoyment sought. This was defined as the point where the error-related events occurred virtually continuously, and muting took place 2 or 3 times in a period of about 30 seconds.

Two values of S/N were recorded for each test, representing the consensus view of the panel of audio engineers.

The results are presented in Tables 14.4 and 14.5, where all S/N values are referred to the significant bandwidth of the Eureka signal, as in Section 14.10.1. The results presented here are the mean values of several tests using different program material.

## *GLOSSARY*

**Common Interleaved Frame (CIF)**    The serial digital output from the main service multiplexer, which is contained in the main service channel part of the transmission frame. It is common to all transmission modes and contains 55,296 bits (i.e., 864 CUs).

**Conditional Access (CA)**    A mechanism by which the user access to service components can be restricted.

**Convolution coding**    The coding procedure that generates redundancy in the transmitted data stream in order to provide ruggedness against transmission distortions.

**Ensemble**    The transmitted signal comprising a set of regularly and closely spaced orthogonal carriers. The ensemble is the entity that is received and processed. In general, it contains program and data services.

**Fast Information Block (FIB)**    A data burst of 256 bits. The sequence of FIBs is carried by the fast information channel. The structure of the FIB is common to all transmission modes.

**Fast Information Channel (FIC)**    A part of the transmission frame (comprising the fast information blocks) that contains the multiplex configuration information, along with optional service information and data service components.

**Main Service Channel (MSC)**    A channel that occupies the major part of the transmission frame and carries all the digital audio service components, along with possible supporting and additional data service components.

**Multiplex Configuration Information (MCI)**    Information defining the configuration of the multiplex. It contains the current (and in the case of an imminent reconfiguration, the forthcoming) details about the services, service components, subchannels, and the linking between these objects. It is carried in the FIC in order that a receiver may interpret this information in advance of the service components carried in the main service channel. It also includes identification of the ensemble itself and a date and time marker.

**Packet mode**   The mode of data transmission in which data are carried in addressable blocks called *packets*. Packets are used to convey MSC data groups within a subchannel.

**Program-Associated Data (PAD)**   Information that is related to the audio data in terms of contents and synchronization. The PAD field is located at the end of the DAB audio frame.

**Single-Frequency Network (SFN)**   A network of DAB transmitters sharing the same radio frequency to achieve a large-area coverage.

**Stream mode**   The mode of data transmission within the Main Service Channel, in which data are carried transparently from source to destination. Data are carried in logical frames.

**Subband**   A subdivision of the audio frequency range. In the audio coding system, 32 subbands of equal bandwidth are used.

**Subchannel**   A part of the main service channel that is individually convolutionally encoded and comprises an integral number of capacity units per common interleaved frame.

**Synchronization channel**   A part of the transmission frame provided to ensure proper time, frequency, and phase synchronization in the receiver.

**Transmission frame**   The actual transmitted frame—specific to the three transmission modes—conveying the synchronization channel, the fast information channel, and the main service channel.

**Transmission mode**   A specific set of transmission parameters (e.g., number of carrier, OFDM symbol duration). Three transmission modes (i.e., I, II, III) are defined to allow the system to be used for different network configurations and a range of operating frequencies.

## *BIBLIOGRAPHY*

1. ETS 300 401: "Radio Broadcasting Systems: Digital Audio Broadcasting (DAB) to Mobile, Portable, and Fixed Receivers," ETSI, Feb. 1995.
2. ITU-R Recommendation BS.774-1: "Service Requirements for Digital Sound Broadcasting to Vehicular, Portable, and Fixed Receivers Using Terrestrial Transmitters in the VHF/UHF Bands."
3. ITU-R Recommendation BO.789-1: "Service Requirements for Digital Sound Broadcasting to Vehicular, Portable, and Fixed Receivers for BSS (Sound) in the Frequency Range 500–3000 MHz."
4. ITU-R Draft Recommendation BS.1114: "Systems for Terrestrial Digital Sound Broadcasting to Vehicular, Portable, and Fixed Receivers in the Frequency Range 30–3000 MHz," June 1995.
5. ITU-R Draft Recommendation BO.1130: "Systems for Digital Sound Broadcasting to Vehicular, Portable, and Fixed Receivers for BSS (Sound) Bands in the Frequency Range 1400–2700 MHz," June 1995.
6. ITU-R Special Publication: "Terrestrial and Satellite Digital Sound Broadcasting to Vehicular, Portable, and Fixed Receivers in the VHF/UHF Bands."
7. EBU, "Technical Bases for T-DAB Services Network Planning and Compatibility with Existing Broadcasting Services," EBU Technical Document, June 1995.
8. Draft EACEM Technical Report No. 09: "Characteristics of DAB Receivers," EACEM, May 1995.

## *ABOUT THE AUTHOR*

Franc Kozamernik graduated in 1972 from the faculty of electrotechnical engineering, University of Ljubljana, Slovenia. He joined Radio-Television Slovenia the same year. For 12 years he was involved in various engineering projects on architectural acoustics, electroacoustics, video engineering, and audio postproduction in television operations.

In 1985 he joined the European Broadcasting Union (EBU). He represented the EBU at the WARC-ORB conferences in 1985 and 1988, and at WARC-92. Since 1989 he has been a member of the Broadcasting Technology Division at the EBU Technical Department in Geneva, Switzerland. He is involved in all digital audio broadcasting matters, audio source coding, and RF aspects of digital video broadcasting by satellite, cable, and terrestrial transmission. He has been involved in the European Telecommunications Standards Institute standardization of many audio and video systems. He was nominated the special rapporteur for the ITU-R Special Publication on DAB and is the founder of the *DAB Newsletter.* He is also the project manager of the EuroDab Forum.

# CHAPTER 14
# DIGITAL AUDIO BROADCASTING

## SECTION 2
## U.S. DIGITAL AUDIO BROADCASTING SYSTEMS

**Skip Pizzi**
*Editor, BE Radio*

## 14.12    A BRIEF HISTORY

The beginnings of DAB in the United States can be traced to August 1986, when public radio station WGBH-FM began simulcasting its audio signal in digital form over its sister station WGBX-TV, a UHF television station. The broadcast used the so-called "F-1" format, which had been developed by Sony as a method of adapting consumer VCRs for recording digital audio. The system converted stereo analog audio to digital audio, then produced a monochrome video output signal representing the digital audio in black or white pixel patterns, intended for recording on any VCR. Consumers with F-1 decoders could feed the received WGBX signal from their VCR's tuner (instead of playing back an F-1–encoded videocassette) to receive the WGBH-FM signal with CD quality.

Although successful in transmitting digital audio to consumer "receivers," the test only underscored the Achilles heel of digital audio systems—their lack of spectrum efficiency relative to analog broadcasts. Here was a 6-MHz TV channel (although only about half of it was being used by the monochrome signal) broadcasting a simple stereo digital audio signal. Meanwhile, in 200 kHz, the analog FM service carried stereo audio plus numerous ancillary services on subcarriers. Clearly, a more efficient transmission system was required.

A breakthrough in this regard occurred under the auspices of the Eureka 147 DAB project, a European consortium charged with development of digital radio broadcasting. One component of the system involved a bit-rate reduction scheme called MASCAM (later updated to MUSICAM) that employed lossy, perceptually based digital signal processing to recode a linear PCM stereo audio signal. Instead of the approximately 1.4-Mb/s data stream produced by the CD format (also used by the F-1 system), Eureka 147 required only 256 kb/s to produce an apparently similar-sounding signal.

The Eureka 147 system then multiplexed several of these compressed digital signals into a wide-band data stream, and applied frequency and time diversity to spread this data across several megahertz of spectrum by QPSK-modulating hundreds of closely spaced, narrowband carriers—the *coded orthogonal frequency division multiplex* (COFDM) technique. This system combined high audio fidelity with resistance to multipath interference, which was the chief nemesis of FM radio (particularly in mobile reception). Moreover, this was achieved with an occupied spectrum per station that approached the efficiency of the FM band. A typical early implementation placed 16 or more audio signals in a 7-MHz-wide channel. Later versions placed five or six channels in a 1.5-MHz-wide channel. The latter's effective spectrum occupancy of 250 to 300 kHz per audio channel was only slightly greater than FM's requirements.

After successful demonstrations of Eureka 147 in Europe in 1988 and 1989, and in Canada during 1990, various proposals to implement Eureka 147 in the United States were launched in the early 1990s. Its requirements for new spectrum and the multiplexing of several channels did not find favor among U.S. broadcasters, however. The hunt was already on for new spectrum for HDTV, and the fiercely independent spirit of U.S. radio station owners was challenged by Eureka 147's need for multiple signals to share a broadcast channel and transmission system.

Meanwhile, other countries moved toward the implementation of Eureka 147, but without total uniformity on its placement in the spectrum. Among different nations, Eureka 147 was proposed for use in the VHF TV band, the UHF TV band, and newly appropriated spectrum for DAB in the L-band (1452–1492 MHz) and S-band (2310–2360 MHz, 2535–2655 MHz). The L- and S-band allocations were agreed upon at the 1992 World Administrative Radio Conference (WARC-92), where these bands were established for both direct-broadcast satellite and "complementary terrestrial" digital audio broadcast transmissions under the title *Broadcast Satellite Service, Sound* (BSS/S).

Although a majority of the world's nations selected the L-band allocation for BSS/S use, several Asian countries (including Japan and the former Soviet states) chose the upper S-band (2535–2655 MHz), whereas the United States chose the lower S-band (2310–2360 MHz). Various nations also chose to interpret the "complementary terrestrial" language of WARC-92 in different ways. For example, whereas Canada will use the L-band for both its terrestrial and (later) satellite DAB applications, the United States plans to use its S-band allocation for satellite DAB only (at this writing). Current directives call for the only U.S. terrestrial origination in the S-band to come from possible ground-based repeaters used to extend satellite DAB coverage into shadowed areas where DBS signals cannot be adequately received—so-called "urban canyons."

The U.S. Federal Communications Commission (FCC) moved throughout the mid-1990s to authorize the lower S-band for *Satellite Digital Audio Radio Service* (S-DARS), under its General Docket 90-357. At this writing, four proponents had applied to provide service in this band (see Table 14.6). Each would offer 30 or more channels of high-quality audio (plus a variety of additional data services) to DAB receivers in homes, vehicles, and personal/portable systems.

On the terrestrial DAB front, U.S. broadcasters' difficulties with Eureka 147 gave rise to a number of proposals for reuse of existing radio broadcast spectrum for DAB. Throughout the early 1990s, various DAB formats of this "in-band" variety were developed and, in some cases, demonstrated to broadcasters. Although it had earlier supported the Eureka 147 format, the National Association of Broadcasters (NAB), the primary trade association for the broadcast industry, switched its backing to the pursuit of an in-band solution in early 1992.

**TABLE 14.6** The Proponents for Direct Satellite DAB Service in the United States under FCC Gen. Docket 90-357[*]

| Proponent | Audio channel | Terrestrial component | Support scheme[a] | System cost |
|---|---|---|---|---|
| American Mobile Radio Corp. | 21[b] | No | Sub+adv[c] | $528M |
| CD Radio | 66 | Yes[d] | Sub[c] | 385M |
| Digital Satellite Broadcasting Corp. | 512[e] | Yes | Sub+adv[c] | 622M |
| Primosphere | 30[b] | No | Adv | 373M |

[*]Including number of audio channels originally proposed, whether or not any terrestrial repeater systems are envisioned, how services will be supported, and the project cost of each system.
[a]Each system's business plan describes whether revenues will come from user subscription fees (sub) or from advertising sales (adv).
[b]Multiple audio fidelity levels offered, from speech-quality mono to CD-quality stereo.
[c]Some program channels may be provided by existing broadcasters.
[d]Terrestrial repeaters for satellite gap-filling also carry multichannel signals from local broadcasters.
[e]Includes national channels plus separate regional channels on spot beams (16 channels per beam).

Shortly thereafter, the Electronic Industries Association (EIA), through its Consumer Electronics Group—now the Consumer Electronics Manufacturers Association (CEMA)—and the NAB, through its National Radio System Committee (NRSC, a joint EIA/NAB standards group) formed a testing body for U.S. DAB format proposals. This test group accepted proposals for both terrestrial and satellite DAB systems for the United States. Eventually, five proponents submitted nine different systems and variations for testing (see Table 14.7).

**TABLE 14.7**   The DAB Formats Submitted to the EIA/NRSC Testing Process in 1994*

| Format | Method | Modes tested | Audio coding |
|--------|--------|--------------|--------------|
| Amati/AT&T | IBOC-FM | 2 | PAC |
| AT&T | IBAC-FM | 1 | PAC |
| Eureka 147 | NB (L-band) | 2 | ISO/MPEG L2 |
| USADR AM | IBOC-AM | 1 | ISO/MPEG L2 |
| USADR FM-1 | IBOC-FM | 1 | ISO/MPEG L2 |
| USADR FM-2 | IBOC-FM | 1 | ISO/MPEG L2 |
| VOA/JPL | NB (S-band) | 1 | PAC |

*Including the technical methodology used, the number of different modes in which the formats were submitted for testing, and the audio-coding algorithm used by each system.

A laboratory testing phase using simulated multipath and other impairments was conducted at the Lewis Research Center, a NASA facility in Cleveland, Ohio. Subjective listening tests held at the Communications Research Center in Ottawa, Ontario, were also included in this phase of the tests. Results of this testing were released in mid-1995. It indicated that many systems had significant technical difficulties, ranging from interference problems to poor audio quality. Systems that performed adequately in the lab tests (including the Eureka 147 format) were tempered by inadequacy or unavailability of spectrum in the United States for their implementation as proposed.

Subsequent field tests were conducted in San Francisco during the summer and fall of 1996, but their value was diminished when all the in-band formats were withdrawn by their proponents from this round of testing. Field-test results for the formats that were tested largely confirmed the laboratory test results.

In the wake of their withdrawals from the tests, in-band DAB proponents AT&T/Lucent Technologies and USADR vowed that their respective systems would undergo further development and improvement. In USADR's case, the resources of consortium member Group W's corporate sibling Westinghouse Wireless Solutions (WWS) were called upon to provide these refinements. WWS reported that the development of an adequate FM-IBOC format seemed possible, although challenging. It further speculated that digital audio bit rates might have to be reduced to 96 kb/s, with little or no auxiliary data capacity.

Meanwhile, another DAB format was proposed by WCRB-FM (a classical commercial station in Boston) and Sanders (a subsidiary of Lockheed/Martin). It used the entire subcarrier spectrum of an FM channel for a single, 40-kHz wide, 10% injected subcarrier, carrying 160 kb/s or more of data. This data could be used for one or more DAB signals (using the AT&T PAC algorithm or projected future algorithms with up to 30:1 compression ratios) plus some auxiliary data. The latter would serve to at least partially replace the data transmission capability lost by the elimination of standard subcarriers (at 67 kHz, 92 kHz, and elsewhere) that implementation of this system would require of an FM station. This system was referred to as "in-band on-carrier" format and conformed completely to existing regulations for the FM band. As such, it was surmised that implementation could be quick and inexpensive. Initial raw-data transmission tests of the system (without FEC or interleaving) were promising, but more specific and relevant tests (such as the effects of multipath reception on perceptually coded digital audio signals) had not yet been conducted with the system as this book went to press.

After the failure of the EIA/NRSC tests to provide complete data on any in-band formats, the NAB proposed that a further round of DAB format tests be organized, which would consider only in-band systems. It would include revised versions of previously tested systems as well as new systems (such as the WCRB/Sanders proposal). At press time, these tests had not yet been designed or scheduled.

The FCC also reacted, albeit somewhat unofficially, to the lack of in-band closure to the EIA/NRSC tests. Staff hinted that the search was on for new spectrum in which to locate U.S. terrestrial DAB, and that the Commission might take a more proactive role in steering the process of DAB development (as it did with advanced television).

In many ways, the path to U.S. terrestrial DAB seems less clear in the late 1990s than it did when the decade began. Nevertheless, the advantages of an "in-band solution" still appeals to U.S. radio broadcasters, and much of the industry still holds out hope for its eventual viability.

## 14.13   THE IN-BAND APPROACH

The difficulty in obtaining new spectrum for DAB and the industry's desire to minimize the upheaval of transition to DAB stimulated the interest in an "in-band" solution. This would allow broadcasters to use existing spectrum allocated to radio broadcasting for new DAB services. It further assumes that any DAB signal would conform to the existing RF mask constraints of radio broadcast channels.

Numerous systems have been proposed. For the most part, they have been divided into two types: systems that would use currently "taboo" channels in the FM band (called *in-band adjacent-channel* [IBAC] *formats*) and systems that would place a station's digital signal within its existing channel, in either the AM or FM band (called *in-band on-channel* [IBOC] *formats*). Either type includes the intention of allowing existing analog services to coexist with the new DAB channels.

A variant called the *in-band replacement channel* (IBRC) foresees a time when analog services will be phased out. Such a service starts out as in IBOC format, and when the analog service is shut down the digital signal takes over the whole channel, allowing increased digital broadcasting capacity.

### 14.13.1   In-Band, On-Channel (IBOC)

For U.S. broadcasters, IBOC has been a sort of holy grail, theoretically providing them with an easy and smooth transition to DAB. From a business perspective, all of the positives of analog broadcasting are retained, with the quality of digital transmission added. In terms of technical reality, however, IBOC development has been difficult.

The narrow bandwidth constraints of existing AM and FM channels that IBOC systems must follow puts tight limits on data rates. There is also no possibility for multiplexing of multiple stations' signals, so these narrow channels are also subject to the same multipath interference problems faced by analog systems. Rather than combating multipath by wideband frequency diversity, IBOC systems are limited to the use of complex coding schemes and adaptive RF equalization.

If that were not challenge enough, IBOC systems must also function in these narrow bandwidths without disturbing the existing analog services with which they share their channels. This places a particular burden on designers because a substantial amount of system complexity must reside in the *receiver*, which implies that consumer cost for conversion to IBOC DAB may be relatively high. Difficult as it may seem, four such formats have been developed, built, and tested—three for the FM band and one for the AM band.

***Amati/AT&T/Lucent Technologies.***    This FM-IBOC system places a DAB signal at the outer-most edges of the FM channel. Although other systems now use a similar approach, the Amati/AT&T system was the first to do so, giving rise to the term "saddlebag IBOC." The DAB carriers of this format are situated so that they begin about 100 kHz above and below the FM channel's center frequency, and end about 180 kHz above and below the center frequency (see Fig. 14.12).

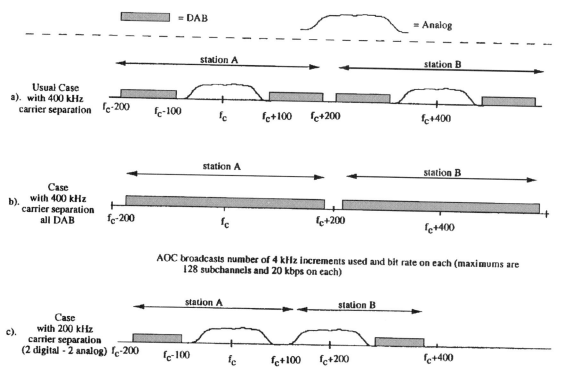

**FIGURE 14.12**    Spectrum plan for "saddlebag IBOC" and IBRC DAB channels (FM band) as proposed by AT&T/Amati.

The system has been proposed in two different forms, one with only a single (upper or lower) sideband and one with symmetrical sidebands on both channel edges. The single-sideband version is designed for use in crowded markets to avoid interference with adjacent channels, but it has correspondingly less capacity and/or lower robustness because it encodes more data into a smaller bandwidth (up to 5 b/s/Hz).

The Amati/AT&T/Lucent system (sometimes simply called "AT&T IBOC") uses a *discrete multitone system,* which is based on the *asymmetrical digital subscriber link* (ADSL) technology proposed for telephone companies' video on-demand services, which Amati also developed. It uses AT&T's *perceptual audio coding* (PAC) algorithm for audio coding and can include a small amount of auxiliary data capacity. The system also includes a robust, narrow-band data channel called the *auxiliary overhead channel* (AOC), which informs the DAB receiver of where in the FM channel any digital information is currently being broadcast.

This provides the system with the ability to dynamically reconfigure a channel. Such capability could be useful in the event that analog FM service were eliminated, allowing the IBOC DAB format to take over the entire channel after the analog carrier is removed (the IBRC approach).

It can also allow multiple services to be carried within the same channel, with the AOC informing the receiver of the channel's current contents.

The PAC algorithm is operated in this system at 128 to160 kb/s. The AOC occupies an additional 8 kb/s, and auxiliary data capacity of up to around 10 kb/s can also be included. Reed-Solomon forward error correction (FEC) is used, followed by convolutional interleaving and an optional trellis encoder. The data signal is then quadrature-encoded onto 20 to 40 carriers of 4-kHz bandwidth each, which are further transformed to as many as 128 carriers occupying up to 550 kHz through an inverse FFT modulator. The signals are then windowed, serialized, synchronized (with training-sequence preambles inserted every 2 ms) and converted to an analog RF signal at a 1.1-MHz IF, which is then up-converted to the broadcast frequency (see Fig. 14.13).

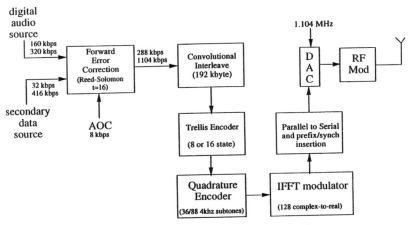

**FIGURE 14.13a**   Block diagram of the AT&T/Amati DAB transmission system.

Ideally, *low-level combining* (i.e., at the exciter stage) of the DAB and analog FM signals are used, but this requires high linearity in the downstream RF amplification, which is not generally provided in existing FM transmitters. (Because analog FM does not require high linearity in its RF amplifiers, these sections typically use class C amplification, often inadequately linear for passage of digital signals.) Therefore, modification of these stages may be required. Alternatively, *high-level combining* can be used, in which two separate cochannel transmitters are used—one high-power FM of standard linearity and one lower-power DAB transmitter of high linearity. The two outputs are then combined through an RF hybrid. Space-diversity receive antennas are optional.

***USA Digital Radio.***   USA Digital Radio (USADR) is a consortium formed in 1991 by three U.S. broadcasting companies: Gannett Broadcasting, CBS Radio, and Group W (Westinghouse Broadcasting). It is the only U.S. DAB format proponent to come from exclusively broadcast roots.

Working with a number of subcontractors having military and secure communications backgrounds, USADR developed a number of different formats, all of which were designed for IBOC applications (for either the AM or FM bands). As this work progressed, the proponent presented a number of public demonstrations of its developing formats at trade shows and elsewhere. USADR then presented three IBOC formats for testing to the EIA/NRSC in 1994—two for the FM band (called simply *FM-1* and *FM-2*) and one for the AM band. These formats are

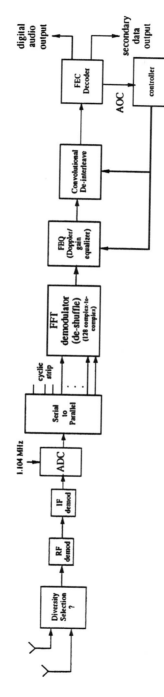

**FIGURE 14.13b**  Block diagram of the AT&T/Amati DAB receiver system.

described below in the forms in which they existed as the EIA/NRSC tests began in 1994. The two FM formats differ primarily in their receiver complexity requirements.

Although USADR's first intent for its FM format was to occupy the entire FM channel with low-level digital information (using de-encryption techniques to extract the data from beneath the analog FM modulation), receiver complexity for such an approach soon ruled it out. The proponent eventually settled on a saddlebag approach, placing its digital data at the edges of the FM channel only. In addition to stereo audio, USADR claimed an auxiliary data capability of 64 kb/s.

The FM-1 system takes a 384-kb/s output from a MUSICAM (ISO/MPEG-1 audio layer II) coder and divides it into 48 separate streams, each of which phase-modulates a narrow-band RF channel. The 48 independently modulated waveforms are then combined into a composite signal, which is converted to the broadcast frequency and mixed with the analog FM signal. The DAB spectrum is distributed in saddlebag form such that it is suppressed in the center of the channel. Adaptive RF equalization and complex filters are employed in receivers, using proprietary DSP for low-cost design of the required high-speed processing.

The FM-2 system uses similar source coding, but employs double-sideband pulse-amplitude modulation (DSB/PAM) with code-division multiple access (CDMA) spread spectrum techniques. Its FEC is at rate 2/3, and a 0.5-s interleaver is used. The proponent claims that the FM-2 system lends itself to a less expensive receiver. Nominal bandwidth of both USADR systems is 300 kHz.

The USADR IBOC AM format is designed to operate within the much narrower channel bandwidths and different propagation characteristics of the AM band. As a result, it has much lower data capacity, providing only 96 kb/s for audio and 2.4 kb/s for auxiliary data. (Auxiliary data capacity is increased to 32 kb/s if no analog AM signal shares the channel.) In operating compatibly on-channel for AM broadcast, the system mirrors its FM counterparts by placing significant energy in the channel "wings" that exist outside the nominal audio bandpass of the AM radio channel ($> \pm 10$ kHz) but within the AM mask, occupying a total of about $\pm 40$ kHz. The DAB data signal is also placed in quadrature with the analog signal, allowing the center of the channel to be used compatibly as well, although less data is placed there than in channel edges. This technique renders USADR AM IBOC incompatible with the C-QUAM AM stereo format, which also uses quadrature modulation for its stereo (L-R) information.

The format's use of 96-kb/s ISO/MPEG layer II joint stereo coding provides it with less than CD-quality audio, but when compared to the typical AM analog audio, the USADR IBOC-AM system's aural quality is substantially superior, providing reasonably quiet, 15-kHz stereo. Error correction is applied at rate 3/4 and interleaving is used. Total data rate is 128 kb/s (or 170 kb/s if no analog AM signal is present in the channel). To put this signal into 40-kHz bandwidth, 32-QAM modulation is used, which can supply the required 4.3 b/s/Hz.

### 14.13.2  In-Band, Adjacent-Channel (IBAC)

The lower power requirements of a digital signal when compared to an analog FM signal make it theoretically possible for a digital broadcast to cover the same service area as an FM station with less transmitter power. This lower-powered signal would then cause less interference to cochannel and adjacent-channel analog FM stations as well. It has therefore been projected that DAB signals might be placed into channels on the FM band where other analog FM signals cannot be operated due to currently established interference criteria. In more technical terms, the protected contours of existing FM stations can be altered due to the differing D/U (desired-to-undesired signal) ratios when comparing digital and analog FM signals.

Typically, this approach assumes that a station's digital signal could be placed in a first-adjacent channel to its FM signal, at a power low enough not to cause interference to the analog FM channel or to neighboring markets' existing cochannel and adjacent-channel analog FM signals. Hence, the name "in-band/adjacent-channel" (IBAC) has been applied to this model.

The obvious advantage is that the signal in this channel would be a digital signal only and would not have to share the channel compatibly with an existing FM analog channel.

A further caveat in this scenario assumes that a station's IBAC transmission site is co-located with its analog site. This guarantees that the two signals will degrade in similar fashion with distance (i.e., their patterns are concentric), which should ensure that any adjacent-channel protection contours will not be violated at any point in the pattern by the lower-power digital signal.

A practical problem is encountered with this method, however. In the most crowded markets and the most densely populated areas of the country (such as the northeast corridor, where neighboring markets are closely spaced and many "short-spaced" exceptions to adjacent-channel reuse exist), there simply may not be adequate spectrum to implement an IBAC format that provides a new adjacent DAB channel for every existing FM channel.

Finally, an IBAC system does not directly address a DAB solution for AM stations. Theoretically, in less crowded markets, some FM channels might be available to use in an IBAC mode for AM stations' DAB signals (any interstitial, "vacant" channel can be used—it need not be truly adjacent to an existing FM station in the market), but this remains an unsettled issue.

***AT&T/Lucent Technologies.***    Although other IBAC formats have been proposed and developed, the sole remaining representative of this format comes from AT&T/Lucent Technologies. It provides from 128 to 160 kb/s of audio data and auxiliary data of about 10 kb/s. Error-correction data takes up an additional 120 to 170 kb/s, implying a channel coding rate of approximately 1/2.

The system employs the AT&T PAC algorithm for audio data compression. Unlike other DAB systems that employ multiple narrow-band carriers, a single wide-band carrier is used, with a high level of error correction relied upon for robustness. A total data rate of 360 kb/s is modulated into a 200-kHz nominal bandwidth via 4PSK, with 1.28-s interleaving plus inner and outer error-correcting codes added to the block error correction included in the PAC subsystem (see Fig. 14.14).

Because the full channel can be used for digital information, data are transmitted at only about 1.8 b/s/Hz. Adaptive RF equalization is employed in the receiver. Space diversity using multiple receive antennas and graceful failure modes are optional.

## 14.14    *SATELLITE DAB SYSTEMS*

Several systems have been proposed for DAB use in the United States that would distribute signals from direct broadcast satellites (DBS). Some of these proposals involve technical formats and hardware, whereas others are essentially proposals for provision of programming services.

### 14.14.1    VOA/JPL

One hardware format proposal comes from the Voice of America (VOA), working in conjunction with NASA's Jet Propulsion Laboratory (JPL). This format was included in the tests conducted by the EIA in 1994 and thereafter. It was also submitted to the International Telecommunications Union (ITU) for consideration in BSS/S discussions, where it has been named "System B." ("System A" is the Eureka 147 system, configured for DBS use.)

The system has been proposed and tested for S-band application, but it is applicable to a wide range of frequency bands, as well as for terrestrial use. As such, it is intended to be an extremely flexible and efficient system of modular design, allowing it to be applied in a variety of ways using inexpensive receivers. (The VOA's initial design philosophy intended to pursue this system for eventual replacement of its network of terrestrial HF ["short-wave"] transmitters worldwide. More recently, the VOA has moved to explore the format as a terrestrial digital HF system,

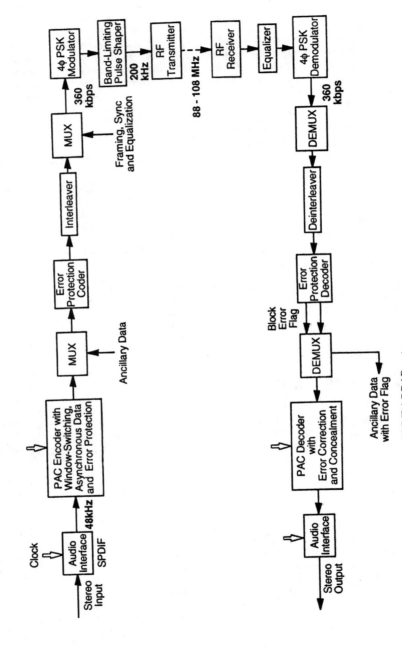

**FIGURE 14.14**  Block diagram of the AT&T IBAC DAB system.

due to reliability and building-penetration problems in a DBS-only delivery mode, especially using geostationary satellites to reach higher latitudes.)

Significant efforts are made by the system to mitigate the impairments encountered in satellite delivery—primarily signal blockage (although multipath can also be addressed by these techniques). This mitigation includes Reed-Solomon coding, convolutional coding (rate 1/2 or 1/3), and interleaving (nominally 0.4 s). AT&T's PAC audio compression algorithm is used (although others can be substituted), operating at data rates from 32 kb/s to 384 kb/s (nominally 160 kb/s), adjustable in 16-kb/s steps. To remain flexible, the system is essentially independent of operating frequency band, specifying only a 70-MHz IF with a ±18-MHz deviation (see Fig. 14.15).

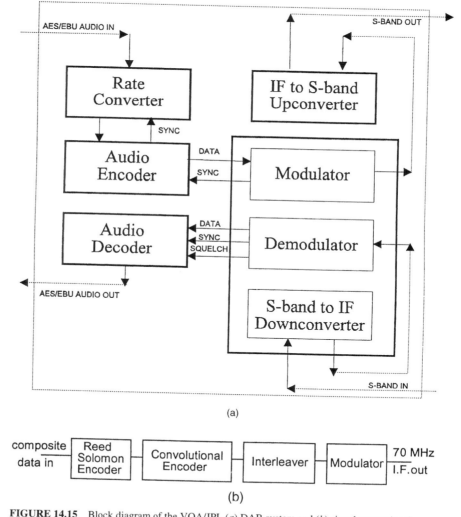

(a)

composite data in → Reed Solomon Encoder → Convolutional Encoder → Interleaver → Modulator → 70 MHz I.F. out

(b)

**FIGURE 14.15**  Block diagram of the VOA/JPL (*a*) DAB system and (*b*) signal processing flow.

### 14.14.2   FCC Docket 90-357 Proposals (S-DARS)

As this book goes to press, four proposals stand ready for action at the FCC, awaiting the licensing of satellite DAB spectrum in the S-band (2310–2360 MHz). These proponents are American Mobile Satellite Corporation, CD Radio, Digital Satellite Broadcast Corporation, and Primosphere Limited Partnership.

All four propose a multichannel audio service of 30 or more channels to be delivered to mobile or fixed receive dishes of very small size and low gain—on the order of a shallow dish 2″ in diameter or a flat plate 4″ or 6″ on a side (Fig. 14.16). It is proposed that these be mounted on (or *in*) the roof of cars or homes. Most channels would provide high-fidelity music, and some proponents include lower-bandwidth channels for voice-based formats. Each proponent requires from 12.5 to 25 MHz of spectrum for their system.

In other ways the four proponents differ in their plans. For example, some expect to support their system by listener subscription (implying addressable receivers and a conditional-access architecture), whereas others plan commercial operation or a mix of the two approaches. Some systems will use two satellites simultaneously (for spatial transmit diversity), whereas others will use one main and one backup. Some do not expect to include terrestrial repeaters (for filling gaps in satellite coverage, particularly in urban areas), but others may. (One proposal would allow carriage of local terrestrial radio signals along with its satellite channels on its terrestrial repeaters.) Perhaps most divergent is one proponent's intention to include small spot beams to target major markets with independent programming streams (Fig. 14.17). All others plan to broadcast the same signals to the entire (contiguous) United States.

The technical systems for space segments are detailed in each format's proposal, including up-link and down-link frequencies, but the actual audio formats are left somewhat unspecified (or with multiple possibilities) on some proposals. This implies that whatever the terrestrial DAB industry has settled on can be implemented into the satellite systems for cost-efficiency in multi-purpose receivers (i.e., satellite and terrestrial DAB in a single radio). Nonaudio data could also be included. For this reason, some of these proposals characterize themselves as "bent-pipe" delivery systems.

The upper portion of the 2310–2360-MHz band appears to interfere with existing aeronautical telemetry uses in Canada, so only the lower half (25 MHz) may be able to be used for satellite transmission. This means that not all proponents will be granted spectrum, since each claims to require a minimum of 12.5 MHz to provide viable service.

### 14.14.3   RadioSat

A final proponent plans to use a portion of the *Mobile Satellite Service* (MSS) spectrum in the L-band for DAB service. MSS is a two-way service, generally expected to be used for satellite telephony. A company called RadioSat has proposed using a portion of this spectrum for radio broadcasting—with each radio receiver having the capability to asymmetrically respond in an interactive fashion by transmitting back to the satellite in the same band. The proponent, and the MSS band itself, have been tied up in protracted litigation for many months, and it is not clear whether this proposal will ever reach fruition.

### 14.14.4   Direct-to-Home (DTH) and Cable-Radio Services

One form of DBS digital radio already exists. It is in the form of the audio-only multichannel services packaged with various DBS-TV services such as DIRECTV, USSB, and Primestar. These services are only receivable under fixed conditions by dishes of at least 18″ diameter—thus the title "direct-to-home." Although they will reach a certain segment of the radio listening audience in this fashion, the growing amount of listening in mobile and office environments will likely not be served by "DTH-DAB."

**FIGURE 14.16**    Prototype antenna for a satellite DAB receiver (*Courtesy of CD Radio*).

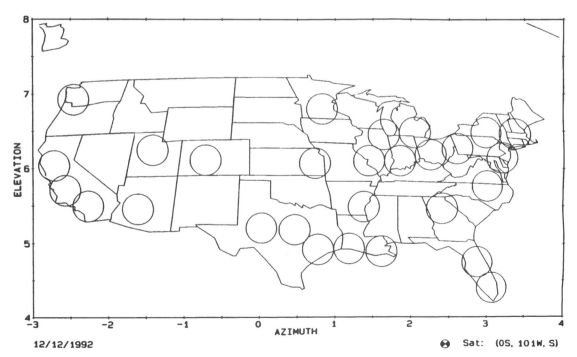

**FIGURE 14.17**    Map showing spot-beam service proposed by one satellite proponent (Digital Satellite Broadcasting Corp.). Each beam would be independently programmable, along with a single large beam that covers the entire continental United States (CONUS).

In these respects they mirror the digital cable audio services that have been available in some parts of the United States since the early 1990s. In fact, much of the DBS audio comes from cable audio content-providers.

These systems utilize a separate set-top decoder that receives the DBS or cable signal and selects the audio channel desired by the listener. Auxiliary text display of the current artist and song title (or other data) can be displayed on an associated TV screen or on an LCD screen built into the set-top box's remote control.

## 14.15  FUTURE PROSPECTS

It is unclear when and what kind of DAB will eventually emerge in the United States. The most likely possibility at this writing (and the one with the most support by U.S. broadcasters) includes a local terrestrial service in the existing radio broadcast band(s), plus a national (and/or regional) satellite service in the S-band. Another scenario envisions both terrestrial and satellite services in the S-band.

A more far-fetched possibility has the United States changing its position on L-band, and moving terrestrial and satellite DAB services there in congruence with the Canadian system that is already under development and early deployment. Finally, an even less likely possibility weaves DAB into future ATV allocations (just as FM sits today in the middle of the VHF-TV band, between channels 6 and 7), such that some freed-up VHF spectrum, or some new UHF spectrum, is dedicated to new DAB terrestrial services.

Current difficulties in testing and policy wrangling virtually guarantee that none of these options will take place in short order, however—probably making the United States one of the last developed nations to actually implement DAB. It is also a virtual certainty that tomorrow's radio receivers will enjoy less worldwide compatibility than today's.

## BIBLIOGRAPHY

R. L. Anglin, Jr., and Pierson, W. Theodore, Jr., Esq., "Application of Digital Satellite Broadcasting Corporation for a DARS Satellite System." Submitted with respect to FCC GEN Docket No. 90-357, Washington, D.C., Dec. 15, 1992.

B. Carlin, "USADR AM DAB System Description." Response to EIA/CEG DAR Subcommittee, Washington, D.C., Feb. 15, 1994.

J. M. Cioffi, "DAB with Auxiliary Overhead Control," Palo Alto, Calif., Jan. 28, 1993.

J. M. Cioffi, and Bingham, J. A. C., "The AT&T/Amati Overlaid DSB System: Maximizing Control and Flexibility," Palo Alto, Calif., Jan. 29, 1993.

E. A. Janning, "USADR System Description—FM2." Response to EIA/CEG DAR Subcommittee, Washington, D.C., Feb. 7, 1994.

Nikil Jayant and Sundberg, C.-E., "AT&T DAR System: General Description and Specific Information." Responses to EIA/CEG DAR Subcommittee, Washington, D.C., 1992–1994.

National Association of Broadcasters, Science and Technology Department, *Understanding DAB: A Guide for Broadcast Managers and Engineers,* NAB, Washington, D.C., 1992 (updated 1995).

Skip Pizzi, *Digital Radio Basics,* Intertec, Overland Park, Kans., 1992.

"USADR FM-1 Description." Response to EIA/CEG DAR Subcommittee, Washington, D.C., Feb. 7, 1994.

Arv Vaisnys, "VOA/JPL DAR System Description." Response to EIA/CEG DAR Subcommittee, Washington, D.C., Jan. 27, 1994.

## ABOUT THE AUTHOR

Skip Pizzi is editor of *BE Radio* magazine and technical editor for *Broadcast Engineering* magazine. He is also the author of *Digital Radio Basics,* which was the first commercially published book on DAB. He is a contributing editor to *The NAB Engineering Handbook* and to several other technical handbooks and textbooks in the field of electronic media technology and electronic journalism. His work has appeared in more than a dozen audio, video, and broadcast trade magazines over the past 15 years.

Pizzi is also a consultant and trainer, specializing in new broadcast technologies. Earlier in his career he spent 13 years on the staff of National Public Radio at its headquarters in Washington, D.C., as an engineer, manager, and trainer.

Pizzi studied physics, international economics, and fine arts at Georgetown University, earning a B.A. degree in 1975. He is a member of the Audio Engineering Society (AES) and the Society of Broadcast Engineers (SBE), is a past chair of the AES's District of Columbia chapter, and serves on the NAB/SBE Broadcast Engineering Conference Advisory Board (which sets the agendas for the two largest technical conferences of the U.S. broadcast industry each year). He is a recipient of the AES Board of Governors Award and the Public Radio PRRO Award.

# CHAPTER 15
# DATA BROADCASTING SYSTEMS

## SECTION 1
## THE RADIO BROADCAST DATA SYSTEM

**Almon H. Clegg**
*CCi*

## 15.1 INTRODUCTION

The idea of broadcasting data along with the radio transmission is quite old. As far back as the early 1970s the National Association of Broadcasters was exploring methods to send a silent digital signal that had encoded in it the station call letters and broadcast frequency. The receiver was to have a display for reading the information.

Also, in the early 1980s a system for broadcasting traffic information was developed called *Automatic Road Information* (ARI). Although it was never popular in the USA, it did have some long-lasting success in Europe and became the nemesis of what is now called *Radio Data Systems* (RDS).

It began in the early 1980s with a strong coordination from the EBU (European Broadcasting Union), a well-recognized authority located in Geneva, Switzerland. The coordinator of the RDS project within the EBU was Dietmar Kopitz, who remains active today as the coordinator of the RDS Forum, a users group composed primarily of experts from Europe and globally oriented companies. The RDS Forum meets at least twice each year and furthers the implementation of RDS technologies and applications.

## 15.2 BEGINNINGS IN THE UNITED STATES

The U.S.A. effort began in earnest in February 1990. The United States Radio Broadcast System Standard was produced by the RBDS subgroup of the National Radio System Committee (NRSC), which is jointly sponsored by the Consumer Electronics Manufacturers Association (CEMA), a sector of the Electronic Industries Association, and the National Association of Broadcasters (NAB).

The approach was to define the standard in a way that would allow usage of the same circuits, especially ICs, and systems as already developed in Europe by the several manufacturers and equipment suppliers. Because most radio manufacturers were already familiar with the technology and had developed receiver circuits for European markets, the NRSC subcommittee recognized that it would help to popularize the technology in the United States if the set makers would not have to design a new, nonstandard product. Also, it was felt that, insofar as possible, the standard should be fundamentally compatible, allowing radio receivers to work in any country of the world. Thus, a new worldwide system standard was being forged.

The standard-setting process was voluntary, with invitations open to anyone with interest in becoming a proponent and contributing to the work. Many meetings of the RBDS subcommittee were held over the next two or three years. Intense activity ensued. Meetings were attended by 30 to 60 persons from the broadcasting community, equipment manufacturers, receiver makers, and government. Early in the standard-setting process it was clear that many parties were interested in the use of the RBDS data channel for other than program- or station-related information, such as paging by RBDS, differential correction signals for Global Positioning System (GPS) receivers, advertising messages on the radio text channel, highway sign control, and traffic control information. In fact, it now appears that the data channel is inadequate to cover all the needs of the industry and an additional subcommittee has been formed for the development of a high-speed (relatively—maybe 9600 baud or more) for additional purposes.

The main function of RBDS remains to unite the broadcaster and the listener through a system of data that would be silently transmitted along with the main baseband carrier signal, which would provide station call letters, program type, information about the music being played, traffic alert, and automatic tracking of national broadcast network programs. These features were the main driving forces and formed the initiative to the working group for writing the initial draft standard language.

The finished document, released in January 1993 at the Winter Consumer Electronics Show, Las Vegas, Nevada, consists of 133 pages and all the charts, diagrams, and group structures to define the system. However, revisions and other guidelines for implementation are being developed on an ongoing basis and will be considered for future publication.

## 15.3 THE BASICS OF RBDS (RDS)

The concept of RDS is quite ordinary. Subcarriers have been employed in FM broadcasting for many years. In the United States the Federal Communications Commission has authorized usage of subcarriers at any frequency in the frequency region from above the stereo difference channel (from about 19 kHz to 53 kHz) to 99 kHz. Some popular subcarrier frequencies in use today are 57 kHz, 66 kHz, 67 kHz, 76 kHz, 92 kHz and 95 kHz. Some subcarriers are used for analog voice or music programs, whereas others are used for data purposes. In-store background music often comes via a subcarrier modulated with an analog music signal. There are many other applications. Of the 5000 or more FM broadcasters in the United States, many broadcast one or more subcarriers. These simultaneous broadcasts are inaudible to those listening to the regular baseband mono or stereo audio program. With care in the modulation levels and filtering of the subcarrier channel, cross-talk into the main channel or adjacent channels is held to a minimum.

RDS is a particular specification for uniform usage of the 57-kHz subcarrier. This carrier is a third harmonic of the 19-kHz tone used to illuminate the stereo light and, to facilitate easy decoding, is held to a close phase relationship with the stereo pilot.

Deviation, or injection level, of the subcarrier is under some control of the broadcaster and is chosen for the best result given the various services to be provided by the transmitter and in conformance with FCC injection requirements. However, the RBDS standard suggests the best compromise is $\pm 2$ kHz, or about 2 to 3 percent modulation injection level.

The subcarrier is amplitude-modulated by the shaped and biphase-coded data signal. The modulation result can be thought of as a form of two-phase-shift-key (psk), with a phase of $\pm 90$ degrees. The data rate of the system is 1187.5 b/s, whereby the basic clock frequency is obtained when dividing the subcarrier (57 kHz) by 48.

The baseband coding is typical of other data transmission systems and is formed in a fashion for easy encoding and decoding. The largest structure consists of 104 bits and is called a "group." Each group comprises 4 blocks of 26 bits each. Each block comprises an information word and a checkword. Each information word contains 16 bits. Each checkword is 10 bits. Figure 15.1 shows the structure. The order of transmission is by blocks 1 through 4, as in Fig. 15.2. Table 15.1 shows the group types and application for each of the groups.

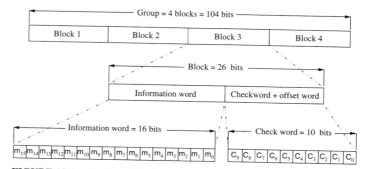

**FIGURE 15.1**    Structure of the baseband coding.

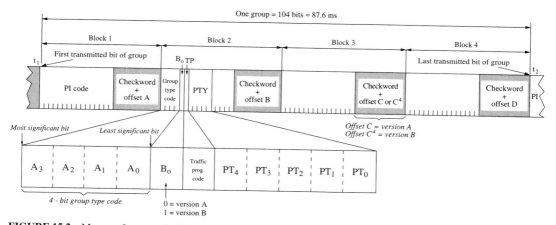

**FIGURE 15.2**    Message format and addressing.

**TABLE 15.1**    Group Types and Applications

| Decimal value | Group type | | | | | Applications |
|---|---|---|---|---|---|---|
| | Binary code | | | | | |
| | $A_3$ | $A_2$ | $A_1$ | $A_0$ | $B_0$ | |
| 0 | 0 | 0 | 0 | 0 | X | Basic tuning and switching information |
| 1 | 0 | 0 | 0 | 1 | X | Program item number and slow labelling codes |
| 2 | 0 | 0 | 1 | 0 | X | Radiotext |
| 3 | 0 | 0 | 1 | 1 | 0 | LN—Location and Navigation |
| 4 | 0 | 1 | 0 | 0 | 0 | Clock-time and date |
| 5 | 0 | 1 | 0 | 1 | X | Transparent data channels (32 channels) |
| 6 | 0 | 1 | 1 | 0 | X | In-house applications |
| 7 | 0 | 1 | 1 | 1 | 0 | Radio paging |
| 8 | 1 | 0 | 0 | 0 | 0 | Reserved for TMC |
| 9 | 1 | 0 | 0 | 1 | 0 | Emergency warning systems |
| 10 | 1 | 0 | 1 | 0 | 0 | Program type name |
| 11–13 | — | — | — | — | — | Undefined |
| 14 | 1 | 1 | 1 | 0 | X | Enhanced other networks information |
| 15 | 1 | 1 | 1 | 1 | X | Fast basic tuning and switching information |

X indicates that value may be "0" (version A) or "1" (version B).

One would do well to obtain a copy of the RBDS standard from either the NAB or EIA for detailed study. This is just for basic understanding of the system; much more information is necessary for product or system design implementation.

## 15.4  DIFFERENCES BETWEEN RBDS AND RDS STANDARDS

The basic structure, group and block layout, modulation, and key applications are the same between the Cenelec EN 10057 RDS standard as used in Europe and the U.S. RBDS standard. Therefore, encoders for broadcast transmitters and ICs for receiver designers are essentially the same. Key differences in the implementation are summarized as follows:

**1.** Because a few hundred FM broadcast transmitters in the United States already employ a multiplex signal for MMBS (a Swedish paging format utilized by CUE Paging of Irvine, California) on a 57-kHz subcarrier, it was necessary to allow for compatibility of operation between the two systems and provide for MMBS stations to support the basic receiver functions. Appendix K of the RBDS standard gives the details of this implementation.

**2.** Because of great interest on the part of the AM broadcasters in the United States for inclusion of the program-type selection process, and because the committee was unable to source a technology that would support RBDS on the AM band, the RBDS standard allows for inclusion in an AM receiver of the ID-Logic system as developed by the PRS Corporation. The system, when employed in an RBDS framework, is called the "In-Receiver Database System" (I-RDS). Essentially, the receiver has a complete database of all AM and FM broadcasters in the United States built into a ROM device. By using a certain bit within the RBDS data stream, the ROM database in an ID-Logic receiver can be updated to include station information changes (such as program-type changes, call letter changes, etc.). See section 4 and appendix L of the RBDS standard for more detailed information.

**3.** There is a placeholder within the standard that allows for the inclusion of an AM RBDS when a technology is found that will support it. The RBDS subcommittee is continually evaluating proposals that may be employed for AM application. To date, a viable technology has not been certified.

**4.** The program-type code (PTY) list is unique to the United States. In Europe and other countries of the world, name titles to characterize the type of music are quite different from the United States. Therefore, the RBDS subcommittee formed a working group to compose a listing of program categories or type descriptors that would be suitable for a listener in setting the search for the type of music preferred. The listing of these categories is given in Table 15.2.

## 15.5  NEXT REVISION OF THE RBDS STANDARD

The RBDS subcommittee, as of this writing, is considering a few basic additions to the standard. It will most likely not be sooner than mid-1997 when the next version of the standard will be available. The basic technology and system parameters will not change; rather the additions will specify two new applications:

**1.** Because a large number of potential users were asking for specific applications, it was decided to use data group 3A as an open data grouping for up to 65,000 vertical applications. The Location and Navigation group (LN) now specified for group 3A will become reassigned to the Open Data Applications (ODA). This approach will provide a long-term future for yet unknown applications. A central agency, yet to be decided but most likely the EBU in Geneva, Switzerland, will act as a clearing house for the ODA codes and will charge a small fee to cover the central

**TABLE 15.2**  Program-Type Codes*

|  | Program type | 8-Character display |
|---|---|---|
| 0. | No program type code | _****_† |
| 1. | News | NEWS |
| 2. | Information | INFORM |
| 3. | Sports | SPORTS |
| 4. | Talk | TALK |
| 5. | Rock | ROCK |
| 6. | Classic rock | CLS_ROCK |
| 7. | Adult hits | ADLT_HIT |
| 8. | Soft rock | SOFT_RCK |
| 9. | Top 40 | TOP_40 |
| 10. | Country | COUNTRY |
| 11. | Oldies | OLDIES |
| 12. | Soft | SOFT |
| 13. | Nostalgia | NOSTALGA |
| 14. | Jazz | JAZZ |
| 15. | Classical | CLASSICL |
| 16. | R&B | R_&_B |
| 17. | Soft R&B | SOFT_R&B |
| 18. | Language | LANGUAGE |
| 19. | Religious music | REL_MUSC |
| 20. | Religious talk | REL_TALK |
| 21. | Personality | PERSNLTY |
| 22. | Public | PUBLIC |
| 23–29. | Spares | |
| 30. | Emergency test | TEST |
| 31. | Emergency | ALERT! |

*It is anticipated that the codes listed above will cover most existing formats and will be capable of handling the future evolution of new formats.
† _ refers to a space in 8-character display.

office costs. The Internet will serve as the vehicle for applications and listing of the various use numbers granted, type of service, service provider, and so on. The Application Identification Codes (AID) will be given to the provider upon completion of the application process. The current ODA protocols that will most likely become the first to be given AID codes are Emergency Alert System (EAS), Differential Global Positioning Systems (DGPS), DAB cross referencing, and Enhanced TMC (Traffic Messaging System). The ODA concept was universally accepted by the standard-setting bodies of the RDS Forum in Europe and the RBDS Subcommittee.

**2.** Differential Global Positioning System (DGPS) provisions. Under a working group led by Dr. Paul Galyean, an appendix will be added that describes the implementation for this feature. It is interesting to note that the differential correction signal sent by RBDS allows for an increase in position accuracy of a typical GPS receiver of more than one order of magnitude. A 50-meter accuracy in position is obtainable with handheld GPS receivers. When differential correction is added, the location accuracy increases to about one meter.

**3.** Recently, the FCC, under legislative mandate, issued a rule-making procedure that replaces the old Emergency Broadcast System with a new, high-technology, digital data–based system. The new system, called the *Emergency Alert System* or EAS, allows for the use of RBDS signals in forwarding and coding various types and kinds of emergency and disaster messages. Sage Alert, along with several others, have started a very large-scale business activity based on RBDS programming. In July 1995 a working group was formed to prepare a document for inclusion within the RBDS standard to accommodate FCC requests. This document is nearly complete and will be included as an ODA in the next issue.

## 15.6   RBDS VS. RDS NOMENCLATURE

Some have sensed confusion over the use of the term *RBDS* in the United States: in all other parts of the world it is called *RDS*. Because of the differences outlined earlier, members of the committee felt that it would be best to distinguish the U.S. standard in order to highlight these differences, thus drawing attention to the differences that would be of serious concern to receiver design engineers and broadcast equipment makers.

However, the RDS logo is specified in the RBDS document, radio receivers use the RDS logo on the front panel, and various companies use RDS only in promoting their products.

Likewise, the term "Smart Radio" is used by many, especially, Denon, for identifying their product. In the commercialization of a product class, it is helpful to have names that the users can identify with. Hence, the committee does not encourage the use of RBDS other than to refer to the U.S. standard document.

Therefore, when one is discussing the technology generally, it is quite correct to refer to it as RDS. When discussing the specifics of implementation as called for in the U.S. standard, it would be proper to refer to the U.S. RBDS standard. The committee does not want to create confusion; it is clear that the basic technology employed is RDS. The encoder manufacturers, the IC makers, and the receiver manufacturers are certain in their understanding of the basic technology. Such an approach provides for the maximum benefit by cost rationalization over the world markets. In fact, if it were necessary for the U.S.-branded products to utilize a different technology or development of unique ICs and logos on the front panel of their sets, it would be highly unlikely that they would expend the development funds to support the businesses.

## 15.7   WHAT IS GOING ON IN THE MARKETPLACE

There are two broad aspects of market applications. The first, which can be referred to as *vertical markets,* are user services such as paging (Panasonic and AXXCESS Global, CUE Paging), DGPS (Differential Corrections, Inc., and ACCQPOINT), advertiser-sponsored information (Coupon Radio), highway sign control (Modulation Sciences), emergency alerting (Sage Alert) and traffic control (Intelligent Traffic Systems [ITS]).

The second market, a horizontal sector, will be a little slower taking off because of the development time and refined engineering that is required. For example, the vertical markets are generally more value-added, smaller manufactured quantities; smaller and more entrepreneurial enterprises; and quicker-to-market technologies. Whereas the horizontal market requires development of ICs, new radio designs, close cooperation with broadcasters (for such things as development of the radiotext channel for music title display), and so forth. But it is the horizontal market that will become the most important driving force as it approaches economy of scale and mass popular appeal.

### 15.7.1   Horizontal Market Features

The features contained on RDS radios of the future will be from among those listed in the glossary. There exist many opportunities for both broadcasters and set makers to customize these features and implement them in unique ways, thus adding to the competitive nature of brand differentiation and market segmentation. For example, Denon, one of the leading manufacturers, who has already placed over 150,000 receivers in the market place, is a maker of high-end audio equipment and emphasizes its ability to quickly add features that distinguish its product from others. Thus, Denon has used this opportunity to be an innovator in the industry. Delco, makers of GM automobile radios for the entire GM line, has also been energetic in providing the technology for their car divisions. The implementation of radiotext is quite different between the Denon and Delco radios. The way in which the user searches for program type is also quite different. See the glossary for definitions of various functions and features of RDS.

## 15.8  MARKET DEVELOPMENT ACTIVITIES

The CEMA (Consumer Electronics Manufacturers Association), with encouragement from the RBDS Subcommittee and many from the industry, has formed a market development advisory group. Along the lines of the RDS Forum, previously explained, it acts as a users group, as an information dissemination point, and as an advisory to the CEMA Audio Division. This group was officially formed in Philadelphia just prior to the Mobile Electronics Show in March 1995. There are several companies actively supporting this movement throughout the United States.

Not unrelated is an offer of CEMA to fund, along with matching funds from the industry, the installation of up to 500 encoders (reaching 85 percent of the country's radio listening audience) in broadcasting markets. This project, which is well underway as of this writing, will generate a total of $3,500,000 for promotional and advertising purposes, in addition to the placement of encoders at radio stations. The 25 markets are outlined by CEMA, and a methodical plan to reach each of them is ongoing.

Similarly, CEMA is meeting with various manufacturers of equipment and receivers to encourage their adoption of RDS into products. Such meetings have taken place already with many of the top brand companies (most of which are currently making radio receivers for sale and distribution in European countries). The corporate sponsors of the development matching fund, as of this writing, are Denon, Delco, Pioneer, and RE America.

## 15.9  PRODUCTS IN THE MARKETPLACE

Already, about 450,000 receivers (not counting pagers, GPS receivers, etc.) are in the market from Denon, Onkyo, Delco, Pioneer, Kenwood, Grundig, and others. They represent only the beginning, if the European experience is any indication of the support that we can expect. In fact, based on the promotional campaign just explained, it would not be surprising to find that the U.S. experience will outshine that of Europe.

A tabulation of products in the market for European countries includes six portable models, 111 home models, and 248 car models. These products come from such manufacturers as Denon, Alpine, Bang & Olufsen, Becker, Blaupunkt, Clarion, Fujitsu-ten, Grundig, Harman Kardon, JVC, Kenwood, Onkyo, Panasonic, Philips, Pioneer, Sanyo-Fisher, Sony, Volvo, and Yamaha.

Also, there are at least six manufacturers or suppliers of RDS encoders for installation by broadcasters. They range in cost from around one thousand to several thousand dollars. Thus, it is not a large investment for a radio station to get "on the air" with RDS; in fact, if the radio station can support the CEMA program, an encoder is given free of charge, in exchange for certain "air time" for promotion.

In order to detect the RDS signal being transmitted over the air, one company, RDS Diagnostix, markets a software decoder that, when installed on a PC, allows the complete visualization of each group, radiotext, and all other pertinent information that a station engineer would want to know. In fact, the same software package can be used for monitoring the signals of other stations in the market to determine PS codes, PTY, radiotext, and all other groups being transmitted by each station.

## 15.10   *CONCLUSION*

Much work has gone into the development of the RDS technology from the greatest engineers and marketing specialists in the world. The success of the technology is without question if one considers the already present number of models and products installed. The market speed-up program of CEMA and cooperation of the NAB will continue to encourage rapid adoption. As large horizontal manufacturers introduce models into the market, other vertical markets will mature. This will lead to a "win-win situation" for all parties: for the broadcaster who will get additional revenue, as well as better station recognition from the listener; for the listener who has a receiver that brings more useful information and entertainment; for the receiver and equipment manufacturer who can introduce more models into the marketplace; and for the advertiser who can focus on dedicated advertising channels through radiotext applications. Also, the fact that RDS can fully support the Emergency Alert System, as newly defined by the FCC, will provide receivers that bring a new level of safety and security to the homes of America. The future of RDS technology, made possible by the adoption of an RBDS standard in the United States, looks healthy, bright, and on an upward path of total acceptance.

## 15.11   *GLOSSARY OF TERMS FOR RDS APPLICATIONS*

**Program identification (PI)**   This information consists of a code that enables a receiver to identify the program. The code is not intended for direct display to the user; rather, it is sort of a program ID to allow distinction from all other programs. This is valuable for "scanning" and "learning" receivers (with or without memory), which are equipped to switch over to an alternate transmitter with the same program (NPR radio affiliates, for example, often broadcast the same program at the same time.)

**Program service (PS)**   This is a text of not more than eight alphanumeric characters coded in conformity with the standard and used to display the service name of the broadcasters. Example of a PS name are "WABC," "WXYZ," "G-100," "HOT93," or "COOL99."

**Program type (PTY)**   This is an identification number to be transmitted with each program; it will identify the "type" or "category" of the program content. There are 31 codes allowed and in the RBDS U.S. standard, 22 categories plus emergency alert have been decided on. For example, PTY "1" is news, and the display on the receiver will show "NEWS"; PTY 15 is for classical music, and the display will show "CLASSICL"—lacking the "A" because only eight characters are allowed (Table 15.2).

**Traffic-program identification (TP)**   This is a digital flag sent by the broadcaster to turn on the lamp or other signal within the receiver to indicate to the user that the broadcaster is giving traffic announcements (when the lamp does not light, it means that the station is not giving traffic announcements at the time).

**Traffic-announcement identification (TA)**   This is an indicator to show when the actual traffic announcement is in progress. It can be used by the receiver to switch from its current mode over to the traffic announcement on the main audio channel, and then back to the previous setting after the traffic announcement concludes.

**Alternative frequencies (AF)**   This list is compiled in memory within the receiver of nearby broadcast stations that carry the same program. This is to allow the motorist who is traveling away from his local station to automatically switch to the next geographically located station for continuation of the same program (such as NPR).

**Location and navigation (LN)**   This feature gives the following information on station location: transmitter's state, city, latitude, and longitude (in the form of a computed grid number).

**Program item number (PIN)**    A number that can be used by a "smart radio" to tune a specific program at a particular time when the program has come on the air. It allows for consumer convenience in receiving special programs.

**Radiotext (RT)**    For receivers suitably equipped with a display, messages could be sent to the listener giving information about the program being broadcast or advertiser-supported messages. A voice synthesizer could be controlled by the radiotext code, for example, to annunciate a certain message in the absence of a display.

**Enhanced other-networks information (EON)**    This feature can be used to update the information stored in a receiver about program services other than the one being received. Alternative frequencies, the PS name, traffic programs, announcement identifications, and other information can be used to provide linkage to such other programs.

**Transparent data channel (TDC)**    These channels could be used to send alphanumeric characters, or other text, or for transmission of computer program data or similar information not intended to display on the user's set.

**In-house application (IH)**    Data used by the station for internal use. For example, IH could be used as a remote switch to turn on/off a remote transmitter site via the studio-to-transmitter link.

**Clock-time and date (CT)**    Coordinated Universal Time (UTC) code can be used to set the clock to the exact time as displayed on the receiver.

**Radio paging (RP)**    The RP feature is intended to provide for numeric and alphanumeric paging via the FM radio transmitter. Already, several international paging operations are in progress. RDS paging receivers are not unlike those already in the marketplace; they are merely operating at a different frequency and from an already installed transmitter.

**Emergency warning systems (EWS)**    This is intended to be used for the coding of audio and digital emergency and alerting messages. Such codes can be coordinated with national emergency and disaster agencies.

**Program-type name (TPYN)**    An alternate name to be displayed after the PTY search if the broadcaster wants to a give more specific music format.

## ABOUT THE AUTHOR

Almon H. Clegg is a consultant to the consumer and multimedia electronics industry on a world-wide basis. He has served as executive vice president and chief operating officer of Denon America, Inc., and has held many positions with Panasonic, Technics, and Matsushita Electric. Clegg received a B.S.E.E. from the University of North Dakota in 1969 and an M.B.A. from Illinois State University in 1972. He was a professor of technology at Illinois Central College, East Peoria, from 1969 to 1974.

Clegg is active in standards-setting bodies of the EIA, NAB, NRSC, ITA, IEC, SAE, and ISO. He is outgoing chairman of the Radio Broadcast Data Service subcommittee of the NRSC (EIA- and NAB-sponsored).

# CHAPTER 15
# DATA BROADCASTING SYSTEMS

## SECTION 2
## TELEVISION DATA BROADCASTING

**Lynn D. Claudy**
*National Association of Broadcasters*

## 15.12  *INTRODUCTION*

The over-the-air television broadcasting medium is well suited for data-broadcasting applications where data are distributed to a large number of recipients from a centralized source. Such an application is called *data broadcasting*. Because of the ubiquity of over-the-air television service, data broadcasting can reach essentially 100 percent of businesses and households in a metropolitan area. This offers the most universal coverage of any available communications medium—exceeding that of the telephone network—and it is available without wires and without incremental transmission costs.

Broadcasting is inherently a point-to-multipoint application. This is a principal advantage of over-the-air broadcasting compared to other media for data broadcasting. Wireline technology transmission costs, on the other hand, increase roughly in proportion to the number of destination sites. This gives great economic advantage to over-the-air data broadcasting.

Immediacy of delivery is another advantage of data delivery using the television broadcast signal. Using broadcast signals, data are simultaneously delivered to all receivers, regardless of the number of receive sites. The difference in delivery time between this architecture and others, such as "broadcast faxing," is profound if the number of receive sites is significant.

Reliability is another strong advantage of television data broadcasting. Broadcast operations are among the most rugged, redundant, and reliable systems in the world, operating at close to 100 percent availability. The reliance of the general public on broadcasting in times of emergency and natural disaster compels the broadcasting system to continue functioning when most other systems cannot.

Finally, the universality of broadcasting is unequaled by any other media. Television is truly ubiquitous. Television penetration is higher than that of any other communications medium, including the telephone network. Data transmission over telephone lines also can be too expensive or not available in rural sites. The incremental cost of adding to the existing telephone network structure can be very significant relative to adding a receive television antenna to access data broadcasts.

Despite the advantages of the over-the-air broadcast medium for data broadcasting, popular applications for high-speed services have not yet emerged. In a 1994 NAB survey of television stations, only 12 percent of stations surveyed reported that they are using part of their vertical blanking interval for data transmission. Although the future for data broadcasting is bright, it is not pervasive at the present time.

## *15.13*   *STANDARDS AND APPLICATIONS*

### 15.13.1   Closed Captioning

A successful and prevalent example of television data broadcasting is the closed-captioning service. Since 1978, FCC rules have reserved vertical blanking interval (VBI) line 21, field 1, and half of field 2, for the transmission of program-related captions. As a result of the U.S. Congress's Television Decoder Circuitry Act of 1990, since July 1993 all television sets manufactured or distributed for sale in the United States with screen diagonals of 13″ or greater must include closed-captioning decoding capability.

The data transmission format of captioning during line 21 on field 1 consists of a clock run-in signal, a start bit, and 16 bits of data corresponding to two bytes of 8 bits each (7 bits ASCII code plus one parity bit). The payload data rate (including parity bits) is 16 bits every 1/30 second or 480 b/s, corresponding to 60 characters per second. Although this is adequate for the captioning application, it is indeed a slow data rate for many other applications. The closed-captioning data signal format is shown in Fig. 15.3.

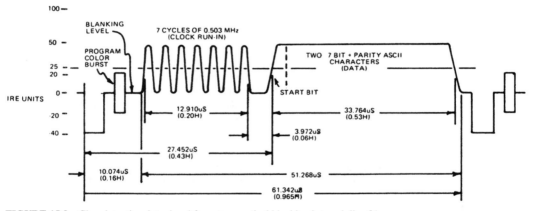

**FIGURE 15.3**   Closed-caption data signal format on vertical blanking interval, line 21.

### 15.13.2   Extended Data Services

The original captioning standard defined two channels of captions which could be interleaved on field 1 (e.g., to implement second-language captioning). However, the captioning industry soon determined that the data capacity of field 1—a maximum of 60 characters per second—is insufficient to carry two complete sets of captions. The industry has stated for many years that the goal of dual-language captioning could only be accomplished by using a second complete field in the VBI.

In 1993 the FCC authorized line 21 field 2 to allow captioning, text, and extended data. Under the auspices of the Electronic Industries Association, a set of services called Extended Data Service (XDS) was developed. XDS is a data-broadcasting scheme that allows broadcasters to send to consumers information about the station and its programming, such as the station's call sign, network affiliation, program title, length of the program, type of show, current time, time left in current program, and other information. The XDS data are interleaved with closed-caption data on line 21 of the TV signal, and its technical specification is documented within EIA Standard 608.

### 15.13.3  Teletext

With the exception of line 21 (reserved for closed captioning) and line 19 (reserved for the ghost-canceling reference signal), lines 10 to 21 of the VBI are available for data transmission. The most popular industry technical standard for VBI data broadcasting is the North American Basic Teletext Standard (NABTS), which is published as an EIA and ANSI (American National Standards Institute) standard, as well as an international specification.

In the NABTS system each line of the VBI can transmit 288 bits per television field, or one packet. The instantaneous transmission bit frequency is 364 times the horizontal scanning rate, or 5.727272 Mb/s. Of this capacity, 64 bits are used for setup and addressing, leaving 224 bits for payload data per line. With 60 fields per second, and assuming 11 available VBI lines per field, and discarding start/stop bits associated with the overhead of RS-232 protocol, a payload data capacity of about 185,000 b/s can be achieved, equating to 83 megabytes per hour.

Because data broadcasting via the over-the-air medium is a one-way path, robust forward error correction (FEC) is needed to guarantee data integrity. The FEC reduces the total capacity available for payload data to slightly over 150 kb/s. However, this allows very close to error-free operation and compares well with wireline-based transmission technologies.

Although teletext has achieved popularity in other countries, it has languished in the United States for the past decade. It is used for various private, business applications, but U.S. consumer interest in teletext applications simply has not developed.

## 15.14  NEW SYSTEMS

In 1993 the National Data Broadcasting Committee was formed by the National Association of Broadcasters (NAB) and the Electronic Industries Association (EIA). The NDBC's purpose is to establish a standardized high-speed data broadcasting system for NTSC television, at data rates higher than can be achieved through standard VBI transmission methods.

Two systems have participated fully in the NDBC and are at the stage of implementing hardware. The systems considered by the NDBC were developed by Digideck, Inc., and Wave-Phore, Inc.

The Digideck system is called "D-Channel," and it inserts a low-level QPSK data signal in the VSB region of a standard NTSC channel. The D-Channel data broadcasting technique uses differentially coherent QPSK modulation (DQPSK) of a carrier located approximately 1 MHz below the NTSC picture carrier, at a power level approximately 30 dB below the picture carrier at peak of sync. The D-Channel transmission is similar to NICAM—an over-the-air broadcast standard now widely used for delivery of stereo digital audio on PAL systems in Europe. However, NICAM inserts the DQPSK carrier above the FM audio carrier of a PAL channel. The spectral profile of the Digideck system is shown in Fig. 15.4. The Digideck system has a total transmitted bit rate of 700 kb/s with a payload data rate of 525 kb/s.

The WavePhore TVT1/4 system inserts a phase-modulated subcarrier located above the NTSC color subcarrier and below the audio carrier, as shown in Fig. 15.5. Unlike the Digideck system, the WavePhore system allows data encoding at baseband video; hence, videotapes can be encoded with the WavePhore signal. However, the data carrying capacity is less than Digideck; total transmitted bit rate is 599 kb/s with a payload data capacity of 300 kb/s.

Because the signals are in the active video area of the NTSC signal, both the Digideck and WavePhore systems cause some visible degradation of the "host" video signal. However, the effect is generally subtle for most types of program material. Based on the results of analysis, laboratory tests, and field tests, the NDBC concluded in December 1996 that the Digideck system is suitable for operating available data broadcasting service.

The FCC began a rule-making proceeding seeking to establish regulations for sending data in the active video portion of the broadcast television signal, as opposed to the vertical blanking

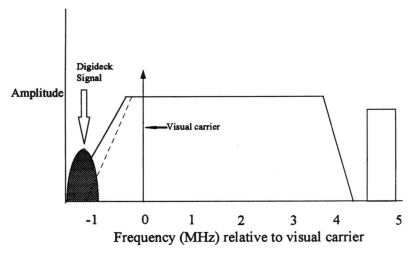

**FIGURE 15.4**   The Digideck signal in the NTSC television signal.

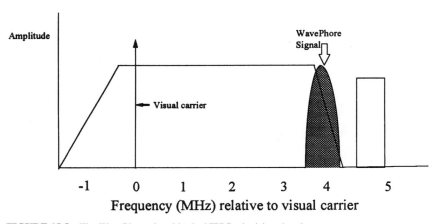

**FIGURE 15.5**   The WavePhore signal in the NTSC television signal.

interval. In this proceeding the FCC is seeking to determine what procedural and substantive rules it should establish regarding the transmission of ancillary digital data within the active video portion of broadcast NTSC signal. In addition, it asks what licensees' responsibilities should be, how to determine the extent to which a data broadcasting signal degrades its host video signal, whether broadcasters have the right to delete the data, and to what extent the Commission's rules should reflect industry standards. In June 1996, based on the work of the NDBC, the Commission authorized the use of both the WavePhore and Digideck systems by television broadcast stations.

## 15.15  OTHER SYSTEMS

Other data-broadcasting proposals are being developed that plan to exploit delivery of data to personal computers. An example is the Intercast system being developed by Intel Corp. along with about a dozen supporting organizations from the television and computer industries. Although the transmission technology is basically that of the conventional teletext system using the VBI, the fusion of television and computers offers compelling applications. By using HTML (hypertext mark-up language), the computer language standard of the Internet, the possibilities of broadcasting to personal computers opens up many possibilities. Examples include in-depth stories on a new topic being broadcast, concert schedules for the artist whose video is playing, and continuous sports statistics for the football game being played. Using personal computers as the target receiving platform, low-cost add-in decoder boards for the Intercast system are feasible. Low-cost decoders, coupled with the processing power of home PCs, may combine to offer applications that can engage the interest of U.S. consumers for higher-rate data broadcast services.

## 15.16  APPLICABLE FCC RULES FOR DATA CARRIAGE BY BROADCASTERS

Excerpts from the FCC rules concerning data transmissions in television broadcasts are as follows.

### 73.646 Telecommunications Service on the Vertical Blanking Interval

**(a)** Telecommunications services permitted on the vertical blanking interval (VBI) and in the visual signal include the transmission of data, processed information, or an other communication in either a digital or analog mode.

**(b)** Telecommunications service on the VBI and in the visual signal is of an ancillary nature and as such is an elective, subsidiary activity. No service guidelines, limitations, or performance standards are applied to it. The kinds of service that may be provided include, but are not limited to, teletext, paging, computer software and bulk data distribution, and aural messages. Such services may be provided on a broadcast, point-to-point, or point-to-multipoint basis.

**(c)** Telecommunications services that are common carrier in nature are subject to common carrier regulation. Licensees operating such services are required to apply to the Commission for the appropriate authorization and to comply with all policies and rules applicable to the particular service.

**(d)** Television licensees are authorized to lease their VBI and visual signal telecommunications facilities to outside parties. In all arrangements entered into with outside parties affecting telecommunications service operation, the licensee or permittee must retain control over all material transmitted in a broadcast mode via the station's facilities, with the right to reject any material that it deems inappropriate or undesirable. The licensee or permittee is also responsible for all aspects of technical operation involving such telecommunications services.

**(e)** The grant or renewal of a TV station license permit will not be furthered or promoted by proposed or past VBI or visual signal telecommunications service operation; the licensee must establish that its broadcast operation serves the public interest wholly apart from such telecommunications service activities. (Violation of rules applicable to VBI and visual signal telecommunications services could, of course, reflect on a licensee's qualifications to hold its license or permit.)

**(f)** TV broadcast stations are authorized to transmit VBI and visual telecommunications service signals during any time period, including portions of the day when normal programming is not broadcast. Such transmissions must be in accordance with the technical provisions of 73.682.

### 73.682 (a) TV Transmission Standards

**(21)** The interval beginning with line 17 and continuing through line 20 of the vertical blanking interval of each field may be used for the transmission of test signals, cue and control signals, and identification signals, subject to the conditions and restrictions set forth below. Test signals may include signals designed to check the performance of the overall transmission system or its individual components. Cue and control signals shall be related to the operation of the TV broadcast station. Identification signals may be transmitted to identify the broadcast material or its source, and the date and time of its origination. Figures 6 and 7 of 73.699 identify the numbered lines referred to in this paragraph.

**(i)** Modulation of the television transmitter by such signals shall be confined to the area between the reference white level and the blanking level, except where test signals include chrominance subcarrier frequencies, in which case positive excursions of chrominance components may exceed reference white, and negative excursions may extend into the synchronizing area. In no case may the modulation excursions produced by test signals extend beyond peak-of-sync, or to zero carrier level.

**(ii)** The use of such signals shall not result in significant degradation of the program transmission of the television broadcast station, nor produce emission outside of the frequency band occupied for normal program transmissions.

**(iii)** Such signals may not be transmitted during the portion of each line devoted to horizontal blanking.

**(iv)** Regardless of other provisions of this paragraph, after June 30, 1994, line 19, in each field, may be used only for the transmission of the ghost-canceling reference signal described in OET Bulletin No. 68. Notwithstanding the modulation limits contained in paragraph (a)(23)(i) of this section, the vertical interval reference signal formerly permitted on line 19 and described in Figure 16 of 73.69 of this part, may be transmitted on any of lines 10-16 without specific Commission authorization, subject to the conditions contained in paragraphs (a)(21)(ii) and (a)(22)(ii) of this section.

**(22)(i)** Line 21, in each field, may be used for the transmission of a program related data signal which, when decoded, provides a visual depiction of information simultaneously being presented on the aural channel (captions). Such data signal shall conform to the format described in Figure 16 of 73.699 and may be transmitted during all periods of regular operation. On a space available basis, line 21 field 2 may also be used for text-mode data and extended data service information.

NOTE: The signals on fields 1 and 2 shall be distinct data streams, for example, to supply captions in different languages or at different reading levels.

**(A)** A decoder test signal consisting of data representing a repeated series of alphanumeric characters may be transmitted at times when no program-related data is being transmitted.

**(B)** The data signal shall be coded using a non-return-to-zero (NRZ) format and shall employ standard ASCII 7 bit plus parity character codes.

NOTE: For more information on data formats and specific data packets, see EIA-608, "Line 21 Data Services for NTSC," available from the Electronic Industries Association.

**(ii)** At times when Line 21 is not being used to transmit a program related data signal, data signals which are not program related may be transmitted, provided: the same data format is used and the information to be displayed is of a broadcast nature.

**(iii)** The use of Line 21 for transmission of other data signals conforming to other formats may be used subject to prior authorization by the Commission.

**(iv)** The data signal shall cause no significant degradation to any portion of the visual signal nor produce emissions outside the authorized television channel.

**(v)** Transmission of visual emergency messages pursuant to 73.1250 shall take precedence and shall be cause for interrupting transmission of data signals permitted under this paragraph.

**(23)** Specific scanning lines in the vertical blanking interval may be used for the purpose of transmitting telecommunications signals in accordance with 73.646, subject to certain conditions:

**(i)** Telecommunications may be transmitted on lines 10-18 and 20, all of field 1 and field 2. Modulation level shall not exceed 70 IRE on lines 10, 11, and 12; and, 80 IRE on lines 13-18 and 20.

**(ii)** No observable degradation may be caused to any portion of the visual or aural signals.

**(iii)** Telecommunications signals must not produce emissions outside the authorized television channel bandwidth. Digital data pulses must be shaped to limit spectral energy to the nominal video baseband.

**(iv)** Transmission of emergency visual messages pursuant to 73.1250 must take precedence over, and shall be cause for interrupting, a service such as teletext that provides a visual depiction of information simultaneously transmitted on the aural channel.

**(v)** A reference pulse for a decoder associated adaptive equalizer filter designed to improve the decoding of telecommunications signals may be inserted on any portion of the vertical blanking interval authorized for data service, in accordance with the signal levels set forth in paragraph (a)(23)(i) of this section.

(**vi**) All lines authorized for telecommunications transmissions may be used for other purposes upon prior approval by the Commission.

(**24**) Licensees and permittees of TV broadcast and low power TV stations may insert non-video data into the active video portion of their TV transmissions, subject to certain conditions:

(**i**) The active video portion of the visual signal begins with line 22 and continues through the end of each field, except it does not include that portion of each line devoted to horizontal blanking. Figures 6 and 7 of Par. 73.699 identify the numbered line referred to in this paragraph.

(**ii**) Inserted non-video data may be used for the purpose of transmitting a telecommunications service in accordance with Par. 73.646. In addition to a telecommunications service, non-video data can be used to enhance the station's broadcast program service or for purposes related to station operations. Signals relating to the operation of TV stations include, but are not limited to program or source identification, relay of broadcast materials to other stations, remote cueing and order messages, and control and telemetry signals for the transmitting system.

(**iii**) A station may only use systems for inserting non-video information that have been approved in advance by the Commission. The criteria for advance approval of systems are as follows:

(**A**) The use of such signals shall not result in significant degradation to any portion of the visual, aural, or program-related data (closed captioning) signals of the television broadcast station.

(**B**) No increase in width of the television broadcast channel (6 MHz) is permitted. Emissions outside the authorized television channel must not exceed the limitations given in Par. 73.687 (E). Interference to reception of television service either of co-channel or adjacent channel stations must not increase over that resulting from the transmission of programming without inserted data.

(**C**) Where required, system receiving or decoding devices must meet the TV interface device provisions of Part 15, Subpart H of this chapter.

(**iv**) No protection from interference of any kind will be afforded to reception of inserted non-video data.

(**v**) Upon request by an authorized representative of the Commission, the licensee of a TV station transmitting encoded programming must make available a receiving decoder to the Commission to carry out its regulatory responsibilities.

## ABOUT THE AUTHOR

Lynn Claudy is senior vice president of science and technology for the National Association of Broadcasters in Washington, D.C. As head of the science and technology department, he is responsible for representation of the NAB in all radio and television technical matters. He joined NAB in 1988 as a staff engineer and served as director of advanced engineering and technology and vice president before assuming his present position in February 1995. Prior to joining NAB, he was employed by Hoppmann Corp., a communications systems integration firm, where he held a variety of technical and management positions. He also was a part-time professor in the physics department at the American University in Washington, D.C., where he taught courses in acoustics and audio technology. Previous to that work, he was a development engineer at Shure Brothers Inc., a manufacturer of audio equipment. He has a B.S. degree from Oberlin College, a B.S.E.E. from Washington University in St. Louis, and an M.S.E.E. degree from the Illinois Institute of Technology.

## BIBLIOGRAPHY

*Television Data Broadcasting Survey Report and Resource Guide,* National Association of Broadcasters, Washington D.C., 1994.

*NAB Engineering Handbook,* National Association of Broadcasters, Washington D.C., 1992.

*D-Channel: A Technique for NTSC-Compatible Data Broadcasting,* Digideck, Inc., Submitted to the National Data Broadcasting Committee, April 1994.

*WavePhore TVT1/4.* Submitted to the National Data Broadcasting Committee, April 1994.

*Code of Federal Regulations,* Title 47, Parts 70 to 79, U.S. Government Printing Office, Washington, D.C., 1994.

# CHAPTER 16
# INTERACTIVE TELEVISION BROADCASTING AND RECEPTION

**Ajith N. Nair**
*Scientific-Atlanta Inc.*

## 16.1   INTRODUCTION

Cable television (CATV) started in the early 1950s. It was the solution for providing television services to areas of the country out of reach of then existing broadcast stations. In those days the typical CATV system had only five channels. The intent was to carry the three major networks. In the 1960s, CATV moved to the urban areas, which were already being served by the broadcast stations. The number of channels increased to 12 or more. The channels carried were mostly broadcast channels. Pay channels started emerging, and these were provided by shipping tapes of movies or special programming from a centralized location to all the CATV head-ends.

The 1970s saw the distribution of the pay-TV channels and "super stations" over satellites. But the number of channels remained from 12 to 20. The 1980s saw a tremendous explosion in all aspects of the industry (with channel capacity reaching 80 in 550-MHz systems) and the emergence of a large number of specialty and cable-only channels. The demand for channel space increased in the 1990s, but the cost of increasing the bandwidth became prohibitive. In the 1990s competition to the CATV systems emerged in the form of direct-to-home satellite services, and changes in the regulatory environment moved toward allowing new entities to provide the entertainment and other services in competition with entrenched CATV operators. Today, some cable systems are going to a bandwidth of 750 and 860 MHz to provide more channels. More and more systems are beginning to make their systems capable of reverse-channel communications necessary for interactive television.

To provide more channels in the available spectrum, data compression techniques have to be used. After several years of research and international committee deliberations, the Moving Pictures Experts Group (MPEG) standard (see Chapter 8) has emerged as the compression standard for the efficient digital delivery of television audio and video programming. There is worldwide consensus on using the MPEG standard for video compression. However, when it comes to the compression of multichannel audio, there are two contending candidates. MPEG uses Musicam as the compression standard, whereas there is a strong push to use Dolby (TM) AC-3 compression in North America.

## *16.2    CABLE PROGRAM ORIGINATION AND DISTRIBUTION*

The main elements of any CATV system (see Chapter 17) are the head-end, the distribution plant, and the subscribers. The head-end is the location where the programs are collected and put together as a single RF multiplex of channels that is delivered to the distribution system.

A current analog head-end consists of the following major elements:

- Integrated receiver demodulators receive channels that are nationally distributed over the satellite. The demodulators receive the RF signal from the satellite transponder in C-band or KU-band through low-noise amplifiers (LNAs) or low-noise block converters (LNBs); select the desired channel, demodulate, and descramble the signals (if scrambled); and provide baseband video and audio signals.

- Signal processors or TV demodulators receive the local channels. The signal processors receive over-the-air channels in the VHF and UHF spectrum, process the signal at an IF frequency, and then remodulate it to the required cable channel carrier frequency for delivery to the subscribers. The signal processors are used when the channel is not scrambled. When the video or audio in the channel has to be scrambled in the baseband domain, a TV demodulator is used. The TV demodulator tunes to any channel in the UHF or VHF spectrum and demodulates the signal to its baseband video and audio signals.

- Scramblers receive the individual channel audio and video and scramble both of the video signals and generate information required for the descrambling at the subscriber end.

- CATV modulators receive the baseband audio and video signal for each channel and modulate them to the appropriate CATV channel carrier frequency.

- Data inserters insert the control data required for the addressable control and management of the subscriber terminals. The control data are usually modulated on a low-bandwidth channel as QPSK data and transmitted along with the program channels. Some systems do not use such an out-of-band data channel. Instead, they use in-band data techniques to control and administer the subscribers. Such systems use data modulation either in the vertical blanking interval (VBI) of the video signal or as amplitude-modulated pulses on the audio carrier.

- Combiners receive the output of the individual signal processors, CATV modulators, and data inserts and then combine them to produce a single broad-band signal output that is sent to the distribution system.

Other elements needed in the head-end are a system manager and a billing system computer. The system manager holds the database of the subscriber units and controls the authorizations and channel allocations. The billing computer has the database of subscribers and information regarding their location, address, credit history, etc., as well as their subscription levels, premium, and pay-per-view purchases. The billing computer generates the billing information for charging the subscribers. Customer service representatives use terminals connected to the billing computer to enter subscriber orders and to manage the subscriber database.

The distribution plant that exists today is predominantly one-way. Large systems have fiber trunks that take the broad-band signal to local neighborhoods. At the neighborhood the signal is converted to electric signals and then distributed over coaxial cables. One-way trunk amplifiers, distribution amplifiers, and line-extenders are used on these distribution systems. Near the subscriber home, taps are used to send the signal the last 100 yards to the home.

The subscriber-premises equipment varies with the type of service that is used. If subscribers need basic cable services only, no special equipment is needed if they have a completely cable-ready TV. If the subscriber does not have a TV that can tune all the cable channels available on the system, then a simple cable TV converter box is needed. This would tune all the channels in the system and then provide that channel signal as either baseband audio and video outputs or on an RF channel output that any TV can tune to (usually channel 3). Subscribers who get premium services and pay-per-view services will need an addressable converter box that can be controlled from the head-end. The latest in advanced analog cable converter boxes provide several new features, such as program guides, virtual channels for information delivery, etc.

## 16.3 CHANGES NEEDED FOR INTERACTIVE DIGITAL DELIVERY

The most extensive and expensive change needed for cable systems to become interactive is a reverse path in the distribution system and accompanying reverse-path amplifiers. It is also necessary to make certain that the whole plant is "clean" in terms of extraneous signal ingress. In a two-way system, noise ingress at any portion of the system will be funneled all the way back and make the system unusable. Digital modulation and transmission of the signals also require that signal-to-noise ratios and signal levels are maintained at higher levels than required for an analog-only system. In the head-end new types of equipment are needed, which are summarized as follows.

### 16.3.1 Real-Time Digital Video Compression Encoders

These encoders receive audio and video baseband signals and digitize and compress them to digital streams. They produce MPEG data streams at data rates that are set by the system controller.

### 16.3.2 Digital Video Servers

Digital video servers store digitally compressed data streams for different kinds of programming (movies as well as other types of information). They can spool out multiple streams of programming simultaneously. The video servers can be used for broadcast applications as well as interactive applications.

### 16.3.3 System Multiplexers

System multiplexers receive multiple streams from real-time encoders, video servers, and digital IRDs and then multiplex them to a single digital stream that can be modulated on a 6- or 8-MHz bandwidth channel.

### 16.3.4 Digital Quadrature Amplitude Modulation (QAM) RF Modulators

Each 6-MHz bandwidth channel can carry 30 or 40 Mb/s of raw data, depending on whether 64 QAM or 256 QAM modulation is used. The digital data stream from the multiplexers is received in the modulators; the stream is processed for forward error correction and then modulated to a 6-MHz bandwidth QAM signal at the chosen frequency.

### 16.3.5 Quadrature Phase-Shift Keying (QPSK) RF Modulators

The control and signaling data for an interactive system may be transmitted on a different, lower-bandwidth frequency channel. The evolving modulation scheme for these signaling channels is quadrature phase-shift keying. These modulators receive the downstream signaling data from the system controllers and QPSK-modulate them at an appropriate frequency.

### 16.3.6   Quadrature Phase-Shift Keying (QPSK) RF Demodulators

The reverse data path for interactive and status-monitoring applications also uses QPSK modulation. The head-end QPSK demodulators receive the upstream QPSK-modulated channels and then demodulate and provide the data to the system controllers.

### 16.3.7   Digital Switches

The real-time encoders, video servers, and other data sources can be connected to a SONET network. The MPEG data from the sources are carried in ATM cells at this level. Digital switches capable of switching high-speed ATM traffic are used for the selection and routing of the proper program data to appropriate cable channel QAM modulators and QPSK transmitters. Occasionally, integrated gateways that support multiple modulators are used.

### 16.3.8   System Controllers

Several powerful computers are needed for the proper functioning of a digital interactive system.

***Connection Managers.***    Connection managers allocate data streams to different frequency QAM modulators and MPEG transport parameters such as program identification numbers (PIDs), and they provide digital switches for routing the different data streams to the correct destinations. The connection manager also allocates frequencies for the forward and reverse signaling channel frequencies for the QPSK modulators and demodulators.

***Session Managers.***    The session manager sets up interactive sessions for the subscriber terminals. They set up, maintain, and tear down the sessions between the home terminal and the service provider. The session manager provides interfaces to the billing system as well as status-monitoring systems.

***Element Managers.***    Element managers are used to control the different network elements of an interactive digital system. These control the QPSK modulators and demodulators, QAM modulators, integrated gateways, etc.

***Conditional Access Managers.***    Conditional access managers are used for the management and control of the system conditional access for the subscribers and service providers. These provide key management for encryption of services.

## 16.4   DELIVERY MECHANISMS

The main delivery mechanism candidates for entertainment and information delivery to users are hybrid fiber coax (HFC), fiber-to-the-curb (FTTC), fiber-to-the-home (FTTH), asymmetric digital subscriber line (ADSL), direct to home satellite (DTH), and multichannel multipoint distribution service (MMDS).

HFC is the system that is most used by the present CATV operators. Some of the new players in the CATV market from the telephony industry are also looking to utilize the HFC networks. HFC uses fiber trunks to carry information to neighborhood nodes and then coaxial cable is used the rest of the way to homes.

The common FTTC architecture uses fiber in the access network all the way up to the curb, to what are called optical neighborhood units (ONU). Then a handful of user homes may be connected with either coaxial cables or copper twisted pairs. A typical distance used for the last leg is 300 yards.

ADSL architecture utilizes special techniques to upgrade the copper twisted-pair networks to support a high bit rate in the downstream direction and lower bit rates in the upstream direction. The downstream data rates supported are 2 Mb/s for 4 to 5 kms and up to 7 Mb/s for shorter distances. The upstream data rates supported are of the order of 600 to 700 kb/s.

The FTTH architecture brings the fiber all the way to the home with a network termination device in the home.

For DTH links, depending on whether C-band or Ku-band transponders are used, data rates of 50 or 90 Mb/s can be obtained per channel.

## 16.5 SPECTRUM USAGE

Figures 16.1 and 16.2 show the spectrum usage for a fully interactive HFC system and an FTTC system, respectively. It is assumed that in both systems the traditional analog channels are available to the subscriber. In the FTTC system only one compressed stream is seen by the user. The selection and switching of the required program/channel occurs in the network. In an HFC network a large number of digital channels are simultaneously present at the input to the ISTT. Some

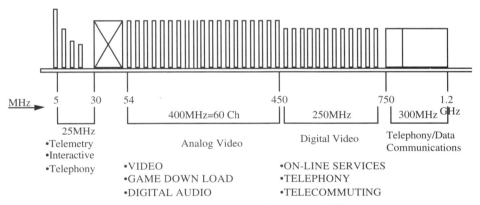

**FIGURE 16.1**    Spectrum usage for a fully interactive hybrid fiber coax (HFC) system.

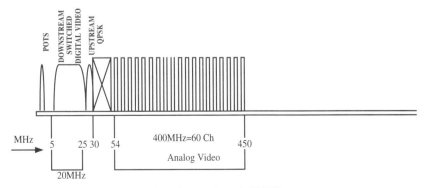

**FIGURE 16.2**    Spectrum usage for a fiber-to-the-curb (FTTC) system.

of these channels may be used for interactive applications, where the data changes depending on the application that the user is running, and other channels could be broadcast channels that everybody can tune to and use.

## 16.6  SET-TOP TERMINAL EVOLUTION

The set-top terminals also evolved along the same lines as the CATV system during the last four decades. There was no need for a set-top terminal in the very early days, because the channels used in the CATV systems were the same used in the television sets of the period. With the advent of premium, or "pay," channels, it became necessary to differentiate those channels and make them not available to those not paying the premium charge. The initial solution was to put these channels in the part of the spectrum that normal TVs of the day could not tune. The televisions caught up with this in the form of "cable-ready TVs." The increase in the number of channels required in the cable systems also played a part in this. The broadcast television systems used the VHF band and the UHF band, whereas CATV used the contiguous spectrum starting around 48 MHz. So the initial set-top or cable boxes were block converters that allowed the user to move the special cable channels to the spectrum that the televisions could tune. When these became too available, some method was needed to scramble the pay channels to make them unavailable to those who did not pay. The first descrambling set-tops were preprogrammed devices in which the information about the channels that the subscriber paid for was stored in a programmable read-only memory (PROM). To get a new pay channel or drop one, the box had to be reprogrammed. This was expensive for the cable operator and inconvenient for the user. This brought about the invention of the "addressable set-tops." These units had more powerful microcontrollers, nonvolatile memory, and descrambling circuitry. These units could be authorized to tune and descramble channels that the user had paid for. The addressability information was transmitted on an independent carrier that the set-top always tuned to. Later developments in the area involved (1) the transmission of the addressability and control data in the vertical blanking interval (VBI) of the video signal and (2) amplitude-modulated pulses on the audio carrier of the channel.

The next big step in the evolution of the set-top architecture was the pay-per-view (PPV) concept. This probably was the first step in interactive television. In the initial PPV scheme, the user had to call the cable operator and ask to buy an event. The operator would then authorize the user's set-top terminal to tune and descramble the channel for the duration of the event. This could be done with a nominal "addressable set-top." This scheme brought problems in terms of billing when a purchased event was not actually viewed and the user wanted his or her money back. There was no way to confirm that the user had actually viewed the program. This problem was overcome with the addition of a telephone modem and additional nonvolatile memory in the set-top terminal. This allowed the cable operator to verify whether or not the program was tuned to. This also allowed for "impulse pay-per-view"(IPPV), in which the user with the right kind of set-top can tune to the PPV channel and view a program for a specified grace period and decide impulsively to buy that event or program. This impulse-buy information is stored in the nonvolatile memory of the set-top terminal and the cable operator can access it through the modem in the unit. Along with PPV and IPPV, the requirement for graphics, or on-screen-display (OSD), also arose to provide user-friendly interaction with the user. The OSD progressed from single-color alphanumeric capability to multicolor bit-mapped graphics. The high-end analog set-tops have most of these features and provide features such as "virtual channels" and *near video-on-demand* (NVOD). NVOD is implemented by starting the same program at specific intervals of, say, 5 to 15 minutes. The set-top will be capable of selecting the closest start and, if the user needs to pause and start again, it catches up on the same segment on another start. Other innovations from the same time frame include improved scrambling techniques and in-band data transmission using the audio carrier as well as the vertical blanking interval(VBI) of the video signal.

Another major change has been the capability for reverse-path communications in the cable plant. This was brought about because of the logistical problems of using the telephone return path for real-time reverse communications. One can imagine what would happen to the telephone system if a large number of cable subscribers decided to buy the first run of a very popular movie. This could bring down the telephone system immediately. The telephone return was used mainly as a store-and-forward system, where the call back is initiated by a command from the cable head-end. This problem was solved by implementing a RF return path through the cable plant itself. However, this approach needed vast improvements in the cable plant and its maintenance. The data rates obtained were in the 20-kb/s range, and the protocols used provided store and forward.

## 16.7   INTERACTIVE SET-TOP TERMINAL

The essential features of an interactive set-top terminal (ISTT) include the support for existing analog cable channels and standard digital channels, as well as a provision for minimum tuning functionality even when the digital network is down. An ISTT should provide an easy and intuitive user interface, using keypads and/or remote controls. It should have a computer interface, and the unit as a whole should be capable of functioning at various signal and interference levels. An ISTT can be said to have two basic elements, as shown in Fig. 16.3. The elements in the network module will depend on the type of access network and delivery mechanism used in the system. The terminal part can be common across most of the network types. However, such a partitioning may cause the complete solution to be very much more expensive than a comparable integrated solution for many of the networks. This higher cost is due to duplication of some of the processing elements in both parts, as well as the difference in complexity of the network-dependent elements. For example, the network elements needed for an HFC system are very much different from the requirements for an ADSL or FTTC system carrying data in ATM.

Figure 16.4 shows a candidate implementation for an ISTT for an HFC system. The shaded blocks represent elements that depend on the network. The unshaded blocks can be common

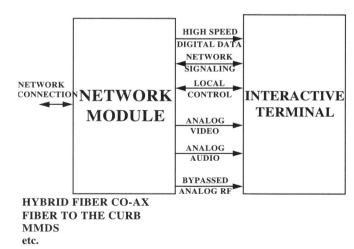

**FIGURE 16.3**   An interactive set-top terminal.

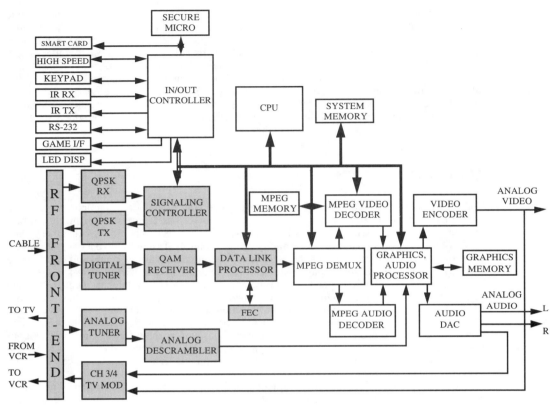

**FIGURE 16.4**    A hybrid fiber coax interactive set-top terminal.

to any kind of network. Figure 16.5 shows a corresponding implementation for an ISTT that would work in an FTTC system.

In the case of the FTTC set-top the digital receiver is a 16-cap demodulator, and the data link processor may be an ATM SAR device. For ADSL the front-end could be line interfaces.

Figure 16.6 shows the RF front-end for an HFC set-top. It consists of a diplex filter that separates the downstream and upstream portions of the RF spectrum. The separated downstream signal then may be amplified to meet the signal level requirements of the subsequent processing elements. The lower part of the spectrum is passed on through a low-pass filter and a splitter to an analog tuner, as well as to a bypass output that could be directly connected to a television set. A VCR may be included in the viewing system as shown. The other output of the splitter that follows the amplifier is passed through another diplex filter that separates the signaling frequencies fed to a QPSK receiver, and the main high-speed digital channels fed to the digital tuner.

The first generation of ISSTs that supported both standard analog cable channels and digital channels used separate tuners. The trend is toward using an integrated tuner that can provide both functions, reducing the overall cost of the unit. An advantage of having independent tuners is that two independent programs are available at any time. The RF tuner output is fed through appropriate AGC circuits to a QAM/VSB demodulator (Fig. 16.7), generating data bits that are

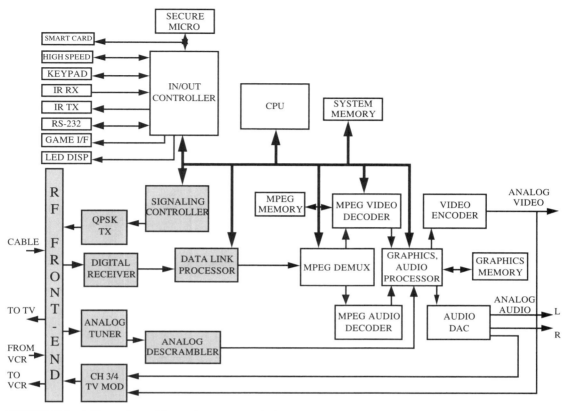

**FIGURE 16.5**   A fiber-to-the-curb interactive set-top terminal.

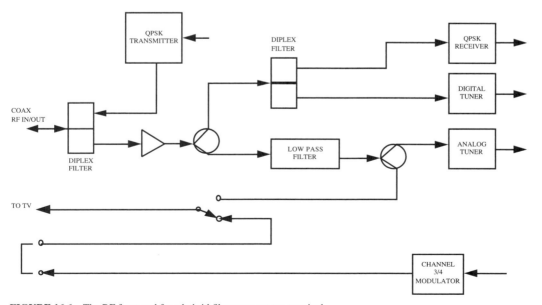

**FIGURE 16.6**   The RF front-end for a hybrid fiber coax set-top terminal.

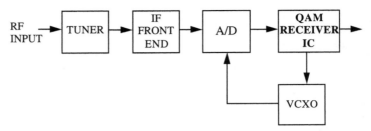

**FIGURE 16.7**   A quadrature amplitude–modulated (QAM) receiver.

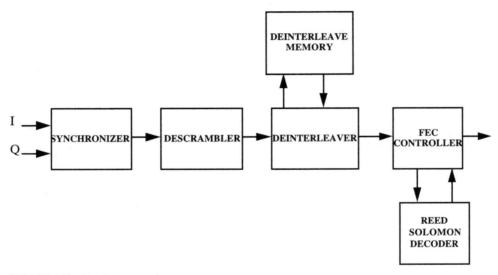

**FIGURE 16.8**   Data link processing.

then processed by the data link processor, which synchronizes, removes the scrambling added for one-zero distribution, de-interleaves the stream, and feeds a forward error correcting element (Fig. 16.8). At this point a transport stream is produced—which could be an MPEG-2 system stream, an ATM stream, or a proprietary transport stream.

## 16.8   MPEG PROCESS

The transport stream carries the main high-speed video and audio compressed data, in-band system data that may be used for the system management, and pure data that could be destined for other applications.

Currently, MPEG-2 and ATM are the competing transport formats. There are proposals for even carrying ATM cells in an MPEG transport stream. MPEG has a very wide following because it is an internationally accepted standard, and a large number of silicon vendors are supporting it with chip designs. Also, it has hooks for conditional access controls, private data, and packet-based encryption. The MPEG standard is being accepted as the overwhelming candidate for compression of both video and audio for the same reasons.

The building blocks required for the MPEG processing are the demux, the video decompressor, and the audio decompressor (Fig. 16.9). The system clock reference is recovered using an SCR VCXO closely coupled with the demux device. Conditional access functions can be implemented in a secure microprocessor, which is also closely coupled to the demux device. The MPEG demux device separates streams based on the program identification numbers (PIDs) and feeds them to the stream-processing elements. The compressed video goes to the video decompressor, audio goes to the audio decompressor, and management data are sent to the system CPU. Auxiliary data can be sent through a high-speed serial interface to a home computer or a high-speed storage element.

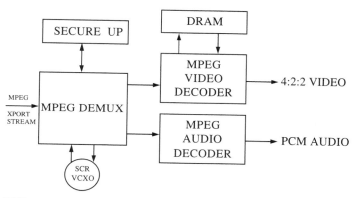

**FIGURE 16.9**    MPEG process.

Several first-generation video decompression devices not only do decoding of MPEG video but also provide low-level graphics capabilities for user interface applications. Most have built-in error concealment. Some also have features such as interpolation and decimation filters built-in to provide different aspect ratios at the output. This makes it easier for some graphics-intensive applications that generate video in non–CCIR-601 aspect ratios to mix and overlay or blend the video and graphics together. One of the main cost concentrations in a digital ISTT is the memory required for the decompression of the MPEG video. A minimum of 16 Mbits of DRAM is required for decoding a main-profile, main-level (MP@ML) video stream at 15 Mb/s.

The audio decoders accept the MPEG stream, decompress, and produce a PCM stream at the original sample rate. The sample rates supported in MPEG are 32, 44.1, and 48 kb/s.

The MPEG transport stream carries time-stamp information on the audio and video streams. These program time stamps (PTS) are used for the synchronization of the video and its associated video, which is very important for lip-sync.

## 16.9   MAIN PROCESSOR AND I/Os

The main processor in the ISTT has to be capable of doing the following functions: overall control of the hardware, low-level drivers, and housekeeping functions; support of an operating system, network connection management, and user interface interactions; and execution of applications. The CPU needs sufficient processing power to support a variety of applications and high-quality graphics. The building blocks associated with the processor are the system DRAM and ROM, graphics controller, and an I/O controller (Fig. 16.10).

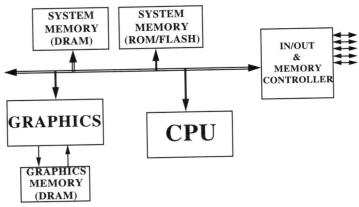

**FIGURE 16.10** Building blocks associated with the main processor.

The traditional inputs and outputs in an ISTT are channel 3 RF output, baseband video, two channels of baseband audio outputs, and the coaxial cable input that connects to the network. The nontraditional I/Os needed in a high-end ISTT are a smart card interface for external conditional access management, keypad for manual control of the unit, LED display, infrared receiver for remote control, an infrared transmitter interface for the use of VCR controls, a general purpose RS-232 I/O for low-speed home computer interactions, game I/O that can support at least two players, and a high-speed data interface such as P1394 for use with a digital VCR, etc.

## 16.10  BACK-END PROCESSING

Figure 16.11 shows the building blocks needed for the back-end video and audio processing. The video from the tuned channel and the graphics generated by the CPU should be brought to the same resolution and aspect ratios and then overlaid or blended together. This composite is then presented to a digital-to-analog video encoder that produces RGB and/or a composite analog video signal that can be fed to a TV or a channel 3/4 RF modulator.

The audio from the tuned channel and audio generated by the CPU are similarly mixed and presented to an oversampling audio DAC, which gives the left and right analog audio signals to drive an amplifier. They may be summed together to produce a single channel to modulate the channel 3/4 RF modulator.

## 16.11  STATE OF TECHNOLOGY

Many of the building blocks required for an ISTT are in the second generation. Integrated tuners capable of supporting both analog channels and digital channels are available. Second-generation silicon available includes single-chip implementations for QAM receivers, integrated MPEG audio and video decompressors with low-level graphics capability, MPEG demux devices with decryption and CPU cores, multimode video encoders, lower-cost Reed-Solomon decoders, and oversampling audio DACs. The current generation of QPSK transmitters and receivers are still discrete. A large amount of development work is going on in the industry to produce single-chip solutions for these as well. The main CPU is still an independent device.

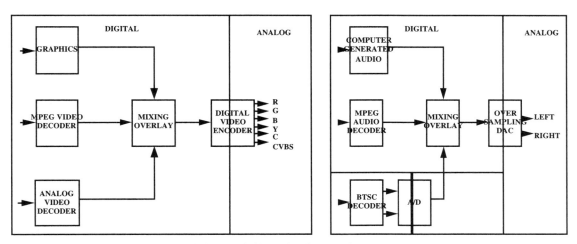

**FIGURE 16.11**    Building blocks needed for back-end video and audio processing.

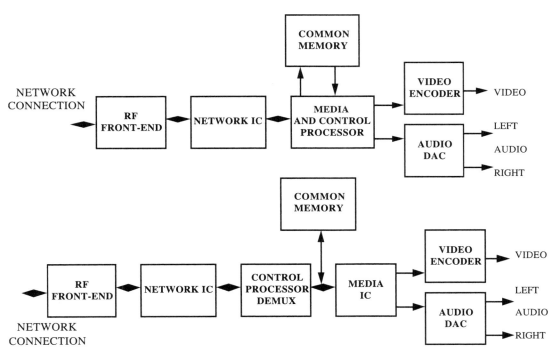

**FIGURE 16.12**    Integrating all the MPEG and graphics processing and the main CPU in one piece of silicon enables use of a common block of memory.

## *16.12  INTEGRATION STRATEGIES*

The only way the cost of interactive set-top terminals can be brought down to fit business models of network and entertainment providers is by integrating more and more functions into a smaller number of integrated circuits and by finding innovative ways to reduce the overall memory requirements. Different manufacturers and IC vendors are tackling this in different ways; the ultimate goal is a "set-top on a chip." On the way to this end there will be several multichip implementations.

One way to reduce the overall memory requirement is to integrate all the MPEG and graphics processing and the main CPU in one piece of silicon and use a common block of memory that will be smaller than the sum if done separately (Fig. 16.12). Another area of integration is to put all the modulation and demodulation processes along with the de-interleaving and forward error correction in one chip that would give a network-dependent device. This would allow for an ISTT with two major devices: a single block of DRAM and the audio and video encoders.

Another integration strategy could be to integrate the control processor and the MPEG demux into one IC, all the media and graphics processing in another IC, and a memory controller design that would allow the use of a common block of memory. Some of these suggested strategies may not be achievable in a single step. However, with the finer geometries that are being developed by silicon foundries these measures are definitely worth trying.

## *ABOUT THE AUTHOR*

Ajith N. Nair is a senior staff engineer at Scientific-Atlanta, Norcross, Ga., where his present assignments include definition of interactive set-top terminal architectures. He has been with Scientific-Atlanta since 1984 and was involved with the design of the company's Home Communication Terminal, used in the Orlando Full Service network trials, and the first MPEG-based compressed television satellite transmission systems. He was also part of the company's team that developed the VQ compression system. Previously, he had designed scramblers and control systems for analog cable TV systems. Nair received a bachelor's degree in electronics and communications engineering from the University of Kerala in 1973 and did two years of postgraduate work at the Queens University at Kingston, Canada.

# CHAPTER 17
# DIGITAL CABLE SYSTEMS

**Paul Moroney, Joe Waltrich, Eric Sprunk, Mark Kolber**
*General Instrument Corporation*

## 17.1  INTRODUCTION

Cable systems have been in existence in the United States and throughout the world for decades, distributing conventional analog television signals. The history and detailed description of cable systems is beyond the scope of this treatment; rather, it will focus only on those aspects that affect digital distribution on the cable plant. Specifically, the cable distribution plant, channel impairments, source origination, security, and access control will be addressed.

Detailed examples will be specific to the 6-MHz RF channel spacing common in North America. However, the concepts and results extend readily to the 7-MHz and 8-MHz channel spacings found in other parts of the world.

## 17.2  CABLE SYSTEM ARCHITECTURE

### 17.2.1  Distribution Plant

The physical cable plant needed to distribute digital CATV signals is not substantially different from the plant designed to carry analog signals. Both the conventional *tree and branch* topology and the newer *hybrid fiber coax* (HFC) topology can be used. In the conventional tree and branch systems, trunk distribution coax supplies signals to neighborhoods, with minimal degradation, and from there the *branches* spread out to streets and ultimately homes. Repeater amplifiers restore signal level and compensate for the significant frequency roll-off of coaxial cable;[1] *taps* provide the ultimate signal to the home via a *drop*. In the newer hybrid fiber coax systems, the trunk distribution coax is replaced by fiber, which improves signal quality, reliability, and flexibility overall.

Channel impairments will have a different effect on digital signals than on analog signals. Most channel impairments caused by the distribution system cause a visible degradation to an analog video signal. A small impairment will usually cause a slight degradation. In digital transmission, with forward error correction coding, minor transmission impairments will normally cause no visible effect on the regenerated video. However, once the impairment or combination of impairments exceeds a critical threshold where the rate and type of symbol errors exceed the capability of the FEC coding, the regenerated video will be severely degraded. Additional impairment will create a complete lack of decoded video.

Table 17.1 gives a comparison of performance generally accepted as excellent for analog channels versus the FCC minimum requirements for cable systems.[2] Digital systems must target the FCC numbers as an operating environment, at a minimum, and consider all other impairments present. Digital signals may even be placed in spectral locations not suitable for analog video. These issues will be detailed in later sections.

**TABLE 17.1** Comparison of Performance for Analog Channels vs. Digital Channels

|  | Excellent analog | FCC minimum requirements |
|---|---|---|
| Gain flatness | ±1 dB | ±2 dB |
| Group delay variation | ±150 ns | Not stated |
| Hum modulation | ±1% max | ±3% |
| Signal-to-noise ratio | 48–50 dB | 43 dB |
| Composite second-order (CSO) | −53 dB | −51 dB |
| Composite triple-beat (CTB) | −53 dB | −51 dB |

## 17.2.2 Cable Frequency Plans

The cable plant's downstream, or forward path, ranges from 54 MHz to 200 MHz for older 23-channel analog systems, and from 54 to 550 MHz for 77-channel analog systems. Newer systems are designed to pass up to 750 or even 1000 MHz. Normally, digital channels are added at the upper end of the spectrum, particularly when a system is being upgraded to extend the upper frequency limit. This allows subscribers equipped with analog receivers to continue receiving services while new digital services are added. Alternatively, analog channels can be deleted and replaced with digital channels.

The downstream frequency ranges are referenced as follows:

Low band       54 to 88 MHz (off-air VHF channels 2 to 6)

Mid band       88 to 174 MHz

High band      174 to 216 MHz (off-air VHF channels 7 to 13)

Super band     216 to 300 MHz

Hyper band     above 300 MHz

Downstream channel assignments typically conform to one of three channelization plans[1]: standard, IRC, or HRC. The standard plan evolved from television broadcast frequency assignments. The IRC (incrementally related carriers) plan is based upon phase locking the carriers to reduce third-order distortions (see Section 17.4.2). The HRC (harmonically related carriers) plan shifts the carriers down in frequency to further reduce distortions.

Many cable systems include an additional "out-of-band" data channel, typically located in the 70 MHz to 110 MHz region of the cable plant spectrum. This data channel provides guaranteed data flow to any set-top equipped with the additional fixed-frequency receiver, regardless of what downstream 6 MHz RF channel that set-top is tuned to. In mixed analog and digital channel systems, this connectivity is especially valuable, so that set-top functionality is not compromised while viewing analog services. Out-of-band channels as low as 14 kb/s and as high as 1.544 Mb/s exist in current North American systems.

The upstream, or return path, is normally located in the 5 to 40 MHz range, also called the *sub band*. Alternatively, a few systems are built to support a "high-end" return path, where the return path is located at the extreme upper end of the frequency band. Upstream capability is largely determined by the amplifiers and filters located in the cable plant.

In both analog and digital systems the return path is used to relay remote broadcasts to the headend and to return billing and other information from the subscriber's set-top. It can also be used for telephony, interactive TV, network monitoring, Internet access, and other two-way applications.

## 17.3  SOURCE ORIGINATION AND HEADEND

In a digital CATV system, the headend has the same basic function of collecting all the signals as in an analog cable system. Because many digital cable systems also contain analog channels, much of the hardware may be similar. In the headend, all the signals that are to be transmitted to the subscribers are collected and converted into the appropriate format. In mixed systems the basic tier channels will usually be sent in analog format, whereas the premium and pay per view channels will be sent in a digital format. This allows for the use of a less expensive set-top, or perhaps no set-top at all for the basic subscribers, reserving the more expensive digital set-tops for the premium subscribers. In all-digital systems, more channels can be transmitted via the cable plant because digital compression can be used on each channel; however, all subscribers will require a digital set-top for each television receiver.

Local off-air signals are received at the headend in analog NTSC format. These are usually part of the basic tier or nonpremium channels, and in a mixed analog/digital CATV system will be sent in an analog format on the cable. This typically requires the use of conventional analog equipment to effect a frequency conversion and a reduction in the audio carrier relative to the video carrier. This can be done with heterodyne processing, which first converts the signal to the standard TV IF frequency, 44 MHz, and then converts it to the desired cable frequency. Alternatively, the demod-remod type system can be used, which demodulates to video and audio baseband and then remodulates onto the desired cable frequency.

In an all-digital CATV system the off-air analog signals must be converted to digital format. Each signal is first demodulated to baseband audio and video using a conventional demodulator. Next the signal is PCM-coded or digitized, usually with 8 bits of resolution, using a sampling rate of 13.5 MHz. This yields a data rate of well over 100 Mb/s and requires excessive bandwidth to transmit on the cable without compression. Thus the data are compressed to typically between 1 and 6 Mb/s using MPEG-2 video compression algorithms. These data are then multiplexed with data from other encoders and with other control data to form a stream of aggregate rate usually between 27 and 40 Mb/s. This bitstream is combined with FEC data and is used to modulate a 44-MHz carrier using 64 QAM or 256 QAM, generating a 6-MHz-wide, 44-MHz IF signal. This QAM signal is then converted to the desired RF frequency for transmission on the cable plant.

Geostationary communications satellites are often used to distribute programming to CATV headends. TV signals can be transmitted via satellite in either analog or digital form and can be carried on the cable in either analog or digital form. Thus there are four signal conversions possible at the headend (Table 17.2).

**TABLE 17.2**  Four Combinations of Signals at the Headend

| Satellite format | Headend equipment | Cable format |
|---|---|---|
| Analog | FM receiver to baseband video to AM modulator/upconverter | Analog |
| Analog | FM receiver to baseband video to encoder to QAM modulator/upconverter | Digital |
| Digital | QPSK receiver to decoder to baseband to AM modulator/upconverter | Analog |
| Digital | QPSK demod to transport stream to QAM modulator/upconverter | Digital |

Analog satellite TV transmissions make use of wideband FM modulation to maximize delivered carrier-to-noise ratio (C/N), given the limited satellite transponder power. The signal path is similar to local off-air broadcasts, except that an appropriate satellite demodulator is used. Some analog satellite signals will also need to be descrambled. If the program is to be sent in analog format over the cable, a conventional AM VSB modulator is used to create an IF and RF signal. If a digital format is to be used, then the audio and video basebands are digitized, compressed, and transmitted as just described.

Many signals that are transmitted via satellite and intended for distribution in CATV systems will be transmitted in an MPEG digital format. Video and audio signals are digitized and compressed before being uplinked to the satellite. This provides more efficient utilization of the satellite transponders and reduces the need for MPEG encoders at each headend. If, however, the signal will be carried on the cable in analog format, the digital signal must first be decoded at the headend to baseband audio and video. This requires a digital satellite receiver, which includes a QPSK (quadrature phase-shift keying) receiver and an MPEG or other decoder, similar to the digital satellite receivers used by consumers.

The last combination, digital on satellite and digital on cable, is more interesting. The signal is not converted to analog or baseband at all but rather kept in digital format throughout. Although the signal received from the satellite and the signal sent out on the cable both contain compressed video bitstreams, there are differences. Modulation format, error correction coding, and access control are typically changed. The headend equipment that performs these changes is called a *transcoder*.

Because the satellite channel is power-limited and the cable channel is more bandwidth-limited, different modulation and error correction codes are used for each. QPSK modulation is used for digital satellite transmission because it is very power-efficient and works well on noisy channels. QPSK uses two carriers 90 degrees apart, one called the *I* (or *in-phase*) *channel* and the other called the *Q* (or *quadrature*) *channel*. Each carrier is modulated using BPSK (binary phase-shift keying) modulation carrying 1 bit per symbol. Together the two quadrature carriers provide 2 bits per QPSK symbol. A 36-MHz-wide transponder can support about 30 million symbols per second, yielding a 60-Mb/s channel rate. Reed-Solomon and convolutional error correction coding consume about 10 percent to 50 percent of these bits. It is also possible to use the I and Q channels separately to carry two independent 30-Mb/s bitstreams using a "split multiplex" mode. This simplifies some of the hardware at the cost of some flexibility in applications.

The use of QPSK modulation combined with coding for error correction provides essentially error-free results with received carrier-to-noise ratios in the range of 1 to 9 dB. This performance is considerably better than analog FM satellite transmission, which requires a received C/N of 10 to 15 dB to achieve video with 50-dB signal-to-noise ratio. The analog system, however, will exhibit a more graceful degradation of performance as the received C/N degrades. Above threshold, the analog video signal-to-noise ratio will degrade approximately 1 dB for every 1 dB reduction of the C/N. A digital system will provide essentially error-free performance over a wide range of received C/N. However, over the range of only a few dB the recovered video will degrade from error free to unusable. Thus, while the digital system provides excellent performance during minor signal degradations such as rain-induced signal fades, it provides little warning of impaired performance before the signal is lost.

At the headend the received bitstreams can be separated into streams that represent individual programs or services. The various services can then be recombined as desired. For example, a particular satellite transponder may carry a multiplex of programs A, B, C, and D, whereas another transponder carries E, F, G, and H. The cable operator may wish to carry programs A, B, F, and G together on one channel and perhaps not carry the others at all. It is important to realize that in a digital CATV system, several services or programs may be multiplexed onto one satellite transponder and onto one physical cable channel. This is transparent to the subscriber. The subscriber can select the desired program by a "virtual channel" number, and the set-top will then select the correct physical channel and also demultiplex the correct bitstream for that particular program. Remultiplexing at the headend allows the operator to map a program onto any cable channel as desired. Each remultiplexed stream carrying the desired programs then modulates a 44-MHz IF carrier with 64 QAM or 256 QAM. The 44-MHz IF carrier, as before, is upconverted to the desired RF frequency on the cable plant.

The satellite transmission link will typically use a different access control mechanism than the cable link. The satellite receiver at the headend will be authorized for reception, will decrypt the satellite services, and may remove the satellite link authorization information. Access control information associated with the particular cable system will then be added back into the streams, if needed, providing the local cable operator access control. Alternatively, it may be desired to control a population of set-tops directly from the satellite uplink facility. In this case the access control information for the set-tops is added to the satellite uplink signal and is passed through the headend to the cable plant.

On the cable plant, 64 QAM, 256 QAM, or other, more bandwidth-efficient modulation is used in place of QPSK; 64-QAM modulation also uses two same-frequency carriers separated by 90 degrees in phase. Each of the I and Q carriers is modulated with 8-level amplitude modulation; each carrier thus encodes 3 bits per modulation interval. Together there are 64 possible combinations, so each 64 QAM modulation symbol carries 6 bits. About 5 million symbols per second can be transmitted in a 6-MHz bandwidth, yielding a channel data rate of about 30 Mb/s. About 10 percent are used for FEC, leaving an information rate of about 27 Mb/s. (Note that the data rate in an 8-MHz RF bandwidth would be correspondingly higher and require a corresponding increase in signal-to-noise for equivalent performance.)

Modulation at 256 QAM is similar except that each of the two carriers (I and Q) is modulated with 16-level amplitude modulation, yielding 256 possible combinations and thus providing 8 bits per symbol. Again, 5 million symbols per second in a 6-MHz bandwidth yield a data rate of 40 Mb/s and an information rate of about 36 Mb/s. The North American cable standard pushes the symbol rate up to 5.36 Ms/s to provide an information rate of 38.8 Mb/s. This rate allows two Advanced Television Systems Committee (ATSC) broadcast A/53 standard channels in a single cable RF channel. (The North American SCTE (Society of Cable Telecommunications Engineers) cable standard references ITU-T Recommendation J.83, Annex B, for modulation and forward error correction).

Modulation at 256 QAM carries more data in a given bandwidth compared to 64 QAM; hence the adjacent constellation points are closer—that is, there is less amplitude and phase difference between the points. Thus a smaller amount of interference or noise can cause detection errors. For the North American cable standard, 256 QAM requires about 6 to 7 dB higher C/N for equivalent error rate performance. Table 17.3 gives examples of digital modulation used in digital CATV.

**TABLE 17.3**   Examples of Digital Modulation Used in Digital CATV

| Usage | Bandwidth, MHz | Bits/symbol | Symbol rate, Ms/s | Transmitted rate, Mb/s | Information rate, MHz |
|-------|----------------|-------------|-------------------|------------------------|-----------------------|
| Satellite link QPSK | 36 | 2 | 29.3 | 58.6 | 27 |
| Satellite link QPSK | 24 | 2 | 19.5 | 39 | 27 |
| Cable 64 QAM | 6 | 6 | 5.06 | 30.4 | 27 |
| Cable 256 QAM | 6 | 8 | 5.36 | 42.9 | 38.8 |

Besides off-air broadcasts and satellite links, there are other sources for video programming at the headend. Many of these are conventional analog sources including local video tape, character generators, and live programming. When these analog sources are to be carried in analog format on the cable, they are processed with conventional analog equipment. If they are to be carried in digital format, they will be locally digitized and then encoded as previously described. The advent of digital video technology has also created new sources for programming. The digital video server is essentially a computer with a large disk drive storage capacity. Compressed video requires about 1 to 10 Mb/s depending on the content and the desired quality. At 5 Mb/s a one-minute advertisement requires 300 megabits, or 37.5 megabytes, of storage, whereas a two-hour feature film would occupy about 4.5 gigabytes. Although this is currently more expensive than other methods of video storage, it offers the tremendous advantage of true random access. Automated advertisement insertion systems make full use of this capability with modest storage requirements. Advertisements are digitized and compressed, and the resulting stream is stored in the video server. The server can be programmed to "play" the ad on cue or at a predetermined time. Video servers with the storage capacity for many movies can be used to provide "video-on-demand," or VOD. This facility allows subscribers to choose what they want to watch when they want to watch it. Pause, rewind, and fast-forward functions are also possible. A less technologically demanding system known as "near video-on-demand" (NVOD) allows subscribers to select from a menu of movies at a selection of different start times. With this system, a selection of popular titles may be offered with staggered running times. As an example, with 40 (virtual) channels dedicated to NVOD and 120-minute films, five top titles can be offered starting every 15 minutes.

Besides delivering video, digital CATV systems also allow entirely new services to be delivered to the subscriber. Various forms of program guides allow the viewer to navigate through the large number of program choices. Because these data are independent of the particular channel or program being viewed at any particular time, the data are sent to the subscriber via an out-of-band channel. The set-top terminal contains a second receiver that is always tuned to the out-of-band channel regardless of which program and in-band channel are in use. Configuration and control information such as the mapping of virtual channels to physical channels and new feature and application code to run on the set-top can also be sent via the out-of-band channel.

Another relatively new service is the delivery of Compact Disc (CD) quality audio programming to the home. Music is selected by programming services that specialize in this field for several dozen audio channels. The computerized programming information is sent to a bank of audio CD players. The audio information, taken directly in digital form from the CD players, is compressed and multiplexed with the other content on the system. Programming information for video games or PCs can also be distributed in a similar fashion.

An important new application for digital CATV is the delivery of Internet access to the home. A server with wideband access to the Internet is either located at or linked to the headend. The downstream data (to the home) for many users is multiplexed onto one or more of the 27 to 39 Mb/s downstream channels, which use 64 QAM or 256 QAM. A cable modem in the home demultiplexes these data and delivers the specific user's data to the PC either directly via the main internal PC bus or through an intermediate interface such as Ethernet. The upstream or return-path data (from the home) is sent either via the cable plant return path or via standard telephone lines. The data flow to and from most Internet users is highly asymmetrical—with much more data flowing to the home and relatively less data flowing from the home. This makes the use of telephone lines for the return path practical. Each click of the mouse represents a few bits to send from the home, but the response is often a large graphical display requiring several hundred thousand bits. The cable plant can also be used to provide the return path. The return path by its very nature consists of many input points being "funneled" into one output point at the headend, creating a buildup of noise and unwanted interference. For this reason, QPSK is typically used for the return path.

## 17.4 TRANSMISSION CHANNEL

When selecting a modulation format for digital transmission, consideration must be given to the characteristics of the transmission channel. An important criterion for systems on which digital transmission will be implemented is that the system should meet FCC specifications for analog cable systems. Nevertheless, a cable system is still subject to channel impairments. Common cable impairments are white noise, distortions, echoes, burst noise, and phase noise in cable headend and terminal equipment.

### 17.4.1 Random Noise

At 64-QAM channel bit rates, carrier-to-noise ratio taken alone is typically adequate on HFC cable systems. FCC requirements state that systems must meet an analog channel C/N ratio of 43 dB (measured in a 4-MHz bandwidth) at the end of the drop. For digital channels, several factors must be taken into account. First, in order to avoid interference into adjacent analog channels due to digital channel induced distortions, most cable system operators plan on running the digital signals at reduced power (typically $-5$ dB to $-10$ dB) relative to peak analog carrier power. Second, digital channel noise bandwidth is about 5 MHz, rather than 4 MHz. Third, analog channel C/N is measured relative to the *peak* analog video signal, whereas digital channel C/N is best measured relative to average signal power. Including all these factors, digital C/N is therefore approximated by the following:

$$(C/N)_D = (C/N)_A - 10 \log(5 \text{ MHz}/4 \text{ MHz}) - \text{Video peak power/Digital average power}$$

$$(17.1)$$

For digital channel power levels at $-10$ dB relative to analog,

$$(C/N)_D = 43 - 1 - 10 = 32 \text{ dB} \tag{17.2}$$

This provides adequate margin over the threshold of SCTE standard digital error correction, which is capable of correcting virtually all 64-QAM errors down to C/N levels of about 21–22 dB. For 64 QAM, this apparent margin proves essential for a variety of reasons, mostly related to distortions and other non-white noise effects. In addition, some cable systems will not meet FCC requirements on all channels, and under all conditions, further eroding the apparent margin. In fact, system planners may add digital channels in portions of the cable spectrum that are inappropriate for analog channels (too noisy, too much roll-off, etc.).

For 256 QAM operation, 6 to 7 dB of additional C/N is required as compared to 64 QAM in white noise. For many cable plants and distortion levels, 256 QAM operation will require digital channel carriage above the $-10$ dB relative level.

### 17.4.2 Distortions

The design objectives of the distribution portion of the cable system involve an adequate level of power to support the attenuation characteristics of the cable as well as to allow energy to be diverted to the subscribers' premises. Because of losses in the distribution system, relatively high power levels are required at the output of amplifiers in this portion of the cable system. The higher the power, the greater the distortion. Line extenders, for example, are operated in nonlinear regions of their transfer characteristics. This gives rise to distortion products such as composite second-order (CSO) and composite triple-beat (CTB) in an analog system.

When an amplifier is operated outside of its purely linear range, its output is approximated by the following equation:

$$e_o(t) = k_1 e_i(t) + k_2 [e_i(t)]^2 + k_3 [e_i(t)]^3 \tag{17.3}$$

where $e_o(t)$ is the amplifier output voltage, $e_i(t)$ is the amplifier input voltage, $k_1$ is the amplifier gain, and $k_2$ and $k_3$ are constants that define the second- and third-order distortion performance of the amplifier.

Because multiplication in the time domain is equivalent to convolution in the frequency domain, the spectrum of the distorted output may be obtained by convolution of the input signal with itself. This yields the following equation:

$$X(f) = k_1 H(f) + k_2 H(f) * H(f) + k_3 H(f) * H(f) * H(f) \tag{17.4}$$

where $X(f)$ is the output spectrum and $H(f)$ is the input spectrum. (The $^*$ denotes convolution). For continuous spectra the function $S(f)$ resulting from a convolution of the variables $H(f)$ and $G(f)$ is given by the following equation:

$$S(f) = \int_{k=-\infty}^{k=\infty} H(k)G(f - k)\, dk \qquad (17.5)$$

or, in discrete form,

$$S(f) = \sum_{k=-\infty}^{k=\infty} H(k)G(f - k) \qquad (17.6)$$

In a mixed-signal system, one must also consider the effect of analog and digital signal interactions. These fall into three categories: analog carriers beating with analog carriers, digital channels beating with digital channels, and analog carriers beating with digital channels. The first case produces an output spectrum containing the original signal plus frequency components that are functions of the sum and differences of the input carrier frequencies (i.e., CSO and CTB).

FCC regulations mandate that analog CSO and CTB levels be no greater than $-51$ dB relative to analog carrier levels. If a digital signal is operated at $-10$ dB relative to the analog carriers, the carrier/interference ratio resulting from analog beat products falling in the digital channel will be 41 dB or greater and should therefore have negligible effect on the digital signal.

Unlike analog CSO and CTB, which generate discrete beat frequencies, digital second- and third- order distortion produces noiselike spectra that, if distortion levels are significant, manifest themselves as an increase in background noise in the analog channels.

An example of a mixed analog/digital cable spectrum is shown in Fig. 17.1. In this example, it is assumed that the digital channels are grouped together at the upper end of the spectrum, because this is where most cable operators are planning to put their digital channels. It is also assumed that all digital channels are at the same power level and that the power level of all the analog carriers is the same, although not necessarily equal to the digital power level.

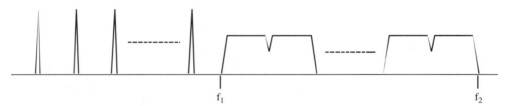

**FIGURE 17.1**   Mixed signal spectrum.

The spectrum of second-order distortion produced by digital/digital interaction is a triangular spectrum given by the following equation.

$$D2_D(f) = 2k_2 P_D^2(f_2 - f_1 - f) \qquad (0 \leq f \leq f_2 - f_1) \qquad (17.7)$$

where $D2_D$ is the second-order distortion due to digital/digital interaction, $k_2$ is the proportionality constant, $P_D$ is the digital signal power spectral density, $f_1$ is the lower frequency limit of the digital channel(s), and $f_2$ is the upper frequency limit of the digital channel(s). The spectrum for second-order digital distortion is shown in Fig. 17.2. If multiple digital channels are grouped together, the second-order digital distortion spectrum will occupy a bandwidth equal to that occupied by the digital channels. Depending on the total digital channel bandwidth, the second-order spectrum may extend into the return path and some of the low-end channels in the system.

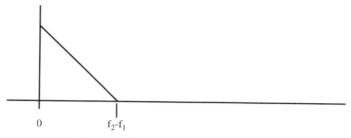

**FIGURE 17.2**  Spectrum of second-order digital distortion.

Third-order digital distortion generates a bell-shaped spectrum that spreads beyond the spectral limits of the digital signal. This spectrum may be expressed as a combination of the following three quadratic equations:

$$D3_D(f) = k_3 P_D^3[1.5f^2 - 3(2f_1 - f_2)f + 1.5(2f_1 - f_2)^2], \quad (2f_1 - f_2 \leq f \leq f_1) \quad (17.8)$$

$$D3_D(f) = k_3 P_D^3[-3f^2 + 3(f_1 + f_2)f + 1.5(f_1^2 + f_2^2 - 4f_1 f_2)], \quad (f_1 \leq f \leq f_2) \quad (17.9)$$

$$D3_D(f) = k_3 P_D^3[1.5f^2 - 3(2f_2 - f_1)f + 1.5(2f_2 - f_1)^2], \quad (f_2 \leq f \leq 2f_2 - f_1) \quad (17.10)$$

where $D3_D$ is the third-order distortion due to digital/digital interaction, and $k_3$ is the proportionality constant.

The third-order digital distortion spectrum is shown in Fig. 17.3. In severe cases this gives rise to visible side lobes, frequently referred to as "spectral regrowth." Third-order distortion increases by 3 dB for every dB increase in digital power level. In addition, if digital signals are placed in a contiguous grouping, the distortion increases by 6 dB for each octave increase in the total digital signal bandwidth.

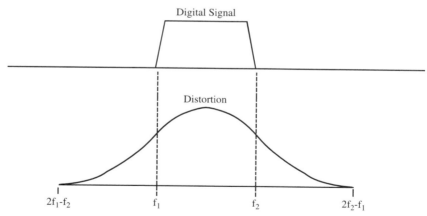

**FIGURE 17.3**  Digital third-order distortion.

Second-order distortion due to analog carriers interacting with the digital signals produces a flat spectrum, which is given by the following equation:

$$D2_A = k_2 P_D P_A(f_2 - f_1)/W_A \quad (17.11)$$

where $D2_A$ is the second-order distortion due to digital/analog interaction, $P_A$ is the analog carrier power, and $W_A$ is the bandwidth occupied by the analog channels.

Third-order digital/analog distortion also produces a flat spectrum given by the following equation:

$$D3_A = k_3 P_D{}^2 P_A (f_2 - f_1)^2 / W_A \tag{17.12}$$

The effect of digital distortion may be minimized by interspersing the digital signals among analog channels rather than by locating them in a contiguous group.

### 17.4.3  Multipath and Echoes

Signal reflections occur throughout the cable plant. These are caused by finite return losses in amplifiers, taps, connectors, etc. on the system itself and in the home by return losses in both active devices such as cable-ready TV receivers, VCRs, and converters, as well as passive components such as splitters and A/B switches.

Delays resulting from reflections of this type are short (on the order of 100 ns to 1 μs) and generally have minimal effect on analog signals. However, their effect on digital signals can be quite pronounced because delays of this order of magnitude are long relative to the symbol period of the digital signal.

The cable trunk does not contribute significantly to microreflections. Reflections generated along the trunk are produced when signals are reflected from amplifier inputs and outputs. The transmitted signal from a given trunk amplifier is reflected due to the return loss of the next amplifier in the cascade and reflected again due to the transmitting amplifier's output mismatch. The power of the echo relative to the transmitted signal is

$$P = -2(R_A + \mu_0 L) \tag{17.13}$$

where $R_A$ = amplifier return loss (dB), assumed identical for amplifier input and output
$\mu_0$ = cable attenuation (dB/ft)
$L$ = amplifier spacing (ft.)

For $N$ amplifiers in cascade, the worst case reflection is produced by the sum of the echoes at the end of the cascade. If all of the amplifiers are equally spaced, all primary echoes will have the same delay and the total reflected power of the $N$th stage of the cascade is essentially $N$ times $P$. For a 20 amplifier cascade, a return loss of 16 dB per amplifier, and a spacing of 2472 feet, the total reflected power vs. frequency is shown in Table 17.4.

**TABLE 17.4**  Total Reflected Power vs. Frequency for a 20-Amplifier Cascade

| Frequency, MHz | $\mu_0$ (dB/ft.) | $P_T$, dB |
|---|---|---|
| 55 | 0.0036 | −37 |
| 300 | 0.0089 | −63 |
| 450 | 0.0122 | −74 |
| 550 | 0.0129 | −83 |

A worst-case situation for reflections on a distribution system would occur in a section containing $N$ equally spaced taps between two line extenders with equal length drops from each tap. In this situation the worst-case echoes occur when the signal is reflected from an adjacent tap back to the previous tap and then passed through the system to the second line extender. For a medium-sized distribution system (about 7 taps between line extenders), the sum of the echoes at the input to the second line extender due to these reflections is about $-28$ dB. Delays are on the order of 115 ns.

Other dominant echoes are those from unterminated drops on the same tap. For a drop length of 125 ft, these reflections are delayed about 300 ns and have the power levels relative to the transmitted signal given in Table 17.5. Different drop lengths will not have a significant effect on the power levels shown in Table 17.5. However, for drop lengths in the range from 75 to 175 ft and cable velocity factors of 82 to 87 percent, the delay of the echoes will vary from about 170 ns to 440 ns.

**TABLE 17.5**  Power Levels Relative to the Transmitted Signal for a Drop Length of 125 ft

| Frequency, MHz | $P_T$, dB |
|---|---|
| 55 | $-24$ |
| 300 | $-30$ |
| 450 | $-32$ |
| 550 | $-33$ |

It is not possible to predict the nature of reflections in every subscriber's home because such installations vary considerably, depending on the number of TV receivers, VCRs, and converters that are connected to the drop. Also, customers may add their own variations to the installation in the form of splitters, cables, connectors, and switches purchased from retail stores.

Splitter performance varies considerably among devices. Worst-case values, based on laboratory tests,[3] have been found to be about $-8$ dB return loss and $-15$ dB interport isolation. Active device return losses also vary widely but, for most equipment, are very near 0 dB when the device is turned off.

When several devices are connected to the output of a splitter, multiple echoes can occur due to multiple reflections between the device input terminals and each splitter output port. Laboratory tests emulating worst-case conditions have yielded the results shown in Table 17.6.

**TABLE 17.6**  Echoes Under Worst-Case Conditions, Relative to Signal Strength

| Delay, ns | Relative Level, dB |
|---|---|
| 20–120 | $-10$ |
| 40–250 | $-20$ |
| 350 | $-26$ |

These echoes can also change dynamically due to channel surfing on a TV connected to the same splitter as the digital consumer terminal.

## 17.4.4  Modulation and Error Correction

Unlike analog signals, digital video signals contain minimal redundant information. Therefore, bit errors caused by transmission impairments such as noise, interference, and multipath can produce catastrophic effects in the recovered picture. Techniques such as forward error correction and adaptive equalization are necessary in order to ensure the reception of good-quality pictures. The manner in which these corrective measures are implemented depends greatly on the digital modulation technique as well as on the characteristics of the transmission medium itself.

Therefore, selection of the correct modulation technique for digital transmission is of paramount importance.

The choice of a digital modulation technique is a compromise between data rate, available bandwidth, and signal robustness. The modulation format for a particular medium must be chosen so as to ensure reliable transmission in the presence of channel impairments that are typical for that medium. Several different techniques have been proposed for digital video transmission over cable. These include 64- and 256-level quadrature amplitude modulation (64 QAM, 256 QAM) as well as 8-level and 16-level vestigial sideband modulation (8 VSB, 16 VSB). Detailed information on modulation techniques may be found in chapter 5 and in the literature.[4,5]

A block diagram of a generic QAM modulator is shown in Fig. 17.4. The input is a serial data stream with error correction overhead bits. This data stream is split into two streams that are digitally filtered and then modulated by in-phase (I) and quadrature (Q) components of the IF carrier, as previously described. The modulated I and Q data streams are then summed to produce the IF output.

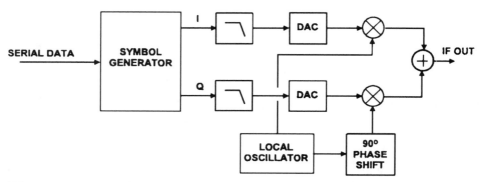

**FIGURE 17.4**    QAM modulator block diagram.

For 64 QAM the serial data stream is assigned to the I and Q components in 3-bit groups; for 256 QAM, 4 bits are assigned to each component carrier phase. Thus 64 QAM permits the transmission of 6 bits in a given symbol period versus 8 bits for 256 QAM.

The IF output of the modulator is upconverted to RF, combined with analog channels, and sent over the cable system. The receiving equipment consists of a tuner that performs RF/IF conversion, a digital demodulator, an adaptive equalizer that corrects for intersymbol interference due to echoes and group delay, and appropriate error correction circuitry.

The digital video compression process removes redundant information from the original digitized picture and codes the remaining information in an efficient manner for transmission. The result is a data stream that is more sensitive to transmission errors. Even the cleanest of transmission channels will cause some data errors and, depending on where they occur, these errors will have varying effects on the reconstructed picture. Unprotected, the viewer may see anything from a few multicolored blocks in the picture to a total loss of picture. Similarly, errors in the audio portion of the encoded data stream can cause anything from a few pops to total loss of audio. Therefore, error detection and correction and error concealment are mandatory for any compressed digital video/audio transmission scheme.

Selection of an appropriate error correction method depends on the nature of the transmission medium, the robustness of the digital modulation technique, and the available bandwidth. The cable channel is bandwidth-limited and therefore an error correction scheme that adds relatively little data overhead must be used. The few errors that cannot be corrected can be masked through error concealment. For example, incorrect portions of a video frame can be copied from a prior frame or predicted. Incorrect audio segments can be muted or predicted.

As the name implies, forward error correction allows the receiver to determine which information bits in a set of received bits are in error and, therefore, to correct the erroneous data. The study of error correction codes has generated a large volume of information on this subject. However, error correction codes may be put into two basic categories; block codes and convolutional codes. A code of each type is often used in tandem to form a concatenated code.

In block codes the data stream is divided into blocks, to which parity bits are appended that relate only to the information contained in the block. A popular form of block coding is that introduced by Reed and Solomon in 1960.[6,7]

An example of a typical error correction process is shown in Fig. 17.5. In this example the serial data stream is first divided into blocks and then Reed-Solomon–encoded. The Reed-Solomon data stream is then interleaved to provide further protection against burst errors. The interleaver output is then convolutionally encoded and sent to the digital modulator.

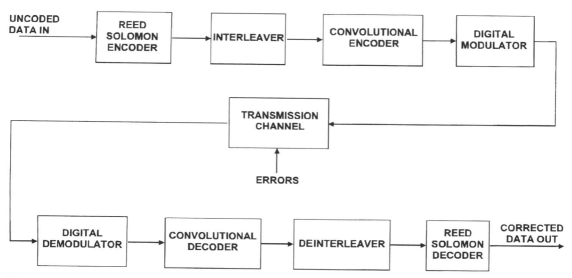

**FIGURE 17.5**    Block diagram of error correction process.

At the receive site, demodulated data, containing errors, is first convolutionally decoded and then deinterleaved, and, finally, Reed-Solomon error correction is performed. The result is a serial data stream that—providing all errors have been corrected—replicates the input data.

A Reed-Solomon encoder first divides the data into blocks of length $k$ symbols. A symbol can be any number of bits, although 8-bit symbols are often used. Each block is then coded by appending a set of $(n - k)$ parity symbols, resulting in a total block length of $n$ symbols. This is known as an $(n, k)$ Reed-Solomon code. The error-correcting capability of the code is given by

$$t = (n - k)/2 \qquad (17.14)$$

where $t$ is the number of errored symbols that the code is capable of correcting.

In order to provide protection against burst errors, the Reed-Solomon data are interleaved prior to transmission. Interleaving is performed by writing data into memory in sequence and reading it out in a different sequence such that the output does not contain contiguous symbols from the same Reed-Solomon block. If a burst error occurs, the corrupted data will be distributed over several different Reed-Solomon blocks, resulting in fewer errors per block and therefore facilitating error correction. Figure 17.6 shows a simple example of a (7, 5) Reed-Solomon code interleaved to a depth of 3.

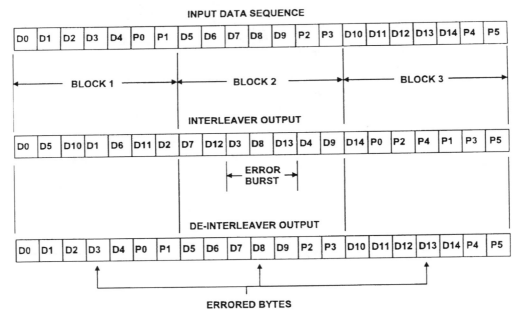

FIGURE 17.6   Reed-Solomon (7,5) code interleaved to depth = 3.

Unlike block codes, convolutional codes do not have a block structure. Instead, the $n$ coded bits are based on $k$ data bits plus the content of the encoder memory. Convolutional encoders are typically implemented by combinations of shift registers and exclusive-OR gates, that is, linear state machines.

A simple example of a convolutional encoder is shown in Fig. 17.7. This encoder consists of a two-stage shift register and three exclusive-OR gates. Because two output bits are produced for each bit shifted into the register, this configuration is known as a rate-1/2 encoder. The rate

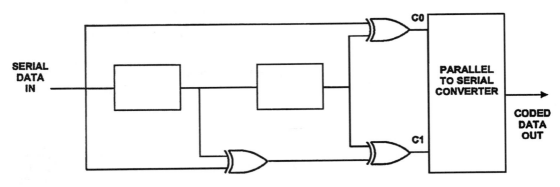

FIGURE 17.7   Convolutional encoder example.

of any convolutional encoder is expressed in terms of the ratio of information bits to code bits. The data overhead of any encoder is given by the following:

$$\text{Percent overhead} = 100(n/k - 1) \qquad (17.15)$$

where $k$ and $n$ refer to the number of information bits and code bits, respectively.

The number of shift register stages in the encoder is known as the constraint length. The example of Fig. 17.7 shows an encoder of constraint length 2. A large variety of convolutional encoders exist and are described in several texts.[6–8]

Because a convolutional encoder has a finite number of shift register stages, only a finite number of input and output combinations are possible. The behavior of convolutional encoders can be described in terms of either state or trellis diagrams. The state of the encoder in Fig. 17.7 is represented by the contents of the two storage elements.

Figure 17.8 shows all the possible input and output combinations for the rate-1/2 coder of Fig. 17.7. From an inspection of these combinations, it can be seen that transitions between certain states are not possible. For example, a transition from state 00 to state 11 cannot occur because shifting a 1 into state 00 causes the contents of the first shift register location to be 0 and not 1.

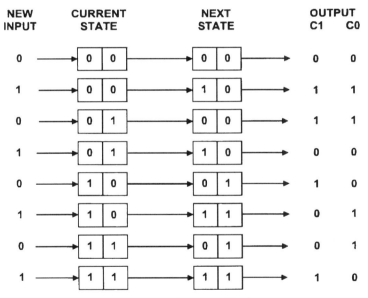

**FIGURE 17.8**    Rate-1/2 encoder input/output combinations.

Another way of showing all possible input/output combinations is in the *trellis diagram* of Fig. 17.9. In this figure each state is shown as a circle and permissible transitions are indicated by arrows. The labels in parentheses on each arrow show the inputs that produce the state change. Those labels without parentheses show the outputs resulting from a particular state change.

The Viterbi algorithm is the most widely used technique for convolutional decoding.[7,8] Viterbi decoding is best visualized in terms of a time expansion of the trellis as shown in Fig. 17.10 for the input sequence 0101100 - - -. The Viterbi algorithm tries to find the path through the trellis that differs in the fewest bit positions from the correct sequence. The algorithm does this by keeping a continually updated score, known as a *branch metric,* on each of the most likely paths

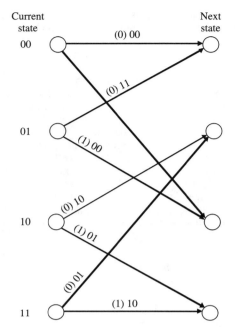

**FIGURE 17.9**   Rate-1/2 encoder trellis diagram.

Input sequence          0101100---

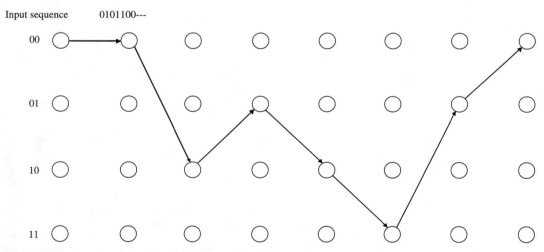

**FIGURE 17.10**   Data sequence through time-expanded trellis.

through the trellis. The path having the lowest score is considered to be the correct path. Those sequences that generate the most likely paths are known as *survivor sequences.*

In practice, convolutional coding is often concatenated with Reed-Solomon coding because the combination of the two processes results in improved coding gain. Coding gain is defined as the reduction in the carrier/noise ratio required to achieve a particular bit error rate for coded versus uncoded data. The concept is illustrated in Fig. 17.11, which shows plots of bit error rate versus C/N ratio for uncoded and coded 64 QAM. In the example shown the coding gain is about 6.8 dB at a bit error rate of $10^{-9}$. Coding gain is always specified at a particular bit error rate.

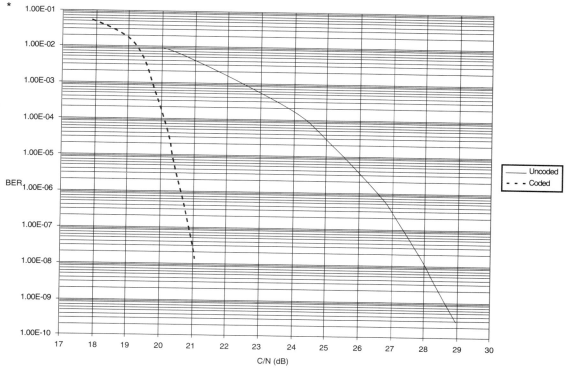

**FIGURE 17.11**    64-QAM coding gain for North American SCTE Standard FEC.

Concatenation of Reed-Solomon and convolutional coding provides excellent error correction because the two codes are complementary in many respects. Reed-Solomon codes generally have lower coding gain than convolutional codes in high-noise environments but are quite good for error detection and, with the addition of interleaving, provide protection against burst errors. Convolutional codes produce higher coding gain but perform poorly in the presence of error bursts. Convolutional codes are extremely easy to adapt to soft decision demodulator outputs, further improving the overall coding gain.[7,10] Thus, a convolutional decoder placed at the demodulator output decodes most of the channel errors, while the following deinterleaver and Reed-Solomon decoder correct the majority of the rest, as well as handle bursts.

## 17.5  CONSUMER PREMISES EQUIPMENT

Customer premises equipment in a digital cable system consists of the digital cable set-top terminal. Figure 17.12 presents a simplified block diagram of the terminal. It includes five major sections: communications, security, transport, decompression, and user interface.

The communications section is often referred to as the "network interface," and has occasionally been proposed as a module external to the rest of the terminal. This block includes the cable tuner, the QAM demodulator, and the forward error correction section. In the case of a terminal with an RF return path, and/or an out-of-band QPSK receiver, those sections are also included. The primary function of the communications section is to deliver reliable information to the terminal, including both error correction and error detection, typically in the form of transport packets. Detected errors are passed to the terminal to support error concealment algorithms.

The security section decrypts transport packets as a function of the user selection and the current authorization state. The security block may be an embedded portion of the terminal, a plug-in smart card or module, or both.

The transport section receives transport packets, filters and demultiplexes them, and establishes a time base for the terminal. Typical digital systems employ MPEG-2 transport, although some ATM-based transport delivery systems are available in the market. MPEG-2 transport is based on 188-byte fixed-length packets, including a 4-byte header, and is specifically designed to support the digital carriage of television services. As such, it includes explicit models describing the multiplexing of compressed video and audio streams, and directory-type streams, into a transport multiplex. ATM transport is based on 53-byte fixed-length packets, including a 5-byte header. The broader application set intended for ATM, including voice, justifies the relatively larger overhead. The transport section supplies demultiplexed video and audio, as well as supporting streams, to the subsequent blocks.

Typical digital cable terminals support MPEG-2, MPEG-1, and/or DigiCipher II video decompression algorithms. Typical audio decompression support includes MUSICAM (MPEG-1 audio) and/or Dolby AC-3 algorithms. In addition to video and audio processing, various types of data service support are usually present, including one or more of the following: asynchronous data channels, isochronous data channels, or IP-type datagram delivery. This block also includes video encoding to the NTSC, PAL, or SECAM standards, and video and audio D/A conversion.

The user interface section includes the most variability in the digital terminals available in the market. Front panel and graphical user interfaces (GUI); support and control of the other sections; and ROM, EPROM, and RAM memory management are present, with varying degrees of capability, depending on the power of the main microprocessor in the terminal, and its total memory configuration.

The digital cable terminal implementation is currently based upon custom, mostly hard-wired VLSI devices to achieve low cost. As such, the terminal is typically able to accept downloaded applications so long as they conform to the constraints of the fixed hardware. In the very near future, cost-effective "soft set-tops" will arrive, where even the video and audio decompression algorithms may be downloaded.

## 17.6  ACCESS CONTROL AND SECURITY

### 17.6.1  Broadcast Security Versus Point to Point

In modern digital systems it is possible to implement a variety of access control and encryption systems ranging from the most primitive up through military-grade levels. This wide range of systems must clearly target a specific application quite narrowly, since there is no general-purpose system that fits all needs. The intended application for the system defines such important parameters as the cost associated with the function, limitations associated with government

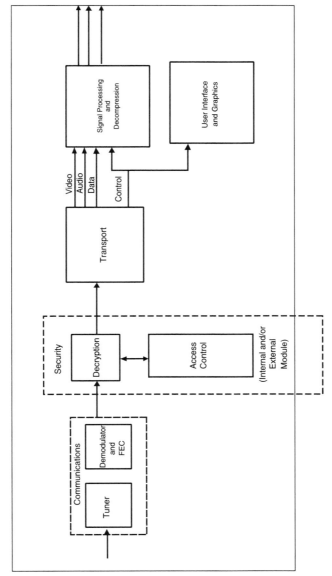

**FIGURE 17.12** Simplified block diagram of a digital cable set-top terminal.

rules for export, the nature of the threat the system is intended to protect against, and the value of the information it is intended to protect.

Different types of protected information have different levels of value. High-value information, such as military information, is usually protected with the most sophisticated security and encryption technologies available, with cost often a subordinate consideration. This type of information could easily be valued in the billions of dollars, and the access control system associated with it is designed accordingly. Other information of a more commercial nature, such as entertainment programming, is less valuable than this but clearly very significant to the future of companies whose revenue depends upon it. Because of this value to its owner, as well as its black market resale value, entertainment information can be valued in the millions of dollars. Personal information specific to individuals usually comprises credit and bank balance information, as well as demand entertainment services such as user-ordered games or videos. Personal information such as this is usually valued less than commercial information or services. Personal information could be valued in the hundreds of thousands of dollars range. One would not use the same access control system for all three of these types of information, since one application or another would either be grossly over- or under-protected.

Focusing on information of direct relevance to the consumer, the highest-security application would certainly be the entertainment information application. This information usually comprises first-run movies such as those shown in a cinema, with the high value being associated with the large number of people who would be willing to pay for the information. A single movie can be worth millions of dollars to the supplier that offers it for sale to a wide audience early in its release.

In a cable TV system this information is transmitted on a physically widespread cable network, with an associated large population of possible buyers. This large potential market could be profitably tapped if the access control and security system protecting the material sent on the cable system could be altered to allow viewing the material for less money than the owners of the system intend. This large potential demand creates substantial value in pirated entertainment materials by creating a viable black market.

Because this illicit financial opportunity is associated with the information being broadcast to numerous possible receivers, the entertainment application is usually referred to as a "broadcast application." The broadcast application for access control and security defines the maximum protection level encountered in the consumer environment. The broadcast application might contain the same information as a "point-to-point" application, but this latter is not usually accessible to a large population on a network. A large population creates demand with corresponding high profit opportunity, so the broadcast application is a potentially lucrative opportunity with high levels of system attack.

The point-to-point applications comprise most obviously what are known as "video on-demand" or VOD systems. A VOD system is equivalent to having a videotape or laser disc player in your cable system, from which specific users on that system may order movies to their own tastes, when they want to see them. Often other nonbroadcast features are available in a VOD system, such as "fast-forward" or "rewind" capabilities. Other point-to-point applications include banking through your cable system, as well as Internet, modem, or telephone access. All of these are heavily personalized applications with limited resale value in the network, which thereby limits the effort unscrupulous individuals will exert to steal the information. This has the effect of lowering the threats the security system must protect against.

With this lessened value due to the personalized nature of the information, point-to-point applications represent the minimum security level commonly encountered by the consumer. The two poles of the consumer security spectrum—broadcast and point-to-point—bound the access control system security at the maximum and minimum levels, respectively. A well-considered design will take this into account, thereby optimizing the security and its associated cost according to the material to be protected. Note that the security spectrum from point-to-point (on the lower security end) to broadcast (on the higher security end) is quite sensitive to the "weak link effect" of security.

This effect is so named because the strength of a chain is always determined by its weakest link. In this context this means that the security level needed for a particular application is always determined by the most valuable information on that network, not by any sort of average or mean. If a system's security level is appropriate for point-to-point applications, and then that system is presumed capable of protecting broadcast applications, then a painful surprise is likely in store.

### 17.6.2 Digital Encryption Basics

Given the definition of a particular access control system application as either point-to-point or broadcast, a wide variety of security techniques are available to add secure functions to the system. Chief among these is digital encryption, which is the key method of obscuring information under the control of one or more secret keys. Unencrypted information is called *clear, cleartext,* or *plaintext;* and encrypted data is called *ciphertext.*

*Block encryption* always operates on a data block of a certain size such as 64, 128, 256, or more bits. *Stream encryption* operates on data a single bit at a time, usually through generation of a *keystream* used to exclusively Or the data to form ciphertext. Stream encryption is sometimes used to encrypt and decrypt higher-speed, higher-volume data, but it is rarely used for protection of messages or other data that is easily handled in block form.

For a data block of $B$ bits, there are $2^B$ possible input data blocks, all of which must have a valid encrypted version. Clearly, there must be $2^B$ possible output encrypted blocks, so the output must be of size $B$ bits (or greater). If the cleartext and ciphertext are the same size, $B$ bits, then the encryption process can be described as the rearrangement of the $2^B$ input data block values into a different order, where there are $2^B!$ possible ways for this rearrangement to be performed. As long as $2^B!$ is acceptably large, then a number of such mappings are available for association with one or more keys, and one nearly has an encryption system. (An example might be for a 64-bit block, which has $2^{64}$ possible mappings. A $10^{21}$-bit key is sufficient to describe all of these, so much smaller keys provide ample choice.)

A good encryption algorithm has to have several properties. To describe these we must introduce some notation, so let $P$ be the clear or unencrypted plaintext, $C$ the encrypted ciphertext, and $K$ the key that maps $P$ to $C$. In this case we are describing a form of encryption known as symmetric encryption, with the more sophisticated *public key* encryption being described later. We denote encryption and decryption by

$$C = E_K[P] \quad \text{and} \quad P = D_K[C] \tag{17.16}$$

where $E$ denotes the encryption operation and $D$ the decryption operation under the key $K$. The algorithm used is stated elsewhere, such as the Data Encryption Standard or DES, the Swiss International Data Encryption Algorithm or IDEA, or even RC4 from RSA Data Security, Inc.

A reasonable encryption algorithm, such as the DES, has the following properties: given $P$ and $K$, it is easy to calculate $C$; given $C$ and $K$, it is easy to calculate $P$; and given $P$ and $C$, it is computationally infeasible to calculate $K$.

The term *computationally infeasible* has a specific meaning here; that is, there is no known way to perform the computation indicated without simply trying out (exhaustively searching) all values of the missing variable $K$ until the correct one is found. This clearly motivates as large a key as possible; one would not want a small number of keys, because they could be exhaustively searched in too short a period of time. Bigger keys are better.

Computational infeasibility implies a certain risk present in encryption algorithms. It is not easy to design a good algorithm, and unless one is very talented a computational shortcut will be created. Such a shortcut might allow the determination of a key from many fewer steps than would be necessary for an exhaustive search, which is the death of an algorithm when it happens. The discovery of such a computational shortcut is generally what is meant when an algorithm is "broken" or "breached." Most algorithms require exhaustive analysis both in the public and private domains for years before there can be reasonable confidence that no such shortcut exists. In the case of the DES, which was adopted as a standard in 1976, some analysis work in the early 1990s provided new insights into the strength of the algorithm but did not find any useful shortcut. Even after 20 years, DES is still secure as far as is known in the public domain.

How big a key must be is determined by the needed level of security in the system (military versus broadcast versus point-to-point) and the technology available to exhaustively try out all keys. If 1 billion keys can be tested per second, one should certainly have many, many more than 1 billion keys possible with the algorithm. In the case of the DES, the key is 56 bits and the number of possible keys is $2^{56} = 7.2 \times 10^{16}$. If 1 million were tested per second, 2283 years are needed to try them all. For many applications this is a perfectly acceptable level of security. In the case of the ANSI standard "triple DES," the key is 112 bits long and the key space is $2^{112} = 5.2 \times 10^{33}$. Even if 1 trillion keys were testable per second, it would take $1.6 \times 10^{17}$ years to try them all, which is long past when the sun will burn out and the next "Big Bang" cycle will begin. (Note that there are strict U.S. export controls on the use of both DES and triple DES.)

Encryption algorithms have other detailed mathematical properties that we will not explore here. Suffice it to say that an algorithm must have a substantially random character, so that someone who is viewing the ciphertext $C$ is not able to determine anything useful from so doing. For example, the number of ones and zeros in the plaintext $P$ should not be a useful indicator of the ones and zeros in ciphertext $C$. Also, the mapping operation for a particular algorithm and key cannot have a notable linear characteristic, because linear equations are solvable with well-known methods. There are many other properties, and the reader is invited to explore these in the literature.[11–13]

### 17.6.3  Public Key Encryption

In the 1970s several pioneers in the cryptography field invented *public key cryptography*. Public key is so named because of its unusual properties that actually allow a key to remain public in many applications without loss of security, and it is these properties that have enabled some amazing applications.

For some notation, we will retain $P$ as plaintext, $C$ as ciphertext, and $K$ as a key. However, we now have two types of key—one that performs an operation and one that performs the inverse of this operation. Sometimes these keys are called the *encrypt* and *decrypt* keys, and this nomenclature is correct in many applications. But it is often possible to use these two keys in either order, so long as both keys are used. For this reason we will refer to them as $K1$ and $K2$, where the following property generally holds:

$$E_{K1}[P] = C1$$
$$E_{K2}[C1] = E_{K2}[E_{K1}[P]] = P \quad \text{and}$$
$$E_{K2}[P] = C2 \tag{17.17}$$
$$E_{K1}[C2] = E_{K1}[E_{K2}[P]] = P \quad \text{so}$$
$$E_{K1}[E_{K2}[P]] = E_{K2}[E_{K1}[P]]$$

The keys are interchangeable, in that each reverses the effect of the other.

A good public key encryption algorithm has the following properties:

**1.** Given $P$ and $K1$, it is easy to determine $C$.
**2.** Given $P$ and $K1$ and $C$, it is hard to determine $K2$.
**3.** Given $C$ and $K2$, it is easy to determine $P$.

**4.** Given $C$ and $K2$ and $P$, it is hard to determine $K1$.

**5.** Given $P$ and $C$, it is hard to determine $K1$ or $K2$.

An interesting encryption system is possible with such an algorithm. An arbitrarily large group of users could all have their own key pairs $K1$ and $K2$, with the $K1$ encryption key listed in a *public key directory* that anyone can access. To send a given user $X$ an encrypted message, the sender simply looks up the $K1$ of $X$ in the public key directory, encrypts the message, and sends it to $X$. Because only $X$ has the matching $K2$, only $X$ can decrypt the message made with $K1$. This fundamental property allows public key encryption to address a very wide range of applications with great efficiency, and more public key applications are introduced each year.

In particular, public key encryption is very appropriate for systems with a large number of users that might each communicate with any other at any time. Encrypted telephone systems or the Internet are the perfect models for public key use, with such uses being developed at this moment. Many point-to-point access control systems are also the perfect domain for public key encryption.

A potential user of public key encryption should be cautious to carefully evaluate the various algorithms available for performance and efficiency. Public key algorithms generally require a large data block size—such as 256, 512, or even 2048 bits—which can affect the minimum size of an encrypted message. (One must transmit all of the encrypted bits to be able to decrypt the message!) Public key algorithms are also several orders of magnitude slower than symmetric algorithms, rendering them unsuitable for high-speed encryption operations. Often, a slower public key algorithm is used to deliver a symmetric encryption key, which is then used for fast encryption.

In broadcast applications there are several ways to address security problems, some involving public key and some not. Symmetric key encryption systems have been successfully used for years in such applications and have seen great success. Public key capabilities are expected to augment the symmetric key systems over the next few years as point-to-point and generic capabilities such as Internet access are introduced.

Finally, public key encryption has another use that is quite valuable—the digital signature. A digital signature involves using a randomizing function (a hash function) to create a *message authentication code* (MAC) that reflects the authenticity of a message. This MAC is then encrypted using the secret decrypt key $K2$ (from the example given). This allows any public user to decrypt the message using the public encrypt key *K1* and confirm the authenticity of the message by confirming the MAC. Digital signatures are an extremely important part of public key technology that will see increasingly widespread use, including use in financial transactions, over the coming years.

### 17.6.4 Access Control System Basics

A primal design decision for an access control system is the level of security (if any) that the system is expected to provide. In some rare cases, there may be no security at all, even though an access control mechanism is in use. Such a system would exert control without the enforcement or protection from attack that comes from security techniques such as encryption—and that can be acceptable in some cases. Most systems require at least some encryption, however.

Once the encryption algorithms and tools are chosen, creating a working access control system involves some additional steps. In all secured cases some information is being transmitted in encrypted form, and one or more receivers are expected to be able to decrypt that information based on some criterion. The most common criterion is when fees have been paid, and the controlling computer has been so informed either by operator input or a connection with a billing computer. Another example criterion might be the geographic location of a receiver. In response to the billing computer, the control computer must send some management information into the system to allow receivers to make the proper decryption decision. Some of this information is sent to specific receivers in the form of an entitlement in an *entitlement management message* (EMM). An *entitlement* is simply the authorization for a receiver to do something, which generally consists of decrypting a signal. Some of this information is sent with the encrypted payload in an *entitlement control message* (ECM). Together, the ECM and EMM must be sufficient to control access to the service or information being protected. Encryption techniques, be they symmetric or public key or large key or small, are usually used in the EMM and ECM to control the dissemination of the key used to encrypt the valuable payload information.

Access to a particular encrypted service can be accomplished in one of two ways, these being direct delivery of the service key used to encrypt that service or tiered delivery of a bit denoting a service. For this discussion the common term *control word* (CW) will be used to denote the service key that encrypts the video or audio or data content.

### 17.6.5   Direct Delivery of Entitlements

To be granted the right to decrypt a service, a specific receiver is simply sent the CW used on that service in an EMM, usually itself encrypted under some secret key unique to that receiver. This direct delivery of entitlements means that the receiver must store one CW for each service it is granted a right to view, which can be troublesome in systems with many possible services.

In this case the ECM contains only basic service identification information as necessary to allow the receiver to determine which key to use. This type of access control is particularly appropriate for VOD or Internet applications.

In short, direct-delivery EMMs deliver control words and CW number designators to specific units, and direct delivery ECMs simply label a service with a CW number to tell a receiver which key to use.

### 17.6.6   Tiered Delivery of Entitlements

Another way to transfer and control entitlements is necessary when the number of services grows into the hundreds or thousands, as some modern digital systems with millions of possible entitlements have done. These systems cannot afford the bandwidth burden of sending that many keys to thousands or millions of users, so a method of tiered delivery is used. Tiered delivery uses fewer keys than direct delivery, but depends upon changing those keys system-wide at a rapid rate to retain security. Tiered delivery involves designating specific bits in an EMM to denote each service. An EMM sent to a single receiver can therefore send as much as one entitlement per bit of tier, which is vastly more efficient than direct delivery. The receiver stores these tiers in a secure location, and refers to them when it tunes to a particular service. The receiver generally is also sent a *group key* in the EMM, which is used as described here.

When a receiver tunes to a particular service, it receives an ECM associated with that service and compares the tiers contained in that ECM to the tiers stored inside the receiver from its previously received EMM. If the receiver contains one or more tier bits that are also in the ECM, it is authorized to use the group key. If the ECM does not contain tier bits in the receiver's EMM tier memory, then that receiver is not authorized and cannot use the group key. The ECM usually also contains a group key encrypted version of the CW used to actually decrypt the service content, where the decryption and use of this service key only takes place if there is a tier bit match as just described.

Obviously, authentication techniques, including the use of digital signatures or public key encryption, are very appropriate for these functions. In fact, these techniques are mandatory to ensure that there is no trivial means to fool a receiver into operating.

Direct delivery of entitlements and tiered delivery of entitlements are, respectively, the least and most efficient forms of entitlement delivery. Tiered delivery is best for systems with a large number of entitlements, where greater accommodation to bandwidth conservation is generally needed, and direct delivery is often best for smaller or simpler systems with fewer entitlements. A system with a medium-sized number of entitlements could conceivably use either method.

## GLOSSARY

**CATV**   Community antenna television.

**C/N**   Carrier/noise ratio.

**CSO**   Composite second-order distortion.

**CTB**   Composite triple-beat.

**FEC**   Forward error correction.

**HFC**   Hybrid fiber coax.

**I**   In-phase component.

**NVOD**   Near video-on-demand.

**Q**   Quadrature component.

**QAM**   Quadrature amplitude modulation.

**SCTE**   Society of Cable Telecommunications Engineers.

**SLM**   Signal level meter.

**VCR**   Videocassette recorder.

**VOD**   Video-on-demand.

**VSB**   Vestigial sideband modulation.

## REFERENCES

1. W. Ciciora, "Cable Television in the United States," *CableLabs Report,* 1995.
2. "Cable Television Service," *Code of Federal Regulations,* Part 76, 1992.
3. J. Waltrich, Glaab, J., Ryba, M., and Muller, M., "The Impact of Microreflections on Digital Transmission over Cable and Associated Media," *NCTA 1992 Technical Papers,* pp 321–336.
4. K. Feher, *Telecommunications Measurements, Analysis and Instrumentation,* Prentice Hall, Englewood Cliffs, N.J., 1987.
5. V. Brugliera, "Digital Modulation and Transmission Technology for Cable Applications," *Communications Technology,* April, 1993.
6. I. S. Reed and Solomon, G., "Polynomial Codes over Certain Finite Fields," *J. Soc. Indust. Appl. Math.,* vol. 8, 1960, pp. 300–304.
7. R. Blahut, *Theory and Practice of Error Control Codes,* Addison-Wesley, 1984.
8. E. Lee and Messerschmitt, D., *Digital Communications,* Kluwer Academic, 1988.
9. A. J. Viterbi, "Error Bounds for Convolutional Codes and an Asymptotically Optimum Decoding Algorithm," *IEEE Transactions on Information Theory,* vol. IT-13, 1967, pp. 260–269.
10. G. Clark and Cain, J., *Error-Correction Coding for Digital Communication,* Plenum Press, 1981.
11. B. Schneier, *Applied Cryptography,* Wiley, New York, 1994.
12. D. Denning, *Cryptography and Data Security,* Addison-Wesley, 1983.
13. D. Davies and Price, W., *Security for Computer Networks,* Wiley, New York, 1989.

## ABOUT THE AUTHORS

Paul Moroney leads advanced development and next generation systems activities at General Instrument, San Diego, Calif., where he is vice president of advanced technology and systems. He joined the company in 1986 and led the systems engineering activities associated with the Video-Cipher and DigiCipher product lines, covering most notably the areas of access control, communications, and video compression. From 1979 through 1986, he worked at M/A-COM Linkabit, San Diego, Calif. He received the B.S., M.S., and Ph.D. degrees from the Massachusetts Institute of Technology, Cambridge, Mass., in 1974, 1977, and 1979, respectively. He co-holds 15 U.S. patents and has various publications to his credit.

Joseph B. Waltrich is manager of digital special products for General Instrument, Hatboro, Pa. He has been employed by the company since 1985, during which time he has worked on digital video processing and transmission of digital video signals. He has been associated with GI's digital video program since 1993 and involved in channel characterization studies for cable and alternative media. He received a B.A. in physics from LaSalle University, Philadelphia, Pa., and an M.S. in physics from the LaSalle University, Philadelphia, Pa., and an M.S. in physics from the University of Notre Dame.

Eric J. Sprunk has led since 1990 the security function at General Instrument through activities involving support of law enforcement in anti–signal theft cases and in the design of several internal and product security systems. One of the primary designers of the Emmy Award–winning DigiCipher II Access Control System, he participates actively in the advancement of security technology and in numerous encryption projects. In the late 1980s he led the Network Test System Division of DSC Technologies, Carlsbad, Calif., following which he joined General Instrument in San Diego, Calif. He graduated from the University of California, San Diego, in 1982 with a Communications Systems BSEE, with an emphasis in mathematics, and then worked on numerous projects in both military and commercial satellite communications and telecommunications. He is the holder of four granted and three pending patent applications.

Marc Kolber is currently working as a systems engineer at the Digital Network Systems Division of General Instrument, where he is participating in the definition and development of products for digital CATV and MMDS systems. Previously he worked at Honeywell Business and Commuter Aviation Systems developing RF communications and navigation equipment for business aircraft, including a satellite-linked digital communications product. Before that, he was employed at Blonder Tongue developing CATV and MATV products. He graduated from the New Jersey Institute of Technology in 1973 with a BSEE degree and received an MSEE degree specializing in communications systems from Arizona State University in 1989.

# CHAPTER 18
# THE VIDEO DIAL-TONE NETWORK

**Ajith Nair**
*Scientific-Atlanta Inc.*

## 18.1 INTRODUCTION

Video dial tone is a new form of video delivery service that is similar to telephone services provided today. When a telephone subscriber picks up the telephone handset, he or she gets a dial tone, indicating a valid connection to the local office of the telephone company and also indicating that the number he or she wants to reach can be dialed. Once the dial tone is received, the telephone company equipment is ready to connect the subscriber to any telephone in the world. Once the connection is established, the subscriber may use the connection for conversation, obtain information from a computer database using a modem connected to the subscriber's computer, use interactive services such as banking and investment inquiries using the phone's push-buttons, or connect to an Internet access provider and browse the World Wide Web.

When you pick up the handset and do not get a dial tone, you know that either the local line is faulty or that the local office has reached the limit of the number of subscribers it can support simultaneously. Current telephone systems are designed with the assumption that, at any one time, only a certain percentage of the subscribers will access the telephone system. The amount of equipment installed by the telephone company will be to support only this average peak demand. When the actual demand exceeds the assumed percentage, you do not get a dial tone and have to wait until another subscriber stops using his or her telephone. With video dial tone the common carriers want to provide this same kind of connectivity and services with video capability as they provide today for audio services.

The voice network of today with its advanced services in addition to voice telephony is accomplished within the limited bandwidth (300 Hz to 3600 Hz) available in the telephone local loop. The local loop of today is copper twisted pair and is limited in bandwidth. A simplistic diagram of the current voice application network is shown in Fig. 18.1. The subscriber is connected by a copper twisted pair all the way to the switch in the first case. The signal is still analog and more or less the same signal that has been used for almost a century. In the second case the copper twisted pair from the subscriber is terminated in a remote terminal (Fig. 18.2). The analog signal is digitized, then the electric signal is converted to an optical signal and then carried to and from the remote terminal to the switch by optical fiber. At the remote terminal more than one

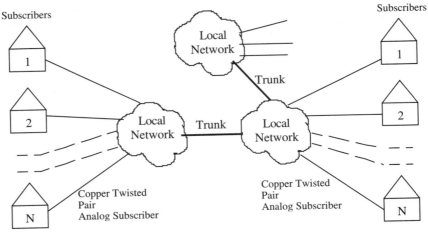

**FIGURE 18.1** Simplistic diagram of the current voice application network.

subscriber's twisted pair may be terminated, and these signals are multiplexed on to the digital carrier on the fiber-optic cable. This second portion of the telephone system is called the digital loop carrier. In the next generation of voice systems the reach of the fiber-optic cables and the digital loop carrier will come closer to the subscriber's premises. This would reduce the length of the copper twisted pair over which the signals have to be carried. An optical network unit would convert the optical signal and then demultiplex the signal to individual subscriber connections for a small neighborhood. This brings the digital subscriber loop very near to the home and, with it, higher bandwidth capability.

The rest of today's telephone network consists of local, national, and international trunks that connect switching centers. The networks are controlled by computers connected to the switches that set up connections and provide operational support for billing and other services.

Video delivery to subscribers today is essentially by broadcast only—by TV stations over the air, by cable TV companies over coaxial cables, or by direct broadcast services over satellites. Figure 18.3 shows a typical cable TV network. The program material at the cable TV head-end is obtained from several sources: national networks, national premium channels over satellite using satellite receivers, local TV stations using TV receivers, and locally generated video. These are modulated on to different TV channel frequencies, combined, and distributed over a coaxial cable. Today's cable systems use analog fiber optics cables for the trunks in a star-star topology. The main characteristic of most of today's cable systems are that they are designed for one-way transmission only.

The intent of the video dial-tone systems is to provide the subscriber with a very much enhanced network with video capability with the same kind of interactive capabilities of the voice telephone system. The services and capabilities that the video dial-tone service is expected to provide are video-on-demand, near video-on-demand, interactive home shopping, interactive home banking, video conferencing, news-on-demand, single-user and multiuser interactive games, program guides, easy-to-use navigators, broadcast services, gaming, Internet, and other data services.

Most of these services are dependent on providing static and moving videos to the subscriber. A fully interactive system opens up immense possibilities with newer applications to enhance the viewing pleasure of the video dial-tone customer. However, video services such as video-on-demand require the capability for real-time video transmissions that require large amounts of bandwidth to the subscriber premises.

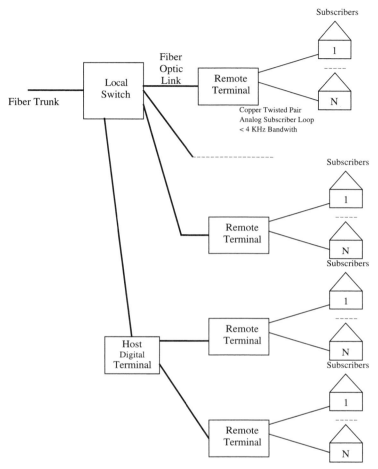

**FIGURE 18.2**    In the voice application network, the copper twisted pair from the subscriber is terminated in a remote terminal.

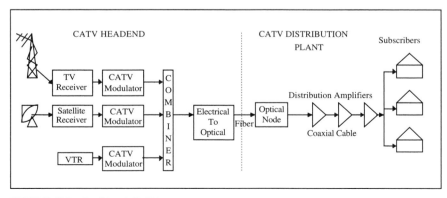

**FIGURE 18.3**    Traditional CATV system.

The essential differences between the current voice telephone network and the cable television network are as follows:

- The telephone system is symmetric in the sense that the forward and reverse bandwidth is the same, whereas the video delivery systems are asymmetric in bandwidth. The bandwidth in the forward direction is very high, and the bandwidth in the reverse direction is either very small or nonexistent.
- The voice telephone bandwidth reaching a customer is only about 4 kHz, whereas the CATV bandwidth is close to 1 GHz.
- The telephone system is real-time–interactive, whereas the cable TV systems are broadcast only.

## 18.2    VIDEO DIAL TONE

The major elements of a video dial-tone system are the home terminal, access network, core network, service providers, and operations support systems (Fig. 18.4).

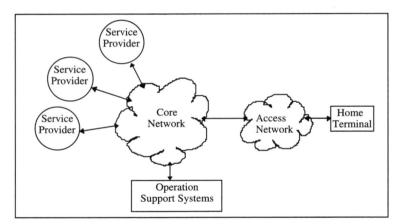

**FIGURE 18.4**    Video dial-tone system.

### 18.2.1    Access Network

The access network provides the connection between the subscriber's home and the local exchange or the core network. The access network carries the control and data information between the home terminal and the core network. There are several choices for an access network for video dial tone. The main types of access network to be considered are hybrid fiber coax (HFC), asymmetric digital subscriber loop (ADSL), fiber-to-the-curb (FTTC), and fiber-to-the-home (FTTH).

HFC for the video dial tone will be an extension of the current CATV system architecture. As mentioned earlier, most of the CATV systems currently in place do not have a return path to provide interactive services. The addition of a sufficiently wide bandwidth return path to the existing HFC plants would provide a good access network for video dial tone. The total bandwidth available in the HFC network is about 1 GHz. A large number of home terminals will be connected to the same coaxial cable and will share the same bandwidth for the downstream and upstream data path. In most cases the HFC network will have regular analog CATV channels coexisting with the video dial-tone service. In this case the total bandwidth available will have to be shared with the CATV system. Special consideration will have to be made for avoiding conflicts in the reverse-path transmissions.

Similarly, ADSL is an extension of the existing copper twisted pair–based voice telephony system. The main modification needed will be to remove filters in place now to restrict the bandwidth to voice only (approximately 4.0 kHz). The ADSL network provides an asymmetrical data rate connection. It provides a higher data rate of 2.0 Mb/s to 7 Mb/s in the forward or homeward direction and a lower data rate of about 600 kb/s in the reverse direction toward the core network. The ADSL provides a direct independent connect for each subscriber to the local exchange or the core network entry point.

In many existing voice telephony systems the extent of fiber reaching out from the central office is increasing. Fiber-optic cables reach to large residential neighborhoods. It will be a natural extension to use this large-bandwidth pipe for video dial tone. The FTTC scheme involves the fiber reaching up to the curb and terminating in an optical neighborhood unit (ONU). The ONU will convert the optical signal to an electric signal for the rest of the network to the home. The rest of the network could be either coaxial cable or twisted pair. If twisted pairs are used for the last leg, then the distance will have to be limited to about 300 yards.

The least common of the access network type is the FTTH network. In this case fiber-optic cable reaches all the way to the subscriber's home. FTTH can provide very large symmetrical data rates in the forward and reverse directions. But the cost of the customer-premises equipment and the installation of the last leg of fiber from the curb to the home is still prohibitive.

### 18.2.2  Core Network

The core network (Fig. 18.5) connects the service provider to the subscriber home through the access network. The core network in the video dial-tone case will be one or many switched networks. In addition to providing the switched interconnection between the home and the service provider, the core network also provides the control and management functions. The main control function provided by the core network is the setting up and tearing down of connections for interactive sessions. It can set up connections for broadcast services also. The network management tasks to be performed by the core network are billing, monitoring, and network configuration management. ATM is emerging as the technique most suited for switching in the core networks for video dial-tone applications.

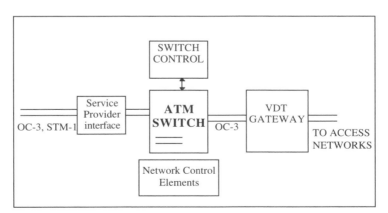

**FIGURE 18.5**  A video dial-tone core network.

The core network could actually be multiple networks spanning a nation or the whole world. The trunks connecting local networks have very large data rate capabilities to support video dial-tone applications. However, when the video dial-tone system spans many core networks, issues of latency and jitter will have to be properly addressed.

### 18.2.3  Network Control System

A comprehensive control system is needed to control and manage the video dial-tone system. The network control system may be implemented as a single entity in the core network or as distributed processes in the core network and service provider computers. The control system takes care of the following functions:

- Verification and authentication of subscriber identity.
- Setting up connections between the subscriber terminal and the core network.
- Allocation of bandwidth for the service or application in the access network.
- Routing the connection through the core network/s to the service provider.
- Maintaining a data base of service providers and the applications available in the system.
- Provide billing services to the subscriber and the service providers.
- In the case of an HFC access network, this will also need to have processes to control the different elements of the system, e.g., QAM modulators, QPSK modulators, and demodulators in the signaling path.

### 18.2.4  Service Provider

The service provider (sometimes appropriately called the *value-added service provider*) is the repository of all the information provided through the video dial-tone system. It is also the server side of the client-server system that can be implemented using the video dial-tone system. The service provider also provides the applications that can be run on the video dial-tone system, which can have multiple service providers. They may be connected to the nearest core network to the access network, as shown in Fig. 18.5, or across multiple core networks. The service providers can be in the same geographical location, across the state, across the country, or anywhere in the world.

The simplest of service provider configurations could be a server and a high-bandwidth connection to the local core network. The server would have the following subsystems: a processor to provide the brains for the whole service provider system, an array of fast-access storage elements, and a communications interface that connects the server to the core network.

The main services that the service provider has to implement are (1) maintaining and updating the database of content that is available from the service provider; (2) providing video, audio, and data streams to applications or users as required; (3) and providing a server gateway function to negotiate for the resources available when a session is being set up. Figure 18.6 shows a simple service provider configuration.

### 18.2.5  The Home Communication Terminal

The home communication terminal is analogous to the telephone in the voice network. The home terminal in a video dial-tone system will require the capabilities shown in Fig. 18.7. The physical layer interfaces with the access network and obtains the high-speed data intended for that terminal/subscriber and transmits the reverse data on to the access network. These interfaces depend on the access network type used. Tables 18.1, 18.2, and 18.3 show access network interfaces for different types of networks used.

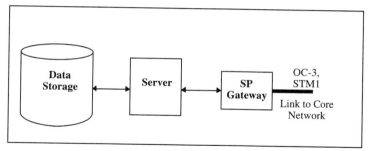

**FIGURE 18.6**   A service provider.

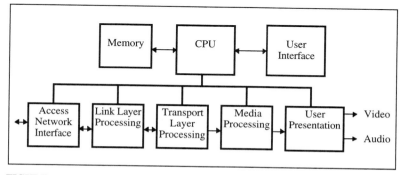

**FIGURE 18.7**   A home terminal for video dial tone.

**TABLE 18.1.**   Access Network Types for Various Interfaced Implementations

| Access network type | Interface implementation |
|---|---|
| Hybrid fiber coax (HFC) | RF tuner |
| Asymmetrical digital subscriber loop (ADSL) | Access arrangement |
| Fiber-to-the-curb (FTTC) | 16-CAP receiver |
| Fiber-to-the-home (FTTH) | Fiber-optic receiver |

**TABLE 18.2.**   Access Network Types for Various Reverse Paths

| Access network type | Reverse path |
|---|---|
| Hybrid fiber coax (HFC) | QPSK transmitter |
| Asymmetrical digital subscriber loop (ADSL) | Access arrangement |
| Fiber-to-the-curb (FTTC) | QPSK transmitter |
| Fiber-to-the-home (FTTH) | Fiber-optic transmitter |

**TABLE 18.3.**   Access Network Types for Various Data Link Processing

| Access network type | Data link processing |
|---|---|
| Hybrid fiber coax (HFC) | QAM demodulation |
| Asymmetrical digital subscriber loop (ADSL) | ATM SAR function |
| Fiber-to-the-curb (FTTC) | ATM SAR function |
| Fiber-to-the-home (FTTH) | ATM SAR function |

The most common transport layer implementation for video applications could be as specified in ISO-13818-1 (MPEG-2 systems layer). However, ATM can also be used to transport the information to the media-processing elements in the terminal.

The main function of the terminal will be to process video and audio streams and present them to the television receiver. These functions will be common across all access network types. (These are described in Chapter 16.)

A video dial-tone terminal to support the applications mentioned earlier will need to have a very powerful microprocessor and graphics acceleration capabilities. The CPU will run software that controls all the hardware, and it could support operating systems that would make third-party development of applications easier.

MPEG has received wide acceptance as the compression standard for good quality video and audio. The media-processing elements in the home terminal will be devices to decompress compressed audio and video streams and mix them with presentation graphics and audio. Graphics-rendering capability in the terminal would provide a very appealing graphical user interface to the subscriber. However, this capability is not a must for providing just the basic features.

The hardware user interfaces for the terminal are infrared remote control, keypad, etc. A high-end terminal could provide full-function keyboards and Ethernet connections for connecting the home terminal to a home computer or a home network. The CPU provides the graphical and audio user interfaces for different applications. The presentation blocks formats and generates audio and video that can be fed to a TV set and/or home entertainment systems.

Some applications such as video conferencing will require the home terminal to have the capability to send audio and video back in to the network. This would necessitate either (1) an external device that can process the audio and video from home and generate compressed streams that can be sent back, or (2) internal resources for compression and formatting.

## ABOUT THE AUTHOR

See About the Author, chapter 16.

# P · A · R · T · 5

# DIGITAL IMAGING PRODUCTS

# CHAPTER 19
# DIGITAL INTERACTIVE MULTIMEDIA PRODUCTS

**David Javelosa**
*Composer/Producer*

## 19.1 INTRODUCTION

By current usage, "multimedia" has become synonymous with "digital" and "interactive." When looking at this generation's consumer products, the emphasis is definitely toward electronic and digital technology. This technology is one born of a relationship between hardware and software. According to the conventional wisdom of this young industry, software drives the hardware. A specific hardware platform is only as good as the "killer app" that runs on it. That desire for a title or program is what inevitably decides the winning hardware technology (see Figs. 19.1 and 19.2).

Multimedia products fall basically into two categories, much like their computer program ancestors. These categories are productivity and entertainment. Although productivity software has been mostly responsible for the growth of desktop computers, entertainment software is rapidly surpassing productivity sales in the computer market. When coin-operated arcade machines, dedicated game machines, and other home players are considered, the whole of interactive entertainment has outgrossed the traditional medium of the feature film industry.

Breaking down consumer interactive entertainment into the two basic categories of dedicated players and personal computer products, it is clear that the overwhelming majority of product sales have been in the dedicated players, or what are referred to as "video games."

## 19.2 WHAT IS A VIDEO GAME PLAYER?

A video game player is a black box consumer electronic product that is designed to entertain. Although they may have several components in common with desktop computers, video game players lack the open architecture and versatility of computers. Video game players are most commonly designed as peripherals to the installed base of the television set, much like the video-cassette recorder. More recently, video game players have been designed with the home theater

Figures and some descriptions are from the book *Sound and Music for Multi-Media: Audio Effects for Interactive Entertainment* by David Javelosa and are used by permission of the author.

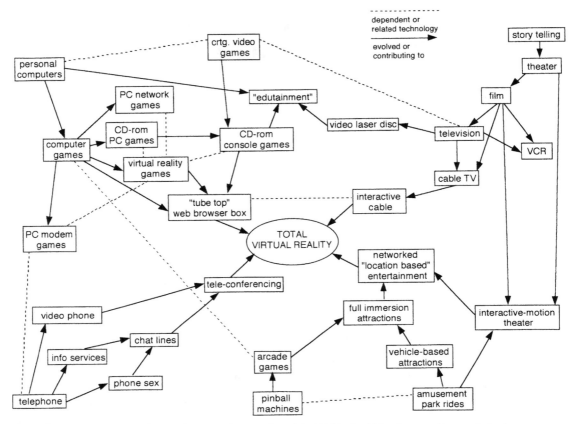

**FIGURE 19.1**   A genealogy of interactive entertainment. (Copyright 1996 by David Javelosa, used by permission.)

concept as part of the ultimate playback system, including a large screen and a powerful stereo or surround-sound system.

A video game is a cartridge or CD-based software product designed to be played on a machine with little instruction for the user. The operating system is completely transparent. The control interface is as simple and unobtrusive as possible. As a closed system, there are few compatibility problems. Bad components are not easily accessible, and generally an entire system is replaced when faulty. The hardware is optimized for the performance specification, manufactured with the most cost-effective parts, and sometimes even sold at a loss to gain market share and platform dominance. Price point to the consumer is a major consideration in this industry.

By contrast, a computer game is a software program designed to run on a multipurpose "productivity" personal computer. The desktop computer is less the central focus of entertainment in the home and, more importantly, not as common a household appliance—yet. The installed base is greatly reflected by the difference in platform cost: a $1500 computer vs. a $100 to $300 game machine. Because of a computer's multiple functionality, there will always be some compromise to the system for doing one thing over another. Game software installation often requires large amounts of computer resources (RAM and hard-drive space), hardware enhancements (CD-ROM, sound card, graphic display), and even unique configurations of both software and hardware to run.

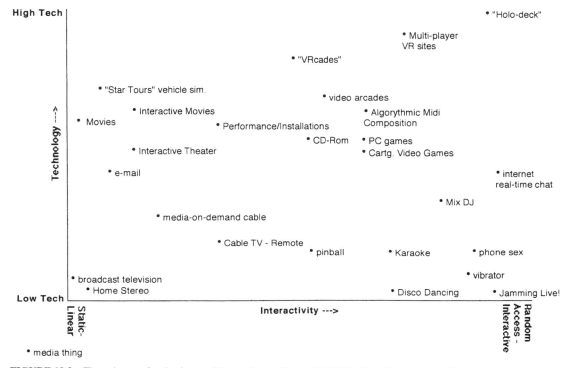

**FIGURE 19.2**   The universe of technology and interactivity. (Copyright 1996 by David Javelosa, used by permission.)

This situation is changing as the technology matures. Multimedia standards in desktop computing are starting to make things easier, and future specifications such as Windows 95 are proposing to make operating more transparent. Still, the computer offers no competition in cost and simplicity to the dedicated "tube-top" game player.

The drawbacks on the video game side seem to focus on the reliance on NTSC display quality. This makes it difficult to use detailed amounts of text, but the product design generally tends toward action orientation. The other major challenge is in development. Because these dedicated machines are designed and released more frequently and with few compatibilities beyond their proprietary technologies, development environments generally comprise exotic, high-end hardware, incomplete development tools, and often redundant, repetitious programming. Unless well-planned from the beginning, development environments rarely mature to a level of standardization until the platform is a few years old and doing well.

Regarding the difference in content between video games and computer games, there are also parallels with the platforms' markets and core audiences. Whereas computer-based products offer a wide range in the education, "edutainment," and self-enrichment areas, the video game product base is for the most part dominated by sports and action games. Because of the non-learning curve of the game machine interface, the term "twitch game" has become associated with this type of game design, even when brought over to the computer-based side.

The result of this action-based dominance has tended to discourage the development of strategy games, simulations, edutainment titles, and other multimedia products for the dedicated game machine. This trend is changing, however, as a result of better machine performance, broader marketing concerns, and a maturing of the core game audience.

## 19.3  *EARLY GAME MACHINES AND CURRENT PORTABLES*

In the beginning, software-based computer games and early game machines were very close in technology and capabilities. The 8-bit processor-based PC, Apple II, Commodore 64, etc., were basic systems capable of simple graphics, minimal sound production, and restricted processing power. Used strictly for game play, these systems were of minor interest. The same technology packaged into a dedicated entertainment box, however, could be optimized to perform quite satisfactorily. The early Atari systems and the enormously successful Nintendo Entertainment System (NES) proved this in a market that had no similar competition. As technology advanced and became more accessible, however, this would change.

### 19.3.1  Eight-Bit Graphics

The first hurdle digital technology had to stride for entertainment purposes was that of graphic presentation. Before the development of bit-map and sprite technology became widespread, most computer games were text-based, either as an abstract, descriptive presentation, or actually using the alphanumeric text characters as graphical components in an 80-column (or less) grid.

Using sprite generation, the game design now could take advantage of an object or character built up from a set of pixels in a grid of much higher resolution. This collection of pixels could be manipulated as one "sprite" across coordinates on the display. Stationary pixel groups or tiles could be repeatedly used as background art with a minimum of memory storage.

The sequencing of several single sprites into the screen location of a character would allow the usage of traditional cell animation techniques in sprite animation. This technique plus the manipulation of its screen coordinates is the basis for most side-scrolling video game character animation. However, even this simple animation technique could have a crucial impact on the processor, affecting game play and speed.

### 19.3.2  The SID Chip

The birth of music and audio synthesis was made possible with the introduction of analog systems composed of various voltage-controlled components, each designed to generate, shape, control, and/or perform sound. The primary elements in a typical configuration of this system would include wave-shape generation (one for each voice), envelope shaping (time-based dynamic shaping), and some interface for controlling note pitch and event triggering. As this relationship was formalized in the commercial synthesizer industry, along with microchip development, a simple synthesizer system could be reproduced in a single digital chip.

With the economics of mass production, these sound chips became available to a number of different applications from toys to alarm clocks and, eventually, game machines and early computer products. One of the most noted home computers to feature such an internal synthesizer was the Commodore 64. This sound interface device, or SID, chip could similarly be found in the Nintendo Entertainment System and later in the portable Atari Lynx. A simpler version (with no wave-shape variation or built-in frequency modulation), known as the *pulse signal generator* (PSG), is featured in the Sega 8-bit machines, the Master System, and Game Gear, and is included in the downwardly compatible 16-bit Genesis.

### 19.3.3  Disk vs. Cartridge

Another major difference between computer-based entertainment and the video game is the delivery medium. The earlier 8- and 16-bit game consoles and a few of the 8-bit home computer systems had titles published on ROM cartridges for a number of reasons. First was the ease of use. Having the software directly attached to the system avoided the need for a user operating system, installations, and file management. Second, cartridges were virtually impossible to duplicate without specialized systems. Third, a user would never be able to accidentally erase the software.

The disadvantages were that there was no writable media attachment and that cartridge delivery was quite a bit more expensive in manufacturing than floppy disks. But because this is a software industry, the added expense of a disk drive on the hardware platform would be prohibitive. Hardware upgrades could also be applied transparent to the user, concealed in the software cartridge.

### 19.3.4   Development Tools

As the game production became larger, production teams would involve specialists in particular fields beyond the programmer's own limited talents in graphics, music, and sound design. Because these specialists would not necessarily have experience in this early digital environment, custom tools would be created by the programming team—sometimes for creating data just for a specific game title or program design.

Because of the low pixel resolution, early sprite design was actually done using graph paper and numbers to represent the colors used in the grid. This would be a multistep process of drawing and coloring the art, converting it to a grid of numbers, and transferring these values to data statements, entered into the program. This process would later be replaced by pen- or mouse-driven graphics programs, but with similarly low resolutions.

Music would be done in a similar fashion, with the $x$-axis representing note occurrences and durations. The values filling in the blocks would be numbers assigned to specific pitches sometimes coupled to a value denoting amplitude or volume. The $y$-axis would be the different voices.

Later, simple utilities would be designed to convert musical instrument digital interface (MIDI) files into editable data text files. MIDI was a much denser format used in professional music gear, and the output would have to be constrained and minimized for game program operation.

The instrument sounds or synthesized effects for each of these tracks would be determined by rows and values assigned to different synthesizer functions. These could include wave shape, envelope functions, and modulation. Some of the first real-time tools were developed for instrument/sound building, allowing manipulation of values sent to the chip, audio feedback, and a display of values on the video output. These values would then be entered as data statements into text files.

### 19.3.5   Rebirth of an Industry

After the rise and fall of Atari with its 2600 and the original killer app, Pong, the home video game market was a virtual wasteland. Several other toy and electronics companies had attempted to make their marks, but each met with a public unwilling to invest in another piece of pricey hardware, yet another incompatible platform, and a limited software title library.

Not until Nintendo released NES did the game world as we now know it come to be. Based on its killer hits Donkey Kong, Zelda, and Mario Brothers, this low-cost, innovative system captured the post-Atari home game market in a single season. It had the right price point, attractive titles, and innovative applications of the 8-bit technology.

## 19.4   SIXTEEN-BIT, CD-ROM, AND BEYOND— THE HEART OF THE INDUSTRY

The home entertainment industry has always been plagued with market dips marking the passing of old technology and the coming of new. As a platform technology matures, the tool set becomes sophisticated and a new generation of talent and creativity arises. Such has been the case with the 16-bit dedicated game platform and the introduction of CD-ROM to personal computers.

### 19.4.1   Sega Genesis

The Sega Genesis is the leading 16-bit game platform, boasting the largest software title base since the original Nintendo system. Based on a Motorola 68000 as the main CPU running at 8 MHz, the Genesis outperformed any game system that came before as well as its later competitor, the Super Nintendo. Optimized for speed side-scrolling graphics, the Genesis took the previously defined video game play metaphors to a new level of velocity.

### 19.4.2   The Next Step Enhancements

The graphic capabilities of the Sega Genesis were not that big a step regarding video output. Basically, the Sega Genesis has an 8-bit video circuit that is able to send multiple layers or video-planes to the screen simultaneously. Each of these v-planes can contain 16 separate colors, organized as a palette, with one of the colors as "null" to allow the v-planes behind to show through. Sprite animation and manipulation is bigger and faster due to the CPU's capability.

The audio chip in the Genesis is a standard six-voice FM synthesizer developed by Yamaha. Each voice consists of four operators or subsystems, each having its own programmable envelope. The four operators can be configured in multiple combinations or algorithms to create a complex variety of sounds. There is also the ability to use different sounds or "patches" on each of the six voices, allowing multitimbral performance of music. One of the voices can also be configured to play low-sample-rate digital audio at a limited number of playback rates. These would be used for sound effects, voice, or percussion sounds.

### 19.4.3   Tool Development

Tool development for the Genesis or any other dedicated entertainment system is the most crucial factor in developing a large catalog of software titles. Tools need to be friendly to the intended users—be they artists, musicians, or game designers. The main purpose of the tool is to bring the most expert talent closer to delivering usable assets for the programmer. A programmer who creates his or her own tools for the content people he or she is working with has the power to customize and optimize this relationship. For instance, the GEMS tool for developing music on the Genesis addressed the functionality of off-the-shelf music sequencing software and, as a result, tapped into a broad, preexisting talent pool.

### 19.4.4   The Development of the Talent Pool

The talent pool for consumer electronic entertainment devices is small but varied. As mentioned earlier, musicians and artists who are already familiar with computer-based tools very easily adapt to the entertainment software development environment. Programmers come from backgrounds as diverse as aerospace, biomed, applications developing, etc. But a surprising amount of game development talent comes from the younger generation postgraduates. Many of these talented programmers get their first jobs in the test departments of larger game companies and have grown into programmers or game producers. Game designers can come from several of the more traditional backgrounds such as toy companies, board games, screenplay writing, etc. Imagination is the premium element here and should not be written off because of seemingly unusual background.

### 19.4.5  Super Nintendo

Similarly, Nintendo countered Sega's 16-bit ante with the Super Nintendo Entertainment System (SNES). Though based on a slower, weaker processor, the SNES development design depended heavily on the various coprocessors in the system. Dedicated hardware for handling video output, as well as graphic scaling and rotation, helped the SNES achieve comparable marketability against the Genesis. Other enhancements such as a 256-color board and an eight-voice sample-based PCM audio chip were considerable steps ahead of the Genesis.

### 19.4.6  Sega CD

The next platform release from Sega was the Sega CD, a peripheral to the original Genesis. Besides introducing CD-ROM technology and CD audio quality, the Sega CD system doubled the main processing power by adding another Motorola 68000 to the array, this one running at 12 MHz. Originally designed to manage the new add-on resources of the Sega CD, the second CPU had limited ability to communicate with the main CPU on the Genesis. The programming strategy called for each processor to handle the duties in its realm and only use the communication bus for staying in sync. The CD side of the system had upgrades similar to that of the SNES. Visual enhancements included dedicated scaling and rotation hardware; CD controller circuits; and an eight-voice, sample-based PCM audio chip. The latter was in addition to the FM synthesizer chip on the Genesis, creating quite a variety of sound generation.

### 19.4.7  CD-I, CD-TV, Peripherals, and Others

Other systems in the market at the time were also 16-bit systems based on desktop counterparts. The Philips CD-I was also 68000-based but suffered from a lack of dedicated software development. As a television-based player system, it could hardly compete with multimedia PCs and Macs running similar titles. The CD-TV was an Amiga development and, for all intents and purposes, was an Amiga in a box bundled with a CD player. Originally designed to tap into the Amiga-based game market, the CD-TV did a small launch in Europe but nothing significant in the United States.

## 19.5  ART AND MUSIC: CROSS-DISCIPLINE DELIVERY

In developing entertainment titles for any of these dedicated home platforms or desktop systems, the concepts of overall design must be comprehensive as well as cohesive. Piecing together a software title from any element evident to the user be it in design, graphics, video, or music detracts from the overall experience or playability. This is the most significant element in the overall success of the electronic game industry.

### 19.5.1  Composition for Film, Video, and CD-ROM

Composing music for multimedia and CD-ROM differs from film and video in that the technology or interactivity allows the art form to exist in many new directions. The main difference is that music for linear media is a passive, sequential experience, whereas music in an interactive product can be driven by the user's actions or the computer's reactions, or generated by the system from random or algorithmic procedures. Also knowing the technology of the music and sound playback system greatly affects the composing process. Whether writing for FM synthesizer, sample-based system, or streamlining CD, each element has its strengths and weaknesses and, depending on the application, must be juggled to perform effectively.

### 19.5.2  Art Direction

One of the most dominating elements in a software title, overall art direction must be complementary to itself and the product. Choice of colors, textures, backgrounds, interfaces, etc. contribute greatly to the success of a title. In some cases, graphics may be the only element that the title will be known for and the main reason for its success.

### 19.5.3  Relating with Game Design

Game design, like so many other elements in the game development process, was originally in the realm of the programmer/designer. With other duties being shared by dedicated professionals, game design became the work of storytellers. Tying the design to the graphic style and audio design is an underrated process in game development: it can sometimes yield the make-it-or-break-it quality that makes a title a hit. Sharing the larger, creative vision is a goal reached by exceptional communication.

### 19.5.4  Work Groups—Dream System Game Design Project

The most effective of small team development can be divided into five roles: producer, art director, writer/designer, audio/musician, and programmer. Recently, because of the overabundance of digital video and animation, the video producer and dedicated animator or 3-D artist began to take a larger role in this group relationship. Communications in this team must be open between every member, and not just in the traditional, vertical organization found in linear media. The producer has the final word on tough decisions, but a good producer drives the strengths of the individual players and is able to coordinate all of the resources into a single, inspired vision. This type of organization is crucial to the game development process.

## 19.6  MIDI AND ANIMATION—THE KEYS TO THE KINGDOM

For the creative side of software production, the two most important skills that have brought the outside talent pool are the knowledge of MIDI specifications and animation techniques.

### 19.6.1  Synthesizer Basics

Understanding the fundamentals of music synthesis is the first step in understanding the need and applications of MIDI. The simple tone module of a sound source, sound modifiers, and control input is the basis for all digital sound source. Manipulation of the ADSR (attack, decay, sustain, release) generators is primary in what gives each synthesized sound its individual characteristics. The control of pitch and expression in a multitimbral setting is where MIDI control comes into play.

### 19.6.2  MIDI Sequencers and SMPTE

MIDI is a low-level digital specification that allows real-time control of music synthesis. As a file format it is time-coded to allow for playback. The spec is based around single, serial statements such as note on, note off, pedal on, pedal off, etc. Each of these statements is coupled with a channel (1–16) and sometimes a pair of values (1–128). It is through this system that all modules or computer-based hardware can be made to respond. The most common spec for synching MIDI or other recorded music on tape to video is the Society of Motion Picture and Television Engineers (SMPTE) spec. This is a time-based striping for video playback that addresses hour: minute: second: frame. Professional MIDI interfaces will support reading and writing SMPTE code connecting MIDI data to videotape.

### 19.6.3 FM and Sample Sound–Editing Tools

FM synthesis requires editing individual "operators" that function like small synthesizers. These are configured in algorithms that allow multiple ways of modulating the frequency output of each operator. The codes for these patch configurations, by comparison, are very small, yet they are capable of describing complex instrumental sounds. The most effective of these editing tools are desktop computer–based and download the data to the synthesizer or target system.

Sound-editing tools are required for preparing digital audio data. Digital audio is described in an $X$-$Y$ axis with $X$ being the amplitude and $Y$ the time. Frequency of sound is recorded in time by a sample rate of twice the desired frequency to be captured, typically 44 kHz, which is twice that of the normal limit of human hearing, around 22 kHz. Data are conserved by reducing the sample rate and thereby reducing the sound quality. The amplitude data are stored in 16-bit values but, to conserve data, they can be captured as 8-bit values, also greatly reducing the quality. The more professional digital audio editing tools allow a graphic interface showing the data as wave shapes with cut-and-paste functions.

### 19.6.4 Video and Animation

The use of animation and video have come a long way in interactive entertainment. The visualization of animated sprites as characters over a scrolling background seems simplistic compared to the ability to stream digital video from CD-ROM. This is accomplished through several graphic images streamed (DMA) from a disc and mapped to video output. Compression issues for this implementation are as simple as reduced screen size, reduced frame rate, reduced color palette, and more complex solutions such as actual compression algorithms dealing with palette and delta state.

## 19.7 THE NOW SYSTEMS: 32- AND 64-BIT

The next generation in home entertainment systems sees a blurring of the dedicated game console and the home desktop computer. Titles are being developed for both platforms and market trends are being projected for both to exist in the same households, each with similar and sometimes overlapping functions.

### 19.7.1 The New Workstations for the New Platforms

Looking at the new high-end graphic work environments such as Sun and Silicon Graphics, a resulting generation of development tools has been developed. Although many of these applications are signature products for their native hardware, porting to other platforms has ensured hardware independence for the future of software development tools. Specifically, SoftImage, a leading 3-D rendering system for the SGI, after being acquired by Microsoft, will be available in a Windows NT version. In many development groups, hardware such as the SGI has been looked at as "one-box" solutions for the total emulation of new game platforms. Economics and the installed base of PCs determined the continuation of external target emulators for most development communities.

### 19.7.2  The Now Generation

The next generation machines again seem to be divided up into two categories: CD-ROM–based and cartridge. These machines take consumer electronics into a level of technology that rivals professional desktop systems and even some high-end workstation capabilities.

### 19.7.3  The Atari Jaguar

The earliest entry into the next-generation arena was the "64-bit" Atari Jaguar. Because most new 32-bit machines would be designed with dual- or multiple-processor configurations, a machine like the Jaguar with two 32-bit CPUs could be considered a 64-bit system. Poor developer relations and a lack of tools and titles doomed the Jaguar to an early grave even though it was a technically superior machine for its time. The legacy of Atari has once more been relegated to the history books, its inventory once again in the hands of discount resellers.

### 19.7.4  The 3DO System

The 3DO system was another early entry of formidable technology and radical business plan, but a poor show in the overall game market. The marketing of this 32-bit, CD-ROM–based machine was that of a multiplayer, attempting to cash in on both the PC-based multimedia market and the console-based video game market. The video quality of a television-based system could not compete with the clarity and sharpness of desktop computer screens. There were a large number of titles developed for the 3DO, but the system itself is destined not to be a leading consumer platform. Both the 3DO and the Philips CD-I systems, incidentally, have been remarketed as training systems for large corporate organizations. Because of the friendly development systems of each (Macromind Director of Macintosh for the 3DO and Amiga for the CD-TV), the player boxes still serve a useful function for distribution of custom interactive presentations.

### 19.7.5  The Sega Saturn and 32X

As the former leader of the pack, Sega had two main strategic entries into the 32-bit arena. The first was the 32X, an add-on "accelerator" to the Genesis/Sega-CD system. This was essentially a competitive move to bring enhanced video output and increased game-processing speed to the installed base of the Genesis. Some games were developed to include the Sega-CD configuration, but by this time the total system was looking very much like the new Saturn.

The Sega Saturn was intended to be the new flagship of the Sega fleet, but because of customer stubbornness to move up to a new nonupwardly compatible machine and because of market confusion about the 32X, a certain amount of steam was already out of this move. Based on a twin Hitachi SH-2 32-bit RISC processor (as was the 32X), the Saturn was engineered from a collection of predesigned multiprocessor, multifunction systems. The audio chip is a new design by Yamaha—the Saturn Custom Sound Processor (SCSP), powered by its own dedicated 68000 with a half a megabyte of work RAM. Also in the array are dedicated video processors for 3-D rendering and the handling of streaming video. A double-speed CD drive tops off the system, and the Sega development community has created several top title and original game products for the system. An early-in-the-year release was thought to give the Saturn a big lead as the new home game console, but because of reasons explained earlier and the anticipated release of the Sony Playstation, Sega was reportedly outsold 10 to 1 by Sony by the end of the following season.

### 19.7.6  The Sony Playstation

Because of Sony's capability as a consumer electronics manufacturing giant, it was possible to design the 32-bit Playstation from Sony's own parts catalog and engineer it to be smaller, lighter, and less expensive to mass produce than the Sega Saturn. With system capabilities nearly identical to the Saturn, the only other obstacle for Sony has been product development of software titles. Many of the third-party publishers and their development groups have planned for both platforms, giving the Playstation as big a title base, initially, as the Saturn. Typically, the specs for these systems include 256 polygons, texture mapping, 16-bit digital audio, 32 voices, built-in digital signal processing for audio processing, and dedicated hardware for 3-D manipulation and compression. Tool support for the Playstation is under direct licensing from Sony Computer Entertainment America and includes stand-alone target emulation systems and Macintosh-based stand-alone audio development tools. The access ports on the Playstation would allow support of (1) modems for use as a dedicated Web-browsing box, and (2) accelerators for MPEG video playback.

### 19.7.7  The Nintendo 64

Just shipped as of this writing, the Nintendo 64 has made a formidable entry into the game market. Launched late in the 32-bit market, it has several seasons of title development to catch up on. One advantage is the much touted processor array codeveloped by 3-D graphics giant Silicon Graphics Inc. Another is the fact that it is cartridge-based. This approach was taken because of the cartridge games' extreme playability due to instant access of data. Also, the elimination of a CD drive lowers the price point of the hardware system, making it easier for first-time consumers. Judging from titles in the release, the machine has the expected amount of 3-D polygon capabilities and 16-bit sound quality, although the latter is severely limited by the storage on the cartridge-based delivery system. Because of the limited success of processor-enhanced SNES titles and the short-lived Virtual Boy portable monochrome 32-bit system, Nintendo will have to make a strong show of the 64 to regain its previous position as a video game leader.

## 19.8  GAME DESIGN AND THE FILM INDUSTRY

One of the biggest buzzes in both the game development community and the entertainment industry in general has been the fusion of interactive games and cinema. Positioning itself to own most of the content, the big film and communication companies are now major players in the game publishing market. Several of these studios have invested heavily in their own internal development groups, attracting several of the veteran game producers, programmers, and designers. The larger companies, however, still must go through the early growing pains that a lot of smaller "garage band" developers have worked out years ago, tending to level out the playing field. The concepts of applying linear cinema stories to interactive systems also has a long way to maturity. What the entertainment industry does hope to do is lend a higher sense of production value to gaming. The game industry, on the other hand, needs to distill the production process for the entertainment business at large in making it aware of the fact that a good game does not need all the bells and whistles to make it a good game.

## 19.9  CONCLUSION

The simplest game design can be elegant to the point of needing only the most minimal graphics and sound. There have been several return-to-basics type titles in development lately that harken a return to the golden age of electronic gaming. What the future will bring is a maturity of both the applications of advanced technologies and the creative processes needed to make them entertain. Many have compared this era in entertainment software development to the early days of film, when sound was first introduced. Even looking at the art form from the metaphors of past art (film games, music recording) may be completely oblivious to a new direction in interactive entertainment that has yet to present itself.

## ABOUT THE AUTHOR

David Javelosa has been involved in interactive entertainment nationwide as a composer, producer, educator, and technologist since the beginning of digital media. He held the position of senior music designer at Sega of America, where he established the music facilities for the Sega Multimedia Studios, as well as composed and designed for several game products, including the Sega-CD, Jurassic Park, and the Sega Channel. Prior to Sega, he was at the Voyager Company, where he produced Schubert's Trout Quintet and did the MIDI design of Microsoft's Multimedia Beethoven. He has independently worked on video game titles, interactive media, commercial recordings, a multimedia opera, and the occasional chamber performance over the past few years. Currently, he writes for *Interactivity* magazine and is involved in production and consulting for a variety of next-generation platforms including Nintendo 64, Sony Playstation, and PC CD-ROM. He has developed projects for clients including Disney Interactive, Microsoft, and Nintendo. He is also the author of *Sound and Music for Multi-Media: Audio Effects for Interactive Entertainment* and serves as audio director for Inscape.

# CHAPTER 20
# DIGITAL VIDEOCASSETTE RECORDERS

**Mikhail Tsinberg**
*Advanced Television Technology Center*
*Toshiba America Consumer Products*

**Kenji Shimoda**
*Multimedia Laboratory*
*Toshiba Corporation*

**Jack Fuhrer**
*Research & Development Division*
*Hitachi America, Ltd.*

## 20.1  INTRODUCTION

This chapter will describe new digital technology that is going to be used for tape recording video and audio signals in the home. Although it may not be obvious why digital is a step forward compared to established analog videocassette recorders (VCRs), such as those using the VHS format that is the standard today, there are several reasons.

Analog VCR picture- and sound-recording quality depends largely on magnetic recording capability—e.g., tape and video head performance—and on the mechanical qualities of the tape deck. To achieve a reasonable cost/performance ratio for mechanical deck accuracy and tape quality, analog VHS VCRs limit recording bandwidth and, therefore, picture resolution, to about 3 MHz for luminance ($Y$) and less than 0.35 MHz for each of the chroma difference ($C_r$, $C_b$) components. This is about half of the broadcast National Television System Committee (NTSC) standard (4.2 MHz for $Y$, 1.5 MHz for $I$, and 0.5 MHz for $Q$). In addition, recorded analog video signals are highly subject to the analog noise inevitable in the analog recording process. The level of such noise will also vary with tape quality, age of the mechanical deck, and recording head conditions.

Digital recording eliminates some of these problems or transfers them from mechanical problems to more predictable and more controllable electronic problems. For example, in a digital VCR the recording bandwidth determines the bit rate as measured in megabits per second (Mb/s). The video resolution is controlled by a compression encoder that reduces the baseband (uncompressed) input video data rate to a recordable bit rate. As will be described in Section 20.3, the recorded resolution of the digital VCR is equal to or better than the broadcast quality standard in standard definition mode (SDTV) and achieves very good results in recorded resolution for extended definition (EDTV) and high-definition (HDTV) modes. In the advanced television (ATV) and digital video broadcasting (DVB) modes (Sections 20.6 and 20.7), digital VCR does not alter the quality of the broadcast signal in any way.

Analog and tape noise do not influence the resultant video and audio quality if the deck system performs within an acceptable error rate. Therefore, the video signal-to-noise ratio is also controlled by the quality of the compression encoder. The resultant resolution capability and noise performance of a digital VCR is of much higher quality than a typical analog VCR. The resolution and noise performance of the encoders also allow fewer losses than analog VCRs during the copying process from one deck to another.

Advances in recording heads and tape technologies has allowed for digital VCRs to use a smaller, 6-mm tape width (compared to 12.3 mm for VHS and 8-mm standards). Therefore, standard size (DV) and a mini DV (Section 20.3) digital cassettes are smaller than a standard 8-mm cassette and are considerably smaller than a VHS cassette.

To summarize, the digital VCR offers five basic advantages to the consumer: higher-resolution standard definition video, extended definition modes, better quality of copy, digital audio, and smaller cassette size. These advantages were understood by broadcasters who have used digital recording for the last 10 years with formats such as D1 and D2. However, only recent advances in recording heads, tape, and compression technology have made it possible to bring digital recording to the consumer market.

## 20.2   HISTORY OF SD, HD, ATV, AND DVB STANDARDIZATION PROCESS

### 20.2.1   First GI/Toshiba HDTV Digital VCR Demonstration

The first publications about digital consumer VCRs appeared in 1989 and 1991.[1,2] In April 1992 Toshiba and General Instrument performed a public demonstration at a meeting of the National Association of Broadcasters (NAB). These two companies demonstrated the world's first digital HD VCR prototype by combining advanced HDTV digital compression technology, developed by General Instruments, with one of the first 8-mm, cassette-based digital VCR decks, developed by Toshiba. The compressed data rate of the GI encoder was 17 Mb/s, and the usable recording data rate of the Toshiba VCR was 24 Mb/s. The digital HD picture quality was free from VTR noises, with greater than standard NTSC resolution, and the digital audio captured the attention of broadcasters and consumers.

### 20.2.2   SD/HD Conference

An HD digital VCR conference was established in September 1993 by most of the companies involved in the consumer VCR business. The VCR standard drafts for SDTV (standard definition TV signals) and HDTV (high-definition TV signals), using high-efficiency compression techniques, were finally approved by the Conference General Meeting in April 1994.

During this time the United States was involved in a new digital TV standard—advanced television (ATV). It made perfect sense to investigate whether an SD or HD digital VCR deck could be used for ATV recording and playback. The data capability of the SD deck (25 Mb/s) was just high enough to enable ATV (the ATV data rate requirement is 19.392 Mb/s) recording.

In 1994 the ATV Working Group (WG) started work to adopt the SD deck for ATV recording. The basic specification of ATV VCR was approved in April 1995. The final specification for ATV VCR is expected to be developed after the Federal Communications Commission (FCC) finalizes the ATV standard, possibly in the middle of 1997.

In Europe in November 1995, digital video broadcasting (DVB) adopted a single-channel recording format based on the ATV VCR format.

To address EDTV standards available in Japan and Europe, the extension SDTV modes were approved in November 1995. Table 20.1 describes the relationships between various modes and required recording bit rates. The standard bit rate (SBR) is 25 Mb/s. As Table 20.1 shows, the SDTV signal can be recorded in standard playback (SP) mode, long playback–narrow track pitch (LP-N) mode or long playback–low bit rate (LP-L) mode. PAL-plus or EDTV-2 signals can be recorded only by using the SDTV/SBR extension mode. The HDTV signal can be recorded only by using 2 · SBR or 50-Mb/s mode. The ATV signal can be recorded only by using SBR at 25 Mb/s. On the other hand, DVB signals can be recorded only by using any of the three modes but with some restricted conditions.

SD/HD recording standards were proposed to the International Electrotechnical Commission (IEC). The HD Digital VCR Conference was dissolved at the end of 1995.

**TABLE 20.1**  DVC Formats for Various Broadcasting Standards

|  | Worldwide | Worldwide | Europe | Japan | United States | Europe |
|---|---|---|---|---|---|---|
| TV signal type/bit rate | HDTV | SDTV | PAL plus | EDTV-2 | ATV | DVB |
| 50 Mb/s (2 · SBR) | Y | N | N | N | N | N |
| 25 Mb/s (SBR) − SP | N | Y | Y* | Y* | Y | Y |
| 25 Mb/s (SBR) − LP − N | N | Y(SDTV/LP − N) | N | N | N | N |
| 25/2 Mb/s (SBR/2) | N | Y(SDTV/LP − L) | N | N | N | Y |
| 25/4 Mb/s (SBR/4) | N | N | N | N | N | Y |

Y   = Draft standard defined.
N   = Draft standard not defined.
Y*  = Optional SDTV mode, not permitted alone.

### 20.2.3  Digital Interface

The digital interface is an important component of the digital VCR standard, especially in the case of digital-broadcasting standards such as ATV and DVB. The digital interface allows interfacing of only the recording part of the deck from ATV- or DVB-capable television. The encoding will be done by the broadcaster and decoding by the ATV or DVB television set. The minimum configuration of digital ATV or DVB VCR would be a digital data recording and playback device with encoding and decoding done by the broadcaster and the TV set, respectively.

The digital VCR conference recommended the IEEE-based p1394 digital interface standard. Full specification of this interface can be found in the IEEE 1394-1995 "High-Performance Serial Bus" document.

## 20.3  BASIC SPECIFICATION OF DVC—6-MM DIGITAL VCR

### 20.3.1  Basic Specification

The basic digital video cassette (DVC) format specifications are the standard definition TV (SDTV) signal and standard play (SP) mode. The SDTV is intended to accommodate current TV standards: NTSC, PAL, and SECAM. All three consist basically of two resolution/frame scanning rates: NTSC (525 lines/frame interlace, 59.94 fields/second) and PAL/SECAM (625 lines/frame interlace, 50 fields/second). The current digital studio (not compressed) standard, ITU-R601, accommodates both 525 and 625 scanning formats. The ITU-R601 standard clock rate is 13.5 MHz for the digital component, 4:2:2, 8 bit per sample/component video signals with a data rate of 216 Mb/s.

The DVC SP mode is used during normal speed playback or recording operations. Table 20.1 indicates the usage of 2 · SBR, SBR SP & LP-N, SBR/2, and SBR/4 recording modes for HDTV, SDTV, PAL plus, EDTV-2, ATV, and DVB systems. The HDTV ATV and DVB recording systems are described respectively in Sections 20.5 through 20.7. The SDTV/SP format specifications are presented in Table 20.2.

**TABLE 20.2**   Main Specifications of DVC

| Television system | SDTV/SP | HDTV |
|---|---|---|
| Tape speed | 18.831 mm/s | 37.594 mm/s |
| Track pitch | 10 μm | 10 μm |
| Minimum wavelength | 0.488 μm | 0.488 μm |
| Areal density | 2.44 $\mu m^2$/bit | 2.44 $\mu m^2$/bit |
| Tape | ME or equivalent | ME or equivalent |
| Video signal | Component: 4:1:1 (525/60); 4:2:0 (625/50) | Component 12:4:0 |
| Encoding parameter | 13.5 MHz, 8 bits | 40.5 MHz, 8 bits |
| Video compression scheme | Intra field/frame DCT | Intra field/frame DCT |
| | Feed forward control | Feed forward control |
| | Adaptive quantization (activity detection) | Adaptive quantization (activity detection) |
| | Modified 2-D Huffman | Modified 2-D Huffman |
| | Compression ratio 1/5 | Compression ratio 1/6.5 |
| Framing | Fixed area | Fixed area |
| Video rate | 24.948 MHz | 49.896 Mb/s |
| Recording rate | 41.85 Mb/s | 83.70 Mb/s |
| Modulation scheme | Scrambled interleaved NRZI | Scrambled interleaved NRZI |
| Error correction mode | Reed-Solomon product code | Reed-Solomon product code |
| Audio signal | 16 bit/48 kHz, 32 kHz, 2 ch or 12 bit/32 kHz, 4 ch | 4 ch or 8 ch |
| Recording time | Standard cassette—4.5 h small cassette—1.0 h | 2 h, 15 min 30 min |

DVC uses azimuth recording similar to that of analog VCRs (VHS, VHS-C, 8-mm). Selected tape width is a quarter-inch, or 6.35 mm. Recently developed metal-evaporated (ME) tape or its equivalents should be used.

This format can accommodate two types of cassettes. One is a DVC standard-size DV cassette and the other is mini DV. The maximum recording times in SDTV/SP mode are 4.5 hours for DV and 1 hour for mini DV. Standard DV cassettes can be used for stationary (desktop) units, whereas the mini DV would be more attractive for camcorder implementations.

Because the maximum recording bit rate is 24.948 Mb/s, input and output video processing and compression encoding and decoding must interface to a display and input video signals in a baseband or noncompressed format. The incoming video signal is processed in 4:1:1 for 525/60 or 4:2:0 for 625/50 digital components, with a sampling rate of 13.5 MHz, 8 bits/sample/component. These input sampling parameters are friendly to the ITU-R601 studio standard. The digital compression system is based on an intra field/frame discrete cosine transform (DCT) with modified two-dimensional (2-D) Huffman coding. This compression technique is cost-effective and, with a compression ratio of 1/5 (see Appendix A), is used to reduce the incoming analog video to a data rate of 24.948 Mb/s.

The audio signal is recorded in a noncompressed format. It is basically digitized with one of the four formats indicated in Table 20.3.

**TABLE 20.3**  Audio Encoding Mode in an Audio Block

| Mode | Sampling frequency | Quantization | Channel |
|------|--------------------|--------------|---------|
| 48k | 48 kHz | 16 bits linear | 1 |
| 44.1k | 44.1 kHz | 16 bits linear | 1 |
| 32k | 32 kHz | 16 bits linear | 1 |
| 32k-2 ch | 32 kHz | 12 bits nonlinear | 2 |

The actual recording rate is 41.85 Mb/s. Additional data—ID data, auxiliary data, subcode data, etc.—are used for protecting the recording data by error correction. In the digital recording, data may endure severe challenges from random as well as large burst errors possibly produced by missing magnetic tape material. Specially designed Reed-Solomon product codes that are used as error correction code (ECC) are generated and added to the compressed data, forming error-corrected data. Subsequently, error-corrected data are scrambled and modulated by the interleaved non-return-to-zero inverted (NRZI) method with one bit insertion in front of a 24-bit-long data pocket.

## 20.3.2  Tape Medium

The tape medium should be able to perform with the following characteristics. The DVC format uses wide-range recording frequencies from pilot signals for tracking with a wavelength of 20 μm to a maximum frequency for video and audio data recording with a wavelength of 0.49 μm. The tape medium should perform equally well in this range of wavelengths.

The tape will also need to perform an overwrite without an actual erasure head. This requirement is necessary because of the small size of the magnetic layer, as well as other requirements such as interchangeability and data rate requirements during still-frame, and lifetime for archival application.

ME tape technology has been proven to satisfy these conditions and was successfully used as a reference during DVC development. The chemical composition of its magnetic layer consists of an obliquely deposited Co-O layer. This tape medium has about 3 dB higher signal output than conventional Hi-8 ME tape. Of course, advanced MP tape can be used. It is less expensive than ME tape, so advanced MP tape was used for the digital VCR.

## 20.3.3  Cassette Specifications

The dimensions of DV and mini DV are shown in Fig. 20.1. In addition, semiconductor memory-in-cassette (MIC) is specified for optional use. MIC can record various information such as cassette type, data content, title of tape, etc. The data from MIC pass through special pin contacts or ID board terminals. Table 20.4 shows various MIC data with respect to the dedicated pin contacts.

In the DVC format there is no DV–to–mini DV cassette adapter such as VHS-C–to–VHS adapters used today for the VHS format. Instead, it is recommended to use DV- and mini DV–compatible mechanical decks if playback of both cassette sizes on the same deck is necessary.

## 20.3.4  Mechanical Deck

The DVC mechanical deck is basically the same as in an analog VCR. Given one tape format in DVC, it is possible to choose several drum diameters and rotation speeds. To accommodate main recording modes such as HDTV, ATV, SDTV, and DVB, a single rotation speed of 9000 rpm is

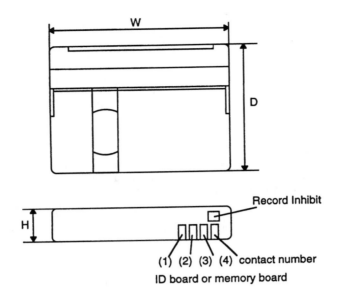

| | Size (mm) (W) X (D) X (H) | Recording time (min) |
|---|---|---|
| DV Cassette | 125 X 78 X 14.6 | 270 |
| mini DV Cassette | 66 X 48 X 12.2 | 60 |

**FIGURE 20.1**    Physical appearance of the digital videocassette.

**TABLE 20.4**    Assignment of the Four Contacts

| Contact number | | Cassette with ID board | | |
|---|---|---|---|---|
| | Assignment | Identification | Resistance value | Detecting resistance range of VCR |
| 1 | Tape thickness | 7.0 μm | Open | > 22 kΩ |
| | | 5.3 μm | 1.80 kΩ ± 0.09 kΩ | 0.8–2.9 kΩ |
| 2 | Tape type | ME | Open | > 22 kΩ |
| | | Reserved | 6.80 kΩ ± 0.34 kΩ | 3.7–13.2 kΩ |
| | | Cleaning | 1.80 kΩ ± 0.09 kΩ | 0.8–2.9 kΩ |
| | | MP | Short | < 0.49 kΩ |
| 3 | Tape grade | Consumer VCR | Open | > 22 kΩ |
| | | Non–consumer VCR | 6.80 kΩ ± 0.34 kΩ | 3.7–13.2 kΩ |
| | | Reserved | 1.80 kΩ ± 0.09 kΩ | 0.8–2.9 kΩ |
| | | Computer | Short | < 0.49 kΩ |
| 4 | GND | — | — | — |

**TABLE 20.5**  Specification of Recommended Scanner

| Dimension | Unit | 525/60 system | 625/50 system |
|---|---|---|---|
| Scanner diameter | (mm) | $\phi 21.7$ | ← |
| Scanner lead angle | deg. (°) | 9.150 | ← |
| Scanner rotation speed | (1/s) | 150/1.001 | 150 |
| Tracks/scanner rotation | — | 2 | ← |
| Effective wrap angle | deg. (°) | 174 | ← |

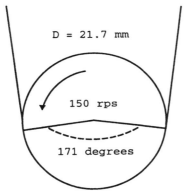

**FIGURE 20.2**  A single rotation speed of 9000 rpm is recommended with a drum diameter of 21.7 mm.

recommended with a drum diameter of 21.7 mm, as shown in Table 20.5 and Fig. 20.2. This rotation speed is five times faster than in current analog VCRs. The mechanical system has been technologically upgraded compared to current analog VCRs in order to accommodate necessary performance characteristics for the amount of noise and vibration and the length of the head life.

Other mechanical system innovations involve making stable movements of a thinner tape and enabling two cassette sizes to be played back on a single mechanical deck.

#### 20.3.5  Tape Format

The DVC SDTV/SP tape format is shown in Fig. 20.3 (view from a magnetic coating side viewpoint) and described in Table 20.6. Each single television frame is recorded on 10 tracks for the 525/60 system and on 12 tracks for the 625/50 system.

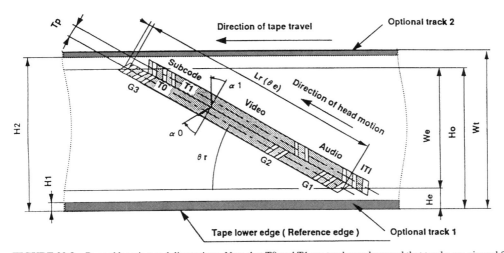

**FIGURE 20.3**  Record location and dimensions. Note that T0 and T1 are track numbers and that tracks are viewed from the magnetic-coding side.

**TABLE 20.6**   Record Location and Dimensions

| Dimensions | Nominal | Tolerance[†] |
|---|---|---|
| Track pitch ($T_p$) | 10.00 μm | Ref. |
| Tape speed ($T_s$) | $A*$ | ±0.5% |
| Track angle ($\theta_r$) | 9.1668° | Ref. |
| Effective track length ($L_r$) | 32.890 mm | ±0.122 mm |
| Tape width ($W_t$) | 6.350 mm | ±0.005 mm |
| Effective area lower edge ($H_e$) | 0.560 mm | ±0.025 mm |
| Effective area upper edge ($H_o$) | 5.800 mm | ±0.045 mm |
| Effective area width ($W_e$) | 5.240 mm | Derived |
| Optional track 1 upper edge ($H1$) | 0.490 mm | Max. |
| Optional track 2 lower edge ($H2$) | 5.920 mm | Min. |
| Azimuth angle ($T_0$) $\alpha 0$ | −20° | ±0.15° |
| Aximuth angle ($T_1$) $\alpha 1$ | +20° | ±0.15° |

$*A = 18.831/1.001$ mm/s for 525/60 system; $A = 18.831$ mm/s for 625/50 system.
[†] Tolerances shall be satisfied under all guaranteed conditions of the recorder. These tolerances shall be measured in the tape's standard environment. This table shows the values for recording the standard video signal.

The SDTV/SP format is subdivided into a four-pilot track sequence pattern. Each track is numbered in an incremental order from the beginning of each television frame, as shown in Fig. 20.4. In a 525/60 mode, one frame is recorded on tracks numbering from 0 to 9. In a 625/50 mode, one frame is recorded on tracks numbering from 0 to 11.

To achieve appropriate tracking performance, a four-pilot track sequence system was adopted. The pilot tracking signals are tracks F0, F1, and F2. These tracks are combined into a *pilot frame,* which corresponds to a television frame of 10 tracks in the 525/60 system. In the 525/60 system there are two types of track 1. In the pilot frame 0, track 1 is track F1, and in the pilot frame 1, track 1 is track F2. These two types of pilot frames alternate with each other in the 525/60 system. In the 625/50 system the situation is simpler. Only the pilot frame 0 type is used.

Figure 20.5 also shows the basic data allocation for this format. The first part of the lower edge on the tape is designated the insert and track information (ITI) sector. This sector is used for two functions regarding the track it is written on—for detecting the position of the editing and for tracking purposes in the insert mode. The second part of this sector is designated for the recording of an uncompressed audio signal. The third part of the sector is the video and audio area for the recording of compressed video and uncompressed audio signals. The last part is the subcode area, which is used for recording the subcode information signal.

As shown in Fig. 20.3, there are three special gaps called G1, G2, and G3 in each track. These gaps allow some tolerance for recording position deviation. Considering the jitter and the skew of the tape, G1 should be smaller than G2 and G2 smaller than G3.

### 20.3.6   Error-Correction Capability

To protect the recorded audio, video, and other data from the errors that are unavoidable in a tape medium, an erasure correction system is used with the following Reed-Solomon product codes:

For audio data, C1 code is (85,77) and C2 code is (14,9) over $GF(2^8)$.

For video data, C1 code is (85,77) and C2 code is (149,138) over $GF(2^8)$.

For subcode data, C1 code is (14,10) over $GF(2^4)$.

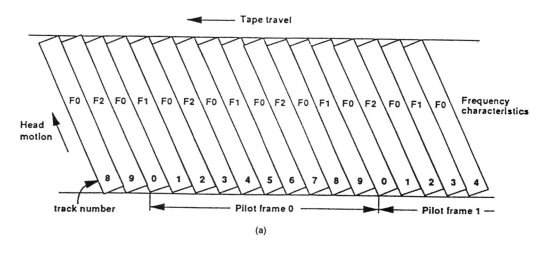

(a)

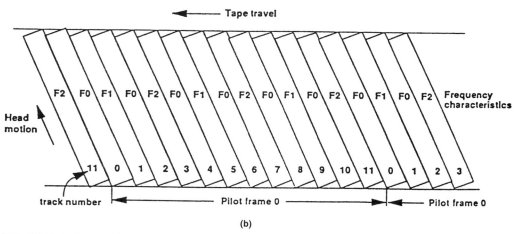

(b)

**FIGURE 20.4**   Frame and tracks (*a*) for the 525/60 system and (*b*) for the 625/50 system.

Theoretically calculated results for corrected video and audio symbol error are shown in comparison with DAT (digital audio system) in Fig. 20.6. Different types of tape damage that produce different kinds of burst errors are shown in Fig. 20.7.

## 20.4   SDTV FORMATTING AND ENCODING METHOD

### 20.4.1   SDTV System Image

Figure 20.8 shows a basic DVC block diagram. The video input signal is based on ITU-RS601 signals. As was described in Section 20.3, this format is based on a 13.5-MHz sampling clock

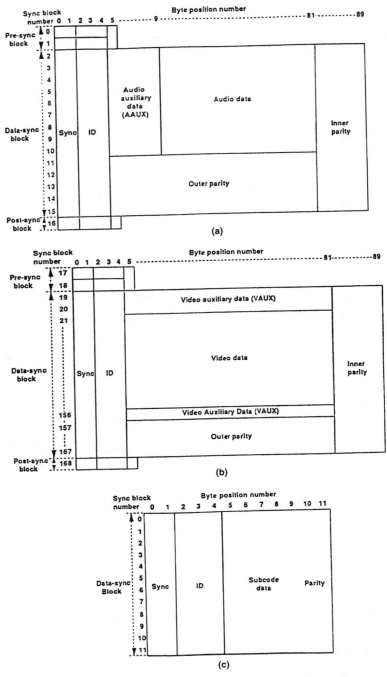

**FIGURE 20.5**    Structure of the sync blocks (*a*) in the audio sector, (*b*) in the video sector, and (*c*) in the subcode sector.

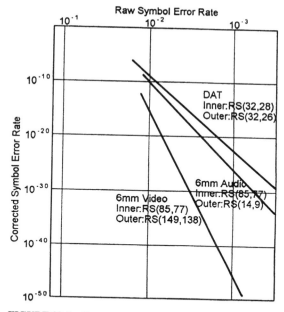

**FIGURE 20.6** Corrected error symbol rate.

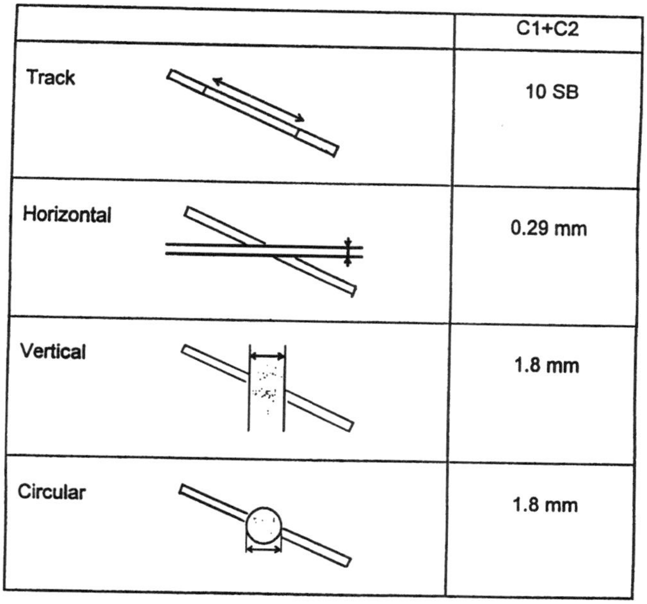

|  |  | C1+C2 |
|---|---|---|
| Track |  | 10 SB |
| Horizontal |  | 0.29 mm |
| Vertical |  | 1.8 mm |
| Circular |  | 1.8 mm |

**FIGURE 20.7** Burst error correction capability. Conditions: C2, 11 bytes erasure correction.

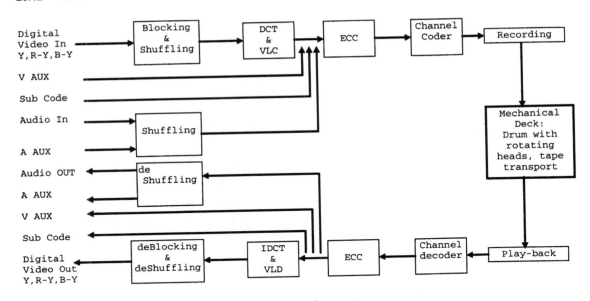

**Abbreviations:**

VLC – Variable Length Coding
VLD – Variable Length Decoding
ECC – Error Correction Coding
A AUX – Audio Auxiliary Data
V AUX – Video Auxiliary Data

**FIGURE 20.8** Digital VCR block diagram.

and 4:2:2 as $Y:C_r:C_b$ luminance/chrominance components sampling ratio. The overall data bit rate is 216 Mb/s. The 4:2:2 input signal is further subsampled to 4:1:1 for 525/60 and 4:2:0 for 625/50, with an overall bit rate of 125 Mb/s.

This subsampling takes place in the blocking-and-shuffling block, which is, in addition, shuffling the data to decrease system error impact. The DCT&VLC block performs video compression from 125 Mb/s to 24.98 Mb/s.

Compressed video data, noncompressed audio, and subcode data are formatted as described in Section 20.3.5. Error-correction codes are applied to audio, video, and subcode data in each ECC block. The channel coder block function is to scramble all input data, add one bit in the beginning of the 24-bit pocket, and modulate with an interleaved NRZI method as described in Section 20.3.1. The functionality of the additional one bit is to generate a pilot tracking signal, generate a dip if necessary, and to keep DC free for each track.

In the playback mode, inverse channel decoder, ECC, IDCT&IVLC, and deblocking/deshuffling operations take place.

## 20.4.2 SD Requirements

The following are DVC system SD requirements:

- Smaller cassette size with long recording and/or playback time
- Better picture quality compared with current analog VCR
- High-performance trick play (fast-forward, fast-reverse) modes

- Home-use editing capability
- Price reduction to achieve low-cost system

All items are related to compression subsystem architecture.

In principal the DVC system has adopted well-known intraframe coding for motion similar to that of the Joint Picture Experts Group (JPEG) (see Chapter 9). Because of frame-by-frame editing and trick (fast-forward and fast-reverse) capability, however, the details of DVC compression architecture are tailored for use in a VCR system. One of the basic reasons specified DVC was chosen is that trick play is not the same as the JPEG requirement, where fast-forward or fast-reverse trick play is needed to transverse each video track, and the compression decoder must be able to play back randomly scanned data from each track. The randomness comes from the fact that, during the trick play, the playback head crosses each of the tracks (recorded by a regular azimuth method) at a different angle than during normal playback. Therefore, the data are available only if the playback head is scanning the recorded track at the same azimuth as the recording head (see Fig. 20.9.)

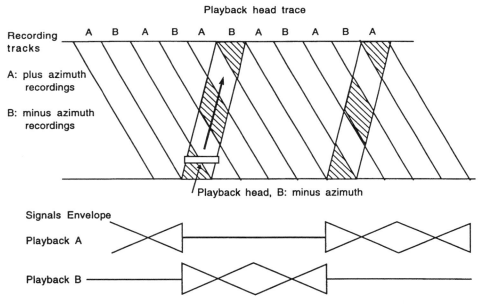

**FIGURE 20.9**  Playback area for trick play.

Intraframe compression allows instant decoding at the time when data are available, which makes the maximum amount of video available during trick play and keeps the trick-play image quality similar to or better than that of an analog VCR trick play. Straightforward application of the interframe compression system for digital VCR in a trick play presents a problem. Randomly recovered data would not be decodable in this case because the decoding would depend on history or previously encoded frames that would not be readily available during the trick play. This problem, however, is addressed and solved in a specific way applicable only in a digital broadcasting environment of ATV, as is described in Sections 20.6 and 20.7. An additional disadvantage of the interframe method for the SD analog-in, analog-out system is the complexity and subsequent cost of implementation of interframe encoder and decoder.

### 20.4.3   Video Compression System and Formatting

The characteristics of DVC compression and formatting are as follows:

- Intra field/frame DCT using motion detection
- Feed/forward bit-rate control
- Adaptive quantization using activity detection
- Modified two-dimensional Huffmann coding
- Special data allocation for trick play

Figure 20.10 shows the basic block implementation of the DVC compression system. The following explanation is done for the example of the 525/60 system. Basic functions performed during DVC compression are as follows.

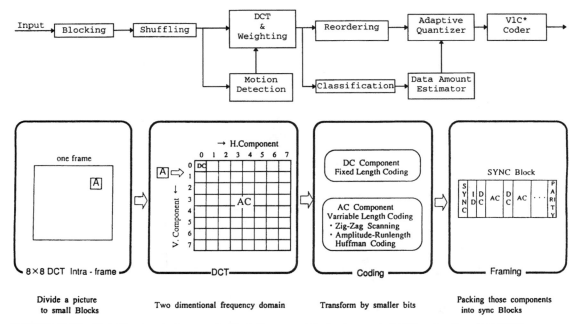

**FIGURE 20.10**   (*Top*) Bit rate reduction encoder block diagram and (*bottom*) concept of DCT system for a digital VCR.

***Block Division.***   In block division the image is subdivided into $8 \times 8$ (8 horizontal pixels by 8 vertical lines) blocks as shown in Fig. 20.11*a, b*.

The macroblock or MB (Fig. 20.11*c*) consists of four $Y$ (luminance) blocks and one each of $C_r$ and $C_b$ blocks ($C_r$ and $C_b$ are representations for R-Y and B-Y, respectively). Each of the $Y$, $C_r$, and $C_b$ blocks occupies the same area in one frame for the purpose of uniformity. For the same reason, every five MBs contain the same number of bits. The superblock (SB) consists of 27 MBs. Each frame consists of 50 SBs.

***Shuffling.***   Shuffling is done to reallocate blocks to reduce the impact of a possible error, as shown in Fig. 20.12. To maintain the fixed data rate without any picture degradation, five MBs'

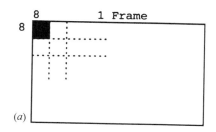

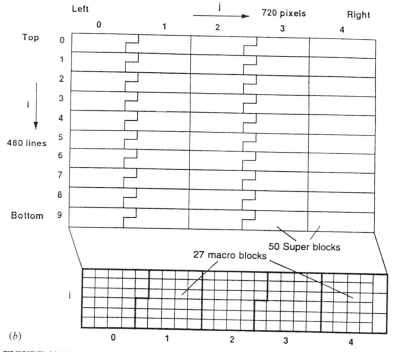

**FIGURE 20.11** DCT block intraframe. (*a*) Pixels of one frame partitioned in blocks of $8 \times 8$ pixels for *Y*, *R-Y*, and *B-Y*. (*b*) Super blocks and macroblocks in a frame on a TV screen for the 525/60 system.

data are combined under a known combination. Each frame's data are divided into five columns horizontally and 10 rows vertically to create an SB. Each SB consists of 27 MBs (0 to 26). There are basically three steps to arrange a 10-track (NTSC 525/60 mode) sequence and to form one picture-frame data area (Fig. 20.12).

- Five MBs including five sync blocks form 135 sync blocks in each SB.
- Five of the SBs are allocated to form the data area on one track.
- One picture area is formed by 10 tracks.

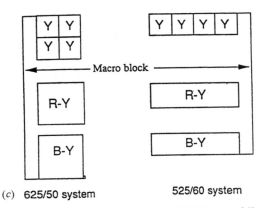

(c)  625/50 system

525/60 system

**FIGURE 20.11** (*Continued*)   (*c*) Four DCT blocks of *Y*, one DCT block of *R-Y*, and one DCT block of *B-Y* are defined as one macroblock.

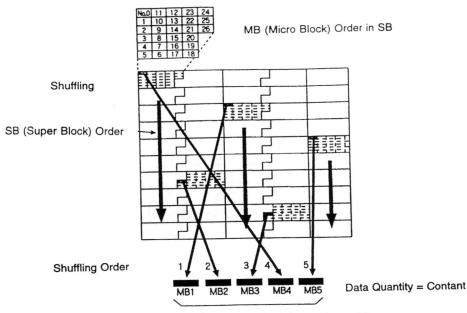

**FIGURE 20.12**   Shuffling is done to reallocate blocks to reduce impact of a possible error.

***DCT Operations.***    Each of the 8 × 8 DCT blocks are transformed into DCT coefficients. The DCT process has two modes, depending on the motion detector decision. One is a static picture mode (no motion) using 8 × 8 DCT (frame DCT) as a regular DCT in a manner similar to a regular JPEG method. In the case of detected motion, the DVC compression encoder uses 4 × 8 DCT (field DCT), as shown in Fig. 20.13. In the case of detected motion, the first-field DCT

2X(4X8) DCT Coefficients

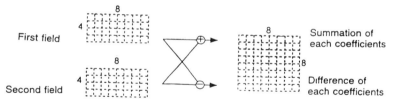

FIGURE 20.13   The digital videocassette has adopted the 4 × 8 DCT process.

coefficients (4 × 8) and second-field DCT coefficients are added and subtracted to form a special 8 × 8 DCT block. The result of the summation is located in the upper side, and the result of subtraction is located in the lower side of the special 8 × 8 DCT block. Figure 20.14 shows the weighting function $W(h, v)$ used in the compression decision-making process that derives from human vision characteristics.

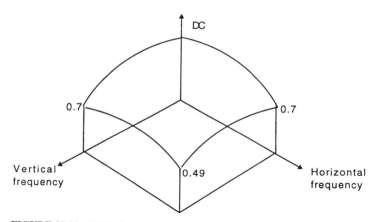

FIGURE 20.14   Weighting factor.

***Motion Detection.***   The two SDTV formats, 525/60 and 625/50, are interlaced. However, basic DVC compression architecture is based on a frame structure. In the case of a quick motion or camera panning, the field structure is more suitable for compression. In order to maintain a constant bit rate and constant (motion/no-motion–independent) recording position on the tape, DVC has adopted the modified 8 × 8 DCT process shown in Fig. 20.13.

***Classification.***   Table 20.7 gives general quantization criteria based on class and area numbers. The main purpose of this block is to select the quantization range according to DCT block activity. There are four classes of classified activity. Class 0 represents the lowest activity, which is the most sensitive for distortion. It uses a small value for its quantizer step. On the other hand classes 1, 2, and 3 represent high activity, which increases from 1 to 3. Specific decisions on quantizer steps for classes 1 to 3 are left as a manufacturer option.

**TABLE 20.7** Quantization Step

| | Class number | | | | Area number | | | |
|---|---|---|---|---|---|---|---|---|
| | 0 | 1 | 2 | 3 | 0 | 1 | 2 | 3 |
| | 15 | | | | 1 | 1 | 1 | 1 |
| | 14 | | | | 1 | 1 | 1 | 1 |
| | 13 | | | | 1 | 1 | 1 | 1 |
| | 12 | 15 | | | 1 | 1 | 1 | 1 |
| | 11 | 14 | | | 1 | 1 | 1 | 1 |
| | 10 | 13 | | 15 | 1 | 1 | 1 | 1 |
| | 9 | 12 | 15 | 14 | 1 | 1 | 1 | 1 |
| Quantization | 8 | 11 | 14 | 13 | 1 | 1 | 1 | 2 |
| number | 7 | 10 | 13 | 12 | 1 | 1 | 2 | 2 |
| (QNO) | 6 | 9 | 12 | 11 | 1 | 1 | 2 | 2 |
| | 5 | 8 | 11 | 10 | 1 | 2 | 2 | 4 |
| | 4 | 7 | 10 | 9 | 1 | 2 | 2 | 4 |
| | 3 | 6 | 9 | 8 | 2 | 2 | 4 | 4 |
| | 2 | 5 | 8 | 7 | 2 | 2 | 4 | 4 |
| | 1 | 4 | 7 | 6 | 2 | 4 | 4 | 8 |
| | 0 | 3 | 6 | 5 | 2 | 4 | 4 | 8 |
| | | 2 | 5 | 4 | 4 | 4 | 8 | 8 |
| | | 1 | 4 | 3 | 4 | 4 | 8 | 8 |
| | | 0 | 3 | 2 | 4 | 8 | 8 | 16 |
| | | | 2 | 1 | 4 | 8 | 8 | 16 |
| | | | 1 | 0 | 8 | 8 | 16 | 16 |
| | | | 0 | | 8 | 8 | 16 | 16 |

***Adaptive Quantization.*** Each of the DCT blocks (8 × 8, or 64 coefficients) is divided into five areas (Fig. 20.15). The first area, the DC coefficient, is recorded as is. The other 63 AC coefficients are divided into four areas from 0 to 3, according to frequency. The quantization step is always a product of the power of two, so hardware implementation of the quantizer is simple. Each area has the same quantization step, indicated by the "estimation of data quantity" block.

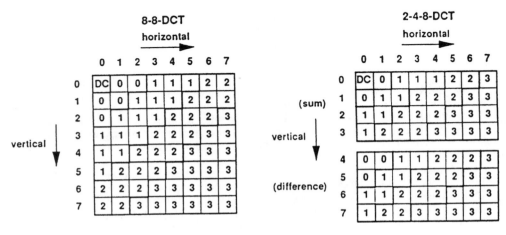

**FIGURE 20.15** Area numbers.

***Estimation of Data Quantity.*** The data rate must be constant in each of the five MB (a,b,c,d,e) blocks (see Fig. 20.16). These blocks are used to estimate feed-forward control. The classification fixed and area numbers are automatically defined by their coefficients. Table 20.7 shows how other criteria are used to decide the quantization number (QNO). To decide the optimum QNO, one of 16 quantizer options is selected before variable length coding (VLC). The largest quantizer (smallest quantization step) is selected to stay within the constant requirement of the bit amount target.

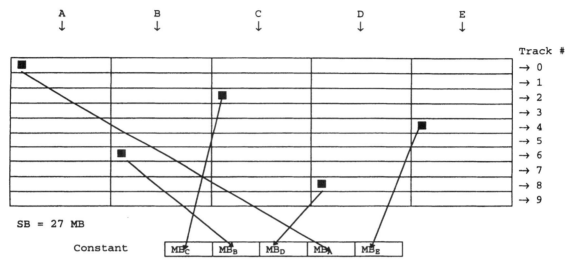

**FIGURE 20.16** Shuffling.

***Variable Length Coding (VLC).*** After quantization, the VLC operation is performed. The DVC VLC system has adopted two-dimensional Huffman code using the number of continuous zero and nonzero values that are located after the continuous zero values.

***Deshuffling and Formatting.*** Deshuffling and formatting operations are performed after the VLC (see Fig. 20.17). Five SBs (sync blocks) are selected from each respective column, but a different row (Fig. 20.16). Shuffling and deshuffling are needed to maintain good picture quality, especially during fast-forward and reverse modes, where large chunks of read data could be missing.

The recording order is from left SB or MB to right for each of the video tracks. Each following SB row is recorded into the next track. Because this is a segmented recording, the recording order is still similar to an analog VCR recording. Figure 20.16 shows data allocation in each of the MBs and SBs for tracks 0 to 9. Figure 20.17 shows data allocation in each of the MBs for the sequence of five MBs and one MB for one of the VCR sync blocks.

## 20.4.4  Audio Processing and Formatting

Table 20.3 describes four available audio modes. A 2-channel (2-ch) mode is available for 48-kHz, 44.1-kHz, and 32-kHz sampling frequencies using 16-bit linear quantization. The first channel of the 2-channel audio signal is recorded in the front of the continuous 5-track sequence.

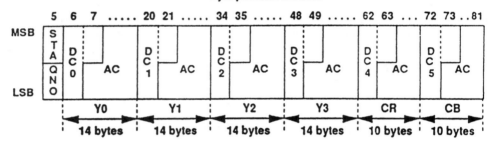

**FIGURE 20.17** (*Top*) Five sync block data and (*bottom*) the arrangement of a compressed macroblock.

The second audio channel is recorded in the back of the same 5-track sequence, as is shown in Fig. 20.18.

A 4-channel (4-ch) mode is available for a 32-kHz sampling frequency and uses 12-bit non-linear quantization. In this case, channel A and channel B are recorded into the channel 1 area. Subsequently, channels C and D are recorded into the channel 2 area.

Audio recording could be done in one of the "lock" or "unlock" modes. The "lock" mode means a tight synchronization between video and audio signals. The predetermined number of audio data could be repeated to create a target data period. "Lock" mode is available only for 48-kHz and 32-kHz sampling frequencies. The "unlock" mode is available for all audio modes. During this mode the amount of audio data can deviate within some restricted size. There is no need to synchronize in such a mode.

### 20.4.5 ITI Sector

The ITI sector is used to maintain stable tracking during editing and for indication of the data format.

### 20.4.6 Subcode

The system data in DVC are recorded in many sections, such as TIA, ITI, AUX in audio, AUX in video, and subcode area, as well as in MIC. MIC recording is optional. The TIA section carries the track information. AUX in audio and video data are used to decode audio and video data.

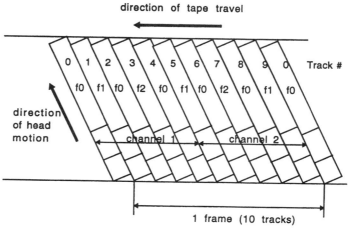

**FIGURE 20.18**    The location of the different pilot signals on the 10-track (525/60) sequence.

A special subcode sector has an ID and subcode data that include absolute track number, index ID, and title/time code data that can be used for a wide variety of purposes. The MIC is used for the table of contents and cassette reference information such as grade of the tape, tape type, tape thickness, etc.

Because there are many different data types in DVC, a special "pack structure" has been adopted (Fig. 20.19). One pack consists of 5 bytes, where the first byte is the header that is used to indicate the content and the other 4 bytes are used to indicate the data.

**FIGURE 20.19**    Pack structure.

### 20.4.7    Modulation and System Features

Figure 20.20 shows the bitstream structure of the 24/25 modulation method. The EB (extra bit) is inserted every 24 bits to generate a three-type tracking pilot signal. Figure 20.21 shows frequency

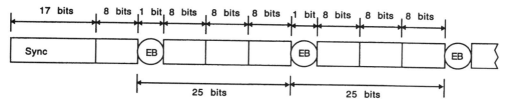

EB:  Extra Bit

**FIGURE 20.20**    Bitstream structure of the 24/25 modulation method.

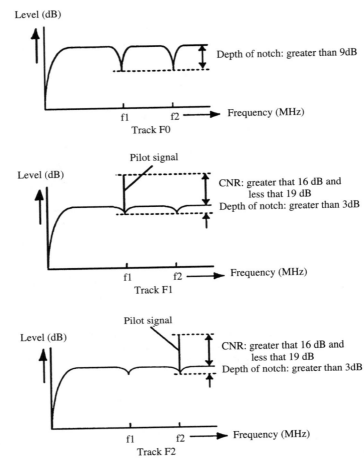

**FIGURE 20.21** Frequency characteristics of three pilot signals: F0, F1, and F2. (a) Note that F1 = Fb/90, F2 = Fb/60, and Fb = the frequency whose period is a time interval of one channel bit; Dc is free; and the resolution bandwidth kHz = Fb (MHz)/20,925 (MHz).

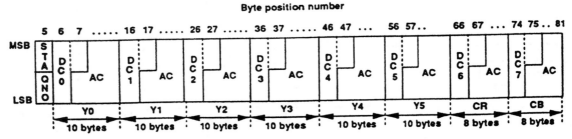

**FIGURE 20.22** The arrangement of a compressed macroblock

characteristics of three pilot signals: F0, F1, and F2. Figure 20.22 shows the location of different pilot signals on the 10-track (525/60) sequence.

## 20.5 HD FORMATTING AND ENCODING METHOD

### 20.5.1 HD System Image and Requirements

DVC formats can also support HDTV (high-definition TV) signal recording. There are two basic HDTV baseband studio standards—1125/60 and 1250/50. The 1125/60 standard will be used in countries now using the SDTV 525/60 standard. The 1250/50 standard will be used in the countries now using the SDTV 625/50 standard. Keeping the same 60- or 50-frame rate is important to minimize the costs of HDTV-SDTV-HDTV conversions. Because the NTSC rate is 59.94 Hz, the U.S. ATV (HDTV) standard can accommodate both 59.94-Hz and 60-Hz refresh rates.

Necessary conditions for the home use of HD-VCR are:

- Commonality with SDTV format
- Same user features as SDTV format
- High HDTV picture quality
- Commonality between 1125/60 and 1250/50 hardware in order to minimize cost to consumers worldwide

The input baseband data bit rate for HDTV is about five times higher than that of SDTV. Therefore, the recording bit rate of HDTV was selected to be 50 Mb/s. This recording bit rate will limit the maximum playback time on the standard DV cassette to 2 hours 15 minutes.

### 20.5.2 Video and Audio Compression System

To address basic parameters of 1125/60 and 1250/50 HDTV (HD) formats, the input sampling frequency was selected to be 40.5 MHz for the luminance signal and 13.5 MHz for the color difference signals. The $Y : C_r : C_b$ sampling ratio is 12:4:0. The compression ratio is 1/6.6, compared with 1/5 for the SDTV (SD) that is 525/60 and 625/50.

To maintain relative SD and HD picture qualities with different compression ratios, the quantization table and area division selections in the HD compression encoder are optimized to use more fine values. Another optional HD compression optimization is the "red detection method," which can improve subjective picture quality based on human visual characteristics, which are more sensitive to the color red.

The HD compression system has specific requirements for block division and shuffling. One HD MB consists of $6 \cdot Y + C_r + C_b$ DCT blocks. As a result, the number of horizontal and vertical MBs is different in comparison to the SD format.

The specifications for DCT and weighting, estimation of data quantity, and variable length coding are the same as in the SD DVC standard. The specifications for classification and motion detection are similar to SD, but optimized for HD.

Adaptive quantization parameters are given in Table 20.8. DCT AC coefficients are divided into eight areas instead of five (SD standard), with a finer quantization step.

In deshuffling and packing, the number of DCT blocks used in one MB is larger than SDTV/SP. This affects the data packing, as shown in Fig. 20.22. The 10-track recording sequence used in the SD 525/60 system is changed to a 20-track sequence for 1125/60 and to a 24-track sequence for 1250/50 systems, shown in Fig. 20.23.

In audio and subcode HD specifications, twice the number of channels are allowed as in SD. All other HD specifications are identical to SD.

**TABLE 20.8**  Quantization Number (QNO)

| Class number | | | | Area number | | | | | | | |
|---|---|---|---|---|---|---|---|---|---|---|---|
| 0 | 1 | 2 | 3 | 0 | 1 | 2 | 3 | 4 | 5 | 6 | 7 |
| 15 | | | | 1 | 1 | 1 | 1 | 2 | 2 | 2 | 2 |
| 14 | | | | 1 | 1 | 1 | 1 | 2 | 2 | 2 | 2 |
| 13 | 15 | | 15 | 1 | 1 | 1 | 1 | 2 | 2 | 2 | 2 |
| 12 | 14 | | 14 | 1 | 1 | 1 | 1 | 2 | 2 | 2 | 2 |
| 11 | 13 | 15 | 13 | 1 | 1 | 1 | 1 | 2 | 2 | 2 | 2 |
| 10 | 12 | 14 | 12 | 1 | 1 | 1 | 2 | 2 | 2 | 2 | 4 |
| 9 | 11 | 13 | 11 | 1 | 1 | 2 | 2 | 2 | 2 | 4 | 4 |
| 8 | 10 | 12 | 10 | 1 | 2 | 2 | 2 | 2 | 4 | 4 | 4 |
| 7 | 9 | 11 | 9 | 2 | 2 | 2 | 2 | 4 | 4 | 4 | 4 |
| 6 | 8 | 10 | 8 | 2 | 2 | 2 | 4 | 4 | 4 | 4 | 8 |
| 5 | 7 | 9 | 7 | 2 | 2 | 4 | 4 | 4 | 4 | 8 | 8 |
| 4 | 6 | 8 | 6 | 2 | 4 | 4 | 4 | 4 | 8 | 8 | 8 |
| 3 | 5 | 7 | 5 | 4 | 4 | 4 | 4 | 8 | 8 | 8 | 16 |
| 2 | 4 | 6 | 4 | 4 | 4 | 4 | 8 | 8 | 8 | 16 | 16 |
| 1 | 3 | 5 | 3 | 4 | 4 | 8 | 8 | 8 | 16 | 16 | 32 |
| 0 | 2 | 4 | 2 | 4 | 8 | 8 | 8 | 16 | 16 | 32 | 32 |
| | 1 | 3 | 1 | 8 | 8 | 8 | 16 | 16 | 32 | 32 | 32 |
| | 0 | 2 | 0 | 8 | 8 | 16 | 16 | 32 | 32 | 32 | 32 |
| | | 1 | | 8 | 16 | 16 | 32 | 32 | 32 | 32 | 32 |
| | | 0 | | 16 | 16 | 32 | 32 | 32 | 32 | 32 | 32 |

## 20.6  ATV

### 20.6.1  ATV System Image

The basic specifications of the ATV (Advanced TV) standard in the United States are defined as shown in Table 20.9. The HDTV video compression format is main level, high profile (MP@HL) as defined in the international compression standard MPEG-2. The ATV standard can also be used to carry SDTV video, defined by the main level, main profile (MP@ML) specifications of the MPEG-2 video standard. In this chapter, however, usage of DVC for the HDTV portion of the ATV standard is described. The audio compression standard is Dolby AC-3. The transmission format is MPEG-2 transport. Total transmitted ATV data rate is 19.39 Mb/s.

The recording of the ATV signal by DVC follows a different route than SD or HD recordings. The ATV signal is highly compressed by high-compression-ratio (factor of approximately 60 for highest HD format) encoders at the broadcasting studios. If decoded and re-encoded in the DVC, the data rate will be larger than the capability of SD or HD DVC decks. To achieve ATV recording, SDTV/SP DVC decks are used in a transparent configuration when DVC is basically recording and playing back data encoded by a broadcaster. In such a transparent system the ATV receiver video and audio decoders are used to display the broadcast program.

A digital transparent DVC system has both advantages and disadvantages. The advantages include the following:

- There is no need to have an ATV decoder/encoder in the DVC itself.
- The recording time is the same as in SDTV/SP VCR.
- The video is of broadcasting quality.

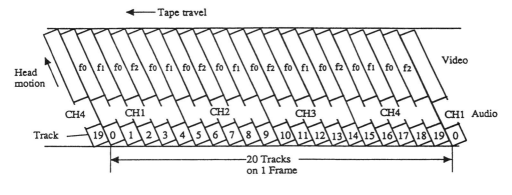

(a) 1125/60 system

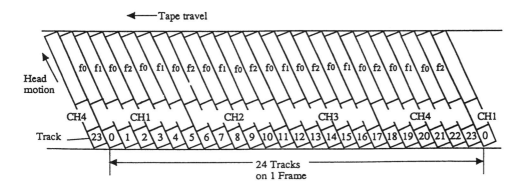

(b) 1250/50 system

**FIGURE 20.23** Frame and track.

Meanwhile, the disadvantages include the following:

- Trick-play picture quality is lower than in SD or HD systems.
- External to DVC encoder is needed for recording of baseband (not-encoded) video.

To achieve trick-play capability, trick-play data are duplicated and spread across many tracks. This method allows the pickup of discretely decodable data from across many tracks during trick-play head scanning. The trick-play data duplication and spreading process does not alter received broadcast data that are recorded without any actual decoding.

The data generated during trick play must be constructed in such a way that the ATV receiver decoder is able to decode and reconstruct the picture. However, the variable length code in the MPEG-2–encoded bitstream can decode only from the start code, for example, the slice start

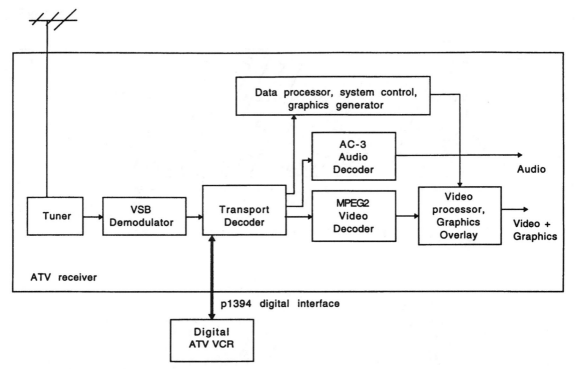

**FIGURE 20.24** Basic ATV receiver/DVCR configuration.

**TABLE 20.9** ATV Specifications

---

ATV Scanning formats:
  HDTV input formats, 16:9 display:
- 1920 pix × 1080 lines interlaced at 30/29.97 Fps
- 1920 pix × 1080 lines progressive at 24/23.98 Fps
- 1920 pix × 1080 lines progressive at 60/59.94 fps
- 1280 pix × 720 lines progressive at 60/59.94 Fps
- 1280 pix × 720 lines progressive at 30/29.97 Fps
- 1280 pix × 720 lines progressive at 24/23.98 Fps

  SDTV formats:
- 704 pix × 480 lines progressive at 60/59.94 Fps, 4:3/16:9 display
- 704 pix × 480 lines progressive at 30/29.97 Fps, 4:3/16:9 display
- 704 pix × 480 lines progressive at 24/23.98 Fps, 4:3/16:9 display
- 704 pix × 480 lines interlaced at 60/59.94 fps, 4:3/16:9 display
- 640 pix × 480 lines progressive at 60/59.94 Fps, 4:3 display
- 640 pix × 480 lines progressive at 30/29.97 Fps, 4:3 display
- 640 pix × 480 lines interlaced at 24/23.98 fps, 4:3 display
- 640 pix × 480 lines interlaced at 60/59.94 Fps, 4:3 display

| | |
|---|---|
| Transport | MPEG2 systems layer, ISO/IEC 13818-1 |
| Video compression | MPEG-2, MP@HL, and MP@ML, ISO/IEC 13818-2 |
| Audio compression | Dolby AC-3 |
| Auxiliary data | Format to be determined by ATSC |
| Data rate | 19.39 Mb/s |

---

Fps—frames per second for progressive scanning
fps—fields per second for interlaced scanning
pix—horizontal pixels
lines—vertical lines
4:3 or 16:9—display aspect ratio

code. Reconstructing the interframe-coded picture is difficult without having the history of the previous frame's data. Therefore, the best candidate for trick-play data is the intradata from the received MP@HL MPEG-2 bitstream. Such intradata does not require previous frame history and can be instantly or discretely decodable during random trick-play pickup. This intradata is duplicated and recorded to special areas of the tape to allow for fast-forward/reverse trick-play mode head scanning.

### 20.6.2 ATV Requirements

As was described earlier, the input data are MPEG-2 transport layer packets delivered from the interface of the ATV receiver (Fig. 20.24). For normal recording and playback, the input data are recorded sequentially in each sync block of DVC. Two transport packets (188 bytes × 2 packets) are mapped into five sync blocks (77 bytes × 5) of the DVC. This method is designed to protect error propagation and to detect the beginning of each transport packet easily. The rest of the packaged recording area is used for many different types of information regarding transport packets, such as time stamp, ID code to discriminate normal or trick-play data, and other data. This specific packaging scheme allows 54 transport packets to be recorded in each video track. Figure 20.25 shows a special 2-to-5 mapping system.

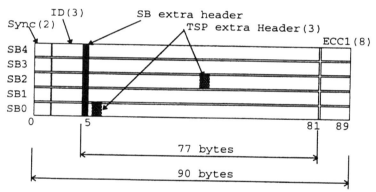

**FIGURE 20.25**   A five-SB unit where TP-1 is the first TS packet in the five-SB unit and TP-2 is the second TS packet in the five-SB unit.

### 20.6.3 Trick-Play Formatting and System Requirements

The ATV, as well as the DVB system, is designed to allow multiple forward and reverse speed depending on the number of trick-play data repetitions. ATV recording is based on a 25-Mb/s mode, as described in Tables 20.10 and 20.11. The basic idea is that each piece of trick-play data has its own recording data. Each trick play is recorded on either plus azimuth or minus azimuth and arranged in a specific sequence. This is basic to the DVC sequence, which consists of four tracks—the same as the tracking pilot signal sequence: F0, F1, F0, and F2.

The basic trick-play recording format is shown in Fig. 20.26. Each of the shaded areas is a trick-play area designated for specific forward and reverse speeds. There are two classes of speeds: TP-L for low speeds and TP-H for high speeds, as shown in Fig. 20.27. The possible variations of trick-play speeds are shown in Table 20.11. To record only the ATV standard, the

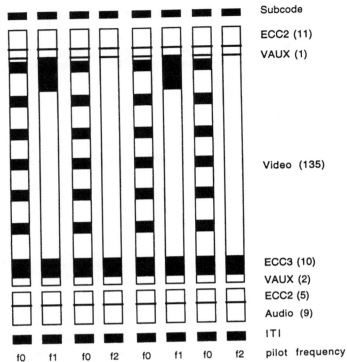

**FIGURE 20.26**    Track format for 25/12.5/6.25-Mb/s mode. ECC3 has 10 SBs in all tracks (21–30); TP-L has 25 SBs in all F1 tracks (131–155); TP-H has 30 SBs in six bursts of five SBs recorded in all F0 tracks (40–44, 62–66, 84–88, 106–110, 128–132, 150–154); VAUX as in SD (19–20, 156).

**TABLE 20.10.**    Number of Repetitions in Trick-Play Data

|        | 25 Mb/s | 12.5 Mb/s | 6.25 Mb/s  |
|--------|---------|-----------|------------|
| TP-H   | 18/36   | 9/18/36   | 5/9/18/36  |
| TP-L   | 2       | 1         | 1          |

25-Mb/s mode is used. The actual (nonrepetitive) compressed video data bit rate varies from mode to mode as well as from the choice of repetitions. The quality of the trick-play picture increases with a higher bit rate as it would for any compressed video. In order to properly align the scanning head with the trick play data during trick play, both speed lock and phase lock are used. Speed lock insures alignment of the scanning head with trajectories of the trick-play data, but not specifically with a particular tracking phase.

The phase lock goes one extra step and places the scanning head on the particular track phase during trick play. To achieve this kind of phase tracking, four pilot signals are used to control the tracking phase. This is one of the reasons that the trick-play data allocation is based on the basic four-track unit F0, F1, F0, F2. The manufacturer is free to decide what kind of speed- and/or phase-lock combinations of TP-L and TP-H should be included in its products.

**TABLE 20.11.**  Relation Between Number of Repetitions, Possible Search Speeds, and Maximum Bit Rate with Speed Lock and Phase Lock

| | | Speed lock | |
|---|---|---|---|
| Mode | Repetitions | Search speeds | Bit rate at speed |
| 25 Mb/s | 36 | $-17.5; -16.5; \ldots; +16.5; +17.5$ | 1.31 Mb/s at 17.5* |
| | 18 | $-8.5; -7.5; \ldots; +7.5; +8.5$ | 1.27 Mb/s at 8.5* |
| 12.5 Mb/s | 36 | $-35; -33; \ldots; +33; +35$ | 1.31 Mb/s at 35* |
| | 18 | $-17; -15; \ldots; +15; +17$ | 1.27 Mb/s at 17* |
| | 9 | $-9; -7; \ldots; +7; +9$ | 1.35 Mb/s at 9* |
| 6.25 Mb/s | 36 | $-70; -66; \ldots; +66; +70$ | 1.31 Mb/s at 70* |
| | 18 | $-34; -30; \ldots; +30; +34$ | 1.27 Mb/s at 34* |
| | 9 | $-18; -14; \ldots; +14; +18$ | 1.35 Mb/s at 18* |
| | 5 | $-10; -6; \ldots; +6; +10$ | 1.35 Mb/s at 10* |

| | | TP-H with phase lock | |
|---|---|---|---|
| Mode | Repetitions | Search speeds | Bit rate at speed |
| 25 Mb/s | 18 | 18 | 2.70 Mb/s at 18* |

| | | TP-L with phase lock | |
|---|---|---|---|
| Mode | Repetitions | Search speeds | Bit rate at speed |
| 25 Mb/s | 2 | $+4$ | 2.25 Mb/s at 4* |
| 12.5 Mb/s | 1 | $(-4; -2; +2) + 4$ | 2.25 Mb/s at 4* |
| 6.25 Mb/s | 1 | $(-8; -4; -2; +2) + 4(+8)$ | 1.12 Mb/s at 4* |

## 20.7   EUROPEAN DVB PROJECT AND A DVB DVCR

Digital video broadcasting has been investigated in Europe under the formal guidance of the European DVB Project since 1993. DVB is a joint project of more than 170 organizations from 21 countries. DVB is establishing consensus standards for digital video-broadcasting signals and equipment. DVB is addressing the requirements of various media, including satellite, cable, and terrestrial broadcast. Several SDTV and HDTV formats are being standardized.

### 20.7.1   DVB DVCR

The DVB DVCR is similar to an ATV DVCR in that it is a bitstream recorder. In other words, a DVB DVCR does not perform analog-to-digital conversion and then compression of analog signals. Instead, it records an incoming already-compressed DVB bitstream and later, on playback, delivers that identical bitstream to an external decoding device.

Because of the broad scope and multiplicity of DVB standards, multiple bit rates must be accommodated by a DVCR that is aimed at the diverse DVB market. The capability of recording and playing at several different bit rates is the major difference between an ATV DVCR and a DVB DVCR.

The DVB standard has adopted the MPEG-2 transport, audio compression, and video compression standards. DVB has added some extensions to the MPEG-2 systems layer.

DVB and MPEG-2 permit the several programs to be multiplexed into one transport stream. However, initially, a DVCR for DVB will record only one program at a time from a multiplexed multiprogram transport stream. This single-program restriction greatly simplifies the DVCR hardware and customer interface.

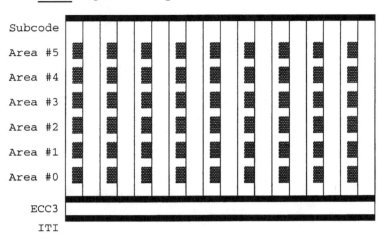

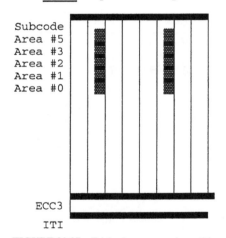

**FIGURE 20.27**   Trick-play areas and repetition cycle.

The standardization of HDTV, including its allowed bit rate, has not yet been completed by DVB. For SDTV broadcasting the maximum bit rates are restricted by MPEG-2 MP@ML to be 15 Mb/s. Many SDTV programs are expected to be broadcast at bit rates considerably less than 15 Mb/s.

In order to minimize unnecessary tape consumption for lower-bit-rate programs, DVB VCRs can operate in any of three modes. The modes are standard bit rate (SBR) at 25 Mbs, SBR/2 at 12.5 Mb/s, and SBR/4 at 6.25 Mb/s (see Table 20.10).

The instantaneous recorded bit rate is the same for all modes. This means that the range of recorded wavelengths is the same for all modes. This greatly simplifies the design of the recording heads. To achieve a recorded bit rate of SBR/2, the head-wheel rotation speed is held constant, whereas the transport speed is divided by two. The recording heads are alternately energized and deenergized on alternating rotations of the head-wheel. A short buffer is required to buffer the data for the head-wheel since it is recording at a SBR rate but only on a 50 percent duty cycle. In this way azimuth recording of tracks is obtained for both SBR and SBR/2 tracks. With this technique, SBR and SBR/2 tracks are the same width (unlike VHS SP and LP mode tracks).

To achieve a recorded bit rate of SBR/4, the head-wheel rotation speed is held constant, whereas the transport speed is divided by four (compared to SBR mode). The recording heads are alternately energized and deenergized. One rotation ON is followed by three rotations OFF and then the four-rotation sequence is repeated. A short buffer is required to buffer the data for the head-wheel since it is recording at an SBR rate but only on a 25 percent duty cycle. In this way azimuth recording of tracks is obtained for both SBR and SBR/4 tracks. With this technique, SBR and SBR/4 tracks are the same width (unlike VHS SP and EP mode tracks).

### 20.7.2  Formatting and System Features

The formatting and system features are basically the same as in ATV. There are extensions to identify the mode. Data areas that are reserved for trick play on an ATV DVCR may optionally be used for normal play in a DVB DVCR. This feature may be important because of the wide variation of bit rates that may occur in DVB. If a signal to be recorded has a bit rate of SBR/$N$ + $\alpha$ for $N = 1, 2, 4$, it may be desirable to operate the recorder at SBR/$N$ and record the extra $\alpha$ bits of normal play data in the area that is normally used for trick-play data.

### 20.7.3  Trick-Play Feature

The track format is shown in Fig. 20.26. It is the same for all modes. Trick play (fast visual scan) on a DVB-DVCR has all of the same characteristics of trick play on an ATV DVCR, plus the added complication of having to deal with three possible recording bit rates (SBR, SBR/2, and SBR/4) at normal speed.

To accomplish trick play of interframe-coded MPEG bitstreams on either an ATV DVCR or a DVB DVCR, specific sync blocks are reserved for trick-play data (Fig. 20.26). During normal playback, data in the trick-play sync blocks are ignored. Conversely, during trick playback mode, data in the normal-play sync blocks are ignored and only data in the trick-play sync blocks are processed. The trick-play data is repeated multiple times, so that, despite the uncertainty of the relative trajectory of the playback heads to the recorded data, it is guaranteed that each unique trick-play sync block is picked up by the playback heads at least once. The trick-play sync blocks are organized into two groups. Sync blocks designated TP-H are sync blocks whose placement is optimized for higher-speed search. TP-L are sync blocks whose placement is optimized for lower-speed search.

TP-H sync blocks are recorded repeatedly in groups of 30 sync blocks consisting of 6 bursts of 5 sync blocks each. TP-H sync blocks are recorded in all F0 tracks. The same 30 sync blocks are repeated in each F0 track for a number of repetitions, as shown in Figure 20.27. Note that there are two repetition options for the SBR speed, three repetition options for the SBR/2 speed, and four repetition options for the SBR/4 speed (see Table 20.10). The maximum search speed and the playback data rate at any given search speed depend on the number of repetitions of the TP-H data. Table 20.11 gives the possible search speeds, the maximum search speeds, and corresponding bit rates during playback as a function of normal-speed playback data rate and the number of repetitions of the TP-H sync block data. The playback data rate at a specific trick playback speed less than the maximum playback speed can be calculated as follows (see Table 20.11):

$$\text{Bit rate at SPEED} = (\text{SPEED/max. speed}) \cdot \text{Bit rate at max. speed}$$

In order to read the trick-play sync blocks successfully during high-speed playback, the capstan servo must meet some additional requirements beyond those required for normal-speed recording and playback. If the capstan servo can achieve precisely the speeds indicated in Table 20.11, then successful trick play can be accomplished. No specific phase relationship is required between the playback head trajectory and the trick-play syncblock pattern. In general, a group of five trick-play sync blocks will not be traversed in a single pass of the playback head. However, because of the specific speeds chosen, any portion of a five-syncblock grouping that is missed in a first pass will be traversed in a subsequent pass of the playback head over a group of five sync blocks containing the same data.

Consider the case of an SBR recording bit rate for normal speed and 18 repetitions of the TP-H data. If the capstan servo can achieve $18 \times$ speed lock and additionally lock to the specific phase, such that a complete group of five TP-H sync blocks is reproduced in a single pass of the heads, then trick play at $18 \times$ at the relatively high trick-play bit rate of 2.70 Mb/s can be obtained. Note that with 36 repetitions, a high search speed of $17.5 \times$ can be obtained with only speed lock. However if the DVCR has phase-lock capability and the 18 repetition mode was recorded, the picture quality of the $18 \times$ playback with 18 repetitions will be much better than the $17.5 \times$ with 36 repetitions. This is because of the much higher data rate that is achieved with the phase-lock servo capability combined with the fewer repetitions of the trick-play data.

A choice of number of repetitions is permitted because of the following trade-off that must be exercised. More repetitions of each unique trick-play sync block permit a higher maximum search speed. However, fewer repetitions of the trick-play sync block data give a higher trick-play playback data rate and corresponding better image quality on lower search speeds.

A servo with phase-lock capability as well as speed-lock capability permits high speed and high data rate as well as low speed and relatively high data rate during trick playback. The penalty is a slightly more costly servo.

TP-L sync blocks are arranged so that they are useful for relatively low search speeds and their use requires phase lock. TP-L data give a much higher data rate during playback at low speeds than does TP-H data. Although TP-H data can also deliver low-playback search speeds, the low-playback data rates achieved are barely acceptable, even for searching. The choices of speeds, repetitions, and playback data rates using TP-L data are as shown in Table 20.11.

An example comparing low-speed use of TP-H and TP-L is as follows. For SBR recording mode (25 Mb/s), $4 \times$ trick-play speed using TP-L with phase lock delivers a playback bit rate of 2.25 Mb/s (from Table 20.11). For SBR recording mode (25 Mb/s), $4 \times$ trick-play speed using TP-H with 18 repetitions with speed lock delivers a playback data rate of $(4.5/8.5) \cdot 1.27 = 0.672$ Mb/s.

## 20.8    OTHER DIGITAL VCR FORMATS

### 20.8.1    Digital VHS

Digital VHS is a digital VCR that uses $1/2''$ tape cassettes that are similar to standard analog VHS cassettes, but with SVHS quality tape. The tape format is roughly analogous to the $1/4''$ DVCR format. Most of the technical details of the DVHS format have not been disclosed as of this writing.

The digital portion of DVHS serves functions similar to those of an ATV or a DVB DVCR rather than an SD-DVCR. A DVHS performs neither compression nor decompression. A DVHS is a "bitstream" recorder/player. It records a compressed bitstream and plays it back at a later time. A principle feature of DVHS is that it is compatible with analog VHS tapes. In addition to its digital capabilities, it can play and/or record analog VHS tapes.

### 20.8.2 DVC-PRO

DVC-PRO is a ruggedized version of the SD-DVCR for professional use. The DVC-PRO track pitch is about 18 μm, 1.8 times the SDTV/SP mode. A new cassette size intermediate between the small and large DVCR cassettes may be used. DVC-PRO uses MP (metal power) tape, the same as DVC.

Recording times for the new M-cassette and the previous L-cassette are 63 min. and 123 min., respectively. The times are reduced because of the wider track pitch, which is used for added ruggedness and more dependable editing. DVC-PRO machine can also play DVC tapes, but the reverse is not true; DVC machines cannot play DVC-PRO tapes.

## 20.9  DIGITAL CAMCORDER

### 20.9.1  System Description

Digital camcorders were introduced in September 1995 by a few companies as a high-end product. Early models contained LSI made in 0.5-μm design rules with particular attention given to low power consumption.

### 20.9.2  Unique Differences Between Digital and Analog Camcorders

Some products have a digital interface using IEEE 1394 to connect other DVCR- and 1394-equipped peripheral devices.

Discussions are ongoing in several venues regarding copyright protection and DVCRs. Pending outcome of these discussions, DVCR camcorders will become available, but DVCR decks for playing prerecorded movies will not yet be available.

## 20.10  CONCLUSION

### 20.10.1  Convergence of PC and TV Consumer Digital Media

Worldwide convergence of TV analog broadcasting to digital is evident by establishing major digital broadcasting standards such as ATV in the United States and DVB in Europe and Japan.

Digital broadcasting provides advantages similar to those of digital VCR—namely, noise-free and high-quality video and audio for the TV viewer. However, there is one more and a very significant change: the information transfer, processing, and storage for digital TV systems and digital computer systems become identical. The implications of such unification are global. Digital TVs equipped with communication hardware and software compatible with computers could be connected to the information superhighway. And digital computers could receive large amounts of digitized information inexpensively through terrestrial, cable, or satellite digital broadcasts. It is probably fair to say that content previously developed for specific media (such as analog TV broadcasting and analog VCR or digital Internet and digital CD-ROM) will start to become more portable throughout all digital delivery media. For example, both a digital TV viewer and a PC user will be able to receive digital TV broadcastings, surf the Internet, and use high-density storage media such as digital video disk (DVD) and digital VCR.

### 20.10.2  Vital Role of Low-Cost, High-Capacity Digital Recording Media

Digital storage media such as the digital VCR will play a vital role in the world of global digital communications. At this time magnetic tape remains the lowest cost per bit digital storage medium. The mechanical deck used for digital VCRs is a very complex mechanical device. However, there is very little difference in basic deck technology used for analog or digital VCRs. The digital deck is a result of a very mature, cost-effective, and reliable technology perfected by years of analog VCR's competition. Both a cost-effective and reliable mechanical deck and a low-cost-per-bit magnetic tape represent a most attractive digital storage choice available today.

## GLOSSARY

**ATSC**   Advanced Television System Committee

**ATV**   Advanced television.

**DCT**   Discrete cosine transform.

**DV**   Digital video.

**DVB**   Digital video broadcasting.

**DVC**   Digital videocassette.

**DVCR**   Digital videocassette recorder.

**ECC**   Error correction code.

**EDTV**   Enhanced-definition television.

**HD**   High definition.

**HD D VCR**   High-definition digital VCR conference.

**HDTV**   High-definition television.

**IDCT**   Inverse discrete cosine transform.

**LP-L**   Long playback–low bit rate.

**LP-N**   Long playback–narrow track pitch.

**SBR**   Standard bit rate.

**SD**   Standard definition.

**SDTV**   Standard definition television.

**SP**   Standard playback.

**VLC**   Variable length coding.

**VLD**   Variable length decoding.

**WG**   Working group.

## REFERENCES

1. C. Yamamitsu, et al., "An experimental study for a home-use digital VCR," *IEEE Transactions on Consumer Electronics,* vol. 35, no. 3, August 1989.
2. M. Yoneda, et al., "An experimental digital VCR with new DCT-based bit-rate reduction system," *IEEE Transactions on Consumer Electronics,* vol. 37, no. 3, August 1991.

## BIBLIOGRAPHY

A. Ide, et al., "[DVC] Standards for Consumer-Use Digital VCRs," National Technical Report, vol. 41, no. 2, Apr. 1995 (in Japanese).

Y. Kubota, et al., *Digital Video Dokuhon,* Ohmsya, Tokyo, 1995 (in Japanese).

"Specifications of Consumer-Use Digital VCRs Using 6.3 mm Magnetic Tape," Blue Book, HD Digital VCR Conference, 1995.

M. Tsinberg, et al., "Data-placement procedure for multi-speed digital VCR," *IEEE Transactions on Consumer Electronics,* vol. 40, no. 3, Aug. 1994.

## ABOUT THE AUTHORS

Mikhail Tsinberg joined Philips Research Laboratories in Briarcliff Manor, N.Y., in 1983 as a member of the research staff, advanced TV department, in the area of HDTV systems development. Subsequently, he was promoted to the position of research department head, advanced TV department responsible for analog and digital HDTV system development and testing as part of the FCC advanced television process. He joined Toshiba America Consumer Products in 1991 as senior research manager responsible for the development of consumer digital technology of the ATV receiver, digital video disk, and digital VCR. He received the M.S.E.E. from Rybinsk Institute of Aviation Technology, Russia, in 1975. He holds 38 U.S. patents.

Kenji Shimoda joined Toshiba Corp. in 1980 as a member of the new videocassette recorder group in the consumer electronics laboratory. He has worked with VCRs such as the 8-mm VCR, digital VCRs, and digital processing systems for error correction and data compression. He is presently the manager of the digital recording and digital processing department of the Multi-Media Engineering Laboratory in Yokohama. His additional responsibilities include development of compression technology. He received the B.S. and M.S. degrees from the University of Osaka Prefecture. He holds 11 patents in the United States and Japan.

Jack Fuhrer joined the David Sarnoff Research Center in 1971 as a member of the technical staff, working on the RCA video disk system. In 1977 he joined RCA's Consumer Electronics Division in Indianapolis, Ind., as manager of signal circuits. He became director of the New Products Laboratory in Consumer Electronics in 1982, where he pioneered a variety of applications of new technology to consumer products. In 1985 Fuhrer returned to the David Sarnoff Research Center as director of television research and led a team in the creation of advanced compatible television. He joined Hitachi America, Ltd., in May 1991 as senior director of the Advanced Television and Systems Laboratory, where he is responsible for developing prototypes and LSI for digital advanced television products. He has a B.S.E.E. and two graduate degrees in electrical engineering, all from M.I.T. He holds 18 U.S. patents.

## APPENDIX A

**1.** Total ITU-R601 bit rate including nonpresentation (not affected by actual image) for total number of 858 pixels per line:

13.5 MHz is the sampling for 525/60 $Y$ signal; 858 horizontal pixels;

525 vertical pixels or lines

Total bit rate for $Y = 858H \cdot 525V \cdot 29.97$ frame/sec. $\cdot$ 8 bits.
Total bit rate for $C_r$ or $C_b = 858/2 \cdot 525 \cdot 29.97 \cdot 8$.
Total for $Y + C_r + C_b = 858 \cdot 525 \cdot 29.97 \cdot 8 \cdot 2 = 216$ Mb/s.

**2.** Effective total bit rate for image affected (720) pixel area with 4:2:2 $Y/C_r/C_b$ luma-chroma ratio:

$$720H \cdot 480V \cdot 29.97 \cdot 8 \cdot 2 = 165.722 \text{ Mb/s}$$

**3.** Effective total bit rate for 4:1:1 ratio:

$$720 \cdot 480 \cdot 29.97 \cdot 8 \cdot 6/4 = 125 \text{ Mb/s}$$

**4.** Necessary compression ratio: $125/24.948 \approx 5$.

# CHAPTER 21
# DIGITAL PHOTOGRAPHY

**Ken Parulski**
*Eastman Kodak Company*

## 21.1 FUNDAMENTALS OF DIGITAL PHOTOGRAPHY

Digital photography is a rapidly growing field that is enhancing, augmenting, or replacing conventional silver halide–based photographic systems in many applications. In digital photographic systems, images may be captured by a solid-state imager in a digital camera, or scanned from photographic film or prints. Image-capable computers in the home, office, or photofinishing lab allow these digital images to be organized, modified, or transmitted to other locations via digital networks. The microprocessors, memory, modems, and displays developed for desktop and portable computers can process, store, transmit, and display color pictures. The image sensors used in digital still cameras use the same fabrication technology as those in camcorders and fax machines. As this technology development produces smaller, faster, and less expensive devices, digital cameras will increase in performance while becoming more affordable.

Digital photography was first applied in professional applications such as graphic arts, where images captured from film-based cameras were scanned and manipulated electronically to correct for the printing process or provide artistic modifications to the image. Digital still cameras first replaced film cameras in applications where immediate access to the image in digital form was more important than obtaining the highest image quality. One such application is photojournalism, for example, sports photography. To meet tight deadlines with late-breaking stories, photojournalists use digital cameras to capture images that can be immediately transmitted by the wire service and distributed to newspapers worldwide.

Lower-cost "point-and-shoot" cameras were first successfully targeted at business and industrial applications. In these applications the images are typically used as "information" to create more powerful presentations or newsletters, or to document building construction or insurance claims. Recently, digital cameras have been targeted directly at the consumer market. For example, digital still cameras with LCD displays allow images to be instantly captured and shared during social occasions, much like "instant" photography. It is unclear, however, whether digital still cameras will replace film cameras in many homes in the foreseeable future. Instead, current digital still cameras may be displaced by future digital camcorders offering the capability to easily capture and organize high-quality still images as well as motion video images.

### 21.1.1 Early Still Video Floppy Based Analog Cameras

During the 1980s, analog electronic still cameras were developed by many Japanese companies.[1] One example is the Sony Mavica camera,[2] first demonstrated in 1981. The cameras used the Still-Video Floppy (SVF) standard, which allows 50 single field or 25 frame images to be recorded on a 5-cm magnetic floppy disk within the camera. The SVF standard uses the NTSC and PAL interlaced scanning standards, though progressive scanning is more appropriate for still cameras. The floppy disk was developed specifically for this application and spins at 3600 rpm in NTSC cameras and 3000 rpm in PAL cameras. Therefore, the same camera could not be used worldwide. SVF cameras use image sensors and recording technology adapted from video camcorders. Analog video processing is used to form a 4.5-MHz luminance signal and a 1.0-MHz line-sequential R-Y and B-Y color signal. The signals are recorded using analog FM recording.

These SVF cameras were used in limited industrial applications but were not successful consumer products because the high-priced cameras offered limited image quality. The horizontal luminance limiting resolution of 350 TV lines was acceptable for TV display, but inadequate for even small-size photographic prints. The analog recording limited the noise and stability of the image. The cameras provided poor connectivity to computers, which required a video frame grabber in order to access the images. In the 1990s, digital still cameras were developed to address these issues.

### 21.1.2  The Photographic Imaging Chain

To better understand the applications of digital photography, it is useful to consider the "photographic imaging chain." Both conventional film and digital cameras are seldom complete systems by themselves. They are normally just one part of an imaging system, or imaging chain.[3] The chain includes functions such as capture, storage, manipulation, transmission, soft copy, and hard copy (Fig. 21.1). Each link in the chain is important, because the quality of the final image, and the happiness of the user, is determined by the "weakest" link in the chain. Each function can be implemented in different ways. For example, a traditional photographic chain begins with a camera loaded with a roll of photographic film. The film is used to both capture the incident light from the scene, and store it as a latent image. The film is developed, and the developed images can be (1) enlarged and cropped by optical printing, (2) enhanced by optical techniques, or (3) combined with other images using complex optical processes or a pair of scissors. The pictures are transmitted by physically moving them from place to place. To provide "soft copy," slide film is used. Hard copies are made by optically printing the color negatives onto photographic paper.

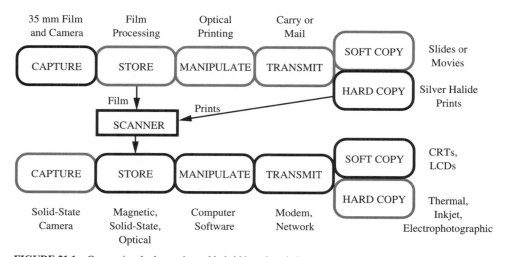

**FIGURE 21.1**  Conventional, electronic, and hybrid imaging chains.

The bottom of the figure shows a second type of imaging chain using only electronic links. The images can be captured using a solid-state sensor in a digital camera, and stored as electric signals on a magnetic hard drive. These digital images can be manipulated in complex ways using computer software and transmitted via computer networks. The soft-copy display may be a computer CRT or an LCD display on a digital camera. A variety of hard-copy output devices are available to print digital images, including color inkjet and thermal printers. In a third type of imaging chain, the film or prints from conventional photography systems can be scanned to create digital images. These "hybrid" systems are described in Section 21.7.

### 21.1.3  Comparing Conventional and Digital Photography Systems

Electronic systems have some significant advantages over film-based systems, as shown in Table 21.1. When the image is captured, it is immediately converted into signals that can be stored, manipulated, and transmitted. There is no need for the time, equipment, and chemicals necessary to develop and print photographic film. The media used to store the images can be erased and reused. Silver-halide photography has advantages in other attributes, however. The cameras are low in cost compared to digital alternatives, and often easier to use. The photofinishing infrastructure allows prints to be quickly and inexpensively produced. Finally, the image quality offered by low-cost film cameras can be approached only by the highest-cost digital alternatives.

**TABLE 21.1.**  Comparison of Digital and Conventional Cameras

| Attribute | Digital still cameras | Conventional film cameras |
|---|---|---|
| Equipment cost | High but decreasing | Low |
| Media | Erasable | Nonerasable but archival |
| Image access | Immediate | Requires chemical processing |
| Ease of use | More complex but flexible | Simple to learn |
| Printing | Requires new equipment | Existing photofinishers |
| Image quality | Depends on camera cost | Excellent |

The image quality required from a digital camera depends on the type of hard copy or soft display used, as well as the size and purpose of the final image. The most significant limitation of digital cameras is their resolution, because the number of photosites in the imager array sets an upper limit on the maximum resolution of the image. The sharpness of the final image is a function of the "acutance" of the photographic system, which depends on the modulation transfer function (MTF) of the camera lens, sensor, and printer, as well as the print size and viewing distance.[4] The quality of images from low-cost digital cameras is limited compared to film-based systems (see Table 21.2), but it is rapidly improving. However, the information-recording

**TABLE 21.2.**  Image Quality Attributes

| Attribute | Digital still camera | Conventional film camera |
|---|---|---|
| Resolution | Limited by image sensor | Excellent |
| Exposure latitude | Limited | Wide for color negative films |
| Sensitivity | Variable | Function of film ISO speed |
| Tone/color fidelity | Adjustable | Function of film, photofinisher |
| Image quality | Function of cost | Excellent |
| Noise and artifacts | Noise, streaks, aliasing | Grain is function of film speed |

capability of silver-halide technology also continues to advance,[5] creating photographic emulsions with increased resolution, enhanced photographic speed, and reduced granularity. This technology is applied in new film-based photography systems such as the advanced photo system. This system allows digital data to be stored on a magnetic coating on the film, providing new capabilities for film-based photography.

## 21.2  DIGITAL CAMERA ARCHITECTURES AND APPLICATIONS

Digital photography is rapidly growing in many diverse applications covering a range of requirements. To meet these needs, digital cameras using a range of image sensors, digital storage devices, and display capabilities have been developed. Some cameras operate as stand-alone systems where the captured image can be instantly viewed, whereas others use a computer to "finish" the image as the first step in a computer-based imaging application.

### 21.2.1  High-Resolution Cameras for Professional Use

High-resolution digital cameras have been developed for professional photographers, whose business depends on taking high-quality pictures. These cameras often use existing 35-mm or medium-format film camera bodies and lenses.[6] This allows the photographers to easily operate the camera and to use their existing equipment. However, the imager's photosensitive area is often smaller than the film area. As a result, the relative magnification of the lens is larger than for 35-mm film. A relay lens can be used to correct the magnification,[7] but this limits the usable f-number range of the lens.

A block diagram of the Kodak Professional DCS 460 digital camera,[8] the first portable 6.3-megapixel digital camera, is shown in Fig. 21.2. The camera uses a Nikon N90 35-mm format single-lens reflex (SLR) camera body and interchangeable lenses. The scene is focused through the camera lens onto the imager. The camera body automatically controls the exposure time and the lens aperture. The image may be focused manually or by using an autofocus lens. The image sensor has 3072 columns and 2048 rows of photosites, providing a total of 6.3 million pixels. The output of the image sensor is amplified and digitized using a linear 12-bit flash A/D controlled by a 5-MHz pixel rate clock. The output of the A/D is processed by logic arrays programmed by the camera firmware, and temporarily stored in DRAM. The digital images are converted to a TIFF (tag image file format) compatible record and stored as a DOS file on a removable PCMCIA (Personal Computer Memory Card Industry Association) type-III hard drive. To download the images, the camera is connected to the SCSI interface of the host computer. A 340-Mbyte hard drive can store over fifty 6.3-megapixel images.

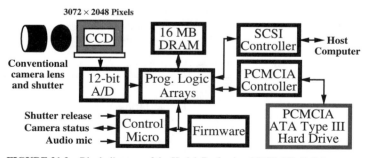

**FIGURE 21.2**   Block diagram of the Kodak Professional DCS 460 digital camera.

Image processing is done entirely in the digital domain, rather than using analog circuits. This minimizes noise and allows sophisticated image processing algorithms to be employed. Some digital processing is performed in the camera, before the image is stored. More complex algorithms are implemented in the host computer as the image is downloaded. A small color "thumbnail" image is stored along with each high-resolution image. When the camera or removable memory is connected to the computer, the thumbnail images are downloaded first to allow the user to select images of interest. This software- and firmware-based processing enables the cameras to use sophisticated image-processing algorithms that can be easily upgraded in the field and customized for special customer requirements.

### 21.2.2  Cameras for Consumer Applications

A wide variety of medium- and low-resolution digital cameras have been developed for business and consumer use. These cameras are less expensive than the digital cameras used by professional photographers, but they offer more limited image quality. They are used to add images to documents or presentations and for entertainment purposes. Most of these cameras include a serial connection to a desktop computer as well as a removable memory card. Some offer immediate viewing of the images by using an attached LCD or by providing an analog video output signal that can be connected to a home TV.

A block diagram of the QV-10 LCD digital camera developed by Casio Computer Co., Ltd., Japan,[9] is shown in Fig. 21.3. This was the first digital camera to incorporate a built-in color LCD image display. The scene is focused through a 5.2-mm camera lens onto the imager, which has 320 columns and 240 rows of photosites to provide a total of 77,000 pixels. The camera automatically controls the sensor exposure time. The lens has a fixed-focus setting. The output of the image sensor is amplified and A/D converted in a first LSI (large-scale integration) IC and then processed by a second LSI to provide an NTSC video signal. This NTSC signal is displayed on a 1.8-inch diagonal TFT (thin film transistor) LCD, which serves as the camera viewfinder.

When the user presses the shutter button, the captured still image is temporarily stored in the 0.5-Mbyte DRAM. The image is compressed by the coprocessor block in the second LSI, which performs $8 \times 8$ block DCT compression and Huffman coding. The compressed images are stored in 2 Mbytes of flash EPROM memory, which can hold up to 96 images. The stored images can be decompressed and viewed on the LCD screen. This enables the images to be immediately shared in social settings, much like instant photos. The images can be downloaded to a host computer via a serial port. However, the limited number of pixels provided by the camera significantly limits the quality of the image on the computer monitor or hard-copy printer.

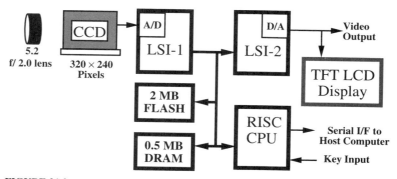

**FIGURE 21.3**   Block diagram of the Casio QV10 digital camera.

## 21.3   *IMAGE SENSORS FOR DIGITAL CAMERAS*

The image sensor is the most important component of a digital camera because it sets the maximum resolution, sensitivity, and signal-to-noise ratio capabilities of the camera. There are many different image sensor architectures, including frame transfer devices,[10] interline devices,[11] frame-interline transfer (FIT) devices,[12] MOS $x$-$y$ addressed devices,[13] and active pixel sensors (APS).[14]

### 21.3.1   Full-Frame Transfer Image Sensors

The image sensors used in most professional digital cameras employ a full-frame architecture,[15] shown in Fig. 21.4. A full-frame imager consists of a parallel, light-sensitive CCD (charge-coupled device) array, also called the "vertical" shift register; an opaque serial or "horizontal" CCD shift register; and an output amplifier. It is similar to a frame transfer CCD, without the separate light-shielded storage array. When the camera shutter is opened, the camera lens focuses the image onto the parallel array, which is precisely positioned in the camera's film plane. The parallel array is composed of discrete photosites, each the same size. When photons of incident light from the scene penetrate into the photosites, the energy of the photons may be absorbed by the silicon. This interaction releases an electron-hole pair from the silicon lattice structure. The electrons are collected into "charge packets" in a "well" created at each photosite. The number of electrons collected is proportional to the number of photons of light hitting that particular photosite. Therefore, the photosites in bright image areas collect a relatively large number of electrons, and those in dark areas collect relatively few electrons.

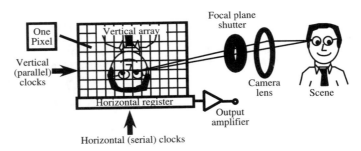

**FIGURE 21.4**   Full-frame CCD image sensor.

In very bright areas of an image (e.g., specular reflections) the number of electrons generated in particular photosites may exceed the charge capacity of the photosite. These excess electrons must be eliminated or they will contaminate nearby photosites, causing the specular reflection to grow into a blob or a streak. Image sensors used in digital cameras normally include a lateral or vertical overflow drain to direct the excess charge into the sensor substrate and prevent it from spilling into neighboring pixels.

The number of electrons collected at each photosite continues to increase as long as the camera shutter remains open. At the end of the exposure time, the mechanical shutter closes to block the light, and the electron charge packets at each photosite are transported to the output amplifier. This is accomplished by applying the proper vertical (parallel) clocks to shift each row of the vertical registers down by one row, so that the bottom row of the array (corresponding to the top of the image) is transferred into the horizontal (serial) register. Next, a series of many horizontal clocks are applied in order to shift the charge packets to the right, toward the output amplifier. This process repeats until all rows are transferred from the parallel register to the serial register, and then to the output amplifier, in order to read out the entire image. The output amplifier provides an output voltage that varies with the number of electrons in each charge packet. This analog voltage is later converted to a digital code value. The digital code value for each pixel—corresponding to each photosite on the CCD—is therefore proportional to the number of electrons collected by the photosite.

In the "true two-phase" CCD shown in Fig. 21.5, each photosite is composed of two slightly overlapping layers of polysilicon. Underneath a portion of each polysilicon layer is a 1.6-micron-wide implanted barrier region that creates a stepped electric potential field. The two layers are electrically isolated from one another by a thin layer of silicon dioxide. Each photosite is separated from its horizontal neighbors by 1.2-micron-wide, vertically oriented, thick layers of silicon dioxide called *channel stops*. The photosite spacing is 9 microns, both vertically and horizontally, providing a "square" sampling grid.

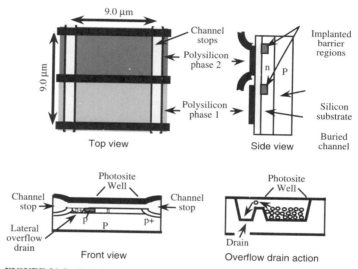

**FIGURE 21.5**   Full-frame photosite.

## 21.3.2   Interline Image Sensors

To eliminate the need for a mechanical shutter and to allow motion images to be captured, an electronic camera can use an interline CCD architecture. Most interline sensors use interlaced readout to conform to video-scanning standards. They have one vertical register stage for every two pixels. For still camera applications a progressive scan architecture is preferred. This requires one vertical register stage for each line of pixels. Figure 21.6 shows the pixel used in a megapixel interline transfer CCD with a progressive scan architecture.[16]

The pixels are 9 microns square, the same as in the full-frame sensor shown in Fig. 21.5. The photodiode is formed using a *pnpn* structure. This allows the n-layer to be fully depleted, eliminating lag and allowing fast electronic shuttering. Electronic shuttering is accomplished by clocking the n-type substrate to a sufficient voltage to sweep the accumulated charge from the photodiodes into the substrate. The photodiode comprises only a 3.4-by-7 micron area of the

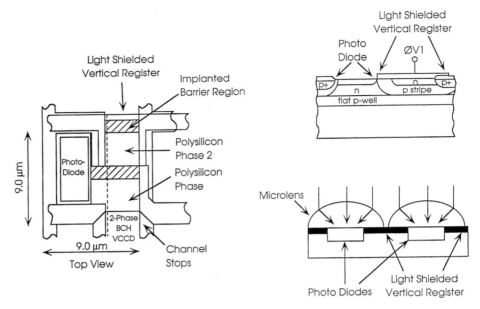

**FIGURE 21.6**    True two-phase charge transfer process.

9-by-9 micron cell. Thus, the sensitivity is reduced by the 30-percent fill factor. To improve the sensitivity, a microlenticular array is fabricated on top of the CCD.[17] These microlenses focus the light away from the opaque light-shield regions and into the center of the photodiodes. This increases the sensitivity by a factor of almost 2.5.

### 21.3.3    Charge Transfer

Charge transfer in CCD imagers can use four-phase, three-phase, pseudo two-phase, true two-phase, or virtual phase clocking, which use a corresponding number of clock lines per pixel. The imagers shown in Figs. 21.5 and 21.6 use two-phase CCD clocking, since there are two polysilicon gates per pixel in both the vertical and horizontal registers. To transfer the electron charge packets, complementary clock voltages are applied to the phase 1 and phase 2 electrodes, as shown in Fig. 21.7. At time $t_1$, the phase 1 voltage is positive (for example, $+6$ V) relative to the phase 2 voltage (for example, $-4$ V). This creates a "potential well" in the buried channel underneath the phase 1 polysilicon layer, because the negatively charged electrons are attracted by the positive electric field under the phase 1 gate. The lower phase 2 voltage creates a potential barrier under the phase 2 polysilicon gate, because the electrons are repelled by the negative electric field. As a result, the electric charge packets are confined under the phase 1 gate. To transfer the charge packets to the right at time $t_2$, the phase 1 voltage is switched negative and the phase 2 voltage is switched positive. At this point, the phase 2 gate becomes a well, whereas the phase 1 gate becomes a barrier, so each charge packet moves to the right. Note that the implanted barrier regions under each polysilicon gate create a "step" barrier, which forces the charge to move to the right instead of the left. Finally, at time $t_3$, the phase 1 and phase 2 voltages are returned to the same levels as at time $t_1$. This causes the charge to again move to the right, into the potential well under the phase 1 gate. Thus, cycling the phase 1 clock negative and then

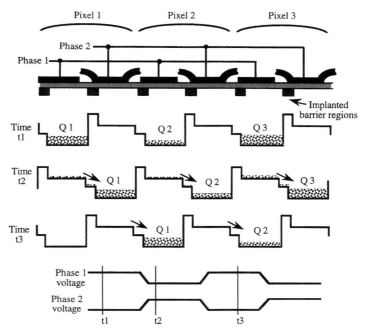

**FIGURE 21.7**   Full-frame photosite.

back positive while phase 2 cycles with the opposite polarity causes the charge packets to all shift one pixel to the right. Imagers using this charge transfer technique are called *charge-coupled devices* because the charge is coupled from beneath one polysilicon gate to the next—like water being poured from one bucket to the next—as the gate voltages are cycled positive and negative.

### 21.3.4  Charge Readout

At the end of the horizontal shift register is an output converter. This structure normally includes a floating diffusion sense node and an output amplifier, as shown in Fig. 21.8. When the phase 1 gate is switched to a negative potential, the charge spills over the output gate (OG) and onto the floating diffusion. The voltage of the floating gate is modified by the transferred charge according to the following equation:

$$\Delta V_{\text{fd}} = \Delta Q/C_{\text{fd}} \tag{21.1}$$

where $\Delta Q$ is the amount of charge transferred and $C_{\text{fd}}$ is the capacitance of the floating diffusion. The floating diffusion is connected to the input of a current amplifier, normally a 2-stage or 3-stage source follower, which provides a CCD output signal capable of driving external electronic circuits. After the charge is transferred to the floating diffusion, the charge is removed by turning on a reset FET (field-effect transistor), which resets the voltage on the floating diffusion to its initial value. To eliminate the thermal noise associated with resetting the floating diffusion, as well as the "1/f" output amplifier noise, *correlated double sampling* (CDS) is often used.[18]

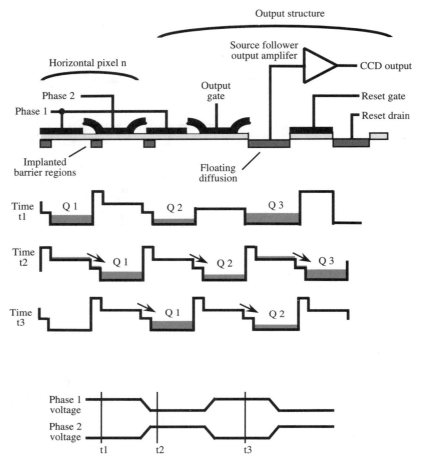

**FIGURE 21.8**   Output structure.

## 21.3.5   Color Separation Methods

Color negative films are coated with separate layers sensitive to red, green, and blue (RGB) light. Unfortunately, silicon-based image sensors have no natural ability to determine the varying amounts of RGB information in the image. There are three methods of capturing color images using a solid-state sensor:

- *Color sequential.*   A color image can be created by taking three successive exposures while switching in optical filters having the desired RGB characteristics. The resulting color image is then formed by combining the three color separation images. The filters may be dichroic or absorptive RGB filters mounted in a color wheel or a tunable LCD filter. The disadvantage of this technique is that the subject must remain stationary during the three exposures, because any subject or camera motion will result in colored edges.

- *Multisensor color.*   A multisensor digital camera typically uses a dichroic prism beam-splitter to separate the light into red, green, and blue components that are focused onto three separate monochrome image sensors.[19] Cameras with a combined red/blue imager,[20] and one or two green imagers, have also been developed. The disadvantages of the multisensor approach are the high cost of the sensors and beam-splitter and the difficulties of maintaining image registration.

- *Color filter arrays (CFAs).*   A single CCD sensor can provide a color image by integrating a mosaic pattern of colors, known as a *color filter array* (CFA), on top of the individual photosites. Many different arrangements of colors are possible.[21] Each photosite is sensitive to only one color spectral band.

### 21.3.6  Single-Chip Color Sensors

Two popular CFA patterns for single chip color sensors are shown in Figs. 21.9 and 21.10. The Bayer pattern[22] has 50 percent green photosites arranged in a checkerboard and alternating lines of red and blue photosites. The complementary mosaic pattern[23] has equal proportions of magenta-(Mg), green-(G), yellow-(Ye), and cyan-(Cy) sensitive photosites arranged in magenta-green and yellow-cyan rows. The position of the yellow-cyan columns is staggered by one pixel on alternate yellow-cyan rows. The Bayer pattern is normally fabricated on a sensor employing progressive scan readout, so that each color pixel is converted to a separate digital value. The complementary mosaic pattern is normally fabricated on an interlaced readout sensor using field integration mode. After the image is integrated, two rows of photosites are summed together in the vertical readout register. For example, in the bottom two rows, the Mg and Ye photosites are summed and the G and Cy photosites are summed. This reduces the vertical image sharpness but increases the signal level.

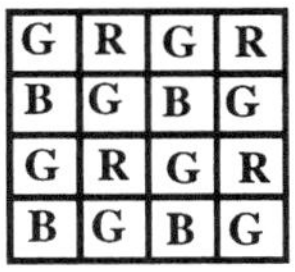

The color photosites sample the image differently, depending on the color. Some properties of the two-dimensional sampling process can be shown using the "Nyquist-domain" plot in Fig. 21.11. This indicates the limiting resolution of a Bayer-pattern color sensor in all directions. It is possible to properly recover scene details that have spatial frequencies falling inside the Nyquist-domain boundary. Scene details falling outside this boundary are aliased to lower spatial frequencies by the color image sensor, causing artifacts in the digital image. When the image falling on the sensor is sampled by the colored photosites, the original image spectrum is replicated around frequencies that are inversely proportional to photosite spacing $d$, as shown in Fig. 21.12. The Nyquist-domain border is the lowest spatial frequency at which these replicated spectra just touch the original spectrum centered at the origin.

The checkerboard arrangement of the green photosites in the Bayer CFA results in a diamond-shaped Nyquist domain for green and smaller rectangular-shaped Nyquist domains

| G | R | G | R |
|---|---|---|---|
| B | G | B | G |
| G | R | G | R |
| B | G | B | G |

**FIGURE 21.9**   Bayer CFA pattern.

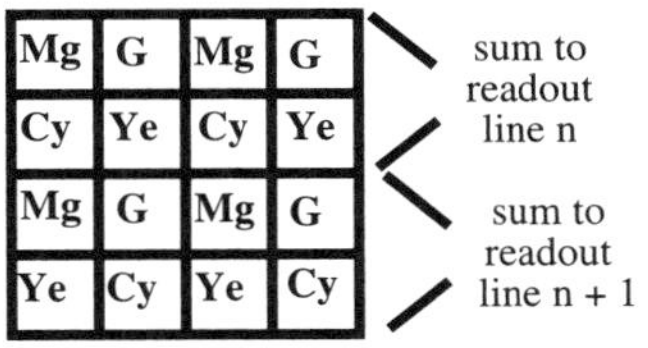

| Mg | G | Mg | G |
|----|---|----|---|
| Cy | Ye | Cy | Ye |
| Mg | G | Mg | G |
| Ye | Cy | Ye | Cy |

sum to readout line n

sum to readout line n + 1

**FIGURE 21.10**   Complementary mosaic CFA pattern.

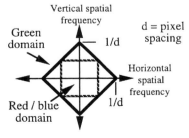

**FIGURE 21.11**   Nyquist domains for Bayer CFA.

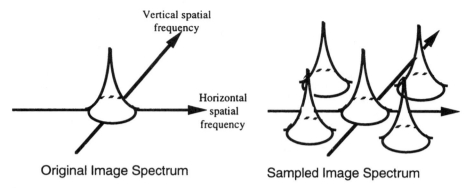

**FIGURE 21.12**  Image spectra.

for red and blue. The RGB Nyquist domains are vertically and horizontally symmetric if the photosite pitch is "square," so the limiting resolution due to the sensor sampling is the same in the vertical and horizontal directions. People are more sensitive to high spatial frequencies in luminance than in chrominance, and luminance is composed primarily of green light. Therefore, the Bayer CFA improves the perceived sharpness of the digital image by allocating more photosites to the green image record.

## 21.4  OPTICS FOR DIGITAL CAMERAS

Many of the optical issues for digital cameras are the same as for conventional film cameras, but there are two key differences. First, imagers normally have a much smaller active area than film, so the lens focal length must be correspondingly shorter to provide the same field of view. Second, the image sensor can create visible aliasing artifacts, particularly with color imagers. Aliasing is normally minimized by optically low-pass filtering the image. To obtain a high-quality image, the lens focus and sensor exposure must be properly adjusted.

### 21.4.1  Lens Formats and Magnification

The magnification of a lens is proportional to the focal length and inversely proportional to the size of the detector. Except for close-up photography, the height of the largest object, $O$, within the field of view captured by a digital camera is given by

$$O = D \times I/F \tag{21.2}$$

where $D$ is the distance from the object to the lens, $I$ is the height of the image sensor, and $F$ is the lens focal length. The sensor height $I$ equals the vertical pixel pitch times the number of active lines on the imager. Smaller sensors require correspondingly smaller-focal-length lenses in order to maintain a normal field of view. For example, a "normal" 50-mm-focal-length lens on a 35-mm SLR film camera, which has an image height of 24 mm, has the same field of view as a 7.5-mm lens on a 1/3-inch format image sensor, which has an image height of 3.6 mm.

The lens normally includes an adjustable aperture or iris that controls the cone angle of light illuminating the image sensor. The lens f-number, $A$, equals $F/d$, where $d$ is the aperture diameter. The sensor illuminance is proportional to $(1/A)^2$, so the sensor illumination with an f-number setting of 2.0 is twice that of an f-number setting of 2.8. Decreasing the aperture diameter provides less light to the sensor but also provides a larger depth of field.

### 21.4.2    Lens Depth of Field and Autofocus Methods

Objects in the scene that are not located at the lens focus distance will be blurred to some extent. Although a slight amount of blurring is acceptable, larger amounts cause a noticeable loss of sharpness in objects located at distances significantly nearer or farther than the lens focus distance. The minimum and maximum distances where the camera is acceptably focused is called the *depth of field*. The depth of field increases for larger lens f-numbers and smaller focal lengths. A point light source at the near or far depth of field limit creates a "circle of confusion," Cc, on the image sensor. The diameter of this circle is the value that causes a "just noticeable" amount of blurring in the final digital image. The proper value of Cc depends on the size and viewing conditions of the output image, which depends on the camera application. In practice, Cc is usually set to a value between 1 and 2 times the pixel pitch.

The hyperfocal distance[24] $H$ of a camera lens can be calculated using the following formula:

$$H = \frac{F^2}{A \times Cc} \tag{21.3}$$

where $F$ is the lens focal length, $A$ is the lens f-number, and Cc is the circle of confusion. If the camera is focused at $H$, all objects at distances between $H/2$ and infinity will be in acceptable focus. Fixed-focus cameras are normally focused at the hyperfocal distance. They are usually designed to work with relatively wide-angle (low-focal-length) lenses in bright-light situations, where the lens f-number can be relatively large. For example, if $F = 8$ mm, $A = 2.8$, and Cc $= 12$ microns, then $H = 1.9$ meters and the depth of field is approximately 3 feet to infinity.

If this camera instead has a zoom lens with a maximum focal length of 20 mm, the hyperfocal distance for the telephoto position is approximately 12 meters. Therefore, an adjustable focus lens is needed in order to focus on objects closer than 20 feet. Zoom lenses normally use an autofocus system to set the focus distance. The near depth of field is equal to $(H \times D)/(H + D - F)$, and the far depth of field is equal to $(H \times D)/(H - D + F)$, where $D$ is the focus distance. There are two main methods of automatically determining $D$, *contrast autofocus* and *correlation autofocus*.

Through-the-lens correlation autofocus methods are used in most 35-mm SLR cameras. These systems use a special optical assembly within the lens to form two images that are asymmetric with respect to the optical axis of the camera lens. The two images are focused side by side onto an autofocus sensor, normally a one-dimensional or "linear" image sensor. The two images are digitally correlated, and the separation between the images provides a measure of the subject defocus. The lens focus motor is then driven to the location providing the separation distance corresponding to the best focus.

Contrast autofocus systems are used in most consumer camcorders. A contrast autofocus method normally uses the image sensor signal to determine if the lens is properly focused. The signal from the central portion of the image is bandpass-filtered, rectified, and either averaged or peak-detected, in order to form a *focus value*. Next, the lens focus position is adjusted slightly, and the focus value from the new position is compared to the value at the previous position. If the value increases, the lens focus is stepped again in the same direction. If not, the lens position is stepped in the opposite direction. This process is continued until the peak focus value is determined. At this point a second bandpass filter with a characteristic more sensitive to fine image details may be used to determine the final focus position. The image may also be divided into multiple zones, with a separate focus value computed for each zone. A "fuzzy logic" approach, as described in Chapter 6, may be used to determine how to set the lens focus position based on the focus values from the various zones.[25] Because contrast autofocus methods are iterative, they require a longer focusing time than correlation methods.

### 21.4.3    Optical Prefilters for Aliasing Suppression

Most single-chip color digital cameras use an optical prefilter to suppress aliasing artifacts. Aliasing occurs when there is significant information in the scene at spatial frequencies above the

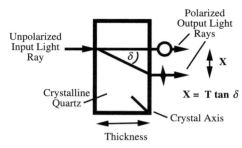

**FIGURE 21.13**   Birefringent quartz antialiasing filter.

Nyquist limit set by the number and arrangement of photosites on the imager. In this situation, high spatial frequencies take on the appearance, or *alias,* of lower spatial frequencies. Aliasing is a particular problem with single-chip color cameras because the different colors are sampled at different spatial locations.

Aliasing can be reduced by using an optical prefilter or "blur" filter, positioned in front of the sensor.[26] The blur filter is typically made of birefringent quartz, with the crystal axis oriented at a 45-degree angle, as shown in Fig. 21.13. In this orientation the birefringent quartz exhibits the double refraction effect. An unpolarized input ray emerges as two polarized output rays. The output ray separation is proportional to the filter's thickness. A 1.5-mm-thick plate will give a separation of about 9 microns—equal to the pixel pitch of the sensors described earlier. Figure 21.13 shows a simple "two-spot" filter, but typical filters use three or more pieces of quartz cemented in a stack.

### 21.4.4   Exposure Determination and Control

Digital cameras include an exposure determination system in order to provide a properly exposed digital image. The exposure of the image sensor depends on the lens f-number, sensor exposure time, scene illumination level, and scene reflectance, as well as many other secondary factors. The fundamental relation between the scene luminance and the sensor exposure is expressed by the following equation:

$$H = \frac{\pi v T \cos^4(\theta) L t F^2}{4A^2 i^2} + H_f = \frac{qLt}{A^2}\left(\frac{F}{i}\right)^2 + H_f \qquad (21.4)$$

where $q = (\pi/4)Tv[\cos^4(\theta)]$
  $A$ = lens f-number
  $F$ = lens focal length (meters)
  $H$ = sensor focal plane exposure (lux-seconds)
  $H_f$ = sensor exposure due to flare light (lux-seconds)
  $i$ = image distance (meters)
  $L$ = scene luminance (candelas per square meter)
  $T$ = transmission factor of the lens
  $t$ = exposure time (seconds)
  $v$ = vignetting factor
  $\theta$ = angle of image point off axis

When the camera is focused on infinity, $H_f \ll H$, $T = 0.90$, $v = 0.98$, and $\theta = 10°$ so that $\cos^4 q = 0.94$; then $q$ is equal to 0.65, so that the equation reduces to the following:

$$H = 0.65L \cdot t/A^2$$

To obtain an acceptable image, the sensor exposure $H$ for all important areas must be less than the saturation level of the image sensor, yet large enough so that the resulting signal level provides an adequate signal-to-noise ratio. Scenes contain a wide range of luminance values. The simplest camera exposure controls use a single photosensor to measure the average luminance in the lower-center portion of the image. This reduces underexposure problems with many outdoor images having bright sky in the top portion of the image. More sophisticated methods use a multisegment light meter or use the image sensor to provide exposure control data. Standard methods have been developed to measure the ISO speed rating of digital cameras.[27]

## 21.5  *IMAGE PROCESSING IN DIGITAL CAMERAS*

Digital photography systems usually utilize extensive digital image processing. Prior to digital processing, the output of the image sensor is typically processed by an analog CDS circuit[28] to reduce noise from the sensor's output amplifier, amplified by a programmable gain amplifier, and A/D converted. In a professional camera a linearly quantized A/D converter having 12 to 16 bits is normally used. In lower-cost consumer cameras, a nonlinear 8-bit A/D is typical. The quantization characteristic uses more bits for low signal levels, because the human visual system has a nonlinear response to the luminances of objects in a scene.[29]

The digital image processing normally includes processing to provide good color and tone reproduction and enhance the sharpness of the image. In single-chip color cameras the processing also includes a reconstruction algorithm to create the necessary color pixel values at each photosite. The processing may also include image compression, typically using the JPEG algorithm described in Chapter 9. The various processing steps may be performed in the camera, or in a host computer as the images are downloaded. The camera processing can be done using a firmware-based processor,[30] or using custom integrated circuits.[31]

### 21.5.1  Color Pixel Reconstruction in Single-Sensor Cameras

To produce a high-quality color image, the color samples from a single-sensor color camera must be processed to provide red, green, and blue values (or luminance and chrominance values) for each pixel, because each CCD photosite can capture only one of the three colors. The process of filling in the "missing" color values is known as *color pixel reconstruction*. The simplest reconstruction algorithms use nonadaptive linear interpolation, where the missing values are always estimated by calculating the average of the vertically or horizontally adjacent sample values of the appropriate color. For example, the missing green values of the Bayer pattern would be estimated using the average of the four values from the adjacent green photosites. This approach is simple to implement but reduces the sharpness of the image while increasing the visibility of false color artifacts due to aliasing of high-spatial-frequency luminance patterns in the scene.

More sophisticated color pixel reconstruction algorithms can provide sharper images with reduced false color artifacts. These algorithms may use template matching[32] or a gradient-based approach[33] to adaptively interpolate the missing color pixel values. In the latter method, vertically and horizontally oriented gradients are calculated in order to deduce whether the missing green pixel lies along a vertical or horizontal image edge. If it does, the missing green value is formed by averaging values along the edge. For example, the horizontally adjacent green pixel values are averaged if the missing green pixel is located along a vertically oriented edge. Once the missing green values have been estimated, the ratios of the red or blue photosite values and the estimated green values can be calculated. These ratios can be linearly interpolated at all of the missing blue and red sample positions. This approach calculates the missing red and blue values by effectively adding high-frequency luminance details from the green record to the sparsely sampled red and blue records.

The color pixel reconstruction method used with the complementary mosaic pattern in Fig. 21.10 normally recovers luminance ($Y$) and color difference signals R-Y and B-Y. Luminance is obtained by low-pass filtering the CCD output pixel values, which contain the sum of a magenta/green line and a yellow/cyan line. The color difference signals are decoded by bandpass filtering the CCD output. This method reduces the sharpness of the luminance because of the low-pass filtering and can cause false color artifacts. More elaborate algorithms exploit the correlation between the luminance and color difference signals to provide a sharper image with reduced false color artifacts.[23]

## 21.5.2   Tone Reproduction and Noise

In conventional CCD image sensors the number of electrons collected by a photosite is linearly related to the illumination level until the maximum capacity, or "saturation," signal level is reached. Beyond this point the output is "clipped" to the saturation value, as shown in Fig. 21.14. The figure also shows the film density–versus–log exposure curve of a typical color negative film. Note that the film curve has a "toe" at low exposure levels, a straight-line region for normal exposure levels, and a gradual "shoulder" for high illumination levels. The straight-line portion of the film curve has a slope or "gamma" of approximately 0.6, whereas the CCD curve has a gamma of 1.0. The higher gamma limits the exposure latitude of the solid-state image sensor. Setting the proper camera exposure level is critical because overexposure will clip the image highlights. Underexposures can be corrected by digitally adjusting the code values of the final image, but this will also increase the noise.

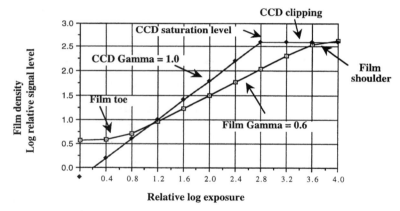

**FIGURE 21.14**   CCD and film *D*–log *E* curves.

Although there are many sources of noise in a digital camera, the most significant is often the dark-current variability of the image sensor. This noise is caused by minute imperfections or impurities in the sensor structure that generate a spatially and temporally varying number of "noise" electrons in each sensor photosite, independent of the sensor illumination level. The signal from each photosite includes electrons resulting from incident photons of light and electrons due to dark current. The number of dark current electrons is proportional to the photosite integration time and an exponential function of the sensor temperature. Therefore, some professional cameras use thermoelectric coolers to reduce the sensor temperature in order to reduce the dark-current noise level.

The average dark-current level can be compensated for by masking off some sensor pixels at the edges of the image sensor, forming an average of these black pixel values, and subtracting this average value from each image pixel value. This ensures that the signal levels of dark objects do not change with varying sensor temperatures or drifting dc circuit levels. The black level–corrected pixel values are normally requantized to provide an output signal that is a nonlinear function of the scene luminance. This allows the image to be represented by 8 or 10 bits while minimizing the visibility of quantization distortion in the final image. The requantization is often called *gamma correction,* since it has a functional form that compensates for the exponential light output–versus–voltage input characteristic of CRT displays. A typical requantizer uses the ITU-R BT.709 optoelectronic conversion function:[34]

$$V = 1.099L^{0.45} - 0.099; \qquad \text{for } 1 \geq L \geq 0.018 \qquad (21.5)$$
$$V = 4.5L; \qquad \text{for } 0.018 > L \geq 0$$

where $L$ is the relative scene luminance and $V$ is the relative output code value. This function is typically implemented on the individual RGB signals using a digital lookup table after the signals have been color corrected. The optoelectronic conversion function can be measured using standard techniques.[35]

### 21.5.3  White Balance and Color Correction

To provide acceptable color reproduction, digital cameras normally utilize white balance and color correction algorithms. White balance requires adjusting the RGB values from the camera to correct for the color temperature of the light source used to illuminate the subject. The amounts of red and blue light in daylight sources are approximately equal, as shown in Fig. 21.15. However, many artificial light sources, such as tungsten light bulbs, provide a much higher proportion of red light than blue light. Images taken using these illuminants must have the blue signal amplified to prevent white objects from appearing unnaturally yellow in the reproduced image.

White balance is performed by amplifying the red and blue signal levels so that they equal the green signal level for neutral (white or gray) objects in the scene. The most foolproof way to

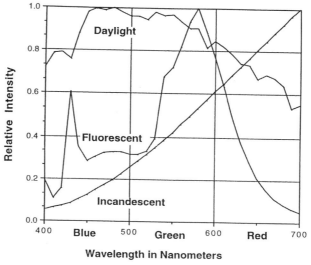

**FIGURE 21.15**  Spectral intensity of various illuminants.

achieve proper white balance is to have the camera user either capture a test image of either a white or gray card held in the scene or select a neutral object in the scene as the image is downloaded. This is appropriate for professional cameras, but for consumer applications an automatic method for determining the illuminant color temperature is necessary. One such method uses a set of photodiodes covered by RGB filters on the front or top of the camera, aimed upward toward the light source. However, this approach can give incorrect results because of reflections from walls or other colored objects. Better results are normally obtained by estimating the proper white balance setting from the image sensor data. The captured image average RGB values are computed for different regions. These average values, along with the absolute scene light level, are used by a scene classifier or fuzzy-logic control that decides on the most likely illuminant color temperature.

After white balancing, the image data can be processed to improve the color reproduction.[36] The red, green, and blue sensitivities of a color CCD have responses that are always positive and typically include some amount of color cross-talk. For example, the blue photosites may have some slight response to red light. On the other hand, the optimum camera sensitivities for the reference primaries, such as those specified by ITU-R BT.709, are negative at some wavelengths. Therefore, a $3 \times 3$ color correction matrix is often used to improve the color reproduction by correcting the camera responsivities properly for the reference display chromaticities. The matrix has the following form:

$$R_o = a_{11}R_i + a_{12}G_i + a_{13}B_i$$
$$G_o = a_{21}R_i + a_{22}G_i + a_{23}B_i$$
$$B_o = a_{31}R_i + a_{32}G_i + a_{33}B_i$$

where $R_i$, $G_i$, and $B_i$ are the linearly quantized white balanced matrix input signals, and $R_o$, $G_o$, and $B_o$ are the color-corrected matrix output signals. The coefficients $a_{ij}$ depend on the responsivities of the RGB signals prior to matrixing, including the optics and color sensor. The off-diagonal matrix terms, such as $a_{12}$, are normally negative values and serve to increase the color saturation of the image.

### 21.5.4   Edge Sharpening

The processing used in digital cameras often includes edge sharpening. This is done to compensate for the image blurring due to the lens, optical antialiasing filter, and CCD aperture, and to provide a subjectively sharper image. The appropriate amount of sharpening depends on the size and distance at which the final image is viewed. An image that appears "crisp" when printed at a small size may appear artificially overenhanced when enlarged.

Edge enhancement can be implemented using a two-dimensional filter kernel, for example:

$$-1/4$$
$$-1/4 \quad 1 \quad -1/4$$
$$-1/4$$

In this case the filter kernel output equals the present luminance pixel value minus the average of the four vertically and horizontally adjacent values. The filter output is zero in uniform areas of the image, but nonzero for edges. More elaborate filter kernels can be used to tailor the frequency response. The filter output may be "cored" by setting small output values to zero, because these small values are often the result of noise. The signal is then amplified and added back to the original luminance signal or to the red, green, and blue signals to increase the image sharpness.

### 21.5.5  Image Compression for Digital Cameras

Digital cameras often employ image compression to allow more images to be stored in the camera's digital memory. Standard JPEG compression is often used. Typically, the fully processed 24-bit-per-pixel RGB data is converted to luminance and subsampled color difference values and then compressed. The average amount of data after compression is often selectable and is normally in the range of 1 to 4 bits per pixel. The advantage of using JPEG compression is that the image data can be decompressed by standard computer software applications if the JPEG image data is stored using a standard image file format.

An alternate approach is to use a compression algorithm optimized for data from one-chip color sensors.[37] This can provide more efficient compression because it operates on the 8-bit-per-pixel CFA data prior to color pixel reconstruction. On playback the images are decompressed and then processed by the color pixel reconstruction, color correction, and other algorithms implemented on a host computer. This approach requires less digital processing in the camera but creates a nonstandard compressed-image record that must be processed after decompression using algorithms appropriate for the particular sensor CFA pattern.

## 21.6  DIGITAL IMAGE STORAGE

Many digital storage technologies may be used in digital cameras. These include ICs, magnetic disks and tape, and optical disks. All of the digital storage technologies and standards used in digital cameras were first developed for high-volume computer or consumer audio applications. Digital photography systems use three different types of storage: (1) an image buffer for temporarily storing data for one or more images during in-camera processing; (2) transfer storage, normally in the form of a removable card or disk, for holding a larger group of images that will be transferred to a computer or other playback device; and (3) long-term storage for hundreds or thousands of digital images. The buffer storage is normally one or more dynamic random access memory (DRAM) ICs. The transfer storage is typically nonvolatile flash EPROM ICs or magnetic hard disks so that the images are retained without drawing power. The long-term storage may be a recordable optical disc or a magnetic hard drive or removable disk cartridge.

### 21.6.1  Solid-State Memory and Memory Cards

Most digital cameras use some amount of IC memory for image buffering. The buffer memory uses volatile storage, such as static or dynamic RAM, which requires continuous battery power to maintain the image data. The memory write speed must match the data rate from the image sensor. Digital cameras often use programmable microprocessors or DSPs that are controlled by firmware stored in ROM or EPROM. The EPROM approach allows the firmware to be upgraded in the field.

Flash EPROM memory is often used for transfer storage in low-resolution digital cameras. One advantage of an IC memory is that it has no moving parts. This allows the cameras to be small and rugged. However, IC memory is more expensive than magnetic memory if a large capacity is required. Therefore, digital cameras that use IC memory for transfer storage always include image compression to reduce the size of the image files. Many flash EPROM devices have a relatively slow write rate, which limits the camera frame rate.

Flash memory ICs may be mounted on a PC board inside the camera to reduce the cost, or on a removable memory card. The latter approach provides a higher camera image storage capacity, limited only by the type and number of cards the user wishes to purchase. It also allows the images to be transferred to any computer or reader with the appropriate card interface. A number of card formats, listed in Table 21.3, have been developed. They include the PCMCIA (Personal

**TABLE 21.3.**   Memory Card Formats

|  | PCMCIA | CompactFlash | MiniatureCard | SSFD |
|---|---|---|---|---|
| Dimensions (mm) | 54 × 86 | 36 × 43 | 33 × 38 | 37 × 45 |
| Thickness (mm) | 3.3, 5.0, 10.5 | 3.3 | 3.5 | 0.76 |
| Connect type | Pin in socket | Pin in socket | Elastomeric | Contact |
| Connector pins | 68 | 50 | 60 | 22 |
| Flash technology | All | All | NOR | NAND |
| Other memories | RAM, hard drive | None | RAM, ROM | None |
| File system | DOS FAT | DOS FAT | DOS FAT | Proprietary |
| Host software | ATA or FTL | ATA | FTL | ATA controller |
| Supply voltage | 5 V or 3.3 V | 5 V or 3.3 V | 5 V or 3.3 V | 5 V or 3.3 V |

Computer Memory Card Industry Association) PC Card, CompactFlash, MiniatureCard, and SSFD (solid-state floppy disk) formats.

### 21.6.2   Magnetic Recording Options

A magnetic recording is created by a varying magnetic field from an electromagnetic head in relative motion with a magnetic storage medium.[38] The medium may be a magnetic hard drive, floppy disk, or magnetic tape. Digital magnetic tape recording—used in the digital camcorders described in Chapter 20—is the least expensive digital storage medium. However, it requires a complex and therefore expensive recording mechanism. Magnetic tape does not offer random access and can be damaged by repeatedly playing the same videotape tracks. The capacity of videotape far exceeds that necessary for digital still photography. Therefore, digital tape recording is not used in still-only digital cameras, although digital video camcorders incorporating special still photography features have been developed.

Magnetic disks offer the advantage of random access and a simple recording mechanism. Standard 3.5″-diameter floppy disks are considered too large and too low in capacity (i.e., 1.44 Mbyte) to be used in digital cameras. Nevertheless, because most computers can read these disks, they are often used to distribute limited-resolution, compressed images.

Magnetic hard drives allow many gigabytes of information to be stored in a relatively small volume. The capacity of a hard drive depends on the head gap, the track spacing, the diameter of the disk, and the number of platters. Hard drives are available in the PCMCIA-ATA type III PC Card format, which are small enough for many digital camera applications. For memory sizes greater than about 10 Mbyte, magnetic PCMCIA hard drives are less expensive than PCMCIA memory cards. Disadvantages of hard drives include the time and power needed to spin up the disk before recording can begin. Hard drives are also less robust than IC memory cards when mishandled by users.

High-capacity 3.5″ Winchester disk drives have been introduced with capacities up to 1.3 gigabytes. The cost per Mbyte for these removable disks is much lower than for PCMCIA format hard drives, but the size is too large for in-camera applications. These drives are attractive for longer-term storage of digital images, however.

### 21.6.3   Optical Recording Options

Most optical discs, such as the CD and DVD discs described in Chapter 12, are created by pressing many duplicate discs from the same master recording. To store images from digital cameras, however, write-once or erasable techniques[39] must be used to create unique discs for each user. Write-once systems use a semiconductor laser to burn holes, form bubbles, or change the phase of a thin layer of material coated on the disc substrate. The recorded "pits" are then read back by a low-power laser to recover the digital image. It is advantageous to design the write-once discs so that they can be read by standard CD-ROM or DVD players. Write-once discs that can be played on conventional CD and CD-ROM drives are called *CD-R discs*. They use a 120-mm-diameter injection-molded polycarbonate substrate, and store up to 600 Mbyte of image data. The Photo CD system described in Section 21.7.1 uses CD-R discs.

Erasable optical discs use phase-change or magneto-optical methods. Phase-change erasable discs use different laser intensities to heat the disc to two different temperatures. Pits are formed at the higher temperature because the material remains in the amorphous state as it cools rapidly, providing low reflectivity. At the lower temperature the material crystallizes as it cools, providing higher reflectivity. One such system stores 256 Mbyte on a 3.5″-diameter disc.

Magneto-optical discs use materials exhibiting the Kerr effect. Data are recorded by heating the material with a laser while applying a modulated magnetic field. The playback signal is obtained by analyzing the polarization of light reflected from the disc as it is scanned with a low-power laser. The 64-mm-diameter Mini Disc (MD), developed for audio applications, uses magneto-optical recording to provide 140 Mbyte per disc.

Recordable CDs are too large for portable camera applications and require too much recording power. They are well-suited for archival storage of images, however. They cannot be inadvertently erased or easily damaged and are very stable. They have high capacity and are less expensive than any other type of digital memory except tape. Most multimedia computers include CD-ROM drives that can be used to read write-once CDs. However, the relatively high cost of CD writers has limited the use of recordable CDs to commercial and photofinishing applications. Erasable discs, such as MD discs, are smaller and more suitable for in-camera recording, since the discs can be erased and re-used. However, there are relatively few readers available.

### 21.6.4    Storage Requirements for Digital Cameras

Table 21.4 lists some of the important parameters for digital storage. The storage capacity is the number of megabytes per unit area or per unit volume. The size of each image file depends on the number of pixels and the compression level. A highly compressed image from a low-resolution digital camera may be as small as 50 kbyte, whereas an uncompressed image from a high-resolution, color sequential camera may exceed 20 Mbyte. In most applications the transfer memory stores from 20 to 100 images, so the required capacity for the transfer storage can range from 1 Mbyte to 2 Gbyte.

The transfer rate is the speed at which image data are stored or retrieved. The storage speed should be less than 5 seconds per image, so the required transfer speed can range between 10 kbyte/s and 4 Mbyte/s. To quickly select desired images from the dozens or hundreds of images

**TABLE 21.4**   Digital Image Storage Attributes

|  | Solid-state | Magnetic | Optical |
|---|---|---|---|
| Types | Flash EPROM<br>RAM | Tape<br>Floppy disk<br>Hard disk | CD-R<br>MD |
| Relative cost | No "drive" cost<br>Low cost for $\leq$ 1 Mbyte<br>Very high for $>$ 10 Mbyte | Medium-priced drives<br>Tape is very low cost<br>Disks are medium cost | High-priced drives<br>Low-cost media |
| Size | Very small for $\leq$ 10 Mbyte |  | Relatively large discs |
| Advantages | No moving parts | Low media cost | Low media cost |
| Disadvantages | Cost | Ruggedness | Recorder cost and size |

stored on the transfer memory, small thumbnail images are often recorded along with the full-resolution image data. The thumbnails can be quickly transferred and displayed as a composite image.

The digital storage must maintain the integrity of the digital information. Digital magnetic and optical-based recording systems always include error detection and correction to handle random and burst errors caused by electrical noise or physical defects or debris. All digital recordings have a finite life, limited not only by the media but also by the availability of playback devices. As a result, digitally archived images may need to be transferred to current storage media every decade or two, or the images may be unrecoverable. One advantage of photographic film is that the stored image is human-readable, so a playback device is always available.

### 21.6.5 Image Formats for Digital Cameras

To gain broad market acceptance, digital cameras must produce images that can be seamlessly transferred into other digital devices for editing, display, transmission, archival storage, and printing. This requires a standard image data format, not just standards for the physical and electrical compatibility of the transfer media. A flexible image format is desirable in order to accommodate a range of storage media, image types and sizes, and optional ancillary information. This nonimage data may include parameters such as the time and date, camera zoom position and focus distance, illumination level, camera calibration data, subject, and copyright owner.

A number of standard image formats have been developed for digital camera applications.[40] These include TIFF/EP (tag image file format for electronic photography),[41] based on the TIFF specification, and Exif (exchangeable image format),[42] based on the JPEG file interchange format. Both of these formats, as well as FlashPix, a recent hierarchical image format employing structured storage,[43] support both uncompressed and JPEG compressed image data. To provide good color rendition, the meaning of the digital color values produced by a camera must be standardized or communicated to the printer and the display. This can be done by using an ICC color profile[44] to unambiguously define the RGB reference primaries, white point, and optoelectronic conversion function of the camera.

## 21.7 HYBRID FILM/DIGITAL PHOTOGRAPHIC SYSTEMS

The strengths of silver halide and electronics can be combined to create hybrid digital photographic systems. This is accomplished by scanning either the film or prints produced by conventional photographic systems. Scanning the original film produces images with higher resolution and wider dynamic range than scanning prints. The scanned images may be stored on many different types of media, including recordable optical discs, magneto-optical discs, and floppy disks. The images may also be stored on an *image server* so that they can be downloaded and viewed on a home computer via networks.

### 21.7.1 Photo CD System

The Kodak Photo CD System, shown in Fig. 21.16, was introduced in 1992 as the first consumer hybrid film/digital photographic system.[45] Conventional cameras and 35-mm film are used to produce images that are developed and printed. The developed film is scanned on a high-resolution scanner, where it is converted to a digital image with 2048 rows and 3072 columns of "square" pixels and a 3:2 image aspect ratio.[46] The data are processed by a computer workstation and stored on a write-once optical disc. Each disc is a unique digital data set recorded with a laser in a CD writer. The images on the disc can be viewed on a home television using a Photo CD or

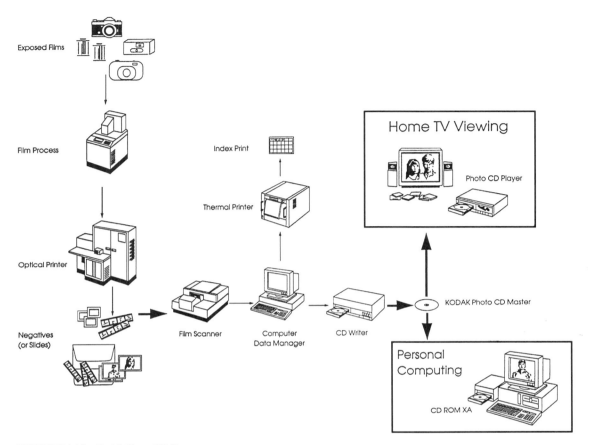

**FIGURE 21.16**    Kodak Photo CD System.

CD-I player, or accessed by a computer using a CD-ROM drive. Computer software packages allow users to manipulate, organize, and transmit the images.

The hierarchical format used with Photo CD master discs has five resolution levels—ranging from 128 to 2048 lines—to allow rapid access for various applications. The Base image has 512 lines, the 4Base image has 1024 lines, and the 16Base image has 2048 lines. The 4Base and 16Base images are compressed using adaptive DPCM compression of the *residual,* obtained by subtracting the up-sampled lower-resolution image, so that the stored file is less than 6 Mbyte per image. The same Photo CD can be played on any computer platform and on TV receivers worldwide using a player with the appropriate TV standard. The optical disc has a capacity of approximately 600 Mbytes. This enables one disc to hold up to 100 "digital negatives" that store essentially all of the image information recorded on each 35-mm film frame.

### 21.7.2   Magnetic Disk–Based Systems

A number of systems have been developed that allow consumers to receive their photographic images on standard 3.5″ computer floppy disks. Because of the limited capacity of these disks, both the number of images per disk and the image quality is limited. These systems normally use an image having $640 \times 480$ or fewer pixels per frame, with aggressive image compression, in order to fit from 12 to 36 images on a disk. Some systems also store an application program on the disk to allow users to view, manipulate, and print their images.

In one system, for example, up to 28 images can be stored on a 1.44-Mbyte floppy disk along with a 500-kbyte application program. Therefore, each image must be compressed to about 30 kbyte. Images are scanned at high resolution, low-pass filtered, and subsampled to obtain $600 \times 400$ pixel images. The images are compressed using JPEG 4:2:0 baseline compression with an average of 1 bit/pixel. Thumbnail images are also stored on the disk so that the disk contents can be quickly previewed.

Another type of hybrid system has been developed using the 64-mm-diameter MD disks originally introduced for music.[47] The disk capacity is 140 Mbytes, with a transfer rate of 150 kbyte/s. The system includes image formats having $480 \times 640$ pixels with a 4:3 image aspect ratio, $480 \times 848$ or $1080 \times 1920$ pixels with a 16:9 aspect ratio, and $1024 \times 1536$ or $2048 \times 3072$ pixels with a 3:2 aspect ratio. The customer selects their desired image format when placing the order. JPEG 4:2:0 baseline compression is used on the main image, and an $80 \times 60$ pixel thumbnail image is also stored along with each image.

### 21.7.3  Network-Based Systems

Many home computers are connected to on-line services and can download and display true color images. A number of hybrid imaging systems transfer digital images to home customers using these networks instead of optical or magnetic disks. In an example system, shown in Fig. 21.17, film images are developed, scanned, and stored as high-resolution files on an image server located at the photofinisher or service provider. Low-resolution versions of these images can be downloaded from the network and viewed by the customer. The customer may then decide which images to print or enlarge, or may e-mail the images to others or place them on their personal home page. With some services the customer can decide how to crop or otherwise modify the

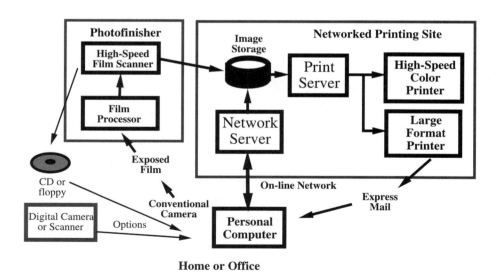

**Home or Office**

**FIGURE 21.17**   Network-based systems.

image prior to printing. Images can also be combined together or merged with "creative" backgrounds.

Some systems allow digital images to be transmitted via the network from home computers to the networked printing site. At this site, color hard-copy printers are used to provide different types and sizes of hard-copy output. These systems can also be used to print digital images from optical or magnetic disk-based hybrid systems, or from digital cameras.

## 21.8  DIGITAL COLOR HARD COPY

Digital photography systems can produce color prints using many technologies.[48] These include thermal wax and dye transfer, color inkjet, color electrophotography, and silver-halide photography methods. Some methods, such as inkjet and thermal wax transfer, can be used in low-cost home printers. Others, such as silver-halide printers, are more appropriate for high-throughput printers used by service providers.

The quality of the final print depends on both the printer hardware and the print media. The media include the inks, dyes, or toner particles used to form the image, as well as the characteristics of the paper or transparency substrate. To produce high-quality prints, the tone reproduction of the print must be carefully controlled and must provide a wide density range. The densities on a print can be produced by three methods:[49] (1) a continuous-tone pixel composed of a single continuous-tone dot, (2) a fixed number of dots having a fixed density level to form a binary halftone, or (3) a smaller number of dots having variable size and therefore variable density levels. The last two methods reduce the perceived resolution of the printed image, because the number of continuous-tone pixels per inch is a small fraction (typically 1/6 or smaller) of the number of dots per inch.

### 21.8.1  Thermal Dye Printers and Media

Most thermal printers use a resistive head to heat a thermal dye or wax donor, as shown in Fig. 21.18. The heads normally have from 100 to 400 resistive elements per inch and print one

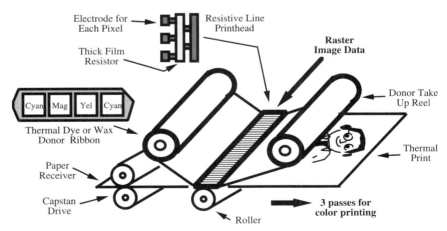

**FIGURE 21.18**  Thermal printer.

column at a time. The head has a continuous thick-film resistor that is locally heated to different temperatures due to the varying currents supplied by each electrode. A servo-controlled capstan drive motor moves the donor web and receiver (paper or transparency) through a three-pass printing process. The donor web is coated with cyan, magenta, and yellow sequential patches, equal in size to the print. A four-pass process, using an additional black donor, is used in some applications. When a thermal sublimation dye is heated, the amount of colorant transferred to the paper is proportional to the amount of heat applied. This allows each pixel (dot) of the print to provide continuous-tone values, which is an important aspect of providing photographic-quality prints.

With a thermal wax donor the wax melts and adheres to the receiver. Either all or none of the dye is transferred. A binary halftone method is needed to produce intermediate density levels, so that each continuous-tone pixel must be represented by a group of binary dots. This limits the range of tones that can be produced and also limits the print resolution while introducing patterns into the image.

After the colorant is transferred to the paper, the paper may be "finished" by heating and pressing the paper to drive the dyes into the receiver. This keeps the dyes from easily transferring from the paper when it is handled or contacted by other materials. Some very high-quality sublimation dye printers use a laser diode, rather than a resistive head, to heat the dye donor.[50]

### 21.8.2  Inkjet Printers

Inkjet printers propel droplets of ink from a reservoir, through the air, and onto the paper, as shown in Fig. 21.19. The ink may be supplied in liquid form or as a solid wax stick that is melted into the reservoir. The ink is piped from the reservoir to an ink chamber, which creates a stream of ink drops from a group of ultrafine nozzles. The stream of ink drops may be modulated by resistive heaters (which expand the liquid in the chambers) or by piezoelectric crystals (which flex to reduce the chamber volume). In color printers, multiple ink chambers are used to supply cyan, magenta, yellow, and, in some systems, black ink. The printhead may be a disposable unit that includes the ink supply, nozzles, and thin-film resistor headers that control the output of each nozzle.

Inkjet printers do not provide true continuous-tone capability, but it is possible for variable-dot-size devices to produce acceptable-quality continuous-tone images. Solid-ink systems dry when the dots contact the paper. The image is then pressed by fuser rollers to improve the surface texture. Liquid systems provide the best image quality when used with specially treated paper, which prevents the ink from running or soaking into the paper.

### 21.8.3  Laser Electrophotographic Printers

Color electrophotographic printers use a four-stage system, with cyan, magenta, yellow, and black toner particles, as shown in Fig. 21.20. A photoconductive belt or drum is first exposed to a corona discharge in order to impart a uniform electrostatic charge. The photoconductor is then illuminated to discharge selective areas and create a latent image. In a printer this is normally done by modulating the output of a gallium arsenide (GaAs) diode laser. As the laser beam scans the photoconductive belt or drum, it modifies the electrostatic charge. Colored toner particles are electrostatically attracted to the charged photoconductor. The toner particles are then transferred to a separate electrostatically charged drum.

The photoconductor belt is cleaned, and the process continues using the next color for a total of four cycles, until the drum contains all four toner colors. The image is then rolled onto the paper, which is heated and pressed to fuse the image. Halftoning methods are normally used to create continuous-tone images. Other methods may be used to form the latent image on the photoconductor, such as a light-emitting diode (LED) printhead assembly incorporating thousands of LED emitters.

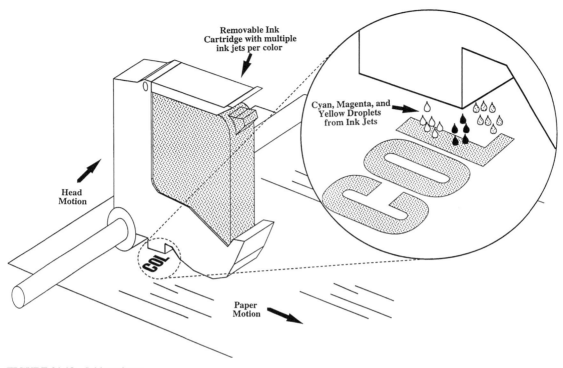

**FIGURE 21.19**  Inkjet printer.

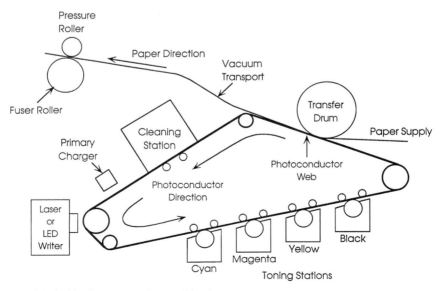

**FIGURE 21.20**  Color electrophotographic printer.

### 21.8.4  Silver-Halide Printers

Most silver-halide printers use standard color photographic paper that is developed in the conventional manner. The printers are typically used by photofinishers and other service providers who require very high throughput. Some printers include the chemical-based paper development process in the printer unit. Others provide a light-tight roll of exposed prints, which are processed using the same high-speed equipment used to develop conventional optical prints.

The photographic paper may be exposed by a cathode ray tube (CRT), as shown in Fig. 21.21, or by a modulated laser or liquid-crystal light valve. In a CRT printer, color sequential exposures are obtained by a color wheel, which has RGB filters specially tuned for the sensitivities of the color paper.

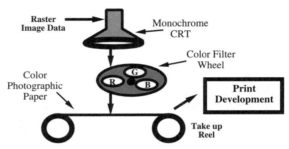

**FIGURE 21.21**   Silver-halide printer.

## GLOSSARY

**Aliasing**   Image artifacts in captured images caused by significant energy in the scene at frequencies higher than the Nyquist limit of the sensor. In single-chip color cameras, aliasing can produce unexpected color patterns in highly detailed monochrome objects.

**Analog-to-digital converter (A/D or ADC)**   A circuit that converts an analog signal, having a continuously varying amplitude, to a digitally quantized representation using binary output signals.

**Charge-coupled device (CCD)**   A type of silicon integrated circuit used to convert light into an electronic signal.

**Color filter array (CFA)**   A mosaic or stripe layer of colored transmissive filters fabricated on top of an imager in order to obtain a color image from a single-image sensor.

**Color pixel reconstruction**   An algorithm that creates a fully populated color image record from the output of a CFA-type sensor by interpolating values for each color at each pixel location.

**Correlated double sampling (CDS)**   A circuit commonly used to process the output signal from a CCD image sensor in order to reduce low-frequency noise components.

**Depth of field**   The difference between the maximum and minimum distances from a camera to objects in a scene that can be captured in acceptably sharp focus.

**Digital camera**   An electronic camera that captures images using a solid-state image sensor, and then outputs a digital signal representing the images or records the images on a digital storage medium.

**Edge enhancement**   A signal-processing operation that accentuates edge details within an image to increase the apparent sharpness. Such operations may also be called *aperture correction, sharpening,* or *peaking.*

**Exposure determination**    A method for setting the appropriate lens aperture and exposure time for a scene to be captured.

**Full-frame imager**    A type of image sensor consisting of a single light-sensitive array of photoelements that also store the image during the sensor readout period.

**Gamma correction**    A signal-processing operation that changes the relative signal levels in order to adjust the image tone reproduction, typically to correct for the nonlinear (nonunity gamma) light output–versus–signal input characteristic of the display. The relationship between the camera light input and the camera output signal level, called the *optoelectronic conversion function* (OCEF), provides the camera's gamma correction curve shape.

**Hybrid photography**    A type of digital photography system in which original scenes are captured using traditional silver-halide–based photography, and the resulting film or prints are scanned to provide digital images that may be displayed, manipulated, transmitted, etc.

**Hyperfocal distance**    The focus distance of a camera lens that offers the greatest depth of field. When a camera is focused at the hyperfocal distance, all objects from half the hyperfocal distance to infinity are within the camera's depth of field.

**Imaging chain**    A flow diagram that indicates all of the components used to produce a final image in a digital photography system.

**Image compression**    A process that alters the way image data are encoded in order to reduce the average size of an image file.

**Image data format**    A specification for storing image data and related information in a digital file. One example is TIFF, tag image file format, which can be used to store various types of monochrome or color bit-mapped images.

**Interline sensor**    A type of image sensor consisting of a two-dimensional array containing light-sensitive photoelements adjacent to light-shielded vertical storage registers.

**Memory card**    A small, thin, removable memory unit, containing digital integrated circuit memory chips, housed in a rugged package.

**Optical prefilter**    An optically transmissive device, such as a stack of birefringent quartz plates, that limits the high-frequency content of an image focused on a solid-state image sensor in order to reduce aliasing.

**Photo CD disc**    A compact disc–recordable (CD-R) optical write-once disc that stores scanned photographic images using the Image Pac image data format.

**Still-video floppy (SVF)**    A standardized recording medium for analog electronic still cameras developed in the early 1980s. SVF cameras use a 2-in.-diameter floppy disk capable of storing either 50 field images or 25 frame images.

**White balance**    A process for adjusting the relative signal levels of the red, green, and blue channels from a camera to correct for the color of the light source illuminating a scene, so that objects that appear to be white in the scene are reproduced as white on the soft-display or hard-copy print.

## *REFERENCES*

1. K. Hanma et al., "An electronic still camera technology," *IEEE Transactions on Consumer Electronics,* vol. CE-30, Aug. 1984.

2. N. Kihara et al., "The electronic still camera: A new concept in photography," *IEEE Transactions on Consumer Electronics,* vol. CE-28, no. 3, Aug. 1982, pp. 325–335.

3. T. E. Whiteley et al., "Synergism: Photography into the 21st century," *Journal Society of Photographic Science Technology, Japan,* vol. 53, no. 2, 1990, pp. 95–105.

4. M. Kriss et al., "Critical technologies for electronic still imaging systems," *SPIE,* vol. 1082, 1989, pp. 157–184.

5. T. Tani, "The present status and future prospects of silver halide photography," *Journal Imaging Science and Technology,* vol. 39, no. 1, Jan.–Feb. 1995, pp. 31–40.

6. T. Jackson and Bell, C., "A 1.3 megapixel resolution portable CCD electronic still camera," *Proceedings SPIE,* vol. 1448, Feb. 1991, pp. 2–12.

7. M. Takemae et al., "The development of SLR-type digital still camera," *Proceedings IS&T 48th Annual Conference,* 1995, pp. 414–417.

8. K. Parulski et al, "Enabling technologies for a family of digital cameras," *Proceedings SPIE,* vol. 2654, Feb. 1996, pp. 156–163.

9. H. Suetaka, "LCD digital camera QV-10," *Proceedings SPSTJ 70th Anniversary Symposium on Fine Imaging,* Tokyo, Oct. 1995, pp. 63–64.

10. A. Theuwissen, *Solid-State Imaging with Charge-Coupled Devices,* Kluwer Academic, Dordrecht, The Netherlands, 1995.

11. W. Steffe et al., "A high performance 190 × 244 CCD area image sensor array," *International Conference on Applications of Charge-Coupled Devices,* 1975, pp. 101–108.

12. L. Thorpe et al., "New advances in CCD imaging," *SMPTE Journal,* May 1988, pp. 378–387.

13. M. Aoki et al., "2/3-inch format MOS single-chip color imager," *ITTT Transactions on Electron Devices,* vol. 29, no. 4, April 1982, pp. 745–750.

14. E. Fossum, "CMOS image sensors: Electronic camera on a chip," *IEEE International Electron Devices Meeting Technical Digest,* Dec. 1995, pp. 1–9.

15. W. Miller et al., "A family of full-frame image sensors for electronic still photography," *IS&T's 47th Annual Conference,* 1994, pp. 649–651.

16. E. Stevens et al., "A 1-megapixel, progressive scan image sensor with antiblooming control and lag free operation," *IEEE Transactions on Electron Devices,* vol. 38, no. 5, 1991, pp. 981–988.

17. A. Weiss et al., "Microlenticular arrays for solid state imagers," *Journal Electrochemical Society,* vol. 133, no. 3, 1996, p. 110C.

18. G. Holst, *CCD Arrays, Cameras, and Displays,* JCD, Winter Park, Fla., 1996, pp. 83–84.

19. E. Edwards et al., "New digital photo-studio camera using two-directional spatial offset-type imager," *Proceedings IS&T's 47th Annual Conference,* 1994, pp. 658–660.

20. Y. Okano et al., "Electronic digital still camera using 3-CCD image sensors," *Proceedings IS&T's 48th Annual Conference,* 1995, pp. 428–432.

21. K. A. Parulski, "Color filters and processing alternatives for one-chip cameras," *IEEE Transactions on Electron Devices,* vol. ED-32, no. 8, Aug. 1985, pp. 1381–1389.

22. P. Dillon et al., "Color imaging system using a single CCD area array," *IEEE Transactions on Electron Devices,* vol. 25, no. 2, Feb. 1978, pp. 102–107.

23. H. Sugiura et al., "False color signal reduction method for single-chip color video cameras," *IEEE Transactions on Consumer Electronics,* vol. 40, no. 2, May 1994, pp. 100–106.

24. M. Shubin, "Lenses: The depth of the field," *Videography,* Feb. 1986.

25. Y. Lee et al., "A fuzzy control processor for automatic focusing," *IEEE Transactions on Consumer Electronics,* vol. 40, no. 2, May 1994, pp. 138–143.

26. J. Greivenkamp, "Color dependent optical prefilter for the suppression of aliasing artifacts," *Applied Optics,* vol. 29, no. 5, Feb. 1990, pp. 676–684.

27. ISO. "Photography—Electronic Still Picture Cameras—Determination of ISO Speed," ISO 12232.

28. M. White et al., "Characterization of surface channel CCD image arrays at low light levels," *IEEE Journal of Solid State Circuits,* vol. SC-9, Feb. 1974, pp. 1–14.

29. W. Schreiber, *Fundamentals of Electronic Imaging Systems* (3rd ed.), Springer-Verlag, New York, 1993.

30. K. Parulski et al., "Digital, still-optimized architecture for electronic photography," *Proceedings IS&T's 47th Annual Conference,* 1994, pp. 665–667.

31. K. Parulski et al., "A digital color CCD imaging system using custom VLSI circuits," *IEEE Transactions on Consumer Electronics,* vol. 35, no. 3, Aug. 1989, pp. 382–389.

32. D. Cok, "Reconstruction of CCD images using template matching," *Proceedings IS&T's 47th Annual Conference,* 1994, pp. 380–384.

33. E. Shimizu et al., "A digital camera using a new compression and interpolation algorithm," *Proceedings IS&T's 49th Annual Conference,* 1996, pp. 268–272.

34. ITU, "Basic Parameter Values for the HDTV Standard for the Studio and for International Programme Exchange," ITU-R BT.709, 1993.

35. ISO, "Photography—Electronic Still Picture Cameras—Methods for Measuring Opto-electronic Conversion Functions (OECFs)," ISO/CD 14524.

36. K. Parulski et al., "High performance digital color video camera," *Journal Electronic Imaging,* vol. 1, Jan. 1992, pp. 35–45.

37. Y. T. Tsai, "Color image compression for single-chip cameras," *IEEE Transactions on Electron Devices,* vol. 38, no. 5, May 1991, pp. 1226–1232.

38. A. Hoagland et al., *Digital Magnetic Recording* (2nd ed.) Wiley, New York, 1991.

39. T. Saimi, "Optical disk scanning technology," in *Optical Scanning,* G. Marshall (ed.), Marcel Dekker, New York, 1991.

40. ISO, "Photography—Electronic Still Picture Cameras—Removable Memory," ISO/CD 12234.

41. K. Parulski et al., "TIFF/EP, a flexible image format for electronic still cameras," *Proceedings IS&T's 48th Annual Conference,* 1995, pp. 425–428.

42. M. Watanabe et al., "An image data file format realized in PC card for digital still camera," *Proceedings International Symposium on Electronic Photography (ISEP),* Köln, Germany, Sept. 1994.

43. J. Milch, "A storage format for resolution independent imaging systems," *Proceedings International Symposium on Electronic Photography (SEP),* Köln, Germany, Sept. 1996.

44. InterColor Consortium, "InterColor Profile Format," ver. 3.0, June 10, 1994.

45. J. Larish, *Photo CD, Quality Photos at Your Fingertips,* Micro Publishing Press, Torrance, Calif., 1993.

46. J. R. Milch, "Photo CD: A system for effective generation of digital images from film," *Proceedings International Symposium on Electronic Photography (ISEP),* Köln, Germany, 1992, pp. 86–90.

47. Y. Ota, "The new digital photography system (Konica Picture MD System)," *Proceedings SPSTJ 70th Anniversary Symposium on Fine Imaging,* Tokyo, Oct. 1995, pp. 93–96.

48. R. Durbeck, *Output Hardcopy Devices,* Academic Press, Orlando, Fla., 1988.

49. S. Ohno, "Pictorial hardcopy for electronic photography," *Proceedings International Symposium on Electronic Photography (SEP),* Köln, Germany, Sept. 1992, pp. 34–38.

50. C. DeBoer, "Laser thermal retained imaging," *Proceedings IS&T's 47th Annual Conference,* 1994, pp. 607–608.

## *ABOUT THE AUTHOR*

Ken Parulski is chief architect for digital camera development at the Eastman Kodak Company in Rochester, N.Y. He joined the physics division of the company's research laboratories after receiving the B.S. and M.S. degrees in electrical engineering from the Massachusetts Institute of Technology in 1980. He has worked at Kodak on the development of digital still cameras, video cameras, film scanners, HDTV, and the Kodak Photo CD system. Parulski chairs the U.S. delegation to the ISO standards group on electronic still picture imaging. He has authored more than 30 technical papers and has given invited lectures on digital photography to audiences in the United States, Europe, and Japan. He has been granted over 50 U.S. patents related to digital photography.

# P · A · R · T · 6

# DIGITAL AUDIO PRODUCTS

# CHAPTER 22
# COMPACT DISC PLAYERS

**Koki Aizawa**
*Pioneer Electronic Corporation*

## 22.1 INTRODUCTION

An example of a block diagram of a Compact Disc (CD) player is shown in Fig. 22.1. Signals recorded on a disc are read by an optical pickup and transformed into an electric signal. The signals corresponding to the laser beam defocusing or out-of-track signals are then separated from the readout signals. The focusing servo enables the laser beam to focus accurately on the recorded pits on the disc. The tracking servo enables the laser beam to trace a track. Tracking error signals are used to generate the carriage servo signals, which are then fed to the servo carriage so that the entire pickup unit can move freely in the radial direction. The servomechanisms are basically the same as those for a laser disc player.

Digital data signals modulated by eight-to-fourteen modulation, or EFM (see Section 22.3.2; also see Chapter 7, Section 7.2.5), are equalized by an RF amplifier to enlarge the eye pattern's opening. They are then converted to binary signals through an automatic threshold control (ATC) circuit. The ATC feeds back low-frequency components (DC to several Hz) to the ATC comparator input utilizing the fact that the digital sum value (DSV) of EFM becomes zero.

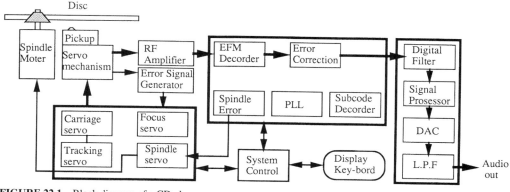

**FIGURE 22.1**    Block diagram of a CD player.

The frame synchronization pattern is separated by the bit clock regenerated in a phase-locked loop (PLL) circuit. The spindle servo control, at the same time, ensures that the disc rotation reaches a targeted linear velocity speed.

After the EFM demodulation, the digital signal passes through various processes such as de-interleaving, error detection, and error correction or concealment. The signal then passes through a digital filter, a digital signal processor, and a digital-to-analog converter (DAC) and, finally, an audio output is filtered out by a reconstruction low-pass filter (LPF). There is no wow and flutter at the audio output signal because the digital signal with jitters is stored in a memory and is then read by a fixed and precise quartz clock.

Random access is fast because the carrier is a disc and the laser readout is a noncontact system and also because the table of contents (TOC) information is provided in the subcode data recorded in the lead-in area of a disc. By utilizing not only the TOC that describes information on the entire contents of a disc but also additional information written along with it pertaining to an individual selection such as a musical number and elapsed time, a specific position can be found instantaneously and eventually displayed.

## 22.2  SERVO TECHNOLOGY

In order to read signals from the disc exactly, it is necessary to keep a very small light spot on the disc and to trace the track by accurately controlling the distance between the disc surface and the objective lens of a pickup.

The focus error signal (proportional to the distance between the disc and the objective lens) and the tracking error signal (proportional to the off-track amount) are detected and fed to the servo system. Many different methods have been developed for obtaining the focus error signal and the tracking error signal.

### 22.2.1  Generation of Focus Error Signals

*Astigmatic Method.*   The astigmatic method of generating focus error signals uses astigmatic aberration caused by the returning light path. Astigmatic aberration is the phenomenon in which

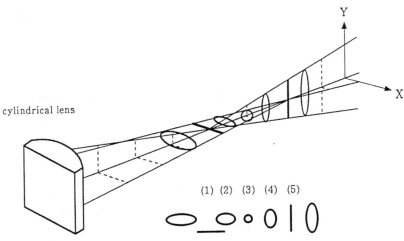

**FIGURE 22.2**   Principle of the astigmatic method of generating focus error signals.

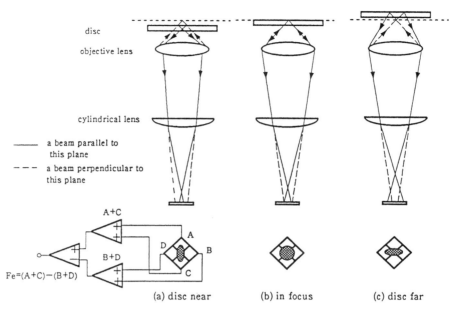

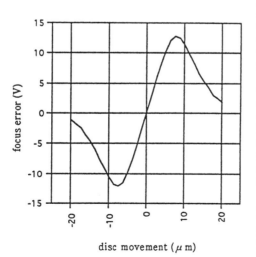

**FIGURE 22.3**    Focus error detection by the astigmatic method.

the focal point in the $x$ direction is formed at a different position from the focal point in the $y$ direction (Fig. 22.2). The signal can be generated by an optical element such as a cylindrical lens or a tilted glass plate. The cross section of the astigmatic light beam changes along the center axis—(1) horizontal line, (2) horizontal ellipse, (3) circle, (4) vertical ellipse, and (5) vertical line—as shown in Fig. 22.2. This illustration also shows the relationship between the vertical disc movement and the light beam shape on the detecting surface. The light beam is detected by a quadrant photo diode (PD) and the focus error signal (Fe) is given by Fe = $(A + C) - (B + D)$ (Fig. 22.3$a$). When the disc is near the focal plane the shape of the light beam becomes a vertical ellipse and Fe > 0. When the disc is far from the focal plane (Fig. 22.3$c$), the beam is a horizontal ellipse and Fe < 0. When the disc is in focus (Fig. 22.3$b$), the beam is a circle and Fe = 0. Consequently, an S-shaped signal curve is obtained as a function of the focal depth (Fig. 22.4).

**FIGURE 22.4**    Focus error curve.

*Knife-Edge Method.*    The knife-edge method uses a knife edge positioned in the return optical path so that the focus error signal is obtained by detecting the movement of the light image on the PD.

The relationship between disc movement and detected beam profile is shown in Fig. 22.5.

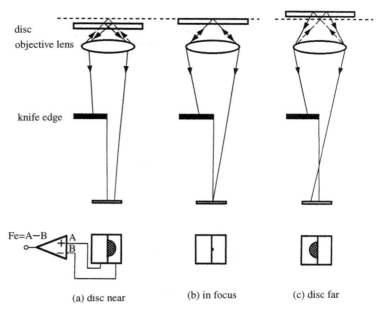

**FIGURE 22.5**   Focus error detection by the knife-edge method.

When the disc is near the focal plane, the shape of the light beam becomes a right-sided semi-circle. When the beam is far from the focal plane, the shape becomes a left-sided semicircle. And when the beam is focused on the disc, the return beam is also focused on the detector surface. Because this beam is detected by the two-division PD (2D-PD), the focus error signal Fe is given by Fe $= A - B$.

When the knife edge is used, a part of the light beam is cut off and the light power decreases. In order to minimize the power loss, a prism or hologram (Fig. 22.6) is applied to separate the light beam. Both of the separated beams are detected by the 2D-PD positioned on them.

***Beam-Size Method.***    The beam-size method uses a detector—for example, a three-division PD (3D-PD)—without modifying the beam shape of the return light in order to get the focus error

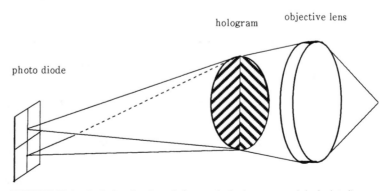

**FIGURE 22.6**   Optical path using a hologram (only the return path is depicted).

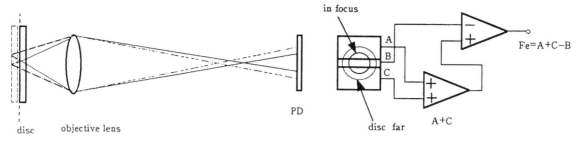

**FIGURE 22.7** Focus error detection by beam-size method.

signal. The beam diameter on the PD changes as the disc moves vertically. The change is detected by the 3D-PD, the pattern of which is shown in Fig. 22.7. The focus error signal is given by $Fe = A + C - B$.

When the beam is just in focus on the disc, the beam diameter is decided as $Fe = 0$. The optical path is as shown in Fig. 22.7. When the beam is far from being in focus on the disc (this means the disc is far from the objective lens), the beam diameter becomes larger and $Fe > 0$. When the disc is near, the beam diameter becomes smaller and $Fe < 0$. An M-shaped focus error curve is procured by this method.

Another example of the beam-size method is shown in Fig. 22.8. In this case the change of beam shape reverses between the front and rear sides of the beam waist, and the beam diameter on PD2 increases as the beam diameter on PD1 decreases. The focus error signal $Fe = (A + C + E) - (B + D + F)$. This can also be expressed as $Fe = (A - D) + (E - B) + (C - F)$. This means that the focus error signal is obtained by subtracting the output of each PD element from that of the corresponding element. Then a stable focus servo can be realized that is not affected by a useless signal in the light spot. An S-shaped focus error curve results.

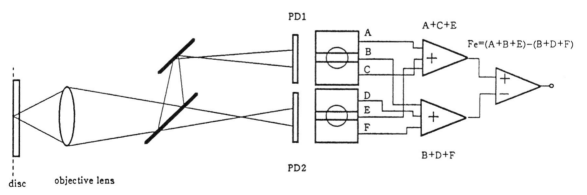

**FIGURE 22.8** Improvement in beam-size method.

## 22.2.2 Generation of Tracking Error Signal

***Three-Spot Method.*** The three-spot method uses sub-beams ($\pm$first-order diffracted waves) generated by the diffraction grating and the main beam (zeroth order diffracted wave) for RF signal detections, to get tracking error signals. The $\pm$first-order beams, separated from the beam by the diffraction grating, are focused on the disc side-by-side at very little distance from the main beam focused position (Fig. 22.9). The return beams from the disc reach the PD

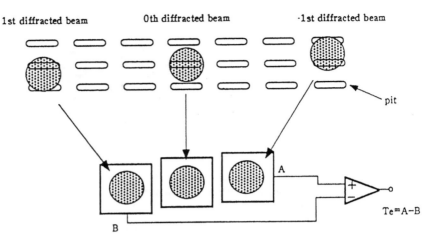

**FIGURE 22.9**   Detection of tracking error.

via almost the same light path as that of the main beam. The PD consists of three parts, each of which is illuminated by one of the three beams, as shown in Fig. 22.8. The tracking error signal Te is given by subtracting each of the sub-beam outputs, or Te $= A - B$.

When the main beam is placed on the track, output of the sub-beam becomes the same and Te $= 0$. When the main beam shifts slightly to the side, the balance between the sub-beam outputs changes and Te $> 0$. When the main beam shifts to the other side, Te $< 0$.

The three-spot method is widely used for CD players because of its reliable movement. But it does have weak points, in that a diffraction grating is needed and the light power of the main beam decreases because of the sub-beams.

***Push-Pull Method.***    In the push-pull method the light intensity distribution of the light that is reflected and defracted by the track differs with respect to the relative position of the track. The light is detected by a 2D-PD placed symmetrically to the center of the track. The light intensity distribution reflected by the track changes depending upon the location of the track (Fig. 22.10). When the center of the beam is placed on the track, or between tracks, the light intensity distribution becomes symmetrical (Fig. 22.10$a$). When the beam is located elsewhere (Fig. 22.10$b$), the light intensity distribution becomes asymmetrical. The light intensity distribution is detected by a PD that has the divided line parallel to the track. The tracking error signal, Te, is obtained by Te $= A - B$.

The push-pull method needs no special optical element for generating the tracking error, and the error signal can be obtained by a very simple structure. But the tracking error signal is affected by such factors as pit depth, disc tilt, and movement of the objective lens. Various corrections are needed to offset these factors in order for this method to be practical.

***Heterodyne Method.***    The heterodyne method overcomes the shortcomings of the push-pull method so that the effect of the tracking error signal due to pit depth, disc tilt, and movement of the objective lens can be decreased. A quadrant-PD is used instead of a 2D-PD (Fig. 22.11$a$). The tracking error signal is generated by the signal processing described later using signals such as S1 $= A + B + C + D$ and S2 $= (A + C) - (B + D)$.

In Fig. 22.12 the relationship between the pit position inside the light spot on the disc and the light intensity distribution on the PD is depicted when the pit depth is $\lambda/4$, where $\lambda$ is the wavelength of light. When the pit enters into the light spot (Fig. 22.12$a$), the tracking error signal

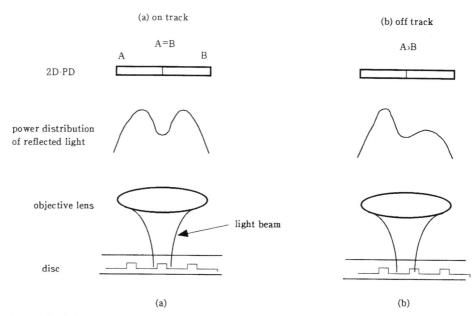

**FIGURE 22.10**    Tracking error detection by the push-pull method.

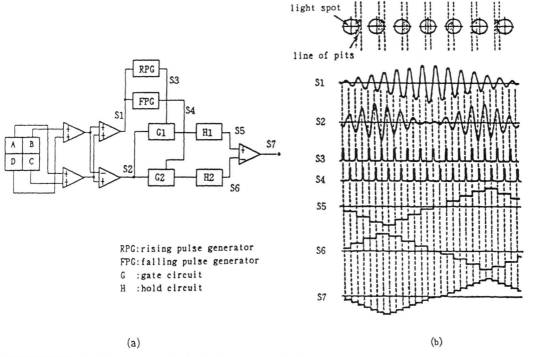

RPG: rising pulse generator
FPG: falling pulse generator
G   : gate circuit
H   : hold circuit

(a)

(b)

**FIGURE 22.11**    Tracking error detection by the heterodyne method.

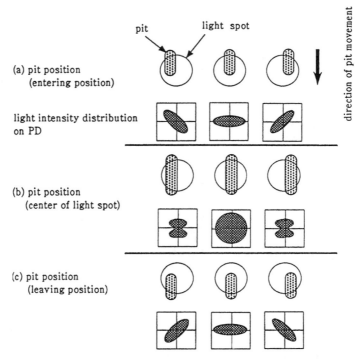

pit    light spot

direction of pit movement

(a) pit position
(entering position)

light intensity distribution
on PD

(b) pit position
(center of light spot)

(c) pit position
(leaving position)

**FIGURE 22.12**    Pit position vs. light intensity distribution on the photo detector
(pit depth $= \lambda/4$).

is proportional to the S2 signal. When the pit leaves the light spot (Fig. 22.12*c*), the tracking error signal is also proportional to the S2 signal. But the phase of the S2 signal in the entering point is different by 180° from that of S2 in the leaving point. The tracking error is obtained by running the signal processing shown in Fig. 22.12*b*.

### 22.2.3  Detection of RF Signal

The playable limitation of linear density is decided by the optical pickup. Figure 22.13 shows the modulation transfer function (MTF) as a function of spatial frequency. This represents the contrast of the reflected light when the disc on which the various frequency signals are recorded is played back. The MTF is proportional to the level of the played back signal.

The spatial frequency $v$ is different from the conventional frequency, which corresponds to the linear density and is expressed in $\text{mm}^{-1}$. MTF decreases linearly as a function of the spatial frequency and, at a cutoff frequency $v_c$ ($v_c = 2\text{NA}/\lambda$, where NA is the numerical aperture of the objective lens and $\lambda$ is the wavelength of the laser used), becomes zero. This formula indicates that a higher linear density can be obtained as NA gets larger and $\lambda$ gets smaller.

In the case of the disc, the linear velocity is greater at the inner side of the disc than at the outer side of the disc when the disc rotates at a specific rotational speed. Therefore, if the same signal is recorded, the linear density differs according to the position from the inner to the outer side of the disc. In order to keep an identical linear density all over the disc, the rotational speed should be decreased at the mastering site as the recording position moves from the inner to the outer side of the disc. This method is called *constant linear velocity* (CLV).

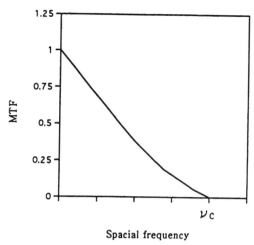

**FIGURE 22.13**  Modulation transfer function as a function of spatial frequency.

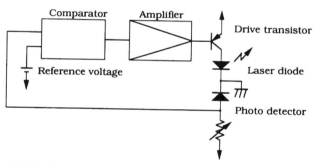

**FIGURE 22.14**  The automatic power control circuit block diagram.

## 22.2.4  Automatic Power Control (APC)

Any semiconductor laser used for a CD player usually has a maximum rating of 5 mW or, in practice, about 3 mW of emitting power. The laser output power should not vary with temperature or temporal change in order to have reliable RF signal readout and stable servo control.

The characteristics of the output power versus current are highly sensitive to temperature, so a simple constant current drive alone cannot hold the emitting power constant. The remedy for this problem is to use a photo detector to detect the laser emitting power directly and to control the laser-driving current so as to keep the laser emitting power at a constant value. A circuit for doing this is called an *automatic power control circuit* (Fig. 22.14).

## 22.2.5  Focus Servo

It is necessary to focus on the reflective layer of a disc in order to read signals embedded on the disc in the form of pits. In general terms, to maintain playback signal quality, the focusing depth should be within the range of $\pm 1$ $\mu$m. Because discs are not necessarily flat during rotation, an actuator mechanism is incorporated in the pickup unit to ensure that the objective lens moves in a direction perpendicular to the disc at all times. Servo control lets the objective lens follow the disc surface. A normal CD has a maximum vertical deviation of $\pm 500$ $\mu$m and a maximum vertical acceleration of 10 m/s$^2$ (Fig. 22.15). In other words, if the target of the servo is within $\pm 1$ $\mu$m, an open-loop DC gain needs to be 54 dB or more, and the loop bandwidth becomes approximately 1 kHz by adding phase compensation to the actuator, which consists of a voice coil like bobbin and a magnetic circuit whose dynamic characteristics are shown in Fig. 22.16. As shown in this illustration, the actuator forms a second-order system.

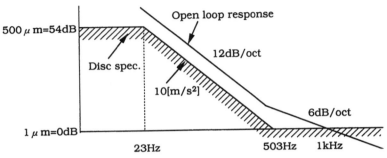

**FIGURE 22.15** Disc specification (focus).

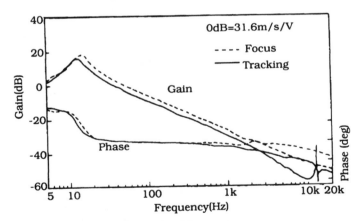

**FIGURE 22.16** Dynamic behavior of the actuator.

### 22.2.6 Tracking Servo

In order for the system to be able to read out pits recorded on a disc in helix form, the laser beam position must be controlled within $\pm 0.1$ $\mu$m with respect to the center of the track pitch of 1.6 $\mu$m. To accomplish this accuracy, it is necessary to provide a tracking servomechanism in the radial direction in the disc. The normal CD has a maximum radial deviation of $\pm 70$ $\mu$m and a maximum radial acceleration of 0.4 m/s$^2$ (Fig. 22.17). To meet these specifications, it is

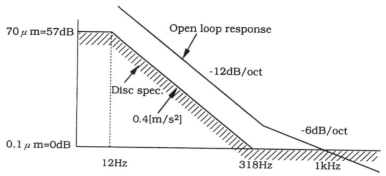

**FIGURE 22.17**    Disc specification (tracking).

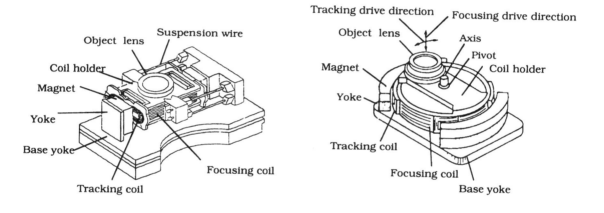

**(1) Wire-suspension Type actuator**

**(2) Sliding and rotary type actuator**

**FIGURE 22.18**    Actuator types.

common practice to use a DC gain of 57 dB or more and approximately 1-kHz bandwidth to take into account phase compensation, playability, and shock resistibility.

Objective lens actuator schemes for tracking include wire-suspension, sliding-and-rotary, and hinge types (Fig. 22.18). Phase compensation is necessary because each of these types shows a second-order system response in terms of dynamic behavior.

## 22.2.7    Carriage Servo

The driving scheme for CD player tracking generally uses an actuator, and the service range is limited to about ±300 μm. This means that the entire playable area of the disc cannot be covered. To solve this problem, a servo is provided so that an entire pickup unit can slide in a radial direction on the disc. Such a servo is called the *carriage servo.* Low frequencies, including DC components from the driving voltage of the tracking actuator, are fed to the driving input as a driving signal for the sliding mechanism. The tracking actuator is placed almost in the center position or the self-standing position.

### 22.2.8  Spindle Servo

The disc rotational speed varies from about 500 rpm to 200 rpm according to the linear velocity. A constant linear velocity (CLV) of 1.2 m/s to 1.4 m/s is used. Frame synchronization (referred to as *sync* hereafter) signals are recorded on the disc. These are reference signals that provide a constant interval in steady rotation.

A phase-locked loop servo works by detecting and feeding to the spindle motor driving circuit the differences in phase and velocity between the frame sync signal and a signal that is derived from a quartz precision clock signal in the CD player. The servo bandwidth depends on the variation of the angular velocity and is generally set at about double the frequency of the eccentricity of the disc at the inner position or about 8 Hz. Playback audio data are free from wow and flutter because they can be pitch-synchronized by the quartz clock.

### 22.2.9  Servo Signal Processing

The servo signal processor circuit controls basic functions of the mechanism inside the CD player. The main functions controlled are focus search and track jump.

*Focus Search.*    When the laser beam is focused on the pit surface, focusing error as a function of focusing depth is an S-shaped curve in which the linear range is only 10–20 μm. This means that a simple closure of a servo loop cannot bring the beam to an in-focus position. The solution is to move the focus actuator in the focal direction with a constant speed and then lock and close the servo loop when the error signal reaches the linear range of the S curve close to the in-focus position. The in-focus position is detected when overall light intensity of the RF signal is obtainable beyond a reference level in order to keep false error signals caused by the disc surface reflection from being misdetected (Fig. 22.19)

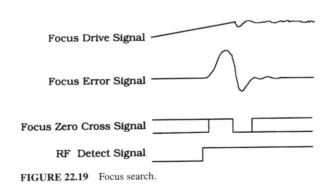

Focus Drive Signal

Focus Error Signal

Focus Zero Cross Signal

RF  Detect Signal

**FIGURE 22.19**   Focus search.

*Track Jump.*    Consider the case where one wants to play a target track other than the one currently being played. The laser beam stops tracking and needs to move across tracks. This action is called *track jump,* which is classified by three categories according to the number of tracks that must be crossed: one-track jump, multitrack jump, and track-count jump. One-track jump literally moves the laser beam by one track or to an adjacent track. Multitrack jump jumps an arbitrary number of tracks from two to about 100. Track-count jump jumps several hundred to several thousand tracks at a time. In practice, actual search to a target track is organized by combining these three jump modes (Fig. 22.20).

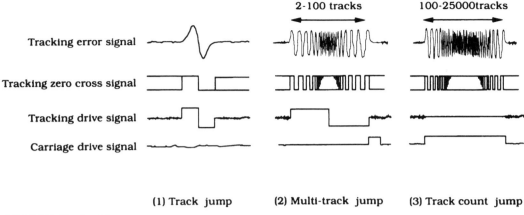

**FIGURE 22.20**   Track jump.

## 22.3   *SIGNAL PROCESSING*

### 22.3.1   RF Signal Demodulation

A series of recorded pits on the disc are translated into an RF signal by means of optoelectronic conversion. The RF signal contains clock components, and, if monitored by an oscilloscope, diamond-shaped waveforms called the *EYE diagram* are observed. The larger the EYE diagram is (wider in both width and height of the waveform), the smaller the jitters become and fewer errors occur in decoding. In this situation the EYE is said to be *open.*

The characteristics of the modulation transfer function (MTF) present no phase change and a constant rate decay in amplitude, as shown in Fig. 22.13. This signifies that the RF signal contains attenuated high-frequency components and, therefore, the EYE usually cannot be open sufficiently without electronic compensation. Such compensation, called *MTF compensation,* consists of a linear equalizer having the reverse characteristics of the attenuated high frequencies. The compensated signal is fed through a comparator, where it is translated into a binary digital signal. At the same time, asymmetry or unbalance of the EYE's opening occurs generated by the pit lengths' dispersion associated with disc mastering or the disc replication process or the performance of the optical pickup. The EYE's opening is adjusted by varying the comparator level making use of the digital sum value (DSV) with zero DC components. This is one of the features of eight-to-fourteen modulation (EFM), to be discussed next (Fig. 22.21).

### 22.3.2   EFM Decoding

The binary playback signal is reconfigured by the 4.3218-MHz clock extracted by the PLL and is then fed to the input of an EFM decoder. The decoder first separates the frame sync signal from the subcode and data. The frame sync signal has a unique pattern with specific polarity reversal intervals that never appear in data patterns. By identifying these patterns, it is possible to decode the subcodes and data that immediately succeed the frame sync signal patterns. The sync pattern is also used as a reference signal for the spindle servo control.

The detection circuit incorporates the sync protection and interpolation functions and takes into account such instances when frame sync patterns are not detected or are misdetected because of a disc imperfection or acquired flaw that results in the sync position being skewed from its expected position.

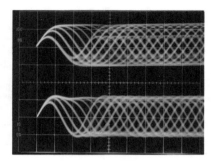

**EYE diagram**

**Upper:without MTF compensation**
**Lower:with MTF compensation**

**FIGURE 22.21**    The RF signal.

Because subcodes and data are recorded with every 8-bit unit encoded to 14 bits, this system is called eight-to-fourteen modulation (EFM). During decoding, the 14 bits revert to 8 bits. EFM-decoded data are stored in random access memory (RAM) and then undergo error detection, error correction, and de-interleaving. Audio data after error correction are then read from the RAM by the quartz clock and relayed to a digital-to-analog converter.

### 22.3.3  Error Correction

The error correction scheme used in the CD system is called cross-interleaved Reed-Solomon code (CIRC; see Chapter 4, Section 4.3.2), which is a combination of interleaving and a double-stage Reed-Solomon code. The minimum length of one stage is five, and each series named C1 or C2 can, respectively, correct up to double errors.

Actual error correction is designed to minimize miscorrection. The system sets a pointer at the position where a C1 series detects an error and references an error detection result by a C2 series so that the error will be corrected only if both results coincide. Correction does not take place if they disagree with each other and data interpolation is made instead.

Error concealment—either the previous word holding or the mean value interpolation—takes place only when C1 and C2 cannot correct errors. The previous word (data) holding functions when more than two consecutive error words continue to happen. Mean-value interpolation works when the preceding and the succeeding words, with respect to the error word, are correct. In both cases audio output is protected from any noise occurrence.

## 22.4  AUDIO

The CD-DA (Compact Disc–Digital Audio) contains digital audio data with a sampling frequency of 44.1 kHz. Quantization is 16 bits linear in the 2s complimentary format. The role of the "audio" part is to transform the digital data into analog signals and output them as purely and faithfully to the original sound source as possible.

### 22.4.1  Digital Filter

A $2^N$ oversampling or an interpolation filter is used for the reproduction of the CD-DA, where $N$ is a natural number. The input to the filter is a 44.1-kHz, 16-bit signal and the output is oversampled by the sampling frequency $2^N \times 44.1$ kHz, whereas the number of bits can be expanded to 18 to 24 according to the performance of a digital-to-analog converter. The filter generally consists of an odd-number-tapped and symmetry finite-impulse response filter with linear phase.

The advantage of oversampling is that it can reduce phase distortion caused by the analog low-pass filter or by passband ripples. The higher sampling frequency suppresses spurious components in low frequencies at the DAC output and thereby reduces the number of orders of an analog reconstruction filter needed to remove the spurious components. The higher the sampling frequency is, the gentler the frequency response or the lower the order of the filter. The requantization noise caused by the digital filtering distributes uniformly, ranging from DC to the Nyquist frequency of the oversampling frequency. Therefore, given the number of bits, the requantization noise within the audio bandwidth due to the rounding at the filter output decreases as the oversampling frequency increases.

The higher number of bits can reduce rounding errors as a result of computation at the digital filter output. However, even if a higher oversampling frequency is used as well as a higher number of bits, quantization noise during real operation cannot be improved from the level provided by 16 bits as long as the data are handed over in their original form to the processor, since data recorded in a CD-DA are 16-bit data.

### 22.4.2 Wide-Range Technology

*Wide Range in the Frequency Domain.* Because the sampling frequency is 44.1 kHz for a CD-DA, frequency components higher than 22.05 kHz (the Nyquist frequency) cannot be recorded or reproduced in principle. However, original sounds—sound events in the real world or instrument sounds—actually contain high-frequency components above 20 kHz, though their levels may be low. Recent studies reveal that the existence of these high frequencies relates to the psychoacoustics of human hearing and influences greatly the sound quality of what is heard. Some of the features of music signals are as follows:

* Original sounds contain frequency components above 20 kHz, though at low levels.
* The nature of the original sounds in the frequency domain are characterized as a function $1/f$ showing amplitude decay with the increasing frequency.
* Music signals are regarded as a set of triangular-like waves having very good transient responses.

A Pioneer proprietary system called *Legato Link conversion* regards these last two items as the salient features of music signals and uses this conclusion as the basis for the arithmetic algorithm for waveform reproduction. Furthermore, because music signals are those that vary instant by instant in time, it is necessary to link the sampling data sequence smoothly as a function of time in order to reproduce natural sound. By applying an algorithm that links data at every 22.7 µs of a CD recording by a smooth function curve that best reproduces spectral components of the original sound, the system estimates original signals and generates the interpolation data. It can then reproduce audio signals above 20 kHz.

Figures 22.22 and 22.23 show characteristics of a conventional D/A conversion and Legato Link conversion, respectively.

*Wide Range in the Level.* One of the technologies used in disc manufacturing is noise shaping. This consists of suppressing the quantization noise in the mid-to-low frequency range, which is acoustically quite audible, and then raising it to the high-frequency range, where it is less audible. This is done when the 20- or 24-bit data stream emanating from the A/D converter in the initial recording is converted to 16 bits for CD-DA.

During disc playback, when the reproducing 16-bit data stream has little or no changes (about 1 least significant bit, or LSB), data less than 1 LSB are generated from data preceding and succeeding itself so that they best fit to a smooth curve. These newly generated data are then

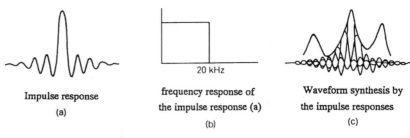

Impulse response

(a)

frequency response of
the impulse response (a)

(b)

20 kHz

Waveform synthesis by
the impulse responses

(c)

**FIGURE 22.22**   Convention digital-to-analog conversion.

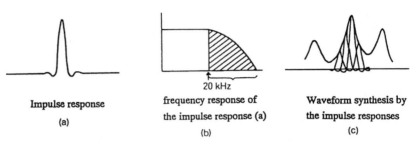

Impulse response

(a)

frequency response of
the impulse response (a)

(b)

20 kHz

Waveform synthesis by
the impulse responses

(c)

**FIGURE 22.23**   Legato Link conversion.

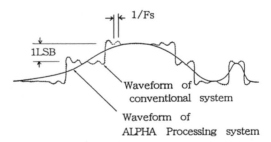

1/Fs

1LSB

Waveform of
conventional system

Waveform of
ALPHA Processing system

**FIGURE 22.24**   Alpha processing digital-to-analog conversion.

synthesized along with 16-bit input data to create 20- to 24-bit output data, which are then fed to
the DAC. An example is given in the case of Denon's ALPHA (adaptive line pattern harmonized
algorithm) processor in Fig. 22.24.

### 22.4.3   DAC

There are two general categories of digital-to-analog converters, or DACs: the multibit DAC,
where information resides in the level of the output signal, and the 1-bit DAC, where the information resides in the time axis.

***Multibit DAC.***    There are any number of ladder resistors or current sources that correspond to
the number of bits. These sources output voltage or current corresponding to the input data at
every sampling timing. Therefore, in a general sense, the larger the number of bits, the higher

the performance. Distortion is limited by the precision of the resistors or current sources. Their precision should be $\leq 2^{-(N+1)}$ in order to acquire $N$ bit precision, where output error is $\leq 2^{-(N+1)}$. As a result, the precision is limited to about 20 or 22 bits at the present time even by trimming or selecting the resistors.

Noise in the DAC distributes uniformly over the frequency range. Careful implementation of the DAC ensures the relatively easy realization of the S/N ratio and dynamic range specifications. Because the DAC does not require specific high clock signals, it is advantageous in terms of electromagnetic interference.

**One-Bit DAC.**   The output of the single-bit-scheme DAC is either a voltage or current waveform of a binary pulse train. By varying the pulse density of the pulse width modulation or pulse density modulation, low-frequency components are embedded in the pulse train. By filtering the low-frequency components by a low-pass filter, the audio frequency output is obtained.

The output from the DAC is sampled by an extremely high frequency compared to the input sampling frequency. Then the difference between the output and input audio signals of the DAC is fed back to the input via a specific transfer function. Through this type of noise shaping, the audible in-band noise can be reduced.

Distortion is easy to manage compared to a multibit DAC scheme because there is theoretically no nonlinear distortion. Therefore, relatively low-cost, high-performance DACs are easily realizable. But because noise shaping makes out-of-band noise greater and high-frequency clocking is needed, any clock jitters may result in deteriorating characteristics.

## 22.5  *CONTROL AND DISPLAY SYSTEM*

### 22.5.1  General Characteristics

After EFM demodulation, one frame (136 $\mu$s) contains subcode information for control and display purposes in 8 bits represented by P, Q, R, S, T, U, V, and W. These eight bits are called the *subcoding symbol*. Ninety-eight subcoding symbols form one subcode block (13.3 ms), as shown in Fig. 22.25. Among the 98 subcoding symbols, the first two are 14-bit sync patterns when EFM is decoded as shown in Fig. 22.25. If the sync patterns are not found in the EFM 8-bit to 14-bit conversion table, the pattern is still detected at EFM decoding. The remaining 96 subcoding symbols are converted back via 14-bit to 8-bit conversion.

### 22.5.2  System Controller

A system controller addresses from subcode Q the signal decoded by the decoder section, displays the music number as well as minute and second, and also controls the servos. In the search mode the system controller moves the pickup unit by activating the tracking servo or carriage servo by an external command to a target address, comparing the current address to the target address.

### 22.5.3  CD Graphics

One application making use of the subcode information is CD graphics that show the lyrics or other information on the display in accordance with the music selection being played. The subcodes used for the display are the 6 bits from R to W out of the 8-bit subcode. They are treated as one symbol. A unit of 6-bit data by 96 symbols is defined as one packet, to which two symbols for synchronization are added. One pack of 24 symbols, obtained from 96 symbols divided by 4, is defined as a minimum unit for one instruction.

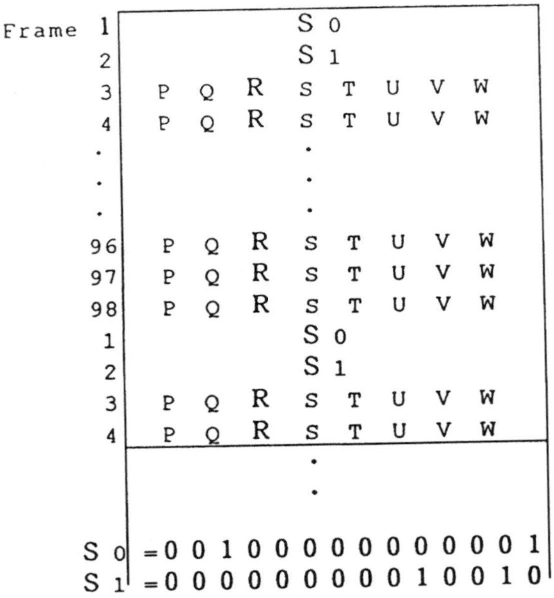

```
Frame  1 ┌─────────────────────────────────┐
       2 │              S 0                │
       3 │ P  Q  R   S  S 1  U   V   W      │
       4 │ P  Q  R   S  T  U   V   W        │
         │ P  Q  R   S  T  U   V   W        │
       · │          ·                      │
       · │          ·                      │
       · │          ·                      │
      96 │ P  Q  R   S  T  U   V   W        │
      97 │ P  Q  R   S  T  U   V   W        │
      98 │ P  Q  R   S  T  U   V   W        │
       1 │              S 0                │
       2 │              S 1                │
       3 │ P  Q  R   S  T  U   V   W        │
       4 │ P  Q  R   S  T  U   V   W        │
         │          ·                      │
         │          ·                      │
         │          ·                      │
         └─────────────────────────────────┘

       S 0 = 0 0 1 0 0 0 0 0 0 0 0 0 0 1
       S 1 = 0 0 0 0 0 0 0 0 0 1 0 0 1 0
```

**FIGURE 22.25**   Subcoding block.

The display screen construction has two modes: a TV graphics mode and a line graphics mode. The TV graphics mode has a screen consisting of 288 horizontal by 192 vertical dots with a maximum of 16 colors specified from 4096 colors. Each RGB color is represented by a 16 gray scale. The line graphics mode has a screen consisting of 288 horizontal by 24 vertical dots with a maximum of eight colors specified. Each RGB color is represented by a 2 gray scale.

The unit represented by horizontal 6 by vertical 12 dots on the screen is called the *one font,* which corresponds to one drawing instruction. Channel numbers from 0 to 15 are prepared for the drawing instruction so that one can select one specific instruction from them. This function helps opt for one desired language from multilingual commentary in the disc, for instance.

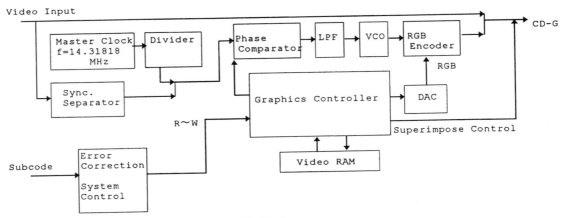

**FIGURE 22.26**   An example of the block diagram of the CD-G system.

Shown in Fig. 22.26 is a block diagram of a system that has added the function enabling superimposed display information on an external video signal. The display is realized by playing back display data synchronized to the external video signal and by high-speed switching of the background video and display signal.

## 22.6  CD PLAYER MECHANISMS

### 22.6.1  Types of CD Changers

Many schemes have been proposed for CD changers. They all fall into one concept, because each one is equipped with a CD container, a CD playback unit, and a conveyor mechanism between the container and the playback unit. From the point of view of how a user loads or changes CDs, there are the following three types of changers:

- The jukebox type has a CD container that holds a few tens to several hundred CDs. The basic design hides the container behind the playback mechanism.
- The magazine type has a dedicated container, or magazine, used as one unit to keep 5 to 12 CDs. The magazine is then inserted into the changer.
- The turret type has several discs set directly on a revolving turret of a changer.

These three types of CD changer have acquired their market positions by making use of individual features to be described later. The jukebox type is used mainly for commercial purposes, and the magazine and turret types in homes.

### 22.6.2  Jukebox Changer

The jukebox changer dates back to the era of disc records and was the first type of CD changer on the market. A representative structure is shown in Fig. 22.27. The disc container consists of a stocker that holds a disc by its circumference and provides many trenches for the discs. The changer incorporates a locking mechanism that prevents a disc from dropping. The layout of the disc container can be made vertical (as shown in Fig. 22.27), horizontal, or even circular. Many ideas have been proposed, but space efficiency and disc capacity make the vertical changer the most popular choice.

Many products have been marketed, moreover, that improve space efficiency by providing plural columns for disc containers. A conveyor moves in the disc-stacking (up and down) direction and stops at a specific position, where it pulls out a disc by clamping onto its circumference. The unit holding the disc is then transferred to the CD playback unit for playback.

The CD playback unit is designed basically in the same way as consumer products except that a brushless motor is used because of the extensive usage. The jukebox is equipped with a sophisticated automated function, and the mechanism and control unit are of a larger scale than in the other two types of changers. With a large stock capacity, music selection and playback can be manipulated by button operation without touching a disc.

### 22.6.3  Magazine Changer

Longer playback time and large capacity can be realized while still maintaining the ease of changing discs by setting multiple discs in a special case, or magazine, which is handled as a unit. The first magazine changer was introduced as a product in 1985. It used a magazine that held six CDs and a single adapter that enabled relatively easy play compared to a conventional single CD.

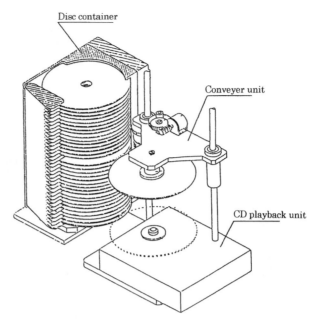

**FIGURE 22.27**    Structure of a jukebox-type changer.

A typical structure is shown in Fig. 22.28. The CD container is the magazine itself, which consists of multiple movable trays for discs to be removed or placed, as well as the casing. The changer main body has a mechanism that can remove the magazine. The disc conveyor mechanism consists of a driving lever that pushes off and puts back the tray and a mechanism for selecting a particular disc by moving the conveyor unit up and down. Another type of structure is one in which the player part is fixed on the base just like that in the jukebox changer. A magazine structure different from the one shown in Fig. 22.28 is also marketed. It has the CD discs inserted directly into the slots of a molded plastic case.

Magazine changers are creating a large market because of their smaller size, the ability to create a personal library of magazines, and the easy setting and exchange of a magazine to and from a changer's main body.

### 22.6.4  Turret Changer

The turret changer is equipped with a rotary tray on which CD discs are placed flat and circumferentially. Disc change takes place by rotating the tray. A patent for a turret changer was applied for in 1984.

A typical turret changer construction is shown in Fig. 22.29. The disc stocker is the rotary tray itself. Five discs can be placed on the tray, which slides toward the front of the playback unit when activated by the loading mechanism so that a user can set or change CDs on and from the tray when it is partially exposed from the player body.

A disc is transferred by rotating the turntable tray. The playback unit is basically identical to that of the conventional CD player in that it is elevated from underneath the tray to clamp the disc. The mechanism can be made at a minimum cost and discs can be changed while one CD is being played back.

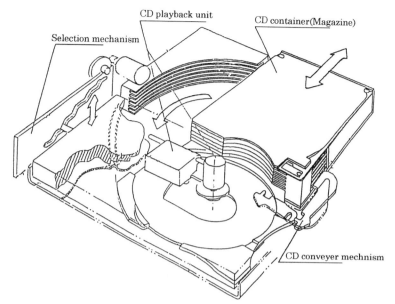

**FIGURE 22.28**    Structure of a magazine-type changer.

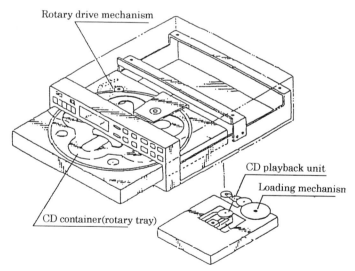

**FIGURE 22.29**    Structure of a turret-type changer.

The most salient feature of the turret changer is that, if a user wants to change all CDs, he or she has to do so one disc at a time. The construction also limits the size of the changer. Despite these shortcomings, the turret changer has the benefits of not requiring any dedicated magazine or accessory, and it gives the lowest-cost solution compared with the other two changer types.

## *22.7  CLD PLAYER*

The CLD player is a player that can reproduce both laser discs (LDs) and CDs using the same pickup, the same digital audio decoding circuit, and almost the same servos. The LD has a multiplexed recorded signal on a single track consisting of an FM video signal, two FM audio signals, and an EFM digital stereo audio signal that is the same format as that of a CD. The reproduced multiplexed signal has a frequency spectrum up to 14 MHz, as shown in Fig. 22.30. Table 22.1 gives the principal differences between LDs and CDs.

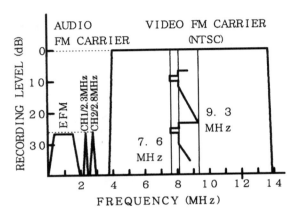

**FIGURE 22.30**   Laser disc signal spectrum.

**TABLE 22.1**   Basic Differences Between CDs and LDs

|       | Diameter | Weight | Rotation speed | Linear velocity | Lead-in |
| ----- | -------- | ------ | -------------- | --------------- | ------- |
| CD    | 12 cm    | 20 g   | 200–500 rpm    | 1.2–1.4 m/s     | 50 $\phi$ mm |
| LD    | 30 cm    | 200 g  | 600–1800 rpm   | 11 m/s          | 110 $\phi$ mm |

### 22.7.1  Tilt Servos

An LD is likely to have a warp of an umbrella shape. The pickup has to be parallel to the disc surface for good signal reproduction even with the existence of warps. The tilt servomechanism makes the good signal reproduction possible.

The tilt sensor detects the angular difference between the pickup and the disc as shown in Fig. 22.31. A light-emitting diode (LED) projects a light beam to the disc. The reflected light beam from the disc is projected onto two photo detectors, $-1$ and $-2$, by an equal amount if the disc is not tilted against the tilt sensors. In other words the pickup does not tilt against the disc surface. If the disc is tilted, the amount of projected light beam on each photo detector will differ.

Figure 22.32 shows a block diagram of the tilt servo system. The tilt error signal (the difference in voltage between the outputs of the two photo detectors) is converted to a digital signal by a mechanism control microprocessor. It processes the digital signal and generates a pulse width–modulated signal to drive a motor that controls the tilt angle of the pickup through a driving circuit to cancel out the angular difference between the disc and the pickup; i.e, it makes the pickup parallel to the disc surface in the radial direction.

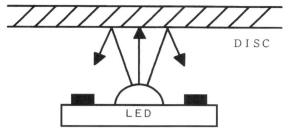

**FIGURE 22.31**  Tilt sensor.

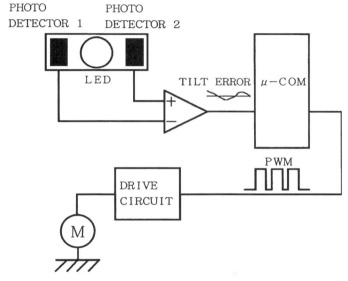

**FIGURE 22.32**  Tilt servo.

### 22.7.2  Reference Shift Method and Digital Time Base Corrector (DTBC)

The horizontal synchronizing pulses of the recorded video signal are aligned in the radial direction in the case of a constant angular velocity (CAV) disc but not in the case of a constant linear velocity (CLV) disc. Therefore, after the pickup has been jumped abruptly from one track to an adjacent track of the CLV disc for a special effect reproduction (e.g., still picture), the reproduced video signal has a discontinuity in the periodicity of the horizontal synchronizing signal. This discontinuity produces a large time-base error in the digital time-base corrector (DTBC) circuit. As a result, a disturbed picture is reproduced on the TV screen. To correct this problem, the reference shift method is used.

Figure 22.33 shows a block diagram of the DTBC, including the reference shift circuits. The time-base error is detected by sampling and holding a certain point of a trapezoidal waveform that is made from reference H, with a sampling pulse that is made from reproduced H (horizontal synchronizing pulse). The sampling pulse is sampled out of the center of the slant portion of the trapezoid during the normal operating condition of the DTBC servo system, as shown in the left portion of Fig. 22.34. At the time of track jump start, the CKH pulse is latched on high level

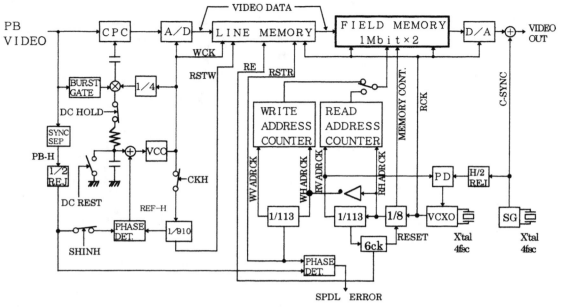

**FIGURE 22.33** Digital time-base corrector.

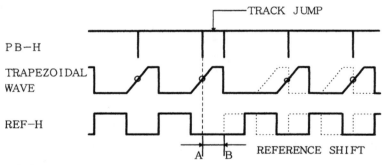

**FIGURE 22.34** Reference shift.

at the timing of reproduced H just before jumping starts, as shown in timing A in Fig. 22.34. This pulse prevents the supply of clock pulses to the 1/910 counter that generates the reference trapezoidal waveform. As a result, the counter stops counting and preserves the counting state of the last count. After the jump is finished, the CKH pulse (which is latched on low level by a first reproduced H as shown in timing B in Fig. 22.34) makes a pass for supplying the clock pulse to the counter. As a result, the counter starts counting from the preserved state. Therefore, the phase of reference H is shifted by the same amount as the discontinuity of the reproduced H. Then, new sample and hold action starts at the center of the slant portion of the trapezoidal waveform. As a result, a large time-base error is not produced. In this way the TBC servo works correctly and continuously even if the reproduced H occurs intermittently upon jumping for the CLV disc.

The reproduced video signal contains tiny timing jitters that are caused by disc eccentricity and vibration of the mechanism. After these jitters are removed by the color phase compensator based on the phase modulation principle, the video signal is converted from analog to digital and stored in a line memory circuit with correct timing supplied by the clock pulses. A voltage-controlled oscillator with a nominal oscillation frequency of 14.318 MHz (which is controlled by the time-base error) generates the clock pulses. As a result, a clean video signal is obtained without any time-base errors and jitters by reading out of the line memory circuit using a stable reference 14.318-MHz clock signal generated by a crystal-controlled oscillator.

### 22.7.3 Signal Processing

***Video Signal Demodulation.*** The RF signal reproduced from a laser disc includes a frequency-modulated video signal at 1.7-MHz frequency deviation (7.6 to 9.3 MHz), and two kinds of FM audio signals at $\pm 100$-kHz frequency deviation on both center frequencies (2.3 and 2.8 MHz), and an EFM digital stereo audio signal (Fig. 22.30).

Figure 22.35 shows the video signal demodulation circuits. The reproduced RF signal has some amplitude variations caused by the fluctuation of spatial frequency characteristics on the reproduction point of a disc. These amplitude variations are corrected by an FM corrector. The video FM carrier signal (which is separated from the RF signal by a bandpass filter of 3.5 MHz to 15 MHz) is demodulated to a video signal by an FM demodulator. Meanwhile, dropouts that cause an abrupt decrease of amplitude of the RF signal (caused by scratches or stains on a disc) and produce noises on a TV screen are detected by a dropout detector. It generates a command signal to control a switch to replace the current video output signal by the previous video output signal. This video signal is generated by delaying the demodulated video signal by just 1-H ($1 \text{ H} = 63.5 \text{ }\mu\text{s}$) period through a delay circuit using a charge coupled device, as shown in Figure 22.35. The video signals on the adjacent horizontal scanning lines have large correlation

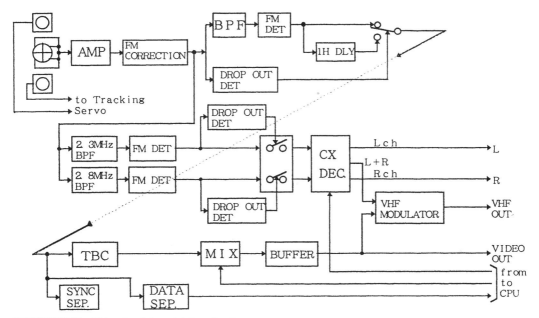

**FIGURE 22.35** Video signal demodulation circuits.

and also the possible occurrence of the dropout in the laser disc is for very short periods of less than a few μs. Therefore, almost all dropouts are corrected and are hardly recognizable on the TV screen. Tiny jitters in the dropout corrected video signal are removed through a jitter correction circuit and various kinds of characters such as frame number, time code, etc. are superimposed on the video signal. Then the video signal is sent to a video output terminal and VHF modulator.

***Y/C Separation Circuit.***   For improving the signal quality (signal-to-noise ratio, for example) of the demodulated video signal, some add-on signal processing is usually applied separately for each of the *Y* (luminance) and *C* (chrominance) signal of the demodulated video signal. A *Y/C* separation circuit is necessary for this purpose. There are various types of *Y/C* separation circuits used for LD players. One type is a 3-line digital comb filter. Its block diagram is shown in Figure 22.36.

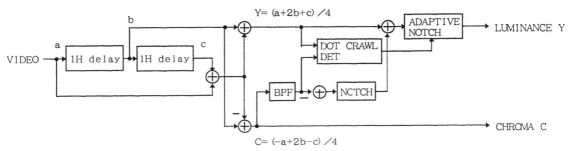

**FIGURE 22.36**   Three-line digital comb filter.

The left part of Fig. 22.36 is a usual 3-line *Y/C* separation comb filter, which has the following faults: dot interference may happen comparatively, and detail of the luminance signal may leak into the chrominance signal and is lost from the luminance itself. In order to improve the dot interference situation, new circuits are added as shown at the right of Fig. 22.36. The dot interference detector monitors both luminance and chrominance signals. When it detects dot interference about to occur, it inserts a notch filter in the luminance channel to prevent the dot interference.

To prevent leakage of the luminance signal into the chrominance signal, signal components of the chrominance channel within the frequency bandwidth of 3.5 to 8 MHz (except for the chrominance signal itself) are put back into the luminance channel to improve the detailed expression of the luminance signal. This brings about improvements that a usual comb filter cannot achieve.

***Digital Audio.***   In the LD recording signal spectrum there is an empty frequency bandwidth below 2 MHz. Fortunately, the recording signal spectrum for CD is below 2 MHz so this is where the CD signal can be recorded (Fig. 22.30). An LD with digital audio is based on this approach and can reproduce the same audio quality as a CD.

The LD reproduction system with digital audio differs from that of a CD in regard to how a reference clock signal is generated for the reproduction of the digital audio signal. In the case of a CD, the reproduced EFM signal is accurately phase-locked to the reference clock signal generated from a crystal oscillator by controlling the disc rotation. In the case of an LD, the disc rotation is controlled with the reproduced video signal, and the reproduced EFM signal is accurately phase-locked to the reference clock signal generated from a crystal oscillator separately. Therefore, the reference clock signal for decoding the CD signal must be phase-locked to the reproduced EFM signal.

The LD with digital surround audio requires an additional digital audio channel together with the CD format digital audio. Table 22.2 shows its format. The digital surround audio signal

**TABLE 22.2**   Laser Disc Format

| Modulation signal | |
| --- | --- |
| Modulation | D-QPSK |
| Carrier frequency | 2.88 MHz |
| Carrier level | −30 dB |
| (to no-modulated video carrier level) | |
| Baseband signal | |
| Recording data rate | 576 kb/s |
| Compression audio data rate | 422 kb/s |
| Error correction | Double Reed-Solomon product code C1:37, 33, 5 C2:36, 32, 5 |

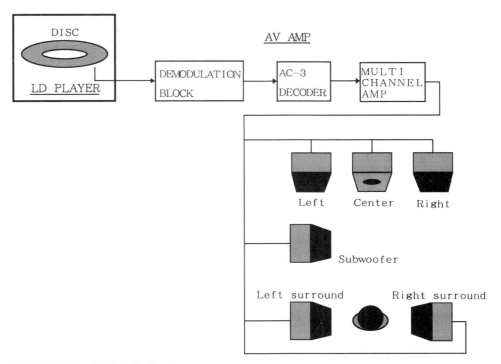

**FIGURE 22.37**   AC-3 playback system.

is compressed by the AC-3 format and is differential quadrature phase shift key (D-QPSK) modulated at a carrier frequency of 2.88 MHz. The modulated signal is then recorded within the frequency bandwidth that was formerly occupied by the FM audio carrier of the channel 2 audio signal (Fig. 22.30). An AD-QPSK–delayed detector and an AC-3 decoder are needed besides the usual LD player to reproduce the surround signal. Figure 22.37 shows a reproduction circuit for AC-3.

### 22.7.4  CLD Player Mechanism

In a CLD player a disc clamp is used to securely clamp different types, sizes, weights, and thicknesses of discs; to cope with the maximum 2700-rpm disc revolution; and to fix a disc accurately to the spindle of the motor with a precision of a few tens of μm in order to minimize the time-base error.

Figure 22.38 shows a clamp mechanism that satisfies these requirements. A disc is centered and placed at the correct position using the attractive force of a magnet by being held between a clamp above and a hub beneath that is fixed to the spindle of the motor. Because the thickness of discs varies drastically depending on the type of disc, an attractive force of a magnet is not enough to hold a disc securely. Constant clamping power is obtained by using a clamp spring, regardless of the disc thickness variation (Fig. 22.39).

An LD differs from a CD in the ability to store information on both sides of the disc. A continuous reproduction mechanism of a double-sided disc was developed to play back entertainment software extending over two sides of the disc without interruption. Figure 22.40 shows a typical mechanism for doing so. It is called an *α-turn mechanism* and has the characteristic of shifting

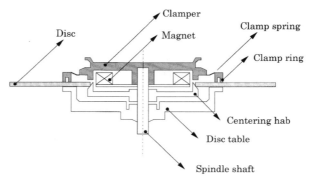

**FIGURE 22.38**   Linear clamp mechanism.

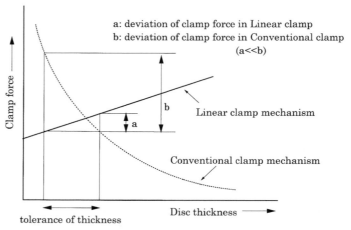

**FIGURE 22.39**   Relationship between disc thickness and clamp force.

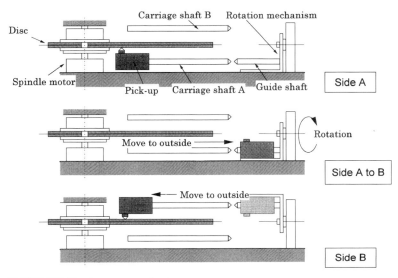

**FIGURE 22.40**   An $\alpha$-turn mechanism.

a pickup from reproduction of side A to side B using a turnover mechanism. Because this mechanism can keep an even characteristic condition between the pickup and disc surface on each side, the reproduced signal from the pickup has the same quality on both sides.

## *ABOUT THE AUTHOR*

Koki Aizawa received the B.E. degree from the Science University of Tokyo, Japan, in 1967. He joined Pioneer Electronic Corp. in that year and was engaged in designing a high-fidelity system. In 1972 he was transferred to the Engineering Research Laboratory, where he was in charge of developing a new digital audio and video disc technology and also a recordable disc technology. Since 1993 he has been working on the development of multimedia products such as DVD at the company's AV & Recording Development Center.

# CHAPTER 23
# DIGITAL COMPACT CASSETTE, MINIDISC

## SECTION 1
## DIGITAL COMPACT CASSETTE

**Henk Hoeve**
*Philips Components, Philips Key Modules*

## 23.1 A BACKWARD-COMPATIBLE DIGITAL AUDIOTAPE SYSTEM

With the advent of the Compact Disc, the limitations of analog audio sources became apparent. Although the analog Compact Cassette (CC) has been a successful audio medium for over 25 years, it is not able to compete in sound quality with the new digital audio system. Nor does it have the facilities for play control, programming, and associated information display.

To overcome these deficiencies, the Digital Compact Cassette (DCC) system was conceived as an extension to CC. The DCC project started in 1989 in the Research Laboratories and the Advanced Development Center of Philips Consumer Electronics in Eindhoven, the Netherlands. The main goals were identified as follows:

- Digital audio recording on an existing tape formulation using a stationary head.
- Compact Disc audio reproduction quality.
- Compatibility with the CC system.

A secondary goal was to include CD-style control and display facilities.

A major problem to overcome was to lower the bit rate without audible loss of sound quality. A breakthrough was needed, and it came with the development of Precision Adaptive Subband Coding (PASC), which, while enabling the transfer of digitally encoded signals to and from tape at a much lower bit rate, nevertheless maintained CD-comparable audio quality. The first DCC recorder for the consumer electronics market was introduced in 1992.

## 23.2   *DESIGN PHILOSOPHY*

DCC was conceived as a combination of the proven CC technology with new digital audio processing and new magnetic head technology.[1] The DCC mechanism is a slightly modified version of the autoreverse CC tape drive. Most of the mechanism modifications relate to the slim-line design of the DCC cassette housing. Digital recording and playback, as well as analog playback, are achieved within one single head using new thin-film technology. The system allows for recording of extra information concerning the music, which can be displayed during playback.

The tape speed in both analog and digital modes is 4.76 cm/s. In principle, the tape used is the same hard-wearing formulation as that used in today's ½-inch domestic video recorders. In the same way as CC, the DCC tape runs in both forward and reverse directions. It is divided into Sector A (forward direction) and sector B (reverse direction), each of which carries a total of nine digital information tracks (Fig. 23.1). DCC is basically an overwrite system and, as such, possesses no erase head.

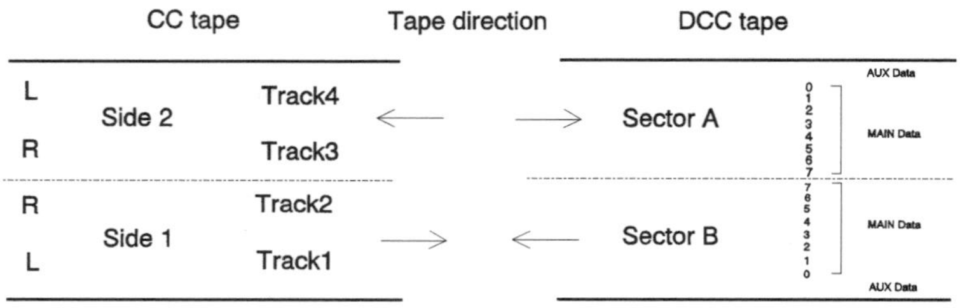

**FIGURE 23.1**   Comparison of CC and DCC tapes.

## 23.3   *THE CASSETTE*

In the years since the CC standard was defined, new concepts, materials, and processes have been developed. Although a definite continuation of the CC, the DCC includes various enhancements made possible by advancing technology. The differences from the CC are shown in Fig. 23.2. Recognition holes are assigned as follows:

| | Cassette type | | |
|---|---|---|---|
| Hole | Use | Blank | Prerecorded |
| 1 | (DCC = open, CC = closed) | 0 | 0 |
| 2 | Erasure protection: (Not protected = open; protected = closed) | O/C | C |
| R | Rear erasure protection: (Not protected = open; protected = closed) | O/C | C |
| 3,4,5 | Tape length indication | Yes | No |
| 6,7,8 | Reserved | No | No |

Thus the DCC is slimmer, with flat sides. By removing the "shoulders" around the tape feed part, the overall thickness has been reduced from 12.1 to 9.6 mm. A slider has been added to protect the tape from exposure to dust and fingerprints and to lock the reels when not in use. Because all DCC mechanisms are autoreverse types, only two holes are provided, for access

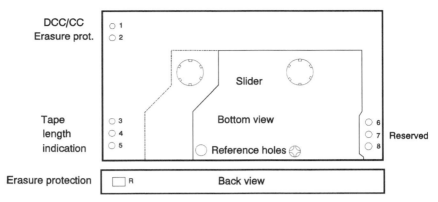

**FIGURE 23.2**   Recognition holes.

to the reel hubs, at the bottom of the cassette. A cross-shaped reference hole is included so that DCC cassettes cannot be inserted into existing CC recorders or players, the corresponding pin in the DCC mechanism still providing the reference height for both cassette types—blank or prerecorded. Additional hole positions in the blank DCC cassette determine the length of the tape. There is also a hole to operate a mechanical switch for erasure protection (only for blank cassettes). Finally, three recognition hole positions are reserved to allow for possible future applications.

## 23.4   THE DCC RECORDER

On one sector of the tape, the digital signals are recorded in nine parallel magnetic tracks, each 185 $\mu$m wide with a track pitch of 195 $\mu$m. The width of the corresponding playback heads is 70 $\mu$m. This 2.5:1 ratio has the advantage that DCC is less sensitive to azimuth errors than CC. Two kinds of data are recorded on the tape: Main data (PASC data and System Information) in eight tracks and Auxiliary data in one track.

### 23.4.1   PASC Data

In the recording mode the analog input (after analog-to-digital conversion) or the digital input (directly) is processed by the PASC encoder.

Two-channel audio signals can be recorded with a sampling frequency of 48, 44.1 (default), or 32 kHz. Optionally, preemphasis (50/15 $\mu$s) may be used. The dynamic range is better than 105 dB, and the total harmonic distortion including noise is less than 0.0025 percent.

### 23.4.2   System Information

The System Information (Sysinfo) is recorded simultaneously with the audio signal in the Main channel. Sysinfo contains music-related data along with information about the status of the Serial Copy Management System (SCMS) for copying protection, as well as the type of recorded tape. Sysinfo cannot be changed by the user except by making a new recording.

Furthermore, for prerecorded tapes, text can be contained in this channel. The purpose of the text is to enable users to obtain information on such matters as the album and music track title, the artist's name, synchronously sung lyrics, and the table of music titles. The text can be made available to the user on a display coupled to the DCC player.

### 23.4.3  Auxiliary Data

The Auxiliary (Aux) data, recorded on a separate track, mainly consists of data to which the DCC player reacts during search and playback modes. Additional recorded information includes several items: absolute time, track time, remaining time, music track numbering, the Table of Contents (TOC), and text information. This additional information enables track start locations to be found in search mode. Aux data can be changed by the user without making a new recording.

### 23.4.4  Encoder-Decoder

The DCC system can be seen as a digital communication system with a transmission path and a reception path (Fig. 23.3). The encoder can be split up into an analog-to-digital converter (for analog inputs), a source encoder, and a channel encoder.[2–5]

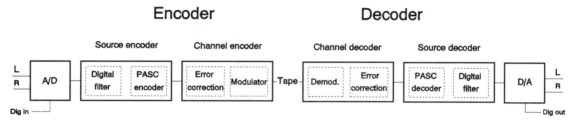

**FIGURE 23.3**    Digital communication system.

In the source encoder a digital filter transforms the baseband audio signal, with a bandwidth of half the sampling frequency ($fs/2$), into 32 subband signals, each with a bandwidth of $fs/64$. Each subband represents a different portion of the baseband signal, such that all the subbands together contain all the information in the signal. So far, no information has been left out.

The PASC encoder, however, now determines which of the subbands contain information perceivable by the human ear. It has been known for decades that our ears do not respond to all audio signal components in the same manner all the time.[6] Phenomena such as temporal and frequency masking have long been successfully used in subband coding systems for telephony to increase the number of simultaneous connections within the finite capacity of the telephone system. In the same way PASC is based on the following phenomena of the human ear:

- One can only hear sound above a certain level called the *hearing threshold* (Fig. 23.4). This depends on the frequency of the sound and differs from one individual to another. In the PASC system a description of the hearing threshold of a very sensitive person is used.
- Loud sounds hide, or mask, softer sounds in their vicinity. They dynamically adapt the threshold of hearing by adding a masking threshold.

PASC codes the audio signal very efficiently by computing the dynamically adapting threshold and transferring only those signal components that extend above the threshold. In addition, the samples transferred are encoded using a floating point notation, which decreases the bit rate dramatically. The output of the source encoder has a constant bit rate of 384 kb/s, independent of the value of the sampling frequency, fs (see Section 23.7, "Bit Rates of DCC").

In the channel encoder the error correction block adds parity symbols to form the error-detecting and -correcting codes that provide immunity against errors related to the recording and playback processes. The error correction system is capable of correcting erroneous symbols during playback to a certain level. Two Reed-Solomon codes ($n, k, d$) are used for the Main data.

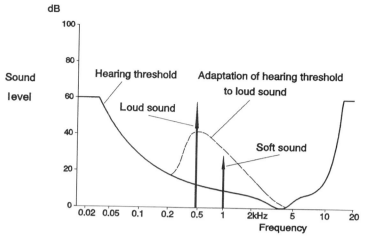

**FIGURE 23.4** Hearing threshold.

These are C1:(24,20,5) and C2:(32,26,7). The Aux data use (24,18,7). Further information on this subject is given in Section 23.5.

Finally, the information is passed through a modulation block, which translates each 8-bit symbol (data or parity) into a 10-bit symbol in a process known as Eight-to-Ten Modulation (ETM) such that the frequency spectrum matches the characteristics of the transmission channel (consisting of the recording head, tape, and playback head).

ETM has a limited run length: maximally, five consecutive channel bits can have the same value. This facilitates easy clock regeneration from the played-back tape signal. It is a DC-free block code with a rate of 8/10. The code design has the special feature that only six Digital Sum Values (DSVs)—the accumulated number of ones in the channel signal minus the accumulated number of zeros—are occupied by the channel bitstream, including the sync pattern.

Each ETM code word consists of the data word and the DSV of the previous code word (DSV old), as shown in Fig. 23.5. The channel information from the modulation block is written onto the tape in eight tracks. Auxiliary information is recorded in a ninth track. Demodulation is free from error propagation, because errors in one channel symbol do not influence the decoding of its successor.

## 23.5 TAPE FORMAT

On the tape the formats of the Main data (PASC and Sysinfo) and the Auxiliary data are similar. The Main data channel will be described first, with reference to Fig. 23.6. For Main data the bit rate is 96 kb/s per track.

The signals recorded on the tape are constructed of self-contained digital audio units known as *tape frames*. Between the tape frames, Inter Frame Gaps (IFGs) of variable length are inserted to accommodate small deviations from the sampling frequency used during recording. The nominal length of an IFG is 64-bit periods, corresponding to about 0.4 percent of the nominal tape-frame length (including IFG) of 16384-bit periods. An IFG carries a signal that alternates in polarity at every bit position.

Each of the Main data tracks carries 32 tape blocks per tape frame. A tape block contains 51 tape symbols of 10 bits. These are generated according to the rules of the 8-to-10 modulation table.

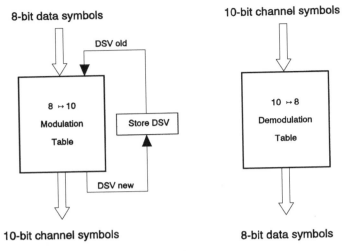

**FIGURE 23.5**   (De)modulation ETM code.

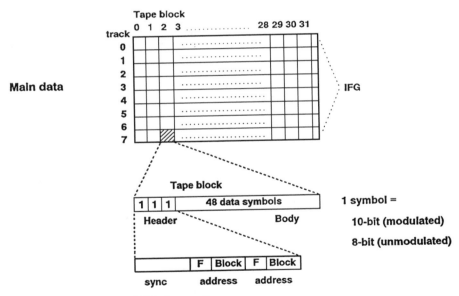

**FIGURE 23.6**   Structure of Main data tape frame.

The first 3 symbols in a tape block constitute a block Header, consisting of a synchronization pattern, a 3-bit frame address, and a 5-bit block address. The synchronization pattern is used to identify the start of a tape block and to find the boundaries of the 10-bit channel symbols. The addresses are needed in the playback process to place the recovered data in the correct RAM location ready for error correction.

The remaining 48 symbols in a tape block (called the *Body*) carry the PASC audio data, the Sysinfo, and the parity symbols for the error detecting and correcting code. A Cross-Interleaved

Reed-Solomon code is employed to protect the data against random and burst errors. During playback, a C1 code word is first evaluated to detect and, if possible, correct errors within a tape block. Because the C1 code may correct at maximum 4 error symbols per tape block (because two code words are interleaved), any errors that have not been corrected are passed to the second (C2) correction phase as erasures, indicating the location of the erroneous symbol but not the error value.

The C2 code can correct a maximum of 6 of these erasure symbols per code word per tape block (with two interleaved code words). The distribution of symbols in a C2 code word is such that an optimal physical spacing is achieved between them, resulting in a "honeycomb" pattern on the tape (Fig. 23.7). The pattern allows the correction of dropouts with diameters up to 1.45 mm. The system can even reconstruct a completely missing track and still have capacity left for correcting small errors in other tracks.

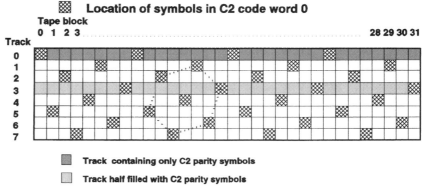

**FIGURE 23.7**    Interleaving of C2 code word.

The PASC symbols are distributed across the tracks in such a way that no failure of the error correction system disrupts consecutive PASC symbols: an uncorrectable C2 code word cannot result in a burst error in the PASC signal. This helps the PASC processor to conceal these errors.

For Auxiliary data the bit rate in the single track is one-eighth that of the Main data bit rate at 12 kb/s. This eases the requirements for error detection and correction (Fig. 23.8). The number of tape blocks is reduced to 4 per frame, and, to improve the data reliability, all the written tape blocks in a tape frame contain the same information.

As for Main data, an Aux data tape block contains a block Header of three tape symbols and a body of 48 tape symbols. The block header consists of PLL data on either side of a synchronization pattern. The PLL data symbols are used to indicate whether or not the adjacent Body is recorded.

The nominal length of an Aux data tape frame IFG becomes 8-bit periods. A single Reed-Solomon code protects the data, with the capacity to correct a maximum of 6 error symbols in a tape block consisting of two code words.

To enable the easy detection of locations on the tape, the Bodies of tape blocks 0 and 2 only are recorded at these locations (Fig. 23.9). On the other hand, to make it possible to find the start of a music track quickly without the need to decode the Auxiliary data, Labelled Frame Markers are also provided. These markers are 16 frames long, and in these frames all Bodies of all blocks are recorded. A marker is detected by analyzing the envelope of the Aux channel signal output. Three kinds of markers exist: (1) *Start Markers,* to indicate the beginning of a music track; (2) *Sector Markers,* which have influence on the tape movement (e.g., Home, Stop,

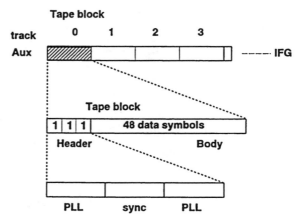

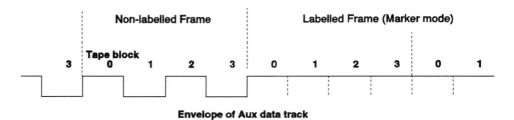

**FIGURE 23.8** Structure of Auxiliary data tape frame.

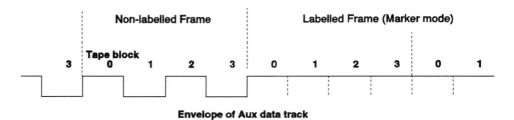

**FIGURE 23.9** Labelling.

and Reverse Markers); and (3) Feature Markers (e.g., Mute, Fade, and Skip Markers). These markers influence the player's behavior while it is playing the music track.

## 23.6 MAIN PARAMETERS OF DCC

| | |
|---|---|
| Playing time (total) | 90 min (120 min max.) |
| Tape speed | 4.76 cm/s |
| Tape width | 3.78 mm |
| Tape thickness | 12 $\mu$m |
| Tape type | Chrome (coercivity 650–750 Oe) |
| Tape protection | Slider |
| Track width | 185 $\mu$m |
| Track pitch | 195 $\mu$m |
| Cassette size | 100.4 × 63.8 × 9.6 mm |
| Cassette | Blank, prerecorded |
| Cassette (record protect) | Recognition holes |

| | |
|---|---|
| Sampling frequency (fs) | 32, 44.1, 48 kHz |
| Quantization | 16 bits/sample, 2's complement |
| Number of audio channels | 2 (stereo or 2-channel mono/sector) |
| Preemphasis | Optional (50/15 $\mu$s) |
| Source coding | PASC |
| Error correction | Reed-Solomon |
|    Main data | C1-RS (24,20,5) |
| | C2-RS (32,26,7) |
|    Aux data | RS (24,18,7) |
| Number of sectors | 2 |
| Number of tracks/direction | |
|    Main data | 8 |
|    Aux data | 1 |
| Channel bit rate/track | |
|    Main data | 96 kb/s |
|    Aux data | 12 kb/s |
| Bit cell length | |
| Main data | 0.496 $\mu$m |
| Aux data | 3.968 $\mu$m |
| Digital I/O | IEC 958 (amended) |

## 23.7  BIT RATES OF DCC

Refer to Fig. 23.10.

**A.** Assuming 16 bits/sample/L,R: $16 \times 2 \times$ fs

| | |
|---|---|
| fs = 32 kHz | 1024 kb/s |
| fs = 44.1 kHz | 1411.2 kb/s |
| fs = 48 kHz | 1536 kb/s |

**B.** Source coding PASC

| | |
|---|---|
| fs = 32, 44.1, 48 kHz | 384 kb/s |

**C.** System information (Sysinfo)

**D.** Auxiliary data

After error correction of main data and Sysinfo:

**E.** C1 + C2 Reed-Solomon

$(24/20) \times (32/26) \times (384 + 6)$:       576 kb/s

After error correction of Aux data:

**F.** Reed-Solomon

$(24/18) \times 6.75$:       9 kb/s

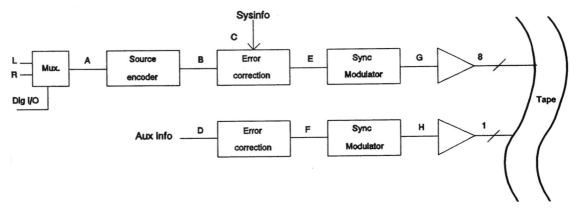

**FIGURE 23.10**   Bit rates of DCC.

**G.** Main data
   Modulation ETM (8–10)
   Synchronization 51/48, IFG 0.4%
   $576 \times (10/8) \times (51/48) \times 1.004$:         768 kb/s
   Each track                                               96 kb/s

**H.** Aux data
   ETM, sync, IFG
   $9 \times (10/8) \times (51/48) \times 1.004$:           12 kb/s

## GLOSSARY

**Aux channel**   The channel containing Auxiliary data, which is recorded on the separate Aux track on the tape.

**Aux data**   Data, recorded in the Aux track on the tape, that contains sync information, search information, and music track–related information.

**Aux data tape frame**   A tape frame in the Aux track consisting of four Aux data tape blocks containing digital data.

**Aux track**   A longitudinal track, positioned at the outer edge of the tape, containing Aux data. An Aux track is defined for both Sector A and Sector B of the tape.

**C1 code word**   Reed-Solomon code word for DCC error correction of the type RS(24,20,5).

**C2 code word**   Reed-Solomon code word for DCC error correction of the type RS(32,26,7).

**CC**   Compact Cassette.

**CD**   Compact Disc.

**Code interleaving**   Merging of two or more independent codes by interleaving their symbols. The purpose is to make the arrangement more robust against perturbation of adjacent symbols.

**Coercivity**   The reverse magnetic field necessary to reduce the magnetic flux to zero after saturation.

**DCC**    Digital Compact Cassette.

**Demodulation**    In DCC the conversion of the 10-bit channel (ETM) symbols as read from the tape to 8-bit data symbols. The conversion is carried out via table lookup.

**Digital I/O**    The digital interface of a DCC recorder/player defined according to the IEC 958 standard.

**Digital sum value (DSV)**    The value obtained when the accumulated number of ones of a code word are subtracted from the accumulated number of zeros. The DSV is used together with the modulation table.

**Eight-to-Ten Modulation (ETM)**    Process in which 8-bit symbols (data or parity) are translated into 10-bit symbols such that the frequency spectrum matches the transmission channel. The conversion is carried out via table lookup.

**Envelope detection**    The odd tape block bodies of the Aux data may be recorded or not, causing a labeled state or nonlabeled state, respectively. When reading the Aux data track, the envelope can be analyzed to detect the state.

**Erasure**    A signal in a received transmission sequence that is reset to a certain value because it is known to be in error.

**Erasure protection hole**    On blank DCC cassettes this hole is defined to prevent the recorder from overwriting an already recorded music track.

**Error correction**    A system in digital communications that is designed to detect and correct errors occurring during transfer of data.

**Frame Address**    Three bits in the header of a tape block in the Main data are coded with the Frame address.

**Hearing threshold**    The intensity level above which one can just detect the presence of any audio signal. The level is frequency-dependent.

**IEC**    International Electrotechnical Commission.

**Inter Frame Gap (IFG)**    Variable-length gap separating the successive Main data tape frames and Aux data tape frames.

**Interleaving**    Mixing data from different sources in a defined and fixed sequence.

**Main data**    The digital information in the Main data tracks.

**Main data tape frame**    The tape frame consisting of 32 tape blocks and adjacent IFG.

**Main data tracks**    The eight tracks on the tape containing the Main data.

**Marker**    A code that is recorded in the Aux channel to invoke automatic player actions. Three kinds of markers exist: Start Markers, Sector Markers, and Feature Markers.

**Modulation**    In DCC the conversion of 8-bit data symbols to 10-bit tape symbols, yielding the advantages of a limited length of strings of subsequent 1s or 0s and an equal average number of 1s and 0s in the code sequence. The modulation is performed via table lookup.

**Parity symbols**    Symbols added to digital data to make error checking and/or correcting possible.

**Precision Adaptive Subband Coding (PASC)**    The audio-coding method used with DCC.

**Preemphasis**    Intensifying the higher frequencies in the frequency spectrum of a signal more than the lower frequencies.

**Reed-Solomon code**    A code defining the error detecting and correcting method.

**Reference pin**    The pin in a cassette deck that fits into the reference hole in a cassette to establish the height reference.

**Sampling frequency**    The frequency of the samples representing the audio signal.

**Sector**    The longitudinal data tracks recorded on DCC tape are divided into two areas: Sector A and Sector B. Each Sector is composed of eight tracks for recording Main data, plus one track for Aux data.

**Serial Copy Management System (SCMS)**    This system specifies which recording may or may not be copied from the source.

**Source coding**    The coding of an information source, mainly to obtain a higher efficiency. In DCC, PASC is the coding system used for coding the audio signal source in a highly compressed way, without loss in sound perception.

**Subband signal**    One of 32 signals coming from a filter bank, which operates on the incoming signal. Each subband signal represents a band-limited portion of the original signal. The 32 signals together fully contain the original information.

**Sync pattern**    A pattern in the header of a Main data tape block and of an Aux data tape block. The 10-bit sync word (unmodulated) can be found in the modulation lookup table.

**Sysinfo**    The 128 bytes per tape frame in the Main data containing tape type and copyright data plus the ITTS text information.

**Table of contents (TOC)**    A table, recorded in the Aux data, allowing the recorder access to music tracks and other points on the tape.

**Tape block**    A subdivision of a tape frame consisting of a group of 51 bytes. A Main data tape frame is divided into 32 tape blocks. An Aux data tape frame is divided into 4 tape blocks.

**Tape frame**    A unit of data consisting of 32 tape blocks for Main data, and 4 tape blocks for Aux data.

**Thin-film head**    The multiple read-write head for DCC is made in thin-film deposition technology.

**Track**    In DCC the term *track* comes with two meanings: a piece of music recorded on the tape, and one of 18 narrow stripes on the tape in which information is written magnetically.

**Two-channel mono mode**    Audio-coding mode that records two mono audio channels—one on L (channel I) and one on R (channel II)—on both Sector A and Sector B of the tape.

## REFERENCES

1. G. C. P. Lokhoff and van der Plas, P. A., "New developments for the Digital Compact Cassette system," *IEEE Transactions on Consumer Electronics,* vol. 39, no. 3, Aug. 1993, pp. 350–355.
2. G. C. P. Lokhoff, "Digital Compact Cassette," *IEEE Transactions on Consumer Electronics,* vol. 37, no. 3, Aug. 1991, pp. 702–706.
3. G. C. P. Lokhoff, Nikkei Electronics, 1991.9.2, (no. 535), pp. 134–141.
4. G. C. P. Lokhoff, "Precison adaptive subband coding," *IEEE Transactions on Consumer Electronics,* vol. 38, no. 4, Nov. 1992, pp. 784–789.
5. A. Hoogendoorn, "Digital Compact Cassette," *Proceedings of the IEEE,* vol. 82, no. 10, Oct. 1994.
6. E. Zwicker and Feldtkeller, R., "Das Ohr als Nachrichtenempfanger," S. Hirsel Verlag, Stuttgart, Germany, 1967.

## ABOUT THE AUTHOR

Henk Hoeve received his electrical engineering degree in information theory from the University of Technology, Eindhoven, the Netherlands, in 1974. He joined the Philips Electronic Component Division in Eindhoven that same year, working on magnetic bubbles. In 1978 he became involved in the development of the Compact Disc system, in the Audio division of Consumer Electronics. Since January 1990 he has been working in the Magnetic Recording Laboratories of Philips Key Modules in the field of digital magnetic recording (Digital Compact Cassette). Hoeve is a member of the IEC committees TC60 and TC84 (merged into TC100) and is secretary of working group WG17, 60A.

# CHAPTER 23
# DIGITAL COMPACT CASSETTE, MINIDISC

## SECTION 2
## MINIDISC

**Tsutomu Imai**
*Sony Corporation*

## 23.8  FORMAT OVERVIEW

The MiniDisc is a 64-mm disc housed in a rigid protective cartridge. It is available in two versions: playback-only MiniDiscs for prerecorded music and recordable MiniDiscs for home recording and playback. The MiniDisc contains 74 minutes of digital sound (the same as a CD), allows fast random access, and is highly resistant to shock during playback.

The playback-only versions are manufactured using the same process as CDs and cannot be rerecorded. The recordable versions are magneto-optical discs that make use of magnetic-field modulation technology to enable recording by the user and may be rerecorded a virtually unlimited number of times. Once recorded, however, the life of the sound data is the same as that of a CD, thanks to a direct overwriting technology developed by Sony. This system performs recording through the use of a laser beam focused on one spot on the disc, and a magnetic field applied on the corresponding spot on the other side. Only through this combination of laser light and a magnetic field can the recordable MiniDisc be recorded or rerecorded.

Table 23.1 gives MiniDisc specifications.

## 23.9  PLAYBACK-ONLY MINIDISCS

Playback-only MiniDiscs are similar to CDs in several ways. The lead-in area is on the inner circumference of the disc, followed by the program area, and the lead-out area on the outer circumference of the disc (Fig. 23.11). In addition, information in the form of pits are created on the disc substrate in the same way as CDs (Fig. 23.12). Moreover, discs can be stamped in mass quantities, like CDs, with an injection-molding machine. The entire front of the disc can be used for visuals and graphics because the shutter opens only on the back side of the cartridge.

**TABLE 23.1** MiniDisc Specifications

| Major specifications | |
|---|---|
| Recording/playback time | 74 minutes (148 minutes mono) |
| Cartridge size | 72 by 68 by 5 mm |
| Disc specifications | |
| Diameter | 64 mm |
| Thickness | 1.2 mm |
| Diameter (center hole) | 11 mm |
| Diameter (beginning of program) | 32 mm |
| Diameter (beginning of lead-in) | 29 mm |
| Track pitch | 1.6 microns |
| Linear velocity | 1.2–1.4 m/s (CLV) |
| **Audio characteristics** | |
| Channels | 2-channel stereo and monaural |
| Frequency range | 5 Hz–20,000 Hz |
| Dynamic range | 105 dB |
| Wow and flutter | Unmeasurable |
| **Signal format** | |
| Sampling frequency | 44.1 kHz |
| Compression system | Adaptive transform acoustic coding |
| Modulation system | Eight-to-fourteen modulation |
| Error correction system | Cross-interleaved Reed-Solomon code |
| **Optical parameters** | |
| Laser wavelength | Standard 780 nm |
| Laser diameter | Standard 0.45 |
| Recording power | 5 mW (max) |
| Recording system | Magnetic field modulation |

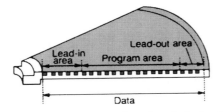

**FIGURE 23.11** Cross section of a playback-only MiniDisc.

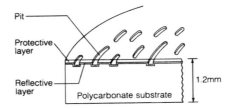

**FIGURE 23.12** Cross-sectional representation of the disc substrate.

## 23.10 RECORDABLE MINIDISCS

Magneto-optical technology is central to the functioning of recordable MiniDiscs. The lead-in area is on the inner circumference of the disc (Fig. 23.13), followed by the user table of contents (UTOC) area, the program area, and the lead-out area on the outer circumference of the disc. Since a magnetic recording head and a laser are used on opposite sides of the disc, the shutter opens on both sides of the disc (Fig. 23.14).

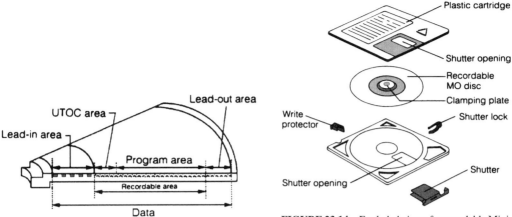

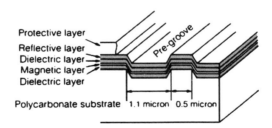

**FIGURE 23.13**   Cross section of a recordable MiniDisc.

**FIGURE 23.14**   Exploded view of a recordable Mini-Disc.

The recordable disc's unique layer structure, along with pre-groove configuration, is shown in Fig. 23.15. This magneto-optical layer construction is engineered to enable magnetic field modulation overwriting and has been proven to endure over one million recording operations without degradation.

Figure 23.16 is a cross-sectional representation of the disc housed in its cartridge. A magnetic clamping plate is mounted in the center of the disc substrate in both the recordable and playback-

**FIGURE 23.15**   Layered construction of a recordable MiniDisc.

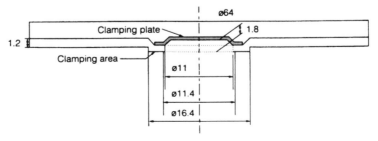

**FIGURE 23.16**   Cross section of disc and clamping plate.

only versions to ensure that the disc is centered when mounted by the drive unit. As in CDs, the center is aligned using the inner circumference edge of the disc substrate as a reference.

The MiniDisc uses a magnetic plate in the center of the disc substrate to stabilize the disc inside the player. Stabilizing discs by clamping from both top and bottom (as in CDs) requires a center opening on both sides of the cartridge, but because it is not necessary for the MiniDisc, the entire front side of the cartridge can be used for the label. Whereas the disc is 1.8 mm thick along the inner circumference, the surface of the disc itself is 1.2 mm, the same as that of CDs.

## 23.11   SIGNAL RECORDING FORMAT

The MiniDisc system uses the eight-to-fourteen modulation scheme in writing data on a disc and the cross-interleaved Reed-Solomon code (CIRC) for error correction. Audio data reduced by adaptive transform acoustic coding are grouped into blocks for recording in a format very similar to the CD-ROM mode 2 standard (Fig. 23.17).

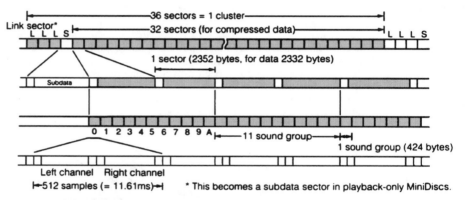

**FIGURE 23.17**   MiniDisc data format configuration.

In the MiniDisc format, 98 frames of a CD make up one sector, which is equivalent to 13.3 ms playback time. CIRC interleave length is 108 frames, equivalent to 14.5 ms, which is longer than one CD-ROM sector. Three sectors must be used as *linking sectors* to record data using CIRC error correction code, and their area is called the *link area*. A link area of more than 108 frames (one sector) must be provided before data are written. Similarly, an area of more than 108 frames must also be provided after the data have been written so that the error correction interleave can finish properly.

If data are written in random places, link areas will be scattered throughout the disc, reducing data utilization efficiency. Data are written only after being grouped into fairly substantial recording units called *clusters*. In the MiniDisc each cluster has 36 sectors, and rewriting is always performed in integer multiples of one cluster. Data to be recorded are temporarily stored in a random access memory before being written on the disc. This RAM also serves as the shock-resistant memory during playback.

The first three sectors of one 36-sector cluster are used as link sectors during recording, with the fourth sector reserved for subdata. In the remaining 32 sectors, the compressed digital data are recorded. When the last sector has been written, error correction data must be written in the first link sector and half of the second sector of the following 36-sector cluster.

In the playback-only disc, data are written in one continuous stroke, obviating the use of three sectors for a link area. Therefore, all of the first four sectors can be used for subdata, such as character information for the incorporation of lyrics or other special text. This also makes it impossible to record the entire contents, including subdata, of a playback-only disc onto a recordable disc.

During ATRAC encoding (Fig. 23.18), the audio data are compressed to one-fifth their original volume and then handled in 424-byte units called *sound groups,* with left and right channels allocated 212 bytes each. A total of 11 of the sound groups are distributed into two sectors. Recorded in the first sector are the left and right channels of five sound groups, and the left channel of a sixth group, whereas the right channel of the sixth group and the left and right channels of another five groups are recorded in the second sector. Each of the two sectors can be expressed as follows:

$$424 \times 5 + 212 \times 1 = 2332 \text{ bytes}$$

In this manner, 11 sound groups are written per every two sectors in each 32-sector cluster. Upon playback, ATRAC decoding restores these data to their original volume and time axis, with one sound group becoming equivalent to 512 samples ($512 \times 16 \times \frac{2}{8} = 2048$ bytes) for both channels, with a playing time of 11.6 ms.

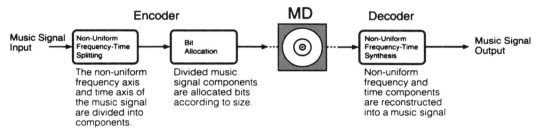

**FIGURE 23.18**    ATRAC operation layout.

## 23.12  *ATRAC DATA COMPRESSION*

When music is recorded onto a MiniDisc it is slightly different from the same music recorded on a CD. This difference is due to increased levels of quantization noise resulting from the MiniDisc's smaller size. If this quantization noise is rendered inaudible, then the sound will be as good as a CD. One ATRAC feature is its ability to make use of psychoacoustic principles (the way humans actually perceive sound) to minimize the audibility of quantization noise.

### 23.12.1  Psychoacoustic Principles

The sensitivity of the ear depends on frequency (see Chapter 3, Section 3.8). It is most sensitive around 4 kHz and least sensitive toward the higher frequencies. A tone at a given power that is audible at 4 kHz might not be audible at another frequency. In general, two tones of equal power but of different frequency will not sound equally loud. For this reason, quantization noise is less audible at some frequencies than at others.

Another important psychoacoustic principle is simultaneous masking. A soft sound can be rendered inaudible by a louder sound, in much the same way that a conversation becomes inaudible when a train goes by. Masking is strongest when the frequencies of the two sounds are close together. This means that quantization noise is less audible at frequencies on, or closely adjacent to, loud tones.

ATRAC makes use of these psychoacoustic principles to adapt the audio signals to the ear's changing sensitivity. It operates in such a way to "hide" quantization noise in frequency regions where there are high signal levels corresponding to a lot of musical activity. This effectively renders it inaudible.

For each block of time, ATRAC analyzes the music signal and determines the sensitivity of each frequency region. The sensitive regions are recorded accurately, with very little quantization noise. The remaining regions are recorded less accurately, but, since they are not too sensitive, the quantization error is hardly noticeable (Fig. 23.19). The result is high-fidelity audio recorded at only one-fifth the bit rate.

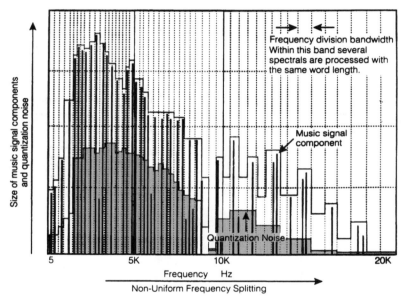

**FIGURE 23.19**   ATRAC effective bit allocation.

ATRAC's frequency and time divisions are shown in Fig. 23.20. Notice that the frequency divisions are not of uniform width. These divisions are based on another important psychoacoustic factor called *critical bands,* in which the ear perceives different frequencies using a series of these critical bands of various widths. Critical bands are essential for understanding the loudness of quantization noise. The width of a critical band increases with frequency. At 100 Hz it is 100 Hz wide, at 1000 Hz it is 160 Hz wide, and at 10,000 Hz it is 2500 Hz wide. Because of this phenomenon, ATRAC has been designed to analyze sounds in nonuniform frequency divisions, with more divisions in the lower frequencies and few in the higher frequencies, to ensure greater accuracy in signal analysis.

ATRAC also features nonuniform time splitting. Because music signals are constantly changing, ATRAC analyzes them in extremely short blocks of time. When the music is changing quickly, as in particularly vivid passages, the sensitivity of the ear also changes quickly, so ATRAC splits time up into short blocks of only 1.45 or 2.9 ms to keep up with the music. When

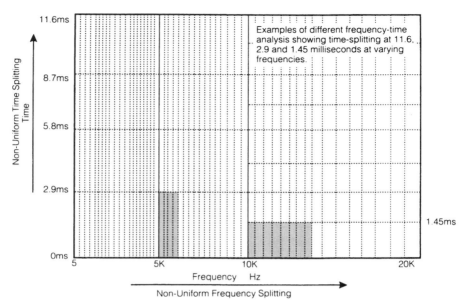

**FIGURE 23.20**    ATRAC nonuniform frequency-time splitting.

the music changes slowly, as in rather stationary passages, the time can be split into longer blocks of up to 11.6 ms, because the sensitivity of the ear also changes more slowly. The use of these longer time blocks enables the use of narrower frequency bands, leading to improved frequency resolution and higher sound quality. ATRAC's active flexibility in using either short or long time blocks is the key to realizing greater efficiency in data compression encoding while controlling quantization noise.

### 23.12.2   Modified Discrete Cosine Transform (MDCT)

In order to achieve nonuniform time and frequency splitting, ATRAC uses a unique combination of filters and transforms (Fig. 23.21). Two splitting filters are used to divide the original signal into three subbands: low (0–5.5 kHz), medium (5.5–11 kHz), and high (11–22 kHz). These signals are then transformed into frequency values by a modified discrete cosine transform (MDCT) operation. Before MDCT is performed, however, the signals are analyzed to determine if they are changing quickly or slowly. As explained previously, if the signal is changing quickly, then the MDCT uses a short time block. If the signal is stationary, a longer block is used for improved frequency resolution.

Once the signal is expressed in terms of frequency, the MDCT values are formed into 52 nonuniform frequency groups. These groups must be requantized to reduce the bit rate, and this is done according to the masking and sensitivity characteristics of each group. A special block-floating algorithm is used to eliminate any wasted high-order bits, such as when a 5-bit value is represented by an 8-bit number. In this way the data word length is reduced top and bottom while avoiding audible degradation of the music.

The decoder reverses the process by first reconstructing the MDCT frequency values back into time values by inverse MDCT operation. Finally, the three subbands are combined to obtain a normal 16-bit digital audio signal.

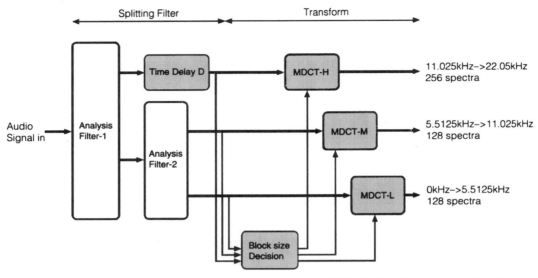

**FIGURE 23.21**  ATRAC frequency analysis circuit construction.

## 23.13  *SHOCK-RESISTANT MEMORY*

The biggest problem in using optical discs in portable applications has been that of skipping due to shock or vibration. This problem has been minimized in the MiniDisc system through the adoption of a high-capacity semiconductor memory system (Fig. 23.22).

The memory acts as a buffer, holding digital data equivalent to about three seconds of playing time before they are sent for conversion into analog signals for playback. If the player is exposed to shock or sudden movement, which jars the pickup from its position on the disc, the semiconductor memory will continue to output digital data to maintain playback. Because the memory itself is a solid-state device, it is unaffected by shock or movement. Since the position of the laser pickup is constantly monitored using address locations which are integral to both recordable and prerecorded MiniDiscs, it can quickly resume position.

One method to achieve efficient operation of the shock-resistant memory is to use digital data compression. Although the pickup reads data from the disc at the rate of 1.4 Mb/s, playback requires a rate of only 0.3 Mb/s. This is possible through the use of ATRAC, as described earlier. If the pickup loses its position, and the flow of data into the 1-Mb memory is interrupted, data will continue to flow out of the memory at the rate of 0.3 Mb/s to enable playback to continue for about 3 s. Once the laser pickup resumes its original position, however, it will read data from the disc at the rate of 1.4 Mb/s and thus replenish the data in the memory in less than 1 s.

During normal operation, signals are read from the disc in intervals because the data pickup rate is about five times faster than is required for playback. Figure 23.23 illustrates how the MiniDisc system reads data from the disc in intervals, while ATRAC decoding reorients the data on the time axis before D/A conversion. In the presence of shock or movement, however, the signal flow is interrupted, and once normal data pickup resumes, the amount of information read in the subsequent interval is increased to cover the time lost when the pickup was out of position. Once the contents of the memory are replenished, however, the pickup resumes reading the signal at regular intervals.

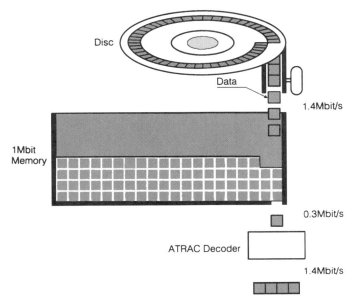

**FIGURE 23.22**   Shock-resistant memory.

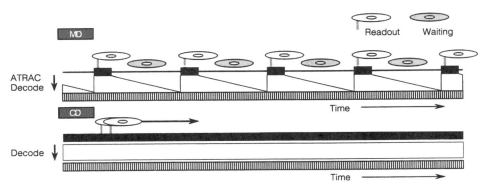

**FIGURE 23.23**   Reading signals in intervals.

## 23.14   *RANDOM ACCESS FUNCTIONING*

The MiniDisc format has been designed for complete random-access functioning. Playback-only discs are recorded like CDs and have addresses for each selection, which enables quick random access. Recordable discs, on the other hand, have special pre-grooves that allow quick and easy access to any point on the disc. In addition, recordable discs also contain a user table-of-contents area, allowing the tracks to be renumbered in any order in a matter of seconds.

The pre-grooves enable tracking and spindle servo control during both recording and playback. As Fig. 23.24 illustrates, the pre-grooves are shifted very slightly to create addresses in 13.3-ms intervals, which allow very stable high-speed random access.

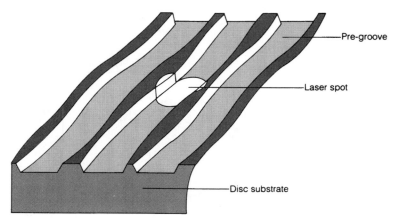

**FIGURE 23.24**    Pre-grooves on a recordable MiniDisc.

Track number addresses (start and finish) are recorded in a user table-of-contents (UTOC) area on the inner circumference of the disc, which is similar to a floppy disk directory. This enables track number editing in seconds, rather than having to wait for the actual performance time to rewrite a track address, as is required in time division systems. Figure 23.25 shows the changes that occur in the UTOC after two common track-editing operations—eliminating an unwanted track and combining two tracks into one.

## 23.15   MAGNETO-OPTICAL RECORDING

Although magneto-optical recording systems that enable recording on a disc one time have existed for some time, a key requirement in the design of the recordable MiniDisc was that it would be recordable over a virtually unlimited number of times. Another requirement was that the hardware be light and compact, and have low power consumption to make portable application practical. Two significant breakthroughs brought about those requirements:

1. The development of a highly stable magnetic disc layer of terbium ferrite cobalt that permits flux reversal using a low-level magnetic field of 6.4 kA/m (about one-third that of conventional magneto-optical discs) to allow the use of a small magnetic head, which produces a weak magnetic field.

2. The development of a magnetic head and related technology that allows virtually instantaneous magnetic flux reversal (in approximately 100 ns) at low power consumption levels. This solves the problem of temperature increases inside the system, whereas low power consumption enables battery-powered operation.

### 23.15.1   Principles of Magneto-Optical Disc Recording

A magneto-optical disc can be rerecorded a virtually unlimited number of times through a method based on the principle of optothermal magnetic recording. A given spot on the magnetic layer is heated with a laser beam to a point above the Curie temperature (which varies depending on the magnetic material involved), which effectively dissipates its magnetization. The disc continues rotation and, once away from the laser beam, the spot begins to cool. A magnetic field is then applied to the spot to orient it to either N or S polarity, corresponding to the 1 and 0 of digital signals.

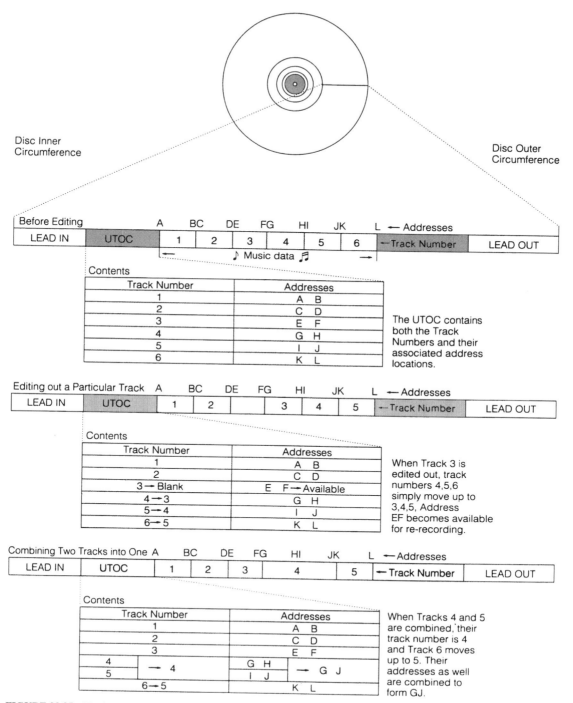

**FIGURE 23.25**  Track renumbering using the MiniDisc user table of contents.

## 23.15.2   Data Rewriting on a Conventional MO Disc

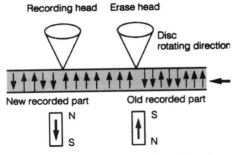

**FIGURE 23.26**   Data rewriting with two lasers in one rotation.

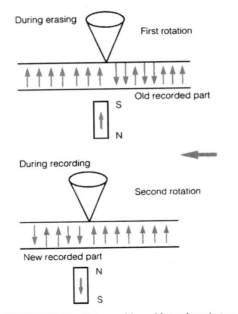

**FIGURE 23.27**   Data rewriting with one laser in two rotations.

On conventional MO discs used for computer data storage, all previously recorded signals must be erased before new data can be recorded. In principle, the track to be rerecorded must be heated (by a laser) while applying a magnetic field to reorient all magnetism in the same direction, effectively erasing it. Previous systems have primarily used two methods to accomplish this:

- Two lasers are used, one for erasing and one for recording, just like a tape recorder with an erase head and a recording head (Fig. 23.26).
- A single laser is used for these two tasks, with old data erased in the first rotation and new data recorded in their place on a second rotation (Fig. 23.27).

The problems encountered in these two approaches were that either two lasers were required, or roughly twice the recording time was required in a system needing a rather complex servomechanism.

### 23.15.3   Magnetic Field Modulation Overwrite System

In order to avoid the large and complex systems necessary with conventional magneto-optical disc technology, MiniDisc uses a magnetic field modulation overwrite (MMO) system that writes new signals over old signals. This system gets its name because it modulates the magnetic field at high speed to create specific magnetic orientations to represent the input signal.

The MMO system uses a magnetic head on one side of the disc and a laser beam on the other side in the corresponding position (Fig. 23.28). With the disc between the magnetic head and the laser, the magnetic head creates a magnetic field, corresponding to the signal, opposite the spot upon which the laser is focused. The laser brings this spot up to the Curie temperature point, which dissipates its previous magnetic orientation. As this spot on the disc moves away from the laser, it cools to below the Curie temperature, and a new magnetic orientation corresponding to the input signal is created by the magnetic head.

Because the size of the magnetic signals recorded on the magneto-optical layer of the disc is controlled by magnetic flux reversal, rather than by switching the laser on and off, the MMO system results in magnetic signals of consistent resolution for greater accuracy. In particular, it is easy to achieve accurate recording of short-wavelength signals, a previously difficult task due to thermal diffusion and other problems. Moreover, the MMO system requires the laser to

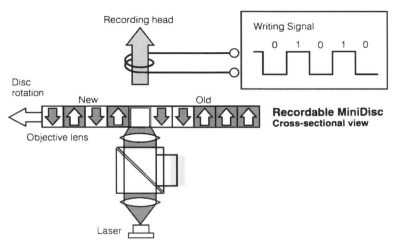

**FIGURE 23.28**    Magnetic field modulation overwrite system.

continually radiate a beam during both recording and playback, which allows a more simplified optical head, contributing to lighter and more compact hardware designs.

## 23.16  MAGNETIC FIELD MODULATION

Recordable MiniDiscs use magnetic field modulation for magneto-optical recording, a technology originally developed for CD-MO, a rewritable CD format. Magnetic field modulation has the following advantages: (1) rewriting capability that enables rerecording; (2) high recording density of 0.6 micron/bit; (3) pit-edge modulation that is ideal for CD-standard EFM signals; (4) less jitter because linear velocity is less critical; (5) wider recording power margin; and (6) superior resistance to disc tilt.

In developing the CD-MO, a magneto-optical disc that can record signals in the same format as CD, Sony set three practical goals: rewriting capability, recording at the same density and linear velocity as CD, and the ability to format addresses on the disc in advance.

Rewriting capability is essential for the continuous recording, in real time, of audio signals onto a previously recorded disc. Although laser modulation recording is used in magneto-optical systems for data storage, it is not applicable for MiniDisc because it erases and records data as separate operations. Instead, magnetic field modulation has been adopted for the recordable MiniDisc.

Achieving the same recording density as CD would be possible if recording could be performed at the same linear velocity of 1.2–1.4 m/s. Only a semiconductor laser with a 780-micron wavelength focused through an NA = 0.45 lens could be used, which would result in a large spot diameter of 0.9 micron. Thus, with the larger light spot, achieving the same recording density as CD was questionable. To enable preformatting of addresses, a slightly wavering pre-groove was adopted. It forms absolute addresses throughout the disc at 13.3-ms intervals.

Using magnetic field modulation results in an erroneous block rate of only 20 per second (compared to 200 per second when recording CD signals by conventional laser modulation) even at a linear velocity of 1.2 m/s (Fig. 23.29). Magnetic field modulation is not only capable of rewriting signals, it is also suitable for recording signals at the same linear velocity.

In magnetic field modulation a semiconductor laser continuously radiates at a power of about 4.5 mW and, when focused on a spot, raises its layer surface temperature to the Curie temperature

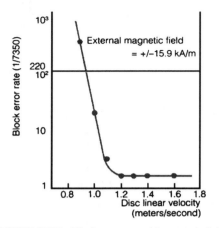

**FIGURE 23.29**   Block error rate with magnetic field modulation.

(about 180 °C). After the laser passes, the temperature drops. The process is then repeated, with the introduction of a magnetic field in one of two different orientations to the spot. Depending upon the orientation, either 1 or 0 is recorded, with the boundary being the isothermal line of the Curie temperature. This determines the shape and length of the recorded 1 and 0 as shown in Fig. 23.30.

If the magnetic field can be inverted quickly enough, it is possible to write patterns at a pitch of only 0.3 micron, even with a laser of 780-micron wavelength, focused through an NA = 0.45 lens. Furthermore, the resulting pattern is highly symmetrical, which is one characteristic advantage of magnetic field modulation.

Laser modulation, on the other hand, records signals by varying the degree of laser power. A magnetic field can only be oriented in one direction, with the area exposed to laser light corresponding to 1 and the unexposed (unrecorded) area corresponding to 0. This results in irregularly

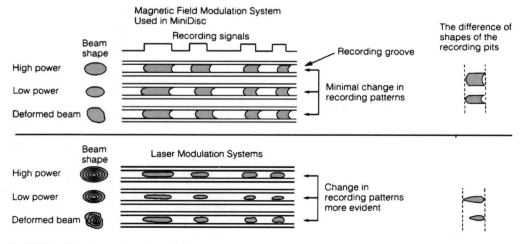

**FIGURE 23.30**   Comparison of recorded patterns.

shaped recording patterns (Fig. 23.30). When attempting to record CD signals with such patterns, increased jitter will result. This occurs because pit length varies in CD signals recorded by EFM, and, when a pattern corresponding to a long pit is recorded at a slow linear velocity of 1.2–1.4 m/s using laser modulation, the latter half will become thicker (Fig. 23.30), creating a shape like a teardrop.

## 23.17  MD PICKUP

Because playback-only MiniDiscs and recordable MiniDiscs are recorded in different ways, they cannot be played back with the same pickup. For this reason the MiniDisc format makes use of an innovative pickup in which both types of discs can be played back on the same equipment. The MD pickup is based on the conventional CD player pickup but has been modified to also contain a polarized beam splitter to detect differences in the light polarization plane direction that represents signals on the recordable disc. Accordingly, the pickup has two types of photodetectors (Figs. 23.31 and 23.32).

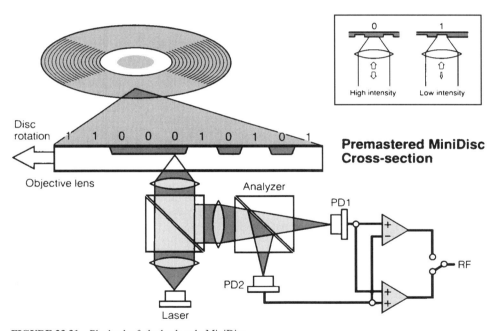

**FIGURE 23.31**  Playback of playback-only MiniDisc.

The playback-only disc is read by focusing a laser beam of approximately 0.5 mW power on a series of pits on the disc. Light reflected directly back indicates the absence of a pit. In the presence of a pit, the light is diffracted, and a lower level is reflected back to the laser. These light-level fluctuations correspond to the 0 and 1 digital signals (Fig. 23.31).

Playback of a recordable MiniDisc is also by a 0.5-mW laser, but the data are picked up in a somewhat different manner. Upon striking a specific portion of the disc, the polarized light will be reflected back along one of two opposing directions, with the polarization plane rotating slightly in a forward or reverse direction in accordance with the direction of the magnetic signal,

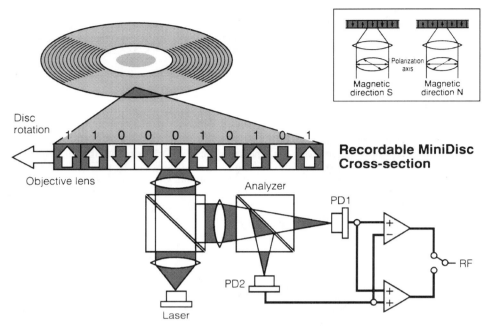

**FIGURE 23.32**   Playback of a recordable MiniDisc.

in a phenomenon known as the *Kerr effect* (Fig. 23.32). The polarization beam splitter varies the distribution ratio of the reflected light to two photodetectors in accordance with the polarization direction. If the reflected light varies in the forward direction, more light is reflected to one photodiode. If reflected light varies in the reverse direction, the other photodiode receives more light. The differences in electrical output of the two photodiodes are thus used to recreate the 1 and 0 digital signals. The fact that light reflected from the disc differs according to magnetic orientation is central to the recording/playback capability of the MiniDisc.

## GLOSSARY

**ATRAC**   The digital data compressing system, developed for the MiniDisc, in which audio signals can be reproduced with only about one-fifth the data normally required for high-fidelity reproduction.

**CD-MO**   A recordable magneto-optical version of the CD.

**CIRC**   An error correction method combining the Reed-Solomon code, which is an error correction code with a high random error correction capability, with an interleave method to convert burst errors into random errors.

**Cluster**   A minimum recording unit in the MiniDisc. One cluster contains 36 sectors.

**CLV (constant linear velocity)**   A pickup must be moved along a track on a disc at a constant linear velocity in order to pick up the same amount of data within a unit of time. CLV in the disc/pickup relationship is maintained by gradually varying disc speed according to pickup location on the disc. The disc rotates slowest at the outer circumference (where the distance covered by the pickup in one disc revolution is greatest) and fastest in the inner circumference (where the distance covered by the pickup in one revolution is shortest).

**Curie temperature**    The temperature at which magnetism of a specific material dissipates. This temperature varies according to the material.

**EFM (eight-to-fourteen modulation)**    The modulation system used to convert 8-bit encoded signals into 14-bit signals to match digital signals to disc transmission characteristics.

**Interleave**    Conversion to rearrange digital signals under a certain protocol to improve error correction capability.

**Kerr effect**    A phenomenon in which the polarization plane of laser light reflected from a material shifts in one of two directions depending upon its "plus" or "minus" magnetic polarization.

**Lead-in area**    The area on a disc before the program area—starting at an inside diameter of 29 mm—in which table-of-contents data are recorded.

**Lead-out area**    The area on a disc, outside the program area, where music signals are not recorded.

**Link sector**    Found only in recordable MiniDiscs, this sector links the beginning and end of a recording. There are three link sectors to one cluster.

**Magnetic field modulation**    The high-speed modulation of a magnetic field in magneto-optical recording to represent an input signal by magnetic orientation.

**Magneto-optical layer**    A thin layer within a recordable MiniDisc on which 1 and 0 signals can be recorded by the reversal of magnetic orientation.

**Masking effect**    A psychoacoustic phenomenon in which certain sounds are rendered inaudible by higher-level sounds on adjacent frequencies.

**MDCT (modified discrete cosine transform)**    A technique to convert time-domain signals into spectral signals for each fixed time block. MDCT ensures highly efficient coding with little connection distortion between blocks because conversion is performed using a special window function for overlapped halves between adjacent blocks.

**Overwrite**    The practice of writing new data over existing unerased data.

**Pre-groove**    A groove molded into recordable MiniDiscs that assists in tracking control. The pre-groove is meandered in a certain pattern in order to create addresses throughout the entire disc.

**Sector**    A unit of data specified for CD-ROM discs and also used for MiniDiscs. One sector is 2352 bytes.

**TOC (Table of contents)**    The generic term for all subdata (track number, playing time, etc.) apart from the audio data.

**UTOC (User table of contents)**    Found only on recordable MiniDiscs, this area contains subdata (track number, etc.) that can be rewritten by the user.

## ABOUT THE AUTHOR

Tsutomu Imai is the general manager of the MiniDisc promotion department at Sony Corp., where he has been employed since 1968. His experience at Sony includes working as a designer of microphones and magnetic recording devices and as product manager for analog record players, Compact Disc players, and digital audiotape products. He studied as an undergraduate at the Musashi Institute of Technology, graduating in 1968. He received an M.S. from the California Institute of Technology in 1975.

# CHAPTER 24
# DIGITAL MUSICAL INSTRUMENTS

**Gordy Carlson**
*Consumer Audio and Communications*
*Motorola Semiconductor Products, Inc.*

## 24.1 INTRODUCTION

*10:00 P.M. New York, N.Y.* Charlie, a young music professional, works to complete a jingle for a Manhattan advertising agency. This project requires a variety of musical textures, including a baseball park organ, a string quartet, and an ending using a standard rock sound. In the past this type of contract would have required several weeks of recording in a professional studio, using a dozen session players. Today, the project is completed in a few days, using one or two keyboardists with access to the latest digital music instruments. Whereas the professional studio would employ equipment worth several hundred thousand dollars, this project will be completed in a fraction of the time and with better fidelity in the young musician's project studio. The equipment used costs only ten thousand dollars.

*9:00 P.M. Austin, Tex.* It's showtime on Sixth Street. The sound of many bands will spill out into the street as onlookers walk by, peering into the windows at the various performers. Trent is just one of those musicians looking for an extra edge to draw in a crowd and get him a contract for a return engagement. His band has three members, each with a keyboard. They will take turns playing different instrument sounds on their keyboards. At one point they will put their electronic keyboards into a synchronized preprogrammed jam while they body surf the crowd for theatrical effect. These wonder instruments cost only a few hundred dollars. Trent's band gets the contract to come back and play again.

*7:00 P.M. Hermosa Beach, Calif.* Paul returns home from his busy day at work. The Los Angeles traffic has been especially brutal. He decides to relax by playing a few CD-ROM games on his computer. In the past, such game sounds were primitive and were played through the tiny built-in speaker. But the newest games use a standard sound format that can be played with consistent CD-quality sound on various computers with various sound cards. These cards are often bundled in with the computer at the time of purchase with minimum added cost to the user. The tensions of the day melt away from Paul as the excellent sound draws him in and makes the game even more engaging.

True stories, real people, real technology. The evolution of music technology over the last decade has been phenomenal. It has changed the way music is created, recorded, and distributed. The basis of this evolution has been driven by the proliferation of personal computers into businesses and homes and the advances in integrated circuits that have enabled low-cost digital audio products. Standards have evolved that permit these technologies to be scalable from professional systems to home consumer products, creating a uniform and consistent quality level, and increasing the available content for all platforms.

This chapter will take a close look at these digital music platforms, from the audio cards used by home computers to the more advanced systems used by music hobbyists and small studio owners. Each system will be evaluated and explained with regard to the technologies and standards used. Because these standards are scalable, the concepts examined in each system become a foundation of the next level system. A music hobbyist could easily find his or her own personal digital music system evolving like the systems outlined in this chapter, depending on budget and needs.

## 24.2    THE "STARTING-OUT" SYSTEM

The basic home digital music system (Fig. 24.1) usually starts out with a personal computer, a reasonable pair of self-powered monitor speakers, a soundcard installed in the PC, and some form of software application that controls the soundcard. Though live music performers may forgo the computer in favor of a keyboard with computer software built in, the concepts involved are the same.

**FIGURE 24.1**    A starting-out digital music system includes a personal computer, a pair of self-powered monitor speakers, a soundcard installed in the computer, and software to control the soundcard.

### 24.2.1    Soundcards

The demand for better supporting audio and sound effects for games was an early market driver for the soundcard industry. Sophisticated users were no longer satisfied with the primitive sounds that the basic computer was capable of providing via its small internal speaker. These sounds were often no better than beeps and clicks, or simple waveform (sine wave and square wave) playback. With the expansion slots available in most PCs, several manufacturers began providing add-on sound boards that provided a more engaging audio experience when used with compatible game software. Game providers were eager to use this new feature as a product differentiator but, in the early stages of the market, found it difficult to provide software that was compatible with the many competing methods to generate better sound. Advances in semiconductor technology drove down the average cost of the soundcard to a point where the average consumer could afford it, and the most economical sound generation methods emerged as standards.

### 24.2.2  Soundblaster

Soundblaster was an early standard that gave the PC an expanded ability to provide better sound by using an expansion card. Soundblaster cards use a sound synthesis method called *FM synthesis*. It uses operators (polyphonic oscillators) to create sounds. By mixing and modulating these oscillators together in various ways, many timbres can be created. The number of operators on an FM synthesizer determines the flexibility and ability of the synthesizer to create complex sounds. Generally, the more operators that are available, the better the sound. A side effect of having more operators is that programming becomes more difficult, but a whole third-party industry has evolved to provide FM sound programming and patches for popular synthesizers, removing the need for users to spend time creating their own sounds. All soundcards come with these preprogrammed sounds that are used as standards for gaming and multimedia.

Early soundcards offered two operators, which could produce a basic quality synthesis. Later soundcards offered four or six operators for even better sound. The more advanced cards today have taken this a few steps further and offer either subtractive synthesis or wavetable synthesis.

Wavetable synthesis uses a ROM resident library of waveform samples as operators. These wavetable samples are then mixed and modulated together to produce a wide variety of sounds. Because the wavetable sample waveforms can be much more complex than the output of a polyphonic oscillator used in FM synthesis, the resulting sound can be of higher quality and more realistic than those produced by FM synthesis.

One limitation of FM synthesis is an inherent inability to produce realistic percussion sounds such as drum kits. To improve the ability to produce percussion, manufacturers began using digitally sampled waveforms of those sounds that were played back through the soundcard's D/A converters in parallel with the FM synthesizer sound.

### 24.2.3  WAV and Sampled Sounds

To overcome the limitations of two-operator FM synthesis, Soundblaster cards turned to digital samples to reproduce certain difficult sounds. Early versions used 8-bit resolution samples with 22-kHz sampling frequency. Later products use 16-bit resolution with 44.1-kHz sampling rates. Although the lower-resolution, lower-sampling-rate boards required less memory for their sounds, the CD quality of the 16-bit systems soon became very desirable to most users.

The initial digital sample players used RAM resident sounds to play back through the soundcard's D/A converter. This method is similar to that used in Macintosh SND files. This RAM-only playback method restricts the playback time of the sample to the amount of available memory in the computer. The digital audio must also compete with other RAM data (graphics and video) for memory and the use of the system bus.

Both the Windows WAV and the Macintosh AIFF formats solve this problem by using a virtual memory scheme to swap data on the fly between RAM and the hard drive. File sizes are limited only by the size of the hard disk drive.

With the ability to store very large sound files of CD-quality audio, the need for better control and editing software emerged. Many users wanted the ability to edit sections of sampled audio for multimedia presentations, video clips, and music recording and production.

The resolution and sampling rate of a soundcard should be taken into consideration with regard to the available hard drive storage and the throughput of the system. In addition to having twice the number of bits per sample, 16-bit boards often use a higher sampling rate of 44.1 kHz. With the higher sample rate and larger sample words, the storage requirements of a 16-bit system can be four times larger than those of a lower-cost 8-bit board that samples at 22 kHz.

Every minute of stereo 16-bit/44.1-kHz audio would require 10 Mbyte of hard disk storage, whereas a minute of a mono 8-bit/22-kHz audio would need only 1.25 Mbyte of storage.

Often the quality in audio is well worth the larger words and the higher sampling rates of the 16-bit boards. A 16-bit/44.1-kHz system can theoretically reproduce CD-quality sound. An 8-bit/22-kHz system's sound would be more typical of a low/medium-quality cassette tape recorder. The difference would be very noticeable on most types of music programs.

### 24.2.4  General MIDI

The ability of multimedia software to have access to a standard set of sounds and instruments on various soundcards and computers is essential to compatibility of software across platforms. *MIDI* stands for Musical Instrument Digital Interface. It is essentially a serial data network over which modern musical instruments and equipment may communicate. MIDI was originally designed as a serial protocol for communicating between electronic musical instruments, but its applications have expanded, bringing musical soundtrack recording and editing to the desktop.

MIDI itself carries no sound. MIDI is a control protocol in which a master device (often a computer) instructs sound-generating devices (soundcards or synthesizers) to turn certain notes on and off upon demand. MIDI can direct these commands to up to 16 different addressable channels. Each channel can be instructed by MIDI to use a predetermined patch (sound or instrument) to create the specific sound required.

Many synthesizers and almost all soundcards are multitimbral, which means they are capable of generating more than one type of sound simultaneously. This is important when the multimedia software or the user-programmed songs need to use several instruments (drums, organ, strings, bass, guitar) playing together for a particular passage.

Synthesizers and soundcards are also specific in the number of voices they can support. A voice is the number of different sonic events that the soundcard can produce at any given moment—not to be confused with the number of patches (multitimbral) that can be played. Voices can be dynamically assigned to different patches as required by the passage being played back. Like human vocalists, voices can sing any part, but the number of voices is the absolute limit of different parts (sonic events) that can happen simultaneously. In general, the more voices possible, the better.

If a cross-platform–compatible multimedia software needs MIDI to play a particular song passage, it is a requirement that the program locations (patches) of these individual sounds (drums, organ, strings, brass, guitar) be consistent across all soundcards or synthesizers. A standard set of patches programmed in identical locations in all soundcards permits cross-platform multimedia audio to work.

The standard set of preprogrammed patches is now known as *general MIDI*. General MIDI is an extension to the MIDI specification that standardizes 128 sound categories and their program locations. It also specifies a minimum of 24 voices. Multimedia software that is general MIDI–compatible should play and sound fairly consistent across all platforms.

In truth, there are significant differences in the quality level of the sound playback and realism of the sounds, dependent upon the method of synthesis used in the individual soundcards. But general MIDI does standardize the name and type of sound that should emanate for identical patch programs on different cards. As the technology improves for all soundcards, the differences in audio quality will diminish over time.

> *Hermosa Beach.* Paul has become intrigued by the capabilities of his computer soundcard. Up to now he has only used prepackaged CD-ROMs that play their soundtracks using the soundcard. For the first time, Paul clicks on the icon for the simple sequencer software that was bundled in with his computer. Using his mouse and keyboard, he enters in the notes for a song melody that he has been humming for the last few days. He adds an accompanying chord pattern and a few drum sounds. He is pleased with his composition. He saves it to hard drive for safe keeping and then plays it later that night for his wife.

### 24.2.5    Music Software

Sophisticated programs available on the market allow control of the soundcard's patches, WAV recording and playback, and recording/edit/playback of the notes performed. Basic featured software is usually included with most soundcards. More advanced software can be purchased separately to enhance the flexibility and programmability of the system.

### 24.2.6    Basic Sequencer Software

Think of a basic MIDI sequencer software package as the music word processor of a MIDI studio. MIDI sequencers enable the entering of notes into a programmed song to be played via a soundcard. The sequencer allows the user to cut and paste sections of the music, copy it, and edit notes that were entered incorrectly. MIDI sequencers also allow the user to change the speed at which the music is played back, and offers programmable access to some of the advanced controls within the synthesizer (volume, pitch blend, modulation).

Many MIDI sequencers do not use standard music notation as their visual interface but offer simpler graphics representations of the notes being played and their durations. Standard music notation sequencers are available for those who desire them, but most musicians find the simpler graphics interfaces easier to use.

Because the soundcards are multitimbral, almost all MIDI sequencers organize their data into separate "tracks" for the individual sounds and instruments being played.

The basic MIDI sequencer included with a soundcard should have the ability to have data entered via the computer keyboard, allow editing/cutting/pasting of selected notes and measures, and allow selection of the various general MIDI standard sound palette for the song being written.

The basic MIDI sequencer should also allow changes in the tempo (beats per minute) of the playback and give the musician enough separate tracks to accommodate the number of different instruments that would be used in any song written. Having many separate tracks allows more flexibility when writing parts for the same instrument. Separate sections of the song can be on separate tracks, to be merged later. Using eight tracks (Fig. 24.2) would be extremely difficult for writing complex passages; 16 tracks would be more useful.

Separately purchased MIDI sequencers can offer an upgrade path to more tracks (sometimes as many as 64–128) and more advanced features. These more elaborate sequencers are not often necessary with the limited palette and programmability of soundcards. They are desirable when upgrading the system with a more capable external synthesizer keyboard.

### 24.2.7    Voice Editing/Patch Librarian Software

Simple soundcards may only support the basic general MIDI set of instruments. These cards would not provide the ability to create new original sounds by adjusting the programming parameters of the soundcard's synthesis circuit. These cards would not need a separate voice editor or patch librarian.

Soundcards with more advanced features allow the musician to create new sounds and store them in unused program areas within the soundcard's instrument (patch) library. These soundcards often come bundled with software that makes this programming easy and flexible.

Many musicians create their own sounds (patches) to differentiate the sound of their work from others. Using creative programming and use of unique sounds results in music that is unique and memorable, a primary goal of many musicians.

A voice editor is the software package that would be used to create these new sounds by manipulating the operators (polyphonic oscillators), the modulation, the filters, the mix, and the envelopes of a synthesizer. Voice-editing software should be easy to use by showing all of the current parameters on the screen at once, with the ability to click and change any parameter and hear the resulting change of timbre immediately.

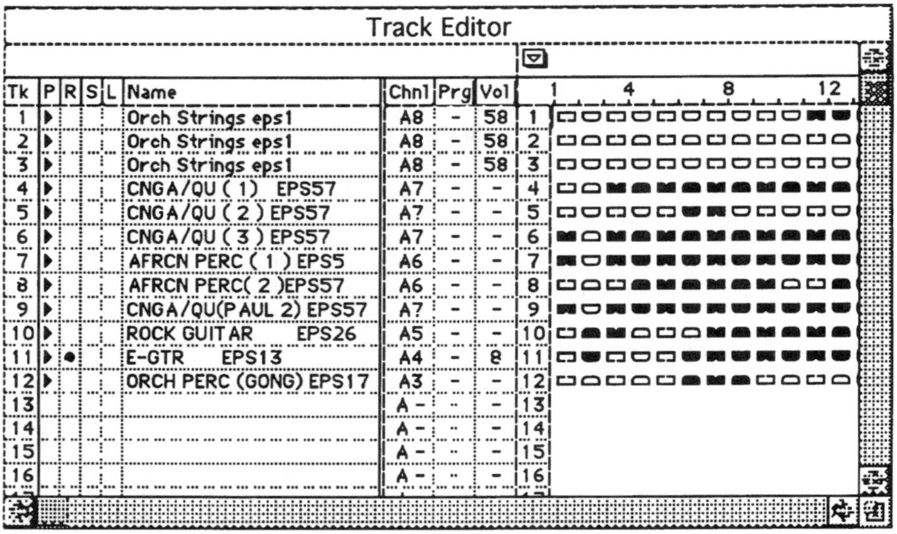

**Track Editor**

| Tk | P | R | S | L | Name | Chnl | Prg | Vol | | 1        4        8        12 |
|----|---|---|---|---|------|------|-----|-----|----|------|
| 1 | ▶ | | | | Orch Strings eps1 | A8 | - | 58 | 1 | |
| 2 | ▶ | | | | Orch Strings eps1 | A8 | - | 58 | 2 | |
| 3 | ▶ | | | | Orch Strings eps1 | A8 | - | 58 | 3 | |
| 4 | ▶ | | | | CNGA/QU ( 1 )    EPS57 | A7 | - | - | 4 | |
| 5 | ▶ | | | | CNGA/QU ( 2 ) EPS57 | A7 | - | - | 5 | |
| 6 | ▶ | | | | CNGA/QU ( 3 ) EPS57 | A7 | - | - | 6 | |
| 7 | ▶ | | | | AFRCN PERC ( 1 ) EPS5 | A6 | - | - | 7 | |
| 8 | ▶ | | | | AFRCN PERC( 2 )EPS57 | A6 | - | - | 8 | |
| 9 | ▶ | | | | CNGA/QU(P AUL 2) EPS57 | A7 | - | - | 9 | |
| 10 | ▶ | | | | ROCK GUITAR       EPS26 | A5 | - | - | 10 | |
| 11 | ▶ | • | | | E-GTR       EPS13 | A4 | - | 8 | 11 | |
| 12 | ▶ | | | | ORCH PERC (GONG) EPS17 | A3 | - | - | 12 | |
| 13 | | | | | | A - | - | - | 13 | |
| 14 | | | | | | A - | - | - | 14 | |
| 15 | | | | | | A - | - | - | 15 | |
| 16 | | | | | | A - | - | - | 16 | |

**FIGURE 24.2**   This MIDI sequencer track editor window shows the instruments used in a song and which MIDI channel they respond to. The solid squares indicate measures that contain MIDI data (notes or controllers) for that instrument. Channels can be selected for play (p), record (r), solo (s), or loop (L).

In deference to the earlier synthesizer—which had dozens of switches and knobs to adjust—some voice editors use graphical pictures of switches and knobs to more easily represent the values currently being used. The rule of "whatever works for you" applies here. It is true that the mechanical graphics can more quickly be read at a glance and offer more ease of use in programming.

When the soundcard or synthesizer has many available user-programmable locations for unique sounds, it often gets difficult to organize, update, back up, and maintain a large library of sounds. Patch librarians make this an easier task. They allow the saving of an entire context of a soundcard's parameter programming as a file on the hard drive. This is very useful for backing up against disaster, or saving a really good sound that you may never want to lose, especially when you need the memory location for something else in the meantime.

Patch librarians that are included with soundcards often have just enough flexibility to work with that particular soundcard, and not much more. More advanced patch librarians are available separately for system expansion with an external synthesizer and need more flexibility.

### 24.2.8   WAV Editors

Many soundcards support 8- or 16-bit digital recording to the computer hard drive. The basic graphic user interface that is used to control this on the computer screen often looks like a standard tape recorder mechanism with play, record, reverse, and fast-forward control buttons (Fig. 24.3).

Once this audio is in digital form on the hard drive, there is almost no limitation on how that sound file can be processed or edited. More advanced functions might require digital signal processors (DSPs) to perform these functions in real time, but the only limitation is the complexity of the software.

WAV editors can often boost or reduce the volume level of the playback, or can be used to cut and paste sections of the sound file. In less complex software utilities, these changes to the sound file are permanent (destructive).

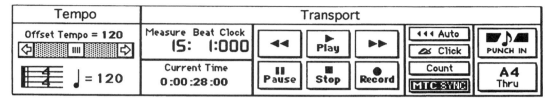

**FIGURE 24.3**    Many MIDI and WAV recorders use windows that resemble consumer audio equipment buttons for control of their record and playback functions.

More sophisticated WAV editors offer nondestructive editing. This type of editing keeps the original recording intact but uses add-on DSP boards to perform the desired function only during playback. The software relies on pointers and other software command hierarchies to define when and where certain processing is used during playback. Using DSPs also enables more complex processing functions such as equalization, digital delays and reverbs, pitch shifting, time compression, and even noise and tape hiss removal for restoring older recordings originally recorded on analog media.

Some of the programmable CD writers also use WAV files as their original data stream for the CD audio. Once recorded onto the hard drive as a WAV file, the applications software package included with the writer inserts the appropriate CD subcodes and time codes to create a compatible CD-ROM file.

### 24.2.9    Soundcard Overview

There is a wide variety of soundcards available in a wide range of prices. The choice of features depends highly on the anticipated uses of the soundcard by the user, as well as the affordability. For low-level enhancement of games, a sound card with FM synthesis is a good choice. For playback of higher complexities, in more advanced games and some multimedia applications, desirable features include FM (4–6 operators) or wavetable synthesis, 8- to 16-bit sampling (WAV compatibility), and compatibility with general MIDI playback. For home musicians and composers a wavetable synthesis synthesizer, 16-bit WAV, general MIDI, ability for custom user synthesizer patch editing and storage, and a MIDI sequencer (playback/recorder/editor) are recommended. Having a soundcard with MIDI IN and OUT ports for connecting to external synthesizers and sound modules also leaves room for system expansion without having to upgrade the soundcard at a later time.

A lot of exciting and creative audio and music work can be experienced with such a system. But for users who learn and master the techniques of soundcards and their control software, there is a road map for system growth. Because the standards used (MIDI, synthesis techniques) are scalable, further steps can be taken to enhance the system and its capabilities while remaining compatible with sounds and compositions already created.

## 24.3    THE "STEPPING-UP" SYSTEM

The beginner's system can provide excellent audio quality for a serious computer gamer and reasonable audio quality for an aspiring composer and musician. The flexibility and programmability of any of the higher featured soundboards allow the composer/musician to write and reproduce fully instrumented pieces due to the multitimbral synthesis feature of these boards. The sounds within the general MIDI sonic palette often satisfy most beginning composers.

One of the limitations that a computer/soundcard musician first notices, and must work around, is the difficulty in entering note data from a computer keyboard or mouse. It is difficult to put into either standard music notation or nonstandard sequencer notation what you hear in your mind. It is much easier to play new ideas on a standard music (piano) keyboard and have the sequencer software record your performance. Later edits and corrections on individual notes are then most easily made using the computer keyboard and mouse (Fig. 24.4). The first area that a computer/soundcard musician considers for upgrade is usually the addition of an external music keyboard to play the notes on more easily.

Theoretically, the musician who has a soundcard (with MIDI IN/OUT ports) installed in his or her computer really needs only a MIDI keyboard controller with no synthesis capabilities or sound generators of its own. But these MIDI-only keyboards are often built to the highest standards as expensive upgrades for very high-end users who have not been satisfied with the "feel" of standard consumer keyboards on most synthesizers. These upgrade keyboards are often much more expensive than a standard keyboard with synthesizer/sound electronics built in.

So most soundcard musicians who upgrade by adding an external keyboard usually buy a complete synthesizer. This is a real benefit, as the synthesizer circuits in the keyboard are often much better than those of the soundcard. With the additional high-quality voices and timbres available in the external synthesizer, the ability to generate better and more complex works of music is greatly enhanced.

By connecting the MIDI IN and OUT ports of the soundcard to the MIDI OUT and IN ports of the external synthesizer, a very integrated system is created. The external keyboard's synthesizer circuits can be driven by the sequencer software within the computer, in parallel with the synthesizer circuits of the soundcard. In addition, any notes performed on the external keyboard can be recorded into the sequencer's memory and later edited with the computer keyboard and mouse.

*Austin, Tex.* Trent and his band have played several successful shows in the last two weeks. The crowds that walk by on Sixth Street often stop by to listen at the club where his band plays. There are many regulars in their audience. Trent wants to try some new songs to keep the regular fans interested in his band, but he feels his keyboard lacks the additional sounds and textures he needs to expand his repertoire. He could buy a whole new keyboard, but he must keep the old one for the benefit of his current song list and the sounds it can produce. Also, using two keyboards would require more space on the already cramped stage. Trent decides to buy an expansion synthesizer module. It is essentially the synthesizer circuits of a modern keyboard without the physical keys. It fits into a standard rack space.

By connecting the MIDI output ports of his current keyboard to the MIDI input ports of his new expansion module, he has added a whole new palette of sounds and textures to his equipment setup. These extra sounds can be played from the keyboard of his original synthesizer. The next night Trent's band plays two new songs they rehearsed earlier. The regulars in the audience enjoy the new songs and sounds they hear.

### 24.3.1  MIDI Physical Interface

Today there are MIDI connectors on keyboards, drum machines, guitars, sound-processing equipment (equalizers, delay units, reverbs), and electronic wind instruments (sax, clarinet). Having a standard interface is important in being able to connect these various pieces of music equipment into an integrated MIDI studio.

In general, all these MIDI devices would be daisy-chained on the same MIDI bus. The protocol aspects of the MIDI data packets route the correct note data to the appropriate destination (synthesizer or soundcard).

### 24.3.2  MIDI Hardware

MIDI is transmitted and received over a MIDI cable a maximum of 50 ft long. It is terminated on either end by two male 5-pin DIN connectors. Most instruments have MIDI IN and OUT ports, as well as a MIDI THRU port. MIDI THRU is included on many keyboards and MIDI devices

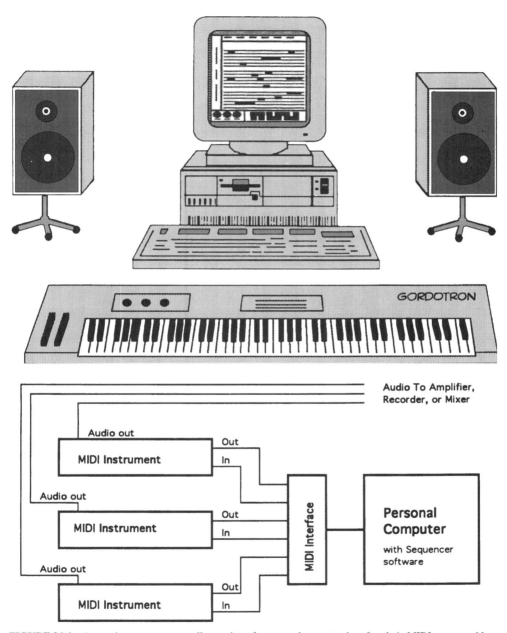

**FIGURE 24.4**   A stepping-up system usually consists of a personal computer interfaced via MIDI to external keyboard synthesizers or sound modules. Using an external keyboard allows the user to "perform" the song on a piano-type keyboard while recording the MIDI data stream using the computer sequence software. Edits and improvements can be completed later with the sequencer's graphical interface on the computer.

to facilitate the daisy-chaining of many devices into one network without having to use extra adaptors or splitters. MIDI THRU provides a direct copy of the MIDI data stream coming into the MIDI IN port.

### 24.3.3  Electrical Specification

MIDI is an asynchronous serial interface. The baud rate is 31.25 kbaud ($\pm 1\%$). There are eight data bits, with one start and one stop bit, for 320 μs per serial byte.

MIDI is a current loop, 5 mA. Logic 0 is current ON. The specification states that the input is to be optoisolated and points out that the Sharp PC-900 and HP 6N138 optoisolators are satisfactory devices. Rise and fall time for the optoisolator should be less than 2 μs.

The MIDI specification uses only three of the five wires for MIDI signals. These are RECEIVE DATA on pin 4, TRANSMIT DATA on pin 5, and GROUND on pin 2. Two wires are reserved for future expansion of the MIDI standard. The (MIDI OUT) transmitter port can be buffered to provide adequate drive if required. There should only be one master in use on a MIDI network, because there is no collision detection/correction implemented in the MIDI protocol.

### 24.3.4  MIDI Addressing

There can be a maximum of 16 channels of information transmitted over one MIDI cable. Each channel is basically a unique address associated with each packet of information. This way, 16 slaves may be chained together on one physical MIDI connection, but each slave can respond to information on a different channel. Because of the multi-timbral nature of many keyboards and soundcards, a single keyboard may use several channels for control of its synthesizer circuits. This allows the musician to create and control many types of sounds from one piece of equipment without having to purchase a complete synthesizer for each sound needed.

### 24.3.5  MIDI Messages

MIDI communicates by using multibyte messages. Multiple bytes are used to denote the channel of the information, the type of information, and the data itself. The most common type of information transmitted and received over MIDI is "note" data—or, more specifically, "what note" is being played by the master, and the duration. But in addition to note on/note off messages, there are many other types of MIDI messages that are used for various reasons in a MIDI system. MIDI messages can be defined within several categories: channel voice messages, channel mode messages, system common messages, and system real-time messages.

Channel voice messages communicate information about the notes being played. These include the ubiquitous note on/note off messages, polyphonic key pressure, overall pressure, controllers, pitch bend, and program change/select.

Channel mode messages are used to denote overall operation of a MIDI synthesizer or device on the network. Some of these mode messages include what information the device should receive. Some devices should receive all notes from all channels (omni on). Other devices are instructed to receive notes only from one channel (omni off). Other commands tell the MIDI device to send the notes received to just one voice (mono on) or to various voices (poly on).

One other very useful channel mode message is "all notes off." Once in a while a note will get stuck on because of a programming error where a "note on" was sent to the keyboard, but a "note off" was somehow lost. Many MIDI sequencers and keyboards have an "all notes off" feature that clears this stuck note.

System-common messages are control commands that apply to all channels. An example is a song position pointer, which communicates to each MIDI instrument how many beats (in sixteenth notes) have elapsed since the beginning of the song. This feature can be used to synchronize the playback of drum patterns and songs programmed into various instruments.

System-exclusive messages are a special type of system common message that are defined by the manufacturer and are not compatible with or understood by instruments from different manufacturers. System-exclusive messages use an ID code that is specifically assigned to one manufacturer. This ID code is used so that other manufacturers' instruments may ignore the special command.

System-exclusive messages are often used by synthesizers to download their patch programs to a computer or mass storage device for archiving. They are also used by sampling keyboards to transfer actual audio sample data to a computer so that waveform-editing software packages can be used to modify the sample.

Advanced patch librarians, voice editors, and waveform editor software products often support many device manufacturers' system-exclusive protocols so that these products can be used by many different musicians. Before downloading a patch or sample, it is necessary to select, or tell the software package, which manufacturer the patch or sample will be downloaded from.

System real-time messages are received by all channels. These messages include a MIDI time clock that sends out a continuous clock signal to which all MIDI devices on the network can synchronize. This MIDI time clock provides 24 ticks per quarter note resolution. Additional system real-time messages include commands sent by the MIDI master for all instruments to start play from the first measure, stop play, and continue. Because the MIDI clock is continuous, all instruments are locked to the time reference prior to any start, stop, or continue command, which results in consistent system operation and synchronization with fewer errors.

### 24.3.6    MIDI Voice Messages

MIDI voice messages are the most common type of MIDI data that are transmitted and received via the MIDI network. Voice messages tell what key is being pressed or released, what MIDI channel the instrument is on, the velocity of the key movement (used to control the dynamics and expression of the note being played), and the pressure used to hold down the key after it has been pressed (for even more expressive control).

MIDI notes have numeric values of 0 to 127. Middle C on the keyboard is equal to note number 60, with lower numbered notes having lower pitches and higher numbered notes being those above middle C. Remember that MIDI can control up to 16 channels of operation, with 127 notes available on each channel.

To allow more expression when playing and recording music on a MIDI network, there are several additional parameters that are sent as MIDI voice messages. When keyboardists play quiet passages, they press the keys slowly. When playing louder, more accented notes, they tend to press down on the key faster. This is the velocity of the key movement. MIDI allows a keyboard to transmit the velocity with which the key is pressed so that the voices responding to that key can alter their volume or timbre to simulate an accented note. Not all keyboards support the transmission of velocity data, and not all patches/voices are programmed to respond to velocity data. A keyboard purchase should be researched carefully to ensure that key velocity is supported. Velocity voice messages can add a great deal of expression to any music passage.

Another voice message is key pressure. This is also used to add expression. Whereas velocity is mostly used as a way to accent or play a note louder, key pressure can be programmed for many different uses. An example of key pressure use would be to have the vibrato of a violin patch deepen when the keys are pressed harder. Another example could be using key pressure to control a filter, so that the sound brightens or mutes the higher frequencies on notes that are pressed harder than others. MIDI supports both polyphonic key pressure where each note is treated individually, or aftertouch key pressure which takes the overall average key pressure of all notes being played, and applies the desired effect to all keys.

The pitch wheel is a small mechanical controller usually placed just to the left of the lowest key on the keyboard. Turning this wheel changes the pitch of the played note by the desired amount. This is used to "bend" notes like a guitarist. The keyboardist plays the melody with the right hand, while the left hand turns the wheel at appropriate moments to change the pitch or bend the notes.

Sequencer software packages would never require the user to program songs in hexadecimal format, but included here is a table (Table 24.1) that shows the data translations that the sequencer is performing. Understanding this table adds to the overall knowledge of how a system works. This is very useful when system problems—such as stuck notes, wrong channels, and mismatch of MIDI functions between different pieces of equipment—need to be worked out.

**TABLE 24.1**   Voice Messages Format

| Status byte | Function | Data bytes |
|---|---|---|
| 0x80–0x8f | Note off | 1 byte pitch, followed by 1 byte velocity |
| 0x90–0x9f | Note on | 1 byte pitch, followed by 1 byte velocity |
| 0xa0–0xaf | Key pressure | 1 byte pitch, 1 byte pressure (after-touch) |
| 0xb0–0xbf | Parameter | 1 byte parameter number, 1 byte setting |
| 0xc0–0xcf | Program | 1 byte program selected |
| 0xd0–0xdf | Channel pressure | 1 byte channel pressure (after-touch) |
| 0xe0–0xef | Pitch wheel | 2 bytes giving a 14-bit value least significant 7 bits first |

### 24.3.7   Channel Mode Messages

MIDI channel mode messages control the way a MIDI master sends data over the 16 available channels and control the way MIDI slave devices receive or ignore MIDI messages that are present on the network. *Omni* mode refers to the ability to receive voice messages on all channels. *Mono* and *poly* modes refer to whether multiple voices are allowed.

*Mode 1: Omni on/poly.* Voice messages are received on all channels and assigned polyphonically. The MIDI slave unit plays any note it gets from any channel up to the maximum number of voices it can handle at one time.

*Mode 2: Omni on, mono.* In this case the slave is a monophonic instrument that will receive notes to play in one voice on all channels.

*Mode 3: Omni off, poly.* The MIDI slave is a polyphonic instrument that will receive voice messages only on the assigned channel (1–16).

*Mode 4: Omni off, mono.* This is an unusual mode to use with keyboards. The receiver can be programmed to recognize several channels but can assign only one voice per channel. This mode is very useful for guitar synthesizers. The guitar would transmit the data from each string to a different MIDI channel. Each channel is monophonic, which helps reduce the effect of false notes and false triggers from the guitar controller. It also allows the electronic guitarist to play multiple instruments simultaneously, one for each string.

To operate in this mode a receiver is supposed to receive one voice per channel. The number of channels recognized will be given by the second data byte or by the maximum number of possible voices if this byte is zero. The set of channels thus defined is a sequential set, starting with the basic channel.

In practice, most keyboards are assigned a basic channel (1–16), over which they receive their data. They ignore any data not addressed to their assigned channel. This allows multiple keyboards to be used simultaneously on a MIDI network, each playing different parts of the song.

Higher-performance keyboards often have the ability to have several channels of sound loaded into memory, with a separate MIDI channel controlling each. These keyboards operate in omni mode but often have a basic channel, with several sequential channels able to respond to the MIDI data and control the "virtual" keyboard sounds. These keyboards have selector switches that allow the user to change which set of sounds the mechanical keyboard is currently controlling. Even in omni mode, the keyboard would ignore MIDI messages for channels it is not using.

Another MIDI mode message either enables or disables local control of the keyboard's voices. With local control on, the voices in a synthesizer respond to the keys being played on that synthesizer. With local control off, they do not. Either way, the MIDI data are transmitted out the MIDI OUT port.

In the example, when Trent the musician purchased an expansion sound module, he needed a way to play its sounds without the voices inside his existing keyboard being heard. In this situation he would turn off local control. Only his expansion module would respond to the notes he played on his keyboard. The sounds in his existing keyboard would stay silent.

The data format of channel mode messages is that of three pieces of data being transmitted (Table 24.2): the channel number; the mode controller number (122–127); and a third byte, which is zero for all controllers except number 122 (local) and 126 (mono mode). This third byte specifies the number of channels to be allocated for mono data, or if local control is on or off.

**TABLE 24.2**    Channel Mode Message Format

| Data | Mode | Third data byte |
| --- | --- | --- |
| 122 | Local control | 0 = local control off, 127 = on |
| 123 | All notes off | 0 |
| 124 | Omni mode off | 0 |
| 125 | Omni mode on | 0 |
| 126 | Monophonic mode | Number of monophonic channels, or 0 |
| 127 | Ployphonic mode | 0 |

### 24.3.8  System Real-Time Messages

System real-time messages (Table 24.3) are received by all devices on the MIDI network. They are used to synchronize all devices to a common clock and permit network-wide commands such as system reset and start/stop song messages to be coordinated among all instruments. Because system real-time messages are to be received by all devices, no channel (1–16) data are needed when sending them.

**TABLE 24.3**    System Real-Time Message Format

| Data | Function |
| --- | --- |
| 0xf8 | Timing clock |
| 0xf9 | Undefined |
| 0xfa | Start |
| 0xfb | Continue |
| 0xfc | Stop |
| 0xfd | Undefined |
| 0xfe | Active sensing |
| 0xff | System reset |

Because of their importance in maintaining a synchronized system timing and commands, system real-time commands have priority over any other type of MIDI message. System reset messages tell each device to return to the parameters it had used as default values after a system power up. The common system timing clock that the MIDI master transmits is 24 clicks per quarter note. This common clock runs continuously. Because it is continuous, all instruments on the MIDI network are able to be in perfect synchronization from the very beginning of any MIDI sequence.

Many MIDI instruments have built-in internal sequencers that offer basic functions such as record and play with a few basic editing modes. Central computer control with a MIDI software package is a much more efficient way to record and edit MIDI songs, but internal sequencers can be very useful, especially when you want to play a sequence in a live show and do not want to have to carry a computer to the gig and set it up.

After completing the detailed programming of songs at home on the computer and software sequencer, the sequence can be downloaded to the internal sequencers in each instrument. But all the individual sequencers in each instrument need not be "conducted" so that they all operate at the same tempo and start and stop their playback in unison. The MIDI timing clock provides the tempo by transmitting the continuous 24 clicks per quarter note.

Other system real-time messages include start, stop, and continue. These commands are transmitted by the MIDI master and are used to tell the instruments to begin playing back their internal sequences at exactly the same time and stop at exactly the same time.

The start message tells each instrument to play its internal sequence from the very first measure. The stop message discontinues the playback. The continue message starts the playback again from the point at which it was last stopped.

Active sensing is a very little used mode. It was intended as a system integrity enhancement for clearing stuck notes. Note data are sent using several voice messages. First the "note on" is sent. When the note is to stop playing, a separate "note off" message is sent. If, for example, the MIDI cable came loose (someone tripped over it) in the middle of a song, several "note off" messages may never be received by the instrument to discontinue the notes that were currently being played. Those notes would be "stuck" notes and would drone on until the system was reset. Active sensing was intended to sense the disconnected MIDI cord and have the sequencer automatically perform an "all notes off" command to resolve stuck notes.

The active-sensing byte is to be sent at least every 300 ms. Its purpose is to implement a time-out mechanism for an instrument to return to a default state. An instrument operates normally if it never times out. If it misses the active-sensing byte, the instrument is usually programmed to turn off its sound circuits.

The problem with the way that active sensing is performed in the MIDI specification is that it significantly increases the MIDI byte traffic on the network. Because system real-time messages have priority over other MIDI messages, this could be a problem during periods of high note and controller activity on the network. Notes could be noticeably delayed during the playback of a sequence. The bandwidth cost of implementing active sensing is more than the solution it offers is worth.

### 24.3.9 System-Exclusive Messages

Manufacturers may want to add extra features to their products that would use the MIDI network as a data media. These manufacturers want the flexibility to specify the best way for these additional features to be implemented. Some of these additional features may include bulk data dumps of the synthesizer patch programs to a computer. There the data can be stored and archived using a voice/patch librarian software package. Another feature for sampling keyboards might be the ability to upload the sample sound from the sampler's memory directly into the computer's memory. There it can be edited using a waveform editor or used as a WAV file for a multimedia application.

These are just two of the many other possible features that manufacturers may want to implement in their equipment (in the way they best determine) and still use the MIDI network for transferring the data and communicating. System-exclusive messages, called *SYSEX* for short,

**TABLE 24.4**  SYSEX Message Format

| Data | Function | Additional bytes |
|------|----------|------------------|
| 0xf0 | System exclusive | Variable length |
| 0xf7 | EOX (terminator) | 0 |

are specified in the MIDI specification (Table 24.4) to give this flexibility for additional or proprietary features for specific manufacturers.

System-exclusive messages have only a few requirements. They must begin with a SYSEX start message and end with an EOX (end of SYSEX) message. In almost all cases, the first data byte is the manufacturer's ID byte. Each manufacturer has reserved a specific identity number for SYSEX messages. Having this ID number in the data stream allows instruments on the MIDI network from other manufacturers to ignore that SYSEX command.

After the manufacturer's ID byte is sent, the number of following bytes sent and received by the two devices communicating is entirely up to the protocol that that manufacturer specifies for its SYSEX features. The only limit is that the data bytes themselves are within the values of 0–127. The EOX message is used to stop the transmission after the variable number of data bytes are sent and received.

### 24.3.10  Sampling Keyboards

The simplest way to reproduce a particular instrument is to actually record that instrument digitally, store it in memory, and play it back when the appropriate key is pressed. Until recently, the problem with this straightforward approach was the very high cost of the large memory arrays required to store these waveforms and the high-speed processors to control the movement of so much data. Technology advances in both DRAMs and processors have now made sampling keyboards affordable to the common consumer.

The design architecture of these keyboards is almost identical to that of PCs, with the exception of advanced video features. The block diagrams of these keyboards include microprocessor, memory, I/O, operating systems, and floppy and hard disk drives. Newer models may also use magneto-optical storage. Digital samples require a lot of memory and storage. Having 16 Mbyte of memory and a 300-Mbyte hard drive built into the keyboard chassis is not uncommon.

With the addition of digital signal processors (DSPs) into the motherboard of the sampling keyboard, many manufacturers are adding signal processing (EQ, digital delays, and reverb effects) to the feature sets of these instruments.

A home MIDI studio that pairs a sampling keyboard with a powerful sequencer and sample (WAV) editing software on a PC can become one of the most versatile and powerful digital audio workstations available to a home musician.

Some of the variables a musician should look for in a sampling keyboard for home use include (1) the size of the internal memory; (2) the ability to upgrade it at a later time; and (3) the type of disk storage, whether built into the sampler or available as an upgrade. A floppy disk is the most economical and might be a reasonable choice starting out. Upgrading to a hard drive can greatly increase the efficiency in quickly loading sounds, storing new ones, and creating macro files of many sounds that can be recalled with one command.

Think of the PC and how having a large hard drive is so much more efficient than reloading programs all the time from floppies. The same analogy applies to a sampling keyboard.

Although most samplers can upload their sounds via the MIDI network to a computer, this is only intended for occasional transfers of data that will be edited on the PC using waveform-editing software. No reasonable system would use this method for routine storage of samples. A hard drive either built into the keyboard or external on an SCSI bus is a wonderful advantage to have.

## 24.3.11 Advanced Sequencers and Wave Editors

The basic MIDI sequencer (Fig. 24.5) included with a soundcard should have the ability to have data entered via the computer keyboard; allow editing, cutting, and pasting of selected notes and measures; and allow selection of the various general MIDI standard sound palette for the song being written. The sequencer should also allow changes in the tempo (beats per minute) of the playback and give the musician enough separate tracks to accommodate the number of different instruments that would be used in any song written.

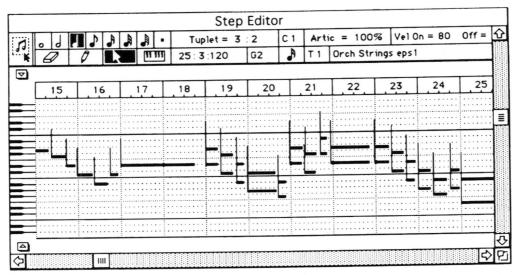

**FIGURE 24.5** Most sequencers use a graphical method of showing the notes instead of conventional music notation. Here the keyboard along the left axis gives a reference to the note pitch. The duration of the note is indicated by its horizontal length. The song measures are the horizontal axis. Notes can be edited with the pencil and eraser tools, and sections of notes can be copied and pasted as needed.

More advanced sequencer software packages available separately offer many more features than the basic sequencers included with computer soundcards. If the intent is to do a lot of song programming, the investment of an advanced sequencer package is well worth the cost.

Some features found in more advanced sequencers include graphical editing, larger number of tracks, quantization/humanize, synchronization, and programmable tempo.

***Graphical Editing.*** The ability to use the computer mouse to select, cut, and paste various types of data can add a considerable amount of efficiency to the work. Being able to use a pencil or eraser icon to add or erase individual notes is also useful.

Even more than note editing, controller (volume and modulation) and pitch shifting editing are much more convenient when using draw techniques. These types of data are more closely packed on a display screen and often represent curves and waveforms. Using a pencil icon to redraw a curve is a lot less painful than individually reentering the data points numerically (Fig. 24.6).

Having the ability to automatically create data value slope changes is useful, too. Imagine a volume fade-out (controller number 7) that on the screen resembles a gently sloping waveform of continuous data points. But suppose a quicker and more dramatic fade-out is desired. Rather than redraw the curve by hand and end up with a free-form, imperfect slope, the area of the fade-

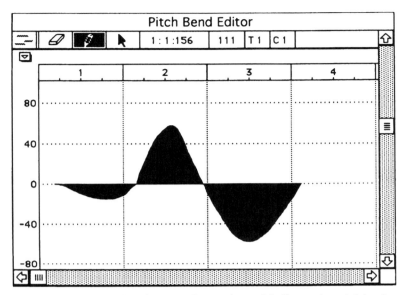

**FIGURE 24.6**   Being able to draw control curves for special effects such as pitch bending is a powerful feature included in advanced sequencers. WAV editors also allow freehand drawing of the audio waveforms for correcting problems such as noise, pops, and glitches.

out could be selected and the sequencer told to recalculate the data values to "change smoothly from volume = 127 (full on) to volume = 0 (full off)." After recalculating, the fade-out curve would be perfect.

***Larger Number of Tracks.***   Basic sequencers may only offer 16 tracks to use with different sections of MIDI data. Better sequencers offer from 64 to 128 tracks. Even though there can only be 16 different MIDI addresses on a MIDI bus, having many available sequencer tracks adds a lot of flexibility in the way work is created and organized. For example, several different versions of an organ solo could be maintained and then just one selected when the final mix is done. The other versions could be kept for future references or for a change of heart in a later mix. The unused tracks can be muted when not in use. Another good use is the ability to record the song in sections without having to punch in and punch out of a single MIDI track. Because drums are a collection of many different sounds and instruments, recording and saving percussion tracks can consume many channels of a MIDI sequencer.

***Quantization/Humanize.***   When recording a MIDI keyboard performance into a sequencer, one of the first things one notices is that no matter how hard you try, your timing will not be perfect. Some notes will fall perfectly onto the beat boundaries; others will be off significantly. This slight imperfection is very desirable. It is the human touch that makes the music engaging and not machinelike in sound. The deviations in timing add expression to the performance. Live drummers often train diligently to be slightly off the beat for certain types of grooves. Sometimes this is called the *swing* of the song. Being slightly behind the beat to add tension is called the *backbeat*. But too much of a good thing can also be undesirable. Being significantly off the beat will sound terrible.

A quantization feature on the sequencer allows the computer to correct the deviations in timing. The software rounds off the note to the nearest fraction of the beat (programmable by the musician). This can be used to tighten up a section of the performance. Care should be taken not to overuse quantization because it makes the song sound very mechanical and machinelike.

The exact opposite of quantization is a humanize feature. This advanced sequencer feature actually adds or subtracts small random deviations to the timing of each note and to the velocity value of the key engagement. This feature is often used after a section of notes are entered into the sequencer by hand using the computer mouse and the sequencer's graphical editing interface. These manually entered notes are usually perfect on each beat. By selecting the section of song and performing a humanize calculation, the notes are given a more "human" feel by adding these timing imperfections.

*Synchronization.*   There are many different ways to time-synchronize various pieces of studio equipment. These include MIDI time code, MIDI song position pointers, FSK (frequency-shift keying), SMPTE (Society of Motion Picture and Television Engineers) time code, and several other manufacturers' proprietary time codes. SMPTE is often used to synchronize MIDI equipment to video and audio tape machines. SMPTE displays its values in hours, minutes, seconds, and frames format. SMPTE is a useful and convenient synchronization method to use and is very scalable to professional studio equipment.

Advanced sequencers should be compatible with the types of time codes that are expected to be encountered. The most common are MTC (MIDI time code), MIDI song position pointer, and SMPTE.

*Programmable Tempo.*   Just as humans are unlikely to play in perfect timing, so too are they unlikely to play songs at a steady tempo. Musicians often fluctuate on various sections of the song to add dynamics and expression. An example would be a slow song that slowly speeds up through several verses, and becomes a fast tempo in the chorus, only to slow down again near the ending for dramatic effect. Being able to program the tempo to different values in different sections of the song adds greatly to the listener's enjoyment.

*Other Advanced Sequencer Features.*   Other sequencer features include auto locator to "jump to" significant sections of the song immediately, rather than having to scroll through from the beginning to find a section to work on or rerecord.

Sometimes songs are written and then found to be out of the vocal range of a singer who wishes to add vocals. A transpose feature solves this problem by shifting the base key of all notes by a desired amount.

If session players are to be hired to play your MIDI compositions, or if you anticipate selling your songs to other musicians for their own uses, you often need to provide a standard music notation printout of the song. Some sequencers have the ability to print out the work in standard sheet music form. Others have upgrade packages that can be purchased separately to provide this function.

With the rapidly evolving developments in multimedia and enhanced full-motion video on PCs, several professional-level sequencer packages have the ability to sync up with a video clip being played on the computer. This sync is often performed at the operating system extension level (e.g., Apple Quicktime). These advanced sequencers need to be compatible with the advanced features of these multimedia computer operating systems. Research this thoroughly if your needs include full-motion video compatibility and sync. Many video software packages will specify and sometimes comarket MIDI packages that work well with their own product.

## 24.3.12   Standard MIDI File Format

Imagine that you are collaborating on a MIDI music project with a friend from a different region of the country. By using general MIDI, you can be assured that the instruments you use will match those of your friend's general MIDI-compatible soundcard or synthesizer. You complete your programming of your parts using your sequencer, save the file, and either copy the file to a floppy and mail the floppy or electronically transfer the file to your friend. So far, so good.

But what if your friend's sequencer software (which may be different from yours) saves and reads its MIDI files differently than your sequencer? What if it uses a different file format that your sequencer cannot read?

The general MIDI standard only denotes that the instrument patches are in the same program areas, not that the MIDI file saved by one sequencer is compatible with the MIDI file format of another.

Early MIDI sequencers had this problem. Sometimes the only way around this was to connect the two MIDI computers together and do a "live" data dump of the sequencer data by playing back the file on one sequencer while the other computer and its sequencer were in record mode—a very primitive solution.

The answer to this problem was the creation of a standard MIDI file format that permits better compatibility across different sequencer packages. Often the default mode of the sequencer is to read its own native files. But most sequencers now allow the import and export of standard MIDI files for compatibility with other sequencers.

Standard MIDI file format does not guarantee compatibility with all the special functions and control features of the sequencers. The file format exchanges a basic level of information about each track, enough to be able to retain the note data, track name, program changes, timing and tempo data, pitch shift, and controllers. Special markers and play lists are not often compatible across sequencers. If you expect to collaborate on MIDI projects with other musicians using different sequencer software, confirm that both of your sequencer packages will at least import/export standard MIDI file format.

## 24.4    THE "SEMIPRO" SYSTEM

The enhancements included with the stepping-up system—the external keyboard, better software for editing synthesizer patches, and sequencing—provide both a better input device for the notes to be played and more and better sounds from the synthesizer's own waveform generators and oscillators. The result of these enhancements is often a repertoire of songs that can be shared with friends, family, and prospective clients. But to share these songs with others, they need to be recorded on a stereo cassette tape. This works fine if the compositions can be played completely by the synthesizer, with no additional layering of sounds from other sources, such as vocals, and acoustic instruments like guitars, drums, brass, or woodwinds. Even though many of these acoustic instruments can be closely approximated by the synthesizer, sometimes there is no substitute for the real thing. Background vocals can often be faked with good sampler patches on a keyboard, but of course the lead vocals must be done live. At this point of the expanding home sound studio, a multitrack recorder and audio mixer become very useful (Fig. 24.7).

### 24.4.1    Multitrack Recorders

Multitrack recorders allow the recording of different parts of a song one at a time. One part can be recorded on one track and then listened to while recording an accompanying part. This allows playing of all different instruments for a song. Each instrument is recorded separately.

Multitrack recorders are to acoustic instruments what sequencers are to multitimbral synthesizers. Having a computer/synthesizer system synchronized to an external multitrack recorder is a powerful combination.

There are many different configurations for multitrack recorders. Systems used in home and semiprofessional systems typically come in 4-, 8-, and 16-track denominations. Professional multitracks used in top studios are usually 24-, 32-, and 48-track sizes.

**FIGURE 24.7**   In a semipro digital music system a multitrack recorder and audio mixer are added.

Analog multitracks can be synchronized with each other to expand the number of available tracks by using optional expansion hardware that uses SMPTE time code to keep each machine locked up at frame accuracy. New digital multitrack recorders often have the master/slave synchronization hardware built into the unit as a standard feature. Most of these units come in eight-track increments and allow slaving of up to 16 units together for a multitrack system as large as 128 tracks.

The synchronization timing signal used to keep these digital modular multitracks working together is much more accurate than the SMPTE standard resolutions of 24 or 30 frames per second. These machines are often synchronized to single sample accuracy (44.1 kHz or 48k samples per second). This type of accuracy is very desirable when synchronizing individual tracks of a song, especially where there are percussion parts on different physical machines. Small timing/sync differences are very noticeable on drum and percussion tracks.

These sync protocols are proprietary and are not universal across equipment from different manufacturers, except with expensive conversion hardware. SMPTE is still a very valid time code to use in a studio due to its status as an open standard and wide availability on many different types of equipment. It also has universal appeal in synchronizing audio to video or movie formats due to its 24 or 30 frame per second time reference.

Most home studios work just fine with 4 to 16 tracks, using either an analog or digital multitrack. These tracks are in addition to the number of tracks/voices that are available for simultaneous use on the soundcard and synthesizer.

### 24.4.2   SMPTE Time Code

SMPTE time code is one developed at NASA for time-coding video data from spacecraft. It has since become a video and music industry standard. SMPTE time code expresses time in hours, minutes, seconds, and frames. The serial digital data that contains the SMPTE time code is recorded on a spare track on the video or audio tape. In a video application this gives each video frame a unique time code. NASA video footage has a running time code displayed in the corner.

There are four types of SMPTE that correspond with the different types of video formats in use around the world: 24 frames per second (fps) for film, 25 fps for European TV, 30 fps for U.S. video, and 30 fps drop frame (a variation of 30 fps).

The real power of SMPTE is in synchronizing audio to video. Because SMPTE is transmitted as a serial data format, many popular computer sequencers and MIDI interfaces will accept SMPTE time code as their synchronizing clock.

> *New York.* Two weeks after Charlie finished his soundtrack for the commercial, the ad agency overnights the draft video footage back for additional audio sweetening. In this commercial a baseball star swings his bat in slow motion and hits a baseball out of the park. The agency wants the sound of the bat's hit to be greatly emphasized. The videotape has SMPTE time code recorded on one of its audio tracks. After viewing the specific tape section several times, frame by frame, Charlie determines the ball connects with the bat at an SMPTE time code of 0 minutes, 23 seconds, and 14 frames.
>
> Charlie loads a special effect sound (a cannon shot) into his sampling keyboard. Using his sequencer, he selects that note to be played exactly when the ball connects with the bat. He starts the videotape, the sequencer locks up with the SMPTE time code on the videotape, and the cannon shot is played at precisely 0 minutes, 23 seconds, and 14 frames. The sound effect is recorded onto the videotape's additional audio track. Charlie sends it back by courier to the ad agency, less than one hour after it arrived at his studio.

This practice is widely used in the advertising commercial industry where music soundtracks and sound effects must be synchronized perfectly to a video event. Film production companies also use SMPTE to trigger synthesizer and sampled sound effects at the appropriate moments in a film score.

Using SMPTE requires giving up one audio track on the tape. The SMPTE time code is recorded (striped) onto this track. As it is played back, the time code is used as a time reference for all slave devices. Because specific time values (hours/minutes/seconds/frames) are striped on the tape, it can be started anywhere and the slave devices will lock up correctly, unlike earlier sync methods which were relative times to the start of the song.

Computers that synchronize to the master SMPTE track simply receive the time code and lock up their sequencer playback with the master clock. This is done in software with no moving parts.

Slaved tape machines receive the master SMPTE clock, compare it to their own SMPTE code, and speed up or slow down the tape to keep the master and slave time codes identical. If the slave's time code is very different from the master's reference, the slave machine will go into fast-forward or reverse mode to bring it in close to the master clock's time reference. Once close, the machines stay locked by varying the motor speed of the slave. This mechanical method of locking up tape machines in synchronization has its limitations.

SMPTE is very beneficial to locking up and synchronizing tape machines and synthesizers when accuracies of $\frac{1}{24}$ or $\frac{1}{30}$ of a second are good enough, such as music to video synchronization. But SMPTE is not accurate enough for locking up multiple audiotape machines with great accuracy. A discrepancy of $\frac{1}{24}$ or $\frac{1}{30}$ of a second can create many problems when tracks that have critical relationships are on different machines. For example, with percussion sounds, such an error can cause the rhythm of the music to sound sloppy. Other effects are frequency cancellation and phasing problems with stereo signals. An example would be a grand piano recorded with two microphones for a stereo image. If the two tracks were recorded on two physically different tape machines that were locked up with SMPTE accuracy, the sound would have phasing, flanging, and frequency problems that would vary continuously with the timing error between the two machines.

To solve this problem, many manufacturers offer proprietary sync interfaces that lock up the machines with single sample accuracy (1/48,000 of a second). These can be used effectively to keep machines in perfect sync. The only problem is that many manufacturers have their own methods that are not universally used with other manufacturers' equipment.

### 24.4.3   MIDI Time Code

SMPTE has become the timing reference standard in the professional audio and video industries. It is also being used more and more in the semiprofessional audio area and is very useful for synchronizing sequencers and computers with external tape machines (video and multitrack audio).

Encoding SMPTE over MIDI allows a person to work with one timing reference throughout the entire system. For example, studio engineers are more familiar with the idea of telling a multitrack recorder to punch in and out of record mode at specific SMPTE times, as opposed to a specific beat in a specific bar. To force a musician or studio engineer to convert back and forth between a SMPTE time and a specific bar number is tedious and should not be necessary.

Some operations are referenced only as SMPTE times, as opposed to beats in a bar. For example, creating audio and sound effects for video requires that certain sounds and sequences be played at specific SMPTE times. There is no other easy way to do this with song position pointers, etc., and, even if there was, it would be an unnatural way for a video or recording engineer to work.

MIDI time code is an absolute timing reference, whereas MIDI clock and song position pointer are relative timing references. In virtually all audio for film/video work, SMPTE is already being used as the main time base, and any musical passages that need to be recorded are usually done by getting a MIDI-based sequencer to start at a predetermined SMPTE time code. In most cases, though, SMPTE is the master timing reference being used.

In order for MIDI-based devices to operate on an absolute time code that is independent of tempo, MIDI time code must be used. Existing devices merely translate SMPTE into MIDI clocks and song position pointers based upon a given tempo. This is not absolute time, but relative time, and all of the SMPTE cue points will change if the tempo changes. The majority of sound effects work for film and video does not involve musical passages with tempos; rather, it involves individual sound effect events that must occur at specific, absolute times, not relative to any tempo.

### 24.4.4   Digital Signal Processors/Effects Units

Soon after the purchase of a multitrack tape recorder (Fig. 24.8), the user will notice that although all the tracks he or she records have good fidelity, everything is "too" clean. All the instruments are there, but the mix may sound a bit flat and one-dimensional. This is because all of the sound stayed in electronic format prior to being recorded to tape. Plugging the synthesizer's audio outputs directly into the multitrack recorder yields the lowest possible noise and distortion, but does not allow the room acoustics to add character to the sound. There is an ambience that is unique to every room and performance hall in the world. This is why some live albums and recording projects very prominently advertise where they were recorded.

Early recording projects tried to replicate these effects in many ways. It has been documented that on Simon and Garfunkel's song "The Boxer," the large reverb sound that enhances the snare drum hit was done by putting a speaker at the bottom of an elevator shaft and suspending a microphone several stories above it. Soon, studios began using mechanical plate reverbs—6-by-4-ft sheets of plate metal—to approximate this effect. An output transducer and a contact microphone were mounted in various positions to create a reverb sound using the resonance of the metal plate. Later, less expensive versions of this effect were made using a metal spring. Many early, classic guitar amplifiers used spring reverbs, which tended to make a noisy twang sound when the amplifier was bumped.

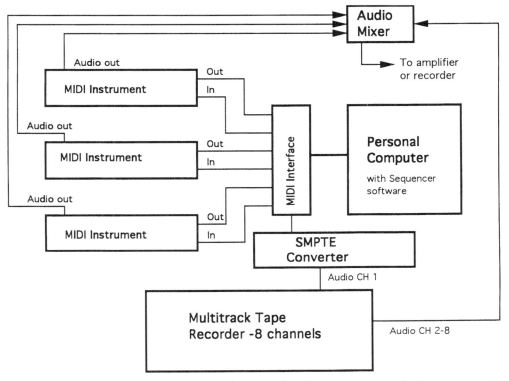

**FIGURE 24.8** Adding a SMPTE writer/reader to the system allows the synchronization of the playback of the MIDI sequencer with audio channels that are already recorded on tape. Because MIDI channels are less expensive than tape channels, this can greatly expand the effectiveness of a small home studio by adding dozens of "virtual" channels of recording capacity using MIDI.

If one listens closely to the original *Woodstock* movie soundtrack, the sound of the amplifier spring reverbs can be heard bouncing and twanging. Apparently someone was dancing nearby on stage and did not notice the aural problem they were creating. Neither did any of the other 400,000 attendees. These sounds were later cleaned up in the rerelease of the original soundtrack by using computer workstations and waveform-editing software.

Digital signal-processing became possible in the late 1970s and early 1980s with the incredible advances in integrated circuit technology and microprocessors. Digital reverb and echo (delay) units were very expensive and were only used in professional studios for several years. But as the electronics revolution continues, the prices of these units fall dramatically. Today they can be purchased for a few hundred dollars.

As DSPs have also become less expensive, other more math-intensive algorithms and features have been added to these units. These other features include pitch shifting, chorusing, flanging, and some sampling capabilities. Some units even do pitch tracking and convert this pitch information to MIDI messages. By pitch-tracking a guitar, any keyboard or synthesizer voices can be controlled by playing on the guitar. These units also have MIDI inputs, which allow the control of functions in real time using computer sequencer software or any other MIDI master. Imagine changing the sound of a synthesizer for the next song in a performance and the effects unit automatically sets itself to the configuration desired for that song.

## 24.5   THE FUTURE OF HOME STUDIOS

The future products and technologies used in home studios will parallel those of the personal computer evolution. One of the first areas to watch will be storage media. Several years ago, analog tape was used almost exclusively in home studios. Today small digital tape multitrack recorders are often used. Tape is a very inexpensive storage medium. An SVHS tape can store almost 5 GB of sound for about $7 per tape.

Hard drive–based multitrack recorders are currently invading professional audio and video studios. Storing audio on a hard drive yields random access, nondestructive editing, and much more control over digital signal processing for effects. Synchronization of the hard disk–stored audio to a video source requires no tape machine lockup, no motor speed adjustments, and no moving parts. As the prices of hard drives continue to fall, and the processing power of PCs continues to rise, many home studios will begin converting to fully tapeless operation (except for backups and for some mix-downs).

Another area on the verge of revolution is the way music is distributed. Today there are several World Wide Web services that allow downloads and distribution of songs to anyone on the network. Emerging artists pay to have their songs and album liner notes available on the Web page. Internet users can download the songs, which are compressed for bandwidth efficiency using MPEG types of algorithms. A song can then be listened to via a multimedia PC and copied to tape if desired. Although, at modem data rates of 14.4 kb or 28.8 kb, these downloads can take several minutes, the coming revolution in bandwidth via broadband networks (cable modems) may soon remove this obstacle.

Two existing Web sites that operate today in this manner include IUMA (the Internet Underground Music Archive) at http://www.iuma.com and Kaleidospace at http://www.kspace.com.

When the logistics of financial transactions over Web browsers are accepted by computer users, one will be able to buy songs from established popular artists using this method. With this change in distribution methods, any Web site can ultimately sign up artists to exclusive or nonexclusive contracts and become a "record company." As a new artist, you will have a new channel to market and distribute your music. In the future this change in the music distribution and promotional channels will dwarf the historical significance of the last great shift from music radio songs to music television videos.

*Hermosa Beach (one year later).*  Paul has expanded his home music system into a small home recording studio with a synthesizer, effects unit, and a multitrack tape recorder. He has enjoyed many evenings of creativity by writing and recording his own songs. He also enjoys programming and sequencing songs from other bands that he hears and likes. Tonight he has finished such a project.

Paul has downloaded a song from an Internet Web site. He likes the song and just for fun wants to try and sequence it with a different arrangement. He opens up the original song file to compare a few measures with his version. He clicks on and reads the album liner notes within the file and learns that this new band is from Austin and the leader's name is Trent.

Paul reaches for his TV remote to turn off the TV before beginning to play the song, but he stops to watch the commercial that is on. It is one of his favorites. It ends with a baseball player swinging at a pitched ball. He connects with his bat. The baseball rockets out of the park like it was shot out of a cannon . . . .

## GLOSSARY

**AIFF files**   A standard file format for storing sound files on Apple Macintosh personal computers.

**Digital signal processing**   In the contest of recording studios, DSPs are the units that alter or change the sound in some desired way. This includes reverb, equalizations, pitch bending, echo effects, and volume processing. In the context of integrated circuits, DSPs are specialized microprocessors that include special math circuits for fast execution of DSP algorithms. DSP integrated circuits are used in music studio DSP equipment.

**FM synthesis**    A type of sound synthesis used in keyboard synthesizers and computer soundcards. FM synthesis uses operators (polyphonic oscillators) to create sounds. By mixing and modulating these oscillators together in various ways, many timbres can be created.

**General MIDI**    A standard set of preprogrammed patches. General MIDI is an extension of the MIDI specification that standardizes 128 sound categories and their program locations. It also specifies a minimum of 24 voices. Multimedia software that is general MIDI-compatible should play and sound fairly consistent across all platforms.

**Humanize**    To introduce minor random variations of timing, duration, and dynamics to a music sequence that was generated by a computer or quantized such that it sounds too perfect or mechanical. Many MIDI sequencers have a humanize tool that introduces these random variations on the desired section of the song. These human imperfections add interpretive emotion to the song and increase its appeal.

**MIDI**    Musical Instrument Digital Interface. A serial data network over which modern musical instruments and equipment may communicate. MIDI itself carries no sound. MIDI is a control protocol in which a master device (often a computer) instructs sound-generating devices (soundcards or synthesizers) to turn certain notes on and off upon demand.

**MIDI addressing**    A maximum of 16 channels of information can be transmitted over one MIDI cable. Each channel is basically a unique address associated with each packet of information. This way, 16 slaves may be chained together on one physical MIDI connection, but each slave can respond to information on a different channel.

**MIDI channel mode messages**    Used to denote overall operation of a MIDI synthesizer or device on the network. Some of these mode messages include what information the device should receive. Some devices should receive all notes from all channels (omni on). Some devices are instructed to receive notes only from one channel (omni off). Other commands tell the MIDI device to send the notes received to just one voice (mono on) or to various voices (poly on).

**MIDI channel voice messages**    They communicate information about the notes being played. These include note on/note off messages, polyphonic key pressure, overall pressure, controllers, pitch bend, and program change/select.

**MIDI messages**    MIDI communicates by using multibyte messages. Multiple bytes are used to denote the channel of the information, the type of information, and the data itself. MIDI message types include channel voice, channel mode, system-common, system real-time, and system-exclusive messages.

**MIDI sequencer**    A computer program that allows one to cut and paste sections of the music, copy it, and edit notes that were entered incorrectly. MIDI sequencers also allow one to change the speed at which the music is played back and offer programmable access to some of the advanced controls within the synthesizer (volume, pitch bend, modulation). A basic MIDI sequencer software package is a music word processor for a MIDI studio.

**MIDI system-common messages**    Control commands that apply to all channels. An example is a song position pointer, which communicates to each MIDI instrument how many beats (in sixteenth notes) have elapsed since the beginning of a song. This feature can be used to synchronize the playback of drum patterns and songs programmed into various instruments.

**MIDI system-exclusive messages**    A special type of system-common message that is defined by the manufacturer and is not compatible with or understood by instruments from different manufacturers. System-exclusive messages use an ID code that is specifically assigned to one manufacturer. This ID code is used so that other manufacturers' instruments may ignore the special command. A system-exclusive message is often used by synthesizers to download their patch programs to a computer or mass storage device for archiving. They are also used by sampling keyboards to transfer actual audio sample data to a computer so that waveform-editing software packages can be used to modify the sample.

**MIDI system real-time messages**    Messages received by all devices on the MIDI network. They are used to (1) synchronize all devices to a common clock, (2) permit network-wide commands such as system reset, and (3) send start/stop song messages to be coordinated among all instruments.

**Monophonic**    The ability of an oscillator to produce only one sound at a time.

**MPEG**    Moving Picture Experts Group. Usually used to refer to digital audio and video compression algorithms. MPEG audio uses a masking algorithm that removes redundant frequencies and quiet sounds that the human ear cannot hear in the context of the total audio output being played. MPEG video uses discrete cosine transforms and motion prediction to approximate adjacent blocks of video pixels and sends/stores only the changes made from one video frame to the next.

**MTC**    MIDI time code. An absolute time reference that is compatible with MIDI message format, and similar to SMPTE time code. Many studios use a SMPTE to MIDI time code converter unit to synchronize the MIDI-based equipment to the SMPTE-based audio and videotape machines.

**Multitimbral**    The ability of a synthesizer or soundcard to play more than one different type of sound (timbre) at the same time, e.g., playing a flute sound at the same time as a piano sound. Most multitimbral synthesizers/soundcards permit playing several different instrument sounds simultaneously.

**Multitrack recorders**    Special analog and digital recorders used in music studios to record songs one at a time to be played back later all at once. Multitrack recorders allow the artist to listen to previous tracks while adding new ones. Multitracks are available in many tape formats and in many track increments. New media developments in multitrack recording include direct to hard disk drive recording, which adds considerable flexibility in sound-editing tasks.

**Operator**    A polyphonic oscillator used in sound synthesis. Often used in the context of the oscillators used in FM synthesis. The number of operators on an FM synthesizer determines the flexibility and capability of the synthesizer to create complex sounds.

**Oscillator**    The basic sound-producing circuit of a synthesizer. An oscillator generates simple waveforms to produce sounds. These simple waveforms are then mixed, filtered, and modulated by other circuits in the synthesizer to create different sounds.

**Patch**    The parameters used to program the synthesizer's operation in creating a sound. This could include types of waveforms used, volume, filtering, tuning, attack, sustain, release, and how several oscillators might be combined to produce the desired tone.

**Polyphonic**    The ability of an oscillator to produce many sounds at a time.

**Quantize**    To align the start times of all played notes either on or very close to the exact beat of the song. Can be used to correct human playing errors or offbeat notes. Subtle quantization is best, because overuse of this feature can result in mechanical-sounding songs. Most sequencers allow subtle quantization to be applied by specifying a random range before or after the exact note that is acceptable, rather than correcting all notes to the exact beat.

**Resolution**    The size of the data word used to store each sample, usually in the range of 8 to 16 bits. One minute of mono 8-bit/22-kHz audio requires 1.25 Mbyte of storage.

**Sampling keyboard**    A keyboard that plays back digital recordings of real instruments to produce its sounds.

**Sampling rate**    The number of times per second that an audio signal is converted to digital format and stored. Compact Discs are sampled at 44.1 kHz. Every minute of stereo 16-bit/44.1-kHz audio would require 10 Mbyte of hard disk storage.

**SMPTE**   Society of Motion Picture and Television Engineers, a standards body that specifies standard interface specifications for many different operations of audio and video production equipment. Most often used as a shortened term for SMPTE time code, an absolute timing reference signal with frame accuracy. It specifies the absolute hour, minute, second, and frame number of a film or video frame and is used to synchronize equipment in an audio or video production studio.

**Synthesizer keyboard**   A keyboard that uses oscillators, operators, or mixed and modulated wave samples to produce its sounds.

**Voice**   Each sound/note that a synthesizer or soundcard can play is a voice. Most synthesizer/soundcards limit the number of voices that can be played at the same time to 20–32 voices. Trying to play more results in either the note not being played or a previous note being cut off before its intended duration is complete.

**WAV files**   A standard file format for storing sound files on Microsoft/IBM type personal computers.

**Wavetable synthesis**   A synthesis technique that uses short samples of complex sounds, instead of oscillators, as its basic building block of creating sounds. The wave samples are then mixed, filtered, and modulated just like an oscillator-based synthesizer. Wavetable synthesis results in better, higher-quality sounds than FM synthesis.

**Woodstock**   The upstate New York artists' community/town that is nowhere near the site of the 1969 Woodstock Music and Arts festival, which was actually held 50 miles away in Bethel, N.Y. The 1994 Woodstock II concert was held somewhat closer in Saugerties, N.Y. The original site is owned by June Gelish and is still an open field accessible to visitors. A small monument marks the location at the corner of Hurd and West Shore Road.

**World Wide Web**   One of the interface protocols of the Internet that simplifies navigation of the network. Several Web sites offer downloads of sound files to distribute songs of new artists. Two existing Web sites that operate today in this manner include IUMA (the Internet Underground Music Archive) at http://www.iuma.com and Kaleidospace at http://www.kspace.com.

## *ABOUT THE AUTHOR*

Gordy Carlson began experimenting with electronic instruments as a teenager in the basement of his childhood home in Jamestown, N.Y. By trying to design a circuit to make his already considerably amplified guitar even louder, he discovered an attraction to electronics.

Several years and several kilowatts of audio later, he graduated from Rochester Institute of Technology with a degree in electrical engineering. His part-time employment during college included professional radio disc jockey/news announcer, video director/producer, and musician in several bands.

Carlson joined Motorola Semiconductor upon graduation in 1982 and has subsequently enjoyed technical sales and marketing assignments supporting consumer audio/video and multimedia customers. He is presently the Northeast Regional Technology Manager. His home recording studio includes much of the equipment described in this chapter and evolved from a "starting-out" system to a "semipro" system as described in the chapter.

# PART · 7

# DIGITAL INFORMATION/ COMMUNICATION PRODUCTS

# CHAPTER 25
# PERSONAL COMPUTERS AND DIGITAL VIDEO/AUDIO

**Chris Day, Jim Seymour**
*AuraVision Corporation*

## 25.1  INTRODUCTION

The personal computer is now a common medium for displaying, capturing, manipulating, and processing video due to advancements in video compression technology and the ability to mix graphics and quality video data. Much of the early adoption of video in PCs occurred in Japan, where consumers use their PCs as television sets, to watch video CDs, and even to play karaoke titles. The limited size of living spaces drives the need for a single product to serve as both PC and entertainment system. Computer-purchasing decisions, therefore, often reflect video quality rather than simply the PC's processor speed, disk capacity, and memory size. With the introduction of DVD drives, high-speed modems, and low-cost TV-tuner solutions, the demand for high-quality video on the PC now looks set to increase rapidly in all markets.

Another key enabler in the increasing adoption of video data as a standard data type in PCs is the operating system (OS) and application support now being offered by Microsoft. Microsoft's Windows 95 operating system now offers support for a wide range of audio-video applications, including "Video for Windows," which allows users to achieve instant video playback. Based on a 32-bit architecture, Windows 95 achieves much greater performance than the previous 16-bit Windows and DOS. It also adds support for MIDI audio. Although Macintosh users might argue that they have enjoyed such seamless integration of audio and video for years, it is clear that the latest operating system developments will help drive the development and adoption of increasingly sophisticated multimedia systems. In June 1995 the Software Publishers Association defined MPC3, a reference standard for multimedia PCs, to include MPEG-1 compatibility and to include support for 16-bit sound with wavetable synthesis.

This chapter deals with the acquisition, processing, and output of video and audio within the personal computer domain. The word *video* generally refers to live-motion video, AVI (audio/video interleaved) files, and MPEG/JPEG files. Video data is usually a clip or a movie that comes from a TV tuner, VCR, camera, Digital Versatile Disc, or MPEG/JPEG decoder. It is of a "natural data" type and in this respect is different from data generated by the computer, such as graphics data. Essential functions of a video capture and display subsystem include video scaling, cropping, filtering, zooming, picture control, color space conversion, audio/video synchronization, color and chroma keying, and scan conversion. The video quality achieved is dependent on a number of attributes and cannot simply be represented by a single parameter, such as the number of bits per pixel in graphics systems.

## 25.2   VIDEO AND AUDIO SOURCES

Video data may enter the computer from a number of sources (Fig. 25.1), including digital camera, TV tuner (NTSC, PAL, SECAM), video CD (MPEG-1), CD-ROM, modem (Internet), LaserDisc player, AVI files, and Digital Versatile Disc (MPEG-2, Dolby AC3).

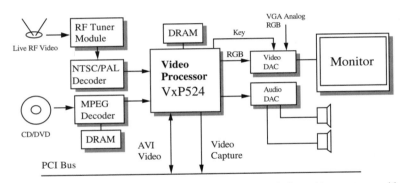

**FIGURE 25.1**   PC multimedia hardware architecture (stand-alone video processor with independent frame buffer).

### 25.2.1   Video Cameras for Personal Computers

Analog video cameras were developed for the television industry to serve the needs of television receivers (as in closed-circuit TV for security) and video recording (as in camcorders). These receivers were analog signal devices by nature, and so an analog video source was required. In the case of camcorders, the video camera "front end" was developed to perform all of the necessary control and processing of the analog signal to generate an industry-standard output video stream. This required a great deal of complexity in the camera circuitry, such as DSPs, microprocessors, ROM, RAM, and analog encoders.

By contrast, a personal computer is by definition a digital device, preferring a digital signal input. It is a powerful and flexible means of adding intelligent control to video cameras without requiring the added circuitry within the camera itself. It also opens up many exciting possibilities for interfaces between the camera and computer, including digital video signals, camera control, and even power. The result can be a smaller, quieter, smarter, and less expensive live video camera source that augments rather than burdens the personal computer.

### 25.2.2   TV-Tuner Applications

The PC-TV–type video board allows the PC to function as a television set. Such boards may also support the capture of TV video, allowing clips of TV programs to be stored and edited. These systems employ a TV tuner to receive the TV broadcast signal and these are typically sold as a single, shielded device for integration onto printed circuit boards. Because of the complex nature of such RF devices it is not usual for computer manufacturers to design their own tuner circuitry, preferring instead to purchase them from companies such as Philips. The output of the TV tuner's video signal is in analog form and is fed into the input of the video decoder. The audio output is typically run to the PC's soundcard.

***Video Decoding.***   The use of a video decoder is necessary to decode the composite NTSC, PAL, or SECAM signals supplied by the TV tuner, to digitize them, and to output them in either RGB or $YC_rC_b$ output formats for further processing or storage within the computer. Video decoders (Table 25.1) are available as single-chip ICs from Brooktree, Motorola, Philips, and other manufacturers. The primary features of a decoder typically include the following:

- Adjustable saturation, contrast, brightness, and hue.
- Video multiplexor to accept inputs from various sources.
- Support for several pixel clock rates.
- Generation of computer video timing signals: h-sync, V-sync, and blanking.

**TABLE 25.1**   Resolutions and Pixel Rates for Video Decoders

|  | Pixel clock rate | Horizontal resolution (total pixels) | Horizontal resolution (active pixels) | Vertical resolution | Application |
|---|---|---|---|---|---|
| NTSC | 12.27 MHz | 780 | 640 | 525 | Square pixel, PC |
|  | 13.5 MHz | 858 | 720 | 525 | CCIR 601 |
|  | 14.32 MHz | 910 | 768 | 525 | Editing |
| PAL | 14.75 MHz | 944 | 768 | 625 | Square pixel, PC |
|  | 13.5 MHz | 864 | 720 | 625 | CCIR 601 |
|  | 17.72 MHz | 1135 | 948 | 625 | Editing |

The line length of a video source is defined as the interval between the midpoints of succeeding horizontal sync pulses. Analog video sources provide line lengths that are not constant, typically varying by as much as a few microseconds. This is true of sources such as TV tuners, VCRs (where tape stretch may provide a worse-case scenario), and even LaserDisc players. PC video subsystems and display monitors, however, require a consistent, fixed number of pixels per line. The decoder must recover the original pixel clock and timing controls from the video signal in a process called *genlocking*.

### 25.2.3   VideoCD and CD-ROM

The MPEG-1 compression algorithm was first developed to allow approximately one hour of play time from a CD-ROM at 1.5 Mb/s. Although the availability of MPEG-1 material is limited outside of Asia, computer manufacturers may support MPEG-1 playback with either a hardware decoder or, increasingly, a software solution. Such software is available from Xing Labs,

Mediamatics, and CompCore, and it offers MPEG-1 decoding at a fraction of the cost of a hardware solution. As microprocessor performance increases, the quality of video decoded using software should increase, furthering the adoption of MPEG-1 compressed video on CD-ROMs. In the meantime, software games developers are continuing to use compression schemes such as Cinepak (Table 25.2), which, though offering more limited compression ratios, can offer full 30 frames per second video without dedicated hardware.

**TABLE 25.2** Compression Technologies Overview

| Compression | Sym/asym | Ratio | Bandwidth | Storage medium | Application |
|---|---|---|---|---|---|
| JPEG | Sym | 15–20:1 | 300 kb/s | Hard disk | Video editing |
| MPEG-1 | Asym | 100:1 | 150 kb/s | CD-ROM | Games/titles |
| P*64 | Sym | 150:1 | 128 kb/s | | Video conferencing |
| Cinepak | Asym | 10:1 | 150 kb/s | CD-ROM | Presentations/games |
| Video 1 | Sym | 8:1 | 150 kb/s | CD-ROM | Presentations/games |
| Indeo (DVI) | Both | 10:1 | 150 kb/s | CD-ROM | Presentations/games |

### 25.2.4 Digital Versatile Disc (DVD)

The Digital Versatile Disc (see Chapter 12) will replace both the CD-ROM drive in the computer environment and the VCR and LaserDisc player in consumer products over time. See Table 25.3 for DVD specifications. The DVD-ROM allows for significantly more data to be stored on a single disc than CD-ROM drives. The adoption of MPEG-2 video compression and Dolby AC-3 audio compression for DVD provides higher quality than available with the Video-CD format, while allowing most movies to be stored on a single disc.

**TABLE 25.3** DVD Specifications

| | |
|---|---|
| Memory capacity | 4.7 Gbytes/side |
| Error Correction | RS-PC (Reed-Solomon Product Code) |
| Data transfer rate | Variable-speed data transfer at an average of 4.69 Mb/s for video and audio |
| Compression | MPEG-2 digital image compression |
| Audio | Dolby AC-3 (5.1 ch), LPCM for NTSC, and MPEG audio LPCM for PAL/SECAM (a maximum of 8 audio channels and 32 subtitle channels can be stored) |
| Running time | 133 minutes per side (average of 4.69 Mb/s data rate) |

## 25.3 VIDEO FORMATS AND COLOR SPACES

Table 25.4 lists the various video formats and their characteristics.

### 25.3.1 Color Spaces

A color space is a mathematical representation of a set of colors. In the personal computer domain there are principally three relevant color space models, RGB, YUV and $YC_rC_b$. All the color spaces can be derived or "color space–converted" from RGB, the color space used widely for computer graphics.

**TABLE 25.4**  Video Formats

| Format | Refresh rate (Hz) | Field resolution | $YC_rC_b$ sampling | Comments |
|---|---|---|---|---|
| QCIF | 10–30 (25) | 176 × 120 (144) | 4:1:1 | Video conferencing |
| CIF | 15–30 (25) | 352 × 240 (288) | 4:1:1 | Video conferencing |
| Square pixel NTSC | 60 | 640 × 240 | 4:1:1 | S-VHS quality |
| Square pixel PAL | 50 | 768 × 288 | 4:1:1 | S-VHS quality |
| CCIR 601 | 60 (50) | 720 × 240 (288) | 4:1:1 | S-VHS quality |
| Square pixel NTSC | 60 | 640 × 240 | 4:2:2 | Broadcast quality |
| Square pixel PAL | 50 | 768 × 288 | 4:2:2 | Broadcast quality |
| CCIR 601 | 60 (50) | 720 × 240 (288) | 4:2:2 | Broadcast quality |

***The RGB Color Space.***  The RGB color space is used in computer graphics frame buffers because CRTs use red, green, and blue phosphors to produce color. Today's graphics chips, therefore, support the RGB color space, and a large number of computer games and software applications have been written for this format. The use of 24 bits per pixel, 8 bits for each color, allows for *true-color* representation of images, whereas 16-bit representations are referred to as *hi-color.* True color provides a color depth that offers realistic color representations of images, such as photographs, including realistic skin tones. Representations using 16 bits per pixel, such as 5:6:5, require less bandwidth and memory to represent colors, but offer reduced color depth.

The RGB color space (Fig. 25.2) is not, however, very efficient when it comes to representing video images. In using RGB it is necessary for each RGB component to require the same (or similar) number of bits to represent a color. Furthermore, if an adjustment to the intensity of a pixel is necessary, then the values for red, green, and blue must be read, recalculated, and stored back in the frame buffer. This is not true of the YUV or $YC_rC_b$ color spaces, where direct access to the intensity ($Y$) information and color difference components ($U$ and $V$) allows simpler calculations and reduced bandwidth requirements.

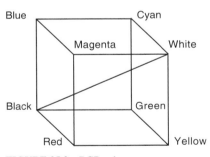

**FIGURE 25.2**  RGB color space.

***The YUV Color Space.***  The YUV color space is used by the NTSC, PAL, and SECAM video standards. The $Y$ value represents luminance, whereas the $U$ and $V$ values represent color difference or "chroma" components. Changes to brightness can be effected by simply adding or subtracting values from the $Y$ component. Contrast changes can be made by multiplying the $Y$ component by some value, and saturation changes are achieved by multiplication of the $U$ and $V$ color difference components.

***The YIQ Color Space.***  The YIQ color space is optionally used by the NTSC composite video standard. The $I$ and $Q$ symbols reflect the transmission modulation method, standing for in-phase and quadrature, respectively.

***The $YC_rC_b$ Color Space.***  The $YC_rC_b$ color space is that used in the CCIR601 digital component video standard. In the $YC_rC_b$ color space, $Y$ remains the component of intensity, as in YUV, but $C_b$ and $C_r$ become the respective components of blue and red. The 4:2:2 and 4:1:1 $YC_rC_b$ formats are used widely in computer video applications. Taking into account the human eye's high sensitivity to luminance data, the YUV 4:1:1 and 4:2:2 formats specify video information

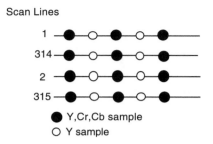

Scan Lines

● Y,Cr,Cb sample
○ Y sample

**FIGURE 25.3**    4:2:2 $YC_rC_b$ format.

by emphasizing the fidelity of luminance while spreading out the chrominance information over time. A 12-bit YUV (8 bits of $Y$ and 4 bits of $UV$) video data provides the same or better quality as a 16-bit RGB picture. In the 4:2:2 $YC_rC_b$ format illustrated in Fig. 25.3, there are two $Y$ samples for each $C_r$ and $C_b$ value. Interpolation is used to provide $C_r$ and $C_b$ values for those samples without them. In the 4:1:1 format there are three Y samples for each $C_r$ and $C_b$.

### 25.3.2  Sampling

The 4:1:1 format requires 12 bits per pixel. Eight bits are required for the $Y$ value and eight for each $U$ and $V$ value, but each $U$ and $V$ value is effectively shared by four pixels, resulting in an average of four bits per pixel. The 4:2:2 format requires 16 bits per pixel, because $U$ and $V$ values are shared between two pixels. The AuraVision Aura1 and Aura2 formats further reduce the number of bits by a factor of two, resulting in six bits for 4:1:1 and eight bits for 4:2:2. The Aura formats are the result of adaptive compression within the video processor, and they allow for reduced storage and bandwidth requirements. Valid ranges for color space components are shown in Table 25.5.

**TABLE 25.5**    Valid Ranges for Color Space Components

| RGB | $R = 0$ to 255 | $G = 0$ to 255 | $B = 0$ to 255 |
|-----|----------------|----------------|----------------|
| YUV | $Y = 0$ to 255 | $U = 0$ to $\pm112$ | $V = 0$ to $\pm157$ |
| YIQ | $Y = 0$ to 255 | $I = 0$ to $\pm152$ | $Q = 0$ to $\pm134$ |
| $YC_rC_b$ | $Y = 16$ to 235 | $C_r = 16$ to 240 | $C_b = 16$ to 240 |

## 25.4  VIDEO PROCESSING

The architecture of a video-processing system can be divided into four modules: the video input pipeline, the video output pipeline, the memory controller, and the host and decoder interface. Although this discussion of video processing is based on AuraVision's VxP500 family (Fig. 25.4), the processes described should apply to video-processing systems in general.

### 25.4.1  Video Input Pipeline

The video input pipeline of an Auravision video processor supports two video inputs, one for MPEG decoders or JPEG codecs and one for live video. This configuration, along with software playback driven through the ISA or PCI bus, enables three different sources of input video. The live video input port should support digitized YUV 4:1:1 and YUV 4:2:2 from a video decoder such as a Philips 7110 or a Brooktree 827. The video decoder also sends control signals to the horizontal sync, vertical sync, and video input clock pins of the video processor. For hardware playback, both MPEG decoder and JPEG codec formats should be supported. The host sends compressed elementary bit streams to the MPEG/JPEG decoder through the host PCI bus or ISA bus via DMA transfers or via programmed I/O. For software playback, video data is fed through the ISA or PCI bus to the input processor. Numerous video formats should be supported, such as

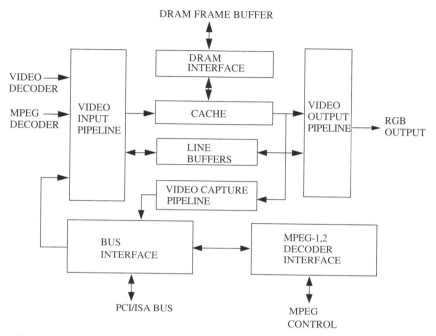

**FIGURE 25.4**   Video processor architecture (AuraVision VxP500 family).

Microsoft's Video for Windows, including RGB 5:6:5, RGB 5:5:5, 8-bit palletized, and YUV 4:2:2. An on-chip palette for VFW 8-bit palletized data color look-up and an interface to AVI formats ensures good-quality, full-screen VFW playback.

The video input pipeline of a video display processor (Fig. 25.5) consists of five major blocks: picture control, RGB-YUV color space conversion, filtering, scaling, and cropping.

***Color Palettes and Color Look-Up Tables.***   Color palettes are used to reduce the number of bits necessary to represent a particular pixel. If the number of bits per pixel is reduced without using a color palette, then the color depth is reduced correspondingly. By employing a palette or *color look-up table,* the components of all the colors for each pixel are removed and are replaced by a number. This number is the index into the look-up table, which stores the relevant color information for the particular pixel. The number of bits necessary to represent the pixel, therefore, simply becomes the number of bits necessary to represent the index into the table.

***Picture Control.***   Unique to the independent frame buffer video subsystem, a video processor may contain on-chip RAM for picture control and AVI palletized data decoding. Integrated color look-up RAM provides picture controls including brightness, saturation, hue, and contrast. This feature is especially important for video capture applications. Controlling the input video for optimum picture quality provides users with "what-you-see-is-what-you-get" video capture. The color look-up RAM also may be programmed as special color maps for pseudocolor special effects and can be used as palette RAM to play back 8-bit palletized VFW data. This improves playback performance and does not require host CPU support. In this configuration the RAM decodes palletized data into RGB 5:6:5 format and transforms it to YUV format through color space conversion.

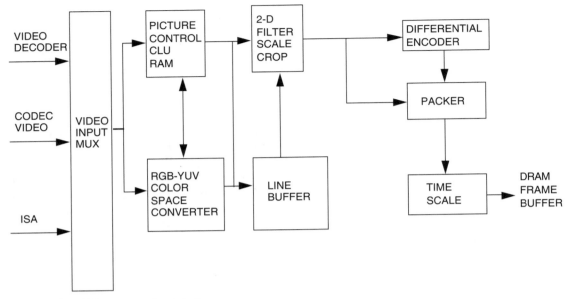

**FIGURE 25.5**    Video processor input pipeline.

***Gamma Correction.***    Gamma correction is necessary because CRT displays produce luminance in a manner that is not a linear function of the input voltage signal. Higher voltage signals (greater luminance) are expanded, whereas lower voltage signals (lower luminance) are compressed. To compensate for this effect it is necessary to *gamma correct* the video signal. The correction involves inversely offsetting the CRT's characteristics by applying a power or *gamma factor* of 2.2 to NTSC signals and of 2.8 for PAL and SECAM. This also has the effect of reducing noise-induced artifacts introduced during transmission.

***Color Space Conversion.***    Most personal computers use the RGB color space, whereas video-processing functions are best handled in the YUV color space. Therefore, although both RGB and YUV color spaces should be supported as input formats, RGB input video must be converted to the YUV color space before any scaling, cropping, or filtering of the input video is performed. This can be done externally; however, most video-processing modules provide an integrated RGB-to-YUV color space converter to reduce external logic and software overhead in video subsystems. All video input formats except YUV are transformed to YUV 4:2:2 for further processing. The color space converter transforms the RGB 5:6:5 and RGB 5:5:5 to YUV 4:2:2 excess 128 format. A YUV video input stream bypasses the color space converter while palletized video input is decoded in picture control RAM to RGB 5:6:5 format and then converted to YUV 4:2:2 by the color space converter.

The following equations may be used to perform color space conversions (R′G′B′ is gamma-corrected):

**R′G′B′ to YUV**

$$Y = 0.299 \times \text{Red} + 0.587 \times \text{Green} + 0.114 \times \text{Blue}$$
$$U = -0.147 \times \text{Red} - 0.289 \times \text{Green} + 0.436 \times \text{Blue}$$
$$V = 0.615 \times \text{Red} - 0.515 \times \text{Green} - 0.1 \times \text{Blue}$$

**YUV to R'G'B'**

$$\text{Red} = Y + 0.000 \times U + 1.14 \times V$$
$$\text{Green} = Y - 0.396 \times U - 0.581 \times V$$
$$\text{Blue} = Y + 2.029 \times U + 0.0 \times V$$

**R'G'B' to YIQ**

$$Y = 0.299 \times \text{Red} + 0.587 \times \text{Green} + 0.114 \times \text{Blue}$$
$$I = 0.596 \times \text{Red} - 0.274 \times \text{Green} - 0.332 \times \text{Blue}$$
$$Q = 0.212 \times \text{Red} - 0.523 \times \text{Green} + 0.311 \times \text{Blue}$$

**YIQ to R'G'B'**

$$R' = Y + 0.956 \times I + 0.620 \times Q$$
$$G' = Y - 0.272 \times I - 0.647 \times Q$$
$$B' = Y - 1.108 \times I - 1.705 \times Q$$

**YUV to YIQ**

$$Y = Y$$
$$I = -0.2676 \times U + 0.7361 \times V$$
$$Q = 0.3869 \times U + 0.4596 \times V$$

**YIQ to YUV**

$$Y = Y$$
$$U = -1.1270 \times I + 1.8050 \times Q$$
$$V = 0.9489 \times I + 0.6561 \times Q$$

**R'G'B' to $YC_bC_r$**

$$Y = \text{Coeff for Red} \times \text{Red} + \text{Coeff for Green} \times \text{Green} + \text{Coeff for Blue} \times \text{Blue}$$
$$C_b = (\text{Blue} - Y)/(2 - 2 \times \text{Coeff for Red}) + Y$$
$$C_r = (\text{Red} - Y)/(2 - 2 \times \text{Coeff for Red})$$

**$YC_bC_r$ to R'G'B'**

$$\text{Red} = C_r \times (2 - 2 \times \text{Coeff for Red}) + Y$$
$$\text{Green} = (Y - \text{Coeff for Blue} \times \text{Blue} - \text{Coeff for Red} \times \text{Red})/ \text{Coeff for Green}$$
$$\text{Blue} = C_b \times (2 - 2 \times \text{Coeff for Blue}) + Y$$

Note that CCIR601 specifies the following coefficients for red, green, and blue: Red: 0.2989; Green: 0.5867; Blue: 0.1144.

***Filtering.***   Essential to the quality display and capture of digital video is the amount of filtering used to remove undesirable noise in a video signal. Most integrated video processors support two-dimensional luma and chroma filtering, which are enabled and disabled independently, to reduce noise and antialiasing artifacts in the video data stream. In better-quality video processors, filtering is accomplished using bilinear interpolation to reduce artifacts and provide high-quality video in a variety of resolutions and window sizes. Other more advanced filtering processes further enhance the displayed image. In more advanced video processors, filtering is accomplished in two stages. First, prefiltering of the video image is conducted before a polyphase filter is applied during the second stage. This dual-stage filtering process notably improves scaled-down video images, especially in high-resolution modes. In addition, an integrated full-length line buffer (minimum $512 \times 768$) provides optimal vertical filtering without costly external line buffers. On less integrated video-processing systems, external line buffers may be required for vertical filtering, or the video may be set to simply pass through. However, this method may introduce video artifacts and noise, resulting in video that is below the standards of even low-end video applications.

***Scaling.***   Scaling a video image reduces the effective resolution of an image displayed on a screen, allowing users to display video in an arbitrarily sized window. When scaling an image, maintaining the proper aspect ratio is of primary importance to prevent image distortion. There are numerous ways to scale an image, and the quality varies greatly depending on the methods used. The simplest method to scale down an image is through pixel decimation, whereby pixels are dropped horizontally and vertically. A line draw algorithm is generally used to determine which pixels not to drop in the horizontal and vertical directions. Although simple to implement, pixel decimation in both the horizontal and vertical directions is not recommended beyond a factor of two because a significant amount of display data is lost. To compensate for the lost display data, many ICs with scaling functionality decimate pixels vertically but perform filtering horizontally. This provides greater scaling flexibility and reduces edgy artifacts caused by simply dropping pixels. Better yet, scaling ICs that perform both horizontal and vertical filtering to scaled video can provide the best image quality with the least amount of unwanted artifacts.

***Aspect Ratios.***   While both computer and TV monitors share the same 4:3 display aspect ratio, the requirements of graphics processing dictate perfect 1:1 square pixels for computer graphics applications. A consumer's video source, after digitization, typically yields slightly rectangular pixels with approximately 12:11 width-to-height pixel ratio. Without using a special square pixel video decoder (digitizer) or compensating for the slight pixel aspect ratio, circular objects displayed on a TV monitor can become slightly elliptical when displayed using a computer graphics display card. To produce a square pixel video output, some aspect ratio correction process must be used. This can be as simple as using a (12.27 MHz NTSC, 14.75 PAL/SECAM) square pixel video decoder or some form of post-decoding pixel processing using downscale filters.

***Interlaced vs. Noninterlaced Video.***   Interlaced video is created by generating two interleaved fields to define a single video frame. Hence, the number of lines in a field are one-half the number of lines in a frame, with odd-numbered lines contained in one field and even-numbered lines contained in the other field.

In an interlaced display, every other scan line belongs to the same field and each field is displayed consecutively—odd, even, odd, even. This provides a sufficient frame rate to prevent flicker in the display. Using two fields, each one providing one-half the frame information and alternately updating the display, the field update rate is 60 fields per second, or 30 Hz. In noninterlaced video, frames are not divided into odd and even fields. Rather, frames are displayed sequentially at a high rate, usually between 70 and 80 Hz to prevent flickering.

Because most computers use a noninterlaced display format, interlaced video must be converted to noninterlaced format prior to being displayed. This may be accomplished in one of three ways: line duplication (Fig. 25.6), line interpolation (Fig. 25.7), and field processing (Fig. 25.8). Line duplication is a process whereby new scan lines are generated by duplicating each real previous scan line. While storing a line into line store A, the previous line store, B, is read out twice (the reading rate being twice the writing rate). The process is then reversed as line store B is written to while line store A is read twice.

Line interpolation creates an intermediate weighted value between two original lines. As in line duplication, the reading rate is twice the writing rate. However, during the read cycle, data is interpolated with data from the previous line store.

Field processing as a method of de-interlacing video merges two consecutive fields to make a single frame of video. Scan lines of a field are merged with scan lines from a previous field, resulting in a single frame. Unlike line duplication or line interpolation, where the vertical scan lines are doubled and vertical resolution is not, field processing doubles the vertical resolution to the full-frame resolution. The drawback of this technique is that motion artifacts are introduced because a moving object is in a different position from one field to the next.

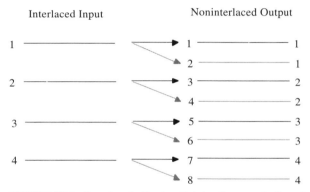

**FIGURE 25.6**  Scan line duplication. Existing lines are duplicated, generating a new line below the existing line.

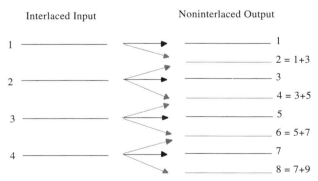

**FIGURE 25.7**  Scan line interpolation. Existing lines are interpolated. Weighted values of existing lines generate a new line above and below the existing lines.

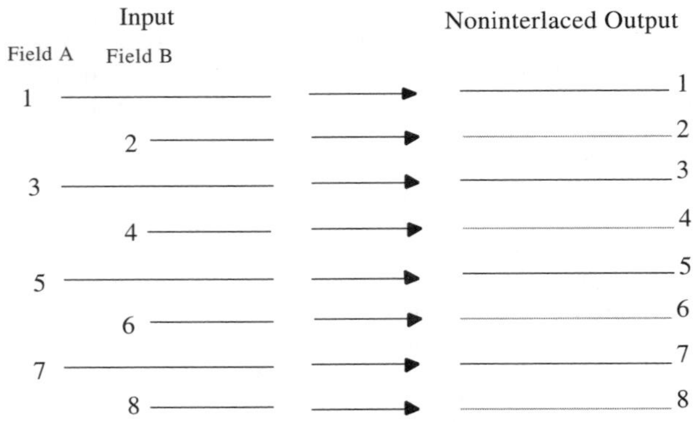

Input                                     Noninterlaced Output

Field A    Field B

**FIGURE 25.8**    Field merging. New lines are generated by using an existing scan line from the next field.

## 25.4.2  Video Output Processing

Because video-processing functions are most effectively performed in the YUV color space, color space conversion must be performed on the processed video. Some highly integrated video processors provide 24-bit color space conversion in the video output region to reduce external logic in video subsystems. The color space converter transforms YUV video data into 24-bit RGB before outputting to an external DAC or TFT flat panel. Applications that require YUV output bypass the color space converter. For maximum output flexibility, a dithering engine (Fig. 25.9) may be required to dither 24-bit RGB to RGB 6:6:6 or RGB 5:6:5 for applications that use a 16- or 18-bit DAC or TFT.

Video-processing functions generally handled in the output region of a video subsystem can include zooming for enlarging the effective resolution of an image displayed on a screen and color space conversion to provide a flexible output. In independent frame buffer architectures the output-processing functions may also include the VGA interface.

In general, there are two methods to zoom output video: pixel duplication or pixel interpolation. Duplicating lines or pixels is easy to implement but introduces blocklike artifacts, a "stairstepping effect," to the video display. Rather than duplicating lines or pixels, pixel interpolation creates intermediate weighted values between two original pixels. Interpolating improves video output quality and reduces display artifacts caused by duplicating pixels and lines. Image quality and cost of implementation of an interpolation process is generally commensurate with the number of taps between two original pixels. For example, in PC video applications a 4-tap filter is generally sufficient to provide good-quality video zoom. However, in high-end video systems used in medical and broadcast editing applications, 30- to 40-tap filters are common.

Many low-end video processors use a combination of pixel duplication and interpolation to zoom video. This reduces the cost of implementation and provides adequate video quality at low resolution modes. In the horizontal direction, a 2- or 4-tap interpolation process is used, whereas the vertical direction relies upon pixel duplication. A more advanced approach is to interpolate in both the horizontal and vertical directions. This algorithm significantly improves video display quality and reduces artifacts caused by duplicating pixels and lines. This is especially effective for applications that require a picture to fill the entire display of a high-resolution screen. In addition, in video capture applications where hard disk and system bandwidth limitations often limit video capture to below $320 \times 240$ at 30 frames per second, horizontal and vertical interpolation is essential. It provides full-screen playback of captured video with minimal image degradation and no blocklike artifacts caused by pixel duplication.

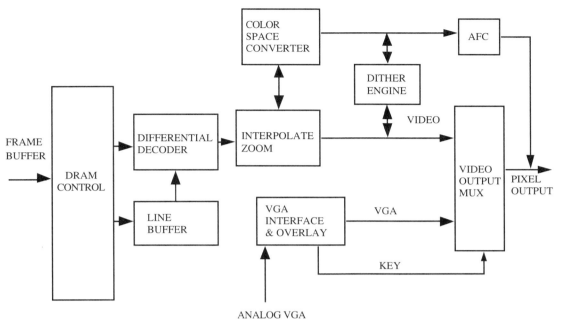

**FIGURE 25.9**   Video output pipeline.

*Video and Graphics Overlay.*   Displaying both video and graphics data simultaneously is achieved by *overlaying,* or *keying,* a visual effect of two pictures superimposed or merged together. A *key* is a value defined on a single pixel basis of the foreground picture in which a portion of background picture is allowed to be seen through. There are two methods of defining an overlay region—windowing and chroma key. The windowing method defines a rectangular region of the background picture to be seen through. The chroma key method detects on a pixel-by-pixel basis a predefined color index in the graphics frame buffer. Wherever the value of pixels in the foreground picture matches the color key, the incoming video is displayed in the color key region. By adjusting the window size and the color key bit-map pattern, an arbitrarily sized and shaped video window may be overlaid on a graphic screen. Pixels that no longer contain the chroma key color display graphics overlaid on the video data.

*Field and Frame Rate Conversion.*   Personal computers and workstations generally display video in a noninterlaced format with refresh rates ranging from 60 to 80 frames per second. This ensures adequate display resolution without excess flickering. Consumer video refresh rates, which are interlaced, are between 25 and 30 frames per second. These different display specifications result in motion distortion and other artifacts when video is displayed without sufficiently compensating for frame rate differences. Unlike computer graphics, where the display rate is much greater than the input rate (20× or more), video input rates are normally closer to display rates (less than 10×). Consequently, when displaying live video on a computer monitor, frame buffer data may be updated in the active region of the display. As a result, two frames of video are displayed simultaneously, "tearing" the video. To prevent unwanted motion distortion, a multiple frame buffer architecture can be used. This configuration enables one buffer to be updated while the second is displayed. The two buffers act like ring FIFO, eliminating motion artifacts in the video display.

### 25.4.3 Memory Interfaces, Modes, and Resolutions

Integrated video-processing ICs contain frame buffer control circuitry to interface the video input and video output display pipelines to the external frame buffer. In higher-end, multifunctional video processors, an additional pipeline may interface to the external frame buffer for video capture applications.

Depending on the IC requirements, video-processing memory systems may use fast-page DRAM, EDO DRAM, or, more costly and increasingly rare, VRAM. Memory requirements for video-processing subsystems depend on the video processor IC and, in the case of shared frame buffer designs, the memory requirement of the graphics portion of the IC. Minimum memory requirements range from 0.25 megabyte of DRAM for lower-resolution displays ($800 \times 600$ or below) to 0.5 megabyte of DRAM for resolutions up $1280 \times 1024$ for ICs such as AuraVision's line of video processors. In a shared–frame buffer design, memory requirements may range from 2 megabytes in low-end displays to 4 megabytes in high-resolution displays.

A video processor with a versatile memory configuration provides numerous benefits for developers designing a video subsystem. By selecting different memory configurations, it is possible to design products for optimum cost and performance. The following three general memory parameters provide developers the most flexible design options: (1) memory bus width and ISA-to-memory interface, (2) video frame buffer partition, and (3) memory speed.

*Memory Bandwidth and Video Resolution.* The bandwidth of the memory bus affects video-processing performance as well as the quality of video data. Memory performance may be increased through pixel interleaving (Table 25.6). The higher the interleave factor is, the greater the available memory bandwidth. For example, an interleave factor of four means that four consecutive pixels are accessed from memory in a single memory cycle. For any given interleave factor, available bandwidth is shared between the input, display, and (depending on the architecture of the video processor) the capture pipeline.

**TABLE 25.6** Example Memory Interleave Factor vs. Video Resolution ($V \times P$ 500)

| Display (graphic) resolution | Memory interleave factor | Available video resolution ($X$ by $Y$) |
|---|---|---|
| $1024 \times 768i$ | 4 | $640 \times 625$ |
| $1024 \times 768i$ | 3 | $480 \times 625$ |
| $1024 \times 768i$ | 2 | $320 \times 625$ |

*Video Frame Buffer Partition.* For motion video display, a video processor's memory architecture should support single- and double-buffer configurations. A buffer is defined in terms of a frame, regardless of whether the incoming video is interlaced or noninterlaced.

For any given memory size, a single-buffer configuration accommodates the largest frame buffer and provides the best memory efficiency for still or slow-transition video. It acts like a ring FIFO, where the input acquires a frame at the input video's timing while the output retrieves data at the graphic's timing. However, a single-frame buffer is not suitable for video that has dramatic contrast transition from frame to frame, because this configuration is susceptible to frequent rolling updates, a visual distortion caused by the display scan line overtaking the input video scan line in an active region of the display. For motion video, a dual-buffer configuration is recommended to eliminate the update line problem. Figure 25.10 compares a moving object using single– and double–frame buffer configuration.

A dual-buffer format partitions memory into two frame buffers. In this configuration, one buffer is updated while the display reads from the second buffer. This is optimal for live video

**Single Frame Buffer**          **Double Frame Buffer**

**FIGURE 25.10**  Examples of single and double frame buffers.

display because the output scan line generally cannot overtake the input display scan line, thereby allowing a full video frame to be displayed without rolling updates or other artifacts. To prevent dual-buffer access interlock, buffer-marking schemes may be implemented to allow both input and output to access the same buffer independently.

*Memory Speed.*    Depending on the design requirements of a given video system, programmable memory parameters allow the use of faster or slower memory. A high-quality video processor will contain a memory control register to program system memory. By adjusting the parameters and the memory clock frequency, different types of memory and performance targets may be met.

### 25.4.4  Video Capture

Advancements in video acquisition, cropping, scaling buffering, and hard disk data transfer rates facilitate high-quality video capture. In addition, flexible video capture and storage mediums, such as the recordable CD-ROM, enable the easy exchange of video as a data type. However, bus bottlenecks and limitations in hard disk transfer rates adversely affect video capture performance (Table 25.7).

**TABLE 25.7**  Capture Speed Requirements—Uncompressed Video

| Resolution | Frames per second | Data transfer speed |
| --- | --- | --- |
| 320 × 240—16-bit color | 30 | 4.6 megabytes/s |
| 320 × 240—16-bit color | 15 | 2.3 megabytes/s |
| 160 × 120—16-bit color | 30 | 1.2 megabytes/s |

Essential to video capture performance is the sustainable data transfer rate of a hard disk. Although peak data transfer rates are often adequate for a short burst of data written to a hard drive, video requires a sustainable rate nearly equivalent to the peak rates of common desktop PC hard disks. Hard disks found in most desktop PCs, slave IDE drives, can generally sustain a capture rate of about 700–900 kilobytes per second. A better performing master IDE drive has a higher data transfer rate of about 1.2–1.4 megabytes per second, whereas drives with data-streaming capability and an SCSI2 interface provide even higher bandwidth with a drive throughput that ranges from 4 to 10 megabytes per second.

Numerous video-processing enhancements and memory buffer configurations can be integrated into a video capture and playback system (Fig. 25.11) to alleviate bus and disk transfer

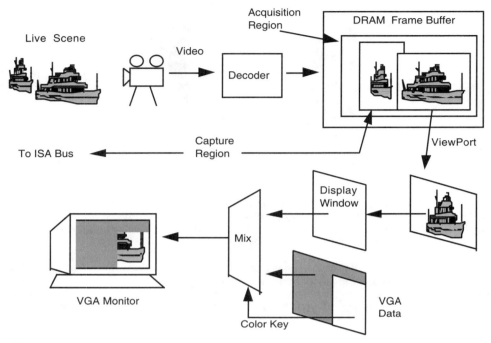

**FIGURE 25.11**    Video capture and display.

rate bandwidth limitations and provide good quality video capture and playback. In addition, the bus-to-memory interface design also affects memory-to-bus throughput and is essential in providing adequate video capture performance.

***Acquiring Video for Capture.***    Video frame buffer setting, filtering, scaling, cropping, and acquiring are interdependent functions. In a normal acquisition sequence, the size and configuration of the video frame buffer is set. In the case of an ISA or PCI input source, the width and height of the ISA window also need to be defined. With decoded video, the incoming horizontal and vertical sync defines the input window size. Next, cropping, scaling, and filtering are set. The cropping function is relative to the incoming video source. An acquisition starting address that is referenced to the frame buffer also needs to be defined (Fig. 25.12). Once this is accomplished, the video processor is ready to acquire input video. During acquisition, filtering, cropping, and scaling are continuously performed.

*Variable-size memory paging* provides a variable frame buffer size and permits the frame buffer address space to be mapped to smaller pages. By splitting the video frame buffer into a number of smaller pages, the video frame buffer is not limited to the extended address space of the bus (1 megabyte and higher).

***Quadruple Buffer Configuration.***    Although a dual-buffer format partitions memory into two frame buffers and is usually adequate for video playback, a quadruple buffer configuration, in addition to solving rolling update problems, is necessary to provide efficient video capture performance. In a double-buffer configuration, the input video may fill one buffer faster than the video processor can write the video data from the second buffer to the system memory.

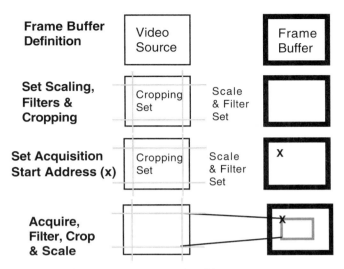

**FIGURE 25.12**    Acquisition conceptual model.

This leaves no room for the video input data frame. The result is that the video frame is dropped or the frame is captured out of sequence. With four video buffers, there are a maximum of three video frames of data available during video capture. This provides enough buffer space between the capture and video update scan lines to support nonuniform data transactions without the loss of video data. In a normal capture sequence, the writing of video data to system memory or from system memory to hard disk is nonuniform due to disk sector seek time and other bus activities. A quadruple buffer configuration resolves this issue by providing two additional buffers to virtually eliminate video data dropping caused by nonuniform data transactions.

***Differential Encoding.***    To enhance video capture performance, a video processor may implement a video data compression function using a differential encoding algorithm. For example, at the beginning of a vertical scan line the YUV color component may be compressed by truncating the least significant bits. Thereafter, each pixel is represented by a differential code of the previous pixel. Decompression of the encoded video data for playback may be performed in software because the input video pipeline can only accept uncompressed video data. Using differential encoding, the buffered video is compressed prior to being sent to the hard disk for storage. This enables more video data to be written to a hard disk and diminishes the frequency of bus bottlenecks.

***Temporal Scaling.***    *Temporal scaling,* also called *time scaling,* provides a solution for video synchronization during periods when full frame rate cannot be supported due to bandwidth and system restrictions. Much in the same way that horizontal scaling drops pixels from an image size of $640 \times 480$ to fit a $320 \times 240$ graphics window, temporal scaling drops fields/frames. For example, when capturing live video for storage, system limitations such as hard disk transfer rates or PC bus bandwidth may limit frame capture rate. If these restrictions limit frame rate to 15 frames per second, temporal-scaling operations enable a system to capture every other frame instead of allowing the hard disk timing restrictions to dictate which frames are captured. This maintains an even distribution of captured frames and alleviates "jerky" effects caused by systems that simply burst in data when the bandwidth becomes available.

***Prescaling and Interpolated Zoom.***    Full resolution ($640 \times 480$) of an incoming raw pixel stream averages over 18 megabytes per second. This far exceeds the sustainable transfer rates of even high-end hard drives. Because of these bus and hard disk bandwidth limitations, video is generally captured using a small capture window. To enhance the display quality of the captured video, hardware zoom on playback can be implemented to enlarge the video window. For example, if a video clip is captured at $320 \times 240$ and displayed at $640 \times 480$, each pixel will need to be expanded eight times. Expanding a video window is best performed using interpolation. Direct pixel replication simply duplicates pixels to increase the effective display resolution. This results in video output with a mosaic effect. On the other hand, an interpolated zoom implements an algorithm that creates intermediate weighted values between two original pixels, thereby expanding the video display size and significantly enhancing the display quality of captured video.

***Direct Video Output Pipeline Access with Zero Wait State Support.***    This memory structure allows independent access to the input video pipeline, output display pipeline, and output capture pipeline. Therefore, within a single frame, video data can be updated, displayed, and captured simultaneously. Video capture throughput is limited only by the bus and its associated mass storage media (i.e., hard disk). Bus access to a video processor can generally be accomplished in three ways: through direct frame buffer random access, through the output capture pipeline, and through the input video pipeline.

***JPEG Codec Interface.***    A direct access compression interface featured in select video processors such as AuraVision's VxP505 enables inexpensive, high-quality JPEG video compression functionality. A codec design provides four essential functions to enhance video capture performance:

1. Compression from live video input (capture) for real-time storage to hard disk
2. Decompression of video data to display overlay (playback)
3. Decompression of image frames back to memory for processing/editing
4. Compression of processed/edited video data back to storage

For JPEG video compression the video processor acquires incoming video from a PAL/NTSC decoder such as the Philips 7111 or the Brooktree BT827. Acquired frames are cropped, scaled, and then stored in the frame buffer to await compression by the JPEG codec. When the system is ready to accept the compressed frame, the JPEG codec inputs the buffered frame from the video processor's capture interface, passing the data to the host via the bus port under the CPU's control. Scaling video images prior to compression reduces the raw pixel rate and allows JPEG compression ratios in the range of 10:1 to 20:1—optimal for a JPEG algorithm.

## 25.5    VIDEO ARCHITECTURES

### 25.5.1    Shared– (Video/Graphics) Frame Buffer and Independent Graphics/Video Buffer Architectures

An essential design task in a video display system is the convergence of graphics and video data (commonly known as *video overlay, video windowing,* or *video/graphic data multiplexing*). Because graphics data are typically generated by the CPU and video data are generated from a storage medium, the two types of data must be combined prior to being displayed. Video and graphic data convergence is a key design issue in multimedia systems because it affects system cost, software requirements, system performance, and overall quality.

There are two common architectures in combining video and graphics. In a shared–frame buffer system, video and graphic data are mixed using a combined graphics and video frame buffer. In an independent graphics/video buffer system, a separate frame buffer is used for video and graphics, and video and graphics data are mixed after the frame buffer stage using an analog mux.

***Shared–Frame Buffer Architecture.***    In a shared–frame buffer implementation, video is sent to the graphics frame buffer and processed with any other data written into the frame buffer. The integrated graphic/video processor processes the data in the frame buffer, and the output is sent to a DAC to generate analog RGB for output to a monitor. Input data composed by the processor can be generated from three independent streams: RGB graphics, RGB or YUV video, and hardware cursor data. The graphics data stream is generated by reading RGB pixel data written to the frame buffer by the graphics controller.

The video data stream is generated by reading pixel data from a separate section of the frame buffer. Data written to this section of the frame buffer is derived from a video source (CPU, digitizer, decoder) in either RGB or YUV format. It may also be RGB data generated by the graphics controller to support moving objects, such as a sprite, commonly used by games programmers. Most processors allow the data to be passed through in its original size or zoomed by an arbitrary amount. In order to be merged with the graphic data, YUV video input data is color space–converted to RGB format. Lastly, to generate the hardware cursor, the cursor data is overlaid on the combined graphics and video streams.

***Advantages of a Shared Frame Buffer.***    Advantages of a shared–frame buffer architecture are that it requires one less frame buffer and one less DAC than an independent frame buffer architecture and, therefore, a low-end video display system may be implemented more inexpensively. Another advantage is that video processing and VGA graphics display are typically implemented on the same board, and video and graphics are mixed in the same buffer. Therefore, no compatibility problems exist between the VGA and video.

***Disadvantages of a Shared Frame Buffer.***    Disadvantages of the shared-buffer architecture involve memory limitations associated with video and graphics sharing the same buffer. In a shared–frame buffer system, FIFOs are used to store video data until there is room in the buffer. Up to an entire scan line may be loaded into the FIFOs. When there is available access to the frame buffer, video data is read out to the buffer. Because video data, VGA data, and other processes require high-bandwidth interfaces to the frame buffer, bus arbitration becomes problematic. Scaling up or zooming the video will increase the bandwidth requirement into the frame buffer up to 400 megabytes per second. This problem can be mitigated somewhat by using more memory, although it negates the cost benefit of a shared–frame buffer solution over an independent frame buffer solution.

> *Case 1:* Shared graphics/video frame buffer (with double frame buffer)
>> Video + graphics (RGB 16) = $1024 \times 768 \times 2 \times 2$ = 4 megabytes
>> Total = 4 megabytes
> *Case 2:* Independent graphics/video frame buffer
>> Video: YUV 4:2:2 NTSC $640 \times 480$ to $1024 \times 768$ = 0.5 megabyte
>> Graphics: RGB 16 = $1024 \times 768 \times 2 \times 1$ = 2 megabytes
>> Total = 2.5 megabytes

Another drawback of a shared–frame buffer implementation is that once written into the frame buffer, video cannot be differentiated from graphics. Therefore, brightness, contrast, hue, saturation, and gamma correction adjustments made to achieve the best video display may adversely affect the graphics display and vice versa.

***Independent–Frame Buffer Architecture.***    In an independent–frame buffer architecture, video processing and VGA graphics display may be implemented on different PC boards. A separate frame buffer is used for video and graphics and they are mixed using an analog mux after the frame buffer stage.

*Advantages of an Independent–Frame Buffer.*    The main advantage of an independent–frame buffer is the ability to handle video and graphics data independently. Because video and graphics are derived from different sources, the ability to adjust the gamma correction, hue, contrast, and saturation independently may be important. Second, although video and graphics share the same memory in a shared–frame buffer implementation, memory requirements for the video portion of an independent–frame buffer configuration can be as low as 0.5-megabyte DRAM to support resolutions up to $1280 \times 1024$. When added to the VGA graphics memory requirement, generally about 2 megabytes, system memory is only 2.5 megabytes. Lastly, larger line buffers are required to support MPEG-2 video data (720 bits in length), which may not be present in integrated graphics/video processors supporting shared–frame buffers.

*Disadvantages of an Independent–Frame Buffer.*    Disadvantages of an independent–frame buffer implementation include added cost and, depending on system architecture, compatibility problems between VGA and video cards. An independent–frame buffer solution requires two sets of memory and two DACs, whereas a shared–frame buffer solution requires one set of memory and a single DAC, integrated into the video/graphics processor. In low-end applications, where full-screen, high-resolution is not a requirement, an independent–frame buffer configuration may be more costly. However, the memory necessary to approach the display quality of an independent–frame buffer configuration may make a shared–frame buffer configuration more costly.

Other drawbacks to the independent–frame buffer architecture are potential compatibility problems between VGA graphics and video cards. Color key information, embedded in the VGA data, is essential for overlaying graphics and video. Because the overlay function takes place on the video board, transferring color key information to the video board has traditionally been accomplished through a common device called a *feature connector.* Although the feature connector provides an essential bridging function for multimedia video overlay products, it also creates major VGA compatibility problems. The feature connector is limited to a speed of 30 to 45 MHz, whereas current computer graphic displays run from 25 to 135 MHz. At speeds beyond its operational maximum, the feature connector fails to carry a clean, solid signal. Hence, many video add-on products are designed to perform at a maximum of 45 MHz. Compatibility problems also exist because the setup and hold times of pixel clocks vary by VGA manufacturer. Thus, designing a video overlay product to be compatible with all VGA subsystems is a complex, time-consuming process. The feature connector also presents other VGA compatibility problems. VGA data can be in 8-bit, 15-bit, 16-bit, and 24-bit format. Because the feature connector's data path is only 8 bits wide, two passes are required for 16-bit data and three passes for 24-bit data. In addition, the pixel clock is doubled for 16-bit data and tripled for 24-bit data. Not only are the pixel frequencies increased, but the video system also must recognize whether it is interfacing one, two, or three passes of graphics data. Many video board designers resolve the compatibility problems between VGA graphics and video and eliminate the feature connector by using a video overlay DAC such as AuraVision's AnP82. Figure 25.13 shows the main components of the AnP82.

An overlay DAC, such as AuraVision's AnP82, detects the color key from the VGA data or chroma key from the video data to provide a smooth overlay function. It has an internal PLL to regenerate the pixel clock for the video controller, thereby eliminating the feature connector. The AnP82 also detects the key information, allowing the host to perform fully automatic video window alignment and automatic color key selection. This simplifies the setup process of the video display system.

### 25.5.2  PCI Video

The development of the PCI bus by an interest group of semiconductor and computer manufacturers, led by Intel, has significantly altered the architectures of computer video subsystems.

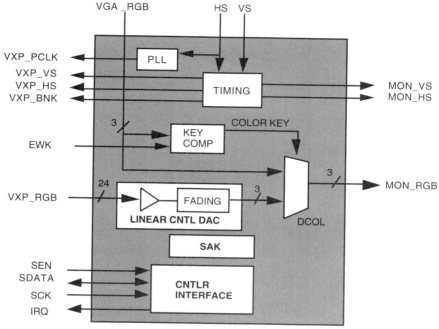

**FIGURE 25.13**   AnP82 block diagram.

The PCI bus has also been adopted by Apple Computer in PowerPC-based Macintosh computers. Prior to the PCI bus, PCs had been based on the ISA bus, with a limitation of 8-megabytes
per second transmission rate. This is insufficient to handle full-scale, full–frame rate interlaced
NTSC video (640 × 480 resolution, 16 bits/pixel, 30 frames per second), which requires 18.4
megabytes per second. It is also barely adequate for scaled CIF video (320 × 240 resolution, 16
bits/pixel, 30 frames per second), which requires 4.6 megabytes per second.

The PCI bus offers a maximum bus burst transfer rate of 132 megabytes per second, enabling
the transfer of video data between (a) video processors and shared graphics/video memory systems, and (b) hard drives. This bandwidth is, however, contended for by hard drives, graphics
devices, and communications subsystems. The ability of a particular device to receive the bus is
therefore very important in calculating available bandwidths and capabilities of the system.

The PCI bus provides high transfer rates through the ability of devices to burst data via burst
mode write transfers. In order to efficiently utilize the PCI bus, data is usually collected at a
source prior to bursting across the bus. This avoids the transfer of small amounts of data, such
as a single word, where the setup information required to accompany the data makes for highly
inefficient bus utilization. The data is usually transferred from a FIFO or from buffer memory,
being timed to take full advantage of the 132-megabytes per second rate available with PCI.

The use of a PCI bridge chip allows devices without PCI interfaces to be connected to the PCI
bus. Such devices might include MPEG decoders or NTSC decoders. PCI bridge chips targeted
at video and audio applications include AuraVision's VxP195, Philips's SAA7145, and Zoran's
ZR36120; they offer a bridge between the typical YUV video bus and the PCI bus. Adding
downscaling functionality to the bridge chip allows PCI bandwidth to be minimized. Not all
the video data being fed into the bridge chip should be scaled down, however. An example of
this is vertical blanking interval (VBI) data that needs to be sent to the system's main memory
for processing by the CPU. The PCI bridge chip must, therefore, be able to send both scaled video

data to a frame buffer and unscaled data to main memory. PCI bridge chips also offer support for serial audio streams and integrate FIFOs necessary to ensure efficient bus utilization.

An alternative approach to the PCI bridge chip is taken with AuraVision's PCI-based VxP524 video processor. This device uses a local frame buffer to support video capture. It also features an output pipeline that can either feed the video port of a graphics chip or a video overlay DAC, such as AuraVision's AnP82. By using a local frame buffer, the possibility of video frame dropping caused by PCI bus contention is eliminated.

### 25.5.3   Graphics Chips and Video Port Architectures

A number of graphics chips are now incorporating a video port on chip, allowing video data to be sent directly to the graphics chip rather than via the ISA or PCI bus. Such an arrangement frees the main system bus from the overhead of handling the video data and avoids latency and the contention issues associated with arbitrating for the bus. The video data must first be sent to the graphics frame buffer and the system must then fetch the data prior to processing, overlay, or other processing.

Figure 25.14 shows one possible video system architecture for an LCD flat panel design utilizing AuraVision's VxP524 PCI-based video processor and a graphics controller with video port. By supporting 24-bit dithering, the VxP524 can provide 16-bit RGB data to 16-bit video ports such as that on Cirrus Logic's Nordic LCD VGA controller.

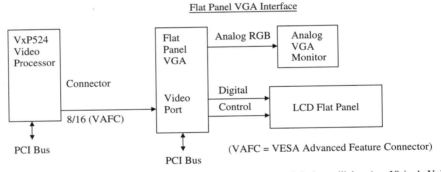

**FIGURE 25.14**   Video system architecture for an LCD flat panel design utilizing AuraVision's VxP524 PCI-based video processor and a graphics controller with a video port.

Another popular example of video support on graphics chips is provided by S3, Inc. S3 provides a local peripheral bus (LPB) on their popular Trio64V+ Integrated Graphics/Video Accelerator, which allows a glueless bidirectional interface to MPEG decoders, such as their own Scenic/MX2. Also offered is an input-only interface to a live video digitizer, such as a Philips 7110 NTSC decoder. These interfaces are collectively called the S3 "Scenic Highway."

The Trio64V+ also incorporates a streams processor (Fig. 25.15), which processes data from the graphics frame buffer, composes it, and outputs the result to the internal DACs for generation of the analog RGB outputs to the monitor. The processor can compose data from up to three independent streams, as follows:

1. **Primary stream.**   RGB graphics data.
2. **Secondary stream.**   RGB or YUV/$YC_bC_r$ video data from another region within the frame buffer.
3. **Hardware cursor.**

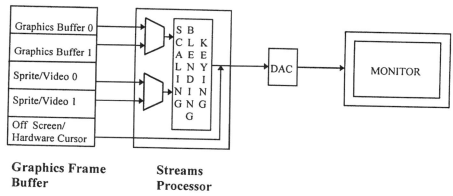

**FIGURE 25.15**   S3 Trio64V+ streams processor.

Regardless of the input formats, the streams processor creates a composite RGB-24 output to the DACs. The secondary stream (video data) is generated by reading pixel data from a section of the frame buffer separate from that used for the primary screen. If these data are in YUV or $YC_rC_b$ format (provided via the LPB), then the data are color space–converted to RGB-24. Before conversion, the data can also be scaled up horizontally and vertically by an arbitrary amount. Horizontal scaling uses filtering for interpolation, whereas vertical scaling is achieved by line replication.

The streams processor can compose a variety of output types from the streams just mentioned:

- Overlay of secondary stream on primary stream in an opaque rectangular window (video window over graphics).

- Overlay of primary stream over secondary stream (graphics window over video).

- Blending of primary and secondary streams on a pixel-by-pixel basis, providing dissolving and fading operations.

- Secondary stream overlaid on the primary stream in an irregular window (e.g., games sprites could be overlaid over graphics). This uses color keying.

- Primary stream overlaid on the secondary stream in an irregular window (graphics text to overlay video). This uses color keying.

## 25.6   SOUNDCARDS AND AUDIO PROCESSING

Personal computer audio is achieving increasingly high fidelity thanks to the adoption of wavetable synthesis for soundcards, MPEG-1 audio for video CD/CD-ROM, and Dolby AC-3 for Digital Video Discs. This section focuses on soundcards that are typical for PC applications. Such cards offer the ability to capture and play back sounds; offer MIDI, SoundBlaster, and Windows Sound System compatibility; and provide both FM and wavetable sound synthesis.

### 25.6.1   Soundcards

Today's soundcards usually feature a single, highly integrated audio IC that performs all wavetable synthesis functions. The audio IC may also integrate IDE CD-ROM interfaces, game interfaces, an ISA bus interface, and a mixer for combining multiple sound sources. FM synthesis may be handled by a dedicated IC or emulated via the wavetable synthesizer and software. The board will also contain a wavetable ROM for storage of the wavetable samples, and a codec (combined A/D converter and D/A converter). Figure 25.16 illustrates the main components of a PC soundcard.

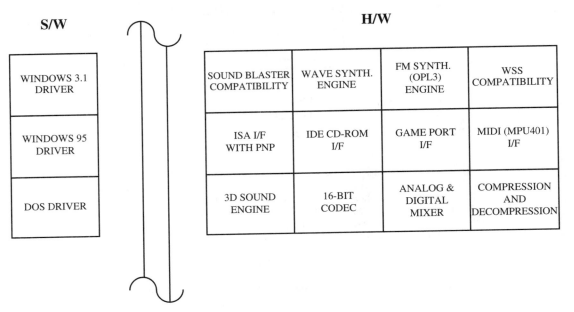

| S/W | | H/W | | | |
|---|---|---|---|---|---|
| WINDOWS 3.1 DRIVER | | SOUND BLASTER COMPATIBILITY | WAVE SYNTH. ENGINE | FM SYNTH. (OPL3) ENGINE | WSS COMPATIBILITY |
| WINDOWS 95 DRIVER | | ISA I/F WITH PNP | IDE CD-ROM I/F | GAME PORT I/F | MIDI (MPU401) I/F |
| DOS DRIVER | | 3D SOUND ENGINE | 16-BIT CODEC | ANALOG & DIGITAL MIXER | COMPRESSION AND DECOMPRESSION |

**FIGURE 25.16**   Soundcard components.

Some soundcards contain a separate FM synthesizer, a controller for game and wave audio, and a wavetable synthesizer. Because all three of these digital subsystems operate at different sample rates, their outputs cannot be mixed digitally without sample frequency conversion. This results in the need for additional DACs and an analog mixing system. The analog mixer is also required for mixing of sound from microphones and other analog sources. For example, in a karaoke application it may be necessary to mix voice with sounds from a video CD.

### 25.6.2  Software Compatibility

SoundBlaster command set compatibility and support for Windows Sound System is frequently required for soundcards in PC applications. The SoundBlaster command set provides a means for software developers to control the soundcard in terms of capturing or playing back sounds and transferring sound data through *direct memory accesses*. The software application is also responsible for setting the sampling rate for the board.

MIDI is the acronym for Musical Instrument Digital Interface, which allows computers, synthesizers, and musical instruments to be interconnected through a standard interface (see Chapter 24). The interface operates at 31.25 kb/s and is asynchronous. The general MIDI sound set provides for 128 different instruments; for example, program change 1 represents an acoustic grand piano and 74 represents a flute. It does not, however, specify how the sounds specified will be reproduced, meaning that the same instrument can sound different if played on soundcards using different synthesis chips. Support for MIDI is offered in MIDIMapper in Windows 95.

The MPU401 interface standard developed by Roland provides a common MIDI interface definition at the hardware level and deals with UART implementations and hand-shaking protocols between host CPU and the sound board. Figure 25.17 highlights the evolution of soundcards, including Creative Labs' SoundBlaster and SoundBlaster Pro.

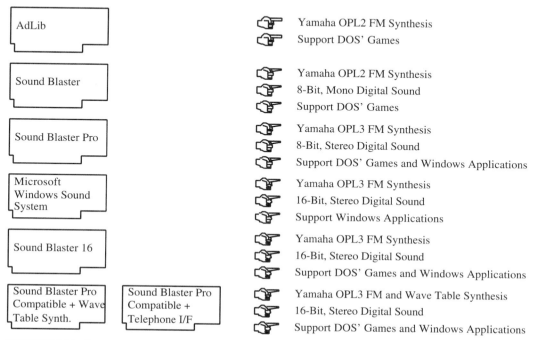

**FIGURE 25.17**    Soundcard evolution.

### 25.6.3  Frequency Modulation

The first sound offered to support games was that provided by a technique called *FM synthesis*. Originally developed at Stanford University and implemented by Yamaha, this technique offers the ability to construct sound by combining several waveforms together. Generally, a periodic signal, or *modulator*, is used to modulate a carrier signal, resulting in an audible change in the sound of the carrier. An FM synthesizer, therefore, requires at least two signal generators, or *operators* (e.g., "OPL3" refers to three operators). Although this can be accomplished with a very cost-effective IC solution, it offers limited fidelity compared with wavetable synthesis. FM synthesis support is now usually offered to maintain backward compatibility with existing software applications.

### 25.6.4  Wavetable Synthesis

The total number of bits necessary to store sound samples is a factor of both the number of bits per sample and the sample frequency rate. According to Nyquist's theorem, the sample rate (fs) has to be greater than or equal to twice the maximum sampled frequency. It is then possible, theoretically, to reconstruct these samples without error. In reality, due to imperfect filtering characteristics and other considerations, the sampling rate is always set at more than twice the highest frequency to be sampled—i.e., 44.1 kHz for CD audio provides up to 20 kHz frequency bandwidth. The number of bits employed for each sample dictates the resolution achievable; i.e., an 8-bit A/D allows for a resolution of 0.02 V for the LSB at 5 V peak to peak, whereas a 16-bit A/D provides 0.000076 V at 5 V peak to peak. Alternatively, quantization error (distortion) = $(1/2)^{N+1}$. For example, 8 bits per sample: $(1/2)^9 = 0.19$ percent and 16 bits per sample = 0.000763 percent.

Using wavetable synthesis, the number of bytes of memory necessary to store the sound of a particular musical instrument is minimized by storing a sample of a note or notes rather than every note in the musical instrument's range. The wavetable synthesis technique offers high-quality sound reproduction because it uses actual digital samples of instruments. The wavetable synthesis engine is then responsible for modifying the pitch of the sound, applying an amplitude envelope, and combining multiple sounds, or "voices," to complete the effect. The greater the number of voices the synthesizer can combine, the greater the realism of the sound for large musical pieces such as orchestral music.

## 25.6.5 Envelopes

Wavetable synthesizers reduce the amount of memory needed to represent a sound by looping sound segments repeatedly rather than storing an entire note's duration. The synthesizer may model an instrument's sound as consisting of four periods (ADSR)—the first is the attack period, the second is the decay period, the third is the sustain period, and the fourth is the release period. The *attack period* is that during which the string is initially plucked or drum hit, and where the dynamics of the sound change very rapidly. The decay is the time during which the sound falls from the peak. The sustain period is that in which the sound decays in amplitude, with less dynamic changes in sound occurring. The release period is the amount of time for the sound to go from sustain to silence. By looping sound segments within an amplitude envelope during playback, the synthesizer can approximate the original sound of an instrument. The envelope models the ADSR periods. The accuracy of such a model depends on the spectral characteristics of the sound remaining fairly constant during the sustain period of the sound. Sounds such as drums may have no sustain period at all. A common amplitude envelope is the ADSR (attack, delay, sustain, release) model, which consists of four linear segments (Fig. 25.18). The wavetable ROMs typically contain both an attack sample and a loop sample. The synthesizer applies the envelope generation function to these samples during playback by modifying the amplitude of the sound loops within its ranges. The length of the loops should be equal to an integral number of the fundamental pitch of the sound being played; otherwise, pitch shifts will occur.

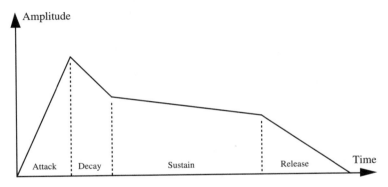

**FIGURE 25.18**    ADSR amplitude envelope.

### 25.6.6  Compression

Some audio ICs use ADPCM (Adaptive Delta Pulse Code Modulation) compression to further reduce the size of sound samples. Delta-PCM allows sample sizes to be reduced by storing only the difference in sample value from one sample to the next instead of the entire sample. ADPCM improves on DPCM by adaptively modifying the range over which difference samples are represented, reducing distortion during large amplitude swings and improving low-level linearity during quieter passages (Fig. 25.19). ADPCM might typically reduce the number of bits per sample from 16 to 4. It achieves such compression ratios at the expense of fidelity and may often be applied only to voice data.

### 25.6.7  Pitch Shifting

Wavetable synthesizers use pitch shifting to allow a number of notes to be generated from a single sample. By accessing the sound samples at different rates during playback, the synthesizer can change the pitch of the note. For example, if the playback rate is twice that of the original, then the frequency is doubled, representing a shift up of one octave. If this playback rate is an integer multiple of the original, then the math is straightforward, but this is not sufficient. For example, to shift a note up by one musical half-step, the playback rate must be 1.05946 that of the original. For this reason, the pitch-shifting increment value must be represented by both an integer value and a fractional value. The resolution of the playback value is therefore determined by the number of bits available to represent the fraction. The interpolation techniques used for video scaling, as previously discussed in this chapter, are also applicable to audio pitch shifting. In the case of pitch shifting, an interpolated value is used when the desired sample value falls between available data points. This provides less distortion than if the nearest available sample is simply used.

When pitch shifting is used during playback, the timbre of the sound may also be audibly affected. Although this may not be noticeable for small shifts in pitch, it is for larger shifts, with the effect on timbre being more noticeable with certain instruments, such as the piano. To avoid these changes in timbre, it is customary to provide more than one sample for a particular instrument. Each sample is then used for pitch shifting only within a particular range or "split." To avoid timbre differences from the high end of one split to the low end of another, digital filtering may be employed to smooth out differences. In other cases the musical instrument's entire range may be sampled to offer maximum fidelity at the expense of memory.

## 25.7  *SOFTWARE*

This section provides an overview of the levels of software needed to support video processing in the Windows PC environment. A diagram of the multimedia systems architecture is shown in Fig. 25.20. Also included in this section is a description of the AuraVision video overlay and capture system architecture, which serves to illustrate an actual implementation running under Windows.

### 25.7.1  Microsoft ActiveMovie 1.00

ActiveMovie is an architecture for processing streams of multimedia data. ActiveMovie software allows users to play back MPEG digital movies, AVI stream files, and audio using either a software MPEG decoder or a hardware MPEG decoding chip (such as C-Cube's CL480). ActiveMovie utilizes video and audio cards that support the Microsoft DIRECTX set of APIs. These DirectX APIs are described in the following sections.

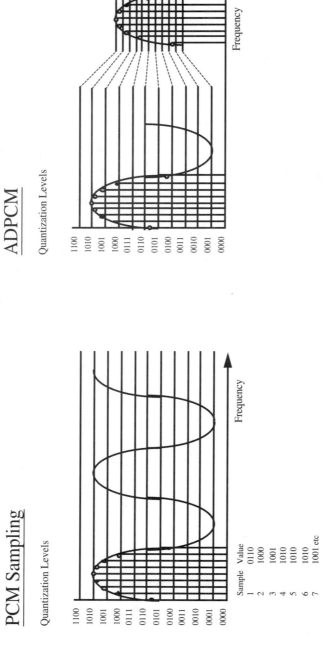

**FIGURE 25.19** Quantization levels for PCM sampling and ADPCM. Using Delta-PCM, the number of bits per sample is reduced by storing only the difference between the current sample and the last, rather than the entire sample value. ADPCM reduces the loss of low-level linearity and reduces dynamic distortions by adaptively modifying the range over which difference (or *delta*) samples are represented. In this example the size of the quantization levels has been reduced to produce more accurate samples.

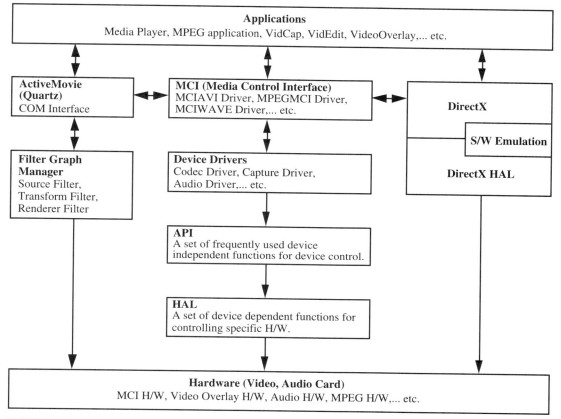

**FIGURE 25.20** Multimedia system architecture.

*DirectDraw.* DirectDraw provides a device-independent way for games and Windows subsystem software (such as 3-D graphic packages or digital video codec) to access display device–dependent features. DirectDraw is a memory manager for video memory. Applications must write to DirectDraw (DDRAW.DLL). This 32-bit DLL provides all the device-independent functionality. DirectDraw HAL is a hardware-dependent layer and provides device-specific code. DirectDraw HEL is a hardware emulation layer that provides software emulation for the requested task if the hardware does not support it or DirectDraw HAL is not present.

*DirectSound.* DirectSound provides audio buffering and effects for audio applications while minimizing CPU overhead.

*DirectVideo.* DirectVideo allows Video for Windows to directly access DirectDraw for fast video acceleration.

*Direct3D.* Direct3D provides a hardware abstraction layer to DirectDraw and a 3-D device driver interface. This allows 3-D hardware to accelerate rendering when present and allows DirectDraw to emulate functions if not. ActiveMovie is also capable of playing back files in audio-video interleave (AVI) format.

### 25.7.2 VFW/AVI Architecture

Microsoft Video for Windows supports the A/V data in the Windows environment. MCI_Level interfaces, such as MCIAVI and MMIO library, handle the A/V streams. From the hardware vendor's perspective, this enrichment of system-level multimedia support presents a clear opportunity to provide value-added products (improved performance, lower cost, etc.) that can be immediately used with a broad range of existing applications. Applications make calls to an API and, as the result of this call, the system sends proper messages to the installable driver's entry point, DRIVERPROC. This then processes the message and calls the appropriate functions in other drivers/Dlls to perform the task. For example, if VIDCAPT opens a video capture device, the system sends the specific video capture device driver—such as VCAPT.DRV(AURAVISION capture driver)—a DRV_OPEN message. The driver's DriverProc function receives and processes this message and opens a video overlay window.

*Windows 3.1.* Drivers are installed through a setup program and/or control panel + driver. They get installed in the system.ini [DRIVER] section, and all of the necessary settings and configurations (IRQ, DMA, . . . , etc.) for the hardware device are written and saved into an appropriate INI file.

*Windows 95 and NT.* Drivers are installed through the Windows registry.

## APPENDIX—AURAVISION VIDEO OVERLAY AND CAPTURE SYSTEM ARCHITECTURE

### *Layer 1: Applications*

SETUP
AVVIDEO
AVCONFIG
Microsoft MediaPlayer, VIDCAP, VIDEDIT

- *SETUP.EXE.* Windows software installation program that is based on the Microsoft setup toolkit (Installshield).
- *AVVIDEO.EXE.* Video application program that is designed to demonstrate live video features of the VxP chips. This application allows the user to capture a single frame in BMP format into a file.
- *AVCONFIG.EXE.* Video configuration program that is used for configuring features of the VxP chips, such as video alignments, color controls, etc.

### *Layer 2*

MCI (Media Control Interface) system layer
Microsoft MCI/AVI system layer
AuraVision MCI driver (DV-MCI )

- *MCI_Level Driver.* Capable of processing AVI data streams (.AVI files). The MCIAVI.DRV driver is provided by the VFW system and is sufficiently general to support custom capture and codec drivers. The MCIMPEG digital video MCI-layer driver is capable of processing MPG file streams.
- *Microsoft ActiveMovie 1.00.* ActiveMovie delivers a consistent streaming architecture for all multimedia data. The Component Object Model (COM) from OLE2 will be used to do the following:

Identify, locate and instantiate each component.

Group APIs and services into interfaces.

Provide common methods for communicating between components.

ActiveMovie components (COM) process streams of data, either providing, transforming, or consuming streams of data. An application program's multimedia task will be performed by a set of connected filters (known as a *filter graph*). Functionality that is presently provided by system-supplied drivers such as MCIAVI video playback driver will instead be provided as a number of discrete components. Capturing, compressing, and decompressing will be done by separate, replaceable, system-supplied filters.

### Layer 3: Codec Driver and Capture Driver

- *Capture driver.* Video for Windows capture driver allows the Windows capture application program, such as VIDCAP, to capture live motion video into an AVI file. It also supports capture of still frames. AVCAPT can capture into the following AVI formats:

  Aura1 compressed: YUV411 (6 bits); Aura2 compressed: YUV422 (8 bits)

  8-bit RGB, 16-bit RGB, 24-bit RGB

- *AVCODEC driver.* Video for Windows compression/decompression driver allows the video for Windows playback application, such as Media Player, to play back AVI files through the video overlay using the features of the VxP chips.

### Layer 4

Video-processing API

Video overlay, compression, and decompression library

*AVWIN.DLL: AVWIN* provides a set of video functions that are used by other drivers and applications in the AuraVision VxP software package. The functions in this library can be called by the applications to control the video controller.

### Layer 5: Hardware Abstraction Layer

Low-level functions (APIs) for setting registers

*AVHAL.DLL.* All of the chip-dependent programming of VxP video registers is done by the HAL functions. The functions in this layer are transparent to the user and should be accessed through the appropriate AVWIN functions (API). For example, to initialize the chip, an application such as AVCONFIG should call AV_Initialize, which is in AVWIN.DLL. AV_Initialize will call HW_Init, which is in HAL.DLL. HW_Init will initialize an appropriate VxP chip (based on the chip type passed on to it from AVWIN.)

### Layer 6: Hardware

AuraVision VxP505, VxP524; Philips 7110, etc.

## GLOSSARY

**Artifacts**    A wide-ranging term involving visible distortion of the video image. This usually refers to the addition of some kind of noise, spots, shimmering effects, crawling dots, etc.

**CCIR601**    The International Consultative Committee recommendation for the digitization of video signals. The recommendation includes the color space conversion from RGB to

$YC_rC_b$, the definition of a 13.5-MHz sample rate, and a horizontal resolution of 720 active pixels.

**Chrominance**    This is the color part of the video signal rather than the luminance part. Often shortened to *chroma.*

**CIF**    Common Interchange Format. Specifies resolutions of $352 \times 288$ or $352 \times 240$. Allows easy interchange of video images between different computers or other video systems.

**Field**    Applies to interlaced systems only. Each TV picture is made up of two fields, the first containing the odd-numbered scan lines and the second containing the even-numbered scan lines. Two fields make up a frame.

**Frame**    A frame is basically one complete still video image consisting of two fields in an interlaced system. The complete frame consists of 525 lines in an NTSC system or 625 lines in a PAL or SECAM system.

**Frame rate**    The number of frames displayed per second. NTSC offers 30 frames per second, resulting in lower flicker than PAL, which has 25 frames per second. Progressive scan displays, such as computer monitors, may offer 75 or 85 frames per second.

**Interlaced**    An interlaced video system uses interleaved fields to create a video frame. By interleaving odd and even scan line fields, interlaced displays avoid the visible flicker caused by the screen's phosphor dimming before line refresh can occur. Interlacing relies on the human eye's ability to blend together the dimming scan line with the newly refreshed adjacent scan lines.

**Interpolation**    Interpolation is the application of a mathematical algorithm to produce missing points between two known points on a curve or straight line. Interpolation can be used when scaling a video stream to create new pixel values between two existing, known pixel values. Interpolation offers greater accuracy than simply repeating pixel values or scan lines, both of which result in "block" or "staircase" effects.

**Luminance**    The black and white part of the video signal.

**NTSC (National Television System Committee)**    This video standard is used in the United States and uses 525 lines per frame, 30 frames per second, and YIQ color space.

**PAL (Phase Alternation Line)**    This video standard is used in most of Europe and uses 625 lines per frame, 25 frames per second, and the YUV color space.

**Resolution**    The number of pixels that make up a picture. The greater the number of pixels is, the greater the detail that it is possible to display. The resolution is usually defined by two numbers representing the horizontal and vertical resolutions; i.e., VHS is $330 \times 220$ resolution, whereas a PC screen may be $1024 \times 768$.

**Scaling**    The ability of a video processor to change the effective resolution of the video image. When scaling down, some information must be discarded, with pixels being dropped and the number of lines being reduced. When scaling up, some information must be "created," perhaps through line replication or interpolation algorithms. The complexity of algorithms chosen to perform scaling greatly affects the achieved video quality in small and large window sizes.

**SECAM (Sequential Couleur Avec Memoire)**    The video standard used in France, as well as in Russia. SECAM is a 625-line, 50-fields-per-second, 2:1 interlaced system that is similar to PAL, but it uses FM modulation for chrominance information.

**Sync**    This signal provides the information needed to position the picture on the display screen. The horizontal sync and vertical sync provide the positional information in the left-right and top-bottom directions, respectively.

**Zooming**    Scaling the video image to make it larger.

## BIBLIOGRAPHY

Jack, K., *Video Demystified: A Handbook for the Digital Engineer,* HighText Publications, Solana Beach, Calif., 1993.

Benson, K. B., *Television Engineering Handbook,* McGraw-Hill, New York, 1986.

CCIR Recommendation 601-2, *Encoding Parameters of Digital Television for Studios,* 1990.

CCIR Recommendations 656, *Interfaces for Digital Component Video Signals in 525-Line and 625-Line Television Systems,* 1986.

*Trio64V+ Integrated Graphics/Video Accelerator User Manual,* S3 Incorporated, July 1995.

*VxP500 User Manual,* AuraVision, 1994.

*VxP524 User Manual,* AuraVision, 1996.

*AnP82 Data Sheet,* AuraVision, 1996.

## ABOUT THE AUTHORS

Chris Day started working at AuraVision in November 1995 as director of marketing. He previously worked for Motorola Semiconductor in California for nine years in a variety of sales and applications positions and, prior to that, was in marketing for Hitachi in London, England. He received his B.Sc. in computer and microprocessor systems in 1984 from Essex University, U.K., and his M.B.A. in 1995 from Santa Clara University, California.

Jim Seymour started working at AuraVision in August 1994 as a marketing engineer. He previously worked at Pacific Gas & Electric Company as a technical writer and has five years experience in marketing and sales in the PC and telecommunications industries. He received the B.S. in 1989 from the University of California, Berkeley.

Thanks must also go to the following people who contributed and helped in the writing of this chapter:

Victor So at AuraVision (Wavetable Synthesis & Audio)

Ellie Abdollahi at AuraVision (Software)

Tommy Lee at AuraVision (Video Processing)

Rudi Wiedemann at Silicon Vision (Digital Camera)

Paul Crossley at S3 (Graphics Chips/Video Ports)

# CHAPTER 26
# PERSONAL DIGITAL ASSISTANTS

**Rhonda Dirvin and Art Miller**
*Motorola, Inc.*

## 26.1  HISTORY

In the early 1990s the public learned that Apple was developing a new class of product: the *personal digital assistant,* or PDA. Small, light, agile, and high-performance, it was rumored to be at the forefront of a new market. The units were portable (consumer and professional) electronic equipment based on new user paradigms that would instantly create a new market. Other developers soon joined the marketing hype: Sharp's Zaurus, Casio's Zoomer, Microsoft's WinPad, General Magic's MagicCap-based consortium, and EO's Message Pad. Most of these were soon identified as high-end organizers with a few interesting new characteristics: graphical user interfaces (GUIs) with liquid crystal displays (LCDs); pen input for character recognition using pressure-sensitive touchpads; and "lightweight" real-time operating systems (RTOSs).

After initial surges in sales, customer demand slackened quickly due to high prices, lack of applications software, poor battery life, large size, slow overall performance, poor cursive handwriting recognition, and lack of synchronization with the desktop computer. PDAs were panned as having instantly created a $0 billion market. Companies such as EO failed, projects such as Microsoft's WinPad were canceled and then redefined, and development staffs were cut back in realization of a long market development cycle. But the PDA had been born.

## 26.2  TYPES OF PDAS

According to a recent report,[1] PDAs can be divided into three general categories: traditional PDAs, personal communicators, and mobile companions. Some subscribe to the theory that there are really more types out there than these few. A PDA can actually be defined as any small, electronic unit that one wants to carry. Typically, its usefulness is attributed to connectivity features, personal applications software, or vertical applications software.

### 26.2.1   Traditional PDAs

Traditional PDAs are described as "pen-based personal organizers [where] wireless connectivity is a secondary or optional feature."[1] This traditional PDA was the entry point for products such as the Apple Newton. One can note that this is evolution at work: advanced user input combined with a graphical user interface. Communications was so secondary that it was always an external piece of electronics. Performance was actually poor, with a low hit rate on the cursive handwriting recognition unless users extensively trained themselves to write like the Newton expected. The operating system was centered around handwriting and user feedback was slow. Other systems of this category, such as Psion's organizer or Casio's Zoomer, were more cautious using screen-based, pop-up keyboards, "chiclet" (small key) physical keyboards, or more robust character recognition. These traditional PDAs were significant for they defined a set of user applications—calendar, address book, etc.—which are now expected as a base set of PDA software referred to as *personal information management software* (PIMS).

### 26.2.2   Personal Communicators

Personal communicators are PDAs specifically designed to operate on wireless networks. Although it should be obvious, it is still worthwhile to remember that a PDA is meant to be portable, so a PDA that is a communicator must connect to some manner of wireless network. Some PDAs are billed as personal communicators even though their only interconnection mechanism is a wireline modem. The user must search for a connection to enable his or her communicator. This is all right for some usage patterns but is not the ideal of a communicator. Note that most personal communicators also contain a wireline modem function, but the wireline modem's primary purpose is to act as a fall-back communications mode in areas where the primary network does not have coverage. Therefore, it needs to be an inexpensive fall-back. Devices fitting this description include the Envoy and Marco, introduced by Motorola, which operate on the ARDIS network, and the Simon cellular phone PDA, introduced by BellSouth.

A short leap of faith will remind us all that personal communicators are also evolutionary. A prime example is that of a two-way pager, now fanning out in the market. With the addition of PIMS functions, such a pager is instantly able to be classified as a personal messenger PDA.

### 26.2.3   Mobile Companions

Mobile companions are described as notebook computers providing "backward compatibility with corporate desktop PC standards."[1] There are at least two examples of this kind of system. One is obvious: a desktop-compatible notebook or subnotebook PC that has a wireless [and wireline backup] network interface either intrinsic to the machine or, most commonly, as a PCMCIA device. The second known example is embodied in the Zenith Data Systems' Cruisepad. This device is actually a duplication of the screen interface on a desktop system. The Cruisepad concept actually lets users unplug the LCD screen on their desktop system and take it with them—usually in a campus (i.e., local) environment. Interface to the screen is via pen—either character, handwriting, virtual keyboard, or other means. Network traffic consists of updates to the user display transmitted to the Cruisepad and command requests from the virtual keyboard to the desktop device. The full services of the desktop system are available to the Cruisepad user without carrying a hard drive, extensive CPU memory, large battery supplies, etc. The components in the Cruisepad consist primarily of the LCD display, a power supply subsystem for portability, a (wireless) communications interface, and a pen input. Compatibility with desktop software is virtually ensured. A graphic of the Cruisepad application is shown in Fig. 26.1.

The rapid proliferation of the World Wide Web and the tools associated with it (e.g., HTML, Hyper Text Markup Language) have greatly expanded the attractiveness of mobile compan-

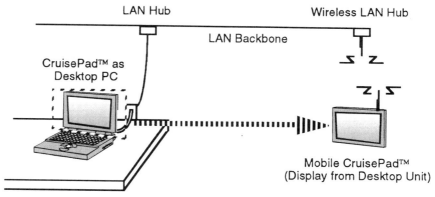

**FIGURE 26.1** Cruisepad™ "wiring" example.

ions. Not only can the user access anything back on his or her desktop, but if the desktop is connected to the World Wide Web through a local-area network, the user can access virtually any public information on the Web. In newer Web applications, Web pages can contain "live code" written in, for example, Sun Computer's Java™, that is then downloaded and interpreted on the local machine. This could let a Java page such as a spreadsheet version of a tax form calculate results from data the user enters locally rather than through further communication of data to and results from the Web server. This distributes the required processing power in the network and keeps Web applications machine-independent.

### 26.2.4 Other PDAs

In order to offer examples of other types of PDAs, it is suggested that digital cameras and cellular phones are also PDAs. Imagine the following scenarios:

• A digital camera is used by an insurance adjuster or a property analyst. The LCD screen on the back is used to either display the current still photo or to act as an interface to PIMS and connectivity software showing the analyst the next appointment, displaying messages from the home office, and allowing electronic uploads along with enhancements such as voice annotation to the home office to accompany reports. Would this be a PDA or a digital camera?

• A cellular phone has PIMS and connectivity added to it along with a graphical user interface. The graphical interface provides not only the ability to access the PDA functions but also acts as an enhanced interface to access information from the cellular provider about additional services available to the cellular phone user. Of course, to keep the phone useful while driving, the added features are accessed through voice input with text-to-voice feedback verifying requests. Again, as in the camera example, this represents a small increase in the cost of the phone but a large increase in the functionality of the portable electronic device that the user already must carry.

## 26.3  COMPONENTS OF A PDA

The block diagram of a typical PDA might look as shown in Fig. 26.2. The "heavy" bus interconnection is for power distribution, whereas the "light" bus interconnection is for signals.

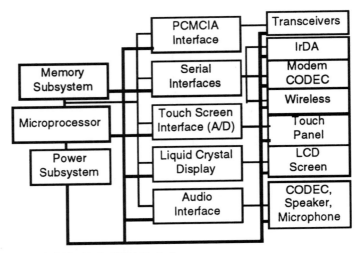

**FIGURE 26.2** Typical PDA block diagram.

### 26.3.1 LCD User Interface

The central point of the system is the LCD display, which is typically a bit-mapped display with gray scale and is either quarter- or eighth-VGA sized. This screen is then covered with a touch-sensitive panel that serves as a user input interface. The touch panel is sampled with analog voltages that are converted (for example, by 10-bit A/D converters) to determine the position of the input device. The input device is usually a plastic pen or nib that provides the pressure touch point but avoids damage to the touch panel. Continuing improvements in touch panels are occurring to improve the transmissivity of the panel, which reduces the light from the LCD. Because of power saving considerations, PDAs typically do not have backlighting; it is therefore important to reflect as much ambient light as possible.

### 26.3.2 Processing Subsystem

Although first-generation PDAs often had *complex instruction set* (CISC) microprocessors, most new PDAs use *reduced instruction set* (RISC) microprocessors specifically designed for low-power, high-performance embedded applications. Often these processors are highly integrated and operate at 3.3 V and lower, providing hundreds of processing "MIPS (millions of instructions per second) per watt." Most such microprocessors are true 32-bit devices, and the memory systems they support are much larger than one would expect from a lightweight RTOS with 2–8 Mbyte of ROM and 256-2 Mbyte of RAM. Often the RAM is static RAM for low power consumption, but it is not unusual to see newer types of DRAM, such as *extended data out* (EDO) DRAM with self-timed refresh and burst modes, used in systems for purposes of lowering cost. Typically, these DRAM structures are used for the most recently executed code, whereas the main programs continue to reside in slower ROM and FLASH memory arrays. Flash memory is often used instead of ROM in order to be able to update systems with new revisions of software as they appear.

Flash is also used for user system data storage to avoid catastrophic loss of information if the battery should fail or be improperly replaced. Flash memory for these purposes is added through PCMCIA ports on the device. The PCMCIA ports are also used for other types of expansion—typically, the addition of communications interfaces such as wireline and wireless modems, pagers, etc. These advanced interfaces require PCMCIA version 2 or higher, whereas memory cards require only PCMCIA version 1 services.

### 26.3.3 Communications Interfaces

Most PDAs also offer at least one serial channel (typically a UART interface) and another serial channel that uses infrared transceivers for short range, line-of-sight communications. In PDAs, infrared has typically been implemented using several incompatible data rates, modulations, frame formats, and message contents. This has changed with the efforts of the Infrared Developers Association, which has produced an IrDA specification that will allow interoperable communications between different manufacturers' systems. IR communication has been used in low-cost organizers as an interchange mechanism to upload information from a desktop computer and as a messaging system between individual units. Schoolchildren have found this second application to be exciting to send notes in class. The real corporate application seems to be evolving into wireless interfacing to utility devices, such as printers. There is also a lot of action in wireless local-area network (LAN) standards, which may use infrared for campus networking of users and utilities, as shown in Fig. 26.3. Note that the wireless LAN units communicate with wireless base stations, which can intercommunicate with each other and with other LAN and wide-area networking (WAN) systems through series of gateways and routers. In some cases the wireless interconnection is very local—such as from a PDA to a printer or workstation through IrDA communications. In this case the communication is not normally routed into these other networks.

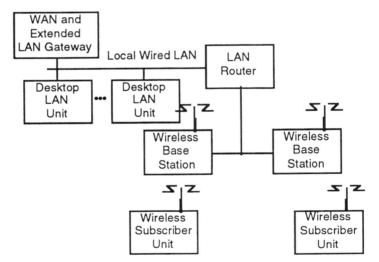

**FIGURE 26.3**  Infrared or wireless LAN.

It is also becoming more common for PDAs to include communications channels that interface to wireline networks (e.g., ISDN or V.32bis modems) or to wireless networks. As noted before, sometimes this is accomplished using PCMCIA interface cards, but now on the market are messaging PDAs in which interconnection into wireless and wireline networks is integrated into the unit because communications is fundamental to the market need being served.

### 26.3.4  Power Subsystem

The final significant block of a PDA is that of the power supply. In order to accomplish a low power objective, PDAs require the usage of low-voltage components whenever possible. Many of these components are not yet available in low-voltage versions, and some of them are analog transceivers or LCD panels requiring various high-level voltage supplies. The challenge of the PDA power supply design is to provide the most efficient conversion of power, usually provided by main and backup batteries, in the most efficient fashion to the various elements of the design. The power supply must also incorporate a number of elements to allow system software in the PDA to turn off portions of the system when not in use and to sense the status of the main and standby power systems in the system. Software is as much a component of the power supply performance of a PDA as the hardware components. PDAs may be turned on by a number of measures including an "on/off" switch and other interrupts, but PDAs always "turn off" through software that selects the proper form of "off" for the situation. A block diagram of a PDA power supply design, indicating control points for software power management, is shown in Fig. 26.4.

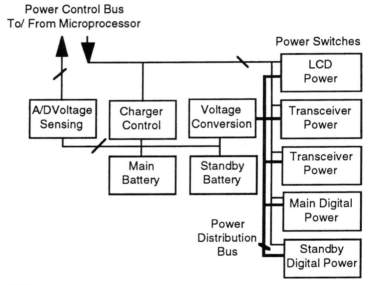

**FIGURE 26.4**    PDA power supply and software control points.

### 26.3.5  Form-Factor Packaging

Although it is not a "block" as such, it is salient to note packaging as a critical component for PDAs. Because of their inherently small form factor by definition, PDAs require compact user interfaces such as LCDs and touch panels. They also require low voltage and compactly packaged components. Packaging is usually of three types, with a fourth type evolving as form factor continues to shrink. The first type of packaging is *surface mount device* (SMD) packaging for capacitors, resistors, connectors, and other passive components. The second type of package is (thin) *quad flat pack* (QFP) packaging for electronic components. When pin counts become very high, the third type of packaging is usually *ball grid array* (BGA). The fourth type of packaging will be an evolution and is already used in some devices and as a submethod for some BGA assembly; this is *direct chip attach* (DCA) using "bumped" die, flipped and mounted onto circuit boards. All of these are important enabling technologies that allow form factors to exceed (be smaller than) the user interface technologies require.

## 26.4    SMALL-FORM-FACTOR (PDA) INPUT

### 26.4.1    Pen Input

The small form factor of a PDA precludes the full-sized keyboard, so there has been a lot of effort on the application of penlike plastic stylus devices and graphical input screens overlying liquid crystal displays (LCDs). Apple glamorized the pen in a cursive handwriting mode, but handwriting input is generally of two types: character or cursive. Character input does not require the high level of processing performance that cursive handwriting does, and it can often appear more accurate even if it is not because the occasional character error can still yield recognizable words. In general, however, character recognition is easier than cursive handwriting because it is usually restricted to specific input fields, so the problem of parsing between characters in a word—the main detractor of accuracy in cursive handwriting—is removed. Another problem with cursive word recognition is that a system can only recognize words that are in its vocabulary, so users often have to train a system over time with the words they use to have any hope of a high hit rate for input. Finally, character recognition is a natural choice for Asian languages with large, character-based language constructs.

A particularly effective method of character input for alphabet-based languages is exemplified by a product named Graffiti. Graffiti requires that the user use a set of single, specific strokes to form a letter. Thus a character begins when the pen is put onto the pressure pad and finished when the pressure is removed. This results in even less ambiguity than in other types of character input boxes; whereas other types of character recognition often require more than one input box. Users must be trained on the character set, but most of the characters are very like the normal character, so the initial "training" often only takes a moment or two and a high confidence level exists after a short period of usage. Part of the character set of Graffiti for the normal alphabet is shown in Fig. 26.5. Punctuation, capitalization, and special characters are formed by combinations of a stroke indicating a command followed by the character. In order to capitalize the next

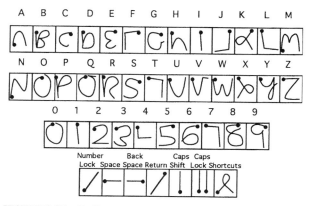

**FIGURE 26.5**    Graffiti characters set. (Copyright by PALM Computing 1994, all rights reserved, used by permission.)

letter, for example, the user makes a vertical upstroke followed by the letter to be capitalized. Two upstrokes puts the input into an all-capitalization mode, etc.

Another method of input is *ink capture.* Ink capture allows the user to, for example, draw maps and figures and have them stored as bit maps. Sometimes specialized software can be used to recognize shapes in order to smooth them and store them in compressed forms or to clean up a drawing for presentation. A rough square drawn by hand would be cleaned up and remembered as a square of a certain size at a certain point on the screen—much more compact to save than a bit map of a screen. Ink capture also could be particularly useful for capturing a signature for functions such as delivery receipt, financial transactions, etc., although this will also require additional certification to be legal in all situations.

### 26.4.2  Keyboard Input

Keyboards are primary input devices for some PDAs such as HP's LX series of devices. They are also common on many organizers such as those from Sharp, Psion, and Casio, but they are often too small to allow any semblance of normal typing input. Pop-up keyboards are also the usual fall-back to pen input PDAs. A pop-up keyboard appears on the screen and allows the pen user to touch the keys of the keyboard with the pen to provide input. This method is highly accurate but has a drawback in requiring much of the screen area for the keyboard image. This would be particularly hampering in a very small-form-factor PDA such as a pager-sized device.

### 26.4.3  Menu and Natural Language Input

Pull-down menus or natural language interfaces are also used to enhance input capability by allowing, for example, a message to quickly be formed from a canned set of words in context. Navigation through the input selection can be through methods such as a pen or by use of cursor keys. One can imagine an allowable set of responses to a message to indicate yes, no, where, when, why or why not, etc. for a two-way pager. Although these responses might be somewhat uncharacteristic of the user's actual word choice and phraseology, the ability to send a response from such a truly small device would override the stilted nature of such methods.

### 26.4.4  Voice Input

Speech recognition requires the same manner of classification that separates pen input into ink capture, character, and word recognition. Speech can be done as isolated characters, isolated words, connected words, or voice capture. Also, as in pen input, speech input is more demanding as the size of the vocabulary increases and if it is speaker-independent (versus requiring user-specific training). Because speech input is inherently hands-free—thereby implying that the user is busy doing something (e.g., driving a car, performing an autopsy)—its usage also tends to imply that visual feedback will be difficult and that audio feedback of results will be important. Therefore, speech input is becoming associated with text-to-speech feedback. Note a possible interaction between a speech input cellular PDA and a driver proceeding down the highway:

Command (User): "Cellular—Dial Carol."

Feedback (Speech-to-text from cellular phone): "Dial Carol Reinhard or David Carroll?"

Command: "Carol Reinhard."

Feedback: "Home, office, or cellular number?"

Command: "Cellular."

Feedback: "Dialing Carol Reinhard on cellular."

For consumer portable electronic devices, speech input is certainly at the technology point where it can work well for unconnected speed with limited vocabulary and speaker independence in a relatively noisy environment. Speech-to-text quality is much better than the game variety speech that many people expect from an electronic device. Voice capture is very good, with advanced algorithms storing significant quantities of speech in products such as digital tapeless answering devices using voice compression algorithms in the 4-kb/s range for compressed voice.

### 26.4.5  Visual Input

Because it is still true that "a picture is worth a thousand words," visual input could be extremely important to communications. Visual input can be either still or motion, and motion can be full motion or slow scan. An earlier scenario for still visual input described a digital camera with communications and PDA features. Pictures could be stored and transmitted with lossless compression methods such as JPEG rather than storing bit maps. Other still visual input could come from markups of transmitted facsimile, scanned-in photographs, or computer-generated graphics. With motion video, ultimately, the "Dick Tracy" wristwatch communicator may actually happen, though it is likely to be belt-mounted initially to support a bigger screen. Near term, video for small devices will likely be some manner of slow scan. For PDA applications the probable forms of video are the H.32P protocol stack, viable now for analog phone line (V.34) rates and higher. New developments in MPEG-4 should further enhance video quality and offer fall-backs to higher-quality video MPEG-1 and MPEG-2 on higher-bandwidth channels.

Before leaving the topic of input and output to PDAs, it is instructive to note that many of the protocols used are asymmetrical, meaning that it takes a different amount of effort to encode (input) than it does to decode (output). Usually, it is more difficult to encode than to decode, which is good for quality images in products such as cable TV distribution, where MPEG encoding is very intensive. MPEG decoding, on the other hand, is less intensive and can be done on a RISC processor in some cases, even for a PDA. There are many estimates for the performance required for such protocols. Table 26.1 gives examples of order-of-magnitude million of operations per second (MOPS) computational and procedural processing requirements for various methods of input and output. The numbers have been checked with a number of expert sources, but are highly dependent upon the processing structures used—integer processor, DSP, task-specific hardware, etc.—as well as the algorithms that are applied to accomplish the procedures.

**TABLE 26.1**  Procedural Computational Requirements

| Input/output algorithm | Encoding MOPS | Decoding MOPS |
| --- | --- | --- |
| MPEG-1/Audio | 25.6 | 21 |
| MPEG-2/Audio | 65 | 48 |
| G.728 audio (video conferencing) | 27 | 17 |
| H.263 video—12.5 frames/second | 24 | In encode figure |
| Character recognition (1 second/character) | 2 | NA |
| Cursive handwriting | 15 | NA |
| Text-to-speech | NA | 2–5 |
| V.SELP, TDMA, 4–8 kb/s audio | 14–18 | In encode figure |
| V.SELP, GSM, half-rate audio | 22 | 3.5 |

## 26.5  NETWORKING FOR PDAS

Networking for PDAs can be broken into a number of categories: wireless or wireline, local- or wide-area, voice channel or data channel, packet or switched data, etc. Within these delineations, there are further breakdowns: wireless can be through radio or infrared and infrared can be at a number of data rates, whereas wireless can be over data channels or primarily voice channels. Rather than cover all of the different possible combinations, networking will be discussed in terms of commonly available interfaces, wide-area wireless networking, and local-area networking. This discussive is not meant to be exhaustive, but it will cover the main features of channel capability and infrastructure that may make particular kinds of networking more or less predominant for PDAs.

### 26.5.1  Commonly Available Interfaces

Nearly all PDAs have available infrared and wireline modem interfaces. Standard infrared interfaces are meant for short-range messaging commonly migrating to the IrDA standard. Although infrared has many advantages, IrDA interfaces are tuned for very-short-range messaging to devices such as printers and desktop computer interfaces. In these applications it offers the benefit of not requiring intermachine cabling and, now that the IrDA standard is gaining multiple platform support, it is becoming reasonable to expect infrared interfaces to be compatible at the physical signaling and protocol levels.

Wireline modem interfaces are extremely common in PDAs because they represent a fallback networking interface available from nearly all environments and with well-understood communications protocols. In areas where wireless coverage is poor, telephones are still common. Wireline modems also provide good data rates and reliable service. This type of messaging is already a success for PDAs in both modem and facsimile formats.

UART and HDLC interfaces are also common, particularly to interface to desktop computers for communications or productivity software updates, and to printers for hard copy. These interfaces are the primary alternatives to infrared interfaces due to their nearly universal availability on a number of devices.

### 26.5.2  Wide-Area Wireless Networking Interfaces—Paging

By far the most common wireless networks in terms of infrastructure and gateways into other networks are cellular and paging networks. Paging networks are primarily one-way, low-bandwidth data networks. They have proven sufficient to provide wide coverage and to support large populations of users, with the primary paging protocol being POCSAG. POCSAG is being replaced by higher-speed protocols to support more users and more intensive alphanumeric messaging. Two higher bandwidth protocols are being promoted—FLEX and ERMES. ERMES appears primarily in Europe, where paging has a relatively low penetration, and FLEX appears throughout the Americas and Asia. Chip sets for these protocols are becoming available, and many PDA vendors are considering the addition of paging interfaces to their system definitions in order to offer users wide-coverage (albeit one-way) communications.

The capability of paging protocols to support intelligent messaging services is expanding rapidly with the introduction of two-way paging protocols, such as REFLEX, which can now acknowledge receipt of messages and provide additional response information. Because of the requirement to add a transmitter function and the additional battery capability required for transmission, two-way pagers are slightly larger than one-way pagers but are still small enough to be easily carried on a belt or in a purse or briefcase. These pagers are being enhanced with graphical screens and touch panels to support the functionality of two-way message creation and display, and it is expected that enhanced communications facilities will be added to two-way paging. Adding software in the form of PIMS capability then can turn a two-way pager into a messaging PDA, or the addition of a two-way paging interface can equivalently turn a PDA into a messaging PDA. The functionality of such devices, along with filtering software in host systems, can provide very satisfactory communications services for short messages.

For those who have experienced wide-area networking through low-bandwidth communications, such as wireline modems and paging services, the comment about filtering deserves some attention. Long messages, messages with attachments, and electronic "junk mail" in the form of broadcast messages able to be read at other times are usually not appropriate for consumption of bandwidth on such communications channels. Such messages should be filtered out either in the paging infrastructure or in a host messaging system such as the user's desktop computer. Fortunately, two-way paging is also being enhanced by higher-speed protocols and additional narrow-band PCS communications channels along with frequency reuse through smaller and smaller communications cell sizes. However, wireless voice and data communications will likely continue to be in a race to provide enough wireless bandwidth for users due to (1) the rapidly expanding number of wireless users, (2) increasing demand for data services with ever-shorter response times, and (3) the eventual addition of multimedia services over wireless. As an indication of the requirement, many independent studies indicate that high-quality multimedia messaging services will require as much as 384 kb/s of channel capacity. Present paging channel capacity is in the low k's of bits per second.

### 26.5.3   Wide-Area Wireless Networking Interfaces—Cellular Voice and Data Channels

Because cellular networks became well established at the same time that notebook computers were becoming widely used in the business market, a natural development achieved some early limited success: the usage of cellular analog voice channels to connect portable computers through the telephone network from cellular to wireline. In order to make this connection viable, the same manner of facsimile and wireline modem protocols that are required for wireline voice channels were applied to cellular voice channels with a new provision—the ability to tolerate the switching handoff that occurs in high-mobility cellular networks. This protocol addition, referred to as *ETC,* was relatively slow to catch on due to its high initial cost and the high cost of cellular connectivity. By the mid-1990s, this had become a popular and reliable method of mobile connection.

With the advent of digital cellular protocols such as Group Special Mobile (GSM) and Japan's Personal Handy Phone (PHP), cellular protocols have provided bandwidth specifically for paging types of functions. In GSM, for example, this is referred to as the short-messaging service (SMS). Methods for using SMS functions in Europe are shown in Fig. 26.6. In all cases, messages are held by the short-messaging center (SMC) until connection with the target phone is made, at which time the availability of the message is provided and the message itself is delivered. The target phone can be a GSM phone or a portable device connected to a GSM phone for cellular devices, as well as computers and other devices connected to the SMC through the

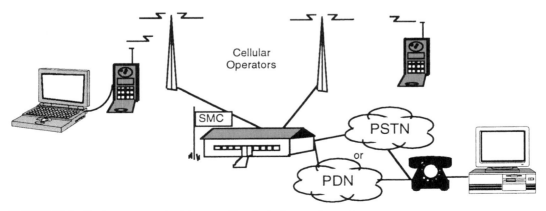

**FIGURE 26.6**   Short-message service interconnection via short-messaging center.

Public Switched Telephone Network (PSTN) or dedicated networking lines through a private or Public Data Network (PDN).

Standards bodies for digital cellular protocols are continuing to develop protocols for enhanced digital data services over their cellular networks. One such method that quickly evolved in the United States is CDPD (cellular digital packet data). CDPD was designed to use the unused bandwidth in the analog cellular network with simple enhancements to the existing analog cellular infrastructure.

### 26.5.4 Wide-Area Wireless Networking Interfaces— Digital Packet Data Networks

Digital packet data networks are the networks such as ARDIS and Mobitex. These networks were originally put together for specialized functions such as the dispatch of service personnel and the tracking and dispatch of delivery vehicles and transcontinental trucks. Providing relatively low bandwidth (9.6 kb/s to 19.2 kb/s), they have very good coverage in their areas of service and have proven to be highly effective in several vertical markets. However, they have not developed a wide following for enhanced services. The volume of users is several orders of magnitude lower than paging and cellular networks, and it may be expected that these specialized networking services will migrate to higher-volume, lower-cost networks such as cellular and paging as they begin to provide packet data services.

### 26.5.5 Local-Area Wireless Networking Interfaces—Spread Spectrum

Spread spectrum technology is usually considered in conjunction with the IEEE 802.11 wireless LAN standard. In wireless LANs the technology can deliver relatively high bandwidth— megabits per second if required—and provide a high degree of frequency reuse through low-power microcells. Spread spectrum is also relatively secure (even if not encrypted) due to the method of allocation of the channel between the user and the network. Combined with the short-range transmission, it has excellent characteristics for campus-wide networks where there is high voice availability and a desire to additionally provide data availability on demand for "locally mobile" users—those attending a meeting, for example. The primary hindrance to IEEE 802.11 is the wide range of options in the yet-evolving standard, which may limit interoperability and, thereby, market penetration.

An evolving alternative to spread spectrum LANs is microcellular systems, or PCS, with installations of low-power base stations that allow cellular phones to operate as cordless phones within range of the low-power personal base station and then as cellular phones when out of range. This type of service is already available with modified program cellular phones and mini-base stations. For PDAs the important point is that wireless data capability is becoming ubiquitous for both campus and wide-area access, so that portable devices can also become wirelessly enabled to a large market of the locally mobile workforce. This ties into information from a BIS study, as shown in Fig. 26.7.

Although this BIS study uses data only for the U.S. workforce and concentrates on the mobile professional, the major nonintuitive result is that most mobile professional workers are actually "local" workers—mobile in the campus or metropolitan area. This large set of professionals is ripe for PDA types of wireless applications in order to increase their productivity in their management and control tasks.

It is also important to note that many of the nonprofessional workforce are also mobile in the same mix and are ripe for PDA types of applications that could help them be more productive with the equipment they operate or services they provide. These applications can be anything from the wireless order-taking in restaurants at Disney World to service dispatching and billing for appliance repair trucks.

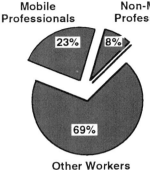

**Of the 23%:**
1. 55% are in the building, away from their desk.
2. 14% are out of the building, on the campus / complex.
3. 18% are out of the campus complex but are in the metropolitan area.
4. 12% are out of the metropolitan area but are in the country.
5. 1% are out of the country.

**Mobile Professionals** 23%    **Non-Mobile Professionals** 8%

69% **Other Workers**

**Total Workers = 118M**
Source: U.S. Bureau of Labor Statistics

**FIGURE 26.7**  BIS study on workforce mobility (BIS Strategic Decision, "Mobility, Computing and Communications: Strategies for Success," Graham Cooper, Chief Executive Officer, 1995).

### 26.5.6  Local-Area Wireless Networking Interfaces—Infrared

Infrared LANs are also being promoted using the same IEEE 802.11 standards. Compared to other forms of wireless network, infrared networks can be truly secure through line-of-sight transmission control, but they also suffer from restrictions in transmission distance due to dispersion of the light beam and the requirement that users and the network interface be properly aligned and in sight. As mentioned before, wireless LANs may become particularly short range for device-to-device interfacing versus becoming ubiquitous network interfaces. Table 26.2 summarizes the data rates and transmission distances promoted by various wireless networks.

**TABLE 26.2**  Wireless Networking Capacity

| Wireless network type | Data rates offered | Connection range |
| --- | --- | --- |
| Paging | 1s kb/s–100s kb/s | Kilometers |
| Packet data | 10s kb/s–100s kb/s | Kilometers |
| Microcellular | 10s kb/s–100s kb/s | 100s of meters |
| Spread spectrum LAN | 1–10 Mb/s | 100s of meters |
| Infrared LAN | 100s kb/s | Meters—line-of-sight |
| Cellular | 10s kb/s–100s kb/s | Kilometers |

## 26.6  SOFTWARE FOR PDAs

Often the most expensive single block in a PDA is the memory system. Memory systems can be expensive because they are fast; extensive; or require high technology such as low voltage, high speed, and low-current drain operation. By the nature of their applications, PDAs can end up fitting all of these "expensive" categories. The performance of the software determines how inexpensive a PDA can be. The techniques used by PDAs in order to keep memory costs low generally fall into the following categories:

• The usage of lightweight, real-time operating systems.
• The usage of lightweight applications such as spreadsheets, calendars, and word processors.

- Translation facilities for desktop connectivity running on desktop machines.

- Usage of translation technologies, intermediate code structures, and data compression technologies to cut storage requirements for nonactive applications.

- Usage of new DRAM technologies such as *extended data out* (EDO) DRAM.

Of these techniques, the usage of newer DRAM technologies is a hardware feature and has already been discussed. The others rely on a software understanding of the market requirements for PDAs.

### 26.6.1   Operating Systems

A good PDA operating system is real-time, multitasking, and small, yet fully featured. Real-time depends on the task at hand. If the major task is to act as an organizer and to work with a buffered keyboard input, real-time operation can live with context switches in the tens and hundreds of milliseconds. If the task is to operate as a communications PDA with organizer functions and the expectation is that user functions and communications functions can cooperate, then context switching needs to be in the low tens to a hundred microseconds. In order to accomplish these fast context switches, the operating system may require the processing element to provide functionality (1) to "lock" certain real-time applications into memories or caches so they can be counted on when task switches occur, and (2) to maintain memory translation in a memory management unit with more extensive translation tables. There may also be a requirement for the operating system to have extensive knowledge of the application being supported.

Multitasking operating systems have the capability to operate in real-time for at least some number of applications. They must guarantee appropriate service to some tasks and operate in an acceptable manner with other active tasks. Relating to desktop and PDA activities for another example, multitasking allows workstations running UNIX to keep alive several applications windows and to perform printing and networking activities at the same time while filling in on assigned networked tasks (e.g., running background batch jobs of routing and placing semiconductor design) with acceptable performance. "Acceptable" here would mean that the communications would never drop a message, the printer may have to wait for the next line of text at times, the windowed tasks would respond with no more than a slight delay (half a second or so) when called by the user, and the background task would only take bandwidth and resources as available. PDAs have to make the same manner of tradeoff—never missing a message, accepting pen input in real time and recognizing it in an acceptable time, and updating a spreadsheet application according to user input. To do all of this, real-time and near-real-time routines in the operating system must often be small with little overhead for the operating system to call and use them.

Small operating systems offer a benefit of keeping cost down by reducing storage memory requirements and by reducing code lengths for tasks, which lowers speed requirements in a device, cutting power drain as well. Operating systems can be kept small by careful attention to the minimum tasks to be accomplished, usually implying an operating system somewhat targeted for the task desired. Operating system effects can also be kept small by being configurable at compile time to only link in the required driver routines and by keeping minimal copies of those routines in the system, for example, by binding a single copy of a driver to the OS, which is then available to all the tasks needing it.

Even so, PDA operating systems must be fully featured because they provide (1) full graphical user interfaces; (2) communications (which must link to other communications systems, thereby implying such connectivity features as TCP/IP communications protocols); (3) advanced user interface features as well as (optional) keyboards and pen-driven pop-up keyboards; (4) advanced power management software; and (5) a range of third-party, independent software vendor (ISV) applications software, etc. Nevertheless, these operating systems cannot demand tens of megabyte footprints in storage and 4 to 8 megabytes of operational RAM; they must be designed to have a 256 kilobyte–4 megabyte footprint in storage and a few 100 kilobyte in operational RAM requirements or they will cost too much in absolute delivered price and in required processor performance.

### 26.6.2  Lightweight Applications

For many of the same reasons that affect the OS, PDAs cannot use a typical desktop application. What is required is that they can take a database from a desktop application, modify it in their format, and then update the main application. This is relatively easy to accomplish for a wide percentage of the PC market through support of two to three each of word-processing packages, spreadsheet packages, and personal information management software (PIMS) packages. PIMSs provide calendars, notepads, currency exchange calculation, name files, search facilities, etc. In order to be able to support such a wide range of desktop connectivity, it is often necessary that several translation facilities be provided for a user's desktop machine. This desktop software would, for example, let the user of an "X" organizer upgrading to a "Y" messaging PDA download his or her PIMS databases to a PC, translate the "X" databases into "Y" format, and then download the databases into their new PDA.

### 26.6.3  Translation Facilities Running on Desktop Machines

This usage of translation facilities running on desktop machines just described then not only keeps the PDA code size down but also gives the PDA manufacturer a common base for obtaining and translating old user databases onto new purchasers' machines. One would hope that manufacturing organizations such as the PDA Forum will eventually be able to develop common database formats that might then give independent software vendors an ability to enhance the features they provide with their PDA applications. Figure 26.8 shows the usage of translation software on desktop applications to convert databases in one PDA's format to that format of another PDA.

### 26.6.4  Usage of Translation Technologies, Intermediate Code Structures, and Data Compression Technologies to Cut Storage Requirements for Nonactive Applications

Several methods are used by various PDA and operating system vendors to support further reductions in code storage requirements. One method is to provide a translation facility that

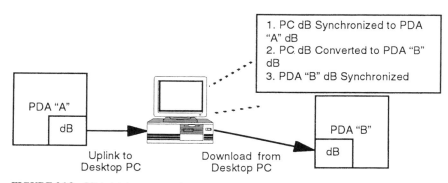

**FIGURE 26.8**  PDA database conversion.

takes applications or code segments and expands them into native processor code for operation. This obviously requires more processor cycles if the code is expanded every time it is run, but this is less onerous if the most recently translated code segments are kept in memory until the least recently used code segment is flushed when operational memory is full and new code segments must be translated. This approach removes the burden of code generation efficiency from the compiler and puts it onto the efficiency of the translator structure, which usually consists of some intermediate code generator (e.g., a C-code–to–intermediate-code generator) and an intermediate-code–to–native-code translator.

Figure 26.9 graphically displays the process of code generation for compiled and translated code. Intermediate code formats can also be significant in allowing processor-independent code distribution where a single, shrink-wrapped "compiled" code version of a product is then fully compiled or interpreted on target machines with multiple processors. A possible form of this is the Virtual Machine code format of Sun's Java™, presently proposed for applications such as Web pages, but also suggested as a binary distribution vehicle, as described here.

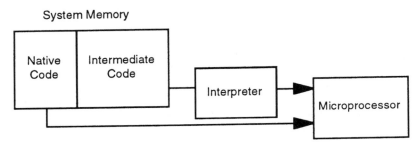

**FIGURE 26.9**   Run-time paths for compiled and translated code.

Data compression is yet another facility that can provide significant code compression for applications. A vendor might, for example, use a simple run-length encoding to compress code. Often, such code compression requires that significant lengths of code be decompressed at a time for application efficiency due to the difficulty in locating starting points for subsequently called segments of code within compressed memory.

## 26.7   APPLICATIONS SOFTWARE—VERTICAL AND HORIZONTAL

Although PDAs can offer good functionality through a base set of applications, their usefulness may actually come from a wide range of vertical and horizontal applications, similar to the way desktop PCs became more useful. A few such applications are described as follows:

A vertical market agricultural PDA that would allow a farmer or rancher to do the following throughout the day:

- Track production figures.
- Call up, wirelessly, futures market events.
- Access databases for farm pests, pesticide dosage, chemical warnings, etc.
- Track sales and cost information.
- Access weather information.
- Access, wirelessly, desktop history for farm animals and crop data.
- Update a desktop PC and databases with information for taxes and other local databases.

A vertical market transportation PDA that would allow a trucker to do the following throughout the day:

- Receive load and off-load updates.
- Call up, wirelessly, requests for service and rates.
- Access databases for road information, weather updates, fuel pricing, etc.
- Track sales and cost information.
- Communicate with families at home.
- Access, wirelessly, company planning calendars for vacation, training, and other events.
- Update a desktop PC and databases with information for taxes and other local databases.

A horizontal market personal transaction PDA that would allow any user to do the following throughout the day:

- Complete and track all financial transactions.
- Call up, wirelessly, requests for financial service and rates.
- Access databases for sale information, product location, services, etc. (perhaps through the World Wide Web).
- Obtain quotes for loan information.
- Communicate and coordinate with other family members.
- Update a desktop PC and databases with information for taxes and other local databases.

Many such applications are becoming possible and cost-effective.

## 26.8   SUMMARY

The PDA market has had a rocky beginning because the functionality and price of devices have not met user expectations. Now the market is developing from existing, valued applications, and it is reasonable to expect that this market will truly emerge while developing new paradigms for form factor, convenience, and user interface functionality. The "killer" application for PDAs can come from any of a number of directions, but it seems obvious that it will come even as the traditional PDA is arising from organizers and becoming ever more functional. PDA developers and marketers no longer expect the market to instantly appear as a totally new market. Along the way, there will be a lot of interesting engineering, which will very likely find its way back to desktop machines.

## GLOSSARY

**BGA (ball grid array)**   Semiconductor package type typically utilized by high-pin-count packages.

**CDPD (cellular digital packet data)**   Protocol used to transfer data over the analog cellular infrastructure.

**DSP (digital signal processor)**   Microprocessor type optimized for signal processing.

**ERMES (European Radio Messaging System)**   A European paging product.

**Flash memory**   A special type of memory that is capable of being written to but does not utilize much power and has a finite limit as to the number of times it can be written to. It is commonly used when changes to an operating system, for example, are anticipated.

**FLEX**   Next-generation paging protocol (following POCSAG) that allows greater density of paging per area.

**GSM (group special mobile)**   Cellular phone standard popular in Europe.

**H.32P** (renamed **H.324**)    International standard for low-bandwidth video conferencing using V.34 wireline modem services.

**HDLC (high-level data link control)**    Low-level protocol used for two-way serial communication.

**IR (infrared)**    Used to wirelessly transfer information or control short distances.

**ISDN (Integrated Services Digital Network)**    A way to transmit both voice and data over the same line. Installed by many of the phone companies throughout the United States but more prevalent in Europe.

**MIPS (millions of instructions per second)**    A common way to compare the performance of microprocessors.

**MOPS (millions of operations per second)**    Standard measure of performance for RISC microprocessors.

**MPEG (Moving Picture Experts Group)**    Standard for video.

**PDA (personal digital assistant)**    A small, portable electronic device. Apple's Newton was introduced as a PDA.

**PDN**    Public Data Network.

**PHP**    Japan's Personal Handy Phone. Cellular communication standard popular in Japan.

**PIMS (personal information management software)**    A standard feature of PDAs, used for calendar, address book, etc.

**POCSAG**    Post Office Communications Special Access Group.

**PSTN**    Public Switched Telephone Network.

**QFP (quad flat pack)**    Semiconductor or integrated circuit package type that does not require the manufacturer to drill through the board (see SMD).

**REFLEX**    Next-generation paging protocol that allows for two-way communication.

**RTOS**    Real-time operating system.

**SMC**    Short-messaging center.

**SMD (surface mount device)**    Semiconductor or integrated circuit package type that does not require the manufacturer to drill through the board.

**SMS**    Short-messaging service.

**UART (universal asynchronous receiver transmitter)**    Common device utilized to transfer serial information.

**V.32bis**    Modem standard that specifies interoperability requirements to communicate at 14.4 kb/s.

**V.34**    Modem standard that specifies interoperability requirements to communication at 28.8 kb/s.

## REFERENCE

1. *PDA's, Personal Communicators and Mobile Companions,* Datacomm Research Co., Wilmette, Ill., 1995.

## ABOUT THE AUTHORS

Rhonda Dirvin is the director of marketing development for personal computing technology for Motorola's Semiconductor Products Sector World Marketing Organization. Previously, she served as the director of marketing of VLSI communication products for Motorola's Data Communications Operations, part of the Microprocessor and Memory Technologies Group.

Dirvin has been instrumental in the development of Motorola's FDDI and data communications technology. She has held the office of secretary as a member of the IEEE 802.4 committee. She has also been part of the ANSI X3T9.5 committee. Dirvin has participated in numerous internal Motorola conferences and presentations. She has spoken at the Map/Top Users Group, Comdex, and various ANSI committee panels. She graduated from Cornell University with a degree in electrical engineering and holds an M.B.A. from the University of Texas, Austin.

Art Miller is the manager for the Portable Systems Operation—a product operation in the Motorola High Performance Embedded Systems Division, formed to develop and support an entire line of integrated microprocessor, tools, and enabling software designed specifically for the PDA and portable electronic market.

He currently oversees the development, market introduction, and support of VLSI for portable electronics devices such as PDAs, advanced cellular phones, pagers, and organizers. Previously, he served as the operations manager in the Data Communication Operation, a part of the Microprocessor and Memory Technologies Group.

Miller has been involved with IEEE 802 for local-area networking standards and the Corporation for Open Systems for networking testing. He graduated from the U.S. Air Force Academy with a B.S. in mathematics and received the M.S.E.E. at MIT. He also holds an M.B.A. from St. Edwards University.

# CHAPTER 27
# DIGITAL/VIDEO TELEPHONY

## SECTION 1
## WIRELESS AND WIRED TELEPHONY

**Bruce H. Benjamin, Eric Boll, David Kolkman, Mark Reinhard**
*Motorola Semiconductor Products*

## 27.1  INTRODUCTION

In retrospect it seems quite natural that digital telephony technology would migrate to the wireless environment. After all, the concept of noise-free communications over a great distance—the fundamental underpinning of digital technology—complements the needs of people who want high-quality communications in a tetherless world. However, wireless communications poses new challenges, resulting in some special requirements for the operation of digital technology, in which the carrier may degrade by 50 dB or more during conversation—something no wired telephony system ever intended to deal with. In addition, wireless system designers must confront the scarcity of radio spectrum when determining the "best" system design, a constraint less imposed by the broad-band nature of wired telephony. This section will study the various systems that have been developed to address the special needs of the untethered communications environment broadly called *wireless*.

## 27.2  WIRELESS TELEPHONY SYSTEMS

The elements necessary to classify a product as both digital and wireless are straightforward (Fig. 27.1). A basic digital wireless communicator contains circuitry to transform the analog audio signal (typically in the 300 Hz–3 kHz range) into a digital representation that is communicated via a radio modem to a similar unit for restoration of the signal to its original analog format. There will be further discussion on this subject later in the chapter. For now, this discussion will center on wireless systems design.

**Digitization Section**

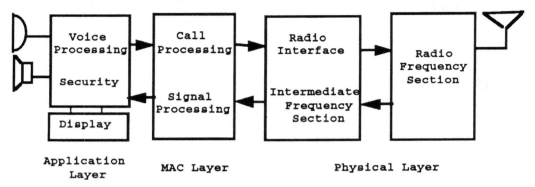

**FIGURE 27.1** Digital wireless communicator block diagram.

When considering digital wireless communications, it is often helpful to divide the various standards into two broad categories: *wide area* (or *high mobility*) and *local area* (or *low mobility*). This distinction is useful because the designers of each type of system had specific applications in mind during development. By understanding the differences, one can come to a deeper appreciation of the nature of wireless telephony.

### 27.2.1 Comparison of Wide-Area and Local-Area Systems

To measure telephone capacity, the concept of the erlang was developed. An *erlang* is a fully loaded telephone circuit. Two calls, each lasting 30 minutes, would constitute one erlang, as would four 15-minute phone calls, etc. To help explain the differences between the various digital wireless telephony systems, however, the concept of density must also be included. Therefore, the various types of systems will be described by their density capabilities, as measured in erlangs/km$^2$.

Wide-area systems have density capabilities of tens to hundreds of erlangs/km$^2$. Local-area systems have densities ranging from several to thousands of erlangs/km$^2$. Intuitively, one can conclude that matching the wireless system operator's needs to the actual system will make the most sense. For example, deploying a wide-area system capable of 50 erlangs/km$^2$ at speeds of 120 km/h into a high-rise building with requirements of 200 erlangs/km$^2$ (where the occupants may move around at speeds of no more than 5 km/h) will leave the system operator with many unconnected calls (from a mismatch between the number of available channels and people seeking a dial tone) and many irate customers. Likewise, deploying a low-mobility system intended to support a neighborhood with 500 erlangs/km$^2$ at "walking around" speeds of 5 km/h into a busy highway corridor, where vehicles often travel at 100 km/h, would likely result in an over-investment by the system operator (from deployment of an excess number of base stations), as well as a large number of dropped calls (from trying to hand off calls from one base to the next at unanticipated speeds). When contemplating the tradeoffs that system designers must make, it is important to consider the target application and therefore the variables that the designer is able to optimize.

To summarize, a low-mobility wireless system is targeted at traffic that moves at pedestrian speeds and is characterized by high(er) density requirements where there are likely to be a large number of users desiring a dial tone at the same time. A wide-area or high-mobility system is one that is targeted at a small(er) number of customers seeking a dial tone at vehicular speeds, sometimes as fast as 200 km/h.

The factors uppermost in the minds of the architects of the wireless standards about to be discussed are as follows:

- *System density.* The number of users likely to seek a dial tone at busy hours in a given locale.
- *Voice quality.* The perceived quality of the communications channel.
- *Security.* The sensitivity of the system operator and users to interdiction of the communications.
- *Spectrum.* The available airspace to support the intended traffic.

***Wide-Area Digital Wireless Telephony.***    The concept of digital wireless telephony was developed in the 1980s and implemented for the first time in a practical, large-scale system in Europe in 1992. The system, known as *GSM* (Global System for Mobile Communications), was the first of many wide-area standards to be developed using digital radio technology. The reason that some digital radio standards have the classification *wide-area* is that the system design is meant to allow a great number of vehicular users to move throughout the service area, seamlessly being passed around the system at speeds up to 200 km/h.

The GSM system architecture embodies several important design features. Some digital wireless features include enhanced voice quality through digital coding, time or code division for spectral efficiency, security for both user and system operator, and the promise for future digital feature enhancements.

As already mentioned, the development of GSM has been followed by the development of a number of other standards with characteristics unique to the country in which they were designed. By studying the GSM communications channel in Fig. 27.2, one can better understand how the system is architected for multiple users. Some of the attributes of this system are a guard band between time slots, interleaving of time slots, and synchronization on a "per slot" basis.

A standard has many components; in the best cases this results in a well-defined and interoperable system. GSM is an example of a well-defined system. To understand how involved the standards activities can become to develop a robust standard that continues to evolve, consider the different ETSI (European Telecommunications Standards Institute) working committees and where the focus of each working group resides (Fig. 27.3).

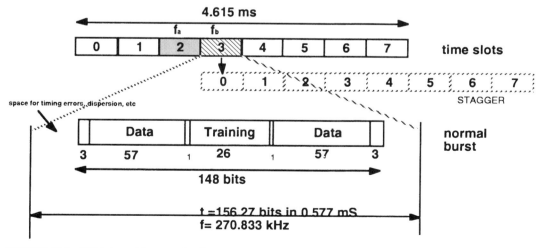

**FIGURE 27.2**    GSM channel frame protocol.

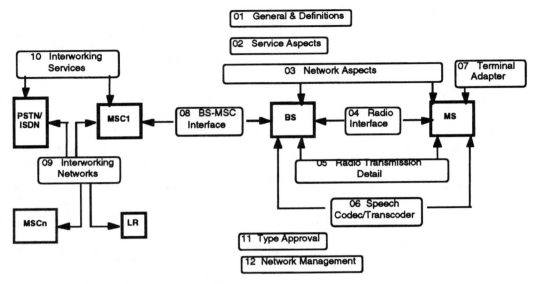

**FIGURE 27.3**    ESTI working committees for GSM.

*Other Systems.*    By way of introduction, IS-54B (which has evolved to IS-136) is a TDMA (*time division multiple-access*) standard similar to GSM. It was developed in the United States, as was IS-95. IS-95 is a form of CDMA (*code division multiple-access*). PDC is the name of the digital cellular standard for Japan and it shares many of the characteristics of the IS-54B system.

A TDMA system (Fig. 27.4) is one in which multiple users gain access to the same channel (same frequency) but at different times. As one might expect, timing on such a system is critical. What is shown in Fig. 27.4 is known as a 6:1 TDMA system, because six users may have access to the same channel in their respective time slot. The pictured channel represents the North

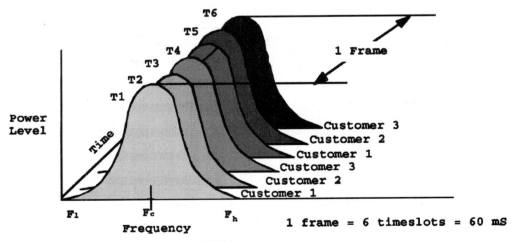

**FIGURE 27.4**    A graphical representation of TDMA.

American Digital Cellular standard IS-54B full-rate system, wherein each user is given two of the available six time slots, enabling the transmission of 8 kb (full-rate) VSELP code. The system was designed so that, at a later date, a half-rate vocoder (voice coder)—requiring roughly one-half as much time to transmit the same amount of vocoded audio information—could be implemented using the same system design.

In contrast to the TDMA approach, CDMA systems operate on the same frequency simultaneously (Fig. 27.5). The concept, first developed for military use to prevent enemy interception of signals, is to spread the intended information over a broad enough bandwidth, and at low energy levels, such that it would be difficult to detect, block, and/or intercept. This spreading technique is a benefit of CDMA in wireless telephony applications and is described as process gain, measured in dB. *Process gain* refers to the improvement in recovered signal performance when compared to a narrow-band signal. In application, each user on the channel is assigned a unique and orthogonal (mathematically unique) code. The digitized voice is mixed with the *spreading code* (i.e., the unique code) to create a new code stream. The new code rate is called the *chip rate* of the channel. For the IS-95 standard the chip rate is given by

$$\text{Data rate} \times \text{Spreading code} = \text{Chip rate}$$

where the data rate is 19.2 kb/s (combined voice/signaling info), the spreading code is 64 bits, and the chip rate is 1.2288 Mchip/s. Although different, each standard was meant to focus on these characteristics. Protocols of some of the best-known wide-area standards are given in Table 27.1.

*Voice Coding.*   Audio to be digitally transmitted is encoded using a number of techniques aimed at reducing the bandwidth required for transmission while minimizing the loss of fidelity

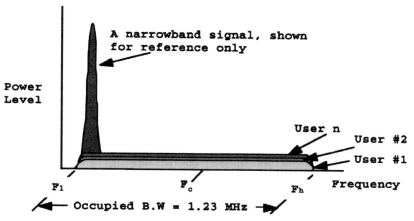

**FIGURE 27.5**   A graphical representation of CDMA.

**TABLE 27.1**   Protocols of Popular Wide-Area Standards

| Standard | Vocoder | Modulation | Access method | Bit rate |
|---|---|---|---|---|
| GSM | 13-kb RPE-LTP | GMSK | 8:1 TDMA/FDM | 270.833 kb/s |
| IS-54B | 8-kb VSELP | $\pi/4$ DQPSK | 3:1 TDMA/FDM | 48.6 kb/s |
| IS-95 | 8-kb QCELP | BPSK/OQPSK | CDMA | 1228.8 kb/s |
| PDC | 8-kb VSELP | $\pi/4$ DQPSK | 3:1 TDMA/FDM | 42 kb/s |

in the process. This is known as *voice coding,* or *vocoding.* The more popular techniques for vocoding are 32-kb ADPCM and Codebook Excited Linear Predictive (CELP). The 32-kb ADPCM is popular because it provides toll-quality audio and 2:1 compression over PCM-encoded audio. CELP coders, on the other hand, break the audio signal down into its components, sending information about pitch and volume to the receiving end for reconstruction.

CELP-type decoders, such as VSELP and RPE-LTP, have found commercial success in digital wireless telephony standards around the world. They provide reasonable fidelity while reducing voice bandwidth up to 16:1 compared to PCM-encoded voice.

The vocoders listed in Table 27.1 differ from each other in the way they digitize the voice and execute the code book search. They also differ somewhat as to how they deal with bit errors—the sort encountered when communicating over a fading RF channel. In addition to those listed, new vocoders are continually being introduced for consideration by the groups responsible for digital wireless standards. Vocoders such as ITU 8 (International Telephony Union, 8 kb) and EFR (Enhanced Full-Rate) are expected to influence the quality of future digital radios.

*Modulation.*    Each of the standards listed in Table 27.1 utilizes a form of modulation that is meant to restrict the occupied bandwidth of the digital carrier while maximizing the number of bits of information per symbol. By reducing occupied bandwidth, more carriers can be placed within a given frequency space, creating another form of spectral efficiency. Two types of modulation are represented: (1) Gaussian-filtered, minimum shift keying for GSM, and (2) binary phase-shift keying for the other listed standards. The model for GMSK modulation is shown in Fig. 27.6.

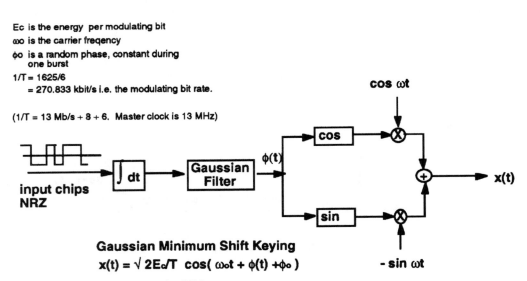

FIGURE 27.6    GMSK modulator for GSM.

*Access Method and Bit Rate.*    Some treatment has already been given to the concept of TDMA and CDMA, which are known as access methods. They describe the manner in which the user is to acquire a portion of an available channel. Taken together with the bit rate of the channel and the protocol, they form a complete description of how to achieve compliance with the standard. The formal name for these elements when taken together is the *air interface.*

***Local-Area Standards.*** Additional standards have been developed for local-area or low-mobility applications. They differ from wide-area standards in several important areas. The vocoder focuses on higher fidelity while still minimizing the need to transmit a large number of bits. Terminal and base transmit power is also lower. Table 27.2 gives local-area protocol features.

**TABLE 27.2**  Local-Area Protocol Features

| Standard | Vocoder | Modulation | Access method | Bit rate |
|----------|---------|------------|---------------|----------|
| CT-2 | 32-kb ADPCM | GFSK | TDD/FDM | 72 kb/s |
| DECT | 32-kb ADPCM | GMSK | 12:1 TDMA/FDM | 1152 kb/s |
| PHS/PACS | 32-kb ADPCM | $\pi/4$ DQPSK | 10:1 TDMA/FDM | 384 kb/s |

Local-area, digital wireless telephony products are meant in many cases to supplement or replace existing phone products, either PBX or CPE based. It is logical then that the system design of local-area systems contain higher-quality vocoders—near or at toll quality—and be low powered to meet the high expectations of long battery life in both active and standby modes.

Although local-area systems are still in their infancy, they offer some of the highest growth opportunities within the wireless marketplace. Taken together, digital wireless communicators—both wide and local area—will greatly extend the productivity and quality of the world's communicating public for decades to come.

## 27.3  WIRED TELEPHONY

Telephones, answering machines, fax machines, or anything that connects to the public network is classified as customer premise equipment (CPE). The latest evolution of CPE has been driven by enhanced calling services in the existing analog telephony system, as well as by the ongoing conversion to digital networks.

### 27.3.1  Analog Services

***Enhanced Calling Services.*** Enhanced calling services such as caller ID utilize in-band transmission for signaling and data—that is, using the existing loop facilities and tip-and-ring connections. Data are transmitted at 1200 baud using phase-coherent FSK modulation; the network employs DTMF tones for signaling (call progress, service requests, etc.). DTMF is a tone-dialing scheme that generates two non–harmonic-related frequencies simultaneously. Eight frequencies have been assigned to the four rows and four columns of a telephone keypad (the fourth column may not be present in all CPEs). Enhanced CPE must therefore include FSK demodulation; UART and data-processing capability; and DTMF decoding, in addition to the standard DTMF generation, speech, and ring/switchhook functions.

Figure 27.7 shows a typical enhanced analog telephone. The tip-and-ring interface splits ring/switchhook, speech, and DTMF/FSK signals as illustrated. This last function enables custom calling features by detecting alerting signals and demodulating the data stream. The microcontroller then processes the data according to predetermined timing and formatting protocols. For example, let us consider caller ID, a complex feature, in more detail.

*Caller ID.* The enhanced service known as *caller ID* delivers data in a standardized packet format for display at the CPE terminal. In its most basic implementation, caller ID provides the

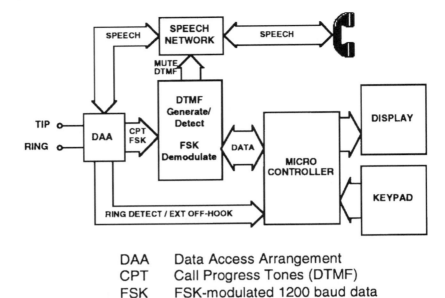

DAA   Data Access Arrangement
CPT   Call Progress Tones (DTMF)
FSK   FSK-modulated 1200 baud data

**FIGURE 27.7**   Block diagram of an enhanced telephone.

time stamp and calling number information normally carried by the public telephone network (single–data message format). Provisions have been made, however, for delivering additional data by appending additional packets, called *parameters,* to the message (multiple–data message format). Caller ID protocols, timing, and signal characteristics are governed by Bellcore document TR-NWT-000030, "Voiceband Data Transmission Interface."

Type I service, also known as *calling line identification delivery* (CLID) transmits data during the four-second interval between the first and second ring signals, as shown in Fig. 27.8. At 0.5 s into the interval, a 250-ms/600-Hz channel seizure tone signals the start of the enhanced calling feature to the CPE. This is followed by 150 ms of "marks." Following this preamble, the CPE must be ready to receive the message. This data transmission must terminate at least 0.5 s before the start of the second ring. A variant of this service is called *message waiting,* which provides on-hook transmission of data in the absence of power ring. Because a ring signal is not available, the CPE must monitor tip/ring continuously for FSK data.

Standard protocols have been defined for off-hook transmission of caller ID messages and type II caller ID (often referred to as *caller identification delivery on call waiting* [CIDCW] or *spontaneous call-waiting identification delivery* [SCWID]). This protocol enhances call-waiting

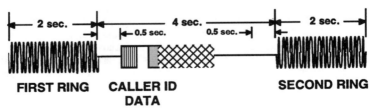

**FIGURE 27.8**   Type I caller ID message timing.

service by displaying information about the new caller. The setup and delivery process follows:

- Far-end (calling) terminal muted.
- Local end receives DTMF *CPE alerting signal* (CAS) tone, a nonstandard DTMF (2125/2750-Hz) signal.
- Local-end handset muted and keypad disabled.
- Local end sends DTMA acknowledge (ACK) tone, confirming it is type II–capable.
- Message transmitted.
- Far end unmuted.
- Local end unmutes handset and enables keypad.

The hand-shaking process confirms that the CPE processes type II caller ID. Other equipment will not respond to the nonstandard CAS tone. If ACK is not received within one second, the far end is unmuted and the call-waiting process continues. The usual call-waiting options are available, including taking the new call, ignoring it, playing a prerecorded message, and adding the party to the existing call.

Type III caller ID provides additional features that enhance the capability of transmitting larger amounts of data. Messages still adhere to the caller ID format, however, requiring the CPE to provide flow control through a series of DTMF progress tones. These functions include message confirmation, next-message request, and repeat transmission of corrupted message.

This protocol is known as the *Analog Display Services Interface* (ADSI). CPE terminals designed to access information through ADSI require significant display capability, and are often marketed as *screen phones*. Applications include white/yellow pages directories, travel assistance, news-on-demand, and community services.

Loop characteristics of Bellcore TR-NWT-000030 are as follows: CO (central office) transmit level $= -13.5 \pm 1$ dBm (ref. 900 $\Omega$) and insertion loss $\leq 14$ dB (residential). Industry experience has shown, however, that insertion loss actually measures 20 to 22 dB in the worst case. Therefore, minimum receiver input sensitivity should be $-36.5$ dBm (ref. 900 $\Omega$) or $-35$ dBm (ref. 600 $\Omega$).

The standard message format is shown in Fig. 27.9. The 1-byte *message type word* in the header, a standard feature of many enhanced calling services, indicates single– or multiple–data message format and identifies the caller ID feature (Table 27.3). The 1-byte *message length word* indicates the size of the entire message in bytes, which may consist of several parameters. In a multiple data message, each parameter carries 1-byte *type* and *length* header words (Table 27.4). A 2s complement check sum is transmitted as the final byte and applies to the entire message.

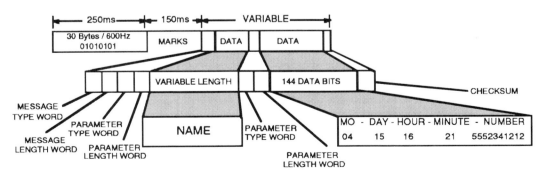

**FIGURE 27.9**   Caller ID message format.

**TABLE 27.3** Caller ID Message Types

| Message type | Binary (decimal) | Bellcore document |
|---|---|---|
| *Single-message data format* | | |
| Calling number delivery (CND) | 0000 0100 (4) | TR-NWT-000031 |
| Message waiting indicator | 0000 0110 (6) | TR-NWT-001401 |
| Message desk information | 0000 1011 (11) | Reserved |
| *Multiple-message data format* | | |
| Call setup | 1000 0000 (128) | |
|     CND (number) | | TR-NWT-000031 |
|     CNAM (name) | | TR-NWAT-001188 |
|     CIDCW (type II) | | TR-NWT-00575 |
| Message waiting notification | 1000 0010 (130) | TR-NWT-001401 |

**TABLE 27.4** Caller ID Parameter Types

| Parameter type | Binary (decimal) | Service |
|---|---|---|
| Time and date | 0000 0001 (1) | Calling number delivery |
| Calling line ID | 0000 0010 (2) | Calling number delivery |
| Dialable DN | 0000 0011 (3) | To be determined |
| Reason for absence of DN | 0000 0100 (4) | Calling number delivery |
| Reason for redirection | 0000 0101 (5) | To be determined |
| Call qualifier | 0000 0110 (6) | Calling number delivery |
| Name | 0000 0111 (7) | Calling name delivery |
| Reason for absence of name | 0000 1000 (8) | Calling name delivery |
| | 0000 1001 (9) | Reserved |
| | 0000 1010 (10) | Reserved |
| Visual indicator | 0000 1011 (11) | Message waiting indicator |
| CPE ID (ADSI) | 0000 1100 (12) | On-hook alerting for feature download, encryption |
| ADSI server display control parameters | 1000 0000 through 1001 0111 (128–151) | ADSI server display control messages (#151 reserved) |
| ADSI feature download parameters | 1000 0000 through 1000 0100 (128–132) | ADSI feature download messages |

Check-sum errors in type I and II caller ID equipment normally result in blank displays. Only in type III is there a mechanism to request retransmission of corrupted messages.

## 27.4 DIGITAL SERVICES

Bandwidth of POTS lines is limited by coils attached to the telephone line. These coils were installed by Bell so that frequencies above 4 kHz could not interfere with the telephone network (4 kHz was chosen as the maximum frequency of the human speaking voice). They can be regarded as 4-kHz lowpass filters.

Modem designers have had to work within the 0–4-kHz frequency range, which effectively limits analog modems to 28.8 kb/s (V.34). In the late 1970s technology advances incorporated the filtering function into central office line cards. The coils were no longer needed on the phone lines, enabling higher data transmission rates.

In an effort to optimize the benefits of increased bandwidth, international organizations developed, in the early 1980s, digital standards known as the Integrated Services Digital Network (ISDN). However, the concept only achieved popularity as the widespread need for higher-speed data communications developed. The basic rate ISDN standard provides for three digital channels over the existing copper pair, as shown in Table 27.5. Basic rate ISDN (BRI) is also known as $2B + D$, which describes its channel configuration.

**TABLE 27.5**   Basic Rate ISDN Channel Definition

| Type | Number | Characteristics |
|------|--------|-----------------|
| B = bearer | 2 | 64 kb/s each; may be aggregated into single virtual 128-kb/s channel; circuit-switched voice/data or packet-switched data |
| D = delta | 1 | 16 kb/s; out-of-band signaling channel; low-speed packet-switched data, max. 9.6 kb/s |

### 27.4.1   Primary Rate ISDN

CCITT standards provide for additional B channels to fill the bandwidth of higher-capacity lines, which are usually of interest to business customers. In North America, a 1.544-Mb/s T1 line can handle 23 B channels plus one 64-kb/s D channel. In Europe a 2.048-Mb/s E1 line can accommodate 30 B channels plus one 64-kb/s D channel. These combinations are referred to as *primary rate ISDN,* or *PRI.*

### 27.4.2   ISDN Reference Model

Network partitioning and interfaces have been standardized in a reference model, a simplified version of which appears in Fig. 27.10. Interfaces known as *reference points* are identified by uppercase letters. They are described in Table 27.6. Table 27.7 defines reference equipment types.

The *network termination* (NT) encompasses the two-wire U-interface–to–four-wire S/T-interface converter (NT1), as well as signaling, addressing, and other control functions (NT2). In the following discussion, *TE* refers to TE1 or to TE2 with terminal adapter. These are indistinguishable at the S/T interface.

### 27.4.3   S/T Interface

The S/T interface provides ISDN access via NT1, PBX, or key system equipment. In North America, where the service provider's responsibility ends at the U-interface, NT1 equipment must be supplied by the customer. This is not necessary in Europe, where the PTT owns the NT1 and provides the S/T as the standard user physical interface. Attributes of the S/T interface include the following:

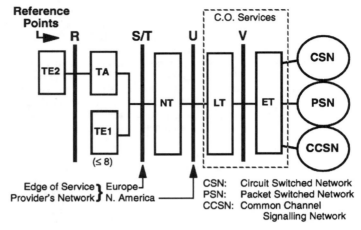

**FIGURE 27.10** Simplified ISDN reference model.

**TABLE 27.6** Simplified ISDN Reference Point Interface Designations

| Type | Description |
| --- | --- |
| V | Proprietary interface within central office |
| U | ANSI T1.601 and ETR 080 standards; two-wire interface up to 18,000 ft; uses 2B1Q line code at 80 kbaud; 160 kb/s (144 kb/s + 16 kb/s overhead) |
| S/T | CCITT (I.430) and ANSI (T1.605/1989) standards; 4-wire interface up to 1 km; point-to-point or point-to-multipoint operation; 192 kb/s (144 kb/s + 48 kb/s overhead) |
| R | Any non-ISDN interface (e.g., RS-232, V.35, etc.) |

**TABLE 27.7** ISDN Reference Equipment

| Type | Equipment | Description |
| --- | --- | --- |
| ET | Exchange termination | Central office hardware |
| LT | Line termination | C.O. switch or remote terminal line card |
| NT | Network termination | CPE connection to network |
| TE1 | Terminal equipment (type 1) | ISDN-compatible terminal |
| TA | Terminal adapter | Interfaces non-ISDN terminal to ISDN network |
| TE2 | Terminal equipment (type 2) | Non–ISDN-compatible terminal |

- Up to eight S/T terminals per subscriber loop.
- Defined by CCITT I.430 and ANSI T1.605 standards.
- Four-wire full duplex data transmission.
- 192-kb/s line rate (48-bit/4-kHz frame containing user data plus other maintenance and framing bits).
- 144-kb/s user data rate ($2 \times 64$ kb/s + $1 \times 16$ kb/s = 2B + D).

- Includes maintenance subchannels.
- Uses pseudoternary line code, as shown in Fig. 27.11.

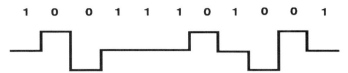

**FIGURE 27.11**  Pseudoternary coding on S/T interface.

ISDN specifications permit point-to-point and point-to-multipoint wiring configurations (subject to the limitation of eight TEs per S/T interface). The specification allows for 6-dB signal loss. This permits maximum NT-to-TE distances to be calculated, as well as TE-to-TE separation for point-to-multipoint bus configurations. These appear in Table 27.8 for 22-gauge wiring. The NY physical interface to CPE is via the RJ-45 connector (wider eight-wire version of the common four-wire RJ-11 jack).

*Information states* have been defined to signal current activity and to allow either TE or NT to activate the other for transmission. Table 27.9 summarizes the definitions. The use of the information states in activation sequences is shown in Fig. 27.12.

The D channel is used by each TE for network signaling. The procedure ensures that when two or more TEs attempt to access the D channel, one (and only one) of the TEs is successful in completing its transmission:

**TABLE 27.8**  S/T Wiring Configurations and Limits (22 AWG)

| Configuration | Max. NT-to-TE distance | Max. TE-to-TE distance |
|---|---|---|
| Point-to-point | 1000 m | N/A |
| Short passive bus (8 TEs max.) | 100–200 m | N/A |
| Extended passive bus (8 TEs max.) | 500 m | 25–50 m |

**TABLE 27.9**  S/T Information States (CCITT I.430/ANSI T1.605)

| State | Definition |
|---|---|
| | TE |
| INFO 0 | No signal transmitted (all 1s) |
| INFO 1 | Wake-up NT (repeated 00111111) |
| INFO 3 | Fully active (transmitting 2B+D) |
| | NT |
| INFO 0 | No signal transmitted (all 1s) |
| INFO 2 | Wake-up TE; respond to TEs INFO 1 |
| INFO 4 | Fully active (transmitting 2B+D) |

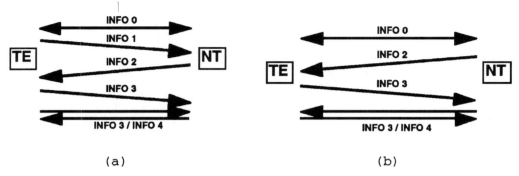

**FIGURE 27.12** S/T interface activation sequences: (*a*) TE initiates activation sequence by sending INFO 1 to NT; (*b*) NT initiates activation sequence by sending INFO 2 to TE.

- Each TE, when not using the D channel, transmits binary 1s (the monitoring state).
- The echo channel (always transmitted from the NT to the TEs) echoes the last D-channel bit received by the NT.
- To gain access to the D channel, a TE must count a programmed number of 1s on the E echo channel before transmitting.
- To avoid collisions, a TE monitors the E echo channel to see if its last transmitted D bit matches the next E echo bit. If these two are the same, the TE continues transmitting; otherwise, it stops transmission and returns to monitoring the E echo channel.

### 27.4.4 U-Interface

The ISDN U-interface is defined over the existing copper pair between exchange and customer. Its attributes include the following:

- Defined by ANSI T1.601 and ETSI ETR 080.
- Two-wire full duplex data transmission with echo canceling.
- Compatible with 99 percent of North American installed bases of nonloaded twisted pair loops, including 18,000-ft (5.5-km) loop length, bridge taps, 1300-ohm loop resistance, wire gauge changes, and 48-dB loss at 40 kHz.
- Transceivers must operate in presence of simultaneous impairments: cross-talk, line imbalances, induced power, jitter.
- 160-kb/s bit rate (including maintenance bits).
- 144-kb/s user data rate ($2 \times 64$ kb/s + $1 \times 16$ kb/s = 2B + D).
- Includes maintenance subchannels.
- 80 kbaud symbol rate using 2B1Q (2-binary, 1-quaternary) line code, as shown in Fig. 27.13.

### 27.4.5 ISDN Implementation

As indicated in Fig. 27.10 (ISDN reference model), service providers differ in their interface to CPE. The North American network ends at the U-interface, whereas European PTTs extend to the S/T interface. Therefore, TE can connect directly (or via terminal adapter) to the Euro-

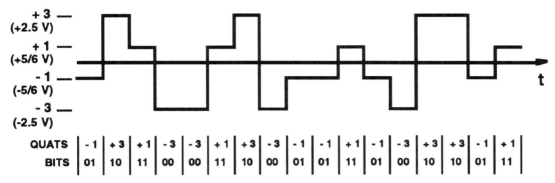

**FIGURE 27.13**    2B1Q coding on U-interface.

**TABLE 27.10**    ISDN CPE Requirements

North America

Customer access point is U-interface.
Phone company drops wire pair at residence.
Phone company does not provide power feed.
Customer must provide for local and emergency power.
Phones are U terminals with integral NT, or S/T terminals
    requiring separate NT1.
Two-wire–four-wire converter.
Residences may require rewiring.
Jurisdiction by state/province as well as country.

Europe

Customer access point is S/T.
U-interface between central office and NT box is proprietary
    to PTT (although defined by ETSI ETR 080).
PTT provides service at S/T four-wire interface.
PTT provides power feed for emergency mode operation.
Terminals must support emergency mode operation.
All phones are S/T terminals.
Residences may require rewiring.
Jurisdiction by country, although pan-European standard exists.

pean network, but the North American network expects to see NT. Table 27.10 summarizes the implementation requirements for ISDN service.

## GLOSSARY

**ACK**    Acknowledge signal (type II and type III handshake).

**ADSI**    Analog Display Services Interface (type III protocol), a subscriber feature that enables transfers and display of larger amounts of data.

**BS**    Base station.

**CAS**   CPE alerting signal (type II and type III handshake).

**Central office (CO)**   A main telephone office, usually within a few miles of a subscriber, that houses switching gear. Commonly capable of handling about 10,000 subscribers.

**CIDCW**   Call identification delivery on call waiting (type II), a subscriber feature that allows for the display of the time, date, number, and other information about the caller to the called party while the called party is off-hook.

**CLASS**   Custom Local-Area Signaling Service, a set of services and enhancements provided to customers that may include CND, CNAM, message waiting, and other features.

**CLID**   Calling line identification delivery (type I), a subscriber feature that allows for the on-hook display of the time, date, number, and other information about the caller to the called party.

**CNAM**   Calling name delivery, a subscriber feature that allows for the display of the time, date, number, and name of the caller to the called party.

**CND**   Calling number delivery, a subscriber feature that allows for the display of the time, date, number, and other possible information about the caller to the called party.

**CPE**   Customer premise equipment, including POTS phones, answering machines, fax machines, or any number of other devices connected to the public network.

**CPT**   Call progress tones, DTMF tones sent while off-hook to signal-enhanced features.

**DAA**   Data Access Arrangement, a CPE interface to network, complying with title 47, part 68 of the Code of Federal Regulations.

**DN**   Directory number.

**DTMF**   Dual-tone multifrequency, a tone-dialing system based on outputting two non–harmonic-related frequencies simultaneously to identify the number dialed or signal call progress. Eight frequencies have been assigned to the four rows and four columns of a typical keypad.

**FSK**   Frequency-shift keying. FSK uses the data stream to modulate a carrier frequency.

**ISDN**   Integrated Services Digital Network, a communications network intended to carry digitized voice and data multiplexed onto the public network.

**Loop**   The loop formed by the two subscriber wires (tip and ring), connected to the telephone at one end and the central office (or PBX) at the other end. Generally, it is a floating system, not referred to ground or ac power.

**LR**   Location register.

**LT**   Line termination, generally CO line card.

**MS**   Mobile station.

**MSC**   Mobile switching center.

**NT**   Network termination, a connection between terminal equipment (TE) and the network.

**NT1**   Converts two-wire U-interface to four-wire S/T interface.

**Off-hook**   The condition in which the telephone is connected to the phone system, permitting loop current to flow. The central office detects the dc current as an indication that the phone is busy.

**On-hook**   The condition in which the telephone's dc path is open and no dc loop current flows. The central office regards an on-hook phone as available for ringing.

**PABX, PAX**   Private Automatic Branch Exchange, a customer-owned, switchable telephone system providing internal and/or external station-to-station dialing.

**POTS**   Plain old telephone service.

**PSTN**   Public Switched Telephone Network.

**Ring**    One of the two wires connecting the central office to CPE. The name derives from the ring portion of the plugs used by operators in older equipment to make the connection. Ring is traditionally negative with respect to tip.

**RJ-11**    Standard four-wire analog phone connector.

**RJ-45**    Standard eight-wire ISDN connector.

**SCWID**    Spontaneous call-waiting identification delivery (type II)—see CIDCW.

**Signaling**    The transmission of control or status information in the form of dedicated bits or channels of information inserted on lines with voice data.

**S/T interface**    Four-wire interface between terminal equipment (TE) and network termination (NT).

**TA**    The terminal adapter, which takes a non-ISDN data stream and converts it to the ISDN data format; used to provide S/T or U-interface connection to type 2 terminal equipment (non–ISDN-compliant).

**TE**    Terminal equipment. Type 1 (TE1) is ISDN-compliant. Type 2 (TE2) requires a terminal adapter to connect to the network.

**Tip**    One of the two wires connecting the central office to CPE. The name derives from the tip of the plugs used by operators in older equipment to make the connection. Tip is traditionally positive with respect to ring.

**U-interface**    A two-wire interface between network termination (NT) and line termination (LT).

**U-TA**    U-interface terminal adapter; converts non-ISDN data stream to ISDN data format; used to provide U-interface connection to type 2 terminal equipment in North America.

## BIBLIOGRAPHY

European standards:

ETSI (European Telecommunications Standards Institute), 06921 Sophia Antipolis, Cedex, France. Phone 33-92-94-4200; fax 33-93-65-4716.

U.S. standards:

Bellcore Customer Service, 60 New England Ave., Piscataway, NJ 08854-4196. Phone 201-699-5800.

TIA (Telecommunications Industry Association), 2001 Pennsylvania Ave., NW, Washington, D.C. 20006-1813. Phone 202-457-5430; fax 202-457-4939.

International standards:

ITU (International Telecommunications Union), Place des Nations, CH-1211, Geneva, Switzerland. Phone 41-22-730-5851.

More details concerning ISDN interfaces, protocols, specifications, and requirements may be obtained from the following documents:

S/T interface:

CCITT I.430 (global standard for BRI at S and T reference points)

ANSI T1.605 (U.S. standard for BRI at S and T reference points)

ETSI 300 012 (European S/T standard)

U-interface:

ANSI T1.601 (U.S. standard for BRI at U reference point)

ETSI ETR 080 (European U standard)

Bellcore TR-TSY-000393 ISDN Basic Access Digital Subscriber Lines

Bellcore TR-TSY-000397 ISDN Basic Access Transport System Requirements

The passive bus arrangement of Table 29.4 uses the LAP-D (link access procedure for the D channel) protocol. It is based on the X.25 LAP-B standard, with an extended 16-bit address field, as specified in CCITT Q.921/I.441. The D-channel signaling protocol, used to control voice and data calls on all channels, is specified in CCITT Q.931/I.451.

## *ABOUT THE AUTHORS*

Bruce H. Benjamin is market development manager of wired communciations at Motorola Semiconductor Products, Phoenix, Ariz., where he is responsible for managing programs to identify and implement system solutions for major telecommunications customers. Previous Motorola assignments included technology development and technical sales to AT&T. Prior to joining Motorola in 1980, he held positions in semiconductor applications, computer design, and radar systems design. He holds a B.S. in physics from the Massachusetts Institute of Technology.

Eric Boll is ISDN applications manager, MOS Digital-Analog Division, Motorola Semiconductor Products, Phoenix, Ariz., where he is responsible for applications support, technical marketing, new product development, and training. Prior to joining Motorola in 1989, he designed naval command and control systems for Lockheed Missiles and Space Co. in Austin, Tex., and point-of-sales systems for Rank Peripherals in Montreal, Canada. He holds a B.S.E.E. from McGill University.

David Kolkman is an applications engineer of wireline telephony, MOS Digital-Analog Division, Motorola Semiconductor Products, Austin, Tex. He joined Motorola in 1988. He has held various assignments in the semiconductor industry for 25 years. He holds an A.S.E.E. from ITT Technical Institute, Fort Wayne, Ind.

Mark Reinhard is director for wireless market development, Motorola Semiconductor Products Sector, Schaumburg, Ill. He has been with Motorola for 20 years, having held positions in the Communications Sector, Cellular Subscriber Group, and Semiconductor Product Sector. He has written several articles on wireless technology. He holds a B.S.E.E. and an M.B.A.

# CHAPTER 27
# DIGITAL/VIDEO TELEPHONY

## SECTION 2
## VIDEO TELEPHONY

**Steve Sperle**
*Motorola, Inc.*

## 27.5   *GENERAL OVERVIEW*

Although video telephony existed in the 1930s and was marketed to the public as early as the 1960s, it has been only recently that companies have been able to create viable businesses by producing video telephony products. The key items that have enabled this market success are the approval and general acceptance of international standards, improved product performance, reduced costs, and greater availability of digital data services.

Previously, the market had been inhibited by the circular problem that demand was constrained due to high cost that was caused by low volume, which was due to the high cost. Fundamental to increasing demand was the creation and acceptance of international standards. This provided a mechanism to not only consolidate the market segments, but also to allow a common technology base to be applied and thereby reduce component cost.

Advances in semiconductor technology have also helped to break the cycle by allowing the price points to decline dramatically and thereby open the technology to larger target markets. It is expected that the development and deployment of international standards will spur the video technology market in a manner similar to the way that they enabled the facsimile market.

## 27.6   *FUNDAMENTAL PRINCIPLES*

Video telephony involves a number of technological disciplines that can yield a high degree of variability in design. Figure 27.14 shows a representative video telephony product.

Many of the design issues for a video phone are similar to those of other consumer communications products. Physical design items such as user interface ergonomics, heat, electromagnetic interference (EMI), radio-frequency interference (RFI), and electrostatic discharge (ESD) must be considered, but they are broadly similar to other consumer communications products. Also, items related to network connectivity and speech compression have the same general principles and methodologies as other consumer communications products.

The uniqueness of video telephony technology tends to revolve around the specifics of how the camera, video compression/decompression (codec), and video display functions are handled. Also, it is the customers' expectations of video quality that is a key driver for their perception of

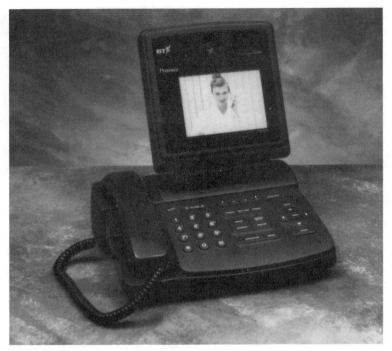

**FIGURE 27.14**   Video phone (courtesy of BT).

**TABLE 27.11**   Video Compression Requirements

| Format | Resolution in pixels × lines | | Bits for color | Frames/ second | Uncompressed source in Mb/s | Available bandwidth in kb/s | Compression required |
|---|---|---|---|---|---|---|---|
| CCIR601 | 720 | 576 | 16 | 30 | 189.8 | 106 | 1791 |
| 4 × CIF | 704 | 576 | 12 | 30 | 139.2 | 106 | 1313 |
| 4 × CIF | 704 | 576 | 12 | 15 | 69.6 | 106 | 657 |
| 4 × CIF | 704 | 576 | 12 | 10 | 46.4 | 106 | 438 |
| 4 × CIF | 704 | 576 | 12 | 7.5 | 34.8 | 106 | 328 |
| CIF | 352 | 288 | 12 | 30 | 34.8 | 106 | 328 |
| CIF | 352 | 288 | 12 | 15 | 17.4 | 106 | 164 |
| CIF | 352 | 288 | 12 | 10 | 11.6 | 106 | 109 |
| CIF | 352 | 288 | 12 | 7.5 | 8.7 | 106 | 82 |
| QCIF | 176 | 144 | 12 | 30 | 8.7 | 106 | 82 |
| QCIF | 176 | 144 | 12 | 15 | 4.4 | 106 | 41 |
| QCIF | 176 | 144 | 12 | 10 | 2.9 | 106 | 27 |
| QCIF | 176 | 144 | 12 | 7.5 | 2.2 | 106 | 21 |
| QCIF | 176 | 144 | 12 | 15 | 4.4 | 20 | 218 |
| QCIF | 176 | 144 | 12 | 10 | 2.9 | 20 | 145 |
| QCIF | 176 | 144 | 12 | 7.5 | 2.2 | 20 | 109 |
| QCIF | 176 | 144 | 12 | 5 | 1.5 | 20 | 73 |
| SQCIF | 128 | 96 | 12 | 15 | 2.1 | 20 | 105 |
| SQCIF | 128 | 96 | 12 | 10 | 1.4 | 20 | 70 |
| SQCIF | 128 | 96 | 12 | 7.5 | 1.1 | 20 | 53 |
| SQCIF | 128 | 96 | 12 | 5 | 0.7 | 20 | 35 |

the value of a video telephony system. This area provides both the greatest opportunity and the greatest challenge.

### 27.6.1  Nature of the Compression Challenge

The fundamental dilemma for video telephony is that user expectations and desires for video are very high, but users also desire inexpensive products. When coupled with the constraint of bandwidth available on public networks, the engineering challenges are extensive. Users have expressed the desire for products with video quality comparable to their television sets while communicating over commonly available networks such as ISDN (Integrated Services Digital Network) or GSTN (General Switched Telephone Network). This represents the desire to transmit the equivalent of almost 190 Mb/s of data while using a 128-kb/s or 28.8-kb/s phone line.

Compression of the data is vital, but because real-time compression and decompression of data by a ratio of around 1800 is not yet commercially feasible, it is also necessary to reduce the amount of video data that is compressed. The most common approaches to reduce the amount of video data that must be transmitted are to reduce the resolution of the image, to reduce the frame rate, and to reduce the number of bits used to represent color. In spite of these reductions, the requirements for compression are still substantial, as Table 27.11 illustrates. The table assumes that the bandwidth available for video on a 128-kb/s ISDN line is 106 kb/s after using 22 kb/s for audio and control. Likewise, it assumes that 20 kb/s of a 28.8-kb/s line is available for video after using 8.8 kb/s for audio and control.

It should be noted that the current state of the technology does not provide the compression ratios that are required to produce the "smooth" motion that users see in 30-frame/s television or 24-frame/s movie theaters. In fact, with compression ratios that are typically under 60, Fig. 27.15 shows that, in most cases, users will experience "jerky" motion at substantially less than 15 frames per second.

The key opportunities for compression are in the spatial domain (by reducing the number of bits required to represent a single frame) and in the time domain (by using motion compensation to reduce the bits required to represent the changes between frames).

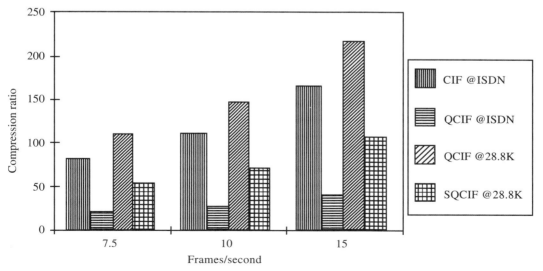

**FIGURE 27.15**  Graph of video compression required for various frame rates, resolutions, and line speeds.

## 27.7  *RELEVANT INTERNATIONAL STANDARDS*

The role of international standards is to provide a mechanism that allows individual vendors to create solutions that will interoperate with solutions from other vendors. The International Telecommunication Union Telecommunication Standardization Sector (referred to as the ITU-T) is the organization that creates international standards, including those related to video conferencing. The standards relevant to video telephony are outlined below.

### 27.7.1  H.320-Series Standards for Video Telephony

The H.320-series is directed toward video telphony using digital networks. Although the initial target for these standards was group video-conferencing systems, the standards are also well suited for consumer ISDN products. Figure 27.16 provides an overview of a generic visual telephone system that is based on the H.320-series recommendations. Table 27.12 provides an overview of the H.320-series standards and their specific titles.

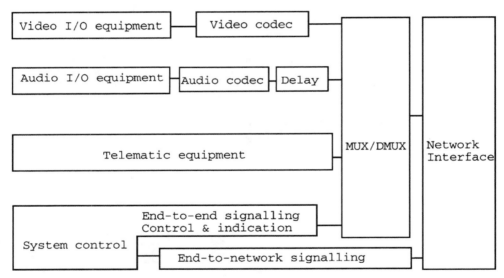

**FIGURE 27.16**   Generic H.320 visual telephone system.

**TABLE 27.12**   H.320-Series Standards

| ID | Title |
|---|---|
| H.320 | Narrow-band visual telephone systems and terminal equipment |
| H.221 | Frame structure for a 64-to-1920-kb/s channel in audiovisual teleservices |
| H.230 | Frame-synchronous control and indication signals for audiovisual systems |
| H.242 | System for establishing communications between audiovisual terminals using digital channels up to 2 Mb/s |
| H.261 | Video codec for audiovisual services at p × 64 |
| G.711 | Pulse code modulations (PCM) of voice frequencies |
| G.722 | 7-kHz audio coding within 64 kb/s |
| G.728 | Coding of speech at 16 kb/s using low-delay code excited linear prediction |

**27.7.2  The H.261 Standard**

The H.261 standard in particular is fundamental to the system because it specifies the video resolutions, the picture organization, and the video compression and decompression transform algorithm (see Chapter 10 for a detailed description). Two video resolutions are specified. The first resolution is the Common Intermediate Format (CIF), which is defined using the luminance sampling structure with 352 pixels per line and 288 lines per picture. The second format is Quarter Common Intermediate Format (QCIF), with a luminance sampling structure of 176 pixels per line and 144 lines per picture. The CIF format is optional, whereas the QCIF format is mandatory.

Also fundamental is the structure for the picture and its subelements. The H.261 video multiplex structure has four layers: picture, group of blocks (GOB), macroblock (MB), and block.

The CIF picture has 12 GOBs, whereas a QCIF picture has three GOBs. Each GOB represents 176 pixels by 48 lines and the GOBs are arranged as indicated in Fig. 27.17. Each GOB is divided into 33 macroblocks. Each MB represents 16 pixels by 16 lines and they are arranged in the GOB as shown in Fig. 27.18. A block is 8 pixels by 8 lines, and blocks are arranged in an MB as shown in Fig. 27.19. This arrangement is summarized in Table 27.13.

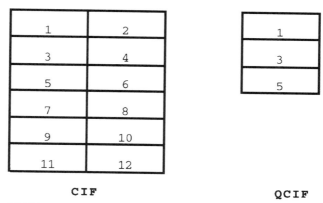

CIF    QCIF

**FIGURE 27.17**    A CIF picture has 12 GOBs (left) and a QCIF picture has 3 GOBs (right).

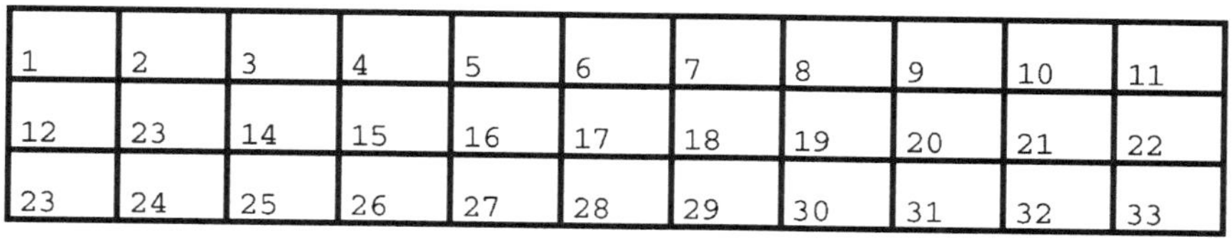

| 1 | 2 | 3 | 4 | 5 | 6 | 7 | 8 | 9 | 10 | 11 |
|---|---|---|---|---|---|---|---|---|----|----|
| 12 | 23 | 14 | 15 | 16 | 17 | 18 | 19 | 20 | 21 | 22 |
| 23 | 24 | 25 | 26 | 27 | 28 | 29 | 30 | 31 | 32 | 33 |

**FIGURE 27.18**    A GOB is divided into 33 macroblocks.

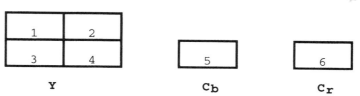

Y    Cb    Cr

**FIGURE 27.19**    A block is 8 pixels by 8 lines, and blocks are arranged in a macroblock.

**TABLE 27.13**  Overview of H.261 Picture Structure

| | | |
|---|---|---|
| Picture | CIF = 352 pixels by 288 lines, arranged as 12 GOBs | QCIF = 176 pixels by 144 lines, arranged as 3 GOBs |
| GOB | 176 pixels by 48 lines arranged as 33 macroblocks | |
| MB | 16 pixels by 16 lines arranged as 6 blocks | |
| Block | 8 pixels by 8 lines | |

The block serves as the fundamental element used during the video compression process. Compression in the spatial domain can be accomplished by reducing the number of bits required to represent the data contained in a block. The first step in the process involves transforming the data into the frequency domain. The algorithm used is the discrete cosine transform (DCT), specified as follows:

$$F(u, v) = \tfrac{1}{4}C(u)C(v)\sum_{x=0}^{7}\sum_{y=0}^{7} f(x, y)\cos[\pi(2x + 1)u/16]\cos[\pi(2y + 1)v/16] \quad (27.1)$$

where $u, v, x, y$ = 0, 1, 2, ..., 7
$\qquad f(x, y)$ = Original $8 \times 8$ block of data
$\qquad F(u, v)$ = Output of $8 \times 8$ DCT values
$\qquad u, v$ = Coordinates in the transform domain
$\qquad x, y$ = Spatial coordinates in the pixel domain
$\qquad C(u)$ = 1/sqrt(2) for $u$ = 0, otherwise 1
$\qquad C(v)$ = 1/sqrt(2) for $v$ = 0, otherwise 1

The inverse transform function is as follows:

$$f(x, y) = \tfrac{1}{4}\sum_{u=0}^{7}\sum_{v=0}^{7} C(u)C(v)F(u, v)\cos[\pi(2x + 1)u/16]\cos[\pi(2y + 1)v/16] \quad (27.2)$$

The transform function does not reduce the amount of data, but it does have the effect of converting most of the data to near-zero values and also concentrating the non-zero data into the upper-left portion of the block. This provides the opportunity for data compression by quantizing, ordering, and then run-length encoding the data. Quantization is a way to limit the set of output values by rounding or truncating the values, as Figs. 27.20 and 27.21 illustrate. In order to take advantage of the concentration of data in the upper-left corner of the block, the order for transmission of the transforms is specified in a zigzag, as indicated in Fig. 27.22.

The next step is to take advantage of the sequences of zero data by grouping the data using run-length encoding. This puts the data in pairs of (number of preceding zeros, non–zero-value). For example, the following sequence,

$$6, 5, 0, -4, 2, -3, 0, 1, 0, 0, 0, -1, 0, 0, 0, 1$$

would be converted to

$$(0, 6), (0, 5), (1, -4), (0, 2), (0, -3), (1, 1), (3, -1), (3, 1)$$

These (run, length) combinations are converted to variable-length codes using a table that is structured to minimize the total number of bits transmitted. Run-length encoding has a significant impact on reducing data transmission requirements where there are long runs of zeros in the data. For example, where a run of 9 zeros followed by a nonzero number would normally require transmission of 10 digits, use of run-length encoding would reduce this to only 2 digits.

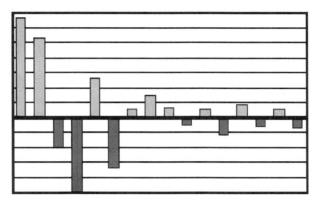

**FIGURE 27.20**    Data prior to quantization.

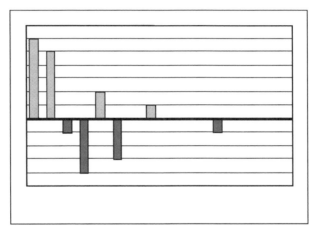

**FIGURE 27.21**    Data after quantization.

| 1  | 2  | 6  | 7  | 15 | 16 | 28 | 29 |
|----|----|----|----|----|----|----|----|
| 3  | 5  | 8  | 14 | 17 | 27 | 30 | 43 |
| 4  | 9  | 13 | 18 | 26 | 31 | 42 | 44 |
| 10 | 12 | 19 | 25 | 32 | 41 | 45 | 54 |
| 11 | 20 | 24 | 33 | 40 | 46 | 53 | 55 |
| 21 | 23 | 34 | 39 | 47 | 52 | 56 | 61 |
| 22 | 35 | 38 | 48 | 51 | 57 | 60 | 62 |
| 36 | 67 | 49 | 50 | 58 | 59 | 63 | 64 |

**FIGURE 27.22**    Order for transmission of transforms is specified in a zigzag.

Another major opportunity for compression is to describe the changes between pictures via motion estimation. Motion is described as a vector that contains horizontal and vertical displacement components and describes the movement of a macroblock. This process can yield significant data compression. This is especially true for macroblocks where there is no movement and the motion vector is zero. Figure 27.23 provides a block diagram of the process.

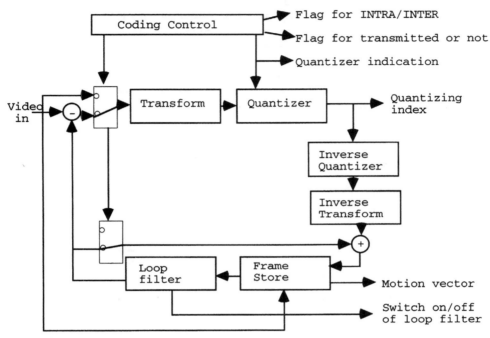

**FIGURE 27.23**  Video processing.

The H.261 standard limits the motion search to a maximum of $\pm 15$ pixels in the horizontal and vertical directions. Motion estimation is a form of pattern matching and is very compute-intensive for an encoder. For example, an exhaustive search requires $5 \times 10^9$ additions or subtractions per second. Motion estimation work must be done by the encoder and can be expensive to implement, but it can also make a significant contribution to picture quality and compression rates. The work for the decoder to implement motion compensation is relatively simple because the operation is to add the motion vectors to a macroblock. The H.261 standard requires that the decoder be able to process motion vectors, but it allows motion estimation to be optional on the encoder. The standard also allows the developer to define the method used by the encoder to create motion vectors and to make the tradeoffs between computational expense and picture quality.

### 27.7.3  H.324-Series Standards for Video Telephony

The H.324-series standards are directed toward video telephony operating over the General Switched Telephone Network (GSTN), but they are based on the same principles as the H.320

standards and include methods for interworking with H.320-based equipment. Figure 27.24 provides a block diagram of an H.324 multimedia system. Table 27.14 provides an overview of the main H.324-series standards and their specific titles.

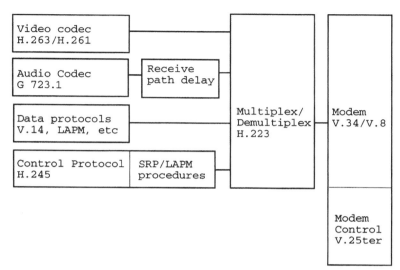

**FIGURE 27.24**   H.324 block diagram.

**TABLE 27.14**   H.324-Series Standards

| ID | Title |
|---|---|
| H.324 | Terminal for low-bit-rate multimedia communication |
| H.245 | Control protocol for multimedia communication |
| H.223 | Multiplexing protocol for low-bit-rate multimedia communication |
| H.263 | Video coding low-bit-rate communication |
| G.723.1 | Dual-rate speech coder for multimedia communications transmitting at 5.3 and 6.3 kb/s |
| V.34 | A modem operating at data signaling rates of up to 28,800 b/s for use on the GSTN and on leased point-to-point two-wire telephone-type circuits |

The H.324-series standards are based on the same principles as the H.320-series standards and are designed to provide interworking with H.320 systems. H.324 does include some new definitions. G.723.1 defines two mandatory audio algorithms (one at 5.3 kb/s and one at 6.3 kb/s). H.263 uses QCIF and CIF image formats as defined by H.261 but also adds definitions for SQCIF, 4CIF, and 16CIF (Table 27.15).

H.263 also provides for some optional modes including unrestricted motion vectors, syntax-based arithmetic coding, advanced prediction mode, and PB-frames. Each of these modes provides some level of video quality improvement, but the developer must decide on whether to include each mode.

**TABLE 27.15** H.263 Video Resolutions

| Format | Resolution | Comments |
|--------|------------|----------|
| SQCIF | 128 × 96 | Mandatory* |
| QCIF | 176 × 144 | Mandatory* |
| CIF | 352 × 288 | Optional |
| 4CIF | 704 × 576 | Optional |
| 16CIF | 1408 × 1152 | Optional |

*Mandatory to decode SQCIF and QCIF, mandatory to encode one of SQCIF or QCIF, optional to encode both.

### 27.7.4 H.323-Series Standards for Video Telephony

The H.323-series standards are directed toward video telephony operating over local-area networks and internet-based communication. The H.323-series incorporates many of the standards from the H.320 and H.324 series. H.323 also includes a gateway and methods for interworking with H.320- and H.324-based equipment (Table 27.16).

**TABLE 27.16** Principal H.323-Series Standards and Their Specific Titles

| ID | Title |
|----|-------|
| H.323 | Visual telephone systems and equipment for local-area networks which provide a nonguaranteed quality of service |
| H.245 | Control Protocol for Multimedia Communication |
| H.225.0 | Media stream packetization and synchronization for visual telephone systems on nonguaranteed quality of service LANs |
| H.261 | Video codec for audiovisual services at $p \times 64$ |
| H.263 | Video coding low-bit-rate communication |
| G.711 | Pulse code modulations (PCM) of voice frequencies |
| G.722 | 7-kHz audio coding within 64 kb/s |
| G.723.1 | Dual-rate speech coder for multimedia communications transmitting at 5.3 and 6.3 kb/s |
| G.728 | Coding of speech at 16 kb/s using low-delay code excited linear prediction |
| G.729 | Speech codec for multimedia telecommunications transmitting at 8/13 kb/s |

### 27.7.5 Other Design Items

In addition to compliance with international standards, the developer must make decisions on support for major components such as cameras, displays, microphones, speakers, and communications interfaces. How each of these items is addressed varies depending on factors such as target application and equipment environment.

In the case where the video phone is installed in a personal computer, it may be possible to take advantage of the existing video, audio, and communications peripherals. General design guidelines for a video phone display include having a viewing distance of at least six times picture height and a 4:3 aspect ratio. Cameras should perform well over a range of illumination between 200 and 4000 lux. Also, cameras with fixed focus are preferred, because auto-focus cameras tend to create excessive picture changes and therefore increase the amount of data that needs to be coded.

In the area of audio, the echo return loss for a handset should be at least 6 dB. For hands-free operation, it will be necessary to implement echo cancellation or echo suppression and to have an echo return loss of over 40 dB.

## *BIBLIOGRAPHY*

J. Angiolilio, Blanchard, H., and Israelski, E., "Video telephony," *AT&T Technical Journal,* May/June 1993, pp. 7–20.

R. Bellman, "I want a Videophone—Now!," *Business Communications Review,* Apr. 1995, pp. 38–41.

S. Early, Kuzma, A., and Dorsey, E., "The VideoPhone 2500—Video telephony on the public switched telephone network," *AT&T Technical Journal,* Jan./Feb. 1993, pp. 22–32.

K. Illgner and Lappe, D., "Mobile multimedia communications in a universal telecommunication network," in *Visual Communications and Image Processing '95,* L. T. Wu (ed.), *Proc. SPIE 2501,* 1995, pp. 1034–1043.

ITU-T Draft Recommendation G.723.1, "Dual rate speech coder for multimedia communications transmitting at 5.3 and 6.3 kb/s," Oct. 1995.

ITU-T Recommendation H.221, "Frame structure for a 64 to 1920 kb/s channel in audiovisual teleservices," Mar. 1993.

ITU-T Draft Recommendation H.223, "Multiplexing protocol for low bitrate multimedia communication," Nov. 1995.

ITU-T Recommendation H.230, "Frame-synchronous control and indication for audiovisual systems," Mar. 1993.

ITU-T Recommendation H.242, "System for establishing communciations between audiovisual terminals using digital channels up to 2 Mb/s," Mar. 1993.

ITU-T Draft Recommendation H.245, "Control protocol for multimedia communication," Nov. 1995.

ITU-T Recommendation H.261, "Video codec for audiovisual services at p × 64 kb," Mar. 1993.

ITU-T Draft Recommendation H.263, "Video coding low bitrate communication," Dec. 1995.

ITU-T Recommendation H.320, "Narrow-band visual telephone systems and terminal equipment," Mar. 1993.

ITU-T Draft Recommendation H.323, "Visual telephone systems and equipment for local area networks which provide a non-guaranteed quality service." May 1996.

ITU-T Draft Recommendation H.324, "Terminal for low bitrate multimedia communications," Nov. 1995.

K. Jack, *Video demystified: A handbook for the digital engineer,* HighText Publications, 1993, ch. 10.

M. Leonard, "Silicon solution merges video, stills, and voice," *Electronic Design,* 2 Apr. 1992, pp. 45–54.

A. Leone, Bellini, A., and Guerrieri, R., "An H.261-compatible fuzzy-controller coder for videophone sequences," *The Third IEEE Conference on Fuzzy Systems,* vol. 1, 1994, pp. 244–248.

F. Mazda, *Telephone Engineer's Reference Book,* Butterworth-Heinemann, 1993, ch. 57.

*Multimedia Device Data DL158,* Motorola, 1995.

S. Okubo, Nishimura, S., and Kato, Y., "International standardization of video codecs for ISDN video conferencing and videophone services," *NTT Review,* vol. 2, no. 3, May 1990.

R. Schaphorst and Bodson, D., "Subjective and objective testing of video teleconferencing/videophone systems," *Globecom '91,* vol. 3, pp. 1771–1775.

J.-C. Schmitt and Eude, G., "Very low bit rate coding for PSTN videotelephony on personal computer—Part 2," *SPIE 2419,* 1995, pp. 486–491.

M. W. Whybray and Ellis, W., "H.263—Video coding recommendation for PSTN videophone and multimedia," *IEE Colloquium (Digest),* vol. N 154, 1995, pp. 6/1–6/9.

## *ABOUT THE AUTHOR*

Steve Sperle is a multimedia strategic marketing manager for Motorola in Austin, Tex. He holds a B.S.E.E., a B.S. in industrial engineering, and an M.B.A. He participates in the International Multimedia Teleconferencing Consortium and is a member of the International Teleconferencing Association. He has served on the advisory board of the DVC Desktop Videoconferencing Technical Seminar and Exhibition and has been a presenter at teleconferencing industry trade shows and meetings.

# P · A · R · T · 8

# DIGITAL APPLIANCES, RESIDENTIAL AUTOMATION

# CHAPTER 28
# FUZZY LOGIC AND MICROPROCESSOR APPLICATIONS IN HOUSEHOLD APPLIANCES

**Adnan Shaout**
*University of Michigan-Dearborn*
*Department of Electrical & Computer Engineering*

## 28.1   INTRODUCTION

In the house of the future, fuzzy logic will have imperceptibly permeated commonly used household appliances. Appliances will automatically adjust to room factors such as the number of people present, temperature, and light levels, or even the cleanliness of the floor. In some cases the appliances will even operate themselves.

Fuzzy logic has helped bring these dreams to the achievable present. This theory has entered many aspects of Japanese life and even some in the United States—for example, automotive applications,[1–6] air- and spacecraft,[7] and even the stock exchange.[1,4,8,9] The concept of the fuzzy-controlled future home has already appeared in Japanese trade shows and households. Numerous appliance applications use fuzzy logic to achieve design goals: the appliance should be simple to operate, have a short development time, and be cost-effective compared to its standard logic counterparts. The designs should also be dynamic, with the ability to adjust to new inputs and different users. Fuzzy logic has allowed designers to achieve all of these goals.

Lotfi Zadeh could not have foreseen the electronic revolution that his obscure fuzzy set theory has produced. However, he predicted fuzzy logic would soon be a part of every appliance: "We'll see appliances rated not on horsepower but on IQ."[10] In Japan the revolution has been so strong that *fuzzy* has become a household word.[11] The Japanese use the term positively to denote intelligence, whereas Americans negatively associate the term with vagueness or uncertainty.

## 28.2   APPLIANCE TECHNOLOGY

Streamlining housework to provide consumers with more free time has always been a design target.[12] Fuzzy logic and neuro-fuzzy logic are being incorporated in Japanese appliances to accomplish this goal. The following sections provide an overview of the microprocessor-based appliance controllers, the fuzzy controllers, and neuro-fuzzy logic.

### 28.2.1 Microprocessor-Based Appliance Controllers

Microprocessor-based household appliances have been available in the consumer market for some time. They have proven to be more safe, convenient, easy and fun to use and program, reliable, and energy-efficient than conventionally controlled home appliances.

A microcomputer can perform certain routine household functions such as menu planning and cooking. Range ovens for household cooking feature digital controls, displays, timers, and microprocessor control. Here a stored program in ROM is used to provide information on the cooking time for various types of food. The program turns the burners or the oven on or off. Touch-control panels and displays for such digitally controlled ranges have been developed. Many of these consumer applications are characterized by large volume and low cost. This is the area of a single-chip microprocessor. These microprocessors are sufficient for simple control functions such as those required by microwave oven controllers. Microprocessors can react much faster and switch with greater precision to variations in the measured process parameters. The microprocessor eliminates electromechanical or hard-wired logic. It also provides more functions and can facilitate reasonable tests. If the user makes an error in selecting controls, the microprocessor can flash a warning and not execute the command. It can also prevent some problems from occurring by detecting them in advance by using continuous records of parameter variations that are saved in microprocessor memory.[13–16]

Refrigerators, sewing machines, dishwashers, food processors, washing machines, dryers, color televisions, microwave ovens, digital clocks, and coffee machines are a small part of the home appliances that are currently controlled by microprocessors. A refrigerator by Whirlpool adjusts its inside temperature when food is added or the kitchen warms up, and a GE refrigerator beeps if the door is left open. Dryers contain moisture sensors that end the cycle when the clothes are dry, and dishwashers alert the users when the drain is clogged and when other mechanical failures occur. A microprocessor has been used in color television sets to provide tuning and programming of the set up to a year in advance. It is easy to provide a digital clock as well as built-in games.[17] New technologies, such as fuzzy technology, are making home appliances smarter and more efficient.

### 28.2.2 The Fuzzy Appliance Controller

Appliances with fuzzy logic controllers provide the consumer with optimum settings that more closely approximate human perceptions and reactions than standard control systems. Products with fuzzy logic monitor user-dictated settings, then automatically set the equipment to function at the user's generally preferred level for a given task.[18] This technology is well suited to making adjustments in temperature, speed, and other control conditions found in a wide variety of consumer products.[19] The following paragraphs briefly outline the design steps for the fuzzy appliance controller. Figure 28.1 summarizes these steps.[8]

The first step in designing a fuzzy controller is to subdivide the input and output variables into their descriptive linguistic terms and then to establish membership functions for each range.[20] A membership function represents each of the fuzzy sets and transforms the crisp real world into the fuzzy view of the real world.[21] These functions provide the appliance with the machine equivalent of perception and judgment.[22] For example, the input Room_Temp may be divided into terms such as Cold, Cool, Warm, Very_Warm, and Hot. Usually, however, a seven-label gradation is used for controllers.[23] Linguistic hedges such as Very, About, or Slightly can be used to narrow or broaden the adjective's definition.[24]

Each membership function identifies the range of input values that correspond to an adjective. Each fuzzy adjective has a region where input values gradually change from being full members to nonmembers.[25,26] This transition corresponds to a change from a one to a zero state.

The system is defined in terms of input and output rules. Generally, a control system rule will consist of two input variables that will combine to direct the output variable.[4,27] These rules may be described in terms of everyday language. For example, a rule for an appliance could be "If

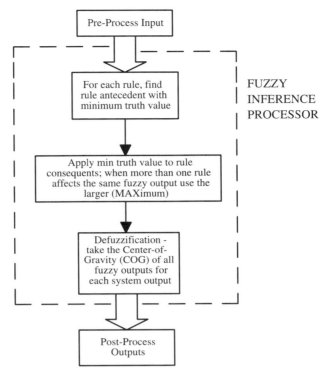

**FIGURE 28.1**    Overall flow of a MCU-based fuzzy logic application.

the wash load is average (cloth quantity) and the fabrics being washed are soft (cloth quality), then the washing time is short."[28]

Finally, the sensor inputs are compared to membership functions, the rules are processed, and then the inputs are combined into a single composite action for each output.[8,23] The output of all the rules is reduced to a single output by taking the logic sum. Next, the defuzzification process takes place. This process changes the fuzzy-inferencing results into output understood by the control system hardware. Although there are several defuzzification methods, the most common method is the *center-of-gravity* (COG) method. COG takes the weighted average of all fuzzy outputs for each system output.[8] This defuzzified result, as it applies to an appliance control system, will give the intake fan speed, the washing time for a laundry load, or the cooking time for a frozen dinner in a microwave oven.

### 28.2.3  Neuro-Fuzzy Logic

Neural networks and fuzzy logic are complementary rather than competing technologies.[7] Neural networks assign relative weights to data that are represented, along with their decision structure, like a biological neural system.[29] A fuzzy set is an extension of traditional set theory that allows grades of membership other than 1 or 0. Fuzzy logic is a digital control methodology that simulates human thinking by incorporating the imprecision inherent in all physical systems.[4] In more general terms, fuzzy logic is a way that electronic devices can react to vague inputs.[30] Fuzziness takes into account the nuances of language and eliminates unnecessary precision.[31]

Many similarities exist between fuzzy logic and neural networks that enhance their relationship. For example, neural networks learn ideas in ways that are similar in form to fuzzy logic's membership functions. This learning capability can be used to master the relative importance between rules and the values of each element in a rule. Another commonality is that fuzzy logic takes a weighted sum of the if-then rules, which is analogous to a neural network action.[32,33]

Neural networks can help fuzzy programmers determine membership functions and logic rules. One method of interaction works by letting the neural net monitor the human expert's reaction to a situation. Models are then built reflecting the relationship between the stimuli and response. Rule structures and membership functions are automatically generated from these models. The neuro-fuzzy controller now matches the knowledge of the human expert using the best qualities of both systems.[7,34] Another method, called the *adaptive fuzzy system,* uses neural networks to learn fuzzy rules.[32]

Taken separately, both neural networks and fuzzy logic have drawbacks. Neural networks learn but do not have an easily discernible structure.[35] To make optimal decisions, fuzzy logic needs a growing amount of data from sensors, which requires time to collect and process.[36] Used together, however, the programmer joins the decision-making capabilities of one with the learning capabilities of the other. The result is a more easily understood rule structure and a processing speed approximately 45 times faster than completely fuzzy systems.[34,36,37]

## 28.3 FUZZY APPLIANCES

The following paragraphs in this section discuss fuzzy logic as applied to various appliances. Table 28.1 summarizes the features of each fuzzy appliance and gives the manufacturer's name when available.

### 28.3.1 Washing Machine and Dryer

These models are equipped with two optical sensors that can sense the quality and quantity of dirt in the wash.[29,38–41] The sensors work by detecting light, passing through a water sample, that is altered by the particles suspended in the solution.[42] The pair of sensors determine the degree of soiling from the wash water's turbidity or soiling speed and also check a stain's dissolving rate to determine whether the stain was caused by oil or mud (mud disperses faster than oil in water). They also discriminate between liquid or powder detergents and meter the amount of detergent required.[12,31,43–45] A load sensor is used to determine the volume of clothes to be washed.[44] The fuzzy controller analyzes the accumulated data then selects the most efficient cleaning method from approximately 600 possible variations of water volume, flow strength, and washing time.[26,29,44,46]

The basic neuro-fuzzy washing machine uses sensors for water temperature, laundry volume (load size), and water level. From these sensors the artificial intelligence unit gathers data on laundry conditions, such as fabric type, then automatically selects one of 250 washing modes.[47] These washing modes optimize water temperature, water level, washing and cycle time, and spin-drying time.[1,27,48,49] Some models also have a feature that schedules unattended washing during the lower power rates that may be charged at night.[50] Other neuro-fuzzy models add the ability to account for detergent type, cloth quality, and water hardness. The machine then chooses from 3800 different operating parameters.[44]

Sharp makes a variation on the basic machine that shoots bubbles into the wash to completely dissolve the detergent. The bubble action is "a combination of air bubbles and swirling water action."[51] The cleaning power of this machine is increased by 20 percent over nonfuzzy machines.[51,52] The machine also claims 30 percent savings in fill and drain times and 70 percent water savings over conventional dual-tub systems.[40]

**TABLE 28.1**    Features of Fuzzy Logic Appliances

| Product | Functions | Manufacturers |
|---|---|---|
| Washing machine | Basic machine programs itself for optimum wash conditions by analyzing water temperature, load size, and water level. Some machines also determine the type of dirt and detergent for ideal wash and detergent dispersal. | Goldstar, Hitachi, Daewoo, Sanyo, Samsung, Matsushita |
| Dryer | Monitors load size, fabric type, and air flow to determine the optimum drying time. | Matsushita |
| Vacuum cleaner | Basic machine correlates optimum suction power and beater-bar speed by sensing the amount of dirt and type of floor. More advanced models analyze the type of dirt for more efficient cleaning. | Matsushita, Toshiba, Hitachi |
| Microwave oven | Monitors temperatures of food and oven and the amount of steam generated. Calculates remaining cooking time. More advanced versions monitor more food and oven attributes to calculate remaining cooking time. Also performs auxiliary food preparation functions. | Toshiba, Sanyo, Hitachi, Sharp |
| Refrigerator | Based on usage, determines the most efficient time to defrost. Uses learned consumer usage patterns for optimum defrost and temperature control. Consumer can "teach" some units when not to create disturbances by making ice. | Kenmore, Sharp, Whirlpool |
| Air conditioner | Senses temperature, humidity, and number of people present and then cools the room accordingly. Defrosts outside unit at the most efficient time. | Mitsubishi, Matsushita, Hitachi, Sharp |
| Dishwasher | Detects the amount of dishes loaded, type of food, and cleanliness level of the dishes and then determines the correct cleaning cycle. | Matsushita, Maytag |
| Rice cooker | Monitors steam, temperature, and volume of rice and calculates the remaining cooking time based on four preprogrammed types of rice. | Matsushita, Sanyo |
| Toaster | Adjusts toasting time based on sensed bread type and previously learned user preferences. | |
| Television | Stabilizes volume based on viewer location. Adjusts picture brightness and contrast in low–ambient-lighting conditions. | Sony, Hitachi, Goldstar, Samsung |
| Shower | Stabilizes water temperature regardless of water pressure. | Matsushita |
| Carpet | Corrects room temperature variations by heating cold floors beneath it. | |

The results, from any of the models, are an output tailored to the required task. The machine will automatically wash durable, highly soiled clothing more thoroughly.[53] For most of the machines the only user input necessary, after loading the laundry, is to press the start button.[10,44,46,53]

The companion dryer uses three heat sensors that monitor load size, fabric type, and hot-air flow. The fuzzy controller determines the optimum drying time and shuts itself off when the contents are dry,[45,54] thus saving time and energy costs.

### 28.3.2  Vacuum Cleaner

The basic fuzzy logic vacuum cleaner uses a single sensor to judge the amount of dust and the floor type.[27,55] By monitoring the change in dust quantity, the controller decides whether the floor is bare (where the dust comes up at once) or covered with thick-pile carpeting (where the dust is gradually released). The 4-bit microprocessor detects the dust by pulsing an infrared light-emitting diode and monitoring the output of a phototransistor. Dust passing between the two components blocks light and changes the output signal.[42,44] On the basis of that data, the fuzzy controller correlates the best suction power and beater-bar speed for each specific job. For example, when a hard floor is detected, the motor and beater-bar are slowed because not much suction is needed.[11,56]

In addition to analyzing the floor type and amount of dust, the neuro-fuzzy version also analyzes the type of dirt. This information is used to adjust both the suction power and brush rotation speed for a 45-percent increase in processing speed.[37,54] Another variation of the basic fuzzy model is the Toshiba vacuum that advertises power steering along with all of its fuzzy features.[57]

Although the efficiency and power savings of these cleaners is greatly increased over conventional vacuums, the user must still press the power button and run the vacuum across the floor. However, in the more advanced unit, the user is not even required to be present. This robot vacuum, which is quiet enough to run at night, navigates a room, cleaning as it travels. It operates approximately 20 minutes per battery charge and, when it runs low on power, it returns to its charging port. After recharging itself, it continues vacuuming where it stopped before recharging. This robot cleaner is actually a cordless vacuum equipped with several sensors including a gyro and tuning-fork structure to control its movements. The gyro and fuzzy logic control the sweeping movements, whereas an ultrasonic sensor detects obstacles, enabling the robot to dodge them. The unit will automatically adjust to a transition from bare floor to carpet and can even be set to clean daily while the owner is on vacation.[58] This time- and labor-saving unit, tentatively called the Home Cleaning Robot, has not yet been released by Matsushita.[42,45,56,59]

### 28.3.3 Microwave Oven

The basic fuzzy-logic microwave oven uses three sensors: infrared, humidity, and ambient temperature. The sensors monitor the temperature of the food and oven cavity as well as the amount of steam emanating from the food.[45] On the basis of these data, the controller calculates the type, size, and weight of the food; whether it is frozen or thawed; whether the oven had been used immediately beforehand; and the degree of doneness. This system results in the most efficient cooking time and usage of cooking conditions (roasting or hot-air blower).[37,45,60] All of the microwaves advertise one-touch operation and use fuzzy logic to simulate a cook's best judgment.[42]

A more advanced model employs eight sensors that monitor, in addition to the previously mentioned attributes, aroma and change in food shape. This unit also has a ceramic grill to emulate a barbecue and, using special attachments, kneads bread dough and mashes potatoes.[10,42]

### 28.3.4 Refrigerator

Kenmore and Whirlpool refrigerator models use fuzzy logic to determine the most energy-efficient time to defrost.[10] The fuzzy controller senses temperature changes and defrosts when necessary, rather than at regular intervals.[41] However, Sharp has taken this application of fuzzy logic much further. One of its models uses a neuro-fuzzy logic control system that learns the consumers' usage patterns for optimum performance.[61]

The Sharp refrigerator memorizes the time and frequency of door and freezer drawer openings. When the usage pattern is learned for each compartment, the control system automatically begins a cooling cycle before heavy traffic periods. This feature minimizes temperature fluctuations in the compartments.[61] Based on the memorized data, the unit also chooses the most appropriate time of day to defrost.[62] An additional feature tells the unit not to make ice at night, which may disturb light sleepers. The consumer pushes a button on the unit before retiring. The system memorizes this time and will repeat this pattern every night.[61]

### 28.3.5 Air Conditioner

Mitsubishi Heavy Industries began production of the first fuzzy air-conditioning system in October 1989. It was based on 50 fuzzy rules and used maximum-product inferencing and centroid defuzzification methods. A thermistor was used to detect room temperature and to control the inverter, compressor valve, fan motor, and heat exchanger. The results from both the simulation and production showed (compared to standard systems) a 20 percent reduction in heating and cooling times, a twofold increase in temperature stability, and an overall power savings of 76 percent for the simulation and production savings of 24 percent for cooling and 17 percent for heating cycles. A contributing factor to the increase in power savings was a reduction in the number of on/off cycles.[35,44]

New models have sensors that evaluate the shape/size of a room and the inside/outside temperatures and humidity levels.[3,11] By using an infrared sensor, the unit also determines the number of people present and cools the room accordingly.[7,10,44,63] These inputs are used by the fuzzy controller to balance the temperature with the power needs of the house, resulting in the greatest possible efficiency.[11,27] Air velocity and direction are adjusted automatically for maximum comfort.[64] Mitsubishi's CS-XG series uses neuro-fuzzy logic to generate 4608 control patterns based on environmental data.[65]

Fuzzy logic also improves the unit's defrosting control. When a room is being kept warm and the temperature outside is low, frost forms on the unit's outside evaporator. Conventional systems defrost at regular intervals regardless of air conditioner demand. However, the fuzzy system evaluates the outdoor and evaporator temperatures and chooses the most efficient interval for defrosting.[7]

A future improvement on this system will encompass all household environmental controls. This system, called *HVAC*, will be a unified programmable command system integrated into the smart house.[62] The HVAC system will automatically recognize a user and adjust the comfort levels accordingly. For example, if the user enjoys a dry environment the system will respond to this preference when the smart house detects the user's presence. These changes will be made without user intervention, as in all fuzzy appliances.[66]

### 28.3.6  Miscellaneous Fuzzy Appliances

*Dishwasher.*    This fuzzy appliance uses two sensors to detect the number of dishes loaded and the amount and general type of food encrusted on the dishes. On the basis of these inputs, the fuzzy controller efficiently varies the soap, water, and cycle time.[11,67] The Maytag model also adjusts for dried food on dishes by tracking the time between loads. Turbidity, optical, and conductivity sensors are used to optimize cycle time and detergent usage.[68] Conventional dishwashers assume the worst case when washing dishes. Because the fuzzy system does not have to make these assumptions, the fuzzy controller should provide 10- to 40-percent water and energy savings.[69]

*Rice Cooker.*    The rice cooker uses three sensors to monitor the steam, the ambient temperature, and the volume of rice. Once a minute, the sensors are checked and the remaining cooking time is calculated.[45,54,64] The unit has four preprogrammed settings for different styles of rice: white, porridge, glutinous (sticky), and mixed. Other features are available that make cooking perfect rice both quick and easy.[54]

The use of fuzzy logic allows the user to fill the pot with the same amount of water each time. The controller changes the steaming method based on the type of rice desired. Without the use of fuzzy logic, the user would be required to change water levels depending on the type of rice desired and the volume of rice in the cooker.[44]

A variation on the rice cooker is in development. It not only cooks rice but seeks a larger market by cooking foods indigenous to many countries. This cooker is said to consume 9 percent less power than conventional designs.[50]

*Toaster.*    A fuzzy-logic toaster is appropriately termed "smart." It adjusts the heat and toasting time depending on the type of bread it senses in the toaster. The user's preferences are also learned and memorized.[11]

***Television Receiver.***   A high-priced fuzzy-logic television receiver is targeted in Japan at wealthy, middle-aged people who love gadgets.[67] The receiver uses fuzzy logic to react to changing conditions and will automatically increase brightness as the room grows darker and increase volume as the user moves farther away from the set.[11,54] Fuzzy TV sets show sharper pictures than traditional sets and automatically adjust contrast, brightness, and color. The TV set also maintains stability across channels, keeping settings constant even when the stations themselves vary the settings.[44]

***Carpet.***   The Japanese also use fuzzy logic in carpets. The carpet has a quick-heating function that adjusts to low room temperatures and warms hard-to-heat floors instantly. The carpet also uses fuzzy logic to adjust to changes in room temperature.[45]

***Shower.***   Fuzzy logic is used in a shower system to keep the water temperature stable even if the water pressure is changing.[3,10,27,64]

## 28.4   FUZZY APPLIANCE DESIGN APPLICATIONS

### 28.4.1   Energy Management System

This section describes existing methods that electric utilities use to reduce their load when the cost to produce energy exceeds the revenue that is produced and when system stability requires it. The benefits in customer satisfaction, the coming information superhighway, and the threats of deregulation in the electric utility business have been the impetus for electric utilities to offer customers a merging of demand-side management and energy management systems. In the future, electric rate schedules will be downloaded to customers periodically to allow them to schedule the use of their electric devices.

The goal of this section is to describe an energy management system for residential applications that makes use of the new two-way infrastructure and fuzzy logic. The inputs and outputs are described for each of these devices, along with how they will be used and generated. In addition to this, a full description is given of how the model will be implemented.

***Direct Load Control.***   Electric utilities have classically charged customers with a constant rate in cents per kilowatt-hour. There are different rates for different customer classes when their service requirements are substantially different. These rates are approved every four or five years by the individual state public service commissions and depend on the cost of service. There is a rate base that is determined by "the used and useful" equipment that the utility uses for generation, transmission, and distribution of electricity. The utility is allowed an agreed-upon rate of return for this rate base plus operation and maintenance costs.

In order to maximize profits, utilities generate, buy, and sell power in the most economical way. These transactions occur in centers known as *power pools*. The Michigan power pool, for example, is in Ann Arbor and is known as the Michigan Electric Power Coordination Center (MEPCC). Typically, arrangements can be made to operate at positive margins, but, during peak load times, inexpensive generation is difficult. To meet high-load demands, the utility must run expensive peaking units and may enter into negative margins for the incremental load. To avoid these conditions, it is desirable for the utility to either reduce the high loads or shift them to off-peak hours. Figure 28.2 shows the ideal case for utility generation.

In the ideal case the load is flat and does not vary with time of day or time of year. This is said to be ideal because, in this case, the utility would build generation to meet the base load and would not need to build, maintain, or operate any extra generation to meet peak loads.

Direct load control benefits the utility by not forcing it to run expensive peaking generation and by not operating with negative margins. Electric utilities began using direct load control radio networks in the mid-1960s and continue using these networks today. In these networks, one-way

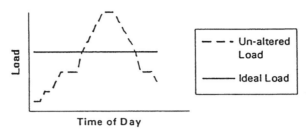

**FIGURE 28.2**  Ideal case for utility generation.

signals go to the customers' homes to cycle off high-load appliances such as air conditioners, water heaters, and heat pumps. In return for the utility's right to curtail power, the customer gets a reduced rate for these appliances.[70]

***Peak Shifting.***    A method for getting closer to the ideal load condition is known as *peak shifting*. It requires customers to modify their behavior by running appliances only during off-peak hours. One method of modifying customer behavior is through real-time pricing, by which a rate schedule is transferred to the customer once per week or once per month and then is tweaked occasionally by the utility during higher-than-expected generation cost periods. Customers choose to pay either the higher rates at peak load times or reduce their load. In either case the utility may not enter into negative margins.

### Implementing Future Energy Management Systems
  *The Problem and the Systems.*    The difficulty of implementing the systems of the last section is in taking a very complex problem of varying cost of generation with load, time of day, time of year, fuel cost, and electrical power system emergency and converting it to one in which a relatively unknowledgeable customer can make some simple choices about comfort and cost of supply. The second part of this problem involves implementing these systems at minimal cost, and minimal discomfort to the consumer. The appliances in the home must be controlled (i.e., turned on, off, reduced, or cycled) via an in-home communication system. Two organizations seem to have risen to the forefront for the in-home network: CEBus (see Chapter 29) and the LONWorks. These two have implemented similar systems in different ways. They use the existing power lines in the home for the communications media. This technology, known as Power Line Carrier (PLC) Communication, has quickly risen to be the chosen media for the home network.
  *CEBus.*    CEBus is a standard developed by the Electronic Industries Association's Consumer Electronics Group (now called the Consumer Electronics Manufacturers Association). The standard specifies transceivers and protocols for home automation.[71] One of the transceiver types specified is a spread spectrum power line carrier transceiver; this is the only one implemented. The Intelon Corporation has implemented the transceiver as a universal asynchronous receiver transmitter (UART) to be associated with a user processor.[72]
  *LONWorks.*    LONWorks has been implemented by the Echelon Corporation with a neuron chip for distributed control and communication, software for operation and network management, and transceivers of several different types. Among the transceivers are three power line carrier (PLC) types: PLT-10, PLT-20, and PLT-30. The PLT-10 is a spread spectrum transceiver in the range of 100 to 450 kHz. This transceiver was originally promoted for North American applications. Inexpensive spread spectrum PLC transceivers have problems with noise sources, and this caused Echelon to switch to the PLT-20 for North America. The PLT-20 and PLT-30 were originally meant for European applications and are narrow-band transceivers. The PLT-20 communicates at about 130 kHz and the PLT-30 at about 90 kHz.[73]
  *Residential Energy Management Equipment.*    Raytheon and Honeywell have developed residential energy management equipment for 120-V, 240-V, and HVAC devices. This equipment

is interfaced to the utilities infrastructure to get pricing information over the power line. The systems are designed to interact with the customer to make simple choices about the maximum price within comfort zones for which given appliances should be run. The customer makes the choices through a user interface with a liquid crystal display and push buttons. The user interface works very similar to a bank ATM. The choices made are relayed to the load nodes around the network. The home equipment currently costs several hundred dollars, but the price is coming down.

***Implementation of Fuzzy Logic in an Energy Management System.***    Figure 28.3 shows the method used to implement a fuzzy-logic–based energy management system. This can be demonstrated with a laptop computer, LONWorks-based PLC-10s, custom electronics to simulate and control electric loads, and four outlets. The communication medium will be PLC and will be implemented through the use of an SLTA (serial LONTalk adapter) with a PLT-10 transceiver being used with the laptop. PLC-10s will be used at different outlets throughout the room. Communication will occur between the outlets, and fuzzy logic algorithms will run both locally and centrally to perform the necessary energy management functions.[73]

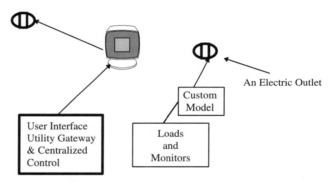

An Electric Outlet

Custom
Model

User Interface
Utility Gateway
& Centralized
Control

Loads
and
Monitors

**FIGURE 28.3**    Implementation for an energy management system.

The laptop has three roles:

- *User interface.*    The consumer of electricity has to have a method of entering his or her preferences for control.
- *Utility gateway.*    The price for electricity will be sent to the users once per week for the customer to make choices.
- *Centralized control.*    Some portions of the control will be performed at a central location, whereas others will be done remotely.

The custom modules have the following roles:

- *Distributed control.*    A certain portion of the control must be distributed, because, if a communication failure occurs, the remote devices actually shedding power must be intelligent enough to return power in these instances.
- *Monitoring.*    Parameters such as inside and outside temperature and humidity will be required throughout the network for the fuzzy-logic algorithms.

The laptop software was written with a package called *Wonderware*. This allowed development of a user interface for the fuzzy algorithms for the energy management system. There are six screens for the system, as follows:

- *Main screen.* Allows the user to input the system variables, move between screens, monitor the fuzzy outputs, and run the lighting or HVAC inputs through the algorithms to modify the fuzzy output.
- *Lighting screen.* Allows the user to input the fuzzy rules in an access database for all variations of the inputs.
- *HVAC screen.* Performs the same function for the HVAC system that the lighting screen does for the lighting system.
- *User interface screen.* Simulates the user interface for the lighting and HVAC. The user can indicate with a slider the desired lighting level for a room and the definitions for hot, cold, etc. for room temperature. The definitions for high, low, etc. for electric rate can also be indicated on this screen.
- *Simulation screen.* Allows the user to test the fuzzy algorithms and measure the energy savings for a typical day (defined in the access database).
- *Membership values.* Allows the developer to monitor the membership values for the fuzzy subsets.

***Equipment Control and Energy Management System Variables.*** The equipment loads that are modeled and implemented in fuzzy logic are lighting loads and clothes-cleaning loads. The inputs and outputs to control these loads in the fuzzy logic algorithms are described in the following.

The variables used for the control of the HVAC and lighting portions, along with their fuzzy linguistic values, are as follows:

| Control Variables | Linguistic Values |
| --- | --- |
| Local occupancy | Present and Absent |
| Electric rate | Low, Moderate, and High |
| Inside temperature | Cold, Cool, Comfortable, Warm, and Hot |
| Outside temperature | Cold, Comfortable, and Hot |
| Time of day | Home, Away, and Sleep |
| Ambient light | Low, Moderate, and High |
| Customer comfort light | Low, Moderate, and High |

The membership functions for these variables are as follows:

*Local occupancy.* This variable is monitored by a sensor in a room to indicate whether someone is present or not. It is a crisp variable that can be a 1 if present and a 0 if absent (Fig. 28.4).

*Time of day.* This variable is crisp and is programmed by the customer to indicate his or her normal pattern of being home, away, or asleep. Only one may be on at a given time (Fig. 28.5).

***Lighting Inputs and Outputs.*** The parameters for lighting control are shown in Fig. 28.6.

***Fuzzy Rules.*** The fuzzy rules for the lighting controller are as follows:

- If Occupancy = Absent, then Light = OFF
- If Time of Day = Away or Sleep, then Light = OFF
- If Ambient Light = High, then Light = OFF
- If Ambient Light = Low and Electric Rate = ! High and Inside Temperature = ! Hot, then Light = Customer Comfort

Electric Rate:

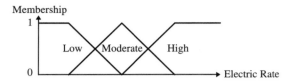

Inside Temperature:

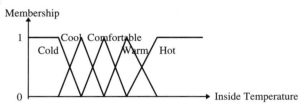

Outside Temperature:

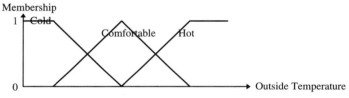

**FIGURE 28.4**    Membership functions for energy management system variables (local occupancy).

Ambient Light:

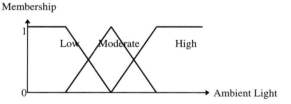

Customer Comfort Light:  This indicates the customer's desired level in a room.

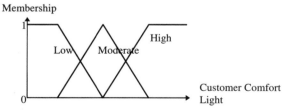

**FIGURE 28.5**    Membership functions for energy management system variables (time of day).

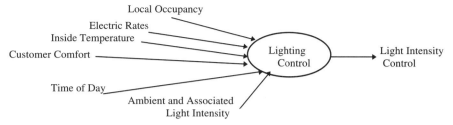

**FIGURE 28.6**   Inputs and output for lighting control.

- If Ambient Light = Moderate and Electric Rate = ! High and Inside Temperature = ! Hot, then Light = Customer Comfort − 1

- If Ambient Light = Low and Electric Rate = ! High and Inside Temperature = Hot, then Light = Customer Comfort − 1

- If Ambient Light = Moderate and Electric Rate = ! High and Inside Temperature = Hot, then Light = Customer Comfort − 2

- If Ambient Light = Low and Electric Rate = High and Inside Temperature = ! Hot, then Light = Customer Comfort − 1

- If Ambient Light = Moderate and Electric Rate = High and Inside Temperature = ! Hot, then Light = Customer Comfort − 2

- If Ambient Light = Low and Electric Rate = High and Inside Temperature = Hot, then Light = Customer Comfort − 2

- If Ambient Light = Moderate and Electric Rate = High and Inside Temperature = Hot, then Light = Customer Comfort − 3

***Simulation Results.***    Two simulations were run for both the lighting and HVAC portions of this system. The first simulation was for a conventional controller implemented in C and run for a typical day. The second simulation was with the computer model just described. The results were as follows:

- The fuzzy controller saved 45-percent more energy than the conventional controller. This is of course dependent on the implementation.

- The user interface for both simulations is simply a wall dial to indicate the customer's desired light level.

- Both controllers were of the same implementation difficulty.

   The fuzzy controller was successfully implemented and demonstrated for lighting and HVAC loads. These concepts could be easily extended to other types of loads (such as washing machines, clothes dryers, dishwashers, etc.).

### 28.4.2   Efficient HVAC Control Using Fuzzy Logic

A method is presented in this section to utilize energy more efficiently in heating and cooling of residential and commercial buildings by taking advantage of off-peak price reduction of electricity; outside air temperature; and other factors such as occupancy, time of day, and time of year. Zoning techniques are also to be taken into account as a method of maximizing efficiency. Also presented in this section is the use of a fuzzy logic controller to receive these inputs, and output control signals to select the most efficient method possible to provide comfortable temperatures all year round.

With ever-increasing energy costs in today's economy, new technologies must be relied upon to provide heating and cooling for homes and commercial buildings in the most efficient way possible. Much emphasis has been placed on the type of HVAC system, such as gas-forced air, heat pumps, and geothermal systems. This section will explore the control aspect of these systems by considering more than a single input (i.e., temperature) and more than a single output (i.e., turn on furnace). Moreover, a fuzzy-logic controller will be utilized to process the inputs and output accordingly using fuzzy relationships and rules.

***Existing Systems.***   Most control schemes for HVAC systems in use today are traditional analog electric circuits[74] that simply sense temperature and close a contact to either begin a heating cycle (for temperatures that are below the set point) or a cooling cycle (for temperatures that are above the set point). Even today's digital controllers merely change the set point as a function of time of day (set back thermostat), with the added input of a clock. If the system were to become more complex—such as taking into account the lag time of the HVAC system or the zone control—the mathematical modeling of the system for control purposes could become more complex. If not done precisely, the system can experience overshoot and undershoot,[74] wasting energy in the process.

***Inputs to an Efficient HVAC Controller.***   In order to develop a more energy-efficient control scheme, however, it is necessary to take into account many more inputs than simply temperature and time of day. For instance, it is not necessary to heat or cool a room that is infrequently occupied to the same degree as a room that is occupied very often. Also, if the air temperature outside the dwelling would assist in satisfying the temperature inside, energy recovery ventilation could be utilized.[75] These inputs, along with others, will be discussed and utilized to develop an efficient HVAC control system.

*Time of Day and Time of Year.*   By knowing the time of day and time of year, a controller could make more intelligent decisions as to whether or not it should begin a heating or cooling cycle. For example, many times during the summer months, there are nights that are relatively cooler than the days. Rather than go into a cooling cycle late in the day, the controller could opt to wait a short period in anticipation of cooler exterior temperatures, and then simply run the blower motor to draw in cool air via the fresh air make-up intake.

Leonard Bachman, an architecture professor from the University of Houston, has recently completed a climate atlas for the state of Texas that divides the state into several regions or "mini-climates" using a psychrometric chart that plots hourly dry-bulb and wet-bulb temperatures, giving a correlation of temperature and humidity.[76] This information could be utilized by the HVAC controller to predict time of day temperature differences.

Time of day could also be utilized in anticipation of occupants leaving the dwelling. For example, if the conditions called for heat, but the controller anticipated an empty building in 15 minutes, it would not initiate a heating cycle.

*Occupancy.*   Occupancy sensing, not only on the building level but on the individual room level, can greatly reduce the need to heat or cool the overall dwelling, or even individual rooms, with the use of duct dampers.[77] Although time of day plays a major part in whether or not a building is occupied, there are many instances, such as weekends, when occupancy sensing could save energy by letting the temperature drift from the set point if no one is present in the building. When individuals are present in a building, occupancy sensing at the room level, along with zone control, could allow unoccupied zone temperatures to stray certain amounts rather than heat or cool empty rooms (provided that temperature information for each zone was input to the controller as well).

Occupancy sensing could be achieved by using optical sensors and determining if the light in the room is on or off.[42] Another method would be to use passive infrared detection which, in many cases, is already present for security systems.

*Zone Temperatures.*   By knowing the temperature in each zone, the controller could satisfy the heating or cooling needs in individual zones by moving heat from hot areas to cool areas by simply running the blower motor and controlling duct-dampers.[77] The controller would be able to position the damper at various positions between fully open and fully closed to more effectively balance the various zone temperatures.

*External Temperature.*   The outside air temperature would have to be an input to the controller for two reasons:

- If the HVAC system were a heat pump, the controller would have to know when to utilize auxiliary heating when the outside temperature falls below about 35°.[78]

- There are occasional conditions when the outside air could be used to satisfy the interior conditions, and the controller would merely have to activate the blower motor, drawing air in from the fresh air make-up intake.

*Current Price of Energy.*   Electricity, like other commodities, varies its price with the level of demand. Typically, the rates fall during nonpeak hours (evenings and weekends) when the load is relatively low. For example, in Chenal Valley, a neighborhood of Little Rock, Ark., residents pay rates that range from 4 cents per kWh at times of low demand to 24 cents per kWh during peak load periods.[79] If this real-time information were available to an HVAC control system, it could make a decision on which method of heating or cooling was currently the most efficient— for example, whether to use gas or electric heat.[79] Even if the rate information were not available on a real-time basis, pricing tables based on time of day could be entered into the controller.

Even air conditioning could take advantage of the cheapest source of energy. Traditionally, air conditioning could only be run by electricity, because the compressor was coupled to an electric motor. However, York International has developed the first natural gas–operated heat pump. The system uses a 5-hp natural gas–driven engine developed by Briggs & Stratton.[80] Their system uses a Honeywell controller to optimize the speed of the engine to match the heating or cooling needs for maximum efficiency. The only electricity needed is for the blower motor.[80]

Rate information could also be used to select levels of comfort. For example, if the rates are at their minimum, the controller would maintain the temperature at 72 °F (cooling); if the rates increased, the controller would then maintain the temperature at 75 °F.[81] If the rates went critically high, the controller would opt not to supply cooling at all until rates decreased again, or, if a gas-powered heat pump were available, it would then take over. Electric utilities would benefit greatly from such a system because load shedding would be automatically carried out by the customer, initiated by the increase in the electric rate.

**Outputs of an Efficient HVAC Controller.**   An efficient HVAC controller would conceivably have the following outputs.

*Heating and Cooling Cycles.*   The controller would initiate the following cycles based on the type of HVAC system: heat pump–cooling, heat pump–heating, central air conditioning, gas–forced air heating, and electric resistive heating.

In the case of the heat pump, it is possible that the controller could control the speed of the compressor motor based on the heating or cooling needs. This could be accomplished by the use of a variable-speed drive unit, which controls the speed of an induction motor by varying the frequency of the voltage source to the motor. This would also allow the use of smaller, multiphase motors.

*Blower Motor.*   Unlike the blower motors of a typical HVAC system, which operate at only two speeds based on whether the system is providing heating or cooling, the controller would be able to vary the speed of the blower over a wide range with the use of a variable-speed drive unit, as mentioned previously. The ability to vary the speed in such a way would provide ultimate efficiency, providing only enough power to the blower to move the air as needed for the current conditions.

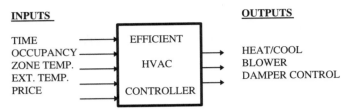

**FIGURE 28.7**   Block diagram of an efficient HVAC controller.

*Zone Control.*   The controller would have the ability to control the flow of air to each and all zones throughout the entire building. This is of course assuming that the hardware exists in the ductwork of the HVAC system. As mentioned earlier, the controller would have the ability to vary the damper in each duct anywhere between fully open and fully closed. Figure 28.7 shows the block diagram of an efficient HVAC controller.

***Implementation of an Efficient HVAC Controller.***   With the number of inputs and outputs involved in this control system, conventional control techniques would rapidly become very complex. Fuzzy control, which has already found its way into washing machines, vacuum cleaners, electric fans, and automobile transmissions,[82] would lend itself very well to the HVAC process, which is governed by essentially very simple rules.[74] In a conventional controller, what is modeled is the system or the process being controlled, whereas in a fuzzy-logic controller, the focus is the human operator's behavior.[82] Fuzzy systems do not need to be precise. The underlying model is not a string of differential equations, but a set of simple rules, such as "if the air is hot, run the motor fast."[74] It is not the intent of this section to provide a thorough design of a fuzzy-logic controller; however, a general overview will be given for the HVAC controller.

*Fuzzy Inputs.*   The first step in laying out the HVAC controller is to define the inputs and develop the fuzzy subsets. As stated earlier, the inputs for this system are time of day/year, occupancy, temperature for each zone, external temperature, and price of energy. The fuzzy subsets for time of day/year could be established as follows: (a) early morning, (b) morning, (c) midday, (d) afternoon, (e) evening, (f) late evening, (g) winter, (h) spring, (i) summer, and (j) fall. Membership functions for each of these subsets would have to be derived as well, defining a membership value in each subset for all values of time of day/year. For example, 0700 hours would have a membership value of possibly 0.9 in the early morning subset; however, it would most likely have a membership value of 0 in the evening subset.

The fuzzy subsets for occupancy could be defined as follows: (a) zone occupied and (b) zone unoccupied. Depending on the accuracy of the occupancy-sensing hardware, this input could possibly be crisp; that is, the membership values would be either a 1 or a 0. The zone temperatures would have subsets defined as follows: (a) cold, (b) slightly cold, (c) comfortable, (d) slightly warm, and (e) warm. Again, membership values would have to be assigned to each temperature in each zone of the building. However, the membership functions would be identical for each zone. Similarly, the input external temperature would have the following subsets: (a) very cold, (b) cold, (c), balmy, (d) hot, and (e) very hot. Finally, the input for price of energy would have the following subsets: (a) cheap, (b) average, and (c) expensive. These subsets would have to be applied to all energy sources available, for example, if the HVAC system had both electric heat and gas heat, there would be two inputs, and fuzzy subsets would be created for each.

*Fuzzy Outputs.*   The outputs of the HVAC controller are (a) heat cycle, (b) cool cycle, (c) blower motor, and (d) damper control. Just as was done with the inputs to the fuzzy-logic controller, the output fuzzy subsets must be defined.

In most cases the heat cycle is either on or off. However, if electric heat is used and it can be "phased on," or if a variable-speed heat pump exists, as was mentioned earlier, then we can define subsets with membership values anywhere between 0 and 1. These subsets could be defined as follows: (a) heat full on, (b) heat medium, (c) heat low, and (d) heat off. Similarly, the cool cycle would have the following subsets: (a) cool full on, (b) cool medium, (c) cool low, and (d) cool off.

Assuming a variable-speed blower motor, the fuzzy subsets for the output blower motor would be (a) blower high, (b) blower medium high, (c) blower medium low, (d) blower low, and (e) blower off. Finally, damper control for each zone would have the following subsets: (a) full open, (b) mid position, and (c) full closed.

*Fuzzy Rule Set.*  This part of the implementation would be most tedious, not because of the complexity of the rules, but because the number of inputs and outputs would require several rules. Again, because this chapter will not address the complete design of the controller, only some examples will be given. A partial rule set for a fuzzy-logic HVAC controller is as follows: (a) if zone N unoccupied and hot, then damper full closed, (b) if zone N hot and external temp. cold, then blower high, (c) if electric price expensive and zone N hot, then cool low, and (d) if time midday and zone N slightly cold, then heat off. Obviously, there are many more rules that have to be stated for this controller to be implemented fully.

***Simulation and Comparison with Classical Control Method.***  A scaled-down fuzzy HVAC controller was implemented in C language by developing a multidimensional array for the rule base, deriving the membership functions for each fuzzy variable, and utilizing the min-average method to determine the output of the controller. Code was also implemented to run a simulation by varying the outside temperature over a 24-hour period and simulating the response of the inside temperature for a particular zone. The output of the HVAC system was fed back into the derivation of the zone temperature, thus providing a fairly accurate simulation of the complete system. Occupancy and electric rate changes were also varied over the period to simulate their impact on the controller.

For comparison purposes a classical controller was also implemented using C language. This controller treated all the fuzzy variables as crisp inputs and then derived the output using the same logic that was implemented in the fuzzy rule base. The classical controller was then simulated using the same parameters as the fuzzy controller.

The following graphs show the results of the simulation. Figure 28.8 simulates an outside temperature swing between 20 °F and 45 °F and shows the effect of the fuzzy controller in the heating mode. The "Home" input goes "hi" when the zone is occupied and the rates are increased for a brief period during the cycle. The membership functions regulate the temperature at 70 °F when the zone is occupied and allow it to drop off to 65 °F when the zone is vacated. Note that the increase in the electric rate has no effect on the heating because the rule base assumes gas heating.

Figure 28.9 shows the results for the classical controller. Note the "ripple" in the inside temperature. This is due to the hysteresis that is designed into the controller to avoid *cycling* of the HVAC system.

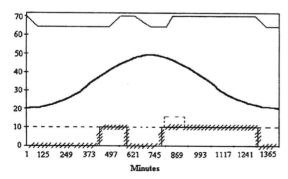

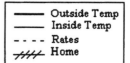

**FIGURE 28.8**  Fuzzy controller (heating).

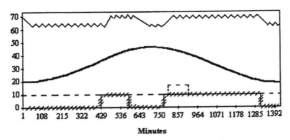

**FIGURE 28.9**   Classical controller (heating).

Figures 28.10 and 28.11 show similar data for a temperature swing between 65 °F and 90 °F and show the effects of the fuzzy-logic control (Fig. 28.10) and the classical control (Fig. 28.11) for cooling. Note that, during the cooling cycle, the temperature is maintained at 70 °F when the zone is occupied and allowed to drift to 75 °F when the zone is vacated. Note also that the electric rate now plays a role in the output of the controller, as can be seen when the rate soars to 15 cents. The ripple in the inside temperature for the classical model is again due to hysteresis.

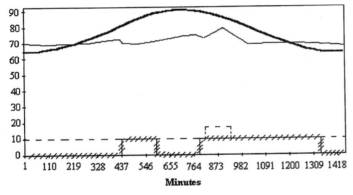
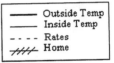

**FIGURE 28.10**   Fuzzy controller (cooling).

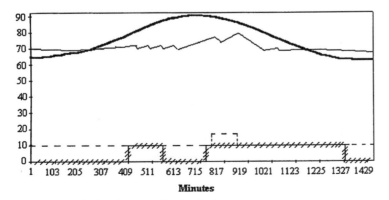

**FIGURE 28.11**   Classical controller (cooling).

**TABLE 28.2**   Energy Usage for Classical and Fuzzy Controls

| Temperature swing | Classical control | Fuzzy control |
|---|---|---|
| 20 °F–45 °F | Heat: 638<br>Cool: 0<br>Blower: 0 | Heat: 644<br>Cool: 0<br>Blower: 0 |
| 65 °F–90 °F | Heat: 0<br>Cool: 206<br>Blower: 99 | Heat: 2<br>Cool: 219<br>Blower: 99 |

This simulation also tracked the energy usage of the HVAC system in terms of heating units, cooling units, and blowing units. The energy usage for both controllers is shown in Table 28.2. As can be seen from the table, the HVAC system, as controlled by the fuzzy controller, used slightly more energy units than that of the classical controller. This is due primarily to the fact that the classical controller allows the temperature to drift by approximately 0.3 °F before initiating another heating or cooling cycle, whereas the fuzzy controller provides optimum regulation.

*State of the Art.*   Apparently, no systems exist that utilize fuzzy logic in HVAC control. This is most likely due to two reasons. First, the hardware needed to implement such a system is currently more expensive than a classical controller. Second, the classical controller performs an adequate job of heating and cooling. The two systems are equally complex as far as implementation; however, the fuzzy controller is easier to fine-tune to the desired results. As the price of fuzzy-logic hardware begins to drop, more applications in the HVAC industry will utilize this form of control.

## 28.5   THE BENEFITS OF FUZZY LOGIC IN APPLIANCES

The commercial importance of fuzzy logic is growing, especially in Japan, where consumers have enthusiastically accepted this technology.[83] In the 1980s the Japanese embraced fuzzy logic as the technology of choice for control systems.[34] The first commercial fuzzy boom was created in 1990, and now the Japanese are preparing for the fifth round and beyond.[38,84] Reportedly, 8 out of every 10 fuzzy applications have been successful.[1] The 12 examples of fuzzy logic presented in this chapter illustrate some of the benefits gained by using this technology. In general, these benefits are simplicity, rapid prototyping, cost efficiency, and flexibility.

### 28.5.1   Simplicity

Fuzzy logic uses preprocessing to turn a large range of values into a small number of membership grades. This action reduces the number of values that must be evaluated by the controller and also reduces the number of rules. The fewer the rules, the faster the fuzzy-logic system can generate output. Most consumer products use less than 20 rules. Typically, a fuzzy system will result in a 10:1 rule reduction, requiring less software and memory to implement the same decision-making capability of conventional methods.[4,85]

### 28.5.2   Rapid Prototyping

This benefit helps the designer react more quickly to market needs. The similarity between human thinking and fuzzy logic facilitates understanding and troubleshooting, which reduces a product's time to market. In a comparison between fuzzy and standard logic, engineers at Rockwell found the fuzzy system resulted in simpler rules, less math, and a more accurate representation of the process. Omron engineers found that fuzzy logic slashed development time by 75 percent.[53]

### 28.5.3  Cost Efficiency

Fuzzy logic exhibits cost efficiency through the reduction in circuitry and in energy savings. With fuzzy implementation, the potential exists to reduce circuitry by a factor of 1000, making simple learning controllers possible.[54] Designers can add advanced features to low-cost microcontrollers. Features that would normally require a 32-bit processor can be implemented in fuzzy logic with a low-cost controller.[86] The consumer experiences these savings in price competition between companies. Energy savings are realized through cycles that are tailored to the specific conditions.[87] For example, the Mitsubishi air conditioner boasts a 24 percent power savings.[35]

### 28.5.4  Flexibility

Flexibility in the appliance could be defined as an adaptation to new inputs and users. The appliances detailed in this chapter definitely illustrate this benefit. Each unit adjusts its output based on constantly changing conditions, such as going from a carpeted to a bare floor. New users are also considered in the fuzzy control system—for example, in the air conditioner that senses the number of people in the room and cools accordingly.[1,10,63]

## 28.6  THE NEGATIVE SIDE

With fuzzy-logic appliances experiencing immense sales in Japan, why aren't they being imported to the United States? Americans would surely be interested in less housework. Reasons vary from the philosophical—the acceptance of imprecision—to the practical—the appliances are structurally different. Physically, Japanese appliances are smaller than their U.S. counterparts; the added cost of shipping a large washing machine overseas would not be profitable. One solution is to release the appliances in the United States, which is exactly what General Electric is considering. General Electric is currently developing a fuzzy-logic dishwasher for release in the United States, although the dishwasher will probably be marketed as "smart" or "intelligent," not fuzzy.[44]

Critics claim that improved sensor technology is the real hero in these appliance applications. Although it is true that better sensors have helped fuzzy logic, they are not the centerpiece of this control technique. All controllers need sensor information to make decisions and have benefited from better and more inexpensive sensors. Fuzzy logic takes advantage of the low degree of precision required for appliance control and turns it into an efficient design. For example, room temperature does not need to be controlled to 0.1 degree. Therefore, an inexpensive, lower-precision sensor can be used with less overhead required than with conventional controllers. Examples such as a washing machine application using 1000 to 2000 rules with a standard PID controller that is implemented with fuzzy logic using 200 rules are not uncommon.[44]

Fuzzy-logic appliances have also been accused of being nothing more than conventional logic using lookup tables and percentages.[44] If some applications have adopted this method and labeled the product as using fuzzy logic, then they are clearly abusing the name recognition that "fuzzy" has in some countries. However, this misuse of the term should not detract from this control methodology. Countless *real* applications of fuzzy logic have been made in household appliances.

## 28.7  THE FUZZY FUTURE

It took two decades for fuzzy logic to transfer from theorists' hands into production. Now engineers worldwide are dreaming up new applications for fuzzy control systems that greatly enhance consumer products.[59] Currently, Japanese companies have the lead in developing fuzzy-logic–based hardware and software, and they have been particularly successful in pushing this technology into consumer products.[19] However, the Europeans are joining the Japanese at the forefront of fuzzy-logic development with major projects underway in France, Germany, and Spain.[33]

Fuzzy logic has been a laboratory toy in the United States since the early 1970s, but the technology has not been widely seen in consumer markets.[88] Recently, however, U.S. businesses have taken an interest in this field. One example is Motorola's Semiconductor Products Sector, which has developed microcontroller technology that makes use of fuzzy logic for automotive applications.

According to some forecasts, the worldwide market for fuzzy-logic semiconductors may total $10–$13 billion by the year 2000.[89] A Goldstar spokesman predicted that "70 to 80 percent of all electric home appliances will use either fuzzy logic or artificial intelligence by 1995."[48]

## GLOSSARY

**Defuzzification**  The process of converting fuzzy outputs into crisp control actions.

**Fuzzification**  The process by which inputs can be transformed into fuzzy subsets.

**Fuzzy adjective**  The fuzzy value of a linguistic variable in the domain of that variable.

**Fuzzy appliance**  The use of a fuzzy controller to control the appliance.

**Fuzzy hardware**  Hardware components that are needed to implement fuzzy logic and control operations.

**Fuzzy inference**  A process of combining conditional (a rule or a relation) and unconditional (a proposition) fuzzy rules or relations.

**Fuzzy logic**  Fuzzy logic deals with propositions that may subjectively ascribe values between zero and one (a closed interval) either in a continuous or a discrete fashion.

**Fuzzy rule**  An association between two or more fuzzy adjectives or two or more fuzzy variables. For example, if temperature is HOT, then pressure is HIGH.

**Fuzzy software**  Software tools that are used to implement and simulate fuzzy logic and control operations.

**Linguistic hedges**  Linguistic terms used to modify fuzzy adjectives. For example, let HOT be a fuzzy adjective; if the linguistic hedge VERY is used to modify HOT, then the fuzzy adjective becomes VERY HOT.

**Linguistic variables**  Fuzzy variables whose value in any one particular instance is a fuzzy subset of a universe of discourse.

**Neural network**  A network that recognizes ill-defined patterns without an explicit set of rules.

**Neuro-fuzzy logic**  The use of both fuzzy logic and neural logic in controlling a process.

**PID controller**  Proportional, integral, and differential controller.

## REFERENCES

1. "Fuzzy logic in Japan: Clearly important and becoming more so," *NTIS Foreign Technology,* vol. 90, no. 20, 1990.

2. L. A. Berardinis, "Building better models with fuzzy logic," *Machine Design,* vol. 64, no. 11, 1992.

3. T. Buckley, "Forecast for a fuzzy future," *Marketing Computers,* Dec. 1990.

4. K. Self, "Designing with fuzzy logic," *IEEE Spectrum,* vol. 27, no. 11, 1990.

5. A. Sprout, "Products to watch," *Fortune,* vol. 124, no. 14, 1991.

6. T. Williams, "Fuzzy logic is anything but fuzzy," *Computer Design,* April 1992.

7. L. A. Berardinis, "Strong bonds link neural and fuzzy computing," *Machine Design,* vol. 64, no. 15, 1992.

8. J. M. Sibigtroth, "Creating fuzzy micros," *Embedded Systems Programming,* vol. 4, no. 12, 1991.

9. F. S. Wong, Wang, P. Z., Goh, T. H., and Quek, B. K., "Fuzzy neural systems for stock selection," *Financial Analysis Journal,* vol. 48, no. 1, 1992.

10. M. Rogers and Hoshai, Y., "The future looks 'fuzzy'," *Newsweek,* vol. 115, no. 22, 1990.

11. T. R. Reid, "The future of electronics looks 'fuzzy': Japanese firms selling computer logic products," *Washington Post,* p. H-1.

12. "Full automatic, electric washing machine using fuzzy theory," *New Technology Japan,* vol. 17, no. 12, 1990.

13. B. B. Brey, *Microprocessor/Hardware Interfacing and Applications,* Merrill, 1984.

14. R. Meadows and Parsons, A. J., *Microprocessors: Essentials, Components, and Systems,* Pitman, 1984.

15. W. V. Subbardo, *Microprocessors Hardware, Software & Design Applications,* Prentice-Hall, Englewood Cliffs, N.J., 1984.

16. R. J. Tocci and Laskowski, L. P., *Microprocessors and Microcomputers: Hardware and Software,* 3rd ed., Prentice-Hall, Englewood Cliffs, N.J., 1987.

17. M. F. Hordeski, *Microprocessors in Industry,* Van Nostrand Reinhold, New York, N.Y., 1984.

18. "Domotechnica '91 emphasizes high-tech environmental protection," *Journal of the Electronics Industry,* vol. 38, no. 4, 1991.

19. S. Loe, "SGS-Thomson launches fuzzy-logic research push," *Electronic World News,* 21 Aug. 1991, p. 1.

20. D. I. Brubaker, "Fuzzy-logic system solved problem," *Engineering Design News,* vol. 37, no. 13, 1992.

21. D. I. Brubaker, "Fuzzy logic basics: Intuitive rules replace complex math," *Engineering Design News,* vol. 37, no. 13, 1992.

22. L. A. Berardinis, "Clear thinking on fuzzy logic," *Machine Design,* vol. 64, no. 11, 1992.

23. F. J. Bartos, "The basics of fuzzy logic," *Control Engineering,* vol. 38, no. 9, 1991.

24. G. B. Cunningham, "Integrating fuzzy logic technology into control systems," presented at AIAA '91, 1991.

25. E. Cox, "The seven noble truths of fuzzy logic," *Computer Design,* Apr. 1992.

26. J. M. Sibigtroth, "Implementing fuzzy expert rules in hardware," *AI Expert,* vol. 7, no. 4, 1992.

27. H. Mitsusada, "Fuzzy logic moving smoothly into the home," *Japanese Economic Journal,* 16 June 1990.

28. A. Sangalli, "Fuzzy logic goes to market," *New Scientist,* vol. 133, no. 1807, 1992.

29. R. K. Jurgen, "Consumer electronics," *IEEE Spectrum,* vol. 28, no. 1, Jan. 1991.

30. "Japanese ministry creates 'fuzzy' research group," *Newsbytes News Network*, 18 Mar. 1991.

31. "Getting fuzzy," *The Economist,* vol. 315, no. 7658, 1990.

32. R. C. Johnson, "Japan clear on fuzzy-neural link," *Electronic Engineering Times,* 20 Aug. 1990, p. 16.

33. R. C. Johnson, "Europe gets into fuzzy logic," *Electronic Engineering Times,* 11 Nov. 1991, p. 31.

34. R. C. Johnson, "Fuzzy logic reawakens," *Electronic Engineering Times,* 21 Jan. 1991, p. 50.

35. T. Imaiida, Hirao, T., Kidokoro, N., Morita, A., and Kato, T., "Development of the fuzzy logic control system for heat pump air conditioners," *Mitsubishi Heavy Industries, Ltd. Technical Review,* vol. 27, no. 3, 1990.

36. "Now neural-fuzzy logic," *Appliance Manufacture,* vol. 40, no. 4, 1992.

37. "Logic calculations is less time," *DIY Week,* 5 Nov. 1991, p. 5.

38. "Fuzzy logic washing machine," *New Materials & Technology Korea,* vol. 4.
39. D. Hulme, "Japan starts fuzzy logic II," *Machine Design,* vol. 64, no. 18, 1992.
40. B. Kosko and Isaka, S., "Fuzzy logic," *Scientific American,* vol. 268, no. 6, 1993.
41. G. Legg, "Microcontrollers embrace fuzzy logic," *Engineering Design News,* vol. 268, no. 19.
42. L. A. Berardinis, "It washes! It rinses! It talks!," *Machine Design,* vol. 63, no. 19, 1991.
43. "Machine decides how to wash your clothes," *Asian Street Journal Weekly,* 25 Dec. 1989, p. 8.
44. D. McNeill and Freiberger, P., *Fuzzy Logic,* Simon & Schuster, New York, New York, 1993.
45. N. C. Remich, Jr., "Fuzzy-logic appliances," *Appliance Manufacture,* vol. 39, no. 4, 1991.
46. S. Dutta, "Fuzzy logic application: Technological and strategic issues," *IEEE Trans. on Engineering Management,* vol. 40, no. 3, 1993.
47. "AI controls automatic washing machine," *Journal of Electronics Industry,* vol. 38, no. 3, 1991.
48. "Fuzzy logic is clearly a trend," *Korea High Tech Review,* vol. 5, no. 11, 1990.
49. K. Hirota, "Fuzzy concept very clear in Japan," *Asahi Evening News,* 20 Aug. 1991, p. 35.
50. R. C. Johnson, "Aptronix launches China fuzzy-logic bid," *Electronic Engineering Times,* no. 791, 1994.
51. "Neuro-fuzzy logic linked to bubble action," *Appliance Manufacture,* vol. 42, no. 5, 1993.
52. "Sharp introduces bubble action washing machine," *Japanese Consumer Electronics Scan,* 8 July 1991.
53. G. Slutsker, "Why fuzzy logic is good business," *Forbes,* vol. 147, no. 10, 1991.
54. N. C. Remich, Jr., "Fuzzy-logic now across most appliance lines," *Appliance Manufacture,* vol. 40, no. 4, 1992.
55. "Vacuum cleaner features a smart analog controller," *Appliance Manufacture,* vol. 43, no. 2, 1995.
56. A. Pargh, "Vacuum cleaners getting smarter," *Design News,* vol. 48, no. 7, 1992.
57. A. Tanzer, "Techie heaven," *Forbes,* vol. 148, no. 6, 1991.
58. Untitled article, *Consumer Electronics,* vol. 30, no. 31, 1990.
59. D. Kaplan and Tinnelly, B., "Fuzzy logic finds a home," *Electronic World News,* 23 July 1991, p. C15.
60. M. Inaba, "Matsushita 'fuzzy logic' m'wave," *HFD,* 8 Oct. 1990, p. 252.
61. "Fuzzy logic refrigerator," *Appliance Manufacture,* vol. 40., no. 2, 1992.
62. R. J. Babyak, "Designing the future," *Appliance Manufacture,* vol. 40, no. 7, 1992.
63. "NEC and Omron to make fuzzy microcomputer," *Newsbytes News Network,* 5 Aug. 1991.
64. R. C. Johnson, "Fuzzy logic escapes the laboratory," *Electronic Engineering Times,* 20 Aug. 1990, p. 16.
65. "Matsushita Electric to market new air conditioner," *Comline Electronics,* 4 Oct. 1990, p. 6.
66. "HVAC next with fuzzy logic and neural networks," *Appliance Manufacture,* vol. 40, no. 4, 1992.
67. "Invisible at home, fuzzy logic crosses the Pacific and bursts out all over," *Computergram International,* no. 1605, 1991.
68. "Dishwasher cleans up with fuzzy logic," *Machine Design,* vol. 67, no. 6, 1995.
69. J. Shandle, "Technology in 1992 ascended to new heights," *Electronic Design,* vol. 40, no. 26, 1992.
70. "Public utility reports view," *Public Utility Reports,* 1991.
71. "Home automation standards," EIA–Consumer Electronics Group, 1993.
72. Product catalog, Intelon Corp.
73. Product catalog, Echelon Corp.
74. "The logic that dares not speak its name," *Economist,* 6 June 1994, p. 89.
75. "Family values," *Popular Science,* Feb. 1995, p. 66.
76. D. Addison, "Comfort zones," *Weatherwise,* June/July 1994, pp. 14–19.
77. M. Morris, "The new American home 1995," *Home Mechanix,* Feb. 1995, pp. 68–71.
78. A. Hingley, "Air-conditioning choices," *Home Mechanix,* Sept. 1994, pp. 16–17.

79. C. Flavin and Lessen, N., "The electricity industry sees the light," *Technology Review,* May/June 1995, pp. 42–49.

80. "First gas heat pump," *Home Mechanix,* Oct. 1994, pp. 28–29.

81. "Giving customers control of their buying decisions produces big savings," *Consumers' Research Magazine,* July 1994, p. 41.

82. G. J. Klir and Schwartz, D. G., "Fuzzy logic flowers in Japan," *IEEE Spectrum,* July 1992, pp. 32–35.

83. D. Dunn, "Fuzzy logic R&D," *Electronic Buyer's News,* 14 Oct. 1991, p. 3.

84. C. R. Johnson, "Making the neural-fuzzy connection," *Electronic Engineering Times,* no. 765, 1993.

85. R. J. Babyak, "To be fuzzy, or not to be fuzzy," *Appliance Manufacture,* vol. 41, 1993.

86. R. Nass, "Fuzzy logic finally gains acceptance in the U.S.," *Electronic Design,* vol. 40, no. 13, 1992.

87. N. C. Remich, Jr., "Ecology drives what's new," *Appliance Manufacture,* vol. 43, no. 2, 1995.

88. "SGS-Thomson is rushing to catch up in embedded fuzzy code," *Computergram International,* no. 1741, 1991.

89. R. Woolnough, "SGS-Thomson to make fuzzy ICs," *Electronic Engineering Times,* 11 Nov. 1991, p. 31.

## ABOUT THE AUTHOR

Adnan K. Shaout is an associate professor in the electrical and computer engineering department at the University of Michigan-Dearborn. At present, he teaches courses in fuzzy engineering applications and computer engineering (hardware and software). His current research is in applications of fuzzy set theory, computer design (hardware and software), computer arithmetics, parallel processing, artificial intelligence and expert systems, image processing, and pattern recognition.

Dr. Shaout has about 13 years experience in teaching and conducting research in electrical and computer engineering at Syracuse University and the University of Michigan-Dearborn. He has published over 60 papers in topics related to electrical and computer engineering. Dr. Shaout obtained his B.Sc., M.Sc., and Ph.D. in computer engineering from Syracuse University, Syracuse, N.Y., in 1982, 1983, and 1987, respectively.

# CHAPTER 29
# THE CONSUMER ELECTRONICS BUS

**Grayson Evans**
*The Training Department*

## 29.1  INTRODUCTION

CEBus (the Consumer Electronics Bus, pronounced "see-bus") is a communications and product interoperability standard designed primarily for consumer products. The goal of CEBus is to allow compliant consumer products to be purchased incrementally by the consumer, brought home, plugged in, and operated together. The standard specifies how products communicate, the media available to use, and what they say to each other. Although residence-oriented, there is nothing that specifically restricts CEBus to residential applications.

The job of writing the standard began in 1984 with a group of consumer electronic member companies of the Electronic Industries Association (EIA). The first version was released as IS-60 (Interim Standard 60) in 1992 for industry review. The standard was revised during 1993 and 1994 and is in the process of release as a joint EIA/ANSI (American National Standards Institute) standard (EIA-600).

CEBus is an open, nonproprietary standard that has been endorsed by dozens of companies in many industry groups that supply products to the home. This includes the audio/video industry, security, climate control, appliance, lighting, computer, and telecommunications companies. The intent of the standard is to allow products from any industry group to control and/or use the resources in products across and within industry groups—to achieve true plug-and-play interoperability.

## 29.2  WHY RESIDENTIAL NETWORKS?

During the 1990s there has been a growing need to solve a related set of communication problems in the home. These problems are the result of the spread of products such as home computers, home automation systems, home theater, energy management systems, and telecommunication products. Traditionally, consumer products are designed to work standing alone. This has perpetuated redundant product functions such as user interfaces, and event scheduling, as well as proprietary wiring. It has made interoperation of products and the distribution of information the business of custom installers. The problem will only get worse with the rise of wide-band service providers.

CEBus solves the problems of integrating products, systems, and services from different manufacturers by establishing the following:

- A standardized, physical product interface so devices can literally plug and play.
- A standardized way to distribute control and broadband data on multiple media.
- A standardized common language for product interoperation.

CEBus allows products to share information such as the time, temperature, occupancy state, status of equipment, etc. This allows centralization of redundant product functions, the removal of the cumbersome user interface from many products, and easy delivery of outside service information directly to products. Products simply "place" information on the network, where it is picked up by products that can use the information to their advantage. Information can originate in the home or from service providers outside the home. One of the first examples of this feature is the delivery of electric rate information by electric utilities. CEBus products such as electric water heaters, dryers, and other high energy users receive the information off the network and can delay operation until rates are lower.

## 29.3   HOW CEBus WORKS

CEBus compliant products work by momentarily *connecting* to another product on the network to perform a sense or control operation. This *virtual connection* is achieved by gaining exclusive use of the medium that connects the two products long enough to transmit a command or request. Once communication is complete, the transmitting devices release the use of the medium. The medium can be the power line wiring, twisted-pair cable, coaxial cable, RF, or infrared (IR). Messages are short, lasting an average of 25 ms, and contain 50 to 300 bits. By keeping messages short, many devices can share the medium without conflict because messages between any two products are relatively infrequent.

To ensure high message delivery reliability, and to ensure that products do not all transmit at the same time, CEBus devices adhere to a strict message protocol. A protocol is a set of rules that define how and when messages are sent, how to recover from transmission errors, the format of messages, etc. The messages contain commands in a common *command language* (CAL) understood by each product. The command language is specifically designed for information sharing and control of residential consumer products. CEBus-compliant products are required to understand a minimum subset of the command language and that portion of the language specific to their product category.

### 29.3.1   Nodes

Any CEBus-compliant device attached to a medium is referred to as a *CEBus node*. A node is any device that can transmit and receive CEBus packets on at least one medium, adheres to the CEBus message protocol, and uses and understands a minimum set of the CAL language.

The minimum CAL language requirements comprise a common set of network management commands plus commands that are specific to the device category. For example, a CEBus light switch must understand the commands to turn on and off the light, but it does not need to understand a command to "turn to channel 13." It must, however, reply that it cannot do "tuning."

### 29.3.2   The CEBus Product Model

All CEBus nodes consist of the three major parts shown in Fig. 29.1. This diagram represents a simple internal model of every node. The parts can be implemented using any combination of hardware or software.

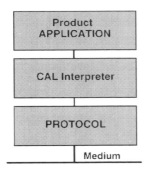

**FIGURE 29.1**   Three major parts of all CEBus nodes.

The protocol is the same in each CEBus compliant device and is responsible for reliable message delivery. The protocol software defines the format of the transmitted packets, the packet delivery services (error detection, message priority, retransmission of lost messages, responses to messages, etc.), and the technique for accessing and transmitting messages on each medium.

The CAL portion of the model is responsible for converting product application events into CAL language messages—or words—that other CEBus nodes understand. Likewise, it interprets received messages and affects product operation.

The *application* represents the specific node application for the product and contains the hardware and software that define the product operation. CEBus uses a "peer-to-peer" connectionless

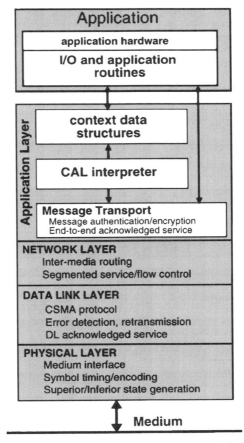

**FIGURE 29.2**   The formal protocol layers of the CEBus model.

service, CSMA/CDCR communication protocol (Carrier Sense Multiple Access with Collision Detection and Collision Resolution). The OSI protocol stack consists of a physical layer, a data link layer, a network layer, and an application layer. Many functions of a transport layer (segmented service, end-to-end acknowledgement) are incorporated into the application and network layers. The protocol service set is a compromise of highly reliable, secure message delivery and easy, low-cost implementation commensurate with consumer products.

Figure 29.2 shows the model in terms of the formal protocol layers. The *I/O and application routines* handle the interface of the device hardware and application software to the *context data structure* and the protocol routines. The application reads and writes the context data structure to reflect product operation. The context data structures serve as the interface from the CEBus network to the product resources. The application can generate messages by writing to the context data structure or by forming a message "manually" and passing it to the *message transport* sublayer (responsible for receiving and generating messages). The context data structure represents a software model of the product's operation to the rest of the network. The *CAL interpreter* translates changes in the context data structure to an appropriate message to another node. Received messages are interpreted and the context data structure is updated accordingly.

Formally, the CAL interpreter and its associated context data structures are part of the protocol stack—the *application layer*—but considering the interpreter to be the protocol application is convenient because its function is unique to CEBus. The CEBus standard treats the CEBus protocol as the physical layer through the message transport layer.

## 29.4  NETWORK COMMUNICATIONS MODELS

Every communications protocol—CEBus included—makes assumptions about the hierarchy of device-to-device communications in terms of network access control and device control. The CEBus protocol uses a peer-to-peer communications model (Fig. 29.3). This means that the protocol is designed to allow any node on any media to communicate with any other node in the home. There is no communication hierarchy or restriction on product-to-product communications over any media. This communications model is essential if CEBus-compatible products are to be installed in the home incrementally. As devices are added to the network, they must be able to communicate with any existing products. There can be no assumptions about the order or the type of devices added to the network.

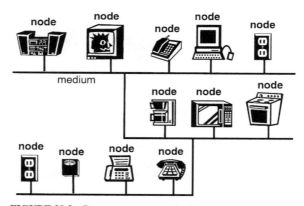

**FIGURE 29.3**  Peer-to-peer communications model.

The communications model also assumes that there is only one CEBus medium shared by all nodes. Though many physical media types are used (twisted-pair, coaxial cable, RF, etc.), they are treated as a single medium, and any message generated by any node will arrive at all other nodes on the medium.

CEBus is a connectionless service protocol. This means that devices can gain access to the network media only long enough to transmit a message and then get off. No "connection" is formed between the two devices to communicate, tying the network up during their "conversation." This is different from a typical connected service network such as the telephone network, in which two *nodes* establish a communication connection and retain exclusive use of the medium until all conversation is complete, then hang up, releasing the network resources for another call.

## 29.5  NETWORK CONTROL MODEL

The protocol supports two control models (Fig. 29.4). A control model defines which nodes can control other nodes on the network. The CEBus protocol assumes peer-to-peer distributed control, allowing any node to control any other node. This is also a result of how CEBus networks are built—incrementally and in random order. Any product can control any other product in the home.

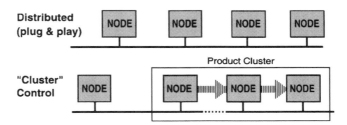

**FIGURE 29.4**    The CEBus protocol supports two control models.

No type of central control device or automation system is necessary in CEBus. The absence of central control was a development goal of the standard. Products may be added at any time by the home owner without having to notify or program a central system.

Although CEBus is not dependent on central control, it does not exclude it or the most common variation—the cluster control model. Cluster control allows one or more nodes to assume the task of controlling several other nodes. This model is employed for system-oriented products in the home (such as HVAC systems, security systems, and lighting control systems) and is used by these systems for normal human control. To set the temperature in the downstairs air-conditioning zone to 69 °F, a user might use an on-screen menu on the living room TV set. The TV set sends a message to the downstairs thermostat requesting a new temperature. The thermostat, in turn, sends (over TP) the control messages required to achieve 69 °F to the HVAC air-handling unit, compressor, motors, etc. The TV set could send messages directly to the HVAC air handler, compressor, motors, etc., but it would not know what to tell these devices. It does not contain any HVAC system control algorithms. Telling the thermostat is easier and lets the thermostat keep track of the HVAC system. The same control scenario applies to the lighting control and security systems.

## 29.6  CEBus REFERENCE ARCHITECTURE AND MEDIA

The CEBus standard establishes a set of physical layer medium specifications to handle all the data communication needs of the home (audio, video, computer data, etc.). It defines the use

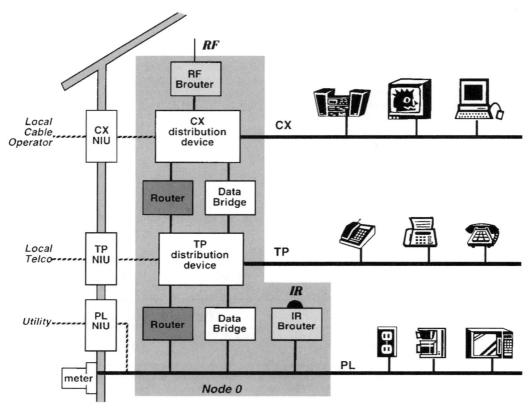

**FIGURE 29.5** The reference architecture for all media supported by CEBus and the interconnection of the media in the home and to outside providers.

of the power line wiring (PL), radio frequency (RF), infrared (IR), and the installation and use of twisted pair (TP) and coaxial cable (CX). Figure 29.5 illustrates the reference architecture, or *topology,* for all media supported by CEBus, as well as the interconnection of the media in the home and to outside service providers.

Figure 29.5 shows all of the media support components that might be used in a typical residential environment for CEBus message communications, as well as maximum utilization of internally and externally supplied wide-bandwidth services. The standard is flexible enough, however, to allow the use of CEBus products in existing homes using the PL, RF, IR, and a majority of existing TP wiring with a minimum of infrastructure support. There is no "initial" network to purchase and install. As additional products are purchased that utilize more of the capability of the media, additional infrastructure support components can be added incrementally.

All media support components are collectively known as *node 0.* The term *node 0* refers to all the physical components necessary to support the various media networks. *Node 0* consists of all routers, brouters, data bridges, and the TP, CX, and PL support hardware necessary for a given installation. The TP network requires a power supply and a distribution/termination device. The CX network requires a distribution device containing block conversion, amplification, and control channel regeneration functions.

Routers connect CEBus messages between two of the wired media (PL, TP, CX). Brouters connect CEBus messages between RF or IR and a wired medium. Data bridges connect any audio, video, or other wide-bandwidth signals between the media.

The TP network architecture is based on TIA-570 (Residential and Light Commercial Premise Wiring standard). Branch runs of four-pair jacketed cable originate at the distribution device and run to each room, where they connect to one or more outlets. Two of the pairs are used for external services; two are reserved for CEBus communications.

The CX network consists of a pair of RG-6 cables that originate at the distribution device and run to each room, where they split to one to four dual outlets. One cable handles delivery of external services (such as cable TV); the other cable is for distribution of in-home-generated control and data.

*Network interface units* (NIUs), typically supplied by an external service provider, provide the interface between the electrical characteristics of the external network and the electrical characteristics of the CEBus network. An NIU may interface any external network medium (coax, twisted pair, RF) to any CEBus medium.

## 29.7  CHANNELS

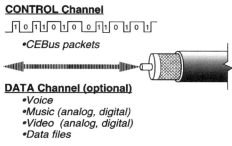

**CONTROL Channel**

•*CEBus packets*

**DATA Channel (optional)**
- •*Voice*
- •*Music (analog, digital)*
- •*Video  (analog, digital)*
- •*Data files*

**FIGURE 29.6**   Each CEBus medium has two communication channels: a required control channel for device communication and a set of optional data channels for distribution of wide-bandwidth signals.

The standard defines two types of communication channels available on each CEBus medium (Fig. 29.6): (1) a required control channel for device message communication, and (2) a set of optional data channels for distribution of audio, video, or any wide-bandwidth signals.

The control channel is used on every medium to transmit and receive CAL messages. The control channel uses a frequency spectrum on each medium that is always available and completely defined by a fixed frequency allocation, amplitude, and encoding method. The control channel is required by all CEBus-compliant products to send and receive messages between nodes.

Data channels make up a reserved frequency space available on some media that may be used by CEBus devices equipped to send and/or receive analog or digital data. The data may be any information (music, computer files, compressed video, voice) as long as it is confined to the frequency and amplitude specified for the medium used. Data channels are currently available on TP and CX, although the standard allows data channels on all media. A VCR that sends its video output to an upstairs TV set over the coaxial medium is a typical example of data channel use. The VCR and TV set use the control channel—exchanging resource allocation messages—to establish the connection, and the data channels are used to send the video, freeing the control channel for other tasks while the data transfer continues on the data channels.

Figure 29.7 illustrates the allocation of the frequency space on one of the two CX coaxial cables. The control channel is assigned the frequency space between 4.5 and 5.5 MHz. This frequency space is reserved for devices attached to CX for sending and receiving messages. Figure 29.7 also shows the frequency allocation for the predefined data channels. The frequency space from 54 to 150 MHz is used for transmission of data in one or more 1.5-MHz channels. These channels are block-converted in *node 0* to the 324-to-420-MHz range for reception by any CX node. The use of data channels is optional.

There are two major differences between the control channel and the data channels:

**1.** While CEBus messages on the control channel are sent and received in packets that are completely defined by the CEBus protocol, the format and content of the data channels are open. A manufacturer can use data channels to send any data that can be transmitted in the frequency allocations and amplitude available for the channel or channels used.

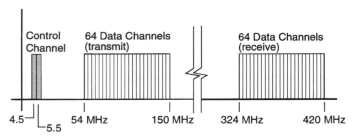

**FIGURE 29.7**   Allocation of frequency space on one of the two coaxial cables.

**2.** As described in the section on the CEBus communications model, the control channel is used for connectionless service only. Devices wait until the control channel is not being used, transmit a message, and relinquish control of the channel. This service is required because there is only one control channel shared by many devices. Each data channel, however, is used for connected service. Devices negotiate for the use of one or more data channels (using messages on the control channel), depending on the bandwidth required for the data to be transmitted. If no other node is using the requested channel(s), the requesting device can use the channel(s) for as long as necessary to transmit the data. For example, transmission of a computer data file might last for several seconds. In the example of the VCR transmitting to the upstairs TV set, the channel might be in use for several hours. There are usually multiple data channels available on each medium so that several devices (64 in the case of CX) can be using data channels simultaneously.

Additional (non-CEBus) frequency space may be available on each medium for use by outside service providers (cable companies, telephone companies, etc.) that share the same medium. This allows products to have access to outside services as well as CEBus control and data channels on the same medium. For example, products such as set-top boxes can access cable company–supplied program material and control messages (pay-per-view video, utility company rates, database access) and CEBus data and control channels all on the same coax cable.

## 29.8   PACKETS AND MESSAGES

Information is transmitted over the control channel in packets of data at about 10 kb/s (regardless of the medium used). Packets contain the necessary "housekeeping" information, such as the address of the originator and destination node, as well as the message (the CAL commands directed to the destination node). A simple analogy can be made between the information contained in a CEBus packet and the information in a typical letter, as shown in Fig. 29.8.

The message contains the packet payload (in the form of a CAL command or reply) that a node wishes to send. The message is typically 4 to 10 bytes—only about one-fourth to one-half of the packet length. The addresses and service bytes make up the remainder of the packet.

For a letter (message) to reach its intended receiver successfully, the letter is sent in an envelope (packet), and the envelope is addressed to a receiver (the TO address). To inform the receiver of the address of the sender (or to allow the letter to be successfully returned if it could not be delivered), the return address (the FROM address) is included on the envelope. Each CEBus product has a unique address, acquired when the product is installed in the home. The address has two parts: (1) a house address (which all products in one home or apartment share), and (2) a device address, unique to the product. When sending letters, the envelope can indicate one or more post office services, such as certified mail, next-day delivery, or return receipt request. Packets contain similar information.

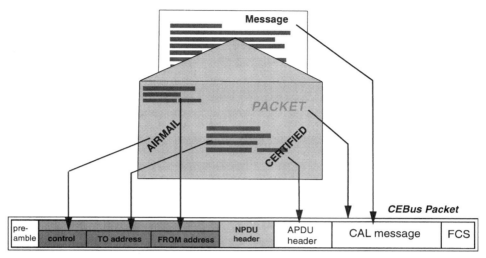

**FIGURE 29.8**    An analogy between the information contained in a CEBus packet and the information in a typical letter.

The data link services (handled by the data link layer) determine network access priority and thus delivery priority, as well as a "return receipt requested" acknowledged service. The network services (handled by the network layer) determine packet routing through the network. The application services (handled by the application layer) determine whether a replay is requested from the receiver and message security. Message *authentication* can be requested in the application services to prevent unauthorized users from accessing or changing information in some products. For example, the electric rate information stored in a CEBus electric meter can be read and changed by the electric utility because it has the correct *authentication keys,* but the information cannot be changed by nonutility devices that do not have the keys.

The *preamble* and *frame check sequence (FCS)* shown at the front end of the packet in Fig. 29.8 are overhead used by the data link layer protocol. The packet preamble, the first byte of the packet, is used for transmission collision detection. The FCS is used for received bit-error detection.

### 29.8.1    Symbol Encoding

When a CEBus device sends a packet, the data in the packet are converted from internal binary form to physical medium symbols. CEBus uses a set of four medium symbols rather than the more common binary symbols used internally in computers:

> 1: Binary one
>
> 0: Binary zero
>
> EOF: End of field—used to separate packet fields
>
> EOP: End of packet—used to identify the end of a transmitted packet

Using four symbols instead of the usual two makes transmission and reception of packets easier. It allows a cheap and easy data compression technique.

All CEBus nodes encode symbols by generating one of two physical medium states. The two states are defined as the *superior state* and the *inferior state*. These terms generically define two electrical conditions of the medium easily detected by a receiving node. The state names imply that the superior state can always override or be detected over the inferior state. The idle state of the medium—the state when no packets are being transmitted—is always the inferior state. Each CEBus medium uses different electrical conditions, appropriate to the medium, to represent the two states. Typical representations of the superior and inferior state on each medium are shown in Fig. 29.9.

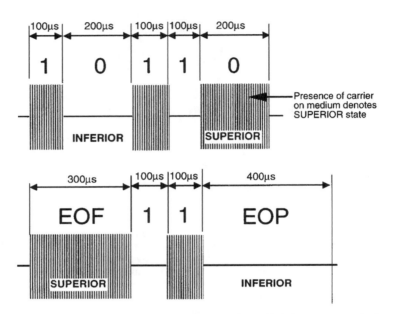

**FIGURE 29.9** Typical representation of the superior and inferior states on each of two physical medium states.

The four symbols are encoded on each medium by using the time the inferior or superior state remains on the media, not by whether the inferior or superior state is used. The 1 symbol is represented by the shortest interval of either the superior or inferior state (100 μs), the 0 is twice the interval of the 1, the EOF is three intervals, and the EOP is four intervals. Note that any symbol can be represented by either a superior or inferior state. The end of one symbol and the start of the next occurs at the transition from one state to another. Therefore, states alternate between superior and inferior for each new transmitted symbol in a packet. Because the idle state of the medium is always the inferior state, the first symbol transmitted in a packet is always the superior state. The packet can end in either state.

The symbol time for the shortest symbol (ONE) is defined as the *unit symbol time* (*UST*) and represents the minimum superior or inferior period. The UST is the basic unit of measure for timing in the protocol software.

The packet data rate is the same for all CEBus media and is stated as 10,000 "one-bits" per second, because a packet containing all 1 symbols would transmit all symbols at 100 μs, or 10,000 b/s. In practice, because each packet contains a mix of the four symbols, the actual data rate varies with each packet, averaging about 8500 b/s. Compression techniques are used in the protocol code to reduce the use of zeros where possible.

The signaling technology used on each medium is tailored for the lowest-cost implementation while providing the highest reliability. PL transceivers use a novel 100–400-kHz spread spectrum carrier technology developed by Intellon Corporation. TP transceivers use a simple 10-kHz,

250-mV carrier, CX transceivers use an equally simple 4.5–5.5-MHz carrier, and IR uses a 100-kHz, 850–1000-nM carrier. RF uses a digitally synthesized spread spectrum carrier centered at 915 MHz.

### 29.8.2  Network Attributes

The symbol-encoding technique used in CEBus allows a uniform packet data rate and encoding on all media in the home. All nodes, regardless of the medium used, transmit packets using the same data rate, the same media states (superior/inferior), the same protocol (CSMA/CDCR), and the same packet format. This allows packets transmitted in one medium to be routed to another medium with a minimum of conversion. The only difference is the electrical representation of the superior and inferior states in each medium. A product that uses the power line can be adapted to use twisted pair by changing only the physical layer transceiver electronics. All other parts of the interface and node software remain the same. This also makes the job of routers and brouters much easier because they need only to convert media states.

## 29.9  *WHAT CEBus PRODUCTS SAY TO EACH OTHER*

The message portion of each packet contains the CAL language command. CAL provides a common language interface so that products know how to communicate with other products without knowing how each specific product operates, who built it, or what features it has.

CAL is responsible for implementing application-level interoperability. To achieve true interoperability on a large scale—to get products to work together in the home—there must be some predefined model of how products operate and a common set of commands to perform the operations. A TV set built by Sony and a TV set built by RCA should operate, from a CEBus standpoint, in the same way. A PC or toaster needs to know how to change the channel on any CEBus-compliant TV set regardless of manufacturer, and without having to consult with the manufacturer. This goal has been achieved by a set of predefined (but extensible) models for all consumer device functions in the home.

### 29.9.1  CAL View of Products

The design of CAL is based on the assumption that all electric appliances and products in the home have a hierarchical structure of common parts or functions, and that the basic operation of the common functions is the same from product to product. CAL treats each product as a collection of one or more of these common parts called *contexts*.

A CEBus TV set, for example, appears as a collection of contexts at a network address in Fig. 29.10. A typical TV set might consist of a video display context, an audio amplifier context, a tuner context, a clock context, a user interface context, etc., depending on the features of the model. CAL defines more than 50 different contexts for everything from lighting, security, and heating/air conditioning to washing and drying. A partial list of contexts is shown in Table 29.1. Each context, regardless of what product it is in, operates in the same way. The audio amplifier in the TV set, stereo receiver, the speaker phone, and the intercom all work alike and "look" alike to the network.

As shown in Fig. 29.11, context consists of one or more objects. An object is a software model of a well-defined *control* or *sensing task*. Objects model the way context functions are normally controlled manually. The volume, bass, and treble analog control objects, as well as the mute binary switch object, represent controls typically found in audio amplifiers. An analog control models controls that can be set to any value from a minimum value to a maximum value. A

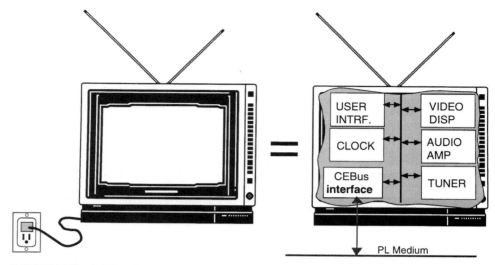

**FIGURE 29.10** A CEBus TV appears as a collection of contexts at a network address.

**TABLE 29.1** CEBus-Implemented Contexts by Context Class and Context Name

| *General* | *Lighting* | *Audio/video* | |
|---|---|---|---|
| 00 Universal | 20 Light sensor | 10 Audio amplifier | |
| 02 User interface | 21 Lighting | 11 Medium transport | |
| 04 Data channel | 22 Lighting zone | 12 Tuner | |
| 05 Time | 23 Light status | 13 Video display | |
| | 24 Lighting zone control | 14 Audio equalizer | |
| | | 15 Camera | |
| | | 17 Switch | |

| *HVAC* | *Utility* | *Security* |
|---|---|---|
| 40 Environmental zone | 50 Utility metering | 60 Security zone/sensor |
| 41 Environmental sensor | 51 Utility sensor | 61 Security system |
| 42 Environmental status | 54 Load center | 62 Security control |
| 43 Environmental zone control | 55 Load center control | 63 Security alarm |
| 44 Environmental zone equipment | 56 Energy control | |
| 45 Environmental system | 57 Energy mangement | |

binary switch object models a two-position (on/off, 0/1) control function. Analog sensors such as temperature sensors, or light-level sensors, are the sensing equivalent of an analog control. They have minimum and maximum values and can assume any value in between.

All objects are defined by a class number (Table 29.2). An object class defines the generic operation of the object. When an object is used in a specific context, it assumes a specific *instantiation* of a function of the context, such as volume or temperature control. CAL uses 26 predetermined objects to model all control functions required for all known consumer/residential products.

Objects, in turn, are defined by a set of *instance variables* (IVs). IVs are like variables in any software program that have a size and a data type. All network operations on contexts are performed by reading and writing object IVs. The IVs that define each object are listed in the

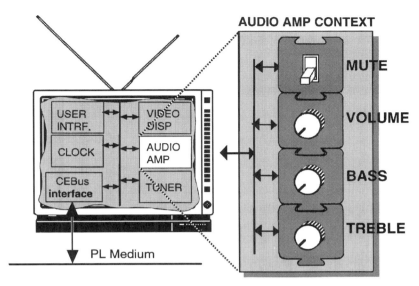

**FIGURE 29.11**   A context consists of one or more objects.

**TABLE 29.2**   CEBus Objects by Object Class and Name

| | |
|---|---|
| 01 Node control | 10 Display |
| 02 Context control | 11 Medium transport |
| 03 Data channel receiver | 13 Dialer |
| 04 Data channel transmitter | 14 Keypad |
| 05 Binary switch | 15 List memory |
| 06 Binary sensor | 16 Data memory |
| 07 Analog control | 17 Motor |
| 08 Analog sensor | 19 Synthesizer/tuner |
| 09 Multiposition switch | 1A Tone generator |
| 0A Multistate sensor | 1C Counter |
| 0B Matrix switch | 1D Clock |
| 0F Meter | |

object tables in the CAL specification. Figure 29.12 shows a typical object table for an analog control object.

The object class and name are given on the top line of the definition. The description section gives the general intended use of the object. This is followed by a listing of all IVs defined for the object class, giving their label, storage class, data type, name, and description. IV labels are an ASCII string of one or more characters. The IV is referenced in a message by its label. Any IV label and name in bold type is required to be supported in the object. Other IVs are optional. The second column shows which IVs are read-only (R) or read/writable (R/W) as referenced from the network, not internally to the product. The third column shows the data type of the IV: b = boolean, n = numeric (integer or real), c = character string, and d = binary data.

Generally, read-only IVs provide information for other nodes on the network about the application of the object. For example, the units_of_measure, step_step, step_rate, min_value, max _value, and default_value IVs in Fig. 29.12 are used to define the characteristics of a particular use of the object. None of these IVs are required, but they may be of benefit to other nodes.

## (07) Analog Control

*Models operation of a continuously variable control such as a knob, slider, or analog setting device. Used to set a variable value for a function over a range of values at a defined resolution.*

| IV | R/W | Type | Name | IV Description |
|----|-----|------|------|----------------|
| U | R | n | units_of_measure | |
| S | R | n | step_size | |
| r | R | n | step_rate | |
| N | R | n | min_value | Minimum value control can be set |
| M | R | n | max_value | Maximum value control can be set |
| D | R | n | default_value | Default value of control on reset |
| C | R/W | n | **current_value** | The current value of control |
| P | R | n | previous_value | |
| R | R/W | d | reporting_condition | Test condition used in reporting |
| H | R/W | d | report_header | Report message body |
| A | R/W | d | report_address | Node address of report message |

**FIGURE 29.12**    A typical object table for an analog control object.

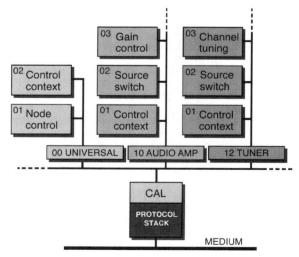

**FIGURE 29.13**    The context and object data structure in a product can be thought of as a tree structure.

### 29.9.2   Context/Object Data Structure

The context and object data structure in a product can be thought of as a tree structure, as shown in Fig. 29.13. Incoming messages are addressed to a specific object in a specific context. Each context is addressed by its class. The objects in each context are addressed by their sequential position in the context, starting at 01. The CAL interpreter locates the object and performs the command in the incoming message on the object IV.

Every CEBus node must contain the *universal context* (00). The universal context does not model any functional system of a product. Rather, it stores the global CEBus housekeeping information about the node in the *node control object* (01). It is the location for global device information, such as the system and node address, the product serial number, and other information that applies to the entire product. IVs in the node control object are provided to simplify product identification on the network, perform address configuration, and determine product capability.

### 29.9.3   Object Implementation

Objects are coded as (1) a set of IV data variables, and (2) application software associated with the variables. Unlike objects in C++ and similar object-oriented languages, CAL object variables are "exposed" to the network through the CAL interpreter, and the application code that executes the function of the object is hidden. IVs are read or written by any node on the network (they can be protected using authentication if necessary). The application code does the same. It checks for changed values and acts accordingly, or updates the values.

### 29.9.4   Object Binding

CEBus objects are divided into three general categories: network output, network input, and network input/output. The categories define whether the object is primarily a sender of messages, a receiver of messages, or both (Fig. 29.14).

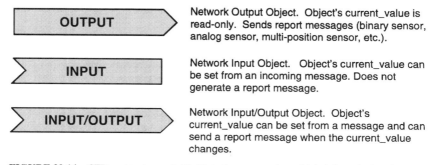

**OUTPUT** — Network Output Object. Object's current_value is read-only. Sends report messages (binary sensor, analog sensor, multi-position sensor, etc.).

**INPUT** — Network Input Object. Object's current_value can be set from an incoming message. Does not generate a report message.

**INPUT/OUTPUT** — Network Input/Output Object. Object's current_value can be set from a message and can send a report message when the current_value changes.

**FIGURE 29.14**   CEBus objects are divided into three categories, which define whether the object is primarily a sender of messages, a receiver of messages, or both.

Object binding means establishing a network correspondence between a *network output object* and one or more *network input objects* for CAL message delivery. Most of the objects in each CEBus context are intended to bind directly with specific objects in other contexts. The binding is expressed in the context interconnect diagram for the contexts in a product category. Figure 29.15 shows a typical example of a context interconnect diagram for two of the eight HVAC contexts. The environmental sensor context is intended to reside in a product that

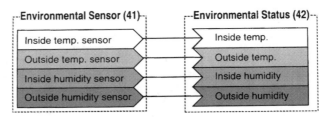

**FIGURE 29.15**    A typical example of a context interconnect diagram for two of the eight HVAC contexts.

measures temperature and/or humidity. The objects bind to a set of corresponding network input objects in the environmental status context intended to be used in products that need or want to know the inside or outside temperature or humidity, such as a thermostat, a TV set, or a PC. The objects in the environmental sensor context send messages reporting a change in temperature or humidity to the environmental status context. This design makes interconnecting products in the field easy because only the system and node address of the device containing the environmental status context need be known.

A more general (and desirable) binding does not require the destination node address. Figure 29.16 illustrates an outside temperature sensor device that contains the environmental sensor context and uses the outside temperature object (03). This product is intended to bind to all environmental status/outside temperature objects in the home. When the outside temperature object sends a message reporting an outside temperature, it uses the broadcast node address (0000) so that it will be received by all nodes in the house.

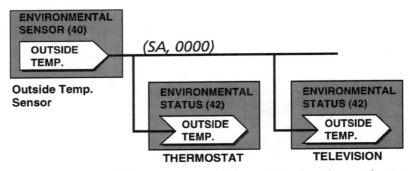

**FIGURE 29.16**    An outside temperature sensor device containing the environmental sensor context uses the outside temperature object (03).

### 29.9.5    CAL Messages

CAL messages are generated from objects using the CAL interpreter or the objects application code. Objects communicate with other objects by setting or reading their IVs. There are two general types of messages: the command message and the response message. The command message is sent to a device to perform an action such as setting or reading the value of an IV. If a command message is sent that generates a return value (such as reading an IV), a response message is returned.

*CAL Message Syntax.*    The expanded portion of the CAL message in Fig. 29.17 shows the syntax of the example CAL message to set the TV channel. The message consists of a <context ID> followed by an <object number> followed by a <method>, optionally followed by an IV, and one or more IV arguments. The <context ID><object number> pair form the destination object *address* for the message within a product. The <> symbols enclose an element identifier and may be made up of one or more simpler elements. The [ ] symbols enclose optional parts.

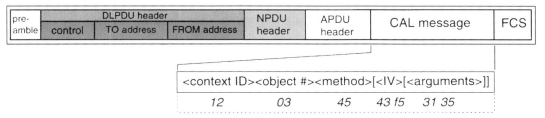

**FIGURE 29.17**    The expanded portion of a CAL message shows the syntax of the CAL message to set a TV channel.

The example message shown in Fig. 29.17 has a context ID of 12 for the tuning context. The object number is 03 for the channel select object. The method is *setValue,* which sets the IV to a value given as the argument. The IV is 43 (ASCII "C"), the current-channel, followed by a delimiter (F5 hex). The argument is the number 15 (ASCII "1," "5"). Numeric values are represented in messages as ASCII characters because representation of a number in a product depends on how it is implemented.

The <context ID><object number> pair forms the destination object address for the message within the node. It identifies a particular object in a particular context in the product.

The method identifies an action to be performed by the CAL interpreter on one or more IVs in an object. Each method operates on a specific data type. Table 29.3 lists a few of the most commonly used methods.

**TABLE 29.3**    Sample Method List

| ID | Name | Arguments |
|----|------|-----------|
| 41 | setOFF | IV |
| 42 | setON | IV |
| 43 | getValue | IV |
| 44 | getArray | IV[■ [offset] ■ count] |
| 45 | setValue | IV value |
| 46 | setArray | IV■ [<offset>] ■ <data> |
| 48 | increment | IV[■ <number>] |
| 4A | decrement | IV[■ <number>] |
| 56 | if | <Boolean> BEGIN <message list> [else clause] END |

*Message Generation.*    Objects send messages to other objects in one of two ways: the object's application code generates a message directly, or the object uses a reporting condition to cause the CAL interpreter to generate a message automatically. The first case is called an *application message*; the second is called a *reporting message*.

Network output and input/output objects contain a set of optional *Reporting IVs*: report_condition, report_header, report_address, and previous_value. These IVs are used as a group by the CAL interpreter to determine when and where to send a message to another node (their function is illustrated in Fig. 29.18):

- Report_condition.   Contains a Boolean expression describing a condition in the object to test whether a report message should be sent. Uses standard CEBus Boolean expression syntax. A typical expression is "C > 85" (or 43 E4 38 35).
- Report_address.   Data variable containing the node and system address of the destination device.
- Report_header.   Data variable containing the CAL message to send to the destination device (less the current _value argument). The message contains the context, object, method, and IV. The IV argument is appended from the current_value.
- Previous_value.   Holds the value of the current_value when the last message was generated. Used by the CAL interpreter to compare against the present value of current_value. It is automatically updated by the CAL interpreter whenever a report is sent.

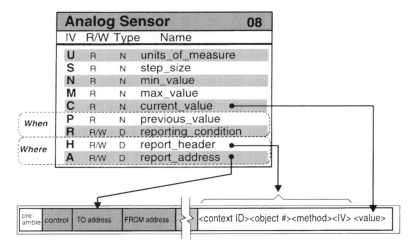

**FIGURE 29.18**   Optional reporting IVs are used by the CAL interpreter to determine when and where to send a message to another node.

Whenever report_condition evaluates true, the message in the report_header (with the object's current_value appended) is sent to the address in report_address to update the value of a target IV. The report_message contains the target context/object address. The advantage of using reporting is that it is field-programmable, because the reporting IVs can be written after product installation.

## 29.10   THE INTEROPERABILITY GOAL

The whole point of the CEBus standard is to allow products to plug and play—to allow the consumer to purchase off-the-shelf appliances, plug them in, and have them provide some function or feature by interoperation between them. This requires two kinds of interoperability: communications interoperability and application interoperability.

Communications interoperability means that products can reliably communicate information between each other. This requires that products adhere to a well-defined protocol and use a communication medium correctly. Application interoperability means that products know how to operate other products correctly, obtain network resources and information, and provide network information and resources across product and industry groups without prior knowledge about what features a particular product has or who built it.

This goal has been met with the CEBus standard. It provides product manufacturers with a totally new way of thinking about product operation by allowing products to network together. Within 10 years, only the most trivial consumer product will be built stand-alone.

## ABOUT THE AUTHOR

Grayson Evans has been actively involved in home automation and communication network development for the past 14 years. He is the founder of the Training Department, a company that specializes in CEBus and related home automation training. He is a product development consultant and a technical consultant to the Electronic Industries Association for the development of the CEBus home automation communication standard. Evans was previously engineering manager at MTI, Inc., product engineering manager at Echelon, and founder and president of Archinetics, Inc., a manufacturer of complete home automation systems. He holds a bachelor's degree in architecture and a masters degree in computer science from the Georgia Institute of Technology.

# THE DIGITAL FUTURE

# CHAPTER 30
# THE DIGITAL HORIZON AND BEYOND

**Roger Kozlowski**
*Motorola Semiconductor Products, Inc.*

## 30.1  INTRODUCTION

Digital consumer electronics in the 1990s and beyond will encompass the true convergence of computing, communication, and consumer technologies to extend the capabilities of present consumer products while creating whole new classes of products.

Attributes of digital consumer design are lower power; single–battery-cell operation; use of embedded microcontrollers; higher-performance CPUs; increased usage of DSP architecture; audio/video decompression techniques; the trend toward "systems on a chip;" and high-level programming language coding. Indeed, the digital revolution is happening today and is being spurred by new technologies and capabilities that were unknown just a few short years ago outside of research and development labs. In this chapter we will examine some of the key enablers for the digital consumer revolution and look ahead to what we might expect in the not-so-distant future.

## 30.2  SEMICONDUCTOR TECHNOLOGY

If any one force can be called a true enabler of this revolution, it is the rapid development of integrated circuit technology. From a few thousand equivalent gates in the late 1970s to the multimillion circuits of today, the designer has been given the capability of working with complete "systems on a chip," which provide the consumer with greater value at ever lower cost.

For over 30 years the semiconductor industry has been the engine that enables countless innovations in analog consumer design. The earliest examples can be found in the first transistorized portable radios, calculators, and watches. In each instance the ability to provide complex circuitry on a single piece of silicon opened new product vistas for the designer. With the transition to new all-digital designs, higher levels of integration are possible through the use of submicron design rules that approach 0.10 micron in size.

Figure 30.1 shows what is known as *Moore's law* for silicon integration over a period of 30 years. On the semilogarithmic $y$-axis is the number of transistors that could be placed on a single silicon die, whereas the linear $x$-axis plots the year in which the product was introduced. The chart shows that chip integration roughly doubles every 18 months. Figure 30.2 shows the cost of an equivalent 1 bit of memory over a 20-year period of time. Taken together, digital

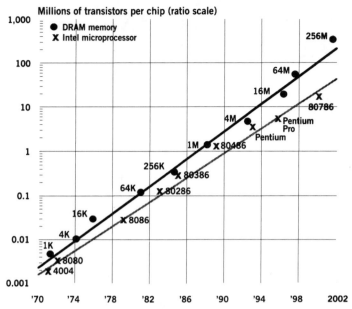

**FIGURE 30.1**  DRAM and microprocessor integration trends (*Source:* VLSI Research).

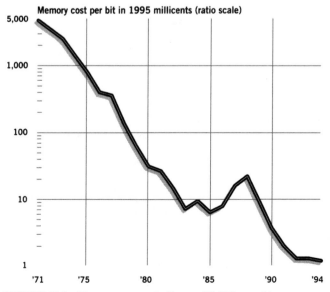

**FIGURE 30.2**  Memory cost trends (*Source:* VLSI Research).

**TABLE 30.1**    Roadmap Projections for Semiconductor Technology

| Year | Smallest feature, μm | Dynamic RAM | | Microprocessors | | | Wiring levels per chip | I/O per chip |
|------|------|------|------|------|------|------|------|------|
| | | Chip size, mm² | Billions of bits per chip | Chip size, mm² | Millions of transistors per cm² | On-chip clock* | | |
| 1995 | 0.35 | 190 | 0.064 | 250 | 4 | 300 | 4–5 | 900 |
| 1998 | 0.25 | 280 | 0.256 | 300 | 7 | 450 | 5 | 1350 |
| 2001 | 0.18 | 420 | 1 | 360 | 13 | 600 | 5–6 | 2000 |
| 2004 | 0.13 | 640 | 4 | 430 | 25 | 800 | 6 | 2600 |
| 2007 | 0.10 | 960 | 16 | 520 | 50 | 1000 | 6–7 | 3600 |
| 2010 | 0.07 | 1400 | 64 | 620 | 90 | 1100 | 7–8 | 4800 |

*High performance.
**Source:**  Semiconductor Industry Association, from Geppert, *IEEE Spectrum,* Jan. 1996, © 1996 IEEE.

consumer product design today is heavily dependent on these charts remaining true in the future with higher levels of integration at a lower cost per transistor than before.

As shown in Table 30.1, the Semiconductor Industry Association predicts that, by the turn of the century, logic ICs will employ upwards of 60 million transistors, six layers of metal, and clock rates over 1 GHz. The critical dimension of a transistor will drop below 0.10 micron, whereas the performance of general purpose microprocessors will climb from 100 million instructions per second (MIPS) to more than 10 billion MIPS.

Of course, many of the circuits that will be used in advanced consumer products will not require such high levels of system integration, nor be solely digital in nature. What the designer will be given, however, is the freedom to explore new algorithms or signal-processing techniques that take advantage of the advances in silicon processing.

## 30.3  ARCHITECTURES FOR DIGITAL CONSUMER ELECTRONICS DESIGN

The advances in silicon integration naturally lead to increasing levels of complexity for embedded microprocessor architectures that can execute more complex systems needs. Table 30.2 is a listing of new consumer products that require some form of processing element, such as a microprocessor, microcontroller, or digital signal processor. In this section, we look at the types of new system designs that use highly integrated microprocessor designs to achieve the promise of the digital era.

**TABLE 30.2**    Embedded Processors in Digital Consumer Applications

| Product category | Microprocessor | Microcontroller | Digital signal processor |
|------|------|------|------|
| Set-top box | X | X | X |
| HDTV | X | X | X |
| Direct broadcast receiver | X | X | X |
| Digital Versatile Disc | X | Optional | X |
| Personal computer | X | X | Optional |
| Video telephony | X | | X |
| Cable modem | X | Optional | X |
| Video game | X | X | Optional |
| Digital audio radio | | X | X |
| Personal digital assistant | X | X | Optional |

### 30.3.1   Microprocessors

Figure 30.3 shows a block diagram of a digital set-top box and Fig. 30.4 shows a block diagram of a Digital Versatile Disc (DVD) player (also see Chapter 12). Both are designed with an embedded microprocessor, but the performance requirements and levels of integration are quite different.

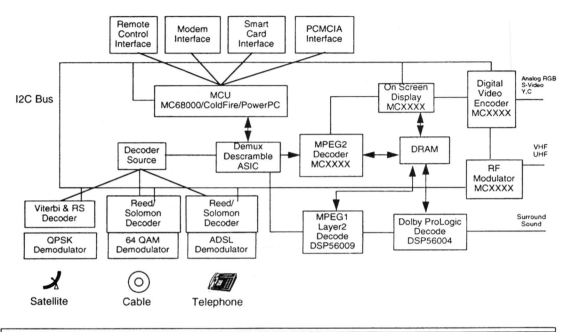

**FIGURE 30.3**   Set-top box block diagram (*Source:* Motorola).

Digital set-top designs use a wide variety of embedded microprocessors depending on the level of graphics, user interface, and communication requirements. In certain high-performance interactive designs, the use of a RISC microprocessor is called for that provides over 100 million instructions per second (MIPS) of processing power for a fast two-way communication link between the user in the home and the service provider's network. In digital set-top designs that are used primarily as one-way broadcast receivers, the processor performance can be in the range of 3–12 MIPS where the embedded microprocessor is used for command, control, transport, and synchronization functions. Specialized processors for audio and video decompression as well as graphics operate under control of the central CPU for the receiver.

DVD requirements are simpler in that the DVD player is not assumed to be connected to a network and the content (disk) is read from an optical disk. In this case the microprocessor handles the command and control of the special processors for audio and video decode as well as acting as a traffic controller for the memory access. In this example the microprocessor used is embedded in an application-specific integrated circuit (ASIC) design that couples the graphics-processing engine. In this way the DVD player minimizes the number of components, thereby also minimizing the delay times for displaying the movie and associated graphics on the TV screen.

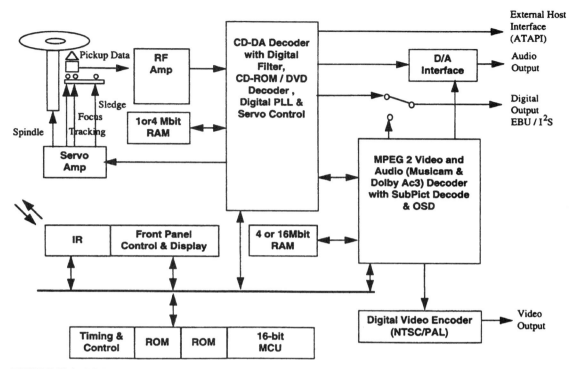

**FIGURE 30.4**   DVD player block diagram.

## 30.3.2   Microprocessor Integration

Figure 30.5 shows the types of ASIC designs possible using microprocessor cores. The designer draws from a library of basic logic and then chooses the type of microprocessor core and supporting peripherals required to be placed on a single IC.

Figure 30.6 shows examples of the types of common analog and digital cells used in a cored-based design, as seen in a floor plan example. The designer must be aware of the interaction of high-speed digital clock signals coupling into low–signal-level analog analog devices when mixed on the same integrated platform.

Figure 30.7 shows the available core functions that can surround microprocessors today. Digital consumer product design in the future will emphasize the need to bring mixed-signal capabilities to the library of functions required for a highly integrated embedded microprocessor design. This will include 9-bit A/D converters capable of working at digital TV rates of 27 MHz all the way to DTV rates of 75 MHz; 8–10-bit D/A converters operating in excess of 125 MHz will be required for future consumer designs. In addition, as the specialized audio and video decoders become a smaller fraction of the total logic budget of the design, the ability to integrate on the same microprocessor die becomes attractive.

By the end of this decade, microprocessor cores will be embedded in dense DRAM products to address the need for portable audio/video DVD players, digital cameras, and digital camcorders where high performance at a low cost is required to make the product small and lightweight for the consumer. These library additions, along with a wide variety of microprocessor cores, will aid the system designer in addressing the tradeoffs between cost, performance, power, size, and ease of software code development.

## Flexibility Of Design - Some Examples

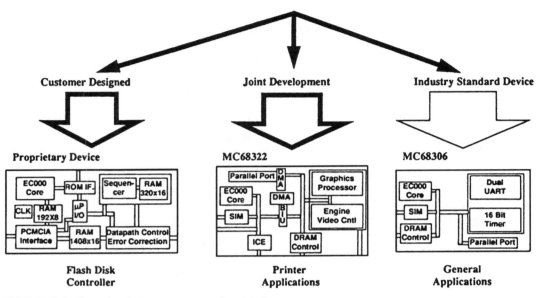

**FIGURE 30.5**   Examples of microprocessor core-based designs.

## Library Functionality

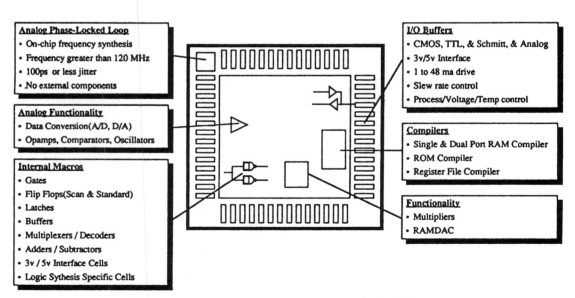

**FIGURE 30.6**   Examples of common standard cells used in digital consumer application design.

## *Memory and Peripherals*

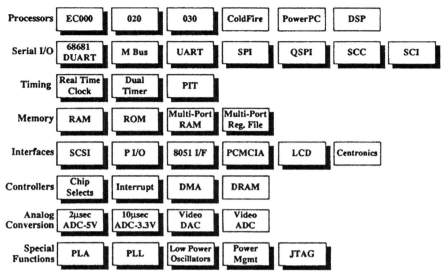

**FIGURE 30.7** Processor architectures compared (*Source:* Motorola Semiconductors).

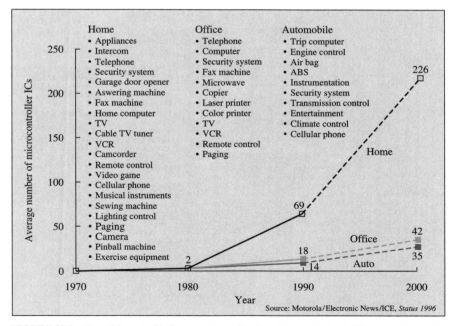

**FIGURE 30.8** Growth in embedded-processing applications (*Source:* Motorola/Electronic News ICE, *Status 1996*).

## 30.4  *MICROCONTROLLERS*

Although microprocessor usage will expand in the next wave of digital consumer applications, one class of processors is already pervasive in digital consumer product use and application—the microcontroller. Whether it be 8- or 16-bit design, the microcontroller is leading the digital consumer revolution by distributing many specialized tasks to low-cost highly integrated microcontrollers. Figure 30.8 shows the numerous applications for embedded microcontrollers.

Embedded microcontrollers are cost-sensitive, with a wide range of on-board peripherals. The designer can take advantage of the integration and low cost of the microcontroller to provide control for radio and television tuners, closed-captioning capabilities in television, system control for camcorders, and conditional access for premium programming. Microcontrollers are stand-alone devices that perform dedicated functions with on-board read-only memory (ROM) and random-access memory (RAM). In addition, special functions such as flash memory, LCD drivers, timers, phase-locked loops, serial and parallel input/output ports, and direct memory access ports make microcontroller design a true single-chip design. Figure 30.9 shows one example of an advanced microcontroller used for security and conditional access consumer applications.

Microcontrollers rarely incorporate the latest advanced processing techniques, relying instead on lowest system cost to provide solutions that range from well below $1 to $10–$15. Because

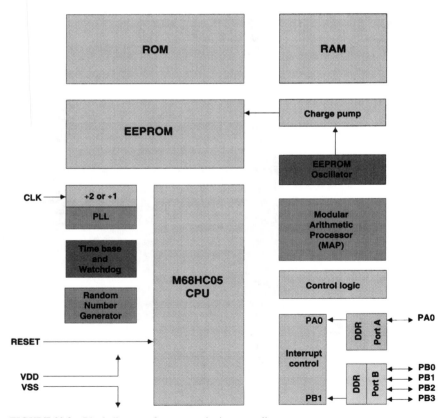

**FIGURE 30.9**   Block diagram of a smartcard microcontroller.

many embedded microcontroller designs do not need a PC or UNIX type of operating system, the software development environment is much simpler as well as portable. The most significant change is in the application of high-level language coding in most consumer designs today and the increased use of a real-time operating system kernel. The next wave of microcontroller designs must tackle the difficult problem of full operation on a single battery cell (1.2 V or less) while providing the rich set of analog and digital capabilities enjoyed by designers today. Special processing techniques to conventional CMOS that will allow operation at well below 1 V are under development. These circuits will enable a new class of high-performance portable products, such as security access for identification and authentication, as well as two-way communications products (e.g., messaging, cellular, digital cordless phones, and digital pocket recorders).

## 30.5  *DIGITAL SIGNAL PROCESSING*

Having been available for nearly 15 years, the digital signal processor (DSP) using a form of Harvard architecture is increasingly providing solutions to advanced digital consumer electronic product designs that require a level of signal processing that is not easily achievable using microprocessors and microcontrollers. Examples include the latest digital audio decoders for MPEG or Dolby AC-3 applications, which require 40 MIPS or more to correctly process the digital bitstream, as well as the next generation of consumer video phone and video teleconferencing applications. DSP techniques offer greater stability; flexibility; and single-clock, multiple-instruction execution for computation.

Figure 30.10 shows that DSP-based processors have continued to follow advanced semiconductor processes. Although DSP continues to follow a similar process road map, the equivalent processing power of DSP designs is moving at an accelerated rate, as shown in Fig. 30.11.

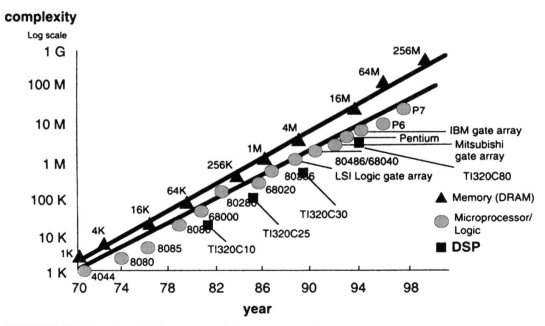

**FIGURE 30.10**  Comparison of DSP, memory, and microprocessor trends (*Source:* Texas Instruments).

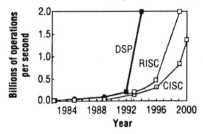

**DSP performance leads RISC, CISC**

**FIGURE 30.11** Comparison of DSP processing power versus CSIC and RISC designs (*Source:* Texas Instruments).

From this, we can see that DSP processors capable of giga-operations per second will be available to the consumer designer. This will allow DSP-based design to encompass many system functions currently done in separate processing units—such as MPEG audio/video decoding, 3-D graphics, speech recognition and synthesis, and high-speed data communications—to be placed in a single DSP integrated circuit. Although microprocessors and microcontrollers continue their steady advancement into consumer design, the DSP is truly the revolutionary enabler that provides the possibility of complex functionality integration.

Table 30.3 compares a number of highly integrated DSP-based designs. These multimedia accelerators or "media engines" bring many of the key technologies required for a consumer product application into a single integrated circuit. Some of these devices are simply combinations of DSP/RISC/graphic controllers interconnected with an on-chip system bus to create a form of acceleration; others use a new, radical architecture called *Very Long Instruction Width* (VLIW). In a VLIW architecture the user has multiple pipelines attached to programmable processor elements to "crunch" the enormous computations required for digital consumer applications. Table 30.4 shows a comparison of the state of the art between the three major CPU architectures.

## 30.6  COMPRESSION TECHNOLOGY

The emergence of digital compression techniques for audio such as Dolby AC-3 digital surround sound and video such as MPEG-1 and MPEG-2 are only the beginning of compression technology (see Chapter 3). Many of today's compression technology relies on the discrete cosine transform (DCT) algorithm, which is able to be executed on DSP- or RISC-based specialized processors. This will change dramatically with the level of integration and performance that will be available to future system engineers. Figure 30.12 shows that the level of compression depends on whether the content is delivered to the consumer in real time (e.g., broadcast television, stored media such as CD-ROM) and on the size of the screen. These factors require different degrees of compression performance and also allow the use of new algorithms that can compress and play back the content in the most cost-effective manner.

Figure 30.13 shows the present landscape of picture quality that can be achieved through the use of software- or hardware-based video-decoding schemes. Software-only decoding will be used increasingly in new consumer products due to lower system cost and because visual artifacts are minimized for smaller screen size or portable products. Software compression will lag behind the hardware capabilities by at least two product generations. As shown in Fig. 30.13, the true frontier for compression technology will be pushed through new hardware architectures (DSP and RISC/DSP combos) and new algorithms that take advantage of this processing power.

Figure 30.14 shows the techniques used today for both reversible (*lossless*) and nonreversible compression (*lossy*). Today, wavelets show the most promise of surpassing the discrete cosine transform as the most widely used compression technique.

**TABLE 30.3**  Comparison of Highly Integrated DSP-Based Designs

| Company | PN | RAMDAC | 2-D GUIX | 3-D Accel | Video Accel | Music synthesis | MPEG-1 decode | MPEG-1 encode | H.320 video conference | Modem data pump | Game port |
|---|---|---|---|---|---|---|---|---|---|---|---|
| Brooktree | BtV2115 | √ | × | — | √ | + | — | — | — | — | — |
| Chromatic Research | Mpact | √ | × | × | × | × | × | × | × | × | × |
| Cirrus Logic | CL-GD5462 | × | × | × | × | √ | √ | — | — | — | — |
| Nvidia | NV-1 | + | × | × | — | × | — | — | — | — | × |
| S3 | Trio64V+ | × | × | × | × | — | √ | — | — | — | — |
| Texas Instruments | TMS320C82 | + | + | × | × | × | × | × | × | × | — |
| Trident | T3D2000 | + | × | × | × | — | √ | — | — | — | — |

× Internal function; √ Part of unified chip set; + External
*Source:* Forward Concepts.

**TABLE 30.4**   Comparison of the State of the Art Between the Three Major CPU Architectures

| Characteristic | Architecture | | |
|---|---|---|---|
| | CISC | RISC | VLIW |
| Picture of five typical instructions □ = 1 byte | □□ □ □□□□ □□□ □□□ | □□□□ □□□□ □□□□ □□□□ □□□□ | □□□□□□□□□□□□□□ □□□□□□□□□□□□□□ □□□□□□□□□□□□□□ □□□□□□□□□□□□□□ □□□□□□□□□□□□□□ |
| Instruction complexity | Varies from simple to complex | One simple operation | Many simple, independent operations |
| Instruction size | Varies | One size, usually 32 bits | One size |
| Instruction format | Field placement varies | Regular, consistent field placement | |
| Registers | Few, sometimes special | Many, general purpose | |
| Memory references | Bundled with operations | Not bundled; load/store architecture | |
| Hardware design focus | Microcoded implementations; one or more pipelines | No microcode; one or more pipelines | Multiple pipelines; no microcode; no complex dispatch logic |

CISC = complex–instruction-set computing; RISC = reduced–instruction-set computing; VLIW = very long instruction word.
*Source:* Geppert, *IEEE Spectrum,* Jan. 1996, © 1996 IEEE.

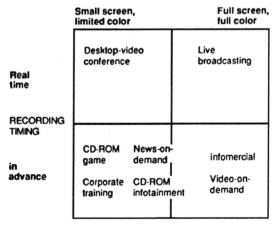

**DIFFERENT APPLICATION REQUIRES
DIFFERENT COMPRESSION PERFORMANCE**

**FIGURE 30.12**   Video compression landscape (*Source:* Motorola).

## 30.7  *SOFTWARE FOR DIGITAL CONSUMER PRODUCTS*

The many capabilities of digital consumer hardware are reliant on the development of reliable, reusable software code—from the simple on-screen user interface seen in television and VCR applications to real-time applications such as digital set-tops, where user interface, 2-D graphics,

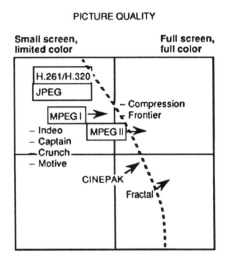

PICTURE QUALITY

communications, and synchronization of the digital bitstream are controlled with a real-time operating system kernel. A number of application drivers and utilities is shown in Fig. 30.15. Software development for many consumer devices continues to be done in assembly code, although users are quickly moving to high-level languages such as C and C++ for digital consumer applications. The modular nature of a high-level language makes possible rapid design cycle times that cannot be achieved using traditional assembly-coding techniques.

An added requirement of many new digital consumer applications is the need for attaching to a network. Certain operating system software and application suites are being developed that address this requirement for networked multimedia. An example from Microware Systems, Inc., is shown in Fig. 30.16. In the diagram the solid lines represent the physical and data link connections. The dashed lines

**FIGURE 30.13**   Relative comparison of compression techniques.

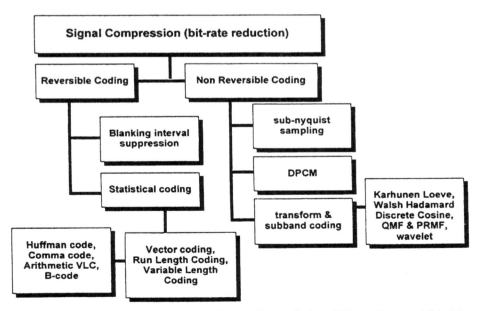

**FIGURE 30.14**   Hierarchy of compression techniques (*Source:* Society of Motion Picture and Television Engineers).

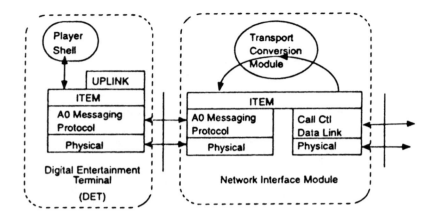

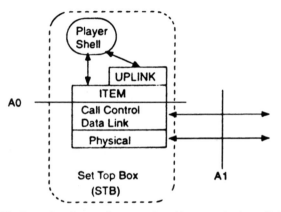

**FIGURE 30.15** Examples of set-top box network architectures: (*top*) type I; (*bottom*) type II (*Source:* Microware Systems, Inc.).

represent the end-to-end connectivity between the digital set-top box and the server. Because of the increasing networking requirements of consumer equipment in the home, future software developments for the digital consumer will continue to borrow techniques from the telecommunications and personal computer industries.

## 30.8 THE FUTURE DIRECTION FOR DIGITAL CONSUMER PLATFORMS

The continued evolution of consumer equipment has brought with it a change in the rate of acceptance by consumers. Figure 30.17 demonstrates the rate of acceptance by consumers of new technology. Whereas black and white television took approximately 20 years to achieve $1 million in sales, personal communicators have achieved this in three years. It is clear that a compression effect is underway as new technologies become available to the consumer. The

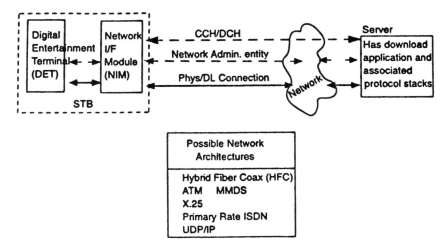

**FIGURE 30.16**  General interactive television network architecture for multimedia delivery (*Source:* Microware Systems, Inc.).

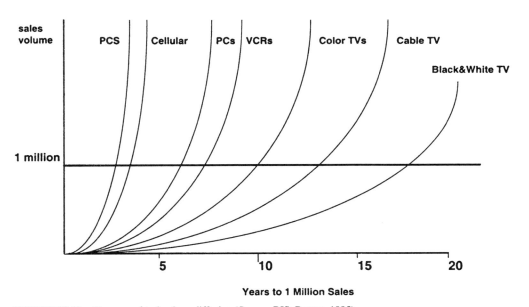

**FIGURE 30.17**  The pace of technology diffusion (*Source:* BIS, Boston, 1995).

transition from analog- to digital-based standards will continue to compress this pattern as the analog C-band satellite, VCR, televisions, and set-tops make way for *direct broadcast receivers* (DBS), recordable DVD, HDTV, and digital set-tops.

Success of new consumer equipment, however, is not ensured. Many product attributes affect the rate of adoption by the consumer. Figure 30.18 shows an analysis of a broad range of consumer equipment. Independent devices are represented by camcorders, portable stereos, and calculators. The success of a product with independent attributes relies solely on a highly specific

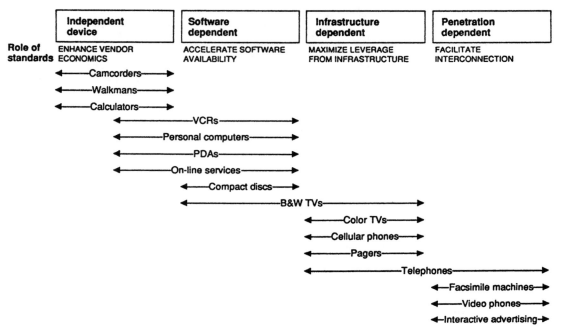

**FIGURE 30.18**   Product attributes affecting rates of adoption.

application at a cost that the consumer is willing to pay. Software-dependent products include VCRs requiring movies as the software content and personal digital assistants (PDAs) with real-time operating systems and software applications. The first generation of PDAs failed to garner consumer acceptance due to a lack of software, whereas VCRs were a success due to a wide range of movies and the ability of consumers to create their own software (home movies on a camcorder).

Increasing complexity of product attributes are shown as infrastructure- and penetration-dependent. Infrastructure dependence offers the longest life cycle of a consumer appliance but is slowed in its deployment as a product due to the high costs of installing a network infrastructure that can utilize the consumer equipment. Penetration dependence is modulated by the role of ubiquitous standards that allow almost universal service, although an infrastructure is assumed to be available.

### 30.8.1   Multimedia Model

A useful way to examine the value to consumers of new equipment and services for multimedia is to look at the interlocking nature of most consumer designs. An HDTV receiver without HDTV programming is a technical statement, not a product. A video telephone standard that doesn't allow for interoperability between callers worldwide—whether the call is made from a personal computer or a consumer video phone—is a fragmented market. Figure 30.19 shows the interlocking relationships for a number of new consumer equipments and the necessary requirements for success. It is important for digital consumer designers to remember that a successful design will take into account what applications are truly useful to the consumer.

| Equipment | Network | Service provider | Info source | Applications |
|---|---|---|---|---|
| Consumer TV players | None | None | CDs, ROMs | Games, movies, training, education |
| Set-top TV converters | Fiber, coax, TP, RF | MSOs/ROBCs | Video servers | Games, movies, shopping, education |
| Video tele.conf. | TP, RF, coax | RBOCs/CSOs, MSOs | Caller/Rcvr | Telephony, business, shopping |
| PC/Workstation | Coax, TP, RF, fiber | RBOCs/source providers | Video servers | Telephony, business, shopping |
| PDAs | RF | RBOCs/CSOs | CDs, Caller/Rcvr | Messaging, business, telephony |
| Digital Cameras | - | - | - | Single- frame video capture |

CSO:    Cellular Subscriber Operator
MSO:    Multiple Service Operators (Cable-Operator)
PDA:    Personal Digital Assistant
RBOC:   Regional Bell Operating Co.'s
RF:     Radio Frequency
TP:     Twisted Pair

**FIGURE 30.19**   Multimedia service network module (*Source:* Motorola).

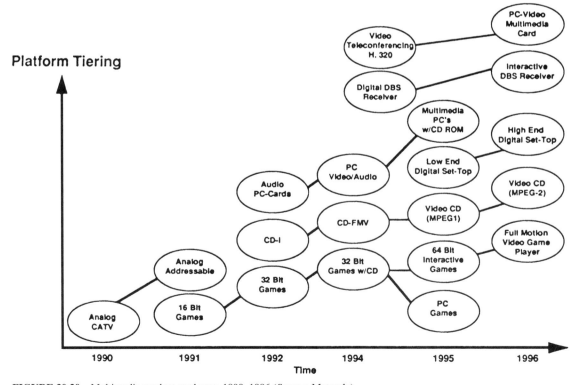

**FIGURE 30.20**   Multimedia product road map, 1990–1996 (*Source:* Motorola).

### 30.8.2  Multimedia Platforms

One way to predict what is on the digital horizon is to look at the recent past of consumer products. Although the rate of adoption of new consumer equipment is clearly evident, it is not always obvious why certain products last longer and others disappear. Figure 30.20 shows the evolution of certain analog/digital consumer products to demonstrate the value of *road mapping* when looking at new product definition. The analog set-top reached a level of acceptance globally beginning in 1990; however, the need for consumers to have a choice of programming led to the development of addressable converters. The 16-bit game machines took center stage beginning in 1991 and have continued to evolve to 3-D arcade graphics with compressed video sequences interspersed into the CD-ROM game title. Digital set-tops begin simply as broadcast receivers, then gradually take on more two-way interactive types of applications that take advantage of the return path of the CATV network. Some new types of platforms, such as DBS, are just starting their road maps with a projection to an expanded role in delivering information to the PC as well as increasing the return-path bandwidth available to the user. Figure 30.21 shows a projection of where some of these platforms may evolve to as well as the incremental developments expected.

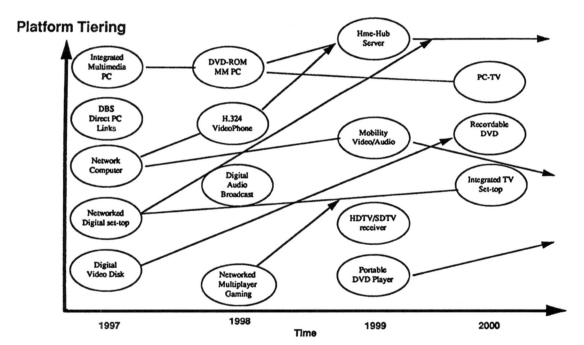

**FIGURE 30.21**    Multimedia product road map, 1997–2000 (*Source:* Motorola).

### 30.8.3  Digital Consumer Electronic Home Networks

Because the new consumer products will likely be digital in nature, using common industry standards (such as MPEG-2 for video, Dolby AC-3 for digital audio, and MPEG-2 for transport), it is possible to think of interconnecting the products through some sort of home network. Figure 30.22 shows one such implementation from the VESA Network Committee. In this example the consumer receives a variety of networked bitstreams representing cable, satellite, terrestrial,

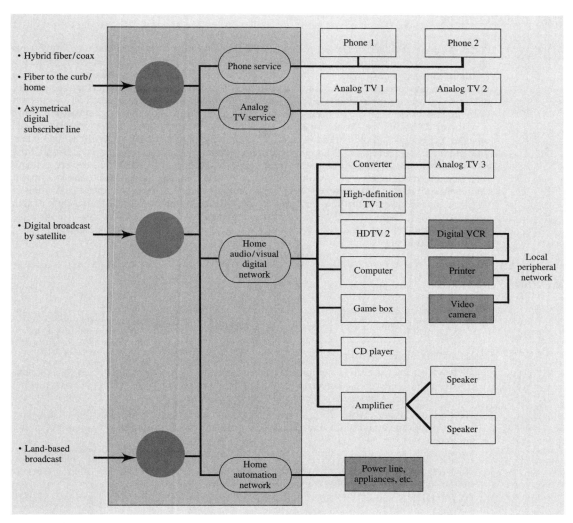

**FIGURE 30.22**   Proposed multimedia home network (*Source:* VESA Home Network Committee, from Braham, *IEEE Spectrum,* Jan. 1996, ©1996 IEEE).

telephony, and fiber-based services. These are in turn converted into baseband signals through a network interface unit, which then "pipes" the information to the *consumer premise equipment* (CPE). Although no standard is yet established for the home network , a number of contenders such as Universal Serial Bus (USB) and P1394—which move data at serial bit rates of up to 100 Mb/s—have been proposed. In addition, the CEBus (see Chapter 29) and Echelon protocols have been suggested for in-home communication and control of security and environmental services. Although the diagram shows a wired connection, it is feasible to consider a wireless connection that makes up the home network. Advantages would include the elimination of additional wiring to the home owner and ease of locating the CPE where the consumer wished. Disadvantages would include the cost of a high-frequency transceiver placed on the CPE to connect to the home network.

## *30.9 CONCLUSION*

Although a convergence of factors is making the digital consumer horizon possible, it is far too soon to say what new products or services will survive as long-term consumer equipment. For example, convergence of products is not always obvious to the consumer who has a toaster, oven, microwave, and grill that, fundamentally, cook food in the same way. Convergence of consumer, computer, and communications will come about only when the consumer perceives a substantially greater value from the combined product than from purchasing the products separately.

Critical decisions for the designer of consumer equipment today involve recognizing (1) whether the product is networked or not, (2) the applications provided, and (3) who the service provider is. In addition, although there is commonality of certain standards such as Dolby AC-3 and MPEG-2, it is equally true that the other vital standards for network transport—user interface and modulation—vary widely across the globe. Making the tradeoff between single-region product design and multistandard product design is complex, with multistandard products adding to the cycle time for new design. Although the factors for a successful product design seem like an impossible task, the opportunity to create useful new products enabled by systems-on-a-chip improvements in digital algorithms and new architectures for implementing product design make this truly the beginning of the digital horizon.

## *GLOSSARY*

**ASIC**   Application-specific integrated circuit, an IC designed for a custom requirement, frequently implemented in a gate array or field programmable array.

**CISC**   Complex instruction set computer, a standard computing approach taken by Intel and MC68000 microprocessors.

**CPE**   Customer premises equipment, a term used to define a class of consumer devices found in the home that connect to a network. Examples include a set-top box, telephone, satellite receiver, and a personal computer with a modem.

**DBS**   Direct broadcast satellite, a digital system for sending/receiving an MPEG-2 transport stream providing audio/video and data services from a stationary satellite to a small receiving antenna.

**DSP**   Digital signal processor, an architecture, based on the Harvard machine, that uses separate data and instruction buses as well as certain instructions optimized for signal processing, such as multiply/accumulate.

**DVD**   Digital Versatile Disc, a high-density CD-ROM technology that provides capacity for full-length MPEG-2 movies as well as data and audio storage and playback.

**MIPS**   Millions of instructions per second; a measure of microprocessor throughput.

**Mixed signal**   The combination of analog and digital circuits on the same semiconductor die.

**RISC**   Reduced instruction set computer, a CPU architecture that optimizes processing speed by using a smaller number of basic machine instructions.

**VLIW**   Variable-length instruction word, a new CPU architecture that relies on a variable-length word size and small parallel processors to provide throughput in terms of billions of operations per second.

## *BIBLIOGRAPHY*

Braham, R., "Consumer electronics," *IEEE Spectrum,* Jan. 1996, pp. 46–50.
Geppert, L., "Solid state," *IEEE Spectrum,* Jan. 1996, pp. 51–55.

Geppert, L., "Semiconductor lithography for the next millennium," *IEEE Spectrum,* April 1996, pp. 33–38.

Gilder, G., "Telecosm, the bandwidth tidal wave," *Forbes ASAP,* 5 Dec. 1994, pp. 163–177.

Kirkpatrick, D., "Riding the real trends in technology," *Fortune,* 10 Feb. 1996, pp. 54–62.

Negroponte, N., *Being Digital,* 1st ed., Vintage Books, 1995.

## ABOUT THE AUTHOR

Roger Kozlowski is the vice president and technical director of multimedia systems for Motorola Semiconductor Products, Inc. His responsibilities have included the development of key technologies and strategies for consumer and multimedia applications over the last six years. His current activities include multimedia standards development; low-cost digital signal processors for consumer/multimedia applications; digital audio applications; and compression technologies for cable, personal computer, and satellite applications.

Kozlowski is currently a member of the IEEE, the IEEE Consumer Society Adcom Board, Society of Cable Engineers, and the Society of Motion Picture and Television Engineers. He received a B.S.E.E. degree from the University of Michigan in 1975 and a Master of Science in Management from Rollins College in 1985. He is a registered professional engineer in the state of Florida.

# INDEX